PHYSICAL METHODS OF CHEMISTRY
Second Edition

Volume VI

DETERMINATION OF THERMODYNAMIC PROPERTIES

PHYSICAL METHODS OF CHEMISTRY

Second Edition

Editors: **Bryant W. Rossiter**
Roger C. Baetzold

PHYSICAL METHODS OF CHEMISTRY

Second Edition

Edited by

BRYANT W. ROSSITER

and

ROGER C. BAETZOLD

Research Laboratories
Eastman Kodak Company
Rochester, New York

Volume VI

DETERMINATION OF THERMODYNAMIC PROPERTIES

A WILEY-INTERSCIENCE PUBLICATION

JOHN WILEY & SONS

New York • Chichester • Brisbane • Toronto • Singapore

In recognition of the importance of preserving what has been
written, it is a policy of John Wiley & Sons, Inc., to have books
of enduring value published in the United States printed on
acid-free paper, and we exert our best efforts to that end.

Library of Congress Cataloging in Publication Data:
(Revised for vol. 6)

Physical methods of chemistry.

 "A Wiley-Interscience publication."
 Vols. 6, 7–   edited by: Bryant W. Rossiter and
Roger C. Baetzold.
 Includes bibliography references and indexes.
 Contents: v. 1. Components of scientific instruments
and applications of computers to chemical research —
v. 2. Electrochemical methods — [etc.] — v. 6.
Determination of thermodynamic properties.
 1. Chemistry—Manipulation. I. Rossiter, Bryant W.,
1931 . II. Hamilton, John F. III. Baetzold,
Roger C.
QD61.P47 1986   542   85-6386

ISBN 0-471-57087-7 (vol. 6)

Printed in the United States of America

10 9 8 7 6 5 4 3 2 1

CONTRIBUTORS

JULIANA BOERIO-GOATES, Department of Chemistry, Brigham Young University, Provo, Utah

*JANE E. CALLANAN, Center for Chemical Engineering, National Institute of Standards and Technology, Boulder, Colorado

†RICHARD S. DAVIS, Automated Production Technology Division, National Institute of Standards and Technology, Gaithersburg, Maryland

PETER J. DUNLOP, Department of Physical and Inorganic Chemistry, The University of Adelaide, Adelaide, South Australia

JEHUDA GREENER, Manufacturing Research & Engineering Organization, Eastman Kodak Company, Rochester, New York

J. REX GOATES, Department of Chemistry, Brigham Young University, Provo, Utah

LEE T. GRADY, The United States Pharmacopeial Convention, Inc., Rockville, Maryland

KENNETH R. HARRIS, Department of Chemistry, The University of New South Wales, Campbell, Australia

REED M. IZATT, Department of Chemistry, Brigham Young University, Provo, Utah

WILLIAM F. KOCH, Inorganic Analytical Research Division, National Institute of Standards and Technology, Gaithersburg, Maryland

JOHN L. OSCARSON, Department of Chemical Engineering, Brigham Young University, Provo, Utah

*Present affiliation: Callanan Associates of Boulder, Boulder, Colorado.
†Present affiliation: Bureau International des Poids et Mesures, Sevres Cedex, France.

v

J. BEVAN OTT, Department of Chemistry, Brigham Young University, Provo, Utah

CHARLES R. TILFORD, Center for Chemical Technology, National Institute of Standards and Technology, Gaithersburg, Maryland

‡DOROTHY K. WYATT, The United States Pharmacopeial Convention, Inc., Rockville, Maryland

DAVID J. YOUNG, School of Chemical Engineering and Industrial Chemistry, University of New South Wales, New South Wales, Australia

‡Present affiliation: Washington Post, Washington, DC.

PREFACE TO PHYSICAL METHODS OF CHEMISTRY

This is a continuation of a series of books started by Dr. Arnold Weissberger in 1945 entitled *Physical Methods of Organic Chemistry*. These books were part of a broader series, *Techniques of Organic Chemistry*, and were designated Volume I of that series. In 1970, *Techniques of Chemistry* became the successor to and the continuation of the *Techniques of Organic Chemistry* series and its companion, *Techniques of Inorganic Chemistry*, reflecting the fact that many of the methods are employed in all branches of chemical sciences and the division into organic and inorganic chemistry had become increasingly artificial. Accordingly, the fourth edition of the series, entitled *Physical Methods of Organic Chemistry*, became *Physical Methods of Chemistry*, Volume I in the new *Techniques* series. The last edition of *Physical Methods of Chemistry* has had wide acceptance, and it is found in most major technical libraries throughout the world. This new edition of *Physical Methods of Chemistry* will consist of twelve volumes and is being published as a self-standing series to reflect its growing importance to chemists worldwide. This series will be designated as the second edition (the first edition, Weissberger and Rossiter, 1970) and will no longer be subsumed within *Techniques of Chemistry*.

This edition heralds profound changes in both the perception and practice of chemistry. The discernible distinctions between chemistry and other related disciplines have continued to shift and blur. Thus, for example, we see changes in response to the needs for chemical understanding in the life sciences. On the other hand, there are areas in which a decade or so ago only a handful of physicists struggled to gain a modicum of understanding but which now are standard tools of chemical research. The advice of many respected colleagues has been invaluable in adjusting the contents of the series to accommodate such changes.

Another significant change is attributable to the explosive rise of computers, integrated electronics, and other "smart" instrumentation. The result is the widespread commercial automation of many chemical methods previously learned with care and practiced laboriously. Faced with this situation, the task of a scientist writing about an experimental method is not straightforward.

Those contributing to *Physical Methods of Chemistry* were urged to adopt as their principal audience intelligent scientists, technically trained but perhaps inexperienced in the topic to be discussed. Such readers would like an introduction to the field together with sufficient information to give a clear understanding of the basic theory and apparatus involved and the appreciation for the value, potential, and limitations of the respective technique.

Frequently, this information is best conveyed by examples of application, and many appear in the series. Except for illustration, however, no attempt is made to offer comprehensive results. Authors have been encouraged to provide ample bibliographies for those who need a more extensive catalog of applications, as well as for those whose goal is to become more expert in a method. This philosophy has also governed the balance of subjects treated with emphasis on the method, not on the results.

Given the space limitations of a series such as this, these guidelines have inevitably resulted in some variance of the detail with which the individual techniques are treated. Indeed, it should be so, depending on the maturity of a technique, its possible variants, the degree to which it has been automated, the complexity of the interpretation, and other such considerations. The contributors, themselves expert in their fields, have exercised their judgment in this regard.

Certain basic principles and techniques have obvious commonality to many specialties. To avoid undue repetition, these have been collected in Volume I. They are useful on their own and serve as reference material for other chapters.

We are deeply sorrowed by the death of our friend and associate, Dr. Arnold Weissberger, whose enduring support and rich inspiration motivated this worthy endeavor through four decades and several editions of publication.

BRYANT W. ROSSITER
JOHN F. HAMILTON

Research Laboratories
Eastman Kodak Company
Rochester, New York
July, 1992

PREFACE

Volume VI of *Physical Methods of Chemistry, Determination of Thermodynamic Properties* deals with important techniques that have been among the mainstay topics over the lifetime of the series. This topical endurance is a tribute to reader demand and to the fundamental value of the individual methods in measuring physical properties of chemical species and their reactions to form new entities of interest to the modern world. World-class contributors give special attention to the presentation of state-of-the-art instrumentation to ensure that the precision, accuracy, ease of use, and presentation of results meet the current requirements of the most sophisticated, as well as the less demanding, user. Often, material is presented in tabular form to aid the reader in making an appropriate selection of instrumentation or measuring devices to meet specific needs. As in earlier editions, the material found in this volume will be of great value in both the laboratory and classroom.

We acknowledge our deep gratitude to the contributors who have spent long hours over the manuscripts for Volume VI; and we welcome two previous contributors, Dr. Peter J. Dunlop and Dr. Lee T. Grady, and several new scholars to the series: Dr. Juliana Boerio-Goates, Dr. Jane E. Callanan, Dr. Richard S. Davis, Dr. Jehuda Greener, Dr. J. Rex Goates, Dr. Kenneth R. Harris, Professor Reed M. Izatt, Dr. William F. Koch, Dr. John L. Oscarson, Professor J. Bevan Ott, Dr. Charles R. Tilford, Dr. Dorothy K. Wyatt, and Professor David J. Young.

We are also extremely grateful to the many colleagues from whom we have sought counsel on the choice of subject matter and contributors. We express our gratitude to Mrs. Ann Nasella for her enthusiastic and skillful editorial assistance. In addition, we heartily thank the specialists whose critical readings of the manuscripts have frequently resulted in the improvements accrued from collective wisdom. For Volume VI these specialists are Dr. R. H. Colby, Dr. K. Liang, Dr. D. J. Massa, and Mr. V. F. Mazzio.

BRYANT W. ROSSITER
ROGER C. BAETZOLD

Rochester, New York
July, 1992

CONTENTS

PHYSICAL METHODS OF CHEMISTRY
Second Edition

Volume VI

DETERMINATION OF THERMODYNAMIC PROPERTIES

Chapter **1**

MASS AND DENSITY DETERMINATIONS†

Richard S. Davis and William F. Koch

†Brand names are given for identification purposes only. Use of brand names neither implies endorsement by the National Institute of Standards and Technology nor assurance that the equipment specified is necessarily the best available.

Physical Methods of Chemistry, Second Edition Volume Six: Determination of Thermodynamic Properties   Edited by Bryant W. Rossiter and Roger C. Baetzold
ISBN 0-471-57087-7 Copyright 1992 by John Wiley & Sons, Inc.

1   INTRODUCTION

The mass m and volume V of a substance determine its density ρ through the relation

$$m/V = \rho \tag{1}$$

Both mass and volume are extensive quantities—they specify how much of the substance is physically present. Density, by contrast, is an intensive quantity, relating to the nature of the substance. Thus, the density of very small amounts of material can, in principle, be determined with high precision. Intensive quantities that correlate with density (such as the index of refraction in transparent liquids) can be used to compare the density of small samples with known reference standards. Where ambiguity is possible, density is sometimes referred to as mass density. Related intensive quantities such as relative density (specific gravity), molar density, and molar volume, which are defined in Section 3, are also amenable to precise determination, even in small samples.

The measurement of mass has been important for millennia because of its usefulness in the trade of precious metals and gemstones. For the Chemist mass measurements are necessary to characterize mixtures; to determine amounts of substance; and, as an intermediate step, to determine density or volume. Measurement of mass changes can elucidate thermally induced reactions, surface interactions, electrolytic processes, and a variety of other phenomena.

Most analytical balances are force transducers; thus, such balances are useful in the determination of forces due to surface tension and magnetic susceptibility.

Measurement of density has a large range of uses in the study of the physical properties of materials. Density measurements are used in industry to control the quality of products as diverse as lubricating oils and beer. The change in density (or volume) upon mixing two fluids tells much about the solvent–solute interactions that have occurred. Relative changes in density as a function of temperature, pressure, and concentration of species are related to the free energy of a solute–solvent system and are thus important in the study of thermodynamics. The same studies help to determine the equations of state of fluids. The measurement of the density of solids is useful to mineralogists, crystallographers, metallurgists, ceramicists, and clinical radiologists, among others.

This brief introduction can only hint at the astonishing variety of analytical uses that have been found for mass and density measurements. This chapter outlines the methods available to the chemist who is confronted with the need for a mass or density determination. The discussion of these techniques makes reference to reviews, monographs, and illustrative reports. Although many examples are cited, we have not reviewed all possible uses of the methods presented below.

2 WEIGHING

Most mass determinations necessary to chemists are carried out on analytical balances [1]. Such balances, capable of high precision, have been available throughout the twentieth century. Nevertheless, the form of these balances has evolved to the point where present models bear little outward resemblance to their ancestors. Principles of construction found in modern devices are reviewed by Erdem [2] and detailed in general references in the Bibliography. Balance development has been motivated by the desire to produce rugged instruments that are convenient to use. Progress has been constrained by the necessity, sometimes difficult to achieve, of maintaining existing levels of accuracy and precision. Many "old-fashioned" balances still exist in chemistry laboratories, and excellent results can be obtained with these instruments. Nevertheless, their relative inconvenience has rendered them unacceptable for the majority of laboratory tasks. There is an inexorable trend toward devices that demand little skill and patience on the part of the operator and that can be controlled by computer when repetitive measurements are necessary.

Even with the newest balances, however, the analyst must be aware of basic operational principles to maximize use of the device's potential and to avoid error. For this reason, we proceed with a review of basic principles of operation of the most common analytical balances. We then discuss other balances that are designed to solve special analytical weighing problems. Finally, we discuss sources of error in weighing and methods for their minimization. Our categorization of balances is somewhat arbitrary because one can find balances that are hybrids of the designs given below.

Unless otherwise specified, we will consider analytical balances as general purpose balances that are designed to operate in ambient laboratory conditions. These balances have capacities up to 1 kg and resolutions of at least $1/10^5$ of capacity, but no better than 1 μg. Balances with better than 1-μg resolution and designed to operate in vacuum or controlled atmospheres are discussed separately in Section 2.5.

2.1   The Balance as Transducer

It is useful to think of balances as transducers; that is, devices that convert a physical quantity of interest to a signal that can be observed. Ideally, a mass transducer produces a readable signal that is proportional to the mass of any object sensed by the transducer. Thus,

$$m = k_t f(z) \tag{2}$$

where f is some function of z, the transducer output (which may be a voltage, a current, the angle of a beam with respect to horizontal, a vertical distance with respect to a fixed point, etc.) and k_t is the ratio $m/f(z)$ at one particular value of m. The ideal mass transducer will have a k_t constant in time. The value of k_t will be unaffected by such perturbing influences as rough handling, changes in laboratory temperature, ground vibration, or the presence of magnetic materials. Finally, the output of the transducer z will be easily resolved to high precision and will be in a form conveniently transferred to a computer or data logger. The functional form of f will be linear. In the rest of Section 2, we will try to be conscious of how closely this paradigm is achieved in analytical balances of various design. For the special case where $f(z) = z$, (2) resembles the equation for a perfect spring. Therefore, by analogy, we refer to dm/dz as the *stiffness* of the transducer. The transducer *sensitivity* is defined as the reciprocal of its stiffness. The smallest mass able to be sensed by the transducer is equal to $(dm/dz)(z_m)$, where z_m is the minimum increment in transducer output that can be resolved *reproducibly*. The quantity $(dm/dz)(z_m)$ has been referred to as the *sensibility* or imprecision of the balance. The reader is cautioned that other nuances in these definitions have been proposed [3]. The ratio of the maximum capacity of a balance to its sensibility is an important figure of merit. This number is sometimes referred to as the load-to-sensibility ratio (LSR). In principle, the sensibility can equal the smallest readable unit of balance output. This is not usually the case for analytical balances, however. As a convenience feature, an adjustment can be made so that $k_t \approx 1$. If this is done, the transducer is said to be *direct reading*.

Most common balances do not sense mass directly. Rather, they respond to *effective* mass or *effective* weight. These important quantities are defined next [see (4)].

2.2   Standard Analytical Balances

The devices considered in this section are designed for general purpose laboratory use at sensibilities above 1 μg.

We first discuss the oldest analytical balances. These first appeared in antiquity with the equal-arm balance and have evolved to the single-pan electronic balances that are widely available from suppliers of equipment for chemistry laboratories.

2.2.1   Equal-Arm Balance

The basic equal-arm balance is shown schematically in Figure 1.1. The principle of operation is well known. Unequal masses suspended from the two arms will produce unequal torques about the central pivot leading to rotation of the beam. In general, the torque exerted by any object X of mass m about a pivot at a distance r is

$$(m - \rho_a V)gr \cos \gamma \tag{3}$$

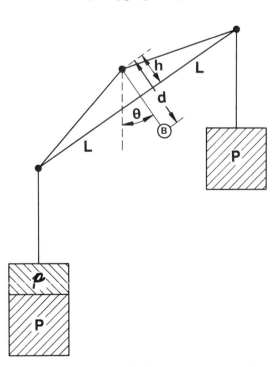

Figure 1.1   Principle of an equal-arm balance. The balance has two arms of length L. The center of mass of the beam is located at B, which is a distance d below the central pivot. A line connecting the end pivots passes a distance h below the central pivot. As long as identical weights P are suspended from each arm, there is no net torque about the central pivot and the beam remains horizontal. An additional weight p causes the beam to rotate through an angle θ, which is determined by the geometry of the beam, its effective mass; and, if h is not zero, by the effective mass of P.

where ρ_a is the ambient density of air, g is the local gravitational acceleration, and γ is the angle between the lever arm from X to the pivot and horizontal. The term involving the volume of X is due to the buoyant effect of air and is simply Archimedes' principle. We note that the term *weight* is defined as $m \cdot g$ [4].† We shall define the quantity $m - \rho_a V$ as *effective mass* and denote it as m^*. The reader is cautioned that other terminology may be found in the literature, often without adequate definition. Making use of (1), we can express the effective mass of X in three useful ways.

$$m^* = m - \rho_a V \tag{4a}$$

$$= V(\rho - \rho_a) \tag{4b}$$

$$= m\left(1 - \frac{\rho_a}{\rho}\right) \tag{4c}$$

Equation (4c) shows the relative importance of air buoyancy. We define effective weight as the effective mass multiplied by the local gravitational acceleration (m^*g). If, instead of air, a body is immersed in a fluid of density ρ_F, the effective mass of the body is again given by (4) with ρ_F replacing ρ_a. Matters relating to air buoyancy are discussed in Section 2.3. Several density determinations presented in Section 3.3 are based on measurements of the effective mass of the same body in two fluids of different density.

Returning to Figure 1.1, the center-of-mass of the beam is below the central pivot for the sake of stability but must be close to the pivot for the sake of sensitivity. The equilibrium angle θ of the beam as a function of the effective mass on the left pan m_p^* is [5]

$$m_p^* = \frac{(m_B^* d + 2m_p^* h)}{L} \tan \theta \tag{5}$$

where m_B^* is the effective mass of the beam, m_p^* is the effective mass of the object suspended on the right, and $m_p^* + m_p^*$ is the effective mass of the object suspended on the left ($m_p^* \ll m_p^*$). The angle θ is measured from the horizontal and is zero when m_p^* is zero. Note the differences between (5) and (2). The most fundamental difference is that this balance is not a mass transducer at all but a *torque* transducer. The torques involved are produced by the earth's gravitational attraction acting on the balance beam and the loads on the beam. Since most analytical balances are used in air under ambient laboratory conditions, the torques are affected by air buoyancy. This is a general feature of analytical balances. A second point worth noting is that even though this balance depends on the force of gravity for its operation, the local acceleration of gravity does not appear in (5) because, as Galileo demonstrated, the acceleration of bodies acted on by earth gravity is independent of chemical composition. Thus g is the same

---

†The word *weight* also denotes an artifact used as a mass standard, for example, in referring to a "set of weights." This usage is, perhaps, unfortunate but nevertheless ubiquitous.

for all the objects in Figure 1.1 and so cancels out of (5). In modern times, this assertion has been tested to very high precision. Any remaining doubts are at a level far below the daily concerns of chemistry. (For unusually precise work one should be aware that the local gravitational acceleration has a vertical gradient of about -0.3 ppm/m at the earth's surface.)

Equation (5) is only valid for a small change in mass on one of the pans; that is, the transducer has a small on-scale range. Consequently an unknown object must, in general, be balanced by an assembly of standards until the difference is small enough to be read on the scale of the balance.

For the balance shown in Figure 1.1 the function $f(z) = \tan(\theta)$ and is, therefore, nonlinear. Approximate linearity is achieved for small angular deflections. If the three pivots are perfectly aligned, the balance sensitivity (defined as $d\theta/dm_p^*$) will not depend on m_p^*. The sensitivity cannot be calculated accurately from (5) because h and d, typically less than 100 μm, are impractical quantities to measure. Rather, sensitivity is determined by measuring the balance response to an additional, small *sensitivity weight* of known mass. The sensitivity can be adjusted by changing the length of d in Figure 1.1, usually by displacing a nut built into the beam directly over the central pivot. A stiff balance with very fine angular resolution may have the same sensibility as a more sensitive balance with crude angular resolution. The former can be made to exhibit a larger on-scale range and has other advantages [6], although nonlinearity may become apparent as the scale is magnified. The balance reading usually is more perturbed by incidental tilting of its base when the on-scale range is large.

Inequality of the arms, not considered in (5), leads to additional complications. Usually, balances of this type have arm equality to a tolerance of 1 : 250,000. This can be improved somewhat, but for best results a transposition scheme is used [7]; that is, a second balance reading is taken with the weights interchanged on the two pans. The average of these two readings is not affected by inequality in the arm lengths. Alternatively, a third weight of unknown value (the tare) can be placed on one pan and the standards and unknown placed in succession on the second pan [7]. This technique is attributed to Borda [8] and bears his name. For determinations of the highest accuracy, it is usual to make repetitive measurements in a symmetric way so that drift of the balance zero point can be eliminated. The usual method for doing this is known as double transposition weighing or, if the Borda technique is used, double substitution weighing [7, 9].

Note that a small change in length of one of the balance arms relative to the other during the course of a measurement can lead to a significant error. If such changes are approximately linear with respect to time, however, their effect can be mitigated by double substitution or double transposition weighing. Geometrical imperfections in the balance also can produce unwanted effects. The interested reader should consult the several detailed studies of equal-arm balances [10–12]. A remarkable balance of this type has been designed by the National Research Laboratory of Metrology (Japan) for the calibration of 1-kg mass standards. Their equal-arm balance has a reported standard deviation of 0.3 μg, a LSR of 3 × 10^{10} [13]. For a beam length of 20 cm this imprecision implies that the two balance arm lengths must track each other to within atomic dimensions. Clearly, such constraints put severe demands on the perfection of the pivots used in these balances. Creating nearly ideal pivots and compensating for the remaining imperfections have been the subject of much research and were dealt with at length in the previous edition of this series [14].

We must abbreviate this discussion and will only point out that mechanical analytical balances have achieved maximum LSR using pivots that are knife-edges bearing on flat surfaces. Knives and flats are often made of sapphire, the knife-edges being honed to an effective radius of less than 1 μm in some cases. Such bearings have very low friction and a well-defined, stable geometry. They also maintain these properties when rotated through a small angle. Their drawbacks include difficulty of manufacture and installation and fragility. The latter necessitates a precise arrestment system to protect the bearing surfaces when weights are placed on the pans or removed. A brief introduction to the other types of pivots available and used in balances is given by Czanderna and Wolsky [3]. All types of pivots can be described by an *effective radius* in analogy to the actual radius of a knife-edge pivot [15]. Generally, this radius is smallest for knife-edges. Although in theory a balance can function with pivots of large effective radius, there are advantages to keeping pivot radius as small as possible [14, 15].

The balance as shown in Figure 1.1 is underdamped; that is, the beam oscillates freely about its equilibrium position with only slight damping due to air resistance and friction in the pivots. The period of these oscillations is related to balance stiffness, again in loose analogy to an undamped spring-mass system. The equilibrium position is usually inferred by noting the position of the turning points through several periods of the oscillation [16]. This is tedious work that can be done more accurately by a machine than a technician [17]. Modern techniques relying on photodiode arrays [18], laser interferometry [19], or measurements of time intervals [13] have been employed for this purpose. Kibble [20] reports gratifying results on the retrofitting of an old equal-arm balance with an electromagnetic servocontrol system of the type discussed below.

The two most inconvenient features of the balance in Figure 1.1 are its small on-scale range and the need to infer equilibrium angle by recording several turning points as the beam oscillates. The problem of oscillation was addressed by P. Curie, who added an air-dashpot mechanism to an existing balance. By proper construction of the dashpot, the balance was critically damped and its equilibrium position read easily [6]. Eddy-current damping produced by the motion of a sheet of pure aluminum (suspended from one end of the beam) through the field of a powerful magnet (mounted on the case) became a popular method as well. To increase the usable range of the balance, a small "rider" weight is added to the beam on some models. This weight can be moved to various positions notched into the top of the beam. Since the same rider will produce different torques when placed in different notches, a single rider can be used to cover a wide range of small increments. Still finer control can be achieved by the addition of a gold chain to one end of the beam of a damped balance. The chain hangs in a catenary suspension, its other end attached to a vertical post. Adjustment of the chain height changes the effective torque applied to the beam. By controlling the chain height, the operator restores the beam to its null position.

A great deal of careful engineering went into two-pan analytical balances. Nevertheless, small errors many times the weighing precision are possible due to the environment in which the balance is kept and to imperfections in the balance construction. These errors were searchingly examined by Corwin [21], who focused his investigations on one representative balance. Macurdy [22] drew attention to the importance of proper thermal conditions within the balance case.

2.2.2   One-Pan Balance, Mechanical

Until the middle of this century, the trend had been to tinker with the basic equal-arm balance in ways that made it more convenient to use, often at the expense of weighing accuracy. A fundamental change in balance design, usually attributed to E. Mettler, shaped the next generation of analytical balances. The new design is illustrated in Figure 1.2. The balance is critically damped by an air-dashpot for ease of inferring the equilibrium position. To circumvent the dependence of balance sensitivity on load, this balance operates at only one nominal load, equal to its weighing capacity. By restricting measurements to Borda weighing (see Section 2.2.1), one pan of the balance can be eliminated and replaced by a counterweight fixed to the beam. This eliminates one set of knife-flat bearings, thereby simplifying the arrestment mechanism, and allows use of a pan whose radius exceeds the arm length of the beam, allowing a more compact case. The beam, in fact, is no longer required to have equal arms. Standard weights are provided along with the balance. These are installed on the suspension above the balance pan and are lifted from the suspension by external controls.

An object to be weighed is placed on the pan, and the calibrated weights above the pan are removed until the equilibrium position of the beam is on scale. The remaining imbalance between the built-in standards and the unknown is read by an optical scale or reticule attached to an end of the beam and projected onto a ground-glass screen fixed to the balance case. This means of projection greatly magnifies small changes in beam angle and allows for a large on-scale range. The scale range is so large in some models (more than a factor of 1000 greater than that of some two-pan balances) that the tangent nonlinearity of (5) can be observed [23]. This source of nonlinearity can be eliminated by manufacturing a compensated reticule.

Although the balance geometry is no longer symmetric about the central knife, certain important symmetries are retained: the unloaded balance is in

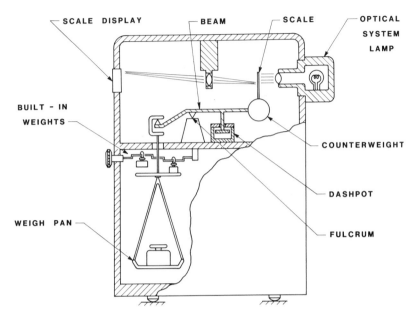

Figure 1.2 One-pan mechanical balance. By turning a knob, the operator removes built-in weights until the object of interest and the built-in weights that have not been removed approximately balance the counterweight. The remaining imbalance causes the beam to rotate as in Figure 1.1. The angle of rotation is related to scale markings, that are projected onto the display [33].

approximate buoyant equilibrium [23]; the large surface area of the pan is compensated by increasing the surface area of the counterweight. These symmetries tend to make the balance zero-reading immune to changes in ambient air density and relative humidity. A comparison of response to changes in humidity of equal-arm balances and single-pan balances was carried out by Almer [24], who found that well-constructed models of either type were acceptable.

Variations on the basic design exist. In one such, the delicate central knife–flat bearing is dispensed with in favor of a more robust torsion fiber. Since the fiber has no bearing surface, it is not influenced by dust. The balance beam is restored to null in much the same way as the catenary chain of a damped equal-arm balance is used to restore the beam to horizontal. By restoration to the null position, the torsion fiber better fulfills its function as a frictionless pivot with well-defined effective radius. The tangent nonlinearity is also avoided. The Berman balance [25] is a torsion balance specially developed for the density determination of small (5–25 mg) mineral samples by the hydrostatic method (Section 3.3).

Detailed theoretical treatments of single-pan balances have been published [23, 26]. A method of testing single-pan mechanical balances for adequate performance is described in [27]. Since these balances typically contain several decades of built-in standard weights, prudent analysts will want to monitor their calibration just as with any set of mass standards. Several schemes of varying complexity have been proposed for this endeavor [13, 28–31], whereby the calibration is carried out without removing the weights from the balance. When balances are used for density determinations, either by hydrostatic weighing or by pycnometry, the built-in weights need not be fully calibrated. It is only necessary that deviations from linearity be known (see below). The schemes given in [30, 31] permit one to obtain the necessary information without recourse to external standards.

The span of the scale, that is, the transducer span, of the balance is calibrated by application of a weight of known value. In the absence of an external standard, the smallest built-in weight can be used. If this weight has a mass m_s and a density ρ_s, then

$$k_t = m_s\left(1 - \frac{\rho_a}{\rho_s}\right)\bigg/ |f(\theta_s)| = m_s^*/|f(\theta_s)| \tag{6}$$

where $f(\theta_s)$ is the change in scale indication upon addition (or subtraction) of the sensitivity weight, and we have set $f(0) = 0$. By adjusting the sensitivity screw on the beam, the value of k_t can be changed so that the span reads the mass value of the added weight. The theoretical value of k_t is qualitatively similar to that shown in (5). Therefore, the value of k_t is subject to small changes due to dimensional instabilities in the balance as well as to gross variations in the ambient air density. The question of whether to apply air buoyancy corrections to scale readings [32] is clarified when one realizes that the scale span is initially adjusted and later calibrated to simulate the response of adding an actual weight to the pan. Since the response of the calibration weight must be corrected for air buoyancy, the simulated response (that is, the scale readings) must also be corrected as shown below in (11a). Linearity of the transducer output between zero and full-scale can be checked with uncalibrated weights [23]. The function f for a particular balance usually remains constant

2.2.3   One-Pan Balance, Servocontrolled

The evolution from the balances presented in the previous section to modern top-loading analytical balances has been reviewed by Schoonover [33]. In its present incarnation, the typical analytical balance is shown schematically in Figure 1.3. The delicate knife–flat bearings are replaced by flat metal springs. This renders the balance rugged enough that no arrestment system is required when objects are added to or removed from the pan. The pan is above the beam for easier access. To make these changes possible, servocontrol by an electromagnetic force cell has been added [34]. The control system shown operates on the same principle as an electromagnetic speaker [20]. Current in the coil mounted on the balance beam travels at right angles to the field produced by the strong magnet fixed to the balance case. Optical sensors

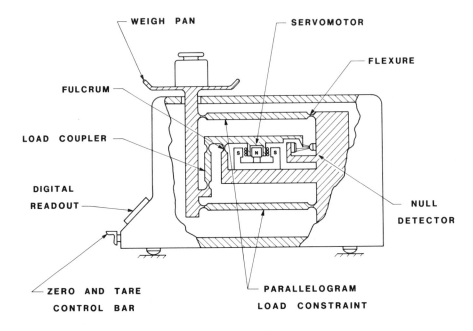

TOP LOADING ELECTRONIC BALANCE

Figure 1.3   A typical electronic balance. Placement of an object onto the pan produces a force that is transmitted by a system of levers to the central pivot (fulcrum). This force is balanced by an electromagnetic force in the servomotor. The balance condition is sensed by an optical detector [33].

provide an error signal when the beam angle changes from the null position. In a standard control circuit, current proportional to the error signal, its time integral, and its time derivative (PID control) is generated to produce a restoring force on the beam. The proportional and differential control make the electromagnetic force generator behave like a critically damped, stiff spring. The integral control increases the stiffness further after the transient oscillation has damped out [35].

The intrinsic, that is, "unservoed", stiffness of the balance and the electromagnetic stiffness produced by the PID control add like springs in parallel. If we neglect transient behavior and recognize that the PID stiffness is far greater than the mechanical stiffness of the balance alone, then

$$m^* = (k'_s/g)f'(\theta) \qquad (7)$$

where m^* is the entire effective mass of the object X placed on the balance pan, not a relatively small difference between the effective mass of X and selected built-in weights. The quantity θ is the difference in equilibrium angle of the beam caused by the addition of X to the pan. The servostiffness k'_s/g depends on the strength of the magnet, the coil geometry, and the gain in the feedback loop of the servocontroller. We show the gravitational acceleration g because the gravitational force generated by X depends on g, while the electromagnetic force generated by the servomotor does not.

Equation (7) is not useful because the values of θ encountered are negligibly small. Instead, one measures the current I in the servocoil and makes use of the following relation:

$$m^* = (k_s/g)f(I) \tag{8}$$

In the usual case, the servofeedback loop has high gain so that g/k_s, the approximate change in current due to a change in mass, depends only on the geometry of the servomotor and the strength of the permanent magnet.

In a well-engineered balance, k_s is stable. Reproducible changes in k_s are compensated by a microprocessor internal to the balance case. For example, in some models the change with temperature in the strength of the permanent magnet is compensated by including a thermistor thermometer on the magnet. The internal microprocessor senses the temperature and applies a predetermined correction. In other models, the thermistor is part of an auxiliary circuit that supplements the strength of the permanent magnet as it is weakened by increasing temperature. In principle, reproducible effects and nonlinearities in the servocontrolled balance can be compensated or corrected by internal data processing before a number is displayed. Even very slow drifts of the balance zero can be sensed and suppressed.

The appearance of the balance pan above the beam deserves an explanation. In the previous designs considered, the balance pan or pans were suspended below the beam. Not shown in Figures 1.1 and 1.2 is a universal pivot in the suspension that allows the pan to swing freely in any direction. Thus, a load placed off-center on a pan causes the pan to swing so that the center of mass returns to a location directly below the end knife. This is an essential feature that compensates for inevitable imperfections in balance construction [36].

The pan location in Figure 1.3 is reminiscent of the shopkeeper's balance invented by Roberval in the seventeenth century [37]. An analysis of Roberval's device shows that placing a load off-center on a balance pan that is not free to realign itself will produce an error except in one special case: if the pan is guided by a perfect parallelogram so that the surface of the pan moves vertically without rotation [37], hence, the appearance of a parallelogram in Figure 1.3. Because the addition of servocontrol can restrict the actual beam travel to less than a micrometer, the parallelogram need only be perfect through this very small distance. Nevertheless, the size of the pan provided on servocontrolled top-loading balance is often limited by the stability of the parallelogram structure that supports it.

The fact that the transducer LSR can be 10^6 or more and is, in principle, dependent on the value of the local gravitational acceleration [See (8)] means that analytical balances of this type must be calibrated at the place of use. (Daily variations about the average g at a particular location are only significant for measurements at an LSR of about 5×10^6 [38].) For this reason, most such balances include one calibrated weight whose nominal mass equals the maximum balance capacity. This weight is used to adjust the span in analogy to what is done on a single-pan mechanical balance. Uncalibrated weights can be used to establish the linearity of the balance between zero and full scale. At least four equally spaced points along the span should be examined.

Although it is still a force transducer, the top-loading electronic balance comes close to fulfilling the design goals given in the introduction to this section. Nevertheless, the balance shown schematically in Figure 1.3 has a few idiosyncracies that must be appreciated by the user [33, 39]:

1. The balance may not work acceptably when weighing magnetic materials. If the balance responds to a ferrous material that is placed near the pan but not on it, clearly there is a problem. One solution is to weigh the ferrous material while it is suspended well below the balance pan (many balances are provided with a hook underneath the case for ease in weighing "below the pan"). Magnetic forces decrease with the distance d between the interacting objects. The attenuation often follows a simple power law d^x, where $1.5 < x < 4$.

2. The balance may be adversely affected by stray electromagnetic radiation at radio frequencies (rf noise).

3. An electronic balance may pose a hazard when operated in explosive atmospheres.

4. The transient response of the balance is optimized for normal air weighing and may exhibit instability if the balance is used for other purposes such as hydrostatic weighing. If hydrostatic weighing is intended, consult the manufacturer before a particular model is chosen. Methods of evaluating single-pan balances have been detailed [38, 39, 40].

Although the catalog of potential problems is fairly long, it is much like the warnings of side effects listed for a miracle drug—the benefits far surpass the drawbacks for the vast majority of users.

The remarkable beam stiffness of servocontrolled balances makes them particularly useful as force transducers for applications such as the measurement of surface tension [41] and magnetic susceptibilities [42]. These measurements are usually carried out using microbalances, which are described in Section 2.5.

2.2.4   One-Pan Balance, Miscellaneous Technologies

There are competing technologies to the servocontrolled analytical balance that may have advantages for users. Although they are not yet in competition with the best servocontrolled balances with regard to LSR, there is a place for balance technologies that are simpler to fabricate (with a consequent reduction in cost). These schemes, which are among those reviewed by Erdem [2], include the strain-gauge load cell (SGLC) and the vibrating string.

The principle of the SGLC is illustrated by considering a cylindrical solid placed upright on one base. A weight placed on the top of the cylinder will then introduce a compressive force. In response to this force, the cylinder height will be reduced slightly and its diameter will be increased slightly. Strain gauges mounted on the cylinder detect these changes in shape with high precision. If the cylinder (or other, more favorable, geometry) is made of a material with good elastic properties, one has produced a transducer that is, quite literally, a very stiff spring. Other load cells based on different principles are also used. Four load cells placed at the corners of a rectangular pan result in a balance that is rugged, has excellent transient response, avoids use of a powerful magnet, and contains no moving parts in the sense that there are no critical bearings or pivots. By careful engineering, the balance can be constructed with a single load cell. Since the device depends on elastic properties of the load cell, limitations due to material creep, hysteresis, nonlinearity, and temperature effects must be overcome.

A second technology exploits the principle that the resonant vibration frequency of a taut string depends on the square root of the tension in the string. Thus, the string is a force transducer whose output variable [z in (2)] is frequency [2, 43].

Finally, we cannot leave this discussion without pointing out that helical springs find use as force transducers in the laboratory. In determinative mineralogy, the familiar Kraus–Jolly helical-spring balance is still used for simple, rapid measurements of specimen densities by the hydrostatic method (Section 3.3). Very sensitive helical springs, although fragile, have been used instead of beam microbalances for special applications such as thermogravimetry [3, 44]. A linear voltage differential transformer (LVDT) can be incorporated with a helical spring to automate the reading of spring extension. Such a device has detected changes in mass of 5 mg on a 5-g load [45].

2.3   Calibration of Weights and Balances

All-purpose laboratory balances are force transducers that respond either to effective mass (m^*) or to effective weight (m^*g). Therefore, a balance cannot be calibrated to read the mass m of all arbitrarily chosen objects (such as helium-filled balloons and platinum crucibles). The best that can be done is to adjust the transducer–balance so that it displays the mass of objects that have one particular density.

2.3.1   Adjustment to a Basis Density

Consider first the balance shown in Figure 1.2. The standard weights internal to the balance are made of the same alloy of density ρ_R. Therefore, when an object X is placed on the balance, the equilibrium condition of the balance is given by:

$$m_X^* = m_D^* + k_t \cdot \theta \tag{9}$$

where m_D^* is the effective mass of the built-in weights that were dialed off to obtain an on-scale value of θ. For simplicity, we assume the functional form of f in (2) is linear. If the scale has been calibrated as described above, then (6) applies, and

$$m_X^* = m_D^* + (m_s^*/\theta_s)\theta \tag{10}$$

where m_s^* is the effective mass of a sensitivity weight of true mass m_s and θ_s is the balance response to addition of the sensitivity weight. For the balance to be direct reading, the sensitivity knob on the beam is adjusted until θ_s as read directly on the projected scale equals the value of m_s. If all these things are done, then (10) becomes

$$m_X \left(1 - \frac{\rho_a}{\rho_X}\right) = (m_D + \theta)\left(1 - \frac{\rho_a}{\rho_R}\right) \tag{11a}$$

where m_x and ρ_x are the mass and density of the object X. The mass of X is thus determined from the balance reading through the relation

$$m_X = (m_D + \theta)(1 + \mathcal{K}) \tag{11b}$$

where

$$\mathcal{K} = \left(1 - \frac{\rho_a}{\rho_R}\right) \bigg/ \left(1 - \frac{\rho_a}{\rho_x}\right) - 1$$

$$\approx \rho_a \left(\frac{1}{\rho_x} - \frac{1}{\rho_R}\right) \tag{12}$$

Some textbooks refer to $\mathcal{K}$ as the *buoyancy correction factor*. Note that (11b) may also be written as

$$m_X = (m_D + \theta)^* \bigg/ \left(1 - \frac{\rho_a}{\rho_x}\right) \tag{11c}$$

where $(m_D + \theta)^*$ is the effective mass of the standards of density ρ_R that are needed to balance the unknown.

Balance and weight manufacturers have chosen $\rho_R = 8000 \text{ kg/m}^3$ as the *basis* density upon which balances will be calibrated at room temperature (20°C). This is the approximate density of the nonmagnetic alloys of stainless steel from which the best weights are now manufactured. Before this convention was adopted, the basis density had been 8390.9 kg/m³ at 20°C (reflecting use of a brass alloy that has a density of 8400 kg/m³ at 0°C and a thermal coefficient of volume expansion of 0.000054/°C between 0 and 20°C). The oldest single-pan mechanical balances were manufactured on the brass basis. Therefore, balances manufactured on the two different density bases can exist side by side in the same laboratory. Since the density of ambient air near sea level is 1.2 kg/m³, this means that direct readings taken on balances manufactured on the two different bases will differ by 7 ppm, that is, 7 mg/kg [9, 26]. The effect of air buoyancy in comparing object of density ρ_X with standards of density 8000 kg/m³ is shown in Figure 1.4.

The electronic balance shown in Figure 1.3 is calibrated as indicated in (8). The mass of an object X is then computed as

$$m_X = I \left(1 - \frac{\rho_a'}{8000} \right) \bigg/ \left(1 - \frac{\rho_a}{\rho_X} \right) \tag{13}$$

where I is the change in balance indication when X is added to the pan. For simplicity, we assume the units in which I is displayed have been chosen to make the balance direct reading in mass (most electronic analytical balances have a calibration mode that accomplishes this task semiautomatically). For instance, a 100-g balance is calibrated so that the balance reading is exactly 100 when a standard of exactly 100 g is put on the pan. It is further assumed in (13) that the

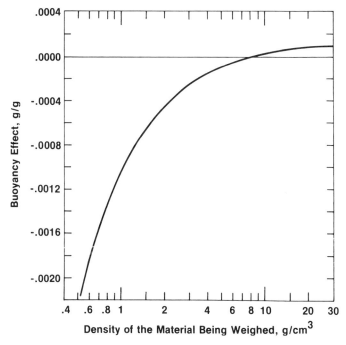

Figure 1.4 The approximate value of the buoyancy effect as a function of the density of the unknown object those mass is to be determined. The mass standards are assumed to have a density of 8000 kg/m³ (8 g/cm³). To correct for this, a buoyancy correction factor must be applied. Also, the magnitude of the effect is the difference between the *apparent mass versus 8000 kg/m³* of a material and its true mass [46].

standard weight used to calibrate the span of the balance has been adjusted on the stainless steel basis. This will be true of any internal weight supplied by the manufacturer. The primed air density is a reminder that, for balances of this type, the air density at the time of calibration belongs in the numerator, while the air density at the time of use belongs in the denominator. When the transducer span is smaller than 10,000 times the sensibility, this is a distinction without a difference. However, in some modern balances the LSR is 10^6 or more. Thus, for critical measurements, it is good practice to recalibrate the span at regular intervals. Since the span calibration in a balance of any design can change slightly over time for a variety of reasons, regular calibration is a good idea.

As can be seen from Figure 1.4, the buoyancy correction factor $\mathcal{K}$ given in (12) is about -0.0001 for a determination of platinum. On the other hand, the weighing of aqueous solutions necessitates a correction factor of order $+0.001$ [46]. Depending on the importance of $\mathcal{K}$ in the final result, the ambient air density may have to be determined at the time of measurement (see Section 2.3.3). In addition, the density (or volume) of X may need to be determined experimentally to achieve the desired accuracy.

Before continuing with means of coping with air density, it is necessary to introduce one last complication regarding standard weights used in calibrating a balance. Materials used to manufacture weights rarely have the exact basis density, ρ_R. Some stainless steel alloys used in the manufacture of weights have a density closer to 7700 than to 8000 kg/m^3. At the time of manufacture, weights are adjusted to *apparent mass* versus 8000 kg/m^3. This means that the weight is made so that, in air of density 1.2 kg/m^3, it will have the same *effective* mass as a standard made of the 8000 basis alloy.

Weight manufacturers traditionally specify their products in terms of apparent mass simply because near sea level, the apparent mass of a weight standard is essentially equal to the reading obtained from a direct-reading balance, uncorrected for air buoyancy; for example, the apparent mass of X is very nearly I in (13). Apparent mass is also referred to as *conventional mass*. Note that Figure 1.4 also represents the difference between the apparent mass and the true mass of a substance.

To clarify these ideas, consider a 100-g weight of density 7770 kg/m^3 manufactured to perfection on the stainless steel basis. The properties of this weight are given in Table 1.1 along with the properties of a similarly perfect weight made of a material with the basis density, 8000 kg/m^3. Three important features seen in Table 1.1 are true in general: (1) The nominal mass equals the apparent mass for perfectly made weights; (2) The apparent mass equals the true mass for weights whose density equals the basis density of 8000 kg/m^3; and (3) Weights having the same apparent mass also have the same effective mass at standard laboratory conditions. This last feature allows the analyst to use the apparent mass and basis density (rather than the true mass and real density) to calculate effective mass in (11c) and all similar equations. Although this is only

Table 1.1   Example of Two 100-g Standard Weights of Different Density[a]

Density ρ (kg/m³)	Nominal Mass (g)	Apparent Mass[b] (g)	True Mass m (g)	Effective Mass[c] m^* (g)
7770	100	100.00000	100.00044	99.8500
8000	100	100.00000	100.00000	99.8500

[a]Each weight is adjusted perfectly to an apparent mass of 100 g with respect to the stainless steel basis.
[b]With respect to standards of density 8000 kg/m³ in air of density 1.2 kg/m³.
[c]In air of density 1.2 kg/m³.

an approximation if the air density is not exactly 1.2 kg/m³, weight specifications [47] require that it be a good approximation for laboratory conditions near sea level. It is safe to say that many users of standard weights and analytical balances are unaware that any such approximation has been made.

That a single weight can be described in terms of its true mass, two types of apparent mass, and effective mass is a source of confusion. Keep in mind that m in (1) is the *true* mass and this is usually the required result of analytical weighing.

When making buoyancy corrections, it is always correct to use the true mass of the standard weight and its true density [46]. It is more convenient, however, and often acceptably accurate to use the apparent value as long as 8000 kg/m³ is also used as the density (8390.9 kg/m³ for weights manufactured on the brass basis). An additional error is incurred by so doing if the air density ρ_a or the density of the standard ρ differ from the conventional values. This error has a magnitude of

$$m(1.2 - \rho_a)\left(\frac{1}{\rho} - \frac{1}{8000}\right),$$

where all densities are in kilograms per cubic meter [9]. Near sea level, this error is usually negligible for routine work. Air density decreases with altitude, however. Thus, laboratories at high elevations must be especially cautious in using an apparent mass convention. In any case, it is wrong to use the value of apparent mass in conjunction with the true density of a standard weight and vice versa.

Unlike the examples given in Table 1.1, the manufacture of actual weights cannot be perfect. Instead, standard weights are certified accurate, to within a specified tolerance [47]. (The tolerance is measured with respect to apparent rather than true mass.) If better accuracy than manufacturing tolerance is needed, then the weight must be calibrated so that its correction to nominal can be known. The cost of such a calibration may exceed that of the weight itself.

2.3.2   Mass Standards

The preceding discussion has assumed the existence of standards calibrated to represent mass in the Système International d'Unités (SI). The SI kilogram is defined by a platinum–iridium artifact that is in the custody of the International Bureau of Weights and Measures (BIPM, for Bureau International des Poids et Mesures) located in Sèvres, France [4]. Mass calibrations within the United States are traceable to the National Institute of Standards and Technology (NIST). In turn, NIST standards are calibrated periodically by the BIPM [48]. Traceability to the SI unit becomes less than routine only for uncertainties below about 0.5 ppm. The needs of most analysts can, therefore, be easily met. Many mass determinations in chemistry, though very precise, are differential; that is, the chemist is concerned with small changes in mass. It is only the final difference that must be known in SI units. Intermediate determinations can be made using uncalibrated (but stable) tares. When analytical balances are used for pycnometric or hydrostatic determinations of density, the SI unit is irrelevant since only linearity is required (see Section 3.3). Similarly, when *percentage by weight* is being determined, any unit of mass can be used. On the other hand, some determinations truly have a critical need for the SI unit. We have been involved in electrochemical measurements of the Faraday constant, a measurement that exemplifies such a need [49, 50].

Commercially produced analytical weights can be bought individually or in sets. The latter contain decades of weights—typically from 100 g to 1 mg. Four weights comprise each full decade. To cover a decade of mass between 10^n and 10^{n+1} kg, the four weights will have nominal values $(m_1, m_2, m_3, m_4) \times 10^n$. Most usually, the m_i take the values 5, 2, 2, 1 (although the 5, 3, 2, 1 set is available in the United States [47] and weight denominations built into commercial balances of the type shown in Figure 1.2 are found in bewildering variety [29]). The best modern analytical weights are made of austenitic stainless steel. This alloy structure is nonmagnetic, but measurable magnetic susceptibilities are usually present in actual material due to imperfections [51, 52].

One expects weight standards to retain their certified mass values through a long period of use. The most stable weights are made of single-piece construction. However, the cost of single-piece weights manufactured to close tolerance is high. Less expensive are two-piece weights in which the lifting knob is screwed onto the body. At the time of manufacture, the apparent mass of this weight can be adjusted easily by adding bits of material to the body cavity before screwing on the knob. Caution must be taken when using two-piece weights so that liquids are prevented from seeping under the knob and into the cavity. In addition, weights of any design must be protected from corrosive atmospheres [53], oils, and dust. Soiled weights must be cleaned with extreme care [7, 54].

2.3.3   Air Density

The air density in the balance must be known to correct the readings of apparent mass obtained to true mass. How accurately air density must be known depends on particular circumstances and can be estimated from (12). In general, ρ_a in a given laboratory depends on the barometric pressure, the ambient temperature, the relative humidity, and, to a small extent, the carbon dioxide level in the atmosphere. The International Committee for Weights and Measures (CIPM, for Comité International des Poids et Mesures) has recommended an equation of state for moist air [55]. This equation is derived from an analysis of the mixing of the constituent gases of air. The relation has been verified experimentally to a level of about 0.01% [56].

The recommended equation of state is cumbersome for everyday use. Jones [57] developed a simplified relation that satisfies most needs in that the uncertainty is dominated by typical errors in determining barometric pressure, temperature, and relative humidity. The relation developed by Jones is

$$\rho_a(\text{kg/m}^3) = \frac{0.0034848}{(t + 273.15)}(P - 0.003796 U \cdot e_s) \tag{14a}$$

$$e_s = 1.7526 \times 10^{11} \exp[-5315.56/(273.15 + t)] \tag{14b}$$

where P is the ambient barometric pressure measured in pascals, t is the ambient temperature in degrees Celsius, and U is the ambient relative humidity in percentages. The calculated quantity e_s is the saturation vapor pressure (in pascals) at temperature t of water in air. The correction for typical changes in carbon dioxide level about the assumed value is so small it is negligible under normal circumstances.

Moist air is approximately an ideal gas. This, coupled with the fact that the relative humidity correction is a second-order effect, allows one to calculate a quick estimate of ρ_a:

$$\rho_a \approx 1.20[(293.15/(273.15 + t)](P/101325) \tag{15}$$

This relation is accurate to better than 1% under laboratory conditions. The pressure P in (14) and (15) is the true barometric pressure, uncorrected to sea level.

The temperature in an analytical laboratory is usually regulated to within a degree or two. Therefore, changes in barometric pressure determine the range of air densities in a particular laboratory. By this reasoning, the air density in a given laboratory should rarely differ from its average value by more than 3%. For many mass determinations, even though a buoyancy correction must be applied, a 3% error in the air buoyancy is not significant. In these cases, it is most convenient to assume an average value for the laboratory. Pontius [58] has produced a table of these values for various locations in the United States.

Many analysts have found it more convenient to measure the air density directly. Suppose m_X is the mass of an object X. This mass can be determined by weighing in a vacuum (i.e., without reliance on an equation of state for air) or at ambient air density the value of which is determined from (14). If the average density of X is lower than other items of interest, then subsequent measurements of m_X^* can be used to infer changes in air density without further recourse to an equation of state. This idea was popularized in the United States by Baxter [59], who manufactured a glass globe for that purpose. Many other schemes have also been used [23, 60–63]. It is convenient to construct a companion weight Y whose density is much greater than that of X, but which matches X closely in mass and surface area. Then

$$\rho_a = -[m_X^* - m_Y^* - (m_X - m_Y)]/(V_X - V_Y) \tag{16}$$

The differences in true mass and volume between X and Y must be known in advance by separate experiment.

A special case of this technique is the weighing of low-density vessels against a tare. As an example, consider a thin-walled, evacuated aluminum gas cylinder X that is to be filled quantitatively with several gases. Buoyancy effects are potentially significant because the volume of the cylinder is relatively large (and difficult to measure). However, if a second, identical cylinder Y is available, the mass m of a gas added to X can be determined as follows:

1. Find the initial difference in mass between X and Y. This will involve a buoyancy correction $\rho_{a1} \cdot \Delta V$, where ΔV is the small difference in volume between X and Y. The comparison can be made by transposition weighing on a two-pan balance or by Borda weighing on a single-pan balance. For simplicity, let us say a one-pan electronic balance is used and the uncorrected difference is I_1.

2. Add a gas charge of mass m to X and again measure the difference between X and Y. The mass m can be computed from these two measurements in obvious fashion. Buoyancy corrections will contain a term $(\rho_{a1} - \rho_{a2}) \cdot \Delta V$, which may be negligibly small, and a term $(I_2 - I_1) \cdot \rho_{a2}/8000$, which is about 0.015% of m. In the example given, one must take care that the change in the volume of X is not significant when pressurized with the added gas. If a tare had not been used, the first term of the correction would grow by a factor of nearly $V_X/\Delta V$. Some chemists prefer transposition weighing on a two-pan balance to Borda weighing on a single-pan balance when tares are involved [64]. There seems, however, to be no great advantage to transposition weighing except that the sensibility of an equal-arm balance is improved by a factor of 2 when transposition rather than Borda weighing is used.

2.4   Accurate Weighing on Analytical Balances

The previous sections discussed ways to correct balance readings for the effects of air buoyancy to calculate the mass of a substance being weighed. Accurate results, however, may also depend on the elimination of perturbing influences. Among these, the most important are inadequate thermal equilibrium, surface instabilities, electrostatic and magnetic effects, vibration, and tilt. Since these considerations become even more important for microbalances, the general reference edited by Czanderna and Wolsky [65] provides a useful introduction.

2.4.1 Thermal Equilibrium

If an object whose mass is to be determined is not in thermal equilibrium with the analytical balance, temperature gradients will generate air currents in the balance case. These may perturb the balance in a systematic way leading to only a small deterioration of the balance standard deviation but to a large error in the reading. Such problems were examined by Blade [66] who demonstrated that the magnitude of the unwanted effect changes with details in geometry of the weighing chamber and the balance pan. A more recent study by Schoonover and Taylor [67] called attention to the geometry and thermal inertia of the objects being weighed. These authors demonstrate that the thermal effects are reproducible on a given balance and so can be made to cancel by proper experimental design.

Poor thermal conditions may also lead to drifts in the balance zero. Such problems are extremely dependent on balance design, but can be qualitatively understood [22, 68].

2.4.2 Surface Adsorption and Evaporation

Generally, the surfaces of objects in air are covered with some amount of adsorbed moisture. Yoshimori [64] discussed methods of desiccating standard reference materials and then determining their mass. He points out that the surface of the weighing container may be unstable to changes in relative humidity, and he gives measurements for several common glasses used. Of the glasses tested, fused silica had the smallest amount of adsorbed surface water, while soft glass had the greatest. Kochsiek [69] gives similar measurements for selected metals. The sample container should be weighed against an identical tare so that surface effects (and buoyant effects) will tend to cancel. If using a single-pan balance, one must determine the difference in mass of the sample bottle and tare by successive readings of the two objects (Borda weighing).

The weighing of filters can pose serious problems with respect to water adsorption. The difficulties arise because of the large surface-to-volume ratio of filters but can be mitigated by choosing filters of poly(vinyl chloride) [70], poly(tetrafluoroethylene), or other nonhygroscopic material.

Volatile substances must be weighed in closed containers. Special syringes have been devised for the accurate weighing to drops of aqueous solutions [71, 72].

2.4.3 Electrostatic and Magnetic Forces

Electrostatic forces may become important in analytical weighing. Although metal surfaces of the balance and its case can be brought to the same electrical potential (most conveniently, that of the earth), nonmetallic objects may present a serious problem. Electrostatic problems are usually anticorrelated with difficulties arising from adsorbed surface water. If such water is not present, a dielectric material will retain static charges more easily. Simple commercial devices can be used to detect the presence of static charges and their magnitude.†

There are three common remedies for electrostatic forces:

1. The relative humidity can be raised artificially within the balance case.
2. The dielectric material can be surrounded by a metal container. Since only the surface is important, the container can be a thin foil. The container forms a Faraday cage so that, if grounded, its surface potential will be at ground regardless of the electric charge on the dielectric material within. The balance case usually has glass windows. These are sometimes coated with a transparent, slightly conductive film, such as tin oxide.
3. The sample can be passed near a radioactive source so that ionized air will discharge the surface of the dielectric. A ^{210}Po alpha source has been found effective [73], although complete discharge may require additional measures [74]. Use of a tesla coil has been advocated in the past, but this "remedy" may actually have the opposite effect [73].

One should be aware that unscreened electric charges external to the balance produce forces on the metallic members of the balance.

Magnetic materials to be weighed may also present a problem. This was mentioned earlier in regard to electromagnetic servocontrolled balances, but problems are often seen in mechanical balances as well. If magnetic interactions are confined to the object being weighed, the pan, and the suspension, there can be no net torque about the main pivot point and hence no problem. Problems do arise, however, when there is a magnetic interaction between the fixed parts of the balance case and the object being weighed (note also that when the object being weighed is not on the balance pan, it is included in the fixed surroundings). In electromagnetic servocontrolled balances, a powerful permanent magnet is a fixed part. This magnet must, therefore, be well shielded by the manufacturer to prevent interactions with a magnetic object on the pan.

The existence of a problem can often be inferred by changing the orientation of the object being weighed. If a rotation of the object on the balance pan, for instance, causes a change in the balance reading, there is undoubtedly a problem. A reorientation of the balance with respect to the room can help determine if there are magnetic interactions between the free balance parts and magnetic materials within the room. Finally, it should be remembered that uniform magnetic fields do not produce a net force on a magnetic material. Thus the earth's magnetic field within a laboratory is seldom a source of concern.

---

†For example, a Precision Electrostatic Locator static detector, ACL, Inc., Elk Grove Village, IL.

2.4.4   Tilt and Vibration

Precise balances may be sensitive to tilts in their support [75, 76]. Because balance pivots have finite effective radii, it is possible to adjust a balance's sensitivity so that the reading is unaffected by tilt. Such an adjustment has been termed *autostatic* [75]. Unfortunately, the autostatic condition is usually at an inconveniently high sensitivity for unservoed balances. Servocontrolled balances can be constructed more easily to be tilt-insensitive [77], but this is not always done. Balances sensitive to tilt must, therefore, be mounted solidly. Basement laboratories are usually preferred since these are more immune to building sway and floor sag. A building usually undergoes slight diurnal changes in level due to differential heating of its exterior. These may affect sensitive measurements that require long-term zero stability [76, 78]. Balances of the type shown in Figure 1.3 are generally much less tilt sensitive than the balance shown in Figure 1.2. Building vibrations may also be troublesome. Excessive vibration can damage the bearings on knife-edge balances and may interfere with the operation of the balances discussed so far. If necessary, vibration isolation must be included in the balance mount, but not at the expense of solidity of the support. A number of schemes for isolating mechanical balances from vibration have been reviewed by Macinante and Waldersee [79]. Similar methods are useful for servocontrolled microbalances [80].

2.5   Microweighing

Mass determinations to better than $1\,\mu g$ and determinations in vacuum or controlled atmospheres constitute a special analytical area. Transducers that operate in this region are referred to as *microbalances*. Some authors use the term ultramicrobalance to distinguish devices with sensibilities smaller than $0.1\,\mu g$. Electrostatic forces, gas motion, radiation pressure, and other perturbing influences may produce effects many times the sensibility of the force transducers involved. Since many of these problems are also seen in critical measurements involving analytical balances, the reader interested in highly accurate analytical weighing should be familiar with experience found using microbalances. Microbalances are used in a very wide variety of applications including thermogravimetry, density measurements, and sorption studies. The measurement of adsorbed gas can be used to determine the surface area of a solid by the Brunauer–Emmett–Teller adsorption isotherm (BET) or other methods [81, 82]. The mass of moisture adsorbed by a hygroscopic material provides a primary measurement of humidity [83]. Advances in microweighing are reported in the *Proceedings of the Conference on Vacuum Microbalance Techniques*, which are published in *Thermochimica Acta*.

We first review the force transducers used in microweighing. We then discuss the quartz-crystal microbalance (QCM) and other transducers that respond directly to mass. Escoubes and co-workers [84] compiled a list of most commercially available microbalances and surveyed instruments suitable for thermogravimetry.

2.5.1   Force Transducers

A range of sensitive force transducers has been developed for microweighing. As early as 1920, Pettersson [85] reported a precision of 0.1 ng using a beam balance carrying a load of 100 mg. The time to perfect and maintain such a fragile instrument severely limits its utility, however. In more recent times, many more tractable balances have been developed. Sensibilities of nanograms are attainable with some of these instruments. All such balances are used in a differential mode, that is, to measure very small mass changes over time.

MAGNETIC SUSPENSION BALANCE

In 1955, Beams and co-workers [86] reported a balance based on a fer-romagnetic "pan" that is held in suspension below an electromagnet. The position of the pan is automatically maintained through a servocontrol circuit. With proper design, the generated force is a linear function of current in the electromagnet. Sensibilities of 0.05 ng were reported. This design strategy is useful when objects must be weighed in corrosive atmospheres. Magnetic suspension also has important application in density determinations (see Section 3.3). Gast [89] developed many interesting and useful microbalances† in which the pan is magnetically suspended from a balance beam [87, 88] (see also Section 3.3, "Method of Gast," and Ref. [89]).

BEAM BALANCE

The form of modern microbalances [90] that rely on a pivoted beam differs somewhat from the analytical balances encountered previously. The micro-balance may have a single pan or two pans. These are typically suspended below the beam, which is servocontrolled to remain horizontal. Most successful microbalances abandon the use of knives as pivots in favor of a torsion pivot or flexure. Long-term zero stability is usually an important consideration in these balances so that differential thermal expansion of the beam must be minimized. Fused silica or fused quartz are often used because of their low coefficient of thermal expansion. (We reserve the term *quartz* for the crystalline form of silica.) Wolsky and co-workers [91] have pointed out, however, that low thermal expansion is only one of several considerations. Also important are the thermal conductivity and, for vacuum applications, emissivity of the beam.

Calibration of the servocurrent in terms of mass is most often accomplished by small mass standards [90]. Eschbach and co-workers [92] have demon-strated a stability of about 1 μg over a period of 10 years in a 1-mg weight used to calibrate a vacuum microbalance. Balances housed in a controlled gaseous atmosphere can be calibrated by observing differences in effective mass as a function of the gas density of two objects of different volume [93]. By a rearrangement of (16),

$$(m_X^* - m_Y^*) = -\rho_g(V_X - V_Y) + (m_X - m_Y) \tag{17a}$$

†Manufactured commercially by Sartorius GmbH, Göttingen, Germany.

where ρ_g is the known density of the gas. A change $\Delta\rho_g$ in gas density will lead to a change in effective mass difference $\Delta(m_X^* - m_Y^*)$, which can be related to the balance sensitivity through (8):

$$\Delta(m_X^* - m_Y^*) = -\Delta\rho_g(V_X - V_Y) = (k_s/g) \cdot \Delta f(I) \qquad (17b)$$

where $\Delta f(I)$ is the resulting change in balance output.

2.5.2 Mass Transducers

QUARTZ-CRYSTAL MICROBALANCE

The most important of the mass transducers is the QCM [94]. The QCM exploits the piezoelectric effect in crystalline quartz. The same effect has long been used as a way to make stable oscillators for radios and, more recently, for computers and wristwatches. In the early days of use, the resonant frequency of crystal oscillators was adjusted by marking one face of the crystal with a graphite pencil [95]. Thus the use of these crystals as mass transducers was foreshadowed. However, it was not until Sauerbrey [96] published a theoretical analysis of the shift in resonance frequency with mass that the QCM became a reality.

Sauerbray considered a quartz crystal undergoing forced oscillation in its fundamental shear mode. The oscillation is driven by thin metal electrodes attached to opposite crystal faces. When the crystal is oscillating in a shear mode, these electrodes move sideways; that is, the thickness of the crystal does not change. If d is the thickness of such a crystal with resonant frequency f_0, then a crystal whose thickness is larger by an amount Δd will resonate at a frequency shifted by Δf according to the relation [96]

$$\Delta f/f_0 = -\Delta d/d \qquad (18)$$

The negative sign indicates that the frequency is reduced as the thickness increases. The right side of (18) can be expressed in terms of the mass change of the quartz crystal:

$$\Delta f/f_0 = -\Delta m_Q/(\rho_Q A d) \qquad (19)$$

where Δm_Q is the mass of quartz in the thickness Δd, ρ_Q is the density of quartz (2650 kg/m^3), and A is the surface area of the electrode. Sauerbray [96] then assumed that if Δd is due to the addition of a uniform solid film of mass Δm, then

$$\Delta f/f_0 = -\Delta m/(\rho_Q A d) \qquad (20)$$

independent of material as long as $\Delta d/d \ll 1$. Since $N \equiv d \cdot f_0 = 1.67 \times 10^3$ Hz$\cdot$m for quartz crystals cut in the so-called AT orientation (20) is more usefully written

$$\Delta m = -N \cdot A \cdot \rho_Q \cdot (\Delta f / f_0^2) \tag{21}$$

or

$$\Delta m = -4.42 \times 10^6 A \cdot (\Delta f / f_0^2) \tag{22}$$

Note that gravitational acceleration g is irrelevant to the preceding analysis. Resonant frequencies of about 10 MHz are convenient to fabricate and frequency resolution of 1 Hz is easily attainable. Thus, a film of mass 5 ng/cm^2 that covers the crystal face uniformly is resolved readily. Higher resolution can be achieved [95]. A more sophisticated theory that takes account of the acoustic impedance of the added material provides quantitative corrections for thick films [97]. Resonance becomes more difficult to maintain as the crystal is increasingly loaded.

The AT-cut crystals are usually preferred because near 20°C they are only affected weakly by temperature variations. At other temperatures, two matched crystals can be used—one of them serving only to compensate for temperature variations. alternatively, two crystals of different cuts can be used for simultaneous measurement of mass and other variables [98].

The QCM has an obvious use in measuring the thickness of vacuum-deposited metallic films. Eschbach and co-workers [92] compare the relative merits of the QCM and a beam microbalance for this purpose. In general, the QCM is rugged and responds quickly. Many uses are found in surface science. Among studies reviewed by Levenson [99] are QCM-based examinations of chemisorption, physical adsorption, and outgassing. By coating one face of the QCM with a film that adsorbs a particular chemical species selectively, one creates a detector for trace amounts of gases. Numerous coatings that have been developed are reviewed by Guilbault [100]. Similarly, an adhesive-coated crystal can be used to detect the mass concentration of aerosols as a function of time. Devices designed for this purpose have detected the impact of a single particle of about 0.01 ng in mass [101]. In the case of particulates, however, limitations of the QCM are important [102]. The chief problems are (1) The theory of operation assumes mass is distributed uniformly over the electrode face. This may not be fulfilled in the case of particle detection, and (2) The sensitivity of the QCM is a function of particle dimension. Advantages are that the QCM gives measurements in real time, is rugged enough for field use, and is relatively inexpensive.

Since the QCM will not resonate if the thickness of a deposited film becomes large, it is somewhat surprising that operation is observed when a QCM is totally immersed in a liquid [103]. Bruckenstein and Shay [104] offer an explanation based on a plausibility argument: Since the QCM operates in a shear mode, it only interacts with a boundary layer of the liquid. From dimensional analysis, the thickness δ of the boundary layer is approximately

$$\delta = (v/f_L)^{1/2} \tag{23}$$

where v is the kinematic viscosity of the liquid and f_L is the oscillator frequency in the liquid. The mass of liquid m_L in the boundary layer is then

$$m_L = A \cdot (\eta_L \rho_L / f_L)^{1/2} \tag{24}$$

where η_L is the absolute viscosity of the liquid and ρ_L is the liquid density. Thus the frequency shift due to the fluid does not depend on fluid height but will depend on viscosity and density, both of which are functions of temperature. A more rigorously derived theory gives qualitatively similar results [105].

By using one electrode of a QCM as the active electrode in an electrolytic cell, Bruckenstein and Shay [104] create what they refer to as an electrolytic quartz crystal microbalance (EQCM). The mass sensitivity of their device was tested by electrolytic deposition of silver and found to agree with Sauerbray's theory for mass deposited in vacuum. The use of EQCMs and related devices in the study of both chemical and biological interfaces has been reviewed by Ward and Buttry [106].

OTHER MASS TRANSDUCERS
An ideal, massless spring with one end fixed and a mass attached to the other is, of course, a force transducer. In a gravitational field

$$m^*g = F = k_t \cdot z$$

where F is the force of the gravitational pull on the suspended mass and z is the spring extension when the mass is added. The undamped spring-mass system will have a resonant frequency

$$f_0 = (1/2\pi)(k_t/m)^{1/2} \tag{25}$$

The mass m appears in (25), but the gravitational acceleration g does not. Thus, by monitoring changes in the resonant frequency of the spring, one can determine changes in m, independent of gravity

$$m = k_t/(4\pi^2 f_0^2) \tag{26a}$$

or,

$$m = k_t'/f_0^2 \tag{26b}$$

in the formalism of (2). An added mass Δm can be determined from the new frequency of oscillation f through the relation

$$\Delta m = k'_t(1/f^2 - 1/f_0^2) \tag{27}$$

The two different schemes—the first gravimetric and the second inertial—are illustrated by a quartz fiber that is attached at one end and free at the other (although such a fiber spring cannot be considered massless in microbalance applications). If the fiber is placed horizontally in a gravitational field, its end will deflect as mass is added. Such cantilever microbalances typically have an imprecision of 1% or lower on samples of about 1 μg [107, 108]. Their observed sensitivity agrees with theory to about 10%. A similar fiber, when operated in the inertial (oscillation) mode, has been used to detect masses from about 0.1 μg to 1 ng [109]. A rugged device based on similar principles is said to follow changes as small as 20 μg in the mass of rigid samples of about 2 g [110].

The tapered element oscillating microbalance (TEOM†) is a commercially available family of devices based on the frequency of oscillation of a tapered tube. The tube is fixed at its wider end and holds a replaceable filter mounted perpendicular to the narrower, vibrating end. Ambient gas is drawn through the filter, trapping particulates. The unloaded filter oscillates at about 200 Hz. Increases in mass lead to decreases in frequency according to (27). Tapered elements can be fabricated to have a range of sensibilities, each with a LSR of 10^5 or more. The TEOM® instrument, like the QCM is suitable for field measurements in real time [111, 112].

Both the QCM and the TEOM instruments can operate in the absence of gravity and in both technologies the sensing surface oscillates side to side (the QCM at megaherz frequencies and the TEOM instrument at humdreds of herz). In both technologies, associated electronics maintain the oscillations and sense their frequency. But the QCM is based on a piezoelectric phenomenon described by (21), while the TEOM instrument is based on a purely mechanical resonance as described by (25).

2.5.3 Other Methods for Determining Small Mass

There are at least two applications for the mass determination of small objects without appeal to a microbalance: (1) in manufacturing standards for use in calibration of specialized microbalances and (2) when the mass to be determined is smaller than the sensibility of available devices. An alternative solution to weighing by transducer relies on dimensional and density measurements. By rearranging (1), the mass of an object X is given by

$$m = \rho \cdot V \tag{28}$$

---

†TEOM is a registered trademark of Rupprecht & Patashnick Company Inc., Voorheesville, New York.

We have noted in Section 1 that the density of small objects can, in principle, be determined to high precision. If the shape of X is regular enough to be measured—either by micrometer or microscope—then the mass of X can be found without reliance on a balance [90]. This technique has been used to find the mass of microspheres (ranging from 60 to 2 ng), which were used in turn to calibrate the sensitivity of a microbalance [109]. Limitations in accuracy can be calculated in obvious fashion by error propagation through (28) [90, 109].

A variation of this technique makes use of a transmission electron microscope (TEM). A TEM micrograph provides simultaneous information on the dimensions and mass per unit area of objects. By including a set of well-characterized objects in the micrograph (such as a range of latex spheres), Bahr and Zeitler[113], were able to determine the dry mass of many biological specimens. Their method was applied successfully to discrete objects that ranged in mass from 10^{-11} to 10^{-18} g. Errors of less than 10% are feasible. Simplifications and improvements in the calibration procedure have been proposed [114, 115].

In favorable cases, optical fluorescence can be used to determine the mass of a particle for use as a calibration standard. The mass of quinine hydrobromide crystals of about 0.1 μg was determined in this way [107].

At the atomic level, molar masses are connected to the SI kilogram through the Avogadro constant. An interesting review of the history and possible future of this connection is given by Deslattes [116].

3   DENSITY DETERMINATIONS

3.1   General Considerations

Although Section 2 was devoted to mass determinations, density was introduced in two places. First, determinations of mass from measurements of apparent mass required corrections involving the ratio of air density to the density of the objects being weighed. Second, at the end of the section, we showed how the mass of an object could be determined by prior knowledge of its density and a computation of volume from dimensional measurements.

In the SI, mass and length are so-called *base units*, which are defined through primary standards. It follows that, since the units of density are (mass)·(length)$^{-3}$, there can be no independent standard of density. Nevertheless, most density determinations do rely on secondary or reference standards. These reference standards (such as distilled water and mercury) have values of density that were determined, at high accuracy, relative to the SI definitions of mass and length. In the sequel we will refer to density *standards* with this understanding.

The reciprocal of density is referred to as specific volume and is denoted by v, where $v = 1/\rho$.

The density of a substance divided by its molar mass M is referred to as the molar density and is denoted by ρ_m. It has units of moles per volume. (*Note*: the molar mass of ^{12}C is defined as $0.012\,kg/mol$.)

The reciprocal of molar density is the molar volume, denoted as v_m, where $v_m = 1/\rho_m = M/\rho$.

Authors may refer simply to density when they mean molar density, number density, or optical density (see Section 3.3). We refer simply to density when we intend mass density as defined by (1). When in doubt, check the dimensions of the quantity given. The term *densitometer* most often refers to a device that measures optical density. For this reason, we will use the term *densimeter* to refer to a device that measures mass density.

3.1.1   The Liter

Reported values of density and related quantities have often been given in units of grams per milliliter. Since 1964, the liter has been defined as

$$1\,L = 1000\,cm^3 = 10^{-3}\,m^3 \quad (\text{exact})$$

so that

$$1\,g/mL = 1\,g/cm^3 = 1000\,kg/m^3 \quad (\text{exact})$$

Between the beginning of the SI in 1901 and 1964, however, the liter was defined as the volume occupied by "one kilogram of pure water, at maximum density and under normal atmospheric pressure." Based on careful experiments carried out prior to 1901 [117], the liter was thus measured to be

$$1\,L = 1000.027\,cm^3$$

so that

$$1\ g/mL = 0.999973\ g/cm^3$$

In 1927, subsequent reanalysis of the same experimental data led to a correction upward by 1 ppm:

$$1\,L = 1000.028\,cm^3$$

so that

$$1\ g/mL = 0.999972\ g/cm^3$$

The relative uncertainty in the conversion factor has been assessed at 4 ppm [118] (see Section 3.2.1 below on water as a standard). Needless to say, these changes in definition have caused considerable confusion in the interpretation of precise data. In what follows, we take the suggestion of the General Conference of Weights and Measures (CGPM) and avoid the use of liters for reporting highly accurate densities.

3.1.2 Relative Density (Specific Gravity)

The importance of water as a readily obtainable density standard leads to the utility of relative densities. The relative density of a substance is the ratio of its density (in any system of units) to the density of water (in the same system of units). Relative density is a dimensionless number. For precise work, one must recognize that densities are a function of temperature. Thus, the relative density (formerly known as specific gravity) must specify the temperature of both the sample X and the water w used as reference:

$$d_{t'}^{t} = \frac{\rho(t)_X}{\rho(t')_w}$$

where t is the temperature at which ρ_X is known and t' is the temperature of the water to which ρ_X is referred. The most commonly used conventions are d_4^{20}, d_{25}^{25}, and d_{20}^{20}, where the superscripts refer to the temperature of the sample and the subscripts refer to the temperature of water, both in degrees Celsius.

Relative densities can be converted to absolute densities, that is, densities in the SI units of kilograms per cubic meter, by multiplying by the density of water at its reference temperature.

3.1.3 Temperature

The volume of a given mass of material, and hence its density, is a function of temperature. A volumetric (or cubic) coefficient of expansion α can be defined as

$$\alpha = \frac{1}{V}\left(\frac{\partial V}{\partial t}\right)_P = -\frac{1}{\rho}\left(\frac{\partial \rho}{\partial t}\right)_P \tag{29a}$$

under conditions of constant pressure. Thus α can be used to estimate the relative change in volume for a 1°C increase in temperature. For solids, the symbol α_L refers to the linear coefficient of expansion. For a solid, the volumetric coefficient defined by (29a) is sometimes denoted by β. For isotropic solids

$$\beta = 3\alpha_L$$

We will adopt the usage of α as a generalized *volume* coefficient for both fluids and solids.

The use of α to describe the thermal expansion of fluids and solids is convenient for theorists but somewhat awkward for experimentalists. Rather, a coefficient α_l defined by

$$\alpha_l = \frac{1}{V_0}\left(\frac{\Delta V}{\Delta t}\right)_P \qquad (29b)$$

is often employed by the latter group. In this definition, V_0 is the volume at a particular temperature (0°C unless otherwise stated). An ideal gas, for example, has $\alpha = 1/T$ and $\alpha_l = 1/T_0$, where T and T_0 are absolute temperatures (in units of kelvin).

If α is constant and thermal expansion is relatively small, the density of a substance as a function of temperature can be found directly by integrating (29a) and recalling the fundamental relation (1):

$$\rho = \rho_0/[1 + \alpha \cdot (t - t_0)] \qquad (30)$$

where the density is known to be ρ_0 at t_0. More usually, a wide temperature range is needed and α cannot be assumed constant with respect to temperature. Then $\alpha \cdot (t - t_0)$ must be replaced by a power series in $(t - t_0)$ as in (45) below.

To obtain the density of a substance at different temperatures, there are two experimental approaches: (1) The density of the substance can be determined at one temperature. Changes in a known volume of the substance can then be measured as a function of temperature by using a volume *dilatometer*. These measurements can be related to changes in density. (2) The density of the substance can be measured throughout the range of interest. In either case, one can fit the measured densities by a polynomial of sufficiently high order in the temperature. The polynomial form is useful for computer storage or for computation of α. Tabulated values of $\rho(t)$ were essential before easy access to computers and are still useful for quick reference.

The density of a material changes discontinuously when it undergoes a so-called first-order change of phase (such as solid to liquid). The lower density of ice with respect to water at 0°C is a familiar example. More subtle changes are often seen in polymers, composite materials, and biochemical systems. Temperature scanning densitometry (TSD) can be used to study these effects as, for example, in [119].

The fact that $\alpha = 0$ for water near 4°C adds to its usefulness as a density standard. From (29a), a coefficient of zero implies an extreme value in the density—in the case of water, a maximum. Near room temperature, the thermal expansion coefficient for water is approximately 3×10^{-4}/°C. For organic liquids, α is considerably larger, often by a factor of 5. Near boiling, α becomes still higher. For an ideal gas, $\alpha = 3.66 \times 10^{-3}$/°C at 0°C. The thermal expansion of mercury (exploited in the design of mercury thermometers and some dilatometers) is about 1.8×10^{-4}/°C. Fused quartz is important in density work because its value of α near room temperature is low (about 1.5×10^{-6}/°C) and is well characterized to high temperatures [120]. Volumetric labware is usually made of borosilicate glass ($\alpha \approx 10^{-5}/°C$).

Special alloys, glasses, and ceramics can be manufactured to have $\alpha = 0$ at some particular temperature (usually room temperature). These can be particularly useful for some applications, but one should be aware that the thermal expansion will not be zero throughout a large temperature range. As an example, in [121] the density of a low-expansion glass is measured over a wide range of temperatures.

The appearance of a density maximum at a finite temperature implies that in a region immediately below this temperature the thermal expansion is "anomalous"—the volume of the sample increases with decreasing temperature. The change in the temperature of maximum density (TMD) for water as a result of interactions with small amounts of added solute is an indication of the intermolecular forces at work [122]. So-called "heavy" water (D_2O) appears to have its TMD near 11°C [123]. Anomalous expansion has been studied in liquid $Hg_{1-x}Cd_xTe$ [124].

3.1.4 Pressure

Materials change their volume in response to applied pressure. The isothermal compressibility κ_T of a substance is given by

$$\kappa_T = -\frac{1}{V}\left(\frac{\partial V}{\partial P}\right)_T = \frac{1}{\rho}\left(\frac{\partial \rho}{\partial P}\right)_T \tag{31}$$

where, as the name implies, the relations apply at constant temperature. Liquids and solids usually exhibit very small isothermal compressibilities so that κ_T can generally be taken to be small and constant between vacuum and several atmospheres pressure. For gases, the term *compressibility factor* is a dimensionless function Z that is a measure of how closely the equation of state approaches the ideal gas equation, as later in (76). The bulk modulus of a material is the reciprocal of κ_T.

Among liquids at room temperature, mercury has an unusually small compressibility ($4 \times 10^{-11}/Pa$; 101,325 Pa = 1 standard atmosphere). The compressibility of water is $4.60 \times 10^{-10}/Pa$ at 20°C [125]. Some organic liquids may have compressibilities four times that of water. Since changes in barometric pressure at a given location may amount to 5 kPa and average barometric pressure falls by more than 10% for each kilometer above sea level, compressibility may become a consideration in high-precision work. The design of experimental apparatus can produce a slight compression of the liquid being studied, for example, in some types of mixing dilatometers [126]. Unless otherwise stated, all pressure-dependent values given in this chapter are assumed to be at standard atmospheric pressure.

Gases (usually air) dissolved in a liquid can greatly increase the compressibility near ambient pressures. Indeed, such problems seem to be present in some measurements of κ_T for water [127].

3.1.5   Impurities

The purity of a substance will, of course, affect its density. Thus, density measurements have been used since ancient times to assay the purity of precious metals. On a more mundane level, chemical impurities, rather than instrumental inadequacies, often limit the accuracy obtainable for the density determination of a given substance. Water is a common impurity in many organic liquids. If a substance (1) and a substance (0) are considered as an ensemble, but not allowed to mix, the average density ρ of the two components taken together is easily calculated: let w_1 be the mass fraction of 1 ($w_1 = m_1/m$) and let w_0 be the mass fraction of 0 ($w_0 = m_0/m = 1 - w_1$). Then,

$$\rho = (m_1 + m_0)/(V_1 + V_0)$$

and

$$\rho = \frac{\rho_1 \rho_0}{\rho_1 + w_1(\rho_0 - \rho_1)} \quad \text{(no mixing)} \tag{32}$$

If the two materials are allowed to mix, generally ρ will change slightly from what is computed in (32) due to interactions between solute and solvent. The difference between ρ calculated from (32) and the actual density on mixing gives useful information about the solute–solvent reactions. This is an important use of density determinations and is discussed in more detail later. However, if (1) is an impurity contained in (0); that is, $w_1 \ll 1$, (32) can provide a good estimate of the effect on ρ.

As an example, let us estimate how much gold (1) is necessary to contaminate a pure sample of mercury (0) to the extent that the density of the mixture will increase by 0.01%: $\rho_1 = 19{,}300 \text{ kg/m}^3$, $\rho_0 = 13{,}500 \text{ kg/m}^3$, $\rho/\rho_0 = 1.0001$. Rearranging (32) to solve for w_1 we find $w_1 = 0.00033$, or 0.033% by mass [128].

GASES DISSOLVED IN LIQUIDS

Gases, such as air, dissolved in liquids will affect the density. For organic liquids, the difference in density when saturated by air or when free of air can be large (0.01%) under normal laboratory conditions. The change in density between degassed and air-saturated water amounts only to several parts per million at room temperature. A review of the extant measurements [129] shows considerable systematic differences among laboratories, indicating the difficulty in preparing air-free water. The usual technique of deaerating a liquid sample involves lowering the atmospheric pressure above its surface. This technique may cause appreciable changes in the density of solutions, however. When accuracy is critical, single-component liquids are vacuum distilled into their container (although this technique can lead to a slight isotopic segregation if the entire sample is not distilled). The solubility of air in water decreases with temperature. Thus water can be deaerated by boiling.

POROSITY IN SOLIDS

As will gas dissolved in a liquid, porosity of a solid will lower the measured bulk density from the intrinsic density of the material. The measured density can, in fact, be seriously in error if a porous sample is mistakenly assumed to be nonporous. On the other hand, the effect of porosity on the density of solids is used as an analytical tool to determine the ratio of pore volume to bulk (that is, total) volume.

Assume a sample has mass m, intrinsic density ρ_0, and a measured bulk density of ρ. The total volume of pores in the sample V_p can be derived directly from the definition of density

$$\rho = m \left/ \left(V_p + \frac{m}{\rho_0} \right) \right.$$

where the mass of evacuated pores is zero. Solving for pore volume

$$V_p = m \left(\frac{1}{\rho} - \frac{1}{\rho_0} \right)$$

or

$$\frac{V_p}{V} = 1 - \frac{\rho}{\rho_0}$$

where V is the bulk volume of the sample. Gas pycnometry, liquid pycnometry, volumetry, ultrasonic velocity measurements, and hydrostatic weighing are used in the determination of porosity ratios.

ISOTOPE EFFECTS

The molar volume of a substance is nearly independent of the isotopic abundances of the constituent atoms. As an example, at 25°C the molar volume of distilled ocean water is 18.069 cm^3/mol, while that of D_2O with the same abundances of oxygen isotopes is 18.133 cm^3/mol [130]. Thus, while the molar mass differs by 11%, the molar volume differs by only 0.4%. (Isotope effects in water will also be discussed in Section 3.2.1.). As a second example, the molar volume of single-crystal silicon has been found constant for various ratios of the three naturally occurring silicon isotopes [116]. The determination of sample density can, therefore, be a sensitive measure of the isotopic mixture in the sample. Virtually all methods capable of high relative precision in density have been used in the determination of isotopic abundance.

The discovery of stable isotopes and, as a consequence, the existence of heavy water led to development of sensitive density techniques for determination of D_2O fractions in natural waters (e.g., 131). The importance of heavy water to development of nuclear energy during and subsequent to World War II added to the assortment of analytical methods based on density variations [132]. These techniques rely on the observation that the H_2O-D_2O system exhibits near-ideal mixing (see Table 1.2 below) despite the difference in molar volume of the pure substances. (Ideal mixing in this case means that the volumes of H_2O and D_2O are exactly additive when a mixture is formed.) One difficulty in determining the deuterium content of water from its density is that there are also three isotopes of oxygen whose relative abundances may vary in water samples. The correction for oxygen isotopes is made by auxiliary methods, for instance by oxygen exchange with carbon dioxide followed by a relatively simple mass spectrographic technique [133, 134]. Once the oxygen difficulty has been attended to, one can show that the mole fraction x of deuterium in a sample of water is

$$x = A(\rho - \rho_0)/[1 - B(\rho - \rho_0)] \tag{33}$$

Table 1.2 Relative Volume Change (in percent) at 25°C Upon Mixing Equimolar Quantities of Liquids[a]

Components	$100(V^E/V_0)_{x=0.5}$
Water, sulfuric acid	-8.4
Water, methanol	-3.3
Water, 2-propanone	-3.1
Water, acetic acid	-3.0
Water, 2-propanol	-1.8
Water, acetonitrile	-1.4
Chloroform, 2-propanone	-0.20
n-Hexane, n-dodecane	-0.13
Methanol, toluene	-0.014
Ethylbenzene, toluene	-0.012
Methanol, benzene	-0.003
Water, heavy water	$+0.0009$
Toluene, benzene	$+0.04$
Chloroform, benzene	$+0.18$
Chloroform, carbon tetrachloride	$+0.18$
Carbon disulfide, carbon tetrachloride	$+0.43$
Cyclohexane, benzene	$+0.66$
Carbon disulfide, chloroform	$+0.67$
2-Propanone, n-decane	$+0.99$
Carbon disulfide, ethyl acetate	$+1.4$

[a]These numbers are taken from various published results. They are meant to be illustrative rather than definitive.

where

$$A = (v_{m,0}/v_{m,1}) \cdot (\rho_1 - \rho_0)^{-1}$$
$$B = (1 - v_{m,0}/v_{m,1}) \cdot (\rho_1 - \rho_0)^{-1}$$

The subscripts 0 and 1 refer to pure H_2O and pure D_2O, respectively. In these equations, ρ is the density of the mixture, ρ_i is the density of the pure species, and $v_{m,i}$ is the molar volume of the pure species, all taken at the same temperature and atmospheric pressure. In the American Society for Testing and Materials (ASTM) method for measuring deuterium concentration [133], (33) is arranged in a more useful form. It is assumed that all densities are measured as ratios to the density of distilled tap water at 25°C. Tap water already has some deuterium in it $[\rho(25)_0/\rho(25)_w = 0.9999984]$ so that (33) becomes

$$x = A'(d_{25}^{25} - 0.999984)/[1 - B'(d_{25}^{25} - 0.999984)]$$

where $A' = 9.2464$ and $B' = 0.0328$. If relative density is measured to 10 ppm, x can be determined with an imprecision of about 0.0002 using the ASTM method [133].

Relative densities are measured by pycnometry in the ASTM method. Other density measuring techniques that have been used to determine deuterium concentration in water include falling drop, gradient column, and various flotation methods. These are all described later. Inherent limitations in the accuracy or convenience of densimetric methods have led to the development of several other ways to analyze deuterium concentration in water. The relative merits of the various techniques and their potential accuracies are reviewed by Babeliowsky and De Bolle [135]. Since publication of this review, a new determinative method based on mass spectrometry and offering high absolute accuracy (6×10^{-6}) has been developed [136].

For a completely ideal isotopic system; that is, the molar volumes are independent of isotope concentration and the volume of any mixture is simply the sum of the volumes of the pure components (33) simplifies to

$$x = (\rho - \rho_0)/(\rho_1 - \rho_0)$$

3.1.6   Partial Molar Volumes and Excess Quantities

The change in volume on mixing of solute and solvent provides important information in such diverse endeavors as the study of the thermodynamics of liquids and the determination of protein structure and molar mass [137, 140].

EXCESS MOLAR VOLUME

The volume of a mixture of two liquids will not equal the sum of the volumes of its pure constituents for a combination of up to five reasons [140]: (1) size differences between solvent and solute molecules, (2) shape differences between solvent and solute molecules, (3) structural changes in the molecular ordering of the liquid, (4) interaction differences between like and unlike molecules, and (5) chemical reactions.

If a solute (1) and solvent (0) are brought together but not allowed to mix, the total volume V of (1) and (0) is

$$V^0 = n_1 v_{m,1} + n_0 v_{m,0} \quad \text{(no mixing)} \tag{34a}$$

where n_1 and n_0 are the number of moles of (1) and (0), respectively. If mixing takes place, the volume changes from V^0 to V. This change can be defined as the *excess volume*,

$$V^E = V - V^0 \quad \text{(mixing)} \tag{34b}$$

It is more convenient to cast (34b) in terms of the excess molar volume v_m^E

$$v_m^E = V^E/(n_1 + n_0)$$

Combining (34a) and (34b),

$$v_m^E = v_m - x v_{m,1} - (1 - x) v_{m,0} \tag{35a}$$

or

$$v_m^E = M/\rho - x M_1/\rho_1 - (1 - x) M_0/\rho_0 \tag{35b}$$

where $M = [x M_1 + (1 - x) M_0]$, $v_{m,i} = M_i/\rho_i$ (i being 0 or 1), ρ is the density of the mixture, M_i is the molar mass of the pure species in units of kilograms per mole, and x is the mole fraction of the solute $[x = n_1/(n_1 + n_0)]$. Values of v_m^E depend on temperature and pressure. Experimental determinations of the excess molar volume can be accomplished in two ways. One can measure the actual volume change upon mixing, using special mixing or batch-type dilatometers. Alternatively, one can measure changes in density using any of several appropriate techniques discussed below. Of these, the vibrating-tube densimeter has become the most popular.

Excess molar volumes are useful as a way of describing the mixture of two miscible liquids throughout the entire range of mixing ratios ($0 \leqslant x \leqslant 1$) [141]. The most thoroughly studied system is that of benzene (C_6H_6) and cyclohexane (C_6H_{12}) at 25°C and 1 atm (1013.25 hPa). This system was proposed by Powell and Swinton [142] as a standard test of excess volume measurements and the suggestion has been taken by most later authors. The excess molar volume for an equimolar mixtures ($x = 0.5$) of benzene and cyclohexane at the standard conditions is 0.652 cm³/mol [140]. For many systems, the excess molar volume is well described by a parabola whose extreme value occurs at $x = 0.5$. Notable asymmetries are found for water–ethanol and water–sulfuric acid systems. The symmetry and even the sign of v_m^E may be a function of temperature.

Approximate values of $(V^E/V^0)_{x=0.5}$ at 25°C and atmospheric pressure are given in Table 1.2. By studying a class of solvents systematically, one can attempt to sort out the five possible contributing effects to the observed excess volumes. As an example, one study looked at mixtures of *n*-octane with a selection of isomeric octanols [143].

An ideal fluid system can be defined as one in which the excess volume of mixing is zero for all mixing ratios because none of the five causes for nonzero excess volume applies. It was once believed that H_2O-D_2O was such a system. Careful measurement, however, reveals that even this system has a measurable (though remarkably small) v_m^E. For an equimolar mixture, the excess molar volume is about 1.8×10^{-4} cm³/mol at 25°C [144]. The excess molar volume has an opposite sign at 4°C suggesting that the H_2O-D_2O system may exhibit ideal behavior somewhere between these two temperatures.

One can define an excess Gibbs energy in analogy to the excess molar volume, such that

$$\left(\frac{\partial G^E}{\partial P}\right)_{t,x} = v_m^E \tag{36}$$

It has been pointed out that determination of the excess molar volume as a function of temperature, pressure, and composition thus yields the information necessary to calculate the pressure dependence of the excess thermodynamic functions [145]. A massive bibliography of excess thermodynamic functions, including excess volume, was assembled by Wisniak and Tamir [146].

APPARENT MOLAR VOLUME AND PARTIAL SPECIFIC VOLUME

Excess volume is a useful quantity for the description of two completely miscible liquids. In many important instances, for example, the dissolution of NaCl in water, the solute is not a liquid. To provide a quantitative description for mixing in such cases, we return to (34). One can amend (34) to account for mixing by defining an apparent molar volume ϕ_v of the solute in solution:

$$\phi_v = (V - n_0 v_{m,0})/n_1 \quad \text{(with mixing)} \tag{37}$$

If we replace ρ_1 with M_1/ϕ_v in (32), that equation will be valid after mixing has taken place and, by rearrangement of terms, one can show that

$$\phi_v = M_1/\rho - (\rho - \rho_0)/(m'\rho\rho_0) \tag{38a}$$

or

$$\phi_v = M_1/\rho_0 - (\rho - \rho_0)/(c\rho_0) \tag{38b}$$

where m' is the solute molality (moles of solute per kilogram of solvent) and c is the molar concentration (moles of solute per cubic meter of solution). The units of ϕ_v are cubic meters per mole. The apparent specific volume is

$$v_\phi = \phi_v/M_1 = [1 - (\rho - \rho_0)/(y\rho_0)]/\rho \tag{39a}$$

or

$$v_\phi = [1 - (\rho - \rho_0)/\delta]/\rho_0 \tag{39b}$$

where y is the ratio of solute mass to solvent mass and δ is the mass (in kilograms) of solute per cubic meter of solution.

Although we present this approach only for the treatment of a single solute, the formalism can be stretched in special cases to provide a useful description of several solutes in the same solution. Millero and co-workers [147] offer an example in their description of volume changes on mixing for the four major sea salts.

The apparent quantities are generally useful for two reasons. First, as an examination of the preceding four equations shows, the apparent quantities are measurable by experiment. Equally important, v_ϕ equals $\bar{v}$, the *partial specific volume* of the solute, for dilute solutions:

$$\bar{v} = \left(\frac{\partial V}{\partial m_1}\right)_{t,P,m_0} = v_\phi + y\left(\frac{\partial v_\phi}{\partial y}\right)_{t,P,m_0} \tag{40}$$

where m_1 is the mass of the solute. The partial specific volume thus represents the infinitesimal change in the solution volume per infinitesimal addition of solute mass. From (39) and (40), the value of ϕ_v/M_1 extrapolated to infinite dilution equals the partial specific volume at infinite dilution. The partial specific volumes of multicomponent systems are additive at infinite dilution. Additivity studies can yield important information on solvent–solute interactions [148].

Partial specific volume measurements of proteins in solution give a variety of useful information regarding molecular interactions, denaturization, hydrolysis, and similar phenomena [149]. In addition, $\bar{v}$ is necessary for molecular weight determinations by ultracentrifuge [150] or small-angle X-ray scattering [151]. The concentration dependence of ϕ_v is usually negligible in protein work.

In the case of aqueous electrolytes, equations of increasing complexity have been proposed for extrapolating ϕ_v to zero molar concentration. The most important are discussed in [137]. Of these, the Masson–Geffken equation is the simplest

$$\phi_v = \phi_v^0 + S_v^* c^{1/2} \tag{41}$$

where S_v^* and ϕ_v^0 are determined by curve fitting. The Masson–Geffken equation therefore predicts that solution densities are described by

$$\rho = \rho_0 + (M_1 - \rho_0\phi_v^0)c - (S_v^*\rho_0)c^{3/2} \tag{42}$$

An improved equation attributed to Redlich and Meyer replaces S_v^* in (41) by a theoretically derived parameter and includes an additional empirical term proportional to c. Data on a wide variety of electrolyte solutions have been compiled in a recent monograph [152]. Nonelectrolytes are usually extrapolated to zero molar concentration by assuming ϕ_v is linear in c.

Partial molar volumes of organic compounds in water have been related to solute–solvent interactions [153]. Several general rules can be inferred from experimental data. For example, ring formation is accompanied by a decrease in volume that is proportional to the number of members in the ring.

Reaction volume can be defined as the difference in partial molar volumes (at infinite dilution) between the products and the reactants of a chemical reaction in solution [154]. Reaction volume can be calculated from the pressure dependence of the equilibrium constant. More to the point of this chapter, it can also be measured by density or dilatometry techniques. A recent series of determinations using a type of dilatometry is reported in [155].

The appearance of the difference term $(\rho - \rho_0)$ in (38) and (39) illustrates the experimental challenge in determining ϕ_v and v_ϕ in dilute solutions: the density ρ of a dilute solution approaches ρ_0 so that precise measurements of each density are necessary to achieve even modest precision in their difference. In the analysis of proteins, for instance, typical requirements imply absolute density measurements of about $0.002\ kg/m^3$ to achieve a 0.15% measurement of v_ϕ [156]. The alternative of dilatometry is, of course, also available for these measurements as discussed below.

3.1.7 Expressions for Concentration

Several expressions for concentration have been introduced in the above equations. We recapitulate them here. For n_1 moles of substance 1 (molar mass M_1) combined with n_0 moles of substance 0 (molar mass M_0):

molar fraction,	$x = n_1/(n_0 + n_1)$
molar concentration,	$c = n_1/1\ m^3$ mixture
molality,	$m' = n_1/1\ kg$ of 0
mass fraction,	$w_1 = 1\ kg$ of $1/1\ kg$ of mixture
mass fraction,	$y = 1\ kg$ of $1/1\ kg$ of 0
mass concentration,	$\delta = 1\ kg$ of $1/1\ m^3$ mixture

The quantities x, m', w, and y are independent of temperature and pressure.

When using quantities c or δ, the temperature and the pressure must be specified. Useful relations are

$$\delta = \rho w_1$$

$$y = M_1 m'$$

$$c = \rho m'/(M_1 m' + 1)$$

$$c = \rho x/[M_1 x + (1 - x)M_0]$$

The units we have chosen were not standard in the past and are still not used universally. For instance, it is common for researchers to report concentration in terms of molarity (moles of solute per liter of solvent) rather than in molar concentration (moles of solute per cubic meter of solvent). There is a factor of 1000 difference between the two. The reader is cautioned that equations such as (38) and (39) will be modified if other units are employed.

3.2   Density Reference Standards

3.2.1   Water

Water has been and remains the most important secondary standard of density. Surprisingly, virtually all the tables and equations of state for liquid water rely heavily on recalculations of two separate determinations whose results were reported in the first decade of this century. There are several questions concerning these measurements, the most important of which are What was the isotopic composition of the water used? and How does the temperature scale used (the Echelle Normale) compare with the present scale (ITS-90)? The impossibility of answering these questions definitively led the International Union of Pure and Applied Chemistry (IUPAC) to call for a modern redetermination of water density [157]. Several laboratories have responded; but, as of this writing, there are unresolved discrepancies in the sixth decimal place.

Thus, the present ambiguities in the published tables for water are only serious for very high-accuracy work. Following Kell [130], one can assert that at a level of 0.1% uncertainty, any water of modest purity will be adequately described by any published table of densities. The same applies at 0.01% uncertainty for pure water at atmospheric pressure. Care must be exercised when one hopes to achieve uncertainties of 10 ppm or less. An uncertainty of 1 ppm is virtually unattainable at this time. One can, however, achieve imprecisions of this level if isotopic variations and dissolved air are taken into account.

Progress has been made in assessing the effects on density of variations in the isotopic abundances of hydrogen and oxygen. It has been shown that the variability of tap waters around the world generally does not exceed 0.007 kg/m^3 [158]. Furthermore, it was found that the isotopic variability in ocean water is an order of magnitude smaller. Specially prepared ocean water, referred to as *standard mean ocean water* or SMOW thus can serve as a less ambiguous standard. A particular lot of this water, with known isotopic abundances, has been set aside by the International Atomic Energy Agency in Vienna. Small aliquots of Vienna SMOW, or V-SMOW, are made available to laboratories with the capability of comparing water samples by mass spectrometry. Density differences between V-SMOW and other natural waters are now known as a function of isotopic abundances [159]. The atomic ratios of $^{18}O/^{16}O$ and of D to H must be known to achieve maximum accuracy. The atomic ratio of $^{17}O/^{16}O$ must also be known, but for natural waters this ratio is assumed perfectly correlated to $^{18}O/^{16}O$ and so need not be measured [158].

A process of multiple distillation can lead to isotopic separation changing the water density by as much as 0.06 kg/m^3 [160]. The purification method of Cox and co-workers [161] is therefore recommended because it does not produce a change in the density between the distilled and feed waters. Even so, evaporation and contamination with atmospheric water can lead to changes after distillation. To achieve relative imprecisions of about 1 ppm, the isotopic abundance at the time of use must be measured [159].

The IUPAC subcommission on Calibration and Test Materials recommends the following equation for the density of SMOW at atmospheric pressure and at a temperature t °C [159]

$$\rho(\text{SMOW})/\text{kg} \cdot \text{m}^{-3} = a_0 + a_1 t + a_2 t^2 + a_3 t^3 + a_4 t^4 + a_5 t^5 \tag{43}$$

where $a_0 = 999.842\,594$, $a_1 = 6.793\,952 \times 10^{-2}$, $a_2 = -9.095\,290 \times 10^{-3}$, $a_3 = 1.001\,685 \times 10^{-4}$, $a_4 = -1.120\,083 \times 10^{-6}$, and $a_5 = 6.536\,332 \times 10^{-9}$. The equation is valid between 0 and 40°C for deaerated water. Temperatures in this formula are assumed to be given in the convention of the International Practical Temperature scale of 1968 (IPTS-68). Thus, small corrections must be applied for temperatures measured in the newly adopted International Temperature Scale of 1990 (ITS-90) [162]. The difficulty in completely deaerating a sample of water or, alternatively, in applying a correction for air saturation is an added problem. It had once been accepted that, at room temperature, the difference in density between air-saturated and deaerated water was well below 1 ppm. Recent work, however, indicates that the difference may actually exceed 3 ppm [129]. The discrepant results, which are reviewed in [129], point to the difficulty in preparing deaerated water.

Most laboratories will not use SMOW but, instead, will use water distilled from a local tap. On the assumption that the isotopic abundance of tap water is similar to that used in the original determinations of water density, IUPAC recommends

$$\rho(\text{SMOW})/\rho(\text{tap}) = 999.975/999.972 \tag{44}$$

The maximum uncertainty in using (43) is estimated to be 0.007 kg/m^3 at 25°C.

A second equation for the density of liquid water is due to Kell [130]. Although similar to (43), the Kell equation is self-consistent from 0 to 100°C and for a wide range of pressures and isotopic abundances. For this reason, Kell's formulation is frequently used for thermodynamic calculations. The Kell equation differs from (43) by no more than 2 ppm throughout the range of the latter's applicability.

The field of oceanography requires an accurate equation of state for seawater. A version of 43 with corrections for nonzero concentration of NaCl has been developed by Millero and Poisson [163].

3.2.2   Silicon

The difficulty in achieving 1 ppm precision in measurements of water density and the impossibility of achieving certainty at this level led Bowman and coworkers [164] of the National Institute of Standards and Technology (NIST) to propose a density scale based on solid objects. As a result of their work, the absolute density (in SI units) was determined for samples of single-crystal silicon ($\rho \sim 2330 \, \text{kg/m}^3$) to an accuracy of about 2 ppm. Unlike water, the density of a semiconductor grade crystal of silicon is stable over time at this level of precision. Such a crystal can be used to standardize water (or, indeed, many other liquids) by hydrostatic weighing as described below. Monoisotopic silicon could thus serve as a density standard. Unfortunately, however, natural silicon has three isotopes whose variable mixture leads to measurable changes in density. Thus each lot of silicon must first be standardized.

For this reason, NIST makes available standardized silicon crystals of 100 and 200 g [165]. The worst-case uncertainty in ρ was estimated at under 10 ppm. The volumetric thermal expansion of silicon is about $7.6 \times 10^{-6}/°C$ at room temperature and is well known throughout a wide range of temperature [166].

3.2.3   Mercury

Liquid mercury [160] has a density of 13545.9 kg/m³ at 20°C and atmospheric pressure. Although it has seven naturally occurring isotopes, variability in the relative density of virgin mercury does not exceed a few parts in 10^6. These properties were established by the remarkable work of A. H. Cook. In two monumental papers [167, 168], Cook reported measurements of the density of mercury in SI units—first by a displacement technique and then by a filling technique. The disagreement between the two methods was only 0.5 ppm. Recently, a technique for comparing the relative density of mercury samples to an imprecision of 0.1 ppm has been developed [169]. Using this method, it was shown that no significant relative changes had occurred between mercury samples that had been standardized by Cook some 30 years previously and stored at the NIST (in the United States) and the Commonwealth Scientific and

Industrial Research Organization)(C.S.I.R.O.) (in Australia) [170].

Several unique properties of mercury recommend its use for density and volume measurements. Its high density and low volatility make it extremely useful for calibrating the volume of transparent capillary tubing. Since the height of the mercury in such a vertical tube can be followed with a cathetometer, knowledge of the mercury volume gives the average cross-sectional area of the capillary. The volume of a mercury drop added or subtracted from the capillary can be found from (1) by weighing the drop and using the known density of the mercury. The capillary to be measured must be scrupulously clean so that the contact angle between the mercury surface and the inner capillary wall will remain constant. Calibration of capillary volume is essential for some methods of precise pycnometry and dilatometry, discussed below.

Mercury also has a remarkably high surface tension (~ 0.48 N/m at room temperature). Thus, under atmospheric pressure, mercury cannot enter pores smaller than about 15 μm. This property can be useful in measuring the bulk density of porous materials by volumetry [171].

Recently, mercury has been used to determine the interior volume of a 3-L sphere to an uncertainty of 0.003 cm^3 and an imprecision of 0.001 cm^3 [170]. In pycnometry, mercury is useful for the determination of the density of UO_2 and ThO_2 pellets [172, 173].

Methods for the purification and storage of mercury are given in [159, 167, 174]. Purified mercury that has been stored for a long period of time under air develops a surface scum. The scum does not indicate contamination of the bulk mercury and can be removed by filtration through a pinhole. Alternatively, bulk mercury can be withdrawn from a tube extending below the air-mercury interface. Traditional containers used in storing mercury for extended periods are soft glass or poly(ethylene). The latter has the advantage of being unbreakable, but pouring mercury from a poly(ethylene) bottle induces electrical charge that may cause the mercury to spatter.

Tabulated values of mercury density as a function of temperature at atmospheric pressure were published by Bigg in 1964. The table is reproduced in [159], where it is corrected to temperature as measured in the IPTS-68 convention. The isothermal compressibility of mercury is approximately 4×10^{-11}/Pa. The density of mercury is most accurately known at 20°C. If θ is the difference in temperature from 20°C, the tabulated values of [159] may be expressed analytically as

$$\rho = 13545.87/(1 + a_1\theta + a_2\theta^2 + a_3\theta^3 + a_4\theta^4) \tag{45}$$

where $a_1 = 181.1417 \times 10^{-6}$, $a_2 = 7.818 \times 10^{-9}$, $a_3 = 3.331 \times 10^{-11}$, and $a_4 = 2.396 \times 10^{-14}$. The equation is valid for $-40 < \theta < 280$, but the uncertainty increases by an order of magnitude at the temperature extremes [159].

3.3   Measurement of the Density of Liquids and Solids

3.3.1   Mass Measurement of a Known Volume

From the fundamental equation (1), the most direct way of determining density is by combining measurements of the mass of a substance with measurements of its volume. Apparatus based on this method are weighed syringes, pycnometers, and vibrating-tube densimeters.

WEIGHED SYRINGES AND PIPETTES

A simple procedure for fast density determinations of small samples of liquid is based on microsyringes [175]. In 2-min time, an accuracy of 1% can be achieved with syringes of 10- or 100-μL capacity. Before use, the volume of the syringe is verified by weighing it first empty and then filled with distilled water. Unknown samples are then introduced and weighed. Buoyancy corrections to the weighings will be less than 0.2% for the lowest density liquids encountered. The small inner diameter of the syringe needle ensures that the evaporation rate of many commonly encountered liquids will be negligible. Viscous liquids cannot be drawn easily into the syringe, however. Accuracy of the method is limited by small bubbles of air entrained with the liquid drawn into the syringe.

A similar technique based on glass pipettes was developed by Alber [176]. Reported accuracies are about 0.1%. Alber made pipettes of two different designs. One has a 100-μL fixed volume and the other has a 1-mm diameter constant bore, which can be filled to any level below its maximum capacity of 20–40 μL. Both pipettes have provision for the ends to be capped to slow evaporation during weighing. To achieve uncertainties of 0.1%, it is necessary to make a rudimentary buoyancy correction. In addition, to minimize surface effects, the pipettes are weighed against a tare having similar shape and surface.

Capillary tubing can be used to find the density of samples whose volumes are of the order of cubic millimeters [177, 178]. Microvolumetric apparatus can be standardized by weighing the amount of liquid of known density necessary to fill or empty the volume of interest. Alternatively, titrimetric methods can be used [179].

PYCNOMETERS

Until the advent of the vibrating-tube densimeter, pycnometry was the most convenient method for achieving high accuracy in the measurement of liquid densities. If solid samples can be introduced to the pycnometer and the remaining volume filled with liquid of known density, the density of the solid can also be determined. Since most analytical laboratories are equipped with a balance having 100-g capacity and 0.1-mg sensibility, liquid samples of 25–50 g can be weighed to a relative precision of better than 25 ppm. The pycnometer must have a mass not exceeding 50 g. These constraints are reasonable for vessels made of laboratory glass. The volume of the pycnometer is a function of temperature. Thus, borosilicate glass ($\alpha \approx 10^{-5}$) or, if necessary, fused silica ($\alpha \approx 1.5 \times 10^{-6}$), is used. Various pycnometer designs have evolved to solve a range of problems. Pycnometry is a mature discipline. The basic considerations could be precisely stated more than 50 years ago [5]. Indeed, this section is largely based on the chapter by Bauer and Lewin [180] in the previous edition of this series.

Hidnert and Peffer [181], in their review of various methods of measuring density list four important features in the design of a precise pycnometer: (1) a form that allows rapid attainment of temperature when placed in a controlled bath, (2) a means of filling without entraining ambient air, (3) a means of filling to a precisely attainable volume, and (4) protection after filling from change in weight due to evaporation or adsorption of moisture.

We now consider the pycnometer designs shown in Figures 1.5 and 1.6. Types A and B and the Lipkin pycnometers are available commercially. Designs with either double capillaries or a wide opening are relatively easy to clean. The latter

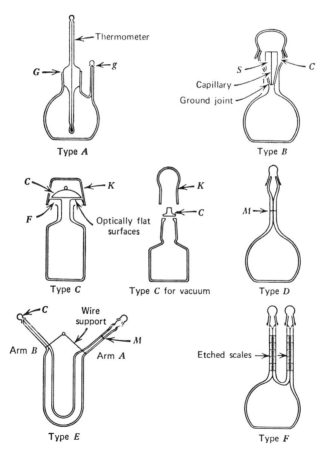

Figure 1.5 A selection of pycnometer types. Reprinted with permission from N. Bauer and S. Z. Lewin, in A. Weissberger and B. W. Rossiter, Eds., *Physical Methods of Chemistry*, Vol. 1, Part 4, Wiley Interscience, New York, 1972, Chap. 2. Copyright © 1972 by John Wiley & Sons, Inc.

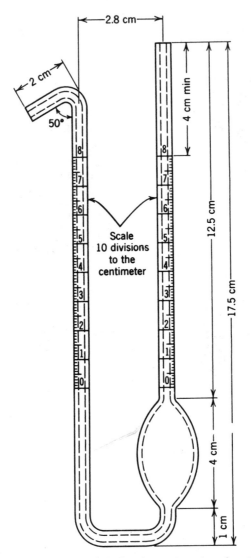

Figure 1.6 Lipkin pycnometer. Standard bulk volumes range from $\frac{1}{4}$ to 10 mL. Reprinted with permission from N. Bauer and S. Z. Lewin, in A. Weissberger and B. W. Rossiter, Eds., *Physical Methods of Chemistry*, Vol. 1, Part 4, Wiley-Interscience, New York, 1972, Chap. 2. Copyright © 1972 by John Wiley & Sons, Inc.

design also allows the pycnometer to be used in the density determination of solid samples (see below).

Type *A* is convenient for rapid determinations to an uncertainty of 0.1% in relative density. The built-in thermometer eliminates the need for a controlled temperature bath. Accuracy is limited by leakage around the joints and the failure of the thermometer plug G to seat reproducibly. A cap g is provided to slow the evaporation of volatile liquids.

Type B is attributed to Weld. Without the cap, it is known as the Gay–Lussac relative density (or specific gravity) bottle. Uncertainties as low as 0.01% are attainable with this design. As with type A, the seating of the plug S gives rise to variability in the volume. Auxiliary apparatus has been devised for seating the plug with the same pressure every time and thus minimizing random error from this source [182]. The design shown in Figure 1.5 provides for a reservoir C to contain an overflow of liquid if the bottle is warmed after it is filled and before it is weighed.

Type C was developed for precision work by Johnston and Adams [183]. They achieved reproducible volumes of $0.2 \, mm^3$, which permits relative density determinations to 5 ppm. The high reproducibility is achieved because the mating surfaces of the cap C and bottle F are optically flat. The open bottle is filled to overflowing with liquid at a known temperature at which point the cap is wrung into place taking care to prevent introduction of air. When properly wrung, the weight of the filled bottle can be supported from the cap. Evaporation through the seal is further slowed by the cover K. It is important that the bottle be filled at a temperature above that of the balance. Otherwise, thermal expansion of the liquid may pop off the cap.

Newkirk [184] used a variation of type C modifed for vacuum in the precise determination of a highly viscous liquid (molasses). The chief problem with measurements involving viscous substances is that they entrain air easily as they are poured into the bottle. The presence of bubbles is disastrous to the determination but difficult to remedy without altering the density of interest. Newkirk found that partial vacuum was superior to heating for the purpose of degassing a sample, and his apparatus represents a convenient method for so doing.

Routine density determinations of viscous samples rely on the Hubbard design [185]. This is a wide-mouthed flask with a special stopper. The stopper has a concave bottom and a small-diameter hole bored vertically through its center. When the stopper is seated into a filled flask, the excess liquid is forced out through the hole. The concavity ensures that air at the surface is also forced through the hole and not trapped within the pycnometer volume. The Fisher pycnometer† operates on roughly the same principle. It is made completely of aluminum, however, and has a cap that is forced into place by a screw mechanism. It is thus particularly suitable for pastes, which require considerable pressure to expel trapped air.

Type D represents a class of pycnometer that is simply a single capillary tube that terminates in a volume of convenient size. Liquid samples are brought to temperature and then adjusted to the mark M. If the capillary bore is 1 mm and the volume is $30 \, cm^3$, fifth-place accuracy is possible. This design does not allow for easy cleaning. In addition, its use is generally limited to pure liquids because

†Fisher Scientific Company, Pittsburgh, PA.

the density of mixtures can change significantly during the filling process (the usual filling technique involves siphoning under vacuum). Mixtures can, however, be measured in a type D pycnometer by use of a special filling apparatus [186]. Use of the pycnometer is made easier if the capillary volume is calibrated as a function of distance from M.

The Bingham pycnometer is similar in design to that shown as type D. The Bingham is not freestanding. It contains an expansion volume above the capillary so that liquid can expand during the weighing process. The filling, cleaning, and use of the Bingham pycnometer are described in ASTM D1217 [187].

A general problem with high-precision pycnometry is that precision exceeds accuracy. Good evidence for this assertion is found in the failure of pycnometric determinations of excess molar volumes to agree among laboratories and with measurements made by other precise techniques [143]. When used in a differential mode, however, pycnometry can give relative differences accurate to the sixth decimal place [129, 188, 189].

A version of type D that substitutes a tube of 1-cm bore for the capillary has been used by Brown and Land [190]. The liquid level is adjusted automatically by overfilling and then withdrawing the excess through a capillary, the end of which is positioned accurately midway up the center of the pycnometer tube. The capillary is then removed and replaced by a cap to retard evaporation of the liquid under test. Relative uncertainty of 0.01% can be achieved.

Types E and F are variations on the Ostwald–Sprengel design, once popular for precision measurements [181]. The design is distinguished by the presence of two capillaries of equal bore. The two openings make this type easy to clean and to fill. A narrow bore helps reduce evaporation while still permitting the volume of liquid to be known accurately. Type F is especially convenient because both capillaries are graduated. The volume of the pycnometer to the bottom of the capillaries is determined with water. To complete the calibration, the pycnometer is then filled with mercury to the top of the tubes. Although the mass of the pycnometer filled with mercury is too large to be weighed conveniently, small quantities of mercury can be withdrawn by syringe and weighed on a suitable balance. The sum of the heights of both columns can therefore be determined accurately as a function of the volume of mercury withdrawn. When used, the pycnometer is first filled with test liquid and weighed. The pycnometer is then equilibrated in a constant temperature bath with provision for reading the liquid levels in the capillaries. Fifth-place accuracy in relative densities is possible. Several nuances in design of the type F pycnometer have been used [191–193]. This pycnometer has also been adapted for use in measuring densities of highly viscous liquids as a function of temperature [194] where a relative uncertainty of 0.04% is claimed. The Ostwald–Sprengel design can also be scaled down to measure sample volumes of 0.1 mL with a relative uncertainty of 0.1% [195].

The Lipkin pycnometer (Figure 1.6) [196] is available commercially.† Its use

† Ace Glass, Inc., Vineland, NJ.

is described in detail in ASTM D941 [197], where techniques for determining volatile liquids are described. Relative accuracies of 0.02% were demonstrated on a device having a bulb volume of only 4.5 mL. This design is particularly useful in determining the thermal expansion of a liquid over a fairly large temperature span. For instance, suppose a liquid having a thermal expansion α of approximately $10^{-3}/°C$ is filled just above the zero marks at room temperature. If the temperature is increased by 15°C, the volume will increase by about 0.07 mL (assuming a bulb of 4.5 mL). If the bore of the pycnometer is 1 mm, the volume increase will raise the liquid level by about 4.5 cm in each arm and is thus accommodated easily. In this way, we see that the Ostwald–Sprengel types of pycnometers can also serve as thermal dilatometers [198]. Determination of the density of a fluid over a considerable range of temperature is an important auxiliary measurement to the determination of solid densities by the temperature of flotation method (discussed later). One must, of course, take into account the thermal expansion of the pycnometer when temperature effects are studied.

Numerous other pycnometer designs have been developed for special uses. Campbell [199] reviewed pycnometers used to determine the densities of liquid metals. Since that review, additional pycnometric devices have been developed for molten metals [124, 200, 201], the last capable of withstanding pressures of 150 MPa. Other devices include a special pycnometer for measuring the bulk density of granular solids [172] and a micropycnometer for use in determining the relative density of 10-mg mineral specimens with an uncertainty of about 5% [202].

Mercury is often used as the filling liquid when determining the density of solid samples. For porous or fine-textured samples, use of kerosene has been recommended because of its ability, given enough time, to penetrate openings of molecular size in an evacuated sample [203]. Ethyl ether has been used at low temperatures. In some special cases, the advantages of a cryogenic liquid may override the inconvenience [204, 205].

Use of Pycnometers for Liquids. Determination of the density of an unknown liquid is done in three steps. The mass of the pycnometer clean and empty m_p is measured. The mass of the pycnometer filled with a reference liquid (usually water) at a known temperature m_s is measured. For pycnometers with graduated capillaries, a range of fillings spanning the capillary volume is made. Finally, the mass of the pycnometer filled with the unknown fluid m_L is measured. The mass of reference liquid m_R in the pycnometer can be calculated from the first two measurements

$$m_R = m_s - m_p$$

The filled volume of the pycnometer V_p is now known in terms of ρ_R, the density of the reference liquid at the temperature of measurement

$$V_p = m_R/\rho_R$$

For a pycnometer with a graduated capillary, a range of volumes is known accurately in terms of the mass and density of the reference liquid. The mass of the unknown liquid m_L can be calculated from measurements one and three. Since this mass is also contained in V_p, the density of the test liquid ρ_L is now known relative to the density of the reference:

$$\rho_L/\rho_R = m_L/m_R \tag{46}$$

Equation (46) is deceptively simple. As discussed in detail in Section 2, mass is not a quantity that can be measured directly on an analytical balance. Instead, it must be inferred from measurements of apparent mass corrected for air buoyancy. In addition, objects of glass or fused silica are difficult to weigh because of adsorbed moisture or induced electrostatic charges. Finally, the requirement of (46) that the unknown liquid is filled exactly to the volume V_p (or, alternatively, that the difference from V_p is known accurately) is difficult to fulfill experimentally. This difficulty arises from instability in the pycnometer volume and from the presence of bubbles. A final difficulty is keeping liquids free from contamination and preventing differential evaporation of mixtures. These considerations also apply to the determination of solids, except that one can avoid the use of liquid mixtures in the pycnometric determination of solid densities.

Let us consider the three measurement steps in detail. First a pycnometer to be used for high-precision work must have a stable volume. Newly formed borosilicate vessels, even when annealed, may have volumes that change significantly with time. In addition, the volume may show hysteresis with thermal cycling. Assuming these problems are negligible, the pycnometer must first be cleaned scrupulously. Chromic acid cleaning solution is regarded as the best cleaner, but good results have also been obtained with detergent. Introduction of the cleaning solution to a type D pycnometer is best done by siphoning under vacuum [187].

The clean pycnometer and all plugs or caps are weighed together. The pycnometer is, however, disassembled to help insure that the air inside the vessel has the same density as the ambient air in the balance. This is an important consideration for precise work. If the pycnometer has been flushed with nitrogen or has been stored in a desiccator, it may be necessary to flush the interior with laboratory air prior to weighing. Weighing techniques detailed in Section 2 are used. In particular, if two identical pycnometers are available, one of them may be used as a tare to eliminate buoyancy corrections involving the glass as well as to minimize surface differences.

The pycnometer is then filled. Operationally, this step is often identical for the calibrating liquid or for the test liquid. Additional precautions may be necessary for volatile or viscous test liquids. For wide-mouthed pycnometers, filling can be done by pipette. Vessels terminating in single capillaries may be filled by a syringe. Care should be taken to degas samples before their use and to avoid entrainment of air bubbles during filling. If a few bubbles appear, they can be withdrawn using the same syringe. Pure liquids can be syphoned into an evacuated pycnometer. Uncomplicated auxiliary devices have been developed for this purpose [168, 170, 188]. Pycnometers having two capillary ports can be filled easily by syphoning liquid under slightly reduced pressure. The Lipkin pycnometer is particularly convenient since one capillary is bent at an acute angle. The end of this capillary can be placed directly into a beaker of the sample liquid.

For accurate work, the pycnometer is then allowed to equilibrate in a well-controlled temperature bath. Forethought should be given as to whether one can allow the liquid in the filled pycnometer to expand during the subsequent weighing process. If expansion is inadvisable, the temperature bath must be maintained slightly above that of the room. For a type B pycnometer, the plug is introduced and excess liquid is forced through the capillary. The excess liquid is then removed carefully from the top of the plug and the expansion area. We can adjust the liquid level in the type D pycnometer to the mark by using a syringe to subtract a small amount of liquid. Ground-glass stoppers are then tightened into place and the pycnometer is removed for weighing.

The exterior of the pycnometer is wiped dry and the vessel is allowed to equilibrate in or next to the balance case. An equilibration time of about $\frac{1}{2}$ h is usually sufficient. The pycnometer is then weighed. For pycnometers whose capillaries are filled to a mark, the gas above the meniscus may be assumed to be laboratory air. This assumption is not justified in work to fifth-place accuracy involving volatile liquids. For a type B pycnometer, one must assume that the small volume inside the cap contains vapor of the volatile liquid at its saturation pressure. The density of this air–vapor mixture can be estimated from the ideal gas laws by assuming that the vapor displaces air

$$\rho_{\text{air} + \text{vapor}} = \frac{P_v M_v}{RT} + \left(\frac{P - P_v}{P}\right)\rho_a \tag{47}$$

where P is the atmospheric pressure, P_v is the saturation vapor pressure of the volatile liquid, T is the absolute temperature, and R is the gas constant. If the pressures are given in ratios to 1 atm (1013.25 h Pa) and T is given in kelvin, then $R = 8.206 \times 10^{-5}$ m$^3 \cdot$ atm/mol $\cdot$ K. The density will be in units of kilograms per cubic meter provided M_v is in kilograms per mole. The ambient air density ρ_a is determined from (13). To apply (47) to pycnometry, first determine the mass of the liquid in the pycnometer under the assumption that the air above the meniscus is the same density as laboratory air. Then add the correction

$$(\rho_a - \rho_{\text{air} + \text{vapor}}) V_g$$

where the volume is that of the cap interior. If V_g is 10% of the pycnometer volume, failure to correct for the vapor above a volatile liquid can cause significant error (for instance, ca. 0.02% for n-pentane at room temperature).

Equations for Liquid Pycnometry. We end this section on liquid pycnometry with the equations that govern the measurements. Each type of pycnometer requires a slightly different treatment, but the general outline is the same. A complete set of equations, including treatment of all known effects is beyond our scope. The interested reader may consult [206] as an example of a sophisticated calibration scheme for type *D* pycnometers. Our simplified relations will generally be adequate, even for work to the fifth decimal place.

We will make the following assumptions: the balance temperature is within a degree or two of the temperature of the controlled bath t_b; a second, identical pycnometer is used as a tare during the weighing (if only one pycnometer is available, one can achieve a similar simplification by weighing the empty pycnometer immediately before being filled with liquid and reweighed so that the air density is the same for both weighings); and weighings are carried out on a single-pan electronic balance that has been calibrated against standards of density 8000 kg/m³.

Step 1. Weigh the empty pycnometer against the tare

$$W_1(1 - \rho_1/8000) = m_t(1 - \rho_1/\rho_p)$$
$$W'_1(1 - \rho_1/8000) = m_p(1 - \rho_1/\rho_p)$$
$$\Delta = m_p - m_t \approx W'_1 - W_1$$

where W_1 is the balance indication with tare, W'_1 is the balance indication with pycnometer, m_t is the mass of the tare, m_p is the mass of the pycnometer, ρ_1 is the air density, and ρ_p is the density of pycnometer glass.

Since the tare is nearly identical in physical properties with the pycnometer being used, the approximation for Δ is justified.

Step 2. Weigh the pycnometer filled with distilled water at t_b. The water is allowed to equilibrate at the balance temperature before weighing. The tare is also weighed.

$$W_2(1 - \rho_2/8000) = m_t(1 - \rho_2/\rho_p)$$
$$W'_2(1 - \rho_2/8000) = m_p(1 - \rho_2/\rho_p) + m_w(1 - \rho_2/\rho_w)$$

where m_w is the mass of the water in the pycnometer and ρ_w is the density of the water at the temperature of the balance. The remaining terms are extended in an obvious way from the relations governing Step 1.

Since the water density appears only in the correction term for air buoyancy, we can just as easily take the density of water at t_b (we have assumed that the balance and bath temperatures are reasonably close together). Making this assumption, one can show that the volume of the filled pycnometer at t_b is

$$V = [(W'_2 - W_2)(1 - \rho_2/8000) - \Delta]/(\rho_{w,0} - \rho_2) \qquad (48)$$

where the subscript 0 indicates the density at the bath temperature.

Step 3. Weigh the pycnometer filled with test liquid at t_b. The liquid is allowed to equilibrate at the balance temperature before weighing. The tare is also weighed.

An equation similar to (48) can be derived for the test liquid, whose density is indicated by a subscript L

$$\rho_{L,0} - \rho_3 = [(W'_3 - W_3)(1 - \rho_3/8000) - \Delta]/V \tag{49}$$

Any needed correction for the gas density over the meniscus of a volatile liquid (see above) is added to the numerator of (49). Combining (48) and (49)

$$\rho_{L,0} = \frac{(W'_3 - W_3)\left(1 - \dfrac{\rho_3}{8000}\right) - \Delta}{(W'_2 - W_2)\left(1 - \dfrac{\rho_2}{8000}\right) - \Delta}(\rho_{w,0} - \rho_2) + \rho_3 \tag{50}$$

Equation (50) makes it clear that air density cannot be ignored for accurate pycnometry. The unknown liquid is determined in terms of the density of water *and* the density of air, the latter typically entering at the 0.1% level. The necessity for air-density corrections at this level is an inherent characteristic of all gravimetric density determinations.

Pycnometric Determinations of Solids

Use of liquid pycnometers. Liquid pycnometers can also be used to determine the density of solids. Samples most often measured by pycnometry are in the form of powders, granules, or fibers. The volume of the pycnometer is first determined experimentally from (48). Then the pycnometer filled with the solid sample to be determined is weighed. Next, enough reference liquid is added to the pycnometer to fill it and the pycnometer is reweighed. The mass of the added liquid divided by its density gives the volume of reference liquid. The difference between this volume and the total volume of the pycnometer gives the volume of the solid sample.

The relative density ρ_s/ρ_R of the solid sample is thus given by

$$\rho_s/\rho_R = m_s/(m_{R,1} - m_{R,2}) \tag{51}$$

where the subscript 1 refers to the mass of reference liquid required to fill an empty pycnometer and the subscript 2 refers to the mass of reference liquid required to complete the filling of a pycnometer that already contains a mass m_s of the solid whose density is to be determined. Equation (51) may be compared with (46), which applied to liquids. Just as (46) became (50) when expressed in terms of observable quantities, (51) must also be amended to be expressed in terms of experimentally accessible quantities. A highly simplified relation that is nevertheless instructive is

$$\rho_s = \frac{m_s^*}{m_{R,1}^* - m_{R,2}^*} (\rho_R - \rho_a) + \rho_a \tag{52}$$

where ρ_a is the air density, assumed the same for all measurements, and the starred quantities are effective masses (see Section 2.2). Ratios of effective masses are the same as ratios of balance readings if all measurements are taken at the same air density. Equation (52) can be compared with (50), which was derived less schematically.

Measuring finely divided samples such as powders or granules is complicated by the necessity of the filling liquid to penetrate the entire volume not occupied by the solid sample. This problem is most acute when using liquids that do not wet the sample. The problem can, however, be turned to advantage in determining the bulk volume of a porous sample. It is even possible to use a very fine dense powder instead of a liquid for measuring bulk densities of porous samples [207]. Of course, one must also find a liquid that does not dissolve or react with the sample. If a sample has finite solubility in a liquid, it may be possible to use a saturated solution as the pycnometric liquid.

Limitations in achieving accurate results on finely divided samples are examined in [208, 209]. An experimental prerequisite to any theoretical analysis, however, is the removal of all air that is trapped inevitably within a powdered sample. The air can be removed before or after the sample is covered with liquid. The most effective technique is to evacuate the sample and then to top off the pycnometer with liquid while still under vacuum. This method has been realized experimentally in many ways [172, 210]. In some situations, it may be more convenient to cover the sample with the reference liquid and then to remove trapped bubbles. Removal can be accomplished by shaking the pycnometer (an ultrasonic bath is useful for this) and then reducing the pressure above the pycnometer by placing it in a vacuum chamber, after which the vessel is topped off and equilibrated to constant temperature [211, 212]. A centrifuge can be useful in the elimination of trapped gas, as described by Russell [213].

Micropycnometers suitable for density determinations of small samples of minerals have been reviewed by May and Marinenko [202]. The volume of several grams of sample can be determined to be better than $0.1 \, mm^3$ under favorable conditions. This corresponds to uncertainities of parts in 10^5 in relative density for a sample of density $3000 \, kg/m^3$. The most serious impediment to achieving high accuracy is the failure of the reference liquid to fill the interstices between solid particles completely. If this problem is significant, the measured density of the solid will be less than its true density.

Volumenometers may be used for routine work to find the bulk density of porous samples by using mercury as the reference fluid [171, 203, 214]. As the name implies, these devices measure the volume of mercury displaced when a sample is introduced to a vessel.

Use of gas pycnometers. Special pycnometers that use gas (often helium) as the filling fluid are available commercially† for the measurement of solid density. Helium is able to penetrate pores of atomic dimension. In addition, some degree of automation can be achieved with gas pycnometry. The disadvantage, of course, is that the density of gas is several orders of magnitude below that of liquids. Therefore, this method is limited to an accuracy of about 0.1–1.0%. The precision varies with the sample volume and the design of the apparatus.

The principle of operation of gas pycnometers assumes that the gas obeys the ideal gas laws. This is a reasonable assumption for helium. Thus, at constant temperature

$$PV = \text{constant} \tag{53}$$

where P is the pressure of a quantity of gas and V is the volume to which it is confined. The various commercial and experimental devices exploit (53), often minimizing errors by differential measurement techniques. References [214–217] describe experimental designs and include a detailed theory of operation. A related device is the gas volumenometer of Hauptmann and Schulze, which is designed for determination of sample volumes of $10 \, \text{mm}^3$ [218].

VIBRATING-TUBE DENSIMETERS

High-precision liquid pycnometry has three serious drawbacks: the technique is tedious and painstaking, the measurements are virtually impossible to automate, and relatively large samples are required. The vibrating-tube densimeter (VTD) is a device that overcomes these objections. The VTD is essentially a pycnometer of type E. It is made of tubing (borosilicate glass, stainless steel, or another suitable material) bent in a V shape, as shown schematically in Figure 1.7. Rather than weigh the device, however, the unclamped part of the tube is forced to vibrate in and out of the plane of the V. The resonant frequency of the tube depends on its total mass m, much as a tuning fork or as the inertial balances described in Section 2.5.2. Since the volume V of the vibrating tube is fixed, the total mass depends on the mass of the tube m_0 and the density ρ of the fluid that fills the tube

$$m = m_0 + V\rho \tag{54}$$

From Section 2.5.2, the resonant frequency of the vibrating mass is

$$f = 1/\tau = (k/4\pi^2 m)^{1/2} \tag{55}$$

†For example, from Micromeritics Instrument Corp., Norcross, GA, and Quantachrome Corp., Syosset, NY.

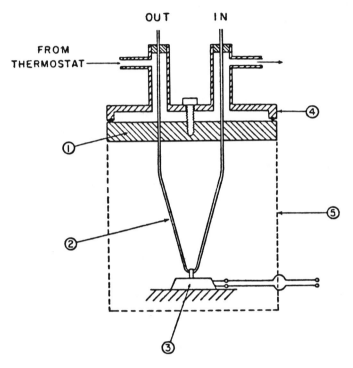

Figure 1.7 A schematic representation of a vibrating-tube densimeter. The stationary ends of the vibrating tube (2) are anchored in a massive metal plate (1). Vibrations of the tube are sensed at (3). Reprinted with permission from P. Picker, E. Tremblay, and C. Jolicoeur, *J. Solution Chem.*, **3** , 377 (1974).

where k is the *spring* constant of the system and τ is the period of oscillation. Based on (54) and (55), the density of an unknown sample can be determined from

$$\rho = A\tau^2 + B \tag{56}$$

In practice, A and B are instrumental constants that must be found by experiment. The VTD was developed by Stabinger and co-workers [219] and is now marketed commercially† Sample size is normally 0.7 mL. A device based on similar principles has been developed by Picker and co-workers [220]‡. It is especially suited to flowing samples and is also sold commercially. The detailed principles of operation and use of the VTD were described in review papers [221]. Instruments can be configured for use at elevated temperature or pressure. A laboratory device having an imprecision of 30 ppm has been

†Anton Paar, USA, Inc., Richmond, VA.
‡Made commercially by Sodev, Inc., Sherbrooke, Quebec, Canada.

constructed for work at temperatures up to 700 K and pressures to 40 MPa (ca. 400 atm) [222]. From (56), it is necessary to calibrate the VTD at a minimum of two points. The most convenient two fixed points are dry air and distilled water. The useful range of most commercial VTDs is 0–3000 kg/m³. Although an imprecision of 1.5×10^{-3} kg/m³ is attainable with a measurement time of under 10 min, absolute accuracy to this level can only be achieved within a narrow range about the calibration points. Thus, the full potential of the instrument is best realized when it is used for differential measurements. Kiyohara and Benson [223] evaluated the accuracy of an Anton Paar VTD when (1) calibrated at two points (with air and water) and (2) calibrated by 10 reference liquids over a range of 740–1050 kg/m³.

Filling the tube to a precise level is not necessary because only the oscillating portion of the tube is sensed. Thus, a flowing system can be measured easily. The major prerequisites are that liquid in the tube contain no bubbles and that thermal equilibrium be attained. Since the surface-to-volume ratio of the tube interior is much larger than for conventional pycnometers, contaminated surfaces must be cleaned scrupulously before subsequent use. In addition, the sample liquid must wet the tube. This is checked in glass tubes by making sure that the meniscus of the liquid is always concave during its injection through the V tube [221]. As with any determinative technique, one must avoid contamination of the sample before it is delivered to the apparatus. An apparatus designed to minimize such problems in the density determination of dialysis samples has been described [224].

It is generally claimed that VTD devices are unaffected by the viscosity of the samples measured. Indeed, a theoretical justification for a viscosity dependence is not obvious. Experimental confirmation of this assertion for a special case is given in [220]. By contrast, the work of Bernhardt and Pauly [225] indicates device-specific errors of about 0.01% at high viscosity (10 mPa · s). The conflicting claims remain unresolved.

Similarly, evidence that one VTD was sensitive to changes in barometric pressure [226] has not been confirmed [223].

Density determinations of powders are possible using the VTD [221]. One must first form a uniform suspension of a known mass of the powder m_s in a known mass of a reference liquid m_R. The density ρ_{sus} of the suspension ρ_{sus} is measured on the VTD as is the density of the pure reference liquid ρ_R. The density of the powder ρ_s is then determined from

$$\rho_s = \frac{m_s \cdot \rho_{sus}}{m_s + m_R(1 - \rho_{sus}/\rho_R)} \qquad (57)$$

The relation (57) is a simple rearrangement of (32) with the powder being treated as an impurity in the reference liquid. A relative imprecision of less than 0.5% should be attainable.

Other geometries than the vibrating tube are, of course, possible. The general terminology *acoustic resonance densitometry* or ARD has been suggested to refer to the entire class of these instruments [227].

3.3.2   Measurement of Effective Mass in a Fluid

We defined effective mass in Section 2.2 by assuming that objects to be weighed are surrounded by an atmosphere of normal air. Archimedes' principle also applies to other fluids including, of course, liquids. Thus the effective mass m^* of a body of mass m in equilibrium with a fluid of density ρ_F is

$$m^* = m(1 - \rho_F/\rho) = m - m \cdot \rho_F/\rho \qquad (58)$$

If the mass is known and the effective mass in the fluid is measured, then prior knowledge of ρ_f determines ρ and, conversely, prior knowledge of ρ determines ρ_F. Devices based on (58) include hydrometers and relative density (specific gravity) balances. General techniques based on (58) include hydrostatic weighing and magnetic suspension.

The special case $m^* = 0$ occurs when $\rho_F = \rho$. Experimentally, this condition can be recognized by the solid neither sinking nor rising in the liquid. Devices exploiting this idea include gradient columns and Cartesian divers. Determination of density from the temperature of flotation is another well-known technique based on this special case. As stated in Section 1, densities of extremely small samples can, in principle, be determined accurately. The technique used, however, must not rely on the measurement of extensive quantities. Flotation techniques fulfill this requirement.

HYDROMETERS

Hydrometers, used to measure relative densities of liquids, are readily available from suppliers of chemical apparatus. Precision to the third decimal place in relative density is attainable. Usually made of glass, most of their volume is in a weighted body, which, when in equilibrium, is submerged in the test liquid. A cylindrical stem with inscribed graduations projects from the top of the body. The hydrometer sinks into the test liquid until the mass of displaced liquid approximately equals the mass of the hydrometer. The level at which this occurs is read on the stem, which is not submerged completely. The exact level to which the hydrometer sinks depends on corrections for the air buoyancy of the unsubmerged stem and for the surface tension σ of the liquid.

At the air–stem–liquid interface, there is a downward force on the hydrometer equal to $\pi d\sigma \cdot \cos \theta$, where d is the diameter of the stem and θ is the angle of contact between the liquid meniscus and the glass stem. The force due to surface tension has the same effect as an additional mass m', where

$$m' = (\pi d\sigma \cdot \cos \theta)/g \qquad (59)$$

Because m' can be an appreciable fraction of the hydrometer mass in precise work, hydrometers must be calibrated for the liquid in which they will be used.

Even then, good reproducibility demands a clean liquid surface and a clean hydrometer stem to ensure reproducible values of σ and θ. An additional inconvenience is that equally spaced graduations on a hydrometer stem correspond to equal volume increments, not to equal density increments. To cancel out the nonlinearity in density, hydrometers are often calibrated in *degrees Baumé* rather than in relative density units. Finally, increased precision in hydrometers is achieved by reducing the stem diameter. Thus, precision comes at the expense of reduced range. For these reasons, use of hydrometers is usually restricted to specialized measurement problems. Details of hydrometer specification and calibration can be found in [228–230].

HYDROSTATIC WEIGHING USING ANALYTICAL BALANCES

General Principles (Type 1). Consider a solid body of mass m, volume V, and density ρ. When weighed in air of density ρ_a, its effective mass m_1^* is

$$m_1^* = V(\rho - \rho_a) \qquad (60)$$

When weighed suspended in a fluid of density ρ_F, its effective mass m_2^* is

$$m_2^* = V(\rho - \rho_F) \qquad (61)$$

The fluid density can then be determined from (60) and (61)

$$\rho_F = \frac{(m_1^* - m_2^*)\rho + m_2^* \rho_a}{m_1^*} \qquad (62)$$

Equation (62) can be rearranged easily to find ρ if ρ_F is known

$$\rho = \frac{m_1^* \rho_F - m_2^* \rho_a}{m_1^* - m_2^*} \qquad (63)$$

One occasionally sees (62) and (63) approximated by setting ρ_a equal to zero. For work to the third decimal place, one must include corrections for air buoyancy. Equations (62) and (63) are useful in estimating sources of error in a hydrostatic determination of density. It is obvious from these equations that if the same instrument is used to determine both values of effective mass, the instrument need only be linear in m^*, but need not be calibrated in terms of a unit of mass.

A good rule of thumb in planning hydrostatic weighings is that a balance imprecision of δm_2^* leads to a relative error in the determination of ρ_F [by (62)] or ρ [by (63)] of about

$$\delta m_2^*/(\rho_F V) \qquad (64)$$

For example, suppose one intends to construct a solid body to be used in hydrostatic measurements with a balance having 0.1-mg precision. (In this application, the solid body is often referred to as a *plummet* or a *sinker*.) The apparatus will be used to compare the relative densities of liquid samples whose nominal relative density is 0.7. The imprecision of the density measurements must be no worse than 50 ppm. Making use of (64), the body must have a minimum volume of about 3 mL. To provide a margin of safety, one might consider 15 mL as a resonable volume. To ensure that the body will sink in a variety of liquids, its relative density should be at least 2. Thus, the useful capacity of the balance need only be 30 g. If the body had a relative density of 8, the capacity of the balance would need to be 120 g. Besides making a lighter demand on balance capacity, a body of minimum density can be suspended from a filer filament than a body whose effective mass in the test fluids is larger (see next subsection).

Experimental techniques. An analytical laboratory balance is conveniently used for measuring the necessary effective masses. Techniques for determining the effective mass in air are given in detail in Section 2. To determine effective mass in a liquid, the sample is suspended from the balance pan or suspension by a fine filament. Since the filament breaches the air–liquid interface, attention must be paid to the submerged volume of the filament and to forces due to surface tension. For rough work, the volume of the submerged filament may simply be kept negligibly small.

For very precise work, considerable pains must be taken to ensure a constant contact angle θ between the surface of the liquid and the filament. The diameter of the filament is also made as small as possible while still maintaining sufficent tensile strength to suspend the sample. Several treatments have been used to improve the constancy of θ for metal filaments. These include: electrodeposition of platinum black onto a platinum (or stainless steel) filament [231], oxidation under partial vacuum of an incandescent nichrome filament [23], and heat treatment of a filament coated with silicone oil [232]. Good results have been obtained with nylon filaments in aqueous solutions [233].

Determination of liquids (Method of Kohlrausch). Determination of liquid densities by hydrostatic weighing has much in common with pycnometry. Achievement of minimum relative imprecision (ca. 10^{-6}) requires attention to thermal equilibrium of the sample, corrections for the volumetric expansion of both sample and plummet, and care in filling the apparatus so that bubbles do not form on the plummet or filament. Hydrostatic determinations usually require a minimum of 25 mL of liquid for precise measurements. Unlike pycnometric determinations, however, there is no maximum limit on the volume of liquid whose density can be determined. A paper by Chappelow and co-workers [234], describing an apparatus for precise density determinations of liquid *n*-octane, gives a good synopsis of important experimental considerations. A similar device has been used for monitoring the density of flowing liquids [235]. Determinations of volatile or hydroscopic liquids require special precautions [236].

Small quantities of liquid can be determined by correspondingly small plummets suspended from a microbalance. Blayzyk and co-workers [78] describe a servocontrolled, two-pan microbalance used for dilatometry measurements on small quanties of biological samples. Several useful design details are given in their report. The apparatus works in a differential mode in that identical plummets, one immersed in a reference liquid and the second in a small volume of the test liquid, are suspended from each arm of the balance. A similar apparatus can be used for general-purpose measurements of liquids and solids, as described by Craubner [237].

A hollow plummet made of invar has been incorporated into a compact apparatus for use over an expanded temperature range [238]. Plummets of stainless steel, tungsten, and molybdenum have been used to determine the densities of liquid metals [239, 240]. Differential measurements between plummets of different volume were employed to cancel the considerable force of surface tension on the suspending filament. Uncertainties to 0.1% were attained.

Special balances, such as the Westphal, incorporate a plummet hanging from one arm. Such balances read directly in relative density. Forziata and co-workers [241] describe in detail the calibration of one such balance and also examine several sources of systematic error.

Determination of solids. The report of Bowman and Schoonover [23] provides a complete analysis of apparatus and techniques required to achieve the maximum accuracy in the density determination of solid samples by hydrostatic weighing. Distilled water is used as the reference density. In subsequent work, Bowman and co-workers [164, 242] extended their techniques to a method based on silicon objects as density standards. When relying on solid standards, the hydrostatic liquid need not be well characterised. Thus Bowman and co-workers [164, 242] were able to use a synthetic fluorocarbon liquid that has some advantages (greater density and chemical inertness, lower surface tension) compared to water.

Major simplifications in technique can be achieved without loss of precision if one is interested only in following small density changes in a solid object [243, 244].

From (64), it is clear that the achievement of a target accuracy places demands on the balance used and the size of the samples measured. These considerations are prerequisites to experimental design. Making use of a commercial balance, Cawthorne and Sinclair [245] devised a hydrostatic system capable of determining densities to a relative precision of 0.05% in samples having a nominal volume of only 4 mm^3. They used a fused silica suspension filament of 20-μm diameter and found that water, with a few drops of wetting agent added to reduce surface tension, gave more reproducible results than organic liquids of higher ρ_F.

Density determination of powdered samples by hydrostatic weighing presents most of the same problems discussed earlier in the section on pycnometry. The primary problem is to free the powder of air and to keep it free as it is submerged in the hydrostatic fluid. Special apparatus for immersion in carbon tetrachloride (CCl_4) have been developed [246, 247]. During evacuation, the CCl_4 is kept frozen in a compartment open to the sample chamber. The system is then closed to the vacuum pump and the CCl_4 is allowed to liquify and distill over to the sample chamber. Carbon tetrachloride was used because of its ability to penetrate small pores [246]. Earlier work suggests that the penetrability of a liquid is correlated with its ability to wet the surface of the particular solid [248].

The bulk volume of a porous body can be measured by hydrostatic weighing provided, of course, that liquid does not enter the pores To exclude liquid, the sample may be coated with a thin layer of wax or poly(styrene). The volume of the coating can be corrected for by measurements of the coating mass and prior knowledge of its density. This technique forms the basis of an ASTM test method [249].

General Principles (Type 2). The hydrostatic weighings, which we have designated type 1, are by far the most commonly used. Nevertheless, in some cases it may be more convenient to use a method that we designated as type 2. This method relies on the fact that if a solid object suspended in a liquid experiences a buoyant force, then the liquid container must experience an equal and opposite force. Consider a single-pan electronic analytical balance. In the first step, determine the effective mass m_1^* of the solid of interest

$$m_1^* = V(\rho - \rho_a) \tag{65}$$

where the meanings of the terms on the right-hand side of the equation are the same as in (60).

In the next step, a beaker of the hydrostatic liquid is placed on the balance pan. Its effective mass is read as m_2^*. Finally, the solid is suspended in the liquid with care taken to avoid touching the walls or bottom of the beaker. Neglecting corrections due to the volume of filament beneath the liquid and surface tension at the liquid–filament–air interface, the new balance reading can be considered a pseudoeffective mass m_3^*

$$m_3^* - m_2^* = V(\rho_F - \rho_a) \tag{66}$$

The relation (66) follows because, in equilibrium, the solid volume V of density ρ can be replaced with fluid of density ρ_F without altering the forces on the balance pan. From (65) and (66)

$$\rho = \frac{m_1^*}{m_3^* - m_2^*}(\rho_F - \rho_a) + \rho_a \tag{67}$$

The type 2 method is particularly convenient for rapid determinations of solid densities to modest uncertainty (0.1–0.01%). Moldover and co-workers [250] used this technique in the density determination of a 15-kg stainless steel sample.

Obvious limitations of the method are the volatility of the liquid and possibility of air adhering to the solid sample. Water with a small amount of wetting agent added has been found to be a suitable liquid. Type 2 determinations are only convenient on balances with high stiffness. Thus, this method was impractical before the advent of electronic balances (see Section 2). The method avoids the inconvenience of elevating the balance so that the object may be suspended below. Hydrostatic determination of types 1 and 2 have their analogs in magnetic suspension densimetry (see next section).

HYDROSTATIC WEIGHING WITHOUT A FILAMENT

For hydrostatic determinations described in the previous section, the solid body or plummet was suspended beneath the surface of a liquid by means of a filament. Either the tensile force on the filament (type 1) or the reaction force on the liquid (type 2) was measured using an analytical balance. Use of a filamentary suspension, however, leads to two major experimental complications: surface tension at the liquid–air interface and difficulty in isolating the sample environment from the laboratory environment.

Submerged Balances. Filamentary suspensions can be eliminated in several ways. Most directly, a modified balance can be submerged completely in the liquid. Commerically available electronic balances were modified for this purpose and are used for density determinations of solids [165, 251] and liquids [252]. A relative standard deviation of better than 0.01% can be attained. Liquid densities have been determined by a float instead of a plummet [253, 254]. The float is attached to one end of a helical spring of fused quartz. The other end of the spring is anchored to the bottom of the vessel, a long thermostatted tube. When the vessel is filled with liquid, the buoyant forces on the float stretch the spring. The technique is very precise but it suffers from the fragility of the spring, the difficulty in its calibration, and the inconvenience of automating the measurements.

A special balance of simple design for use submerged in fluids has been developed by Keramati and Wolgemuth [255]. Relative uncertainties below 0.2% were obtained at fluid pressures to 17 MPa.

Magnetic Buoys

Method of Lamb and Lee. If magnetic material is imbedded in an inert solid volume (the buoy), which is immersed in a fluid of unknown density ρ_F, the buoy can be maintained in equilibrium by adjustment of a magnetic field gradient. The solenoid producing the field can be conveniently placed outside the chamber containing the fluid. This idea was first exploited by Lamb and Lee [256]. Their work evidenced a remarkably low imprecision (ca. 0.1 ppm) for their time (1913) or, indeed, for our own. Lamb and Lee constructed a buoy whose volume was approximately 235 mL and whose density was just below 1000 kg/m³. The buoy, which would otherwise float in the test liquid, was held down by application of a magnetic field produced in a coil external to the liquid container. The buoy was constructed in such a way that it touched the bottom of the liquid container at a single point. The magnetic force produced by the electric current in the external coil was then slowly reduced while the contact point of the buoy was observed through a telescope. The current in the coil when contact was broken thus could be correlated with the density of the fluid. A modern version of the Lamb and Lee apparatus was developed by Millero [257], who paid particular attention to calibration of the device. Apparatus based on the same principle were designed for density measurements at high pressure [258], for general use over a wide density range [259], and for determinations of excess volume [260].

Methods of Beams and Clarke. In 1962, Beams and Clarke [261] presented two methods using magnetically suspended buoys for the determination of liquid densities. Method 1 is basically a Lamb and Lee device except that the buoy is maintained at a fixed position in the liquid by servocontrolling the current in the external coil. Thus a means for calibrating the apparatus in terms of electric current must be found. In Method 2, a cell containing the liquid and the buoy is suspended from a conventional analytical balance. Four balance readings are required corresponding to the following loads: the container alone, the container plus buoy, the container plus buoy plus liquid, the container plus liquid with the buoy magnetically suspended in the liquid. Method 2 does not rely at all on calibration of magnetic forces.

Devices based on Method 1 have proved particularly useful for measurements of the partial specific volumes of proteins. The reason is that the method is capable of low imprecision (1 ppm) in determinations of small volumes of liquid (0.3 mL). These and similar measurements are reviewed in [262]. Haynes [263] also reviewed magnetic densimeters and introduced useful design innovations. Devices tailored for special applications include: a magnetic densimeter capable of inferring excess volumes to an accuracy of 0.2 nL in samples of 0.1 mL [264]; a densimeter with a standard deviation of under 0.02% that operates at cryogenic temperatures and pressures to 5 MPa [265]; apparatus achieving a relative standard deviation of 0.15 ppm, used to determine density changes in distilled water due to dissolved air [266].

Method 2 of Beams and Clarke has found application in measuring the density of fluids over wide ranges of temperature and pressure. Successful devices used for density measurements of both liquids and vapors were developed at the National Physical Laboratory, in England [267, 268]. In the first apparatus, a buoy whose density is less than the test liquid is held in place by the action of a solenoid situated below. Uncertainties of about 50 ppm are attained at temperatures to 150°C and pressures to 0.7 MPa on sample volumes of 6 mL. For measurement of vapor densities, the buoy is necessarily more dense than the test fluid. Thus, in the second device, the buoy is suspended by the action of a solenoid placed above it.

Method of Gast. In another variation of method, workers at the NIST made an apparatus in which a solenoid hangs below a conventional electronic balance, and the fluid chamber, containing a magnetic buoy, rests firmly on the laboratory floor [269]. Servocontrol of the solenoid current keeps the buoy in stable levitation within the chamber. Fluids under pressures up to 15 MPa have been measured with a relative uncertainty of 0.1%.

Although quite different in construction, the principle of this device is identical to that of a balance with a magnetically suspended pan, as pioneered by Gast [89]. In Gast's design, the balance looks conventional with the single exception that the pan is coupled to the suspension by electromagnetic servocontrol. This allows the pan to be hermetically sealed in a nonmagnetic enclosure. An apparatus that resembles the Gast unit was used by Kleinrahm and Wagner [270] to measure the density of methane from the triple point to the critical point. A relative uncertainty of 0.02% or better was attained.

One can construct an apparatus of this type [271] without the need to servocontrol the balance pan. The technique, attributed to Kerl, requires only permanent magnets and thus achieves significant experimental simplicity. In place of servocontrol, the chamber containing the magnetic buoy is moved vertically with respect to an attractive magnet suspended from the balance arm. Initially, the chamber is close to the balance pan so that the buoy is held down by magnetic forces. Within the chamber, the buoy is free to move vertically between two stops. The necessary data are obtained from a record of the balance output as the chamber is raised: at the instant the buoy jumps between the stops, the magnetic force must just balance the effective weight of the buoy. The latter quantity can be inferred easily from the balance output. As with the other apparatus described in this subsection, the magnetic force law need not be known. Kerl's method exhibited an accuracy of 0.01% in the measurement of liquids at standard conditions.

FLOTATION METHODS

When the density of a body of arbitrary volume equals the density of a fluid in which it is immersed, the body will be in equilibrium. This is the principle of flotation methods. The volume of the body can, in principle, be any size large enough to be seen. As with the hydrostatic weighing methods given earlier, a solid of known density can be used to determine the density of a fluid and vice versa. Flotation methods, however, generally require the fluid to be transparent so that the float can be observed visually.

Although the necessary apparatus is generally uncomplicated, measurements are often slow and tedious. Use of gradient columns and ultracentrifugation gives convenience at the expense of additional equipment. Determination of solid densities is complicated by the absence of clear liquids with relative densities greater than 4. Suitable high-density liquids are, in general, highly toxic. Although dense solids can be determined by attaching them to a float of known volume, this technique reinstates a limitation on sample volume. Synthetic ferrofluids offer the possibility of creating a liquid with arbitrarily large and easily controlled pseudodensity. The various techniques listed in this paragraph are now discussed briefly.

Narrow-Range Instruments. In this section we discuss flotation devices designed to study a single sample. Assume that a float of mass m has a volume

$$V = V_0(1 + a\Delta t - k\Delta P) \tag{68}$$

and density

$$\rho = m/V$$

where Δt is the temperature above $20°C$ and ΔP is the pressure above 1 atm. For simplicity, we assume a and k to be constants equal to α and κ_T, respectively. We wish the float to be in equilibrium with a liquid of density

$$\rho_L = \rho_{0,L}/(1 + a'\Delta t - k'\Delta P) \tag{69}$$

Retgers' method (solids). To determine the density of the float, one method is to work at atmospheric pressure with the sample precisely controlled at some convenient temperature. Two liquids, whose densities bracket ρ are mixed to form a solution having a density approximately equal to ρ. (Commonly used liquids are given in Table 1.3.) If the solid sinks in the solution, a small amount of the more dense liquid is added until it floats. Less dense liquid is added if the solid floats. Eventually, one reaches a condition in which the solid will sink or

Table 1.3 Useful Liquids for Flotation Methods

System	Density Range[a] (kg/m³)
Methanol–benzyl alcohol	800–920
2-Propanol–water	790–1000
2-Propanol–diethylene glycol	790–1100
Ethanol–carbon tetrachloride	790–1590
2-Propanone–methylene iodide	790–3200
Toluene–carbon tetrachloride	870–1590
Water–sodium bromide	1000–1410
Water–calcium nitrate	1000–1600
Isopropyl salicylate–*sym*-tetrabromoethane	1100–2900
Carbon tetrachloride–trimethylene dibromide	1600–1990
Trimethylene dibromide–ethylene bromide	1990–2180
Ethylene Bromide–bromoform	2180–2900
sym-Tetrabromoethane–methylene iodide	2900–3200
Aqueous salts[b]	1000–2490
Thallium malonate–formate (Clerici solution)[b]	1000–4050

[a]The density range is approximate for room temperature.
[b]R. P. Cargille Laboratories, Inc., Cedar Grove, NJ.

float with the addition of a very small amount of liquid. At this point, the density of the solid is bracketed. The density of the liquid mixture can then be determined by any other convenient technique. This technique, attributed to Retgers, is described in the review by Wulff and Heigl [272]. These authors were able to achieve a relative imprecision of 50 ppm for crystals of 100 µg. Since organic liquids are used, this level of precision requires temperature stability to 10 mK. The method of Wulf and Heigl is tedious and thus seldom used.

One of the most serious problems with the method as applied to small samples is the length of time necessary to determine with certainty whether the solid will ultimately rise or sink in the liquid. The force driving the solid is equal to $(\rho - \rho_L) V g$, where once again g is the gravitational acceleration. As ρ_L is made closer to ρ, the force tends toward zero and it takes increasingly longer to determine the sign of $(\rho - \rho_L)$. However, if instead of relying on gravity, one places the sample cell in a centrifuge, g is replaced by the centrifugal acceleration, which is much greater. Profitable use of a centrifuge was first demonstrated by Bernal and Crowfoot [273]. The topic is treated extensively in a review by Oster and Yamamoto [274].

Temperature method (*solids*). Rather than adjust the density of the liquid by changing its composition at constant temperature, it is more convenient to change the controlled temperature until the solid body is in buoyant equilibrium with the liquid at atmospheric pressure. From (68) and (69), equilibrium will occur at

$$\Delta t = \left(1 - \frac{m}{V_0 \rho_{0,L}} \right) \frac{1}{(a' - a)} \tag{70}$$

The liquid expansion α' is generally an order of magnitude larger than α and thus can be used to estimate Δt in (70). The liquid density must be known or measured at the equilibrium temperature. In addition, the thermal expansion of the solid must also be known or determined.

It is convenient to do these auxiliary measurements in the same vessel used to find the flotation equilibrium. Two methods of accomplishing this are Andreae's technique, by which the flotation temperature is found in a pycnometer [275]; and a technique due to Oosterhout, by which small samples are floated in one arm of a capillary U tube after which the density of the liquid is determined by the balanced column method (see below) [276].

The measurements of Hutchison and Johnston [277] on the density of lithium fluoride crystals remain a good introduction to techniques of flotation by temperature variation. Many experimental details can be found in [278]. Flotation methods, although appropriate for small samples, are by no means limited to them [279].

Solids too dense to be floated in conventional liquids can be attached to low-density floats. Equation (57) then applies at flotation equilibrium, where ρ_{sus} now indicates the density of the liquid (equal to the float–sample combination) and the subscript R now refers to the float alone. From (57), it is clear that equilibrium may be achieved by varying the mass of the float, the mass of the sample, or the density of the liquid. The density of the float may also be varied, but this is inconvenient because the density of the float must be known and, therefore, measured whenever it is changed. In addition to these considerations, precision and accuracy are determined by the ratio of float volume to sample volume, knowledge of the float density, and measurement of the masses involved. Relatively simple apparatus are described in [280–283]. A thorough treatment of a device sensitive to a relative change of less than 1 ppm in the density of the liquid has been given by Spaepen [284]. By using the device, Spaepen showed that the density of samples of fine platinum wire was less than the bulk density.

Cartesian diver method (solids). An alternative to adjustment of the temperature of flotation at constant pressure is adjustment of pressure at constant temperature. This follows directly from (68) and (69). Devices exploiting this principle are often known as *Cartesian divers*. Since liquids and solids are nearly incompressible, the sample is usually attached to a compressible float that determines the sensitivity and range of the device with respect to changes in pressure. Despite experimental difficulties, the technique has the great advantage that pressure changes at the liquid surface are transmitted almost instantaneously (as opposed to temperature changes, which propagate by diffusion). Because the pressure in a liquid column has a hydrostatic head component that decreases with height, equilibrium is unstable. This problem is not, however, serious. Temperature must be controlled because the relation (70) is still in effect when pressure equilibrium is attained.

Accuracy and precision depend on many experimental variables. The analysis of Haller and Calcamuggio [285] gives a good introduction to experimental considerations. The imprecision of their device corresponds to about 10 μg in the mass of a sample. Bowman and Schoonover [286], making several important innovations, lowered their imprecision to a remarkable 0.2 μg. They were able to measure the relative densities of 2-g crystals of silicon to a relative imprecision of 0.2 ppm. Their accuracy was limited by available density standards. Instrumentation advances should now make possible the addition of automatic servocontrol of the pressure, thereby simplifying the taking of data.

One thus uses the cartesian diver simply as another type of immersed balance. It follows that an incompressible pycnometer of small volume may be weighed to high precision on such a device. Gilfillan and Polanyi [287] constructed an apparatus for comparing small amounts of liquid (5 mm^3) with a standard of nearly equal density. Lauder [288] used a similar device to study changes in water density as a function of dissolved gas.

Ferrofluid method (solids). Ferrofluids are stable colloidal suspensions of magnetite particles in various liquids. They have remarkable properties because of their maintenance of liquid behavior even when magnetized strongly. Of interest here is a property known as *buoyant levitation of the first kind* by which a nonmagnetic object can be made to experience an additional buoyantlike force proportional to a magnetic field gradient in the vertical direction [289]. When properly adjusted, the field can levitate a nonmagnetic sample of arbitrary shape and density. This phenomenon has been exploited by Hughes and Birnie [290], who were able to float a variety of samples from quartz to platinum in the same fluid. The relative accuracy of the Hughes–Birnie apparatus appears to be about 2% on samples as small as 4 μg. Sample size was limited by the ability to observe the specimen with the aid of a microscope. Several ferrofluids are available commercially.†

Temperature method (liquids). The determination of liquid density by a narrow-range instrument relies on the temperature of flotation of an auxiliary buoy of known density. This technique is also referred to as the *isopycnic method.* Equations (68) and (69) still apply. The technique has been used to study changes in water density, most notably in the work of Emeléus and co-workers and Spaepen [131, 284]. Both papers report low relative imprecision (< 1 ppm). For both apparatus it was noted that, near buoyant equilibrium, the velocity of the float is a highly reproducible linear function of the departure of the temperature from the true flotation temperature. This feature was exploited to reduce the observation time. Kozdon [291] used a flotation apparatus to conduct precision densimetric measurements of partial molar volumes in dilute binary liquids.

As with the magnetic float, temperature of flotation can be used to determine the density of small samples (ca. 0.3 mL) of liquid. Experimental considerations in making and using microfloats are given in [292]. The simplicity of the apparatus with respect to Beams and Clarke devices can make the isopycnic technique an attractive alternative as pointed out by Hunter [293], who describes a method suitable for determining partial specific volumes of proteins. Her apparatus, though simple, is sensitive to a change in density of 0.002 kg/m^3 in aqueous solutions.

Wide-Range Instruments. Instruments based on the Retgers method or the temperature of flotation are time-consuming to operate and not easily auto-mated. Use of centrifugation is one technique, described above, for speeding determinations. We now discuss other approaches. The first is based on creation of a continuous vertical density gradient in a liquid column. A sample whose density lies somewhere within the range of densities of the column can be determined with relative ease. This technique is applicable to both liquid and solid samples. A second technique, specific to dense solids, relies on a float whose buoyancy is made to be a linear function of height.

Gradient columns. If the density of a column of liquid varies with height in a known, stable fashion, the density of any sample coming to equilibrium within the column is determined to (68) and (69). The approach to equilibrium may be slow, but can proceed unattended. Many samples of slightly different density can

†Ferrofluidics Corporation, Nashua, NH.

be determined simultaneously. Gradient columns [274] are usually produced in one of two ways: (1) at constant temperature, two miscible liquids of different density are made to form a stable concentration gradient; and (2) a thermal gradient is produced along a column containing a single liquid or a homogeneous liquid mixture. There is an obvious trade-off between the density range of the column and the precision with which the density of a sample within that range can be determined.

A practical lower limit to the size of samples that can be determined, unaided by centrifugation, is imposed by the time necessary to attain the equilibrium condition. Thus, to achieve a relative precision of 1 ppm, samples should have dimensions larger than $30-60\,\mu m$ [294].

The concentration-gradient technique was first used by Linderstrøm–Lang [295] for relative density determinations of small drops of biological liquids. He also recognized the general utility of the technique for other applications. A similar device can be used to measure partial specific volumes [296], density of deuterium-enriched water [297], and densities of protein crystals [298, 299]. Various methods of forming concentration gradients have been described (e.g., [296, 297, 300, 301]). The last reference gives a theoretical treatment of density profile and stability of concentration columns. A properly maintained column can, in fact, be stable throughout several weeks of use. Gradient column apparatus is available commercially for determination of liquids† and solids.‡ Use of a column for the density determination of plastics is described in [302]. Rose and co-workers [303] measured changes in the density of UO_2 fragments after irradiation. They attach the fragment to a poly(tetrafluoroethylene) (PTFE) carrier whose mass is adjusted until the ensemble density is within range of the column densities.

Apparatus based on a thermally induced density gradient are described in [304] and [305]. A theoretical treatment is given in [294].

When used to determine densities of liquid drops, the density gradient is usually calibrated by reference drops of known density. It is obvious that both unknowns and references must be immiscible in the column liquid. The gradients of columns formed in a centrifuge are usually measured optically, taking advantage of the correlation of index of refraction with density (see below). The column can also be calibrated by a selection of solids of known density. Acquiring such solid standards to sufficient accuracy can be a difficulty requiring elaborate auxiliary apparatus [302, 305], although solid standards of specified density are available commercially†. Columns formed by thermal gradient can be calibrated through direct measurement of the temperature profile [304]. In favorable cases, it may be possible to monitor a concentration gradient by radioactive labeling of one component [306].

Gradients can be established by using an ultracentrifuge, as reviewed by Oster and Yamamoto [274]. Since that review, other centrifuge techniques were

†Anderson Laboratories, Inc., Forth Worth, TX.
‡Techne, Inc., Princeton, NJ.
†R. P. Cargille Laboratories, Inc., Cedar Grove, NJ. Techne, Inc., Princeton, NJ.

developed for determinations of microparticles as small as bacteria. A method due to Lange relies on a H_2O-D_2O gradient [307]. Another interesting technique, described by Mächtle [308], exploits a colloidal dispersion of PERCOLL in water. Particles of PERCOLL (PVP-coated SiO_2) have a range of sizes up to 35 nm and so produce a dynamic sedimentation gradient when ultracentrifuged. Microparticles to be determined form a turbidity band within the gradient, which is established in a matter of minutes.

Bowman float. A second wide-range device, suitable for the density determination of small solid samples of arbitrary density, was suggested by Bowman and first built by Franklin and Spal [309]. Two immiscible liquids are layered in a column. The liquids are separated by a meniscus, the more dense liquid being at the bottom. The apparatus is completed by a glass float in the upper liquid. A wire is attached below the float and extends through the meniscus. Test samples can be attached to the bottom of the wire. The sample produces a downward force on the float that is equal to the sample's effective weight in the more dense liquid. This force causes the float to lower its position by an amount determined by the cross-sectional area of the wire and the difference in density of the two liquids. Franklin and Spal's apparatus had a precision of 6 μg for samples with an effective mass of 17 mg in the lower liquid. Schoonover [310] built a similar device that is somewhat easier to use. A major advantage of the Schoonover design is that the device is calibrated using solid samples of known density—for example, bits of silicon and germanium.

3.3.3   Other Methods

Although most precise methods of determining density rely on the principles of pycnometry or buoyancy, several other techniques are available. We now examine the most important of these.

TERMINAL VELOCITY

If a sphere of radius r and density ρ falls at terminal velocity through a liquid, then its velocity v is given by

$$v = (kr^2/\eta)(\rho - \rho_F) \tag{71}$$

where η is the viscosity of the liquid, ρ_F is its density, and k is a constant. Barbour and Hamilton [311] recognized that an apparatus exploiting (71) could be designed to measure densities of small drops of liquid. Their technique, known as the *falling drop method*, is to measure the time t taken by the drop to fall a fixed vertical distance d in the liquid. Then

$$\rho = (\eta d/kr^2) \cdot (1/t) + \rho_F \tag{72}$$

By also measuring a drop of identical size and known density, one can determine $(\eta d/kr^2)$ without knowledge of the actual radius or viscosity. A second reference drop of different density will eliminate the need for prior knowledge of ρ_F. The technique is capable of precision to the sixth decimal place on sample volumes less than 10 mm^3, although many precautions are necessary to achieve both this precision and a comparable absolute accuracy [312]. Among the experimental parameters that must be well controlled are timing precision, temperature, and drop radius.

To achieve good timing precision and to avoid shape distortions, the drop should fall slowly. From (72), it is seen that this constraint limits the range of densities ρ, which can be determined with a single reference liquid. A related problem is finding a suitable liquid, immiscible with the sample that has the desired density. The review of Cohn [313] is instructive, especially regarding the determination of density variations in water samples. A slightly different apparatus is described fully by Hallaba and co-workers [314]. Hoiberg [315] describes a simple apparatus, good to 0.1%, for the determination of oils

Cap [316] presented a theoretical analysis of a microtechnique based on the speed with which the unknown liquid, contained in a capillary tube, is displaced by a reference liquid of lower density.

A time-of-fall technique developed by Roy [317] can be used to determine the densities of small solid samples of arbitrary shape. There is no maximum density limit to the solids. To eliminate shape effects, the unknown must be dropped in two separate liquids of different, known density. The viscosity of these fluids must be high so that the fall time will be long enough to be measured precisely. In analogy to the falling drop method, a small sample of known density is also dropped in the two liquids to account for the difference in their viscosities. (There is, even so, an advantage to choosing liquids with approximately equal viscosities.) Roy's data indicate the errors in the method are less than 0.4% for samples ranging in density from 2,650 to $10,500 \text{ kg/m}^3$ and having linear dimensions of about 1 mm. If two additional known solids are measured using the same apparatus, one avoids the need for measurement of the liquid densities.

PRESSURE HEADS

Balanced Column. Assume a tube is placed vertically into a reservoir of an incompressible liquid whose density is to be determined. If a partial vacuum is created in the tube, the liquid will rise by an amount

$$h = \Delta P/(\rho g) \tag{73}$$

where ΔP is the pressure differential between the top and bottom of the liquid column, g is the local gravitational acceleration, and ρ is the density of the liquid. This is the well-known principle behind manometers and mercury barometers. A column of reference liquid of known density ρ_F, subjected to the identical pressure differential, will rise to a height h_F. The density ρ is now readily calculated

$$\rho = \rho_F \left(\frac{h_F}{h}\right) \tag{74}$$

This technique is sometimes referred to as *Hare's method*. Using a balanced column, Frivold [318] was able to achieve a remarkably low imprecision (0.2 ppm) in the comparison of density between air-saturated and air-free water.

Small samples may be determined by this method. Ciochina [319] was able to measure 2-mL samples of volatile liquids to an imprecision of 0.05% using a relatively simple apparatus. An apparatus described by Blank and Willard [320] in their review of microdensity determinations can measure density to an uncertainty of less than 0.5% in samples of less than 1 mL. Gouverneur and Van Dijk [321] review previous balanced column methods and advance their own apparatus, designed for small samples (0.2 mL) of petroleum products. A very simple setup is described by van Oosterhout [276] in conjunction with determinations of small solid samples by the temperature of flotation method (see above).

If two different liquids are filled to equal height in separate columns, the difference in pressure as measured at the bottom of each column is proportional to the column height and the difference in density between the two liquids. This follows from (73). A sensitive differential instrument based on this principle has been used by Blair and Quinn [322] to study the effect of dissolved gas on the density of a liquid.

Maximum Bubble Pressure. Suppose one end of a narrow tube is lowered to a depth h beneath the surface of a reservoir of liquid whose density is to be determined. From (73), if the top end of the tube is pressurized by an amount ΔP, liquid will be forced all the way to the bottom of the tube. If the pressure is further increased by enough to overcome surface tension, a gas bubble will be created. The overpressure required to form a bubble is proportional to σ/r, where σ is the surface tension of the liquid and r is the bubble radius; thus the maximum pressure occurs at the creation of a bubble. Since σ is independent of h, the extra pressure required to initiate a bubble can be eliminated from the density calculation by taking the difference in the maximum bubble pressure between two immersion depths. If the required pressure is determined by a manomometer, one sees the conceptual similarity between this method and the balanced column method described above. The maximum bubble-pressure method is used frequently for melts and slags where high viscosity and temperature preclude other techniques. A typical application is described by Gaskell and co-workers [323], who achieved uncertainties of less than 1%.

OPTICAL TECHNIQUES

The index of refraction n of a pure, colorless fluid is related to its molar volume v_m through the Lorentz–Lorenz equation

$$\mathscr{R} = \left(\frac{n^2 - 1}{n^2 + 2}\right) v_m \tag{75}$$

where $\mathscr{R}$ is the molar refraction, related to the polarizability of a molecule in the fluid [324]. There are many cases where $\mathscr{R}$ is nearly constant, even for liquids.

For pure fluids (especially gases), nonideal behavior of $\mathscr{R}$ can be conveniently described by a power expansion in ρ_m [325].

When two pure fluids are mixed at constant temperature and pressure, the molar refraction of the mixture is approximately equal to the contribution of the individual components weighted by their respective mole fraction. An excess molar refraction can be defined and measured in analogy to excess molar volume [326]. Tables of the molar refraction of binary liquids as a function of component concentration, if available, can be used to determine density from measurements of index of refraction. Clever use of the refractive index–density correlation is made in the design of a fiberoptic "hydrometer" for determining the density of a battery electrolyte [327].

Even if nonideal behavior is ignored, one can often achieve sufficient accuracy. Zamvil and co-workers [328] found, for instance, that such an approach produces errors smaller than 0.5%, which is sufficient for use with their temperature-of-flotation apparatus.

Optical techniques are used traditionally to quantify the concentration gradient in centrifuge measurements. The schlieren method, until recently the most popular, is sensitive to the first derivative of the refractive index. A good description of this technique is given by Oster and Yamamoto [274]. Additional optical methods—Rayleigh fringe displacement and optical density—are now in frequent use. The Rayleigh method measures n directly. Optical density methods measure the absorption of a particular frequency of light (usually in the ultraviolet) as a result of transit through the fluid. Since the technique relies on absorption at a convenient optical frequency, it is obviously not limited to clear liquids. These optical detection schemes are reviewed briefly by Walton [329]. Alber [176] discussed the use of schlieren methods for density determinations of small samples. An experimental example of the correlation of relative density with optical density for bacterial suspensions is given in [227].

MISCELLANEOUS TECHNIQUES

Virtually any physical phenomenon that owes a part of its theoretical description to the density of a substance can be made into an analytical technique for density. The most generally useful of these phenomena have been described earlier. However, the list is by no means exhausted. In this section, we catalog a selection of additional techniques. They are chosen to reflect diverse possibilities. Recent references were selected where possible. These refer in turn to the fundamental work in each area.

1. Liquid densities can be related to the oscillation amplitude of a submerged disk whose plane is oriented parallel to the surface. The disk is forced to vibrate at right angles to its plane [330]. Density is determined by the amplitude of oscillation at the driving frequency. This method differs from acoustic resonance techniques, as exemplified by the VTD, which relate density to a resonant frequency. The oscillation–amplitude technique is recommended by its developers for determinations in fused salts, liquid metals, and slags.

2. Gamma-ray attenuation can probe the density distribution within compacted powders [331]. A similar method with reduced spacial resolution finds industrial application in monitoring the density of liquids and slurries as they flow through piping. If the γ radiation is produced by positron annihilation, use of coincidence detection offers improved accuracy [332].

3. Compton scattering can be used to determine the absolute density of bone *in vivo* [333].

4. Ultrasonic velocity measurements in liquids, plastics, and sintered metals can be related quantitatively to density [334–336].

5. The densities of gas–liquid systems at high pressure and temperature can be measured by a nuclear spin projection technique [337]. The method relies on a special NMR apparatus.

6. Techniques based on sedimentation field-flow fractionation can be used to determine densities (and, simultaneously, particle sizes) of microparticles [338]. An imprecision as low as 0.1% is possible.

7. Dielectric measurements of fluids can be correlated with density by using the molecular Clausius–Mossotti formula. This is analogous to the Lorentz–Lorenz relation at optical frequencies. Liu and Chow [339] used this technique to measure the densities of cryogenic liquids.

3.4   Dilatometry

Dilatometers directly or indirectly measure the change in volume of a sample as a function of a parameter of interest—temperature, for example. From (1), the volume change can be inferred indirectly from the mass of the sample and the change in its density. Most of the techniques described earlier can thus be used. Indeed, several of the references already cited describe devices configured as dilatometers. In this section, we discuss apparatus that directly measure the volume change of a sample. We divide the instruments into four major applications: thermal processes, volumes of mixing (excess volumes), partial molar volumes, and rates of reaction.

3.4.1   Thermal Processes

For reasons described in Section 3.1, it is often necessary to find the change in sample volume as a function of temperature. A simple dialtometer for this work is described in detail by Bekkedahl [340]. The basic principle is that of a liquid-in-glass thermometer. The sample occupies as much of the volume of the thermometer bulb as is practical and a confining liquid of known properties (usually mercury) fills the remaining volume to the lower part of a calibrated capillary tube. The thermal behavior of the sample can be found by subtracting known thermal corrections for the glass and confining liquid from the total observed volume change. Apparatus operating on the same principle can be configured to work at high pressure [146, 341]. Readings of the capillary can be automated by use of a LVDT [342] or by an electrically conducting wire dipping into the mercury [343].

A second design, used for small samples of biological liquids, confines the sample to the interior of a precision bellows [344]. Expansion of the bellows is measured by an LVDT. Differential measurements between two bellows can be measured optically with a relative imprecisions of 0.02 μL [345].

In the case of isotropic solids, changes in length of the sample can be measured and related easily to the change in volume as in the example of [121]. Linear variable differential transformer-based measurements are used widely in industry to measure the thermal characteristics of ceramics and powder compacts. One such commercial device has been modified by De Jonghe and Rahaman [346] to study the dilation properties of samples under stress.

A problem with measuring either the height of mercury in a capillary or the extension of a bellows is that a single apparatus cannot combine high precision with wide range. Three remedies are available. First, the volume of confining liquid or the volume of the reservoir can be readjusted to bring the dilatometer back on scale [347].

A second alternative method relies on the mass of mercury expelled from the dilatometer to determine the change in total volume. Such a device is described by Gibson and Loeffler [348]. The drawback to this solution is added intricacy of experimental technique.

A third alternative is the clever suggestion of Fortier and co-workers [349]. They point out that instead of reading the height reached by the containing liquid in a capillary, one can obtain the necessary dilatation data by measuring the rate of volume flow of the containing liquid past a point in the capillary. If the temperature change of the sample is known as a function of time, then thermal expansion of the sample can be determined from the measured flow rate. A device based on this principle lends itself to automation. Fortier and co-workers [349] achieved a detection limit of 10^{-9} L/s.

3.4.2   Volume of Mixing

EXCESS VOLUME

Dilatometers suitable for excess volume measurements were reviewed thoroughly by Handa and Benson [140]. Direct-measuring devices come in two types— batch and continuous.

Batch dilatometers can be extremely simple in design. A single filling of a batch dilatometer yields the excess volume for a single mole fraction of solvent. A widely used design is that of Duncan and co-workers [350]. Measured quantities (0.2–0.8 g) of the two liquids to be mixed are separated by a mercury barrier in the base of the dilatometer. The free surface of the mercury reaches the bottom of a capillary stem that connects to the base. The liquids are then mixed by rocking the dilatometer. The resulting change in height of the mercury meniscus is related to the excess volume. One drawback of the batch-type dilatometer is that many fillings are required to obtain sufficient data to fit an empirical curve of excess volume as a function of solvent mole fraction. By using capillaries of various bores, uncertainty in excess volume determinations can be limited to 0.5% for excess volumes as small as 1 μL [350]. A batch dilatometer constructed of readily available parts and designed for use at high pressure (400 MPa) measures volume changes to an uncertainty of less than 0.1 μL [351].

Dilution dilatometers allow the addition of arbitrary increments of solvent to the solute up to a mole fraction of at least 0.5. Thus all mole fractions can be measured in sufficient detail during two runs of data taking. A successful instrument of this type was developed by Stokes and co-workers [352] and later improved by Bottomley and Scott [353]. Kumaran and McGlashan [354] introduced further refinements, and their design has been widely adopted by other investigators (see Figure 1.8).

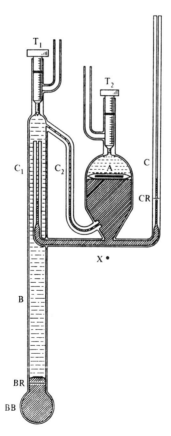

Figure 1.8 Dilutation dilatometer developed by Kumaran and McGlashan. Mercury, represented by diagonal lines, is used to separate the diluent in *B* from the liquid in *A*. When the apparatus is rotated counterclockwise, mercury flows from C_1 into *B*. This displaces an equal volume of diluent through C_2 into *A*. After the apparatus is restored to its original orientation, measurement of the heights of the mercury minisci can be used to determine the change in volume upon mixing. Reprinted with permission from M. K. Kumaran and M. L. McGlashan, *J. Chem. Thermodyn.*, **9**, 259 (1977).

PARTIAL MOLAR VOLUME

A dilution dilatometer based on designs similar to those used for excess volume measurements [353, 354] can also be used to extrapolate apparent molar volumes of aqueous electrolytes to low concentration [355]. Other designs may offer advantages, for example [356, 357]. The case of gases dissolved in a solvent liquid requires specialized dilatometers as, for example, the device described in [358].

A batch dilatometer developed by Linderstrøm–Lang and Lanz [359] has provided a wealth of useful data on biological materials. Details of design and use are given in a review by Katz [360].

OTHER USES

Dilatometers have also been used to follow chemical reactions as a function of time, as in the design of Tong and Olson [361]. It is, of course, essential that the volume of products differs from the volume of reactants. Interferences that also affect volume, such as temperature changes, must be controlled [361]. A practical example of a type of reaction dilatometry comes from the need to know the shrinkage of material used for dental fillings. Many modern dental fillings are composite resins that polymerize upon photoactivation. A dilatometer designed to follow the course of such a reaction is shown in Figure 1.9 [362]. Shrinkage can be tracked with an imprecision of 10^{-5} cm^3 in a small (0.1-cm^3) sample. The sample chamber is made by modifying a spherical glass joint. The height of the mercury column is followed by an LVDT.

Reaction volumes can be measured by dilatometry and used to study the pressure dependence of ionic equilibria [154, 155].

3.5   Measurement of the Density of Gases

The density ρ of a gas may be described by the following equation of state

$$\rho = MP/ZRT \tag{76}$$

where P is the gas pressure, M is the molar mass, T is the absolute temperature, and R is the universal gas constant. The compressibility factor Z is equal to 1 if the gas is "ideal." Gas behavior approaches ideality as the pressure is reduced. Depending on which four quantities on the right-hand side of (76) are known, measurement of the gas density can be used to determine the fifth quantity. Gas density measurements are important in determining molar mass and compressibility factors, as well as for controlling industrial processes. In the case of a binary gas, knowledge of the molar mass can be used to determine the mole fraction x of a given component since

$$M = M_1 x + M_2 \cdot (1 - x) \tag{77}$$

Here M_1 is the molar mass of the component whose mole fraction is x, and M_2 is the molar mass of the second component gas. Obviously, sensitivity is achieved

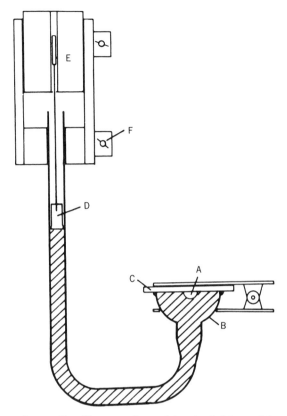

Figure 1.9 Automatic recording dilatometer for studying the shrinkage of dental material [362].

only to the extent that the two molar masses differ. The compressibility factor may not be known for a gas whose molecular weight is to be determined. If such is the case, one can extrapolate data over a range of pressures to find ρ at $P = 0$. A simple apparatus for the determination of the vapor density of a volatile liquid is described [363]: One measures the volume of a known mass of gas at a known temperature and a known pressure. In another simple apparatus [364], which can be used to find the molar mass of a gas, one measures the decrease in mass of a constant volume of gas as its pressure is reduced along an isotherm.

Some of the preceding techniques are suitable for measurement of gas density over extended ranges of temperature and pressure. For example, magnetic float measurements [268] and index-of-refraction measurements [325] have both been used effectively. Traditional methods, including the Burnett apparatus, are nicely reviewed by Ellington and Eakin [365].

In Section 2.5 we showed how the sensitivity of a microbalance could be determined by means of the buoyancy difference between bodies of different volume in a gas of known density. The same apparatus might then be used to

Table 1.4   General Guide to the Most Common Methods of Measuring Densities[a]

Sample	Amount	Precision	Method
Solid			
single piece	Large	High	Hydrostatic weighing
		Low	Volumenometry
			Gas pycnometry
	Small	High	Temperature of flotation
			Cartesian diver
			Retgers' method
		Moderate	Gradient column
			Bowman float
			Terminal velocity
		Low	Hydrostatic weighing
			Liquid pycnometry
			Levitation in ferrofluid
powder	Large	High	Hydrostatic weighing
			Liquid pycnometry
		Moderate	Gas pycnometry
	Small	Moderate	Gradient column in centrifuge
			Vibrating-tube densimeter
Liquid	Large	High	Vibrating-tube densimeter
			Pycnometry
			Hydrostatic weighing
			Lamb and Lee method [256]
			Beams and Clarke methods [261]
			Gast method [89]
			balanced column
		Moderate	Kohlrausch method
		Low	Hydrometer
			Maximum bubble pressure
			Weighed syringe
	Small	High	Vibrating-tube densimeter
			Beams and Clarke methods [261]
			Temperature of flotation
			Terminal velocity
		Moderate	Pycnometry
			Gradient column
			Balanced column
			Weighed capillary
Gas	Large		Buoyancy balance
			Weighed vessel
			Magnetic buoy

[a]The listing is not exhaustive and is necessarily vague. Most high-precision techniques are adapted easily to lower precision.

find the density of an atmosphere of unknown gas relative to the known gas. Thus, microbalances that act as force transducers can be configured for the study of gas density. A servocontrolled gas-density balance designed by Gast and Gebauer [366] is an updated and simplified version of earlier devices such as that of Johnson and Nash [367]. The latter instrument has an imprecision of 0.001 kg/m^3 with good zero stability.

A gas density balance of radically different design has been used as a detector for gas chromatography (see, for instance, [368–370]). The principle of operation, though based on flowing gas, is reminiscent of the balanced column technique for liquids.

3.6 Summary

As with all general methods that are adapted to satisfy diverse requirements, density measurement techniques do not lend themselves entirely to a tidy categorization. For the most part, we have ordered the techniques we have discussed by their underlying physical principles, which we believe is the most logical way to have proceeded. For quick reference, however, we summarize in Table 1.4, which of the many techniques for density determinations are most commonly used. As indicated in this table, the choice depends on the size sample available and the precision required.

Acknowledgments

Sections 2 and 3 are basically revisions and updates of the chapters on mass and density determinations in the 1972 edition of this series. We owe a debt to the authors of that edition for the basic structure and scholarship of these sections. We also thank numerous colleagues, authors of technical papers cited, and manufacturers of laboratory equipment for their cooperation.

References

1. M. Kochsiek, Ed., *Handbuch des Wägens*, Vieweg, Braunschweig, 1985.

2. U. Erdem, *J. Phys. E*, **15**, 857 (1982).

3. A. W. Czanderna and S. P. Wolsky, "Introduction and Microbalance Review," in A. W. Czanderna and S. P. Wolsky, Eds., *Microweighing in Vacuum and Controlled Environments*, Elsevier, Amsterdam, 1980, Chap. 1.

4. D. T. Goldman and R. J. Bell, *Natl. Bur. Stand. (U.S.) Spec. Publ. 330* (1986).

5. R. Glazebrook, Ed., *A Dictionary of Applied Physics*, Vol. 3, Macmillan, New York, 1923; P. Smith, "Balances," *A Dictionary of Applied Physics*, Vol. 3, Macmillan, New York, 1950.

6. P. Curie, *Oeuvres de Pierre Curie*, Soc. Francaise de Phys., Paris, 1908, pp. 517–548.

7. J. K. Taylor and H. V. Oppermann, *Natl. Bur. Stand. (U.S.) Handb. 145* (1986).

8. H. R. Jennemann, *Fresenius' Z. Anal. Chem.*, **291**, 1 (1978).

9. K. B. Jaeger and R. S. Davis, *Natl. Bur. Stand. (U.S.) Spec. Publ. 700-1* (1984).

10. M. Thiesen, *Trav. Mem. Bur. Int. Poids Mes.*, **5**, 3 (1886).

11. W. Felgentraeger, *Feine Waagen, Wagungen und Gewichte*, 2nd ed., Springer, Berlin, 1932.

12. A. J. F. Metherell and C. C. Speake, *Metrologia*, **19**, 109 (1983).

13. Y. Kobayashi, Y. Nezu, K. Uchikawa, S. Ikeda, and H. Yano, *Bull. Natl. Res. Lab. Metrol. Tokyo*, **35**, 7 (1986).

14. A. H. Corwin, "Weighing," in A. Weissberger and B. W. Rossiter, Eds., *Physical Methods of Chemistry*, Vol. 1, Part 4, Wiley–Interscience, New York, 1972, Chap. 1.

15. C. J. van Duyn, C. H. Massen, and J. A. Poulis, "Errors in Weighing Caused by Combining Different Types of Restoring Forces," in C. Eyraud and M. Escoubes, Eds., *Progress in Microbalance Techniques*, Vol. 3, Heyden, London, 1975, Chap. 5.

16. N. Bignell, *Metrologia*, **19**, 127 (1983).

17. H. A. Bowman and L. B. Macurdy, *J. Res. Natl. Bur. Stand. (U.S.)*, **63C**, 91 (1959).

18. G. G. Luther and W. R. Towler, "Redetermination of the Newtonian Gravitational Constant 'G'," *Proceedings of the Conference on Precision Measurement and Fundamental Constants II*, NIST, Gaithersburg, MD, June 8–12, 1981, in B. N. Taylor and W. D. Phillips, Eds., *National Bureau of Standards Special Publication No. 617*, NIST, Washington, DC, 1984, p. 573.

19. M. Kochsiek, R. Kruger, and H. Kunzmann, *Feinwerk. Mess.*, **85**, 86 (1977).

20. B. P. Kibble, *Metrologia*, **11**, 1 (1975).

21. A. H. Corwin, *Ind. Eng. Chem. Anal. Ed.*, **16**, 258 (1944).

22. L. B. Macurdy, *J. Res. Natl. Bur. Stand. (U.S.)*, **68C**, 135 (1964).

23. H. A. Bowman and R. M. Schoonover, with Appendix by M. Jones, *J. Res. Natl. Bur. Stand. (U.S.)*, **71C**, 179 (1967).

24. H. E. Almer, *J. Res. Natl. Bur. Stand. (U.S.)*, **64C**, 281 (1960).

25. H. Berman, *Am. Mineral.*, **24**, 434 (1939).

26. M. E. Cage and R. S. Davis, *J. Res. Natl. Bur. Stand. (U.S.)*, **87**, 23 (1982).

27. ASTM Standard E319, *Standard Practice for the Evaluation of Single-Pan Mechanical Balances*, American Society for Testing and Materials, Philadelphia, PA, most current printing.

28. T. W. Lashof and L. B. Macurdy, *Anal. Chem.*, **26**, 707 (1954).

29. J. W. Humphries, *Aust. J. Appl. Sci.*, **11**, 360 (1960).

30. D. F. Swinehart, *Anal. Lett.*, **10**, 1123 (1977).

31. K. Uchikawa, *Bull. Natl. Res. Lab. Metrol. Tokyo*, **33**, 32 (1984).

32. M. R. Winward, E. M. Woolley, and E. A. Butler, *Anal. Chem.*, **49**, 2126 (1977).

33. R. M. Schoonover, *Anal. Chem.*, **54**, 973A (1982).

34. H. K. P. Neubert, *Instrument Transducers*, Clarendon, Oxford, 1975.

35. H. Nakamura, *Jpn. J. Appl. Phys.*, **17**, 1397 (1978).

36. H. A. Bowman and L. B. Macurdy, *J. Res. Natl. Bur. Stand. (U.S.)*, **64C**, 277 (1960).

37. P. Smith, "Weighing Machines," *A Dictionary of Applied Physics*, Vol. 3, Macmillan, New York, 1950; D. M. Consodine, Ed., "Roberval Principle," *Van Nostrand's Scientific Encyclopedia, Sixth Edition*, Van Nostrand Reinhold, New York, 1982.

38. I. M. Longman, *J. Geophys. Res.*, **64**, 2351 (1959).

39. D. B. Prowse, *The Calibration of Balances*, Commonwealth Scientific and Industrial Research Organization (Australia), Melbourne, 1985.

40. ASTM Standard E898, *Standard Method of Testing Top-Loading, Direct-Reading Laboratory Scales and Balances*, American Society for Testing and Materials, Philadelphia, PA, most current printing.

41. J. Richlin, *Rev. Sci. Instrum.*, **56**, 476 (1985).

42. B. Dellby and H. E. Ekstrom, *J. Phys. E*, **4**, 342 (1971).

43. V. S. Ashanin, A. A. Stepanov, and A. V. Klyuchkin, *Zavod. Lab.*, **51**, 56 (1985).

44. J. McBain and A. Bakr, *J. Am. Chem. Soc.*, **48**, 690 (1926).

45. H. S. Tung and D. J. Friedland, *J. Phys. E*, **20**, 140 (1987).

46. R. M. Schoonover and F. E. Jones, *Anal. Chem.*, **53**, 900 (1981).

47. ASTM Standard E617, *Standard Specification for Laboratory Weights and Precision Mass Standards*, American Society for Testing and Materials, Philadelphia, PA, most current printing; *OIML Rec. No. 20*, International Organization of Legal Metrology, Paris, 1972.

48. R. S. Davis, *J. Res. Natl. Bur. Stand. (U.S.)*, **90**, 263 (1985).

49. W. F. Koch, W. C. Hoyle, and H. Diehl, *Talanta*, **22**, 717 (1975).

50. V. E. Bower and R. S. Davis, *J. Res. Natl. Bur. Stand. (U.S.)*, **85**, 175 (1980).

51. M. Kochsiek, *Wägen u. Dosieren*, **9**, 4 (1978).

52. P. H. Bigg and F. H. Burch, *Br. J. Appl. Phys.*, **2**, 126 (1951).

53. W. Bich, A. Cappa, M. Plassa, A. Torino, *Proceedings of the 8th Conference of the IMEKO Technical Committee TC3*, Krakow, Poland, 1980, p. 33.

54. A. Bonhoure, *Microtecnic*, **6**, 151 (1952).

55. P. Giacomo, *Metrologia*, **18**, 33 (1982).

56. R. Balhorn, "Berücksichtigung der Luftdichte durch Wägung beim Massevergleich," in M. Kochsiek, Ed., *Massebestimmung hoher Genauigkeit, Me-60*, PTB (FRG), Braunschweig, 1984, p. 65.

57. F. Jones, *J. Res. Natl. Bur. Stand. (U.S.)*, **83**, 419 (1978).

58. P. E. Pontius, *Natl. Bur. Stand. (U.S.), Monogr. 133* (1974).

59. G. P. Baxter, *J. Am. Chem. Soc.*, **43**, 1317 (1921).

60. I. Brulmans and H. L. Eschbach, *Rev. Sci. Instrum.*, **41**, 1680 (1970).

61. S. E. Toropin and V. C. Snegov, *Izmer. Tekh.*, **12**, 75 (1975).

62. W. F. Koch, R. S. Davis, and V. E. Bower, *J. Res. Natl. Bur. Stand. (U.S.)*, **83**, 407 (1978).

63. Y. Nezu and Y. Kobayashi, *Proceedings of the 10th Conference of IMEKO Technical Committee TC3*, Kobe, Japan, 1984, p. 105.

64. T. Yoshimori, *Talanta*, **22**, 827 (1975).

65. A. W. Czanderna and S. P. Wolsky, Eds., *Microweighing in Vacuum and Controlled Environments*, Elsevier, Amsterdam, 1980.

66. E. Blade, *Ind. Eng. Chem., Anal. Ed.*, **12**, 330 (1943).

67. R. M. Schoonover and J. E. Taylor, *IEEE Trans. Instrum. Meas.*, **IM-35**, 418 (1986).

68. J. W. Schürmann, C. H. Massen, and J. A. Poulis, "The Influence of Free Gas Convection on Balance Readings," in T. Gast and E. Robens, Eds., *Progress in Vacuum Microbalance Techniques*, Vol. 1, Heyden, London, 1972, p. 189.

69. M. Kochsiek, *Metrologia*, **18**, 153 (1982).

70. H. H. Perkins, Jr., *Text. Res. J.*, **45**, 25 (1975).

71. G. C. Lowenthal and V. Page, *Anal. Chem.*, **42**, 815 (1970).

72. H. A. Wyllie, *Anal. Chem.*, **58**, 3269 (1986).

73. R. E. Hawley and C. J. Williams, *J. Vac. Sci. Technol.*, **11**, 419 (1974).

74. D. R. Engelbrecht, T. A. Cahill, and P. J. Feeney, *J. Air Pollut. Control Assoc.*, **30**, 391 (1980).

75. A. E. Conrady, *Proc. R. Soc. London Ser. A*, **101**, 211 (1921).

76. L. D. Palmer, *J. Phys. E*, **13**, 920 (1980).

77. C. C. Speake, *Proc. R. Soc. London Ser. A*, **414A**, 333 (1987).

78. J. F. Blayzyk, D. L. Melchior, and J. M. Stein, *Anal. Biochem.*, **68**, 586 (1975).

79. J. A. Macinante and J. Waldersee, *J. Sci. Instrum.*, **41**, 1 (1964).

80. J. A. Rooney, *Ultrasound Med. Biol.*, **1**, 13 (1973).

81. S. Brunauer, P. H. Emmett, and E. Teller, *J. Am. Chem. Soc.*, **60**, 309 (1938).

82. E. Robens, "Physical Adsorption Studies With the Vacuum Microbalance," in A. W. Czanderna and S. P. Wolsky, Eds., *Microweighing in Vacuum and Controlled Environments*, Elsevier, Amsterdam, 1980, Chap. 4.

83. A. G. Forton and J. C. C. Day, *Thermochim. Acta*, **103**, 75 (1986).

84. M. Escoubes, C. Eyraud, and E. Robens, *Thermochim. Acta*, **82**, 15 and 23 (1984).

85. O. Pettersson, *Proc. Phys. Soc.*, **32**, 209 (1920).

86. J. W. Beams, C. W. Hulburt, W. E. Lotz, Jr., and R. M. Montague, Jr., *Rev. Sci. Instrum.*, **26**, 1181 (1955).

87. T. Gast, *Kunststoffe*, **34**, 117 (1944).

88. T. Gast, in *Proceedings of the 10th Conference of IMEKO Technical Committee T C3*, Kobe, Japan, 1984, p. 273.

89. T. Gast, *Naturwissenschaften*, **56**, 434 (1969).

90. R. L. Schwoebel, "Beam Microbalance Design, Construction and Operation," in A. W. Czanderna and S. P. Wolsky, Eds., *Microweighing in Vacuum and Controlled Environments*, Elsevier, Amsterdam, 1980, Chap. 2.

91. S. P. Wolsky, E. J. Zdanuk, C. H. Massen, and J. A. Poulis, "The Advantage of Using a Combination of Different Materials for Balance Beams," in A. W. Czanderna, Ed., *Vacuum Microbalance Techniques*, Vol. 6, Plenum, New York, 1967, p. 37.

92. H. L. Eschbach, I. V. Mitchell, and E. Louwerix, *Thermochim. Acta*, **51**, 33 (1981).

93. A. W. Czanderna and J. M. Honig, *Anal. Chem.*, **29**, 1206 (1957).

94. C. Lu and A. W. Czanderna, Eds., *Applications of Piezoelectric Quartz Crystal Microbalances*, Elsevier, Amsterdam, 1984.

95. A. W. Warner and C. D. Stockbridge, "Mass and Thermal Measurement With

Resonating Crystalline Quartz," in R. F. Walker, Ed., *Vacuum Microbalance Techniques*, Vol. 2, Plenum, New York, 1962, p. 71.

96. G. Sauerbrey, *Z. Phys.*, **155**, 206 (1958).

97. C. S. Lu and O. Lewis, *J. Appl. Phys.*, **43**, 4385 (1972).

98. E. P. EerNisse, *J. Vac. Sci. Technol.*, **12**, 564 (1975); E. P. EerNisse, "Stress Effects in Quartz Crystal Microbalances," in C. Lu and A. W. Czanderna, Eds., *Applications of Piezoelectric Quartz Crystal Microbalances*, Elsevier, Amsterdam, 1984, Chap. 4.

99. L. L. Levenson, "Applications of Quartz Crystal Microbalances in Surface Science," in C. Lu and A. W. Czanderna, Eds., *Applications of Piezoelectric Quartz Crystal Microbalances*, Elsevier, Amsterdam, 1984, Chap. 6.

100. G. G. Guilbault, "Applications of Quartz Crystal Microbalances in Analytical Chemistry," in C. Lu and A. W. Czanderna, Eds., *Applications of Piezoelectric Quartz Crystal Microbalances*, Elsevier, Amsterdam, 1984, Chap. 8.

101. R. L. Chuan, "Particulate Mass Measurement by Piezoelectric Crystal," in W. Cassatt and R. S. Maddock, Eds., *National Bureau of Standards Special Publication No. 412*, NIST, Washington, DC, 1974, p. 137.

102. M. H. Ho, "Applications of Quartz Crystal Microbalances in Aerosol Mass Measurement," in C. Lu and A. W. Czanderna, Eds., *Applications of Piezoelectric Quartz Crystal Microbalances*, Elsevier, Amsterdam, 1984, Chap. 10.

103. T. Nomura and O. Hattori, *Anal. Chim. Acta*, **115**, 323 (1980).

104. S. Bruckenstein and M. Shay, *Electrochim. Acta*, **30**, 1295 (1985).

105. K. K. Kanazawa and J. G. Gordon, *Anal. Chem.*, **57**, 1770 (1985).

106. M. D. Ward and D. A. Buttry, *Science*, **249**, 1000 (1990).

107. S. L. Bonting and B. R. Mayron, *Microchem. J.*, **5**, 31 (1961).

108. J. Skelly, *Rev. Sci. Instrum.*, **38**, 985 (1967).

109. H. Patashnick and C. L. Hemenway, *Rev. Sci. Instrum.*, **40**, 1008 (1969).

110. D. R. Andrews, *J. Phys. E*, **16**, 803 (1983).

111. J. C. F. Wang, H. Patashnick, and G. Rupprecht, *J. Air Pollut. Control Assoc.*, **30**, 1018 (1980).

112. J. C. F. Wang, B. F. Kee, D. W. Linkins, and R. W. Lynch, *J. Air Pollut. Control Assoc.* **33**, 1172 (1983).

113. G. F. Bahr and E. Zeitler, *Lab. Invest.*, **14(2)**, 955 (1965).

114. M. K. Lamvik, *Ultramicroscopy*, **1**, 187 (1976).

115. P. W. J. Linders, A. L. H. Stols, and A. M. Stadhouders, *J. Microsc.*, **130(1)**, 85 (1983).

116. R. D. Deslattes, *Annu. Rev. Phys. Chem.*, **31**, 435 (1980).

117. P. Smith, "Volume," *A Dictionary of Applied Physics*, Vol. 3, Macmillan, New York, 1950.

118. E. R. Cohen, K. M. Crowe, and J. W. M. Dumond, *The Fundamental Constants of Physics*, Interscience, New York, 1957, p. 5.

119. K. Lohner, P. Laggner, and J. H. Freer, *J. Solution Chem.*, **15**, 189 (1986).

120. A. H. Cook, *Br. J. Appl. Phys.*, **7**, 285 (1956); National Bureau of Standards (U.S.), *Standard Reference Material No. 739*, NIST, Washington, DC, 1971.

121. R. B. Roberts, *J. Phys. E*, **14**, 1386 (1981).

122. J. E. Garrod and T. M. Herrington, *J. Phys. Chem.*, **74**, 363 (1970).

123. J. Brulmans, J. Verdonck, and H. L. Eschbach, *Z. Naturforsch.*, **30A**, 107 (1975).

124. D. Chandra and L. R. Holland, *J. Vac. Sci. Technol. A*, **1**, 1620 (1983).

125. G. S. Kell, *J. Chem. Eng. Data*, **20**, 97 (1975).

126. F. Tanaka, O. Kiyohara, P. J. D'Arcy, and G. C. Benson, *Can. J. Chem.*, **53**, 2262 (1975).

127. A. J. T. Hayward, *J. Phys. D*, **4**, 938 (1971).

128. A. H. Cook, *Br. J. Appl. Phys.*, **15**, 1111 (1964).

129. H. Watanabe and K. Iizuka, *Metrologia*, **21**, 19 (1985).

130. G. S. Kell, *J. Phys. Chem. Ref. Data*, **6**, 1109 (1977).

131. H. J. Eméleus, F. W. James, A. King, T. G. Pearson, R. H. Purcell, and V. A. Briscoe, *J. Chem. Soc. London*, **1934**, 1207 (1934).

132. I. Kirshenbaum, *Physical Properties and Analysis of Heavy Water*, National Nuclear Energy Series, Division III, Vol. 4A, McGraw-Hill, New York, 1951.

133. ASTM Standard D2184, *Standard Method of Testing Deuterium Oxide*, American Society for Testing and Materials, Philadelphia, PA, most current printing.

134. D. Staschewski, *Ber. Bunsenges. Phys. Chem.*, **73**, 59 (1969).

135. T. Babeliowsky and W. E. De Bolle, *Metrologia*, **10** 129 (1974).

136. W. M. Thurston and M. W. D. James, *Anal. Chem.*, **56**, 386 (1984).

137. F. J. Millero, *Chem. Rev.*, **71**, 147 (1971); "The Partial Molal Volumes of Electrolytes in Aqueous Solutions," in R. A. Horne, Ed., *Water and Aqueous Solutions*, Interscience, New York, 1972, Chap. 13, p. 519.

138. D. W. Kupke, "Density and Volume Change Measurements," in S. J. Leach, Ed., *Physical Principles and Techniques of Protein Chemistry, Part C*, Academic, New York, 1973, Chap. 17, p. 1.

139. R. Battino, *Chem. Rev.*, **71**, 5 (1971).

140. Y. P. Handa and G. C. Benson, *Fluid Phase Equil.*, **3**, 185 (1979).

141. J. S. Rowlinson and F. L. Swinton, *Liquids and Liquid Mixtures*, Butterworth, London, 1982.

142. R. J. Powell and F. L. Swinton, *J. Chem. Eng. Data*, **13**, 260 (1968).

143. A. W. Awwad and R. A. Pethrick, *J. Chem. Soc. Faraday Trans. 1*, **78**, 3203 (1982).

144. G. A. Bottomley and R. L. Scott, *Aust. J. Chem.*, **29**, 427 (1976).

145. P. Engels and G. M. Schneider, *Ber. Bunsenges. Phys. Chem.*, **76**, 1239 (1972).

146. J. Wisniak and A. Tamir, *Mixing and Excess Thermodynamic Properties*, Elsevier, Amsterdam, 1978; supplements, 1982 and 1986.

147. F. J. Millero, L. M. Connaughton, F. Vinokurova, and P. V. Chetirkin, *J. Solution Chem.*, **14**, 837 (1985).

148. N. Nishimura, T. Tanaka, and T. Motoyama, *Can. J. Chem.*, **65**, 2248 (1987).

149. E. J. Cohn, T. L. McMeekin, J. T. Edsall, and M. H. Blanchard, *J. Am. Chem. Soc.*, **56**, 784 (1934).

150. H. Fujita, *Foundations of Ultracentrifugal Analysis*, Interscience, New York, 1975.

151. H. Pessen, T. F. Kumosinski, and S. N. Timasheff, "Small-Angle X-Ray Scatter-

ing," in C. H. W. Hirs and S. N. Timasheff, Eds., *Methods of Enzymology 27*, Academic, New York, 1973, p. 151.

152. O. Sohnel and P. Novotny, *Densities of Aqueous Solutions of Inorganic Substances*, Elsevier, Amsterdam, 1985.

153. J. T. Edward, P. G. Farrell, and F. Shahidi, *J. Chem. Soc. Faraday Trans. 1*, **73**, 705 and 715 (1977).

154. T. Asano and W. J. Le Noble, *Chem. Rev.*, **78**, 407 (1978).

155. S. Srivastava, M. J. DeCicco, E. Kuo, and W. J. Le Noble, *J. Solution Chem.*, **13**, 663 (1984).

156. O. Kratky, H. Leopold, and H. Stabinger, "The Determination of the Partial Specific Volume of Proteins by the Mechanical Oscillator Technique," in C. H. Wires and S. N. Timasheff, Eds., *Methods of Enzymology*, Vol. 27, Academic, New York, 1973, p. 98.

157. IUPAC, Commission I.4, *Recommendation for Redetermination of the Absolute Density of Water*, International Union of Pure and Applied Chemistry, Oxford, England, 1974.

158. M. Menaché, C. Beauverger, and G. Girard, *Ann. Hydrograph.*, 5th Ser., **6(750)**, 37 (1978).

159. G. Girard, "Density," in K. N. Marsh, Ed., *Recommended Reference Materials for the Realization of Physicochemical Properties*, Blackwell, Oxford, 1987, Chap. 2.

160. G. N. Lewis and R. E. Cornish, *J. Am. Chem. Soc.*, **55**, 2616 (1933).

161. R. A. Cox, M. J. McCartney, and F. Culkin, *Deep-Sea Res.*, **15**, 319 (1968).

162. B. W. Mangum, *J. Res. Natl. Inst. Stand. Technol.*, **95**, 69 (1990).

163. F. J. Millero and A. Poisson, *Deep-Sea Res.*, **28A**, 625 (1981).

164. H. A. Bowman, R. M. Schoonover, and C. L. Carroll, *J. Res. Natl. Bur. Stand. (U.S.)*, **78A**, 13 (1973).

165. R. S. Davis, *Metrologia*, **18**, 193 (1982).

166. C. A. Swenson, *J. Chem. Ref. Data*, **12**, 179 (1982).

167. A. H. Cook and N. W. B. Stone, *Philos. Trans. R. Soc. London Ser. A*, **250**, 279 (1957).

168. A. H. Cook, *Philos. Trans. R. Soc. London Ser. A*, **254**, 125 (1961).

169. J. B. Patterson and D. B. Prowse, *Metrologia*, **21**, 107 (1985).

170. J. B. Patterson and D. B. Prowse, *Metrologia*, **25**, 121 (1988).

171. R. Snel, *J. Phys. E*, **17**, 342 (1984).

172. S. Yamagishi and Y. Takahashi, *Meas. Sci. Technol.* **3**, 270 (1992).

173. B. Marlet, Report CEA-R-4954, 1978, available from International Atomic Energy Agency, Vienna.

174. C. L. Gordon and E. Wichers, *Ann. N.Y. Acad. Sci.*, **65**, 369 (1957).

175. J. E. Burroughs and C. P. Goodrich, *Anal. Chem.*, **46**, 1614 (1974).

176. H. K. Alber, *Ind. Eng. Chem. Anal. Ed.*, **12**, 764 (1940).

177. H. Schreiner, *Mikrochemie*, **38**, 273 (1951).

178. A. Haack and G. Wieser, *Mikrochim. Acta*, **1954**, 117.

179. G. Marinenko and J. K. Taylor, *J. Res. Natl. Bur. Stand. (U.S.)*, **70C**, 1 (1966).

180. N. Bauer and S. Z. Lewin, "Determination of Density," in A. Weissberger and B. W. Rossiter, Eds., *Physical Methods of Chemistry*, Vol. 1, Part 4, Wiley-Interscience, New York, 1972, Chap. 2.

181. P. Hidnert and E. L. Peffer, *Natl. Bur. Stand. (U.S.) Circ.*, **487**, 1950.

182. J. Reilly and W. N. Rae, *Physico-Chemical Methods*, 3rd ed., Vol. I, Van Nostrand, Princeton, NJ, 1939, p. 477.

183. J. Johnston and L. H. Adams, *J. Am. Chem. Soc.*, **34**, 563 (1912).

184. W. B. Newkirk, *Tech. Pap. Bur. Stand. (U.S.)*, **161**, 1 (1920).

185. ASTM Standard D1963, *Standard Test Method for Specific Gravity of Drying Oils, Varnishes, Resins, and Related Materials at 25/25°C*; ASTM Standard D70, *Standard Test Method for Specific Gravity and Density of Semi-Solid Bituminous Materials*, American Society for Testing and Materials, Philadelphia, PA, most current printings.

186. N. R. Amnuil, I. N. Pleskach, and V. V. Egorov, *Zh. Prikl. Khim.*, **44**, 2755 (1971).

187. ASTM Standard D1217, *Standard Test Method for Density and Relative Density (Specific Gravity) of Liquids by Bingham Pycnometer*, American Society for Testing and Materials, Philadelphia, PA, most current printing.

188. E. W. Washburn and E. R. Smith, *J. Res. Natl. Bur. Stand. (U.S.)*, **12**, 305 (1934).

189. E. R. Smith and M. Wojciechoeski, *Rocz. Chem.*, **16**, 104 (1936).

190. S. A. Brown and J. E. Land, *J. Chem. Educ.*, **33**, 72 (1956).

191. T. Shedlovsky and A. S. Brown, *J. Am. Chem. Soc.*, **56**, 1066 (1934).

192. G. R. Robertson, *Ind. Eng. Chem. Anal. Ed.*, **11**, 464 (1939).

193. J. P. Wibaut, H. Hoog, S. L. Langedijk, J. Overhoff, and J. Smittenberg, *Recl. Trav. Chim. Pays-Bas.*, **58**, 329 (1939).

194. P. P. Pugachevich and V. M. Zatkovetskii, *Zavod. Lab.*, **33**, 837 (1967).

195. S. Marantz and G. T. Armstrong, *Chemist-Analyst*, **55**, 114 (1966).

196. M. R. Lipkin, J. A. Davison, W. T. Harvey, and S. S. Kurtz, Jr., *Ind. Eng. Chem. Anal. Ed.*, **16**, 55 (1944).

197. ASTM Standard D941, *Standard Test Method for Relative Density (Specific Gravity) by Lipkin Bicapillary Pycnometer*, American Society for Testing and Materials, Philadelphia, PA, most current printing.

198. R. A. Phillips, D. D. Van Slyke, P. B. Hamilton, V. P. Dole, K. Emerson, Jr., and R. M. Archibald, *J. Biol. Chem.*, **183**, 305 (1950).

199. J. Campbell, *J. Phys. E*, **13**, 627 (1980).

200. E. Mathiak, W. Nistler, W. Waschkowski, and L. Koester, *Z. Metallkd.*, **74**, 793 (1983).

201. R. Barrue and J. C. Perron, *Rev. Sci. Instrum.*, **52**, 1536 (1981).

202. I. May and J. Marinenko, *Am. Mineral.*, **51**, 931 (1966).

203. J. N. Bohra, *Fibre Sci. Technol.*, **9**, 315 (1976).

204. T. Nozaki and M. Shimoji, *Trans. Faraday Soc.*, **65**, 1489 (1969).

205. B. Moses, *J. Catal.*, **7**, 290 (1967).

206. A. Kozdon, *Pomiary, Automa., Kontrola*, **17**, 250 (1971).

207. B. Buczek and D. Geldart, *Powder Technol.*, **45**, 173 (1986).

208. M. J. Knight and P. N. Rowe, *Chem. Eng. Sci.*, **35**, 997 (1980).

209. S. K. Kesavan, *J. Mater. Sci. Lett.*, **5**, 497 (1986).

210. M. Vanka and M. Kaclik, *Chem. Listy*, **74**, 550 (1980); W. B. Jepson, *J. Sci. Instrum.*, **36**, 319 (1959).

211. E. L. Gooden, *Ind. Eng. Chem. Anal. Ed.*, **15**, 578 (1943).

212. W. Cheng, *Anal. Chem.*, **57**, 2409 (1985).

213. W. W. Russell, *Ind. Eng. Chem. Anal. Ed.*, **9**, 592 (1937).

214. ASTM Standard C493, *Bulk Density and Porosity of Granular Refractory Materials by Mercury Displacement*, American Society for Testing and Materials, Philadelphia, PA, most current printing.

215. P. L. Fortucci and V. D. Meyer, *J. Vac. Sci. Technol.*, **16**, 963 (1979).

216. I. Suzuki, *Rev. Sci. Instrum.*, **54**, 868 (1983).

217. J. H. Petropoulos, K. Tsimillis, C. Savvakis, and V. Havredaki, *J. Phys. E*, **16**, 1112 (1983).

218. H. Hauptmann and G. E. R. Schulze, *Z. Phys. Chem.*, **A171**, 36 (1934).

219. H. Stabinger, O. Kratky, and H. Leopold, *Monatsh. Chem.*, **98**, 436 (1967); O. Kratky, H. Leopold, and H. Stabinger, *Z. Angew. Phys.*, **4**, 273 (1969).

220. P. Picker, E. Tremblay, and C. Jolicoeur, *J. Solution Chem.*, **3**, 377 (1974).

221. J. P. Elder, "Density Measurements by the Mechanical Oscillator," in C. H. W. Hirs and S. N. Timasheff, Eds., *Methods of Enzymology 61*, Academic, New York, 1979, p. 12.

222. H. J. Albert and R. H. Wood, *Rev. Sci. Instrum.*, **55**, 589 (1984).

223. O. Kiyohara and G. C. Benson, *Can. J. Chem.*, **56**, 2803 (1978).

224. P. Wells, *Anal. Biochem.*, **130**, 189 (1983).

225. J. Bernhardt and H. Pauly, *J. Phys. Chem.*, **84**, 158 (1980).

226. J. Francois, R. Clement, and E. Franta, *C. R. Acad. Sci. Ser. C*, **273**, 1577 (1971).

227. B. C. Blake-Coleman, D. J. Clarke, M. R. Calder, and S. C. Moody, *Biotechnol. Bioeng.*, **28**, 1241 (1986).

228. P. Smith, "Hydrometers," *A Dictionary of Applied Physics*, Vol. 3, Macmillan, New York, 1950; F. W. Cuckow, *J. Soc. Chem. Ind. London*, **68**, 44 (1949).

229. H. A. Bowman and W. H. Gallagher, *J. Res. Natl. Bur. Stand. (U.S.)*, **73C**, 57 (1969).

230. J. B. Patterson, *Water Density Measurement Using Hydrometers, CSIRO Division of Applied Physics (Australia) Technical Memorandum*, No. 37, 1986.

231. I. Henins, *J. Res. Natl. Bur. Stand. (U.S.)*, **68A**, 529 (1964).

232. H. Watanabe, Y. Okamoto, and K. Iizuka, *Oyo Butsuri*, **48**, 133 (1979).

233. J. E. Desnoyers and M. Arel, *Can. J. Chem.*, **45**, 359 (1967).

234. C. C. Chappelow, P. S. Snyder, and J. Winnick, *J. Chem. Eng. Data*, **16**, 440 (1971).

235. N. Fornstedt and J. Porath, *J. Chromatogr.*, **42**, 376 (1969).

236. G. V. Schulz, *Z. Phys. Chem.*, **40B**, 151 (1938).

237. H. Craubner, *Rev. Sci. Instrum.*, **57**, 2817 (1986).

238. J. Jadzyn and J. Malecki, *Rocz. Chem.*, **48**, 531 (1974).

239. S. Hiemstra, D. Prins, G. Gabrielse, and J. B. van Zyveld, *Phys. Chem. Liq.*, **6**, 271 (1977).

240. J. Seerveld, S. van Till, and J. B. van Zyveld, *J. Chem. Phys.*, **79**, 3597 (1983).

241. A. F. Forziati, B. J. Mair, and F. D. Rossini, *J. Res. Natl. Bur. Stand. (U.S.)*, **35**, 513 (1945).

242. H. A. Bowman, R. M. Schoonover, and C. L. Carroll, *Metrologia*, **10**, 117 (1974).

243. G. A. Bell, *Aust. J. Appl. Sci.*, **9**, 236 (1958).

244. R. T. Ratcliffe, *Br. J. Appl. Phys.*, **16**, 1193 (1965).

245. C. Cawthorne and W. D. J. Sinclair, *J. Phys. E*, **5**, 531 (1972).

246. W. W. Barker, *J. Appl. Crystallogr.*, **5**, 433 (1972).

247. Z. Kluz and I. Waclawska, *Rocz. Chem.*, **49**, 839 (1975).

248. J. L. Culbertson and A. Dunbar, *J. Am. Chem. Soc.*, **59**, 306 (1937).

249. ASTM Standard D1188, *Standard Test Method for Bulk Specific Gravity and Density of Compacted Bituminous Mixtures Using Paraffin-Coated Specimens*, American Society for Testing and Materials, Philadelphia, PA, most current printing.

250. M. R. Moldover, J. P. M. Trusler, T. J. Edwards, J. B. Mehl, and R. S. Davis, *J. Res. Natl. Bur. Stand. (U.S.)*, **93**, 85 (1988).

251. R. M. Schoonover and R. S. Davis, in *Proceedings of the 8th Conference of the IMEKO Technical Committee T C3*, Krakow, Poland, 1980, p. 23.

252. R. M. Schoonover, in *Preprints of the 10th Conference of the IMEKO Technical Committee T C3*, Kobe, Japan, 1984, Preprint 5-3.

253. G. H. Wagner, G. C. Bailey, and W. G. Eversole, *Ind. Eng. Chem. Anal. Ed.*, **14**, 129 (1942).

254. G. L. Gaines, Jr. and C. P. Rutkowski, *Rev. Sci. Instrum.*, **29**, 509 (1958).

255. B. Keramati and C. H. Wolgemuth, *Rev. Sci. Instrum.*, **46**, 1573 (1975).

256. A. B. Lamb and R. E. Lee, *J. Am. Chem. Soc.*, **35**, 1666 (1913).

257. F. J. Millero, Jr., *Rev. Sci. Instrum.*, **38**, 1441 (1967).

258. F. J. Millero, J. H. Knox, and R. T. Emmet, *J. Solution Chem.*, **1**, 173 (1972).

259. B. N. Barman and Z. Rahim, *Rev. Sci. Instrum.*, **48**, 1695 (1977).

260. I. A Weeks and G. C. Benson, *J. Chem. Thermodyn.*, **5**, 107 (1973).

261. J. W. Beams and A. M. Clarke, *Rev. Sci. Instrum.*, **33**, 750 (1962).

262. D. W. Kupke and J. W. Beams, "Magnetic Densimetry: Partial Specific Volume and Other Applications," in C. H. W. Hirs and S. N. Timasheff, Eds., *Methods of Enzymology*, Vol. 26, Academic, New York, 1972, p. 74; D. W. Kupke and T. H. Crouch, "Magmetic Suspension: Density–Volume, Viscosity, and Osmotic Pressure," in C. H. W. Hirs and S. N. Timasheff, Eds., *Methods of Enzymology*, Vol. 48, Academic New York, 1978, p. 29.

263. W. M. Haynes, *Rev. Sci. Instrum.*, **48**, 39 (1977).

264. D. W. Kupke, *Anal. Biochem.*, **158**, 463 (1986).

265. W. M. Haynes, M. J. Hiza, and N. V. Frederick, *Rev. Sci. Instrum.*, **47**, 1237 (1976).

266. N. Bignell, *J. Phys. E*, **15**, 378 (1982).

267. J. L. Hales, *J. Phys. E*, **3**, 855 (1970).

268. J. L. Hales and H. A. Gundry, *J. Phys. E*, **16**, 91 (1983).

269. R. Masui, W. M. Haynes, R. F. Chang, H. A. Davis, and J. M. H. Levelt Sengers, *Rev. Sci. Instrum.*, **55**, 1132 (1984).

270. R. Kleinrahm and W. Wagner, *Chem. Ing. Technol.*, **57**, 520 (1985).

271. H. Bettin, F. Spieweck and H. Toth, *Meas. Sci. Technol.* **2**, 1036 (1991).

272. P. Wulff and A. Heigl, *Z. Phys. Chem.*, **153A**, 187 (1931).

273. J. D. Bernal and D. Crowfoot, *Nature (London)*, **134**, 809 (1934).

274. G. Oster and M. Yamamoto, *Chem. Rev.*, **63**, 257 (1963); G. Oster, *Sci. Am.*, **213**, 70 (1965).

275. J. L. Andreae, *Z. Phys. Chem.*, **82**, 109 (1913).

276. G. W. van Oosterhout, *Philips Tech. Rev.*, **28**, 30 (1967).

277. C. A. Hutchison and H. L. Johnston, *J. Am. Chem. Soc.*, **62**, 3165 (1940); H. L. Johnston and C. A. Hutchison, *J. Chem. Phys.*, **8**, 869 (1940).

278. ASTM Standard C729, *Standard Test Method for Density of Glass by the Sink-Float Comparator*, American Society for Testing and Materials, Philadelphia, PA, most current printing.

279. P. Seyfried, R. Balhorn, M. Kochsiek, A. F. Kozdon, H.-J. Rademacher, H. Wagenbreth, A. M. Peuto, and A. Sacconni, *IEEE Trans. Instrum. Meas.*, **IM-36**, 161 (1987).

280. E. L. Simon, *Anal. Chem.*, **35**, 407 (1963).

281. M. Donnelly and R. Hayes, *Powder Metall.*, **15**, 11 (1972).

282. A. F. Naumov, N. A. Mamaev, and Z. N. Pudkova, *Zavod. Lab.*, **39**, 292 (1973).

283. A. Mitsuo and K. Shiba, Report JAERI-M-167, 1984, International Atomic Energy Commission, Vienna.

284. J. Spaepen, *Meded. K. Vlaam, Acad. Wet. Belg.*, **19**, No. 5 (1957); *Chem. Abstr.*, **52**, No. 15170e (1958).

285. W. K. Haller and G. L. Calcamuggio, *Rev. Sci. Instrum.*, **26**, 1064 (1955).

286. H. A. Bowman and R. M. Schoonover, *J. Res. Natl. Bur. Stand. (U.S.)*, **69C**, 217 (1965).

287. E. S. Gilfillan, Jr., and M. Polanyi, *Z. Phys. Chem.*, **166A**, 254 (1933).

288. I. Lauder, *Aust. J. Chem.*, **12**, 32 and 40 (1959); H. Watanabe and K. Iizuka, *Metrologia*, **21**, 19 (1985).

289. R. E. Rosensweig, *AIAA J.*, **4**, 1751 (1966).

290. J. M. Hughes and R. W. Birnie, *Am. Mineral.*, **65**, 396 (1980).

291. A. Kozdon, *Proceedings of the 1st International Conference on Calorimetry and Thermodynamics*, Warsaw, 1969, p. 831.

292. M. Randall and B. Longtin, *Ind. Eng. Chem. Anal. Ed.*, **11**, 44 (1939).

293. M. J. Hunter, "Partial Specific Volume Measurements Using the Glass Diver," in C. H. W. Hirs and S. N. Timasheff, Eds., *Methods of Enzymology*, Vol. 48, Academic, New York, 1978, p. 23.

294. G. Andreev and M. Hartmanova, *Cryst. Res. Technol.*, **21**, K74 (1986).

295. K. Linderstrøm-Lang, *Nature (London)*, **137**, 713 (1937).

296. J. D. Sakura and F. J. Reithel, "Partial Specific Volume Measurements by the Density Gradient Column Method," in C. H. W. Hirs and S. N. Timasheff, Eds., *Methods of Enzymology*, Vol. 26, Academic, New York, 1972, p. 107.

297. C. Anifisen, "The Determination of Deuterium in the Gradient Tube," in D. W. Wilson, A. O. C. Nier, and S. P. Reimann, Eds., *Preparation and Measurement of Isotopic Tracers*, J. W. Edwards, Ann Arbor, MI, 1946, p. 61.

298. B. W. Low and F. M. Richards, *J. Am. Chem. Soc.*, **74**, 1660 (1952); **76**, 2511 (1954).

299. E. W. Westbrook, "Crystal Density Measurements Using Aqueous Ficoll Solutions," in H. W. Wyckoff, C. H. W. Hirs, and S. N. Timasheff, Eds., *Methods of Enzymology*, Vol. 114, Academic, New York, 1985, p. 187.

300. J. M. H. Fortuin, *J. Polym. Sci.*, **44**, 505 (1960).

301. N. Payne and C. E. Stephenson, *ASTM Mater. Res. Stand.*, **4**, 3 (1964).

302. ASTM Standard D1505, *Standard Test Method for Density of Plastics by the Density Gradient Technique*, American Society for Testing and Materials, Philadelphia, PA, most current printing.

303. K. S. B. Rose, J. Williams, and G. Potts, *J. Nucl. Mater.*, **51**, 195 (1974).

304. J. Pelsmaekers and S. Amelinckx, *Rev. Sci. Instrum.*, **32**, 828 (1961).

305. H. Moret, *Rev. Sci. Instrum.*, **32**, 1157 (1961).

306. L. Kjlleén and H. Pertoft, *Anal. Biochem.*, **88**, 283 (1978).

307. H. Lange, *Colloid Polym. Sci.*, **258**, 1077 (1980).

308. W. Mächtle, *Colloid Polym. Sci.*, **262**, 270 (1984).

309. A. D. Franklin and R. Spal, *Rev. Sci. Instrum.*, **42**, 1827 (1971).

310. R. M. Schoonover, *J. Res. Natl. Bur. Stand. (U.S.)*, **87**, 197 (1982); R. S. Davis, p. 207.

311. H. G. Barbour and W. F. Hamilton, *Am. J. Physiol.*, **69**, 625 (1926).

312. V. J. Frilette and J. Hanle, *Ind. Eng. Chem., Anal. Ed.*, **19**, 984 (1947).

313. M. Cohn, "The Falling Drop Method for the Determination of Deuterium," in D. W. Wilson, A. O. C. Nier, and S. P. Reimann, Eds., *Preparation and Measurement of Isotopic Tracers*, J. W. Edwards, Ann Arbor, MI, 1946, p. 51.

314. E. Hallaba, E. Abdel-Wahab, A. Islam, and M. Hamza, *Z. Anal. Chem.*, **247**, 283 (1969).

315. A. J. Hoiberg, *Ind. Eng. Chem., Anal. Ed.*, **14**, 323 (1942).

316. F. Cap, *Mikrochem.*, **33**, 195 (1947).

317. A. S. Roy, *Anal. Chem.*, **33**, 1426 (1961); **45**, 1921 (1973).

318. O. E. Frivold, *Phys. Z.*, **21**, 529 (1920).

319. J. Ciochina, *Z. Anal. Chem.*, **98**, 416 (1934).

320. E. W. Blank and M. L. Willard, *J. Chem. Educ.*, **10**, 109 (1933).

321. P. Gouverneur and H. Van Dijk, *Anal. Chim. Acta*, **7**, 512 (1952).

322. L. M. Blair and J. A. Quinn, *Rev. Sci. Instrum.*, **39**, 75 (1968).

323. D. R. Gaskell, A. McLean, and R. G. Ward, *Trans. Faraday Soc.*, **65**, 1498 (1969).

324. C. J. F. Bottcher and P. Bordewijk, *Theory of Electric Polarization*, Vol. 2, Elsevier, Amsterdam, 1978, p. 290.

325. T. Bose, J. M. St.-Arnaud, H. J. Achtermann, and R. Scharf, *Rev. Sci. Instrum.*, **57**, 26 (1986).

326. T. M. Aminabhavi, R. C. Patel, E. S. Jayadevappa, and B. R. Prasad, *J. Chem. Eng. Data*, **27**, 50 (1982).

327. J. Arikawa, T. Tejima, and S. Kimura, *J. Elec. Eng. (Jpn.)*, **22**, 78 (1985); A. L. Harmer, in *Proceedings of the 1st International Conference on Optical Fibre Sensors*, London, England, 1983, p. 104.

328. S. Zamvil, P. Pludow, and A. F. Fucaloro, *J. Appl. Crystallogr.*, **11**, 163 (1978).

329. A. G. Walton, *Polypeptides and Protein Structure*, Elsevier, New York, 1981, p. 250.

330. R. V. Chernov and B. V. Yakovlev, *Z. Fiz. Khim.*, **49**, 2448 (1975).

331. B. Charlton and J. M. Newton, *Powder Technol.*, **41**, 123 (1985).

332. H. Heusala and R. Myllyla, *IEEE Trans. Nucl. Sci.*, **NS-30**, 398 (1983).

333. E. S. Garnett, T. J. Kennett, D. B. Kenyon, and C. E. Webber, *Radiology*, **106**, 209 (1973).

334. C. A. Swoboda, D. R. Fredrickson, S. D. Gabelnick, P. H. Cannon, F. Hornstra, N. P. Yao, K. A. Phan, and M. K. Singleterry, *IEEE Trans. Sonics Ultrason.*, **30**, 69 (1983).

335. L. Piche, A. Hamel, and P. Y. Kelly, "Automatic Device for Density Measurement in Polyethylene," in J. Vachopoulos, Ed., *Proceedings of the Symposium on Quantitative Characteristics of Plastic and Rubber*, 1984, p. 134; through *Chem. Abstr.*, **102**, No. 114289 (1985).

336. J. P. Panakkal, J. K. Ghosh, and P. R. Roy, *J. Phys. D*, **17**, 1792 (1984).

337. S. Shigezo and E. Yamada, *Rev. Sci. Instrum.*, **56**, 1220 (1985).

338. J. J. Kirkland and W. W. Yau, *Anal. Chem.*, **55**, 2165 (1983).

339. F. F. Liu and S. W. H. Chow, *Rev. Sci. Instrum.*, **58**, 1917 (1987).

340. N. Bekkedahl, *J. Res. Natl. Bur. Stand. (U.S.)*, **42**, 145 (1949).

341. P. S. Z. Rogers, D. J. Bradley, and K. S. Pitzer, *J. Chem. Eng. Data*, **27**, 47 (1982).

342. L. H. Tung, *J. Polym. Sci. Part A*, **5(2)**, 391 (1967).

343. P. Bernatchez and D. Goutier, *Rev. Sci. Instrum.*, **44**, 1790 (1973).

344. J. E. Rothman, D. L. Melchior, and H. J. Morowitz, *Rev. Sci. Instrum.*, **43**, 743 (1972).

345. D. A. Wilkinson and J. F. Nagle, *Anal. Biochem.*, **84**, 263 (1978).

346. L. C. De Jonghe and M. N. Rahaman, *Rev. Sci. Instrum.*, **55**, 2007 (1984).

347. E. S. R. Gopal, R. Ramachandra, M. V. Lele, P. Chandrasekhar, and N. Nagarjan, *J. Ind. Instrum. Sci.*, **56**, 193 (1974).

348. R. E. Gibson and O. H. Loeffler, *J. Am. Chem. Soc.*, **61**, 2515 (1939).

349. J.-L. Fortier, M.-A. Simard, P. Picker, and C. Jolicoeur, *Rev. Sci. Instrum.*, **50**, 1474 (1979); C. Jolicoeur, *Methods of Biochemical Analysis*, **27**, 171 (1981).

350. W. A. Duncan, J. P. Sheridan, and F. L. Swinton, *Trans. Faraday Soc.*, **62**, 1090 (1966).

351. O. Bozdag, M. S. Chaudhry, and J. A. Lamb, *Rev. Sci. Instrum.*, **55**, 427 (1984).

352. R. H. Stokes, B. J. Levien, and K. N. Marsh, *J. Chem. Thermodyn.*, **2**, 43 (1970).

353. G. A. Bottomley and R. L. Scott, *J. Chem. Thermodyn.*, **6**, 973 (1974).

354. M. K. Kumaran and M. L. McGlashan, *J. Chem. Thermodyn.*, **9**, 259 (1977).

355. G. A. Bottomly, L. G. Glossop, and W. P. Staunton, *Aust. J. Chem.*, **32**, 699 (1979).

356. S. J. Swarin and D. J. Curran, *J. Chem. Thermodyn.*, **10**, 381 (1978).

357. B. S. Lark and K. Bala, *Indian J. Chem.*, **20A**, 1163 (1981).

358. Y. P. Handa, P. J. D'Arcy, and G. Benson, *Fluid Phase Equil.*, **8**, 181 (1982).

359. K. Linderstrøm-Lang and H. Lanz, *C. R. Trav. Lab. Carlsberg*, **21**, 315 (1938).

360. S. Katz, "Dilatometry," in C. H. W. Hirs and S. N. Timasheff, Eds., *Methods of Enzymology*, Vol. 26, Academic, New York, 1972, p. 395.

361. L. K. J. Tong and A. R. Olson, *J. Am. Chem. Soc.*, **65**, 1704 (1943).

362. R. W. Penn, *Dent. Mater.*, **2**, 78 (1986).

363. K. L. Uglum, L. M. Carson, and R. V. Riley, *J. Chem. Educ.*, **42**, 559 (1965).

364. G. M. Bodner and L. J. Magginnis, *J. Chem. Educ.*, **62**, 435 (1985).

365. R. T. Ellington and B. E. Eakin, *Chem. Eng. Prog.*, **59**, 80 (1963).

366. T. Gast and K.-P. Gebauer, *Thermochim. Acta*, **51**, 1 (1981).

367. W. Johnson and L. K. Nash, *Rev. Sci. Instrum.*, **22**, 240 (1951).

368. A. G. Nerheim, *Anal. Chem.*, **35**, 1640 (1963).

369. R. W. Swingle, II, *J. Chromatogr. Sci.*, **12**, 1 (1974).

370. F. Kiran, *J. Chromatogr.*, **280**, 201 (1983).

Bibliography: General References for Mass (M) and Density (D) Measurements

Alber, H. K., "Systematic Qualitative Organic Microanalysis. Determination of Specific Gravity," *Ind. Eng. Chem., Anal. Ed.*, **12**, 764 (1940) (D).

Bauer, N., "Density and Specific Gravity," in H. F. Mark, J. J. McKetta, Jr., and D. F. Othmer, Eds., *Encyclopedia of Chemical Technology*, Vol. 6, 2nd ed., Wiley Interscience, New York, 1965 (D).

Blank, E. W., and M. L. Willard, "Micro-density Determination of Solids and Liquids," *J. Chem. Educ.*, **10**, 109 (1933) (D).

Czanderna, A. W., and S. P. Wolsky, Eds., *Microweighing in Vacuum and Controlled Environments*, Elsevier, Amsterdam, 1980 (M).

Elder, J. P., "Density Measurements by the Mechanical Oscillator," in C. H. W. Hirs and S. N. Timasheff, Eds., *Methods of Enzymology*, Vol. 61, Academic, New York, 1979, p. 12 (D).

Erdem, U., "Force and Weight Measurement," *J. Phys. E*, **15**, 857 (1982) (M).

Felgentraeger, W., *Feine Waagen, Wägungen und Gewichte*, 2nd ed., Springer, Berlin, 1932 (M).

Girard, G., "Density," in K. N. Marsh, Ed., Recommended Reference Materials for the Realization of Physicochemical Properties, Blackwell, Oxford, 1987, Chap. 2.

Glazebrook, R., *A Dictionary of Applied Physics*, Vol. 3, Macmillan, New York, 1922; P. Smith, *A Dictionary of Applied Physics*, Vol. 3, Macmillan, New York, 1950 (M, D).

Handa, Y. P., and G. C. Benson, "Volume Changes on Mixing Two Liquids: A Review of the Experimental Techniques and the Literature Data," *Fluid Phase Equil.*, **3**, 185 (1979) (D).

Hidnert, P., and E. L. Peffer, "Density of Solids and Liquids," *Natl. Bur. Stand. (U.S.) Circ.*, No. 487, NIST, Washington, DC, 1950 (D).

Kirshenbaum, I., *Physical Properties and Analysis of Heavy Water*, National Nuclear Energy Series, Division III, Vol. 4A, McGraw Hill, New York, 1951 (D).

Kochsiek, M., Ed., *Handbuch des Wägens*, Vieweg, Braunschweig, 1985 (M).

Lu, C., and A. W. Czanderna, Eds., *Applications of Piezoelectric Quartz Crystal Microbalances*, Elsevier, Amsterdam, 1984 (M).

Ma, T. S., and V. Horak, *Microscale Manipulations in Chemistry*, Wiley-Interscience, New York, 1976 (M, D).

Mason, B., "The Determination of the Density of Solids," *Geol. Föeren. Stockholm Föerh.*, **66**, 27 (1944) (D).

Pratten, N. A., "The Precise Measurement of the Density of Small Samples," *J. Mater. Sci.*, **16**, 1737 (1981) (D).

Rowlinson, J. S., and F. L. Swinton, *Liquids and Liquid Mixtures*, Butterworth, London, 1982 (D).

Schoonover, R. M., "A Look at the Electronic Analytical Balance," *Anal. Chem.*, **54**, 973A (1982) (M).

Wulff, P., and A. Heigl, "Methods of Density Determination of Solid Substances," *Z. Phys. Chem.*, **153A**, 187 (1931) (D).

Chapter **2**

PRESSURE AND VACUUM MEASUREMENTS

Charles R. Tilford

Physical Methods of Chemistry, Second Edition Volume Six: Determination of Thermodynamic Properties   Edited by Bryant W. Rossiter and Roger C. Baetzold
ISBN 0-471-57087-7   Copyright 1992 by John Wiley & Sons, Inc.

1   INTRODUCTION

The art and science of pressure measurement can be fairly said to date from the middle of the seventeenth century. Evangelisti Torricelli is generally credited with the discovery that a glass tube filled with mercury can be used to measure atmospheric pressure. Torricelli's discovery was quickly put to use in a number of significant scientific investigations, but widespread use awaited the Industrial Revolution. During that time a variety of pressure measurement techniques, most of them for "high" pressures, were developed, along with the harnessing of steam power and the emergence of the science of thermodynamics. Their scope was further broadened at the beginning of this century to support the development of the light bulb and the vacuum tube amplifier. The post-World War II surge in research and high-technology manufacturing spurred a rapid increase in the volume and variety of pressure measurements, particularly for very low pressures or vacuum where the availability of new technologies stimulated the rapid development of entirely new industries.

Modern society depends on a diverse complex of sophisticated manufacturing processes and research activities, most of which require pressure and/or vacuum measurements. The accuracy of these measurements can have major effects on the validity of results, product quality, energy efficiency, and, in many cases, the safe operation of a process. Present-day industrial processes require pressure measurements over about 17 decades; research and development projects extend this range two or more decades higher and lower. The applications, accuracy requirements, limitations on the measurement process, and available instrumentation vary widely over this range. At the upper end of the range the pressures that can be generated and measured are limited by the ultimate strengths of materials, while at the lower end they are limited by surface interactions and the release of gas from vacuum chamber walls. Around atmospheric pressure the state-of-the-art accuracy is a few parts per million (ppm). At the high- and low-pressure extremes it is a factor of 2 or 3; indeed, in some cases one must be satisfied to know which decade the pressure is in. Around atmospheric pressure a wide variety of commercial instrumentation, with various strengths and weaknesses, is available. As the pressure becomes higher and lower, the choices are more limited and eventually the measurement becomes a major experiment in itself. For these reasons this discussion is of necessity somewhat limited. It is confined to the *industrial pressure* range, 10^{-8}–10^9 Pa (10^{-10} torr to 150,000 psi). In all cases the accuracy obtained depends not only on the instrumentation and calibration protocols used, but also on the operator's level of understanding of the measurement process. Therefore, although much of this discussion concerns instrumentation, there is also an emphasis on problems generic to all measurements rather than those that are specific to certain instruments. Only static measurements are discussed; dynamic or time-varying pressure measurements present increasingly severe problems as the frequency range increases, and are well beyond the scope of this discussion. An attempt is made to provide guidelines to the range of accuracies that can be obtained at different pressures, and the basic operating principles are described for many of the available gauges and electromechanical pressure transducers. However, it is not possible to provide detailed performance data for the large variety of available commercial pressure instrumentation. The discussion of the higher pressure measurements includes the primary standards that are maintained in many industrial laboratories and often used for calibration or end-use measurements. As the pressure decreases into the vacuum area, primary standards become increasingly complex, to the point that they are maintained almost exclusively in national standards laboratories. The discussion in this portion of the range focuses more on the end-use instrumentation. The extent of the discussion is biased toward the low-pressure end because of the rapid expansion of vacuum technology over the past few decades and the consequent need for improved vacuum measurements.

As noted, the performance of instrumentation is only one of the factors that affect measurement accuracy. This is particularly true for vacuum measurements. As the pressure is reduced three or more orders of magnitude below atmospheric, pressure uniformity is increasingly difficult to maintain. At very low pressures order of magnitude pressure differences can exist in laboratory-scale systems, local pressures may be largely determined by chemical reactions or adsorption–desorption phenomena, measuring instrumentation can significantly perturb the pressure, and most instruments are sensitive to the gas species. Obtaining accurate measurements in these cases requires that we understand the entire apparatus, and steps must be taken to ensure that the gauge is measuring the pressure of interest.

2 DEFINITIONS

2.1 Pressure and Related Quantities

Hydrostatic pressure, P is generally defined as force F per unit area A. This is thermodynamically equivalent to the negative volume derivative of the Helmholtz free energy E at constant temperature T

$$P = F/A \tag{1}$$

or

$$P = -(\partial E/\partial V)_T \tag{2}$$

These definitions illustrate an important point: Pressure is a mechanical quantity and the basic limitations of the devices that measure pressure are mechanical. Although electrical measurement techniques, many of them quite sophisticated, can be used to enhance the mechanical measurements required and greatly improve the ease of the measurement, the basic limitations of the measurement process, such as response time, susceptibility to environmental changes, and instability or failure modes, are largely determined by the mechanical portion of the instrument.

In solids or high-viscosity liquids significant shear forces can exist and the pressure will be neither isotropic nor hydrostatic. This situation is very much the norm at the higher pressures that lie beyond the range of this chapter. In these situations the pressure is more properly defined as the negative average of the diagonal components of the stress tensor and it can vary significantly over volumes of interest.

At the other extreme, very low pressures, density or molecular flux is generally of more relevance than the pressure, although measurements are usually made in terms of pressure units. For a rarefied gas in thermal equilibrium the pressure and density n are related by the ideal gas law

$$P = nkT \tag{3}$$

where k is Boltzmann's constant, T is the absolute temperature, and n is the density in molecules per unit volume. Similarly, if v_i is the flux of molecules with mass m_i striking a surface of unit area per unit time

$$P_i = \sqrt{2\pi m_i kT}\, v_i \tag{4}$$

where P_i is the partial pressure due to molecules with a mass m_i. For mixtures the total pressure would be the sum of the contributions from different molecular species.

As can be the case with very high pressures, very low pressures may not be isotropic, in which case it is necessary to speak of directional dependent molecular fluxes or pressures. The density, molecular flux, and molecular velocities can also have large spatial variations in a laboratory-size vacuum system, particularly near pumps and heated or cooled parts of the apparatus, for example, near hot filaments or cryogenic shrouds. Under these conditions special techniques must be used to measure the molecular flux as a function of direction and location, or the apparatus must be designed to minimize these variations so that the pressure or density measured by the vacuum gauge is characteristic of the region where it is desirable to know these quantities.

The thermodynamic definition of pressure is that of the absolute pressure, with which this discussion is primarily concerned. The measurement techniques discussed here, however, are also applicable to differential pressure, or the difference between two absolute pressures. Differential pressure measurements are often required in industrial process control, most notably in flow measurements. As the absolute or *line* pressure increases in comparison to the differential pressure, unique problems can arise. A special case of differential measurement is gauge pressures. These are differential measurements referred to atmospheric pressure, and many lower accuracy *quasiabsolute* measurements are actually gauge measurements. In fact, it may well be argued that more gauge measurements are made than any other kind. The variability of atmospheric pressure can limit the accuracy of these measurements if they are used to determine absolute pressures.

2.2 Pressure Ranges

In this discussion reference is made to somewhat arbitrarily defined pressure ranges:

Range	Pressure (Pa)	Conventional Units
Very high pressure	$10^8 - 10^{11}$	$1 - 1000$ kbar
High pressure	$10^5 - 10^8$	$15 - 15,000$ psi
Atmospheric	$10^2 - 10^5$	1 torr–1 atm
Low vacuum	$10^{-1} - 10^2$	$10^{-3} - 1$ torr
High vacuum (HV)	$10^{-4} - 10^{-1}$	$10^{-6} - 10^{-3}$ torr
Very high vacuum	$10^{-7} - 10^{-4}$	$10^{-9} - 10^{-6}$ torr
Ultrahigh vacuum (UHV)	$10^{-10} - 10^{-7}$	$10^{-12} - 10^{-9}$ torr

Several of these terms can have common usages that differ somewhat from those defined above; for example, high vacuum is often used for all pressures below 10^{-1} Pa. Reference is also often made to the *transition* regime or range. For most systems this is the higher pressure end of the high-vacuum range and the bottom to middle of the low-vacuum range (10^{-2}–10 Pa). Over this range the mean free path of molecules is comparable to the dimensions of laboratory apparatus, and the molecular dynamics are in a transition from domination by molecule–wall collisions (molecular flow) at the lower pressures, to domination by molecule–molecule collisions (viscous flow) at higher pressures. This range is of special interest because in this range the change in molecular dynamics complicates the design and operation of both primary standards and vacuum gauges, a notable complication being the thermal transpiration effect discussed in Section 6.3. These complications are of some concern since many important industrial processes, for example, sputtering and freeze drying, take place in this pressure range.

2.3 Measurement Accuracy

Pressure measurements, as with all measurements, have a meaning only to the extent that the accuracy of the measurement can be evaluated. A difficult task in any case, accuracy evaluation or error estimation is made all the more difficult by the prevalence of ambiguous or contradictory definitions of the terms used. The same terms are often used in specifications for commercial instruments, where an even wider range of definitions and a sometimes suspicious vagueness can be found. Several common terms are defined below. These definitions are not unique; the Compilation of ASTM Standard Definitions [1] alone lists 17 different definitions for *accuracy*. The definitions presented here, however, are in general accord with those found in [1] and common metrology usage.

Accuracy.   The closeness of agreement between a measured value and the "true" or "accepted" value. Since the true value can never be known exactly, accuracy in this sense can only be estimated. When using accuracy most users specify the error or uncertainty; for example, if a measured value is 99% of the accepted value, it is said to have an accuracy of 1%. Accuracy is often specified for a measuring instrument, in which case it refers to the uncertainty of the measurements made with that instrument.

Confidence Level.   The probability that the error of a measurement will not exceed the specified uncertainty. It may be specified as a percentage or as a multiple of sigma or standard deviations. As an example, if the errors are assumed to be normally distributed and the uncertainty is specified as "2% at the 3 sigma level," this means that there is a 99.7% probability that the measured value is within 2% of the true value. Unfortunately, confidence levels are often not specified and there is no accepted standard confidence level. Some workers or manufacturers may state uncertainties at the 1 sigma level, others at 2, 3, or some other multiplier. Still others may believe or imply that errors can never exceed the specified bounds.

Error.   The difference between a measured value and the true or accepted value. Since the true value is never known exactly, the error in this sense can only be estimated.

Linearity.   More properly termed *nonlinearity*, linearity is the deviation of an instrument output from a linear dependence on the input stimulus. It is often equivalent to the maximum deviation of the output from a linear function determined by the outputs at minimum and maximum values of the input.

Precision.   The degree of agreement among independent measurements of a quantity. In some cases this term is used for the standard deviation of repeated measurements, in which cases it should be more properly called *imprecision*.

Random Error.   The unpredictable variation of repeated measurements of a quantity. It is often characterized by an imprecision or standard deviation.

Repeatability.   The closeness of agreement of measurements made under the same conditions. It typically implies the measurements were taken over a short period of time, although *long-term repeatability* may be specified.

Reproducibility.   The closeness of agreement of measurements taken under different conditions. The conditions may include different environments, laboratories, operators, or apparatus. Reproducibility may also imply measurements that are repeated over long periods of time.

Sensitivity.   The minimum detectable change in a measured quantity, or, the change in output divided by the change in input. Generally, but not always, the context will make apparent which of these two different meanings is intended.

Traceability.   The documentation of the relationship between the accuracy or uncertainty of an instrument or measurement and a recognized standard. For example, "Traceable to NBS" implies that the specified uncertainty or accuracy has been established by documented comparisons or calibrations relative to standards maintained by the National Bureau of Standards (now the National Institute of Standards and Technology). Although this term is often used in contractual or legal documents the rigor of its implementation varies widely from one case to another.

Uncertainty.   The estimated bounds of the error of a measurement. Uncertainty may also be ascribed to the instrument making the measurements. It should include a specified confidence level, but it often does not.

3   PRESSURE UNITS

The seemingly innocuous subject of units is actually the cause of significant pressure measurement errors in a surprising number of instances. Not only are a large number of pressure units found in regular use, but many of them are defined in an incomplete or ambiguous manner; subtle conditions that are part of the definition are not specified or are misunderstood. The force-per-unit-area pressure units, of which the pascal or newton per square meter (Pa or N/m^2) is the internationally accepted unit (the SI unit or Système International d'Unités), and the pound per square inch (psi) is a non-SI but widely used example, are unambiguously defined. Other units, such as the torr, bar, or atmosphere, are now effectively defined in terms of the pascal, although, in the case of the atmosphere, there are at least two different definitions of the atmosphere, *normal* and *technical*, in common use. Also, it should be noted that "standard conditions," implying a standard pressure and temperature, are not uniquely defined. The pressure is generally 1 atmosphere, although which atmosphere is usually not specified; and various standard temperatures, ranging from 0 °C to "room temperature," are used. What is standard is that almost every user of this term is quite confident that his temperature and pressure are the "standard" and that other workers understand this.

Major pressure measurement problems are caused by the widespread use of pressure units expressed in terms of manometer liquid-column lengths, such as *millimeters of mercury, microns* (micrometers of mercury), and *inches of water.* Unfortunately, to derive a pressure from the height of a manometer column, or to convert from a "manometer" pressure unit to a true thermodynamic pressure unit, the density of the manometer liquid and the acceleration of gravity must be known. For example, to convert inches of mercury to pascals or pounds per square inch requires that the acceleration of gravity, the temperature, and the pressure of the mercury all be specified (the density of mercury depends on both the temperature and the pressure). However, while it is fairly uniform to assume "standard" gravity (9.80665 m/s²), there is no standard or convention for the density of the fluid. Thus, we find inches of mercury commonly used to refer to a pressure unit assuming the density of mercury at 32°F, as well as to a different unit assuming the density of mercury at 60°F. A misunderstanding as to which definition is being used will introduce an error of 0.28%, significant for even routine industrial measurements. Smaller errors can be caused by a failure to specify the pressure of the mercury. For example, because of the compressibility of mercury, the pressure equivalence of 1 in. of mercury will change by 4 ppm if the ambient pressure of the mercury is alternately assumed to be 0 or 101 kPa. The possibilities for confusion are compounded when different temperature scales, discussed in Section 4.1.3, are considered. Because of changes in scales, the density of mercury at 20°C also depends on which temperature scale the 20°C is determined.

The conversion factors presented in Table 2.1 for mercury manometer units, and the related torr, are calculated using mercury density values at 1 atm, based on the International Practical Temperature Scale of 1968. Because of the ambiguities in the definition and use of these units, the conversion factors have been deliberately truncated to the number of digits presented, and they should *not* be changed to allow for different temperature scales, as this will only greatly increase the already ample confusion. Similarly, units based on a column of water are uncertain because of ambiguities in the density value used. Additional conversion factors for a variety of units can be found in [2, 3]. To avoid difficulties with units the user is strongly encouraged to use force-per-unit-area units, or, at a minimum, specify clearly the definition and conversion factor to pascals if "manometer" units are used.

Table 2.1 Conversions Factors to Pascals[a]

Unit	Conversion Factor
1 psi	6.894757×10^3
1 normal atmosphere	1.013250×10^5
1 technical atmosphere	9.806650×10^4
1 torr	1.33322×10^2
1 bar	1 (exact) $\times 10^5$
1 mm Hg (0°C)	1.33322×10^2
1 in. Hg (32°F)	3.38638×10^3
1 in. H_2O (39.2°F)	2.49082×10^2

[a]To convert the specified units to pascals multiply by the conversion factors.

4   PRESSURE STANDARDS

Measurement standards are of two different types: primary and secondary. Primary standards are those that permit a measurement of a quantity in terms of other known measurement units with a predictable uncertainty. Primary standards are not calibrated against another standard of the same unit, although parts of a primary standard may be calibrated against standards of other units; for example, a primary pressure standard is not calibrated against another pressure standard, but component parts may require calibration against standards of length and temperature. The design of a primary standard must be based on well-established physical laws relating the quantity to be measured to the component measurements; for example, the relationship must be known between the pressure applied to a standard and the measured quantities: length, temperature, mass, time, and so on. The uncertainty analysis of a primary standard requires a thorough understanding of the operation of the standard, including its susceptibility to environmental influences and a knowledge of the uncertainties of the component measurements. Common examples of primary pressure standards are liquid column manometers, piston gauges, and McLeod gauges.

Secondary or transfer standards are calibrated against a primary standard of the same unit and depend on a predictable stability to maintain their accuracy. In theory, any instrument with a well-evaluated stability can serve as a transfer standard; in practice, the term is generally reserved for instruments of exceptional stability. The uncertainty of a secondary standard will always be greater than that of the primary standard against which it is calibrated, the difference being due to the random errors of the calibration; the short-term random errors of the secondary standard; and the instabilities caused by time, environmental parameters, or operator technique. Some instrument types may be used as either primary or secondary standards, for example, piston gauges. The operating theory of other types may not be well enough understood to permit their use as primary standards, but they may possess the stability required for a transfer standard. It is neither impossible nor uncommon for some transfer standards to have a stability superior to the primary standards against which they are calibrated. In these cases the transfer standard can be used to check the stability of the primary standard. Repeated calibrations of the transfer or check standard are monitored to determine whether the differences between the primary and check standard are within expected bounds. If excessive changes are detected, this may indicate a malfunction of the primary and/or check standards, an undetected environmental perturbation, or an underestimation of the inherent instability of either standard.

Discussed below are commonly used pressure or vacuum primary standards: liquid-column manometers, piston gauges, McLeod gauges, volume or Knudsen expansion systems, and orifice-flow vacuum standards. Liquid-column manometers and piston gauges are discussed in more detail because of their widespread use.

4.1   Liquid-Column Manometers

Liquid-column manometers are widely used as primary standards for the calibration of other instruments and also for many end-use measurements. They are capable of the most accurate measurements, and can be the cheapest and simplest pressure measurement device to construct. Unfortunately, these two attributes are not found in the same instrument. The basic principle is illustrated in Figure 2.1. A pressure P applied to the right-hand liquid surface displaces the liquid, which generates a differential pressure ΔP determined by the liquid density ρ, the displaced height h, and the gravitational acceleration g. When the applied pressure and the displaced liquid are in equilibrium, allowing for a nonzero reference pressure P_{ref} on the left column, the absolute pressure P is given by

$$P = \rho g h + P_{ref} \tag{5}$$

Manometers are widely used for the measurement of atmospheric pressure and the calibration of other instruments, such as aircraft altimeters that are used for this same purpose. Therefore, a great many manometers have a full-scale range of about 100 kPa, although lower range instruments are not uncommon and commercial manometers with ranges up to 350 kPa are found in many

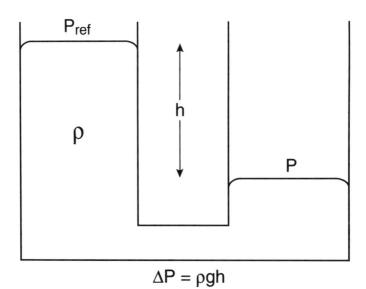

$$\Delta P = \rho g h$$

Figure 2.1   Schematic representation of a basic liquid column manometer. The height h is measured along the vertical, ρ is the average density of the liquid in all nonhorizontal parts of the manometer, and g is the local acceleration due to gravity; P_{ref} is the reference pressure and for absolute pressures it should be "zero"; otherwise, the measured pressure will be a differential pressure with respect to P_{ref}.

industrial standards laboratories. Special purpose manometers have been built for higher pressures [4, 5], although generally piston gauges, discussed in Section 4.2, are used for the higher pressure measurements. Manometers are used for both differential and absolute measurements. For absolute measurements the higher quality instruments use a vacuum pump, generally a two-stage mechanical pump, to maintain a "zero" reference pressure. However, large numbers of barometers, used exclusively to measure atmospheric pressure, rely on a vacuum sealed in the closed manometer tube at the time it was filled with mercury.

Manometers may be as simple as a plastic U tube mounted next to a vertical ruler. However, as the desired accuracy increases so does the complexity and cost, although the complexity and cost of some commercial instruments may be due as much or more to a desire for measurement convenience as to an effort to secure measurement accuracy. The differences between various types of manometers arise largely in the techniques used to determine the location of the liquid surfaces and to measure the differential height between them. This is particularly true in the low-range instruments used for vacuum or small differential measurements where the useful range of the manometer may be limited by the resolution of the height measurement. However, the accuracy of state-of-the-art instruments near atmospheric pressure, and the resolution of state-of-the-art low-range instruments is probably constrained as much by the stability and knowledge of the liquid temperature as by any other factor. High-accuracy manometers of the type found in national standards laboratories have been reviewed by Guildner and Terrien [6] and low-range manometers for use in low vacuum and differential measurements have been reviewed by Ruthberg [7] and Peggs [8].

Manometers utilize a variety of liquids, but most manometers with any pretext to high accuracy use mercury. Mercury is readily purified, is chemically stable, has a highly reproducible density with a small temperature coefficient (for a liquid), and possesses a modest vapor pressure. Many of the properties of mercury are discussed in a very comprehensive review by Wilkinson [9]. For very low pressure measurements, low-vapor pressure, low-density (compared to mercury) oils can be used to minimize the reference pressure and obtain an improved pressure resolution. However, the density of some oils may vary by several percent, depending on composition, and the density of even highly purified oils may vary more than 0.01% from lot to lot [10]. The appreciable amount of gas absorbed in oils can hinder the attainment of low reference pressures.

Overall, good commercial mercury manometry can attain uncertainties as low as 25–30 ppm of the measured pressure near atmospheric pressure. Attempts are being made to produce commercial instruments with uncertainties as low as 10 ppm, and the specialized manometers in national standards laboratories are designed to attain uncertainties as low as a few parts per million [6]. Manometers used for low-range vacuum measurements are designed to optimize length resolution. Commercial versions of a low-range instrument [11] using precision micrometer screws for the length measurements can achieve a length imprecision of about 10^{-3} mm, depending on the skill of the operator. Figure 2.2 shows a low-range calibration system that includes two manometers of this type, one employing mercury, the other oil. Full-scale range of the mercury manometer is 13 kPa. The lower density and vapor pressure of the oil allows a better lower pressure limit than can be achieved by the mercury manometer. Specialized instruments using optical [12, 13] or ultrasonic [14, 15] length measurement techniques can achieve a length imprecision of about 10^{-5} mm, further reducing the low-pressure limit. However, most manometers have much larger uncertainties and the accuracy of results obtained with a given type of manometer will vary widely with environmental conditions, maintenance of the instrument, and, in many cases, the skill of the operator. Some appreciation of the factors limiting accuracy can be gained by considering the individual terms in (5): gravity, height, density, and reference pressure.

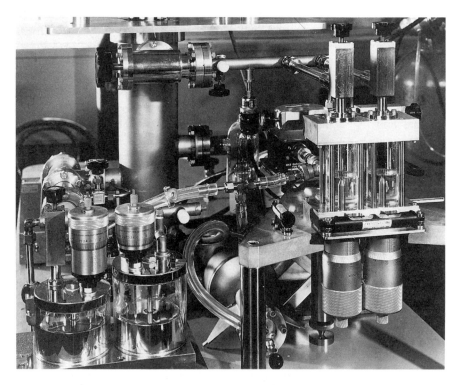

Figure 2.2   Calibration system employing two low-range liquid-column manometers that use micrometer screws to measure the displaced height. The manometer on the left is filled with mercury and the one on the right is filled with a low-vapor pressure oil. The ends of the micrometer screws are sharpened to a point and the liquid surfaces are located by adjusting the screws until the points just "dimple" the surface.

4.1.1   Gravity

The local acceleration of gravity at the site of the instrument g has nothing to do with so-called *standard* gravity. Without going to geographic extremes, g can vary by 0.1% within the continental United States. The acceleration of gravity, in meters per seconds squared (m/s^2), can be calculated from a theoretical formula [16] based on an ellipsoid approximating the earth's shape. Corrected for local deviations from the ellipsoid, the acceleration g is given by

$$g = 9.78032(1 + 5.2789 \times 10^{-3} \sin^2 \phi + 2.35 \times 10^{-5} \sin^4 \phi)$$
$$- 3.086 \times 10^{-6} h + 1.19 \times 10^{-6} h' \tag{6}$$

where ϕ is the latitude, h is the elevation in meters above mean sea level, and h' is the difference in elevation in meters from the average terrain elevation for the surrounding 100-km radius area. The uncertainty in g, calculated from this equation, will typically vary between 10 and 100 ppm depending on the extent of local geophysical anomalies. Improved accuracies can be provided by interpolating between measured reference points,[†] typical uncertainties being a few parts per million. Measured values can be obtained with uncertainties less than 0.1 ppm, although account must be taken of tidal variations at the 0.3-ppm level. Many institutions, such as national geodetic surveys, petroleum exploration firms, and geophysical research organizations, have the capability to make gravity measurements at field locations.

4.1.2   Height

The diversity of techniques used to locate and measure the heights of the liquid surfaces is too great for a comprehensive discussion. The most common techniques involve a visual comparison of the location of the surface with an adjacent length scale. This comparison may be aided by a sighting ring that can be slid up or down a glass manometer tube until the bottom of the ring is coincident with the top of the meniscus. A manometer using this technique to measure the height of one of the columns is illustrated in Figure 2.3. An attached vernier transfers the location to an etched length scale. Theodolites are also used to transfer the location of the meniscus horizontally to an adjacent scale. Cathetometers are similarly used but with the scale built into the cathetometer. Some automated mercury manometers use the meniscus to interrupt the light between a bulb and a photocell. The bulb–photocell combination, a servosystem, and length measuring mechanism are used to track and measure the height of the meniscus. Some short-range instruments and limited-range barometers locate the surface by observing the slight indentation when a pointed tip,

[†]Interpolated values of g and estimated errors for a specified location (latitude, longitude, and elevation) can be obtained in the United States for a fee from the National Geodetic Information Center (N/CG174), National Oceanic and Atmospheric Administration, Rockville, MD 20852, telephone: (301) 443-8623.

Figure 2.3 Commercial *single-cistern* mercury manometer using a sighting ring and micrometer screw to measure the column heights. Pressure is applied to the large diameter cistern; the height of the mercury in the cistern is measured using the micrometer screw mounted on top of the cistern. The reference pressure is maintained in the smaller diameter glass tube; the height of the mercury in the tube is measured using the sighting ring around the tube, the attached vernier, and the length scale to the left of the tube. The sighting ring rests on a stack of gauge blocks used to calibrate the length scale.

attached to a length scale, just touches the surface [11]. Two such instruments are illustrated in Figure 2.2. The precision of the visual techniques usually depends on the visual acuity of the operator. Attempts are generally made to enhance the operator's vision by special lighting techniques. At the best, imprecisions of about 10^{-3} mm can be achieved with careful design and practiced skill. The accuracy of both the visual and the automated techniques can be compromised by irregularities in the glass tubes.

Some high-performance, and high-cost, commercial mercury manometers effectively eliminate the problems due to capillary depression (see below) and operator skill by locating the mercury menisci in large-diameter (7–10-cm) cisterns with a capacitor plate above the mercury surface, as shown in Figure 2.4. The mercury in the cisterns is connected by a small flexible tube, and a

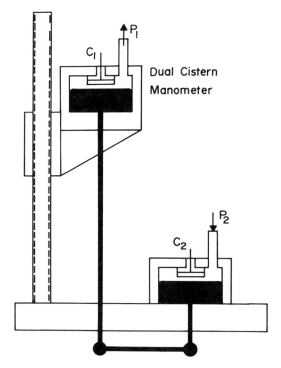

Figure 2.4   *Dual-cistern* type manometer: P_1 is the reference pressure and P_2 is the applied pressure. The height of the mercury in the cisterns is measured using capacitor plates C_1 and C_2. The upper cistern is supported by a movable carriage that is adjusted as the pressure changes so that the height of the mercury in each cistern remains constant.

servomechanism adjusts the height of the reference–pressure cistern so that, as the pressure is varied, the mercury surfaces remain a fixed distance below the capacitor plates. The variable height of the moveable cistern is measured by a calibrated lead screw or a laser interferometer. The original of this design used gauge blocks to measure the cistern height, as described in detail in [17].

Since mercury has a highly reflective surface, it has occurred to many investigators to use the mercury surfaces as mirrors in a laser-illuminated Michelson interferometer, providing a high-accuracy and high-resolution measurement of the displaced height of the mercury column. Unfortunately, mercury is mercurial, and even with good vibration isolation a free mercury surface has standing and random surface waves with amplitudes of micrometers, which greatly complicates the interferometry. Similarly, placing cube-corner reflectors on floats in the mercury is complicated by capillary forces that perturb the height of the floats, and it is difficult to keep the floats centered in the manometer tubes. Several successful high-accuracy manometers have been built that employ a variety of techniques to overcome these problems [15, 18–22]. However, each has required a major effort, and the use of a laser interferometer for manometry is not the undergraduate laboratory exercise it may appear to be at first glance.

Users should be aware that, for computational convenience, the length scales of some commercial manometers are not calibrated in true length units, although they are marked as millimeters or inches. In these instruments the length scales have been adjusted from true length by the ratio of local to standard gravity, and, for some temperature-controlled manometers, for the ratio of the operating temperature to 0°C. If these instruments are used without correction at another location or at a different temperature, significant errors can result.

The accuracy of the length-measuring apparatus is not the only factor limiting the accuracy of the height measurement. The surface of any confined liquid will be displaced by capillary forces. These forces vary with surface conditions at the liquid–tube interface, so that the capillary depression or elevation can vary significantly with time and conditions of use. Since [5] assumes that h is the displaced height of a free surface, these capillary variations can cause additional errors. This is a particularly difficult problem with mercury. Even with carefully cleaned glass tubes and very pure mercury it is possible to observe a wide range of contact angles at the mercury–glass interface, including inverted menisci where the contact between the mercury and the glass is higher than the center of the meniscus. Capillary depression correction tables for mercury in glass have been developed [23], with the correction given as a function of the tube diameter and the height of the meniscus. However, determination of the meniscus height is difficult and the capillary corrections are of questionable reliability. The best practice is to use tubes large enough that the maximum capillary deviations are negligible. A conservative upper bound for possible capillary variations can be calculated from expressions for the meniscus shape of a sessile drop [24]. The depression of the center of a drop of specified diameter, relative to the surface of an infinitely large drop, corresponds to the capillary depression in a tube of the same diameter if the mercury–tube contact angle is 180°. This is an upper bound for possible capillary distortions. As can be seen from Table 2.2, for tube diameters above 10 mm the maximum capillary depression decreases by an order of magnitude for each 10-mm increase in tube diameter. Using small tube diameters in mercury manometers is a false economy.

Since (5) requires the measured height, the length measurement must be made along the vertical axis. Verticality errors depend on the square of the misalignment from the vertical. Therefore, alignment to within 1 mrad, which is not a difficult task, will keep this uncertainty below 1 ppm. Tilt of the manometer during a measurement will cause apparent length changes that introduce errors to the height measurement. Special multiple-column manometer designs [14] can compensate for this effect, but a stable mechanical foundation is highly desirable.

The limited precision with which the surfaces of the liquid can be located will cause random errors in the height measurements. The magnitude can be

Table 2.2   Maximum Mercury Capillary Depressions[a]

Tube Diameter (mm)	Maximum Depression (μm)
10	555
15	161
20	46
25	13
30	3.7
35	1.0
40	2.8×10^{-1}
50	2.0×10^{-2}
60	1.5×10^{-3}
70	1.0×10^{-4}
80	7.3×10^{-6}
90	5.1×10^{-7}
100	3.6×10^{-8}

[a]Values are for mercury with a surface tension of 450 dyn·cm.

evaluated from repeated measurements with zero applied differential pressure. These measurements will be most meaningful if they are repeated several times after the surfaces have been displaced from their rest positions and allowed to return to equilibrium with zero applied differential pressure.

4.1.3   Density

As previously mentioned, many fluids are used for manometry, but mercury possesses unique properties that have made it the most widely used manometric liquid. However, measuring the absolute density of a sample of mercury at the parts per million level is a major experiment, and the results of only a few such experiments have been published. Recently, comparative measurements have been made with imprecisions of less than 0.01 ppm [25]. This technique, and a reference sample with a known density, are used to determine absolute densities of submitted samples on a for-fee basis.[†] More common is the use of published reference values for the density of mercury. Several factors must be considered, however, in evaluating the uncertainty of such a reference value when it is used for a specific sample.

To use a reference density value with any confidence the mercury must be pure. Mercury can be purified at the parts per billion level, but particular attention must be paid to the removal of base metals impurities, such as copper or lead. The most commonly used purification technique, vacuum distillation,

[†]For further information contact the CSIRO Division of Applied Physics, PO Box 218, Lindfield NSW 2070, Australia.

readily removes noble metal impurities, such as gold or platinum; however, the more volatile base metals can distill over with the mercury. Unfortunately, many commercial mercury processors only vacuum distill. The base metals can be removed by distillation in air or by washing the mercury with nitric acid. Some impurities are more effectively removed by washing with a potassium or sodium hydroxide solution. Various cleaning techniques are discussed and referenced by Wilkinson [9]. Exposed to air, very pure mercury will develop a visible film on the surface with time. Sensitive measurements can detect this film within a period of days, over longer periods it will become visible to the eye; but, even after periods of years it will still be thin and 'patchy." However, the development of a thicker "scum" indicates the presence of base metal impurities. With high levels of contamination a readily apparent scum will develop in a matter of days. In general, mercury that has been acid-washed and vacuum distilled, or distilled in air, and that retains a bright surface, will have impurities well below the parts per million level.

Even with very pure mercury the possibility still exists of significant density differences from one sample to another due to isotopic abundance differences. Six different isotopes contribute between 7 and 30% to the natural abundance of mercury. Samples from different mines can have different isotopic abundances, and distillation can cause significant isotopic fractionation. Mass spectrometric measurements have yet to be published with an accuracy that will allow density differences to be determined at the parts per million level. However, an analysis of the densities of 13 different samples determined by three different experiments [25–27] found a standard deviation, due to sample differences, of 0.7 ppm. This indicates that with a high degree of confidence, near 99%, the densities of different samples can be expected to be within ±2 ppm of the mean density of mercury.

Of the limited number of absolute density measurements, the best known are those of Cook [26]. Cook measured seven different samples and averaged data from five of the samples to obtain a reference value widely used as *the* density of mercury. Unfortunately, Cook's analysis of his systematic uncertainties is very brief and somewhat cryptic. Comparison with a more recent set of absolute measurements [27], with a more thorough uncertainty analysis, indicates that a reasonable systematic uncertainty for Cook's measurements is 0.3 ppm (1σ). Thus, we would predict that, within three standard deviations the density of a sample of pure mercury is within ±3.1 ppm of Cook's reference value.

Since mercury expands with temperature and is compressible, its density depends on both temperature and pressure. Therefore, reference values are specified for a particular temperature and pressure, most commonly 20°C and 1 atm (101 kPa). Densities at other temperatures and pressures can be calculated using the thermal expansivity values of Beattie and co-workers [28] and the compressibility values of Grindley and Lind [29]. For most manometry applications the pressure correction is relatively small; the fractional correction is 4.0×10^{-11} Pa^{-1}. Generally, the temperature correction is much larger, and the situation is somewhat confused by changes in the internationally accepted temperature scale, the most recent of which occurred in 1990.

Realization of thermodynamic temperatures at the millikelvin level is very difficult. Therefore, these experiments are carried out at only a few laboratories, and the results are combined in an agreed-upon temperature scale that is considered to be the best available approximation to true thermodynamic temperatures. Periodically, improvements in the determination of thermodynamic temperatures prompt a revision of the accepted temperature scale. This, of course, does not imply any change in thermodynamic properties, but it does mean a change in the values that are assigned to them. For example, Cook determined the density of mercury using thermometers calibrated on the International Temperature Scale of 1948 (ITS-48), and specified his mean value at 20°C on ITS-48. In 1968 ITS-48 was replaced by the International Practical Temperature Scale of 1968 (IPTS-68). Since most thermometers are now calibrated on IPTS-68, and very few ITS-48 calibrations are still in use, Cook's value is generally corrected to the value corresponding to 20°C on IPTS-68. At the time of this writing most thermometers have been calibrated on IPTS-68, but this scale has recently been replaced by the ITS-90, and as time passes an increasing fraction of thermometers will be calibrated on ITS-90.

Between -200 and 630°C the differences between temperatures measured on ITS-90 and IPTS-68, T_{90} and T_{68}, are given in degree Celsius by

$$
\begin{aligned}
T_{90} - T_{68} = {} & -0.148759 \times (T_{90}/630) - 0.267408 \times (T_{90}/630)^2 \\
& + 1.080760 \times (T_{90}/630)^3 + 1.269056 \times (T_{90}/630)^4 - 4.089591 \\
& \times (T_{90}/630)^5 - 1.871251 \times (T_{90}/630)^6 + 7.438081 \times (T_{90}/630)^7 \\
& - 3.536296 \times (T_{90}/630)^8
\end{aligned}
\tag{7}
$$

Using Cook's density reference value and the thermal expansivity of Beattie and co-workers, the density of mercury at 101-kPa pressure is given between 0 and 40°C by

$$
\begin{aligned}
\rho_{68}(\text{kg/m}^3) = {} & 13{,}545.866 \times [1 - 1.8115 \times 10^{-4}(T_{68} - 20) \\
& + 2.5 \times 10^{-8}(T_{68} - 20)^2]
\end{aligned}
\tag{8a}
$$

for thermometers calibrated on IPTS-68 and

$$
\begin{aligned}
\rho_{90}(\text{kg/m}^3) = {} & 13{,}545.854 \times [1 - 1.812 \times 10^{-4}(T_{90} - 20) \\
& + 2.5 \times 10^{-8}(T_{90} - 20)^2]
\end{aligned}
\tag{8b}
$$

for thermometers calibrated on ITS-90.

Because of the uncertainty in the thermal expansivity values, the uncertainty

Table 2.3 Temperature Scale Differences, Mercury Density at 101 kPa, Mercury Volume Thermal Expansivity, and Vapor Pressure of Mercury

T ($°C$)	$T_{90} - T_{68}$ (mK)	ρ_{68} (kg/m^3)	ρ_{90} (kg/m^3)	β (ppm/K)	P (mPa)
0	0	13,595.08	13,595.08	181.5	26.8
5	−1.2	13,582.75	13,582.75	181.4	43.6
10	−2.4	13,570.44	13,570.43	181.3	69.7
15	−3.7	13,558.14	13,558.13	181.2	110
20	−5.0	13,545.87	13,545.85	181.2	170
25	−6.3	13,533.61	13,533.59	181.1	259
30	−7.6	13,521.36	13,521.34	181.0	390
35	−8.9	13,509.13	13,509.11	180.9	579
40	−10.2	13,496.92	13,496.90	180.8	849

in the density will linearly increase from ± 3.1 ppm at 20°C, to ± 4.1 ppm at the extreme temperatures.

These same data can be used to derive the values in Table 2.3. Also included are differences between ITS-90 and IPTS-68, the volume thermal expansivity of mercury [28], and the mercury vapor pressure [30]. To use the density values in Table 2.3 it is necessary to determine whether the thermometer has been calibrated on IPTS-68 or ITS-90, and then use the corresponding density values, ρ_{68} or ρ_{90}. The expansivity can be used to interpolate between entries. There is no significant difference between the thermal expansivity and vapor pressure values on the two temperature scales.

As noted, the density values in (8a), (8b), and Table 2.3 are for a pressure of 101 kPa. For manometry one should correct the density to the mean of the applied and reference pressures $(P + P_{ref})/2$ by using

$$(1/\rho)(\partial\rho/\partial P)_T = 4 \times 10^{-11} \text{ Pa}^{-1} \tag{9}$$

Thus, for a manometer measuring an absolute pressure of 100 kPa the densities given by (8a), (8b), or Table 2.3 should be reduced by 2 ppm to obtain the value at the mean pressure, 50 kPa.

Despite the impressive accuracies attained in the determination of the properties of mercury, the uncertainty in the density is often the limiting factor in high-accuracy manometry. This is because the average density of the liquid in all vertical portions of the manometer is required. Since the temperature of the liquid varies with location and time, and the thermometer(s) used to determine the temperature can have different response times and may not be suitably located or adequate in number to obtain a good spatial average, a significant error may exist in the indicated average temperature of the mercury. This error can be minimized by controlling the temperature, minimizing spatial gradients, and recognizing that even changing the pressure of the gas will change the temperature (adiabatic heating and cooling, discussed in Section 6.1).

4.1.4 Reference Pressure

Manometers inherently measure differential pressures. To determine an absolute pressure the pressure on the reference or low-pressure side of the manometer must either be known, or be so small that it can be neglected. Thus, the reference pressure is an additional, and in some cases, significant source of error for absolute pressure measurements. The vapor pressure of mercury is low enough to be a negligible correction for most measurements in the atmospheric pressure range, although for the most accurate measurements corrections must be made. Reference [30] presents an equation for the saturated mercury vapor pressure derived from a fit to three different sets of experimental data; tabulated values are presented in Table 2.3. Unfortunately, the reference pressure may differ significantly from the mercury vapor pressure. As mentioned, some barometers used for the routine measurement of atmospheric pressure have a sealed reference pressure at the top of the mercury column. If the glass tube is carefully cleaned and properly filled this can give low and stable reference pressures over periods of many years, but significant errors can result if these conditions are not met. Higher accuracy instruments use a pumped reference vacuum, generally employing a mechanical vacuum pump. For best results a two-stage pump should be used. It should be connected to the manometer by clean tubing, preferably metal or glass, with a minimum of rubber or plastic tubing, and the pump should preferably be operated continuously, or, at a minimum, for 24 h before use of the manometer. Adequacy of the vacuum should be checked by a thermocouple or similar vacuum gauge. These precautions are particularly advised when uncertainties of 10–20 Pa or less are required. If uncertainties of 1–2 Pa or less are required, the mechanical pump should be supplemented by a diffusion pump, preferably a mercury diffusion pump. If the pumping system includes a chilled baffle, the pressure can actually be reduced below the saturated vapor pressure of mercury, and for high-accuracy or low-pressure measurements the reference pressure should be monitored by a capacitance diaphragm gauge (discussed in Section 5.1) or a similar device. For measurements in the low-vacuum range, low-vapor pressure oils have been used as manometer fluids. Diffusion pump fluids are available with specified room temperature vapor pressures in the micropascal range; however, unless properly outgassed, the reference pressures can be several orders of magnitude higher than this.

The total uncertainty of a manometer is the sum of both random and systematic errors. The magnitude of the random errors, due to factors such as the imprecision in locating the liquid surfaces and measuring the height differences, variations in capillary effects, and variations in temperature gradients, can be evaluated experimentally by making repeated readings of the indicated pressure with zero applied differential pressure. If these zero readings are made over a period of time and after the liquid has repeatedly been displaced and allowed to return to "zero," the repeatability of the readings (characterized by a standard deviation) should be characteristic of the random errors of the measurement. The evaluation of systematic errors requires a theoretical model of possible error sources and estimates of their magnitudes, and it is much more a matter of judgment. Confidence in the uncertainty analysis can be increased greatly if the manometer is compared with another pressure instrument of comparable or better accuracy.

Manometers are most generally used as primary standards. While they could be calibrated against another standard of superior accuracy and used as a transfer standard, this in general has dubious advantages for mercury manometers. The performance of a manometer depends on environmental conditions and operator skill, as well as on the length and temperature measurements, and these factors can vary considerably. Therefore, at the time of use the performance of the manometer can differ significantly from that at the time of calibration. In short, the reproducibility of a manometer is generally not significantly better than its total uncertainty. It is much better to calibrate the thermometers and the length scale, maintain a stable and uniform temperature environment, and evaluate the manometer as a primary standard. One exception is the calibration of barometers with a sealed reference vacuum. In this case the calibration primarily serves to verify that the vacuum, which is not accessible for independent measurement, has not deteriorated. However, as with any standard, it is a good practice to compare a manometer periodically with another pressure standard to determine if the agreement between the two standards is in reasonable accord with the combined uncertainties of the standards.

4.2   Piston Gauges

The piston gauge, also known as a deadweight tester or pressure balance, is shown schematically in Figure 2.5. The force generated by a pressure P, applied to the bottom of a carefully fabricated piston–cylinder combination with effective area A_{eff}, is balanced by the weight F of the free-floating piston and associated weights, and the reference pressure P_{ref} acting on the effective area

$$P = F/A_{\text{eff}} + P_{\text{ref}} \tag{10}$$

Piston gauges can be operated with either oil or gas as the pressure fluid. Oil-operated gauges are widely used in the high- and very-high-pressure ranges; gas gauges are used primarily in the atmospheric pressure range and the lower part of the high-pressure range. Commercial oil gauges are available with ranges up to 1.3 GPa (200,000 psi), and special purpose gauges have been built for operation as high as 2.6 GPa [31–33]. Since piston gauges measure, or more properly, generate differential pressures, and their operation generally requires frequent access to the weights, they are typically used for the measurement of

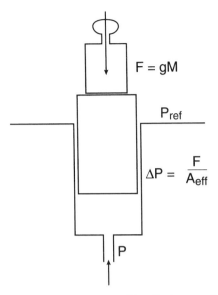

Figure 2.5 Schematic illustration of a piston gauge. The effective area A_{eff} is very nearly the mean of the piston and the cylinder cross-sectional areas. The pressure applied to the effective area is balanced by a stack of weights attached to the piston. The piston is rotated with respect to the cylinder to relieve friction. Typical clearances between the piston and cylinder are 1 or 2 μm for a good quality gas-operated gauge, twice this for an oil-operated gauge.

gauge pressures—differential pressures with an atmospheric reference. This is true of virtually all oil-operated gauges, but gas-operated gauges are used for both gauge and absolute pressures. In the latter case the outside of the piston–cylinder assembly and the weights are surrounded by an evacuated bell jar, as can be seen in Figure 2.6. In all high-quality gauges the force balancing the pressure is generated by the mass of a set of weights (including the piston mass) supported by the free-floating piston. High accuracy requires that contact and friction between the piston and cylinder be minimized. For most piston gauges this is achieved by rotating the piston and weight stack with respect to the cylinder so that viscous forces will center the piston within the cylinder and maintain a lubricating film between the two surfaces. In most gauges the lubricating film is provided by a slow flow of the pressure fluid escaping from the high pressure at the bottom of the piston to the lower pressure at the top. Proper operation requires high-quality pistons and cylinders; both pieces must be straight and round with good surface finish, and the clearance between the piston and cylinder must be small, typically a few micrometers or less. These qualities are particularly important for gas-operated, gas-lubricated gauges, where proper operation further requires careful cleaning of the piston and cylinder and the use of gases free of oil or particulate contamination. A few gas-operated gauges maintain a lubricating film of oil between the piston and cylinder. For all piston gauges the piston and weights will be freely floating, or

Figure 2.6 Gas-operated piston gauge equipped for operation in the absolute mode. The piston and cylinder assembly is located under the stack of weights and is not visible. For absolute measurements the piston and weights are enclosed under a bell jar that can be evacuated through the large port on the left side of the housing. The pressure to be measured is applied to the lower pressure fitting, which connects to the bottom of the piston–cylinder assembly. The base includes a drive mechanism to allow rotation of the piston and weights when the bell jar is evacuated. Three different piston and cylinder combinations provide full-scale ranges of 5, 50, and 500 psi (33, 330, and 3300 kPa).

balanced, only for the discrete pressure that balances the force on the piston; and if the pressure changes, the weights must be adjusted accordingly. This greatly restricts the ability of the piston gauge to follow changing pressures, particularly for gas gauges operating in the absolute mode, where the weights are located inside an evacuated bell jar. References [34] and [35] discuss both the theory and operation of piston gauges; [35] is particularly suitable both for the first-time user and the specialist.

The uncertainty of the pressures generated by a piston gauge are determined by the uncertainties of the acceleration of gravity, the mass of the weights, the effective area, and, for absolute mode operation, the reference pressure. As discussed in Section 4.1.1, the local acceleration of gravity can be determined with uncertainties as small as a fraction of 1 ppm. A competent mass laboratory typically can determine the masses with an uncertainty of less than a few parts per million. In most cases the critical factor limiting the accuracy of a piston gauge is the uncertainty of the effective area. To a good approximation the effective area is the mean of the geometric cross sections of the piston and the cylinder. The accurate determination of the effective area of a primary standard piston gauge is a difficult process. It is generally limited by imperfections in the geometries of the piston and cylinder and by uncertainties in the dimensional measurements. Stated uncertainties typically vary from 20 to 100 ppm, where the lower values are for large diameter (up to 35 mm), low-pressure, gas-operated gauges. As discussed below, further reductions in the low-pressure uncertainties are limited by our lack of detailed understanding of the momentum transfer between the gas and the cylindrical surface of the piston. At high pressures, above about 1 MPa, distortion of the piston and cylinder introduce additional uncertainties, the magnitude of the problem increasing with pressure. Not only will the effective area change with pressure, but the operation of most gauges will also deteriorate since there will be an increased loss of pressurizing fluid through the enlarged annular space between the piston and cylinder, to the point that pressure equilibrium and free floating of the piston cannot be maintained.

Several approaches have been taken to minimize or compensate for the high-pressure distortion. These are illustrated in Figure 2.7. One is to use heavy-walled cylinders in the common *simple* piston gauge. Simple gauges are generally used for pressures lower than 100 MPa, but a commercial simple gauge is available with a range of 700 MPa. The cylinder can also be constructed so that the pressure is applied to the outside of the cylinder as well as to the inside. Depending on the dimensions, the annular space of such a *reentrant* piston gauge can actually decrease with increasing pressure. The reentrant

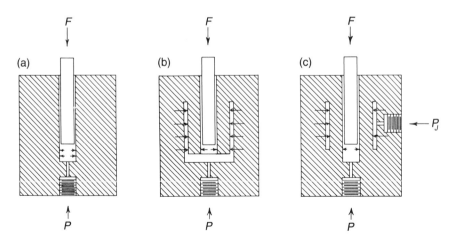

Figure 2.7 Cylinder configurations used in simple, reentrant, and controlled-clearance piston gauges. The operating range of the simple gauge (*a*) will be limited by expansion of the cylinder with increasing pressure. The reentrant design (*b*) decreases the distortion of the cylinder. Depending on the details of the design, a reentrant cylinder will either expand or contract with increasing pressure. The distortion of the cylinder in a controlled-clearance gauge (*c*) can be controlled independently by varying the jacket pressure P_j.

technique is used in commercial gauges operating at up to 300 MPa. However, while the reentrant design maintains satisfactory operation of the gauge at higher pressures, the calculation of the changes with pressure of the effective area is even more difficult than for a simple cylinder. The *controlled-clearance* gauge employs an independent jacket pressure on the outside of the cylinder to control the clearance between the piston and cylinder. This not only permits useful operation of the gauge to pressures above 1 GPa, it also permits, through a self-calibration procedure discussed in [34], a more accurate determination of the effective area than is possible with either simple or reentrant gauges. For this reason controlled clearance gauges have also been used as low-pressure, gas-operated primary standards. Both oil- and gas-operated controlled clearance gauges are available commercially, but their complexity of operation generally restricts their use to standards laboratories. Figure 2.8 shows a high-pressure controlled-clearance piston gauge.

Piston gauges are most widely used as transfer standards, in which case they are calibrated against either a mercury manometer, or another piston gauge

Figure 2.8 Controlled-clearance piston gauge with a range of 200,000 psi (1.3 GPa). The piston and cylinder assembly are located in the housing at the top of the apparatus. The jacket pressure is applied through the pressure line connected to the side of the housing. The presure to be measured is applied through the line connected to the bottom of the housing. The weights are suspended beneath the piston–cylinder housing by a yoke that is attached to the piston at the top of the housing. A bearing and drive motor allow rotation of the piston while the weights remain stationary. The pressure vessel at the upper right of the apparatus encloses a high-pressure transducer.

standard. The piston gauge is very sensitive and stable and much less susceptible to environmental and operator variables than are liquid–column manometers. Good quality gas-operated gauges will achieve imprecisions in the pressure of a few parts per million or less and can be expected to have irreproducibilities of 10 ppm or better. Piston gauge temperature coefficients are an order of magnitude smaller than those of a mercury manometer, and the small size of the gauge further simplifies the temperature-measurement problem. The primary operator skills required are the ability to clean the piston and cylinder, handle the weights with proper care, and control the pressure. Absolute mode measurements require the maintenance of a reference pressure that is low enough that the error in its measurement does not contribute a significant error to the measured pressure. This reference pressure is limited primarily by the gas flowing between the piston and cylinder. For a typical gas-operated gauge a two-stage mechanical vacuum pump with a nominal capacity of 160 L/min will suffice to maintain a pressure below 5 Pa (5×10^{-2} torr) after 15-min pumping time, and the addition of a 2-in. diffusion pump will maintain the pressure at a few tenths of a pascal. It should be noted that at least one piston gauge manufacturer recommends that the reference pressure be maintained at about 1 Pa. This significantly increases the possibility of errors in the absolute pressure due to the measurement of the reference pressure. Our experience has been that we can successfully operate gauges produced by that manufacturer, and others, with reference pressures below 0.1 Pa. With but a few exceptions, most gauges operate better in the absolute mode, with low-reference pressures, than they do in the gauge mode.

As noted, the necessity to balance changing pressures by changing weights restricts the applicability of piston gauges. However, their exceptional stability makes them ideal as calibration reference and check standards. An example of the latter application is the periodic comparison of a gas-operated piston gauge with a mercury manometer. Changes with time of the measured effective area probably indicate a problem with the mercury manometer. Piston gauges can be used to measure changing pressures over a restricted range by combining them with a low-range differential pressure transducer. In this case the piston gauge maintains a stable reference pressure on one side of the differential pressure transducer, and the transducer measures small changes in pressure about that reference pressure. The low-pressure limit of a piston gauge is determined by the weight of the piston. As an example, the lowest pressure attainable with a widely used 50-psi range piston gauge is 10 kPa (1.5 psi, 76 torr). However, operation of piston gauges becomes more difficult at low-mass loadings, and they generally are not used much below 10% of the full range.

Relatively large errors can occur during operation in the gauge mode because of the rotation of the weights. As noted earlier, the piston and weights are generally rotated with respect to the cylinder to center the piston within the cylinder and to minimize friction. Generally, the piston and weights are spun up to some initial rotation rate and then allowed to coast freely until mechanical contact between the piston and cylinder occurs and rotation stops. The minimum rotation rate varies between different gauges from less than 0.1 to about 2 Hz. Most commercial piston gauge bases include a mechanism for rotating the piston and weights. In some cases the initial rotation rate may be as high as 16 Hz. When operating in the gauge mode, the rotation of the weights will generate an aerodynamic force that depends on the diameter of the weights and the square of the rotation rate [36 and 37, p. 105]. The error in the generated pressure can exceed 100 ppm. This error can be minimized by operating at low-rotation rates and using weights with an outer diameter near 80 mm.

Evaluation of the systematic uncertainties of a primary standard is as much an art as a science. Anticipating and assigning magnitudes to all possible sources of error is very difficult. Therefore, it is always desirable to compare standards to see if the measurements agree to within the estimated uncertainties, particularly if the standards employ different measurement techniques and are therefore susceptible to different types of errors. Since primary standard piston gauges and mercury manometers are of quite different design, it is useful to compare them and determine whether the results lie within the combined uncertainties. A few such studies have been made, using gas-operated piston gauges whose areas have been determined by dimensional measurements. At the National Research Council, Ottawa [38], the measurements were made for one piston gauge at about 100 kPa in both the absolute and gauge modes. Measurements at the National Physical Laboratory, Teddington, UK [39], were made at a nominal absolute pressure of 100 kPa with two gases (helium and nitrogen) for four different gauges. In both cases the results were within the combined uncertainties; however, there was some evidence of systematic differences dependent on operating gas or mode of operation.

Recent work at NIST (National Institute of Standards and Technology) has found definite evidence that the effective area of gas-operated piston gauges depends on the gas; mode of operation; and, at low pressures where distortion is insignificant, pressure. Figure 2.9 shows the results of the calibration of a gas-operated piston gauge, with a nominal 2-cm-diameter piston cylinder, against an ultrasonic interferometer manometer [37, 37a]. Data were taken in both the absolute and differential (gauge) modes, and with different gases: helium, nitrogen, and argon. As can be seen in Figure 2.9, at higher pressures there is a systematic difference of about 6 ppm between the absolute and gauge mode effective areas. The increase in the absolute-mode effective area with decreasing pressure has also been observed for other, although not all, gauges. For other gauges the change is as large as 25 ppm. The results in Figure 2.9 for the three different gases are industinguishable; but for other gauges the effective area has differed for different gases by as much as 30 ppm [40]. The geometric cross sections of the piston and cylinder clearly do not depend on the operating gas, and mechanical distortions of these parts over this pressure range are less than 1 ppm. Therefore, the observed changes in the pressure, mode of operation, and operating gas are not explained by current piston gauge theory and indicate that approximating the effective area as the mean of the geometric cross sections of the piston and cylinder is probably not valid if uncertainties of less than 20 or

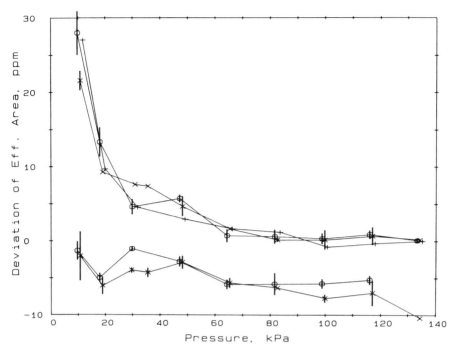

Figure 2.9 The effective area of a 2-cm-diameter, gas-operated piston gauge determined in the absolute and gauge modes for nitrogen (×), argon (+), and helium (○) using an ultrasonic interferometer manometer. The absolute-mode data are plotted at the top and the gauge-mode data, at the bottom. The differences between the gauge and absolute mode effective areas and the pressure dependence of the absolute mode data are not predicted by present piston gauge theory.

30 ppm are desired. The reasons for these changes in the effective area, and in particular why the magnitude of the changes vary so much from gauge to gauge, are not understood at this time, but they clearly are caused by changes in the interaction between gas molecules in the annular space and the piston. Although these systematic effects limit the uncertainty of the piston gauge as a primary standard, it is apparent from the data in Figure 2.9, which includes the random errors of both the manometer and the piston gauge, that piston gauges make excellent transfer standards. Changes in the measured effective area over the 2 months of this experiment were less than a few parts per million. Similar data taken with other gauges over periods of several years indicate changes of less than 2–3 ppm.

4.3   McLeod Gauges

McLeod gauges measure absolute pressures in the low- and high-vacuum ranges and have a distinguished history in vacuum science, having been an essential part of many notable experiments. They also have been used to make a great many very bad measurements.

The construction and accurate use of a McLeod gauge requires a high degree of skill and attention to detail. The principle of the McLeod gauge is very straightforward and applicable throughout the transition range. A known volume of gas is trapped; it is compressed to a much smaller volume using a mercury piston, and the pressure in the smaller volume is measured with a mercury manometer. The original pressure is derived from the pressure in the small volume using the ideal gas law and the known ratio of the volumes. The simplicity of this concept conceals several practical difficulties. The ideal gas law is not applicable if condensable gases are present. Reasonable values of pressure multiplication and the constraints of apparatus size requires very small final volumes, which are difficult to measure. Furthermore, the small-diameter capillaries (a few mm to fractions of a mm) used for the final volume include one arm of the mercury manometer, and capillary effects are very large. The measurement is static and the apparatus has a high surface-to-volume ratio so that outgassing or adsorption can generate significant errors. Mercury vapor "streaming" from the McLeod gauge to cold traps elsewhere in the vacuum system can generate significant pressure gradients between the gauge and the region where the pressure must be known.

McLeod gauges have been used successfully with uncertainties of a few percent down to 10^{-4} Pa (10^{-6} torr), but the novice considering their use is advised to consult the literature for an appreciation of what must be done to use them properly. McLeod gauges are reviewed in [7] and [41]. More recent work is described in [42], and examples of the care that must be taken to obtain reliable results with McLeod gauges can be found in [43–46].

4.4   Volume Expansion Systems

Volume expansion systems, also known as *static*, *Knudsen*, or *series expansion systems*, are the complement of McLeod gauges. A small volume of gas at a known "high" pressure is expanded into a larger volume, generating a lower pressure that can be calculated from the measured volume ratio and the equation of state of the gas. To attain still lower pressures this process can be repeated by additional expansions into still larger volumes, or by reevacuating the large volume and then again expanding the gas remaining in the high-pressure volume into the now empty low-pressure volume. This gives rise to the term *series expansion*. As with McLeod gauges, these devices are simple in concept and the physics of the measurement process is independent of the pressure, so that they can be used in the transition range, as well as at higher and lower pressures. Volume expansion systems have been built and are in regular use in national standards laboratories for the generation of absolute pressures down to the very high vacuum range (10^{-6} Pa). They can achieve uncertainties of the order of 1% in the high-vacuum range for inert gases (e.g., nitrogen and rare gases). This type of standard is reviewed and discussed in detail in Poulter's review [41], with particular reference to the system first described in [45].

Descriptions of systems at two other national standards laboratories are in [47–49]. This type of standard is also discussed in [7] and [50].

The theory of volume expanders is so simple that it appears rather straightforward to generate arbitrarily small pressures by use of the series expansion technique. The number of such systems that have been designed on paper and described at conferences rivals or exceeds the number of proposals for high-accuracy mercury manometers employing laser interferometers. However, since these systems utilize a fixed amount of gas, increasing care must be taken as the pressure is lowered to avoid leaks and outgassing that will add gas, and corrections must be made for the pumping or removal of gas by ion gauges and surface adsorption. The very high vacuum systems have been constructed of specially treated stainless steel or glass to minimize or eliminate the bulk permeation of hydrogen from stainless steel, and their operation is limited to inert gases. To achieve low pressures, starting with measurably high pressures (typically, $100–10^5$ Pa), very large volume ratios must be used. These are difficult to measure and the attendant errors are multiplied in each step of the series expansion process, so that the final generated pressure may have a very large uncertainty from this source alone. Practical limitations require that large-volume-ratio systems employ a very small volume (typically a few cubic centimeters or less), and this may change with time or operation due to deformation of mechanical components, such as bellows or valve sealing materials (including those in all-metal valves). Volume expansion systems can be a reasonable and accurate way of generating pressures in the low-vacuum and upper end of the high-vacuum ranges. Anyone contemplating using this technique for lower pressures is advised to review the experiences of those that have built, operated, and carefully evaluated such systems, such as are presented in [41, 47, 48].

4.5 Orifice-Flow Systems

Orifice-flow systems, also known as *dynamic expanders* or *conductance-limited expanders*, are illustrated in Figure 2.10. These generate a known pressure difference by passing a measured flow of gas through a known conductance, quite analogous to generating a known voltage by passing a measured current through a known resistance. If the pressure on the downstream side of the conductance can be maintained at a low enough value either to be neglected or to be measured with a small error, then the absolute pressure on the upstream side can be determined. Since these devices operate with a continual flow of gas, the problems due to leaks, outgassing, and gauge interactions are significantly reduced compared to static expansion devices. Thus, dynamic systems are more suitable for operation at lower pressures or with reactive gases. In addition, the pressure can be changed to a new stable value much faster with these devices than with a static system, and they can maintain a stable pressure for prolonged periods of time. However, the value of the conductance can be reliably

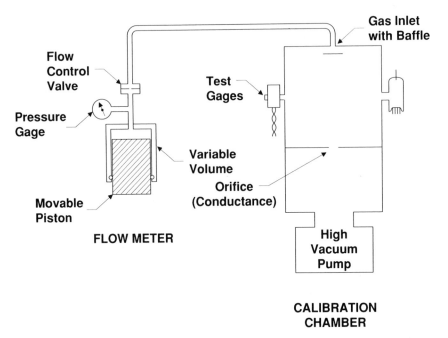

Figure 2.10   Schematic of an orifice-flow primary vacuum standard. Gas flows from the flowmeter into the upper part of the calibration chamber, where the test gauges are attached, and is pumped out through the orifice. Knowing the conductance of the orifice, the flow, and the pressure in the lower part of the calibration chamber, the absolute pressure in the upper part can be calculated. The flow can be measured by monitoring the pressure in the flowmeter and decreasing the flowmeter gas volume at a rate that will maintain a constant pressure. This is accomplished by driving a servo-controlled movable piston into the flowmeter gas volume.

calculated only in the molecular flow regime, generally restricting the upper end of the range of these devices to 10^{-1} Pa (10^{-3} torr) or below, although a few extended range devices have been built that operate in the transition range [51]. Orifice-flow standards are used for vacuum calibrations at both national standards laboratories [49, 52, 53] and industrial standards laboratories. In some cases the operating ranges have been extended down into the UHV range by the use of cryocondensation pumps [54, 55]. Further references to these standards can be found in [41] and [56], and the performance of an orifice-flow system is compared with a volume expansion system in [57].

The fabrication of an orifice such that the conductance can be calculated accurately requires considerable care, the maintenance of a low pressure below the conductance requires a large stable pumping capacity, and the measurement of the small flows required is not trivial. It seems that each designer of an orifice-flow standard seeks a different orifice design; but typically the orifices are a fraction to a few centimeters in diameter, and are fabricated either as "thin" orifices out of shim stock or with a geometry that allows an accurate calculation of edge corrections [52]. The conductance of these orifices requires a pumping speed below the orifice of a few tenths to 1 m^3/s in order to maintain the pressure below the orifice to a few percent of the generated pressure. The flows that must be measured are well below the range of conventional flow standards, and considerable effort has gone into addressing this problem. The constant-pressure flowmeter, illustrated in Figure 2.10, can achieve uncertainties of a few percent for flows as small as 10^{-11} mol/s (10^{-7} std cm^3/s) [58]. In this type of flowmeter the gas flows into the calibration chamber, at a rate determined by a control valve, from a small volume. The pressure in this volume is monitored continually, and the size of the volume is decreased at a rate that will keep the pressure constant. One way to achieve the volume decrease is to advance a piston into the volume through a sliding seal. The flow can be calculated from the pressure, temperature, and rate of volume decrease. Other techniques to measure the flow have been reviewed by Peggs [59]. These include constant-volume flowmeters, in which the flow is calculated from the rate of pressure decrease as the flows out of a small known volume. This has the disadvantage that the flow and the pressure in the vacuum chamber also decrease with the pressure in the flowmeter, but constant-volume flowmeters are in general simpler to build than the constant-pressure type. Variable-pressure flow elements, typically sintered plugs, have also been used. They require calibration as a function of the upstream pressure against a constant-pressure or constant-volume flowmeter. If the conductance of the flow element is stable and they are operated in the molecular flow regime, the flow can be calculated from the measured upstream pressure. Some orifice-flow standards also use flow-division schemes to extend the effective lower range of the flowmeter and, therefore, the range of pressures that can be generated.

5   PRESSURE GAUGES AND TRANSDUCERS

Pressure standards, even those that measure rather than generate a pressure, are designed to optimize accuracy, generally at the expense of low cost, speed, and ease of use. Therefore, most measurements are made with pressure gauges or transducers that must be calibrated against a pressure standard and then rely on their reproducibility and stability to maintain the accuracy of their calibration. Typically, although the relationship between the output of such devices and the applied pressure may be quite stable, it cannot be derived with adequate accuracy from a fundamental theory. As an example, the deformation of a coiled metal tube as a function of pressure applied to the inside of the tube may be reproducible to 0.1% or better, but a calculation of the deformation, based on properties of the material and mechanical deformation theory, will have a much larger uncertainty. Therefore, measurements with such a gauge will have reasonable accuracy only if the gauge is calibrated against a pressure standard. It is possible for the stability and precision of a gauge or transducer to exceed that of the standard against which it is calibrated, but it can never be more accurate.

Pressure gauges can be divided into two broad types: those that are sensitive to the force generated by the pressure (mechanical deflection gauges), and those that are sensitive to gas density. The latter are found almost exclusively in the vacuum ranges where (3) can be used to convert from pressure to density. The diversity of techniques used in different pressure gauges and the range of performance observed is so great, particularly in the atmospheric range, that it is impossible to describe them in any detail. Several general comments can be made, however, about the different types of gauges, their generic strengths and weaknesses, and techniques for evaluating their performance.

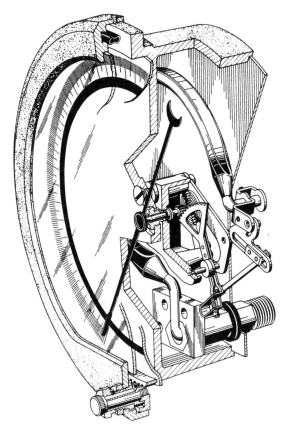

Figure 2.11   Bourdon tube pressure gauge with mechanical readout mechanism. Adjustments in the linkage between the Bourdon tube and the pointer allow compensation for zero, span, and linearity differences.

5.1   Mechanical Deflection Gauges

Figure 2.11 illustrates the familiar Bourdon tube pressure gauge and can serve as the basis for a general discussion of the basic principles and limitations of pressure gauges. Pressure applied to the inside of the tube causes the tube to "uncoil." The motion of the tip of the tube is amplified mechanically and transmitted to a pointer and scale. Gauges of this type are produced with a wide variety of full-scale pressures, level of performance, and cost. The best of them exhibit both ingenious design and quality craftsmanship, and this type of gauge is a cost-effective solution for many pressure measurement needs. For a number of reasons, however, the performance of even the best gauges is limited to an uncertainty of about 0.1%. The precision of the measurement is limited by visual detection of the pointer location and the repeatability of the mechanical linkage. The repeatability of the gauge will also be limited by inelastic behavior of the Bourdon tube, which will depend on the material properties and the level of strain induced by the pressure. Using a more compliant tube may improve the sensitivity, but result in less reproducible behavior. Temperature changes will alter the gauge reading because of thermal expansion of the components and temperature dependence of the elastic coefficients (thermoelastic coefficients). Temperature changes are caused not only by changes in the ambient temperature, but also by the adiabatic heating and cooling induced by pressure changes, as discussed in Section 6.1.

A number of techniques, most employing electronics, can and have been used to reduce these problems. These techniques are incorporated in a wide variety of commercial electromechanical pressure transducers. The mechanical linkage, pointer, and dial in Figure 2.11 can be replaced with an optical lever. The deflection of the tube end can be detected electronically and the output displayed digitally. Or, the electronic signal from the tube deflection detector can be used with a feedback circuit to maintain the tube end in a null position, limiting the strain of the tube and consequent inelastic behavior. In this case the magnitude of the feedback signal, rather than the deflection of the tube, constitutes the gauge output. Since electronic position detection techniques are available that are far more sensitive than mechanical techniques, a stiffer tube, operating with lower strains and closer to pure elastic behavior, can be used. Materials with improved elastic behavior, such as fused quartz, can be used for the sensing element. These changes may cause other problems, however. For example, although the thermal expansion of fused quartz is relatively small, its thermoelastic coefficients are rather large, and fused quartz gauges are correspondingly more temperature sensitive. Temperature sensitivity can be reduced by temperature controlling the sensor or temperature compensating the output.

These same techniques have also been used with sensor configurations other than Bourdon tubes, for example, diaphragms. Diaphragm gauges, where one side of the diaphragm is evacuated or a differential pressure is applied across the diaphragm, are available with capacitive, inductive, or strain gauge sensing of the diaphragm position. The capacitance type is illustrated in Figure 2.12. When

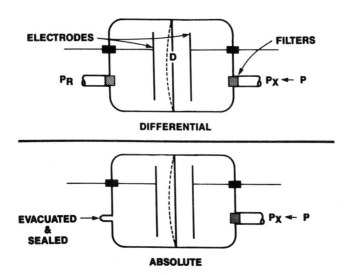

Figure 2.12 Schematic of absolute and differential capacitance diaphragm pressure sensors. Applied pressure P_X causes the diaphragm D to deflect, as shown by the dashed line. This in turn causes a change in the capacitance between each of the electrodes and the diaphragm. This change is measured by external circuitry and converted to equivalent pressure units.

strain gauges are used, the sensor may be either bonded to the diaphragm or implanted in the diaphragm material. Gauges are available with diaphragms made of a variety of materials, including: Ni-Span C, Inconel, titanium alloys, 400-series stainless steels, beryllium copper, fused quartz, and silicon, to name but part of what is undoubtedly a longer list. Other sensor types may be advantageous for different applications. Increased sensitivity can be achieved by using a sensor in the form of a capsule or bellows, which are more compliant than a Bourdon tube or diaphragm. At higher pressures the limitations of material strength can be minimized by using a sensor that is placed in compressive rather than tensile stress. One such transducer used in the very high pressure range employs a coaxial capacitor, with one capacitor plate deposited on a fused quartz rod and the other deposited on the inside of a surrounding fused quartz cylinder. The cylinder and rod are fused together at the ends and pressure is applied to the outside of the cylinder, minimizing tensile stress and taking advantage of the superior compressive strength of the quartz.

The techniques discussed so far rely on measuring the deflection or strain of a sensing element. An alternate approach is to measure the change in frequency of a resonant structure as the applied pressure changes the stress in the structure. Examples include cylindrical cans employing a feedback circuit to sustain a resonant radial oscillation. Pressure applied to the inside of the can changes the resonant frequency in a reproducible manner. However, because of viscous damping, the frequency also depends on the pressurizing gas. The force generated by the pressure applied to a diaphragm can be used to tension an oscillating wire, with corresponding changes in the oscillation frequency of the wire as the pressure changes. A high-pressure sensor applies the pressure to the outside of a structure fabricated from a single crystal of quartz. Electrodes plated on an internal part of the structure are used to excite and measure the frequency of the resonant quartz structure as a function of the pressure. A different quartz resonant-frequency sensor design is used in the atmospheric and high-pressure ranges; the pressure applied to a bellows generates a force that is transmitted to a single-crystal quartz oscillator using a lever and fulcrum mechanism machined out of the same block of quartz as the oscillator. In these single-crystal quartz sensors the crystal is cut to minimize the temperature coefficient. Unfortunately, the crystal orientation for zero temperature coefficient varies with stress, or pressure, so that the sensor will still have a pressure-dependent temperature coefficient.

The performance of a pressure gauge or transducer will depend on the design, materials, manufacturing techniques, and operating environment. Subtle changes in these factors can sometimes cause the performance to vary significantly from one gauge to another of the same design and mnaufacturer. The influence of environmental parameters will depend in part on the effectiveness of the design, for example, the adequacy of temperature control or compensation, if used. This makes it difficult to make a priori predictions of performance based on the design type; different implementations of the same technology can have quite different results. Keeping this in mind, the following discussion should be regarded as only a general guideline.

As noted, the best Bourdon-type gauge with mechanical readout can be expected to have an imprecision and long-term instability of about 0.1%; the performance of less-expensive instruments of this type can be one or two orders of magnitude poorer. High-quality capsule gauges with mechanical readouts, used in the atmospheric range, may exhibit instabilities significantly smaller than 0.1%. The use of force-balance techniques with metal Bourdon tubes will significantly improve their performance below the 0.1% level, as will the use of fused quartz for the tube material. Force-balance fused quartz Bourdon tube gauges can achieve instabilities of about 0.01% over periods of a year or more. This level of performance can also be observed for several atmospheric range transducers employing resonant structures. However, even of the types that can be found with this superior level of performance, not all units will demonstrate a consistent level of performance.

Other factors can limit the performance of high-quality electromechanical sensors. Some of the strain or stress measurement techniques used are highly nonlinear and/or temperature dependent, and careful attention must be paid to the statistical treatment of calibration data to derive the optimum relationship between the applied pressure and the output. For at least one commercial instrument 16 calibration constants must be determined. The failure of some manufacturers or users to consider this problem has resulted in cases where the transducer performance is significantly worse than what the intrinsic capabilities of the instrument will allow. Even where the manufacturer has addressed this problem with an extensive calibration as a function of pressure and temperature, and a careful statistical treatment of the data, the problem remains for the user to verify, at some future time, that the calibration is still valid in the temperature and pressure ranges over which it will be used. The larger the number of parameters required for proper gauge performance, the larger the amount of data required for initial calibration and subsequent verification. The performance of the gauge for some applications may also be limited by secondary design features. Many of the atmospheric range, high-performance gauge types have been designed for aircraft altimetry applications. While demanding of pressure measurement accuracy, this application does not require the chemical resistance or low leak rates required by many chemical-process or vacuum applications. Some, but not all, high-accuracy sensors are designed to be compatible with leak-free systems and/or aggressive chemicals. Some gauge

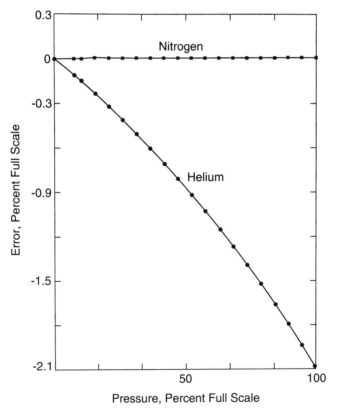

Figure 2.13 Gas density dependence of a vibrating-can pressure transducer. The transducer was calibrated with nitrogen, and can repeat its calibration at the 0.01% level over periods of years. However, if it is operated with helium, the indicated pressure will be in error by about 2% of the reading; when the transducer is operated with air, the error will be about a factor of 4 smaller. Some commercial gauges of this type include an automatic compensation for the air–nitrogen difference.

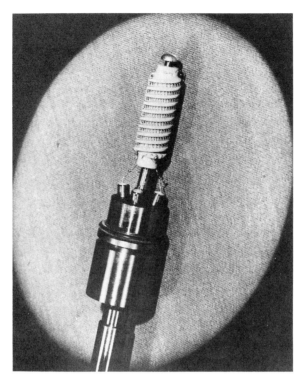

Figure 2.14   Manganin pressure sensor used in the very high pressure range. The resistance of the coil of wire at the bottom of the figure depends on pressure. The assembly attached to the coil includes electrical feedthroughs required for the resistance measurement, and seals to a high-pressure vessel of the type seen in the upper right side of Figure 2.8.

types are sensitive to the density of the pressurizing gas. Data for a vibrating-can-type transducer are shown in Figure 2.13. Although this type of gauge can achieve long-term instabilities of the order of 0.01%, significant errors will occur if the gauge is used with a gas different from the one with which it was calibrated. Some types of gauges may be sensitive to changes in power supply voltage. Proper power supply design can readily avoid this problem, but it does exist with some instruments. Some transducers can exhibit hysteresis after thermal cycling (such as occurs during shipment in an unheated aircraft cargo hold). This is particularly bad if it occurs during shipment between a calibration laboratory and the user.

The mechanical aspects of some pressure gauges may not be entirely obvious. In the high-pressure and very high pressure ranges a number of gauges rely on electrical property changes of a bulk material. The sensor of one such gauge, illustrated in Figure 2.14, is a coil of Manganin wire, a nickel–manganese–copper alloy. Pressure is determined from changes in the reistance of the wire. While these do not involve a macroscopic mechanical deflection, and the gauges may appear to be strictly electrical instruments, the electrical changes are caused by pressure-induced changes in the interatomic spacing. The performance of this type of gauge is limited, just as is a Bourdon-tube gauge, by temperature dependencies and inelastic behavior.

The adverse effects of temperature dependencies and an interesting solution to this problem are illustrated by the dielectric or capacitance high-pressure gauge constructed by Andeen and co-workers [60]. This gauge employs the change of capacitance with pressure of a parallel plate capacitor with a CaF_2 solid dielectric. Use of three-terminal capacitance measurement techniques allows very good sensitivity; and repeatability of the pressure measurement, 0.01% or better for pressures above 100 MPa, is a significant improvement over the performance of Manganin gauges, discussed later in this section. However, the capacitance gauge has a large temperature coefficient, and to achieve the best pressure reproducibility the temperature must be controlled to within a few tenths of a millikelvin, a difficult task in any case, considerably complicated by the adiabatic heating and cooling encountered in a high-pressure apparatus. Colwell [61] made two significant improvements in the design of this gauge. Instead of using a single dielectric material, he measured the ratio of two capacitors employing different dielectric materials with similar temperature coefficients but different pressure coefficients. Both capacitors are contained in the pressure vessel so their temperatures track one another closely. Colwell found the best material combinations to be single crystals of calcite ($CaCO_3$) cut parallel to and perpendicular to the c axis, and arsenic trisulfide (As_2S_3, a glass) and BGO ($Bi_{12}GeO_{20}$). He also was able to reduce the adiabatic heating and cooling problem significantly (see Section 6.1) by minimizing the amount of pressure fluid in the pressure vessel and balancing the adiabatic heating and cooling of different parts of the system to achieve close to zero net change. Instabilities of the gauge could be reduced to a few kilopascals and seemed to be determined primarily by material inhomogeneities.

In selecting a commercial gauge the user should first consult the manufacturer's specifications, keeping in mind that some manufacturers are more optimistic than others in projecting their product's performance and that the meaning of even commonly used terms may vary from one specification to another. Additional considerations are the thoroughness of the specifications and the extent to which they are substantiated by supporting material, and whatever is known of the manufacturer's reputation or the observed performance of a given type of instrument. However, even if experience with several units of a type is available, allowance should be made for unit-to-unit variations, and users are advised to monitor the performance of each gauge used.

One factor to be examined is the stability of the instrument zero. Our experience shows that for several types of high-performance gauges, instabilities are dominated by zero shifts, with changes of the calibration constants (linear and higher order terms in the functional relationship between pressure and gauge output) being much less of a problem. As an example, apart from zero drifts, the response of some force-balance quartz Bourdon-tube gauges will be stable to within 0.01% of the reading, over the upper 50–90% of the full range, for periods of years. For some of these gauges, however, the zero drift can amount to 0.01% of the full-scale pressure over a few days time. Therefore, the magnitude of the zero instabilities for a particular gauge should be evaluated so that zero corrections can be made at appropriate intervals. The zero stability can be evaluated by applying "zero" pressure to the gauge and recording its output as a function of time, temperature, and any other parameter of concern, such as power line voltage. For a differential gauge this requires the application of the same pressure to both sides of the gauge, for an absolute gauge a vacuum must be applied so that the gauge effectively sees "zero" absolute pressure. For an electromechanical transducer this type of test requires little more than a strip chart recorder and possibly a vacuum pump. The same technique can be used to make necessary zero corrections. With zero applied pressure the instrument zero control (if so equipped) is adjusted so that zero pressure is indicated, or the indicated pressure is recorded for later correction of the data.

Determining the accuracy of the calibration constants requires a pressure standard with an adequate uncertainty. If an appropriate pressure standard is not available, the calibration stability can be evaluated by periodic comparisons with another stable pressure gauge or some stable natural phenomena, for example, the vapor pressure of a pure material at a known temperature. The use of multiple sensors, even if of the same type, will help detect random shifts in the gauges and determine what degree of conficence should be placed in the reading of any one gauge.

As more zero and calibration stability data are gathered for a gauge, it will be possible to better predict how often the zero should be checked and how often the gauge should be recalibrated to maintain a desired level of accuracy. Historical data of this type are invaluable in evaluating the expected performance of gauge types as well as that of the individual instruments being monitored.

Unfortunately, performance data of this type are seldom published in readily accessible sources, and much of what is available is published by manufacturers for their own products. In many cases the manufacturer has conducted and reported the tests in a scrupulous manner; in other cases, there may be reason to suspect that "good" news is given more weight than "bad" news. Most of the published independent data for mechanical pressure gauges is for types used at the extreme ranges: very high pressures and low vacuums. The expense and complexity of primary standards in these extreme ranges results in the availability of relatively few primary standards and an increased reliance on the use of gauges as transfer standards. This has prompted different national standards laboratories to undertake studies of the performance of gauges that might be used as transfer standards.

Peggs [62] reviewed the performance of different high-pressure gauges, including several types employing electrical property changes. The most widely used such gauge is the Manganin gauge, illustrated in Figure 2.14. Typically, a Manganin gauge includes a 100-Ω coil of Manganin wire immersed in the pressure fluid, and a temperature-compensating resistor mounted outside the pressure vessel. The resistivity of the Manganin changes as a function of pressure. The better gauges use four-terminal resistors, and care is taken to minimize strains in the resistor winding. An initial "seasoning," consisting of thermal and pressure cycling, will generally improve the performance. Several laboratories have investigated the performance of these gauges, and Molinar and co-workers [63, 64] published results indicating that if precautions are taken, it is possible to achieve long-term instabilities in the measured pressure of 0.04–0.05% between 200 and 1200 MPa.

At the other extreme, low pressures or vacuums, capacitance diaphragm gauges or capacitance manometers are widely used as transfer standards. This type of gauge, illustrated in Figure 2.12, employs a thin stretched metal diaphragm, typically several centimeters in diameter. Deflection of the diaphragm by the applied pressure is detected by measuring the capacitance between the diaphragm and the nearby capacitor plates attached to the gauge housing. The use of thin diaphragms and the high sensitivity to length changes of the capacitance technique allow a pressure sensitivity that is superior to most, if not all, other mechanical deflection pressure gauges. These gauges may have pressure access to both sides of the diaphragm for differential measurements or a reference vacuum on one side for absolute measurements. Most of the gauges are temperature controlled for improved stability, although some use temperature compensation. Several variations of this design are marketed. Performance tests of these gauges show not only characteristic features of the type, but also significant variations from one unit to another.

The experience at NIST [65] indicates that zero instabilities for low-range gauges (1 and 10 torr full scale) are typically 10^{-2} Pa (10^{-4} torr) over a day's time, but there is a difference of three orders of magnitude between the performance of individual gauges, some performing much better, some much worse than this. Figure 2.15 is an example of zero stability data. As can be seen, changes in the gauge output tend to be highly correlated with ambient temperature changes, although these particular gauges are temperature controlled. Also evident in the figure are abrupt changes in the zero, apparently due to instabilities in the electronics or mechanical stresses caused by abrupt changes in temperature. The zero stability of this type of gauge is superior to most others, and is achieved by the design, choice of materials, and manufacturing techniques. The large range of zero instabilities observed for different gauges is probably indicative of subtle differences in material properties or strains induced during manufacturing. If these gauges are used in the high-vacuum range, optimum performance will require; ambient temperature control; frequent zero checks; and, if possible, selection of better than average units.

Evaluation of calibration stability is, as noted, a more difficult task. Figure 2.16 illustrates changes in the calibration of 4 capacitance diaphragm gauges over 11 years. As can be seen, the level of instability differs from one gauge to another, and, for two of the gauges, appears to improve with time. Evaluation of less extensive calibration data for 23 gauges of this same type [65] found the

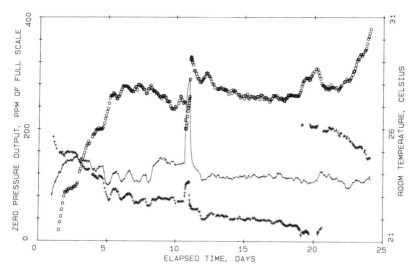

Figure 2.15 Stability as a function of time and ambient temperature with "zero" applied pressure for two different low-range capacitance diaphragm gauges. Indicated pressure outputs for the two gauges are plotted with an ○ and an +; the solid line is room temperature. The time scale starts when the gauges are turned on; the initial trend in the gauge outputs is a warm-up effect. The obvious correlation between temperature and gauge output indicates incomplete temperature control or compensation of the gauges. The discontinuous changes for one gauge at around day 20 are probably due to problems with its electronics.

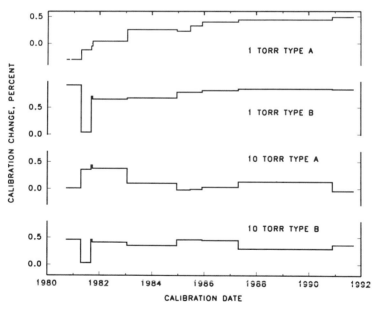

Figure 2.16 Change in calibration with time for four different 1- and 10-torr-range capacitance diaphragm gauges from two different manufacturers, *A* and *B*. The gauges were recalibrated at the times indicated by the vertical discontinuities in the data. The behavior of the instruments between calibrations is unknown; the horizontal data lines merely connect successive calibrations.

143

instability of the calibration factor over 1 yr to be typically about 0.4%. However, changes for some gauges were only 0.01 or 0.02%, while for others they were as large as 3%. For this type of gauge the changes appear to occur as abrupt shifts rather than a steady drift. Most of the gauges were monitored and used under laboratory conditions, although the question remains in some cases as to whether or not the change may be due to abnormal use of the gauge. These results are very similar to those obtained by other national standards laboratories [66–68]. While some gauges were found to be more stable, and others less so, typical long-term instabilities in laboratory use are about 0.4%.

These data indicate the selection of capacitance diaphragm gauges for applications in the low- and high-vacuum ranges requiring uncertainties better than a few percent. Above 1–10 kPa, other transducer types exhibit better calibration stability and might be preferred for measurement applications requiring uncertainties better than a few tenths of a percent; or the same level of performance can be achieved by transducers that are much less expensive than the capacitance diaphragm gauges. Other selection factors may also be important, however. The materials and construction techniques used in capacitance diaphragm gauges are not only compatible with UHV systems, they can also be used with chemicals that might be dangerous to use with other transducer types. Capacitance diaphragm gauge readings are effectively independent of the gas species, a virtue not shared by the density-sensitive gauges used in the low- and high-vacuum ranges. Note that when elevated-temperature gauges, such as the capacitance diaphragm gauges, are used in the low- and high-vacuum ranges, significant errors can result if thermal transpiration, discussed in Section 6.3, is not taken into account.

5.2   Density Gauges

Gauges that respond to gas density and to some species-dependent molecular property, rather than to a mechanical force, are widely used for vacuum measurements. The measurement result is generally given in terms of pressure units, and the gauges are sometimes calibrated against pressure standards. As can be seen from (3), the relationship between density and pressure is temperature dependent and a correction is required if the gauge is used at a temperature different from the calibration temperature. For most measurements this correction is small—a 3 K temperature change will cause a 1% error—and is generally ignored except for the most accurate measurements or cases where temperatures may differ significantly from normal room temperature. Published literature on the performance of several different types of density gauges is reviewed in [69].

As discussed in Section 1, in the ranges where most density gauges are used a number of factors, in addition to instrument performance, can contribute to measurement errors. It is important to examine the entire system to determine whether the pressure gauge is measuring the pressure or density of interest and to be aware of the possible interactions between the gas(es) and the gauge.

5.2.1   Thermal Conductivity Gauges

A relatively inexpensive alternative to capacitance diaphragm gauges for gauging in the low-vacuum range is the thermal conductivity gauge, common types being *thermocouple*, *thermistor*, or *Pirani* gauges. These gauges determine the pressure as a function of the gaseous conductance between a heated element and a temperature sensor within the gauge. Although they have been used at the 1% level [70], for most applications these gauges have much larger uncertainties. They are nonlinear, and the indicated pressure is a function of the gas species and cleanliness of the gauge. Because of their low cost, small size, and sensitivity they are widely used in the low-vacuum range for routine measurements where accuracy is not critical. Capacitance diaphragm gauges should be used for more critical applications. Extended range thermal conductivity gauges are available for use up to atmospheric pressure, although care should be exercised in their use at higher pressures. If the gauge is used with a gas, for example, argon, with a lower thermal conductivity than the air or nitrogen for which the gauge is usually calibrated, the gauge will always indicate a pressure below atmospheric, even though the actual pressure may be much higher. The consequences, particularly in a glass vacuum system, can be disruptive.

5.2.2   Ionization Gauges

The lack of adequate sensitivity limits the useful range of even the best mechanical deflection gauges to the upper part of the high-vacuum range. For measurements at lower pressures a different technique must be used; the solution usually employed is the ionization or ion gauge. Ion gauges employ electron-impact ionization to ionize gas-phase molecules. To a first approximation, the rate of ionization is proportional to the gas pressure, or more properly, gas density. Arguably, the ion gauge is the most widely used vacuum instrumentation, so it will be discussed in some detail. Ion gauges can be categorized into two types, cold and hot cathode, depending on how the ionizing electrons are generated; both types are discussed by Peacock [71].

In cold-cathode gauges the ionizing electrons are generated by a high-voltage (typically, 1000–3000 V) discharge between the cathode and anode. The electron discharge is confined by crossed magnetic and electric fields and amplified by secondary electrons emitted when an energetic electron ionizes a molecule. The discharge current increases with increasing pressure until the electron mean free path decreases to a length so short that the electrons cannot be accelerated between collisions to energies above the ionization threshold. The measured discharge current can be used as a measure of the pressure from the UHV up to the low-vacuum range. Two different gauge types, the *Penning* and *inverted magnetron*, are commonly used.

The simple geometry, rugged construction, and absence of a hot filament that are characteristic of cold-cathode gauges allow them to operate under conditions that would quickly destroy a hot-cathode gauge, such as exposure to atmospheric pressures. In addition, their low-power consumption and the absence of the heat and light generated by a hot filament make them uniquely suited for certain applications. Unfortunately, they also have a number of undesirable features. The discharge current is a distinctly nonlinear function of the pressure, with characteristics depending on the magnetic field and pressure range [71]. The discharge characteristics are not always stable, which can lead to abrupt changes in pressure readings. The discharge can also generate a significant ion pumping speed. For these reasons, with the exception of specialized gauge types developed for the UHV range, most cold-cathode gauges are used for relatively inaccurate measurement of pressures in the high- and low-vacuum ranges, and they are often used in applications where the gauge may be exposed to "dirty" gases or to pressures above their normal operating range. Relatively little effort has been made to evaluate their accuracy, but errors of a factor of 2 or more are believed to be typical for most cold-cathode gauges.

Most cold-cathode gauges are the Penning type, but recently two new cold-cathode inverted magnetron gauges have been commercially introduced. Both employ microprocessors to linearize the output, are available in bakeable versions, and have operating ranges extending from the UHV to the upper end of the high-vacuum range. Both gauges employ design features intended to stabilize the discharge characteristics and improve the accuracy of the indicated pressure. Limited testing of several units of each type at NIST has found that over a period of several months the errors in measured nitrogen pressures, due to residual nonlinearities and hysteresis with pressure, and changes with time, are typically within $\pm 20\%$ for pressures below 10^{-2} Pa. At low pressures relative sensitivities for other gases were found to be about the same as those for hot-cathode gauges, which are discussed below. Near the top of their operating range, however, about 0.1 Pa, the readings for other gases could be in error by as much as an order of magnitude. After longer periods of use, for pressures below about 10^{-6} Pa, these gases may read low by as much as an order of magnitude.

The hot-filament or hot-cathode ion gauge, which is the more widely used type of ion gauge, is illustrated in Figure 2.17. The earliest version of this gauge was the *conventional triode*, basically an outgrowth of the triode vacuum amplifier tube. The filament is heated to maintain an electron emission current, typically 1 mA, that is accelerated by a bias voltage, typically 150 V, between the filament (cathode) and the grid (anode). The ions produced by electron–molecule collisions are attracted to the collector, which is maintained at a negative bias with respect to the grid and filament. The pressure is determined from the ratio of the measured ion current I_C, corrected for a residual current I_R (discussed below), to the measured electron emission current I_e

$$P = (I_C - I_R)/I_e S \qquad (11)$$

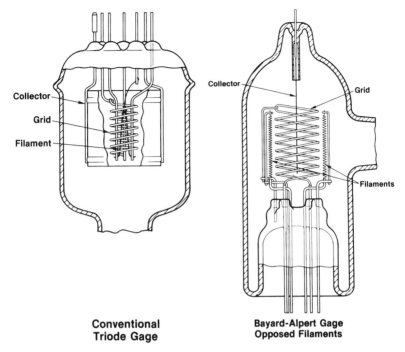

Collector

Grid

Filament

Collector

Grid

Filaments

**Conventional
Triode Gage**

**Bayard-Alpert Gage
Opposed Filaments**

Figure 2.17   Schematics of conventional triode and Bayard–Alpert hot-cathode or hot-filament tubulated ionization gauges.

To a first approximation the sensitivity S is a constant. In fact, it depends on a number of parameters; it can vary with gas species, bias voltages, electron emission current, and the geometry and surface condition of the electrodes. Above 10^{-3}–10^{-1} Pa a number of effects combine to make the ionization process nonlinear, causing the sensitivity to vary with pressure culminating in a rapid decrease in the sensitivity with increasing pressure. This establishes an upper pressure limit for the gauge that varies with gauge design; some gauges have been designed to raise the upper limit of their operation, but for most gauges this upper limit is near the lower end of the transition range from viscous to molecular flow (10^{-1} Pa).

A lower limit to hot-cathode ion gauge usefulness is established by a pressure-independent residual current, I_R in (11). A major contribution to this current is caused by the generation of soft X rays when electrons strike the grid. If these X rays in turn strike the collector, photoelectrons are ejected. Since the emission of an electron cannot be distinguished from the arrival of a positive ion, this *X-ray current* appears as a pressure-independent collector current. Ion currents at lower pressure are masked by this residual current. The large cylindrical collector used in conventional triode gauges will intercept a large fraction of the X rays, and the residual current in these gauges typically corresponds to pressures of about 10^{-6} Pa (10^{-8} torr). This limit presented a barrier to the extension of vacuum science and technology to lower pressures until Bayard and Alpert's development of the inverted triode geometry, commonly known as the *Bayard–Alpert* or *BA gauge*. As seen in Figure 2.17, the BA gauge has the filament outside the grid and a thin wire collector in the center. The thin wire intercepts only a small fraction of the X rays generated at the grid, but electrostatic forces maintain a high ion collection efficiency. This effectively extends the lower limit of the ion gauge down in pressure by two orders of magnitude (to around 10^{-8} Pa). Other designs to further reduce the effect of the residual current have included the *extractor gauge*, in which the collector is moved outside the grid volume and shielded from the X rays. The positive ions are extracted from the grid volume to the collector by electrostatic fields, and in some variants, a magnetic field as well. This can reduce the residual current by about another two orders of magnitude below that found in BA gauges. An alternate approach is to use an additional electrode near the collector in a BA gauge to modulate the collection efficiency for ions. In theory, the residual currents in such a *modulation gauge* are not influenced by the modulating field and the residual current can be determined from the difference of the collector currents when the modulator is alternately maintained at collector and grid potentials.

Significant contributions to the residual current can also be generated by ions desorbed from parts of the gauge structure, particularly the grid, by electron collisions. These electron-stimulated desorption currents will depend very much on the history of the gauge and can be a major problem when the gauge is used with gases that absorb at room temperature.

Operation of the gauge requires power supplies to maintain constant bias voltages and regulate the filament power to maintain a constant emission current. An ammeter or electrometer is needed to measure the collector current, ranging from $10\,\mu A$ at the highest pressures, to $1\,pA$ at about 10^{-8} Pa. The combination of power supplies and ammeter is generally called a controller. Controller designs are relatively straightforward and their operation should introduce negligible errors, although the measurement of currents below $1\,pA$ is not trivial. Unfortunately, in our experience, this is not the case with some commercial gauge controllers. Apparently, the excitement of digital displays and computer interfaces causes some designers to forget or ignore the fundamentals, so that in some instruments bias voltages and emission currents are not sufficiently regulated and collector currents are poorly measured. Therefore, the user is advised to make independent checks of the electrical operation of the controller.

The major determinant of ion gauge performance is the gauge structure or "tube." The relation of the ion current to the emission current depends not only on the pressure or gas density, but also on the ionization probability of the different molecules in the gas, bias voltages, emission current, electrode geometry, and electrode surface conditions. These will change with gas species, pressure, and time. In other words, the gauge sensitivity depends on a great many factors, the effects of many of which cannot be predicted adequately from theory, so that the behavior of ion gauges is best determined by experiment. The results of some such experiments are summarized below.

Although the emphasis is on UHV measurements, Weston's review [72] includes a great deal of information on general ion gauge theory and different designs. Nash [73] discusses many of the practical aspects of ion gauge use and performance. Much experimental ion gauge work was carried out in the 1950s and 1960s; most of that work is summarized in [56]. The rapid growth over the last two decades of industrial and scientific applications of vacuum technology has prompted a renewed interest in vacuum gauging focused largely on commercial instrumentation. Summarizing this work is complicated by the variety of instrumentation in use and sometimes conflicting results.

Of the types commonly used in the United States, we found [74] the best overall performance in the high- and very-high-vacuum ranges, in terms of linearity, stability, and constancy of sensitivity from one unit to another, to be obtained from a common glass-tubulated BA design with tungsten filaments, in particular, a design with two filaments (only one is used at a time) 180° apart about the central collector, as illustrated in Figure 2.17. Variations of sensitivity with time in a benign environment for a sample of this type of gauge were found to be relatively small [75], about 6% over 10,000 h of operation, while variations of sensitivity from gauge to gauge for this type were generally less than ± 10%. Conventional triode gauges with tungsten filaments, of the type illustrated in Figure 2.17, showed similar unit-to-unit variations, but varied about twice as much with time. Another study has found similar levels of instabilities for a slightly different triode gauge [76]. However, the stability of BA gauges with different filaments, thoria-coated iridium, were found to be inferior to conventional triode gauges [77]. Thorium oxide is a low work function material that can achieve a given emission current at a significantly lower temperature than a tungsten surface, and thoria-coated iridium filaments can withstand limited exposure to atmospheric air (they are sometimes marketed as "burnout proof"). However, the stability of gauges with thoria-coated filaments is generally inferior to that of gauges with tungsten filaments. Other studies [67, 73], of gauges of the same types, but with somewhat different construction than those illustrated in Figure 2.17, have observed larger instabilities; in one case, BA gauges have been found to change by as much as 25% over short periods of time [78]. In general, instabilities may be due to geometry changes [76], and/or electrode surface changes [79].

In our experience the performance of some types of commercial gauges is less predictable than that observed for the conventional triode and conventional tubulated BA gauges illustrated in Figure 2.17. Nude BA gauges, illustrated in Figure 2.18, have the same basic electrode structure as tubulated BA gauges, but are constructed on a flange and designed to be mounted directly in the vacuum system without a surrounding enclosure or tube. They were originally designed for use in the UHV range, but are often used in the high-vacuum range. They have been found to have sensitivities varying from 40 to 130% of the specified

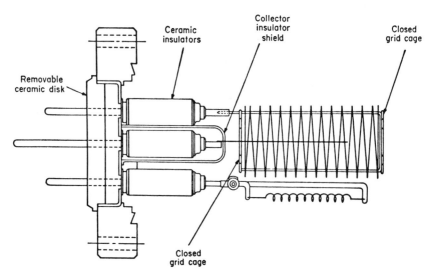

Figure 2.18   Schematic of a nude Bayard–Alpert gauge. The gauge is constructed on a stainless steel and ceramic flange, which is mounted directly to the vacuum chamber. This provides a much higher degree of mechanical integrity than a glass tube.

value [74], and significant changes in the sensitivity with pressure can occur for pressures as low as 10^{-3} Pa. As illustrated in Figure 2.19, the nude gauges will typically have a maximum in the sensitivity at some pressure near 10^{-3} Pa and then, a rapid drop in sensitivity with increasing pressures. This may be due to the closed grid generally used to enhance the sensitivity of these gauges [80]. The sensitivities shown in the figure for tubulated BA gauges are relatively constant over a wider pressure range, which is typical for this type of gauge. It is not known to what extent the data in the figure are typical of extractor and modulated gauges. Another variant of the BA gauge, designed for use at pressures as high as the low-vacuum range, was observed to exhibit sensitivities as low as one-half that specified [81]. They also appeared prone to unstable operation.

Since the development of the BA gauge in 1958 a considerable amount of work has been done on UHV gauging. Particle accelerator and storage ring design and operation and the recent development of semiconductor production processes operating in the UHV stimulated renewed interest in this field. Several reviews addressing the particular problems of UHV gauging have been published [72, 82–84], and additional work is described in the proceedings of a workshop on UHV gauging [85]. Results of particular interest include the observation that the residual currents of different BA gauges of the same design can vary by an order of magnitude [86, 87], and that significant modulation of the residual current can occur in at least one type of modulator gauge [86],

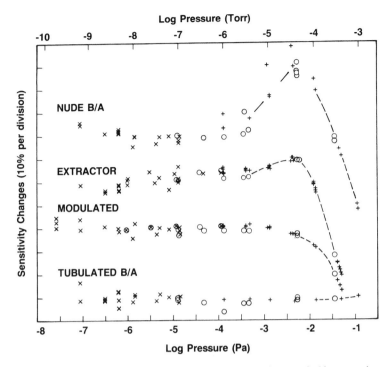

Figure 2.19 Sensitivities of tubulated, nude, and modulated Bayard–Alpert, and extractor ionization gauges. The data were obtained over several months using an orifice-flow primary standard [52] operating in two different modes (0, X) and a calibrated spinning rotor gauge, discussed in Section 5.2.3 (+). The lines connecting points have been added for visual clarity.

rendering worthless, for this particular gauge design, the modulation technique of determining residual currents.

When the gas is not specified, ion gauge sensitivities are generally quoted for nitrogen, although in some cases the argon sensitivity may be quoted. The sensitivities for other gases, if given, are often given in terms of the *relative sensitivity*, the ratio of the sensitivity for the specified gas to that for nitrogen or argon. This is done with the hope or expectation that relative sensitivities for a specific gas will be a constant for different gauges and/or that relative sensitivities can be correlated with a known property of the gas, such as the number of valence electrons, first ionization energies, or the electron-ionization cross section. Holanda's review [88] of older published relative sensitivities found variations greater than a factor of 2 in the published values for some gases, and the closest correlation was found to be between relative sensitivity and electron-ionization cross section. Significant variations can also be found between relative sensitivities in later published work [56, 77, 89–91]. Some of this variation is undoubtedly due to experimental errors in determining gauge

Table 2.4 Ion Gauge Sensitivities Relative to Nitrogen, $S_r{}^a$

Gas	S_r	Gas	S_r
He	0.17	CH_4	1.6
Ne	0.28	C_2H_2	0.6
Ar	1.3	C_2H_6	2.6
Kr	1.9	C_3H_4	1.3
Xe	2.7	C_3H_6	1.8
H_2	0.4	C_3H_8	3.4
O_2	0.9	C_4H_8	2.1
CO	1.0	C_4H_{10}	4.0
CO_2	1.4	C_6H_6	3.8
NH_3	0.6	$C_6H_5CH_3$	6.8

[a]Relative sensitivities for different ion gauges can differ by as much as 30% from these values. Relative sensitivities for partial pressure analyzers can differ by as much as an order of magnitude.

sensitivities, but some is also due to real differences in the relative sensitivities for different gauge designs and variations in the relative sensitivities with gauge potentials and pressure. Table 2.4 presents "typical" relative sensitivities for a number of gases for BA gauges, obtained in our laboratory [56] or selected from other published work [77, 89–91]. For some gauges these values can be in error by 10–30%, with even larger errors possible at the high-pressure end of the high-vacuum range. These values can also be used for conventional triode gauges, and our limited experience indicates that they are also applicable to cold-cathode gauges. However, they should not be presumed to apply to partial pressure or residual gas analyzers, discussed in Section 5.2.4. If more accurate values are needed the gauge to be used should be calibrated with the gas to be used, using a vacuum standard with a predictable performance for that gas.

As noted in Section 1, a number of problems complicate vacuum measurements. One such problem with hot-filament ion gauges is dissociation of gases at the hot filament or reactions of the gases with the filament material and included impurities. In an effort to minimize these effects, filaments have been coated with low-work-function materials, thorium oxide being the most common, so that adequate electron emissions can be obtained with lower temperatures. Unfortunately, as previously noted, gauges equipped with coated filaments tend to be less stable than the same gauge type equipped with tungsten filaments. We believe that part of the problem with coated filaments may be the tendency for the adherence of the coating to the heated substrate to change with time, causing varying electron emission patterns, with consequent sensitivity changes. This defect may be reduced by different filament designs. In addition, we have found that when operated with water, thoria-coated filaments produce significant quantities of several reaction products, principally, hydrogen, carbon monoxide, carbon dioxide, and oxygen. Cold-cathode gauges produce quantities of these products comparable to a thoria-coated filament operating at 1 mA emission, while tungsten filaments, operated at the same emission current produce somewhat smaller quantities.

Another major vacuum measurement problem is nonisotropic and nonuniform pressure and flux distributions. Major problems of this type can be encountered in vacuum systems with cryopanels and/or solar simulators, such as space simulation chambers. The difficulties and some solutions encountered in measuring pressures in space simulation chambers are discussed in detail in [92] and [93]. To measure flux asymmetries gauges have been designed with highly directional characteristics [94].

The available data indicate that, under optimum conditions and after calibration, it is possible to make ion gauge measurements with uncertainties approaching a few percent. However, not all gauge types perform equally well, and for critical applications it is wise to calibrate frequently, if possible, and/or to use multiple gauges that can be cross-checked for stability. A more important concern is that the laboratory results cited here have been obtained under carefully controlled conditions and much poorer results can be expected if any of a number of factors are overlooked. Chief among these are cleanliness of the gauge and vacuum system, pressure equilibrium between the gauge and the system, and orientation of the gauge when temperature or flow asymmetries exist. As examples, order-of-magnitude pressure differences can exist within a laboratory-scale vacuum system when it is pumped down after exposure to atmospheric air. During pumpdown in the high-vacuum range there are probably more water molecules on the vessel wall than there are molecules of any kind in the gas phase. Outgassing of this material can maintain large pressure gradients between the gauge and critical parts of the system. Contamination of gauges with hydrocarbons or overheating of the electrodes can change the sensitivity by large factors. Young [89] found changes of a factor of 4 or more in ion-gauge sensitivities with time when ion gauges were operated with hydrocarbon gases. He convincingly argues that this is due to the buildup of carbonaceous deposits on the collector. These high-resistivity deposits can significantly alter electrical potentials within the gauge. Hydrocarbons are a major source of contamination in some vacuum systems. Depending on its history and the gas in the system, an ion gauge can be either a source or a sink for gas, and the pressure in a tubulated gauge can differ significantly from that in the rest of the vacuum system, particularly if the gauge is attached to the system by a restricted conductance. Tubulated gauges should be purchased with tubulation at least 2 cm or 1 in. in diameter and attached as directly as possible to the vacuum system. Even in a clean and properly conditioned system, large errors can result if the gas composition is not known or controlled. It is apparent from the values presented in Table 2.4 that a small amount of a high molecular weight hydrocarbon can give erroneously high-pressure readings, particularly if one is attempting to measure the pressure of helium or hydrogen. Partial pressure analyzers, briefly discussed in Section 5.2.4, can be quite useful in detecting and analyzing such problems.

5.2.3   Molecular Drag Gauges

Several high-vacuum gauges have been proposed and a few built that rely on the exchange of momentum between gas molecules and some mechanical structure. One such gauge [95, 96] is now available commercially. Variously referred to as a spinning rotor gauge, molecular drag gauge, or viscosity gauge, it is shown in Figure 2.20. The sensor is a 4.5-mm or $\frac{3}{16}$-in.-diameter magnetic steel ball, which is contained in a 1-cm-diameter, thin-walled extension of the vacuum system (thimble). The ball is levitated magnetically by a combination of permanent and electromagnets contained in a suspension head mounted on the exterior of the thimble. A differential transformer position sensor and feedback circuit maintain the vertical position of the ball. A set of drive coils and an inductive drive circuit are used to rotate the ball up to about 400 Hz. Once up to speed, the ball is allowed to coast, and its rate of rotation is determined by timing the signal induced in a set of pickup coils by the rotating component of the ball's magnetic moment. Collisions with gas molecules generate a *molecular drag*, which causes the ball to slow down, the rate of slowing depending on the ball's moment of inertia, the molecular weight of the gas, and the gas density or pressure. If the

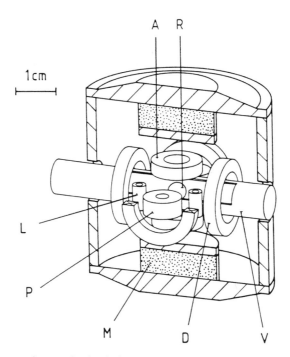

Figure 2.20   Sensor and suspension head of a commercial molecular drag high-vacuum gauge. The bearing ball (sensor) R is located in an extension of the vacuum system V. Permanent magnets M and electromagnets A levitate the ball. Horizontal stability is provided by coils L. Coils D are used to spin the ball up to about 400 Hz. Pickup coils P sense the rotation of the ball.

other factors are known, the density or pressure can be determined from the rate of slowing. The imprecision of this gauge is determined largely by difficulties in measuring the rate of change of the rotational speed and mechanical and thermal noise. With a reasonable signal strength, a quiet mechanical environment, several minutes measuring time, and a statistical procedure that uses multiple zero crossing of the timing signal [96], the rotation period can be determined with about a 1-μs imprecision, with a corresponding imprecision in the equivalent pressure of between 10^{-6} and 10^{-5} Pa. The molecular drag will vary linearly with pressure to better than 1% for pressures below 10^{-1} Pa. Above this pressure, corrections must be made for viscous drag effects.

To obtain optimum performance from the spinning rotor gauge, several departures from this simple model must be taken into account. Asymmetries in the magnetic suspension field will induce eddy currents in the rotating ball. Similarly, the rotating component of the ball's magnetic moment will induce eddy currents in surrounding metallic components. These combine to generate a pressure-independent slowing of the ball, or *residual drag* (RD). The rate of slowing due to molecular drag will depend on the average momentum transfer during a molecule–ball collision. This will depend on the gas molecular weight, the tangential momentum accommodation coefficient, and the ball's surface roughness. The last two factors are generally combined as the *effective accommodation coefficient* σ_{eff}, which is effectively the calibration constant for the gauge. Taking these factors into account, the pressure is given by

$$P = \sqrt{8kT/\pi m}\,(a\rho\pi/10\sigma_{eff})\cdot(-\dot{\omega}/\omega - RD) \qquad (12)$$

where k is Boltzmann's constant, T is the absolute temperature of the gas, m is the gas molecular mass, a is the radius of the ball, ρ is the density of the ball, and $\dot{\omega}$ is the measured rate of change of the ball's rotation rate ω. These factors are readily determined, assuming that the gas composition is known or controlled, except for the effective accommodation coefficient and the residual drag.

The residual drag, which is typically equivalent to a nitrogen pressure between 10^{-4} and 10^{-3} Pa, is experimentally determined by measuring the rate of slowing of the ball at "zero" pressure, that is, any pressure below the desired imprecision of the pressure measurement. The commercial gauge units then automatically compensate for this residual drag or *offset* in subsequent measurements. The residual drag depends on a number of factors; changes in these factors must be minimized to stabilize the residual drag, and several precautions must be taken to determine accurate values. Temperature changes will cause changes in the ball's moment of inertia and rotation rate. These changes will cause errors in the determination of the residual drag, as well as errors in later pressure measurements. Fortunately, since the ball is magnetically suspended in a high vacuum, it has a long thermal time constant, and typical short-term laboratory temperature variations are highly attenuated and have a small effect [97]. However, the inductive drive circuit will cause significant heating of the ball and the surrounding thimble, and the ball's temperature will not stabilize for 4–7 h after it is spun to its operating frequency [98]. Attempts are being made to reduce this effect by redesign of the suspension head and drive circuit and the use of ball materials with a small thermal expansion coefficient [99]. If the apparatus is tilted, the ball will move slightly with respect to the suspension field, changing the residual drag [100]. Orientation of the gauge must be maintained within 1°, which is not a difficult task in a laboratory environment. Competition between inertial and magnetic forces may cause the axis of rotation of the ball to change slightly as it slows down, causing a change in residual drag [101]. For some balls this effect is large enough that corrections must be made to the residual drag as a function of the frequency. Even if these effects are minimized, it is necessary to monitor the stability of the residual drag by periodic redeterminations. In particular, if possible the residual drag or offset should be determined just before measurements are made at the lower end of the range.

A ball that is smooth on the atomic scale and with total momentum accommodation will have an effective accommodation coefficient of 1. For real "smooth" balls, that is, bearing balls as fabricated with a shiny surface, the effective accommodation coefficient for nitrogen can vary from ball to ball as shown in Figure 2.21. Effective accommodation coefficients less than 1 are caused by elastic collisions with no momentum exchange; values over 1 are due to the increased tangential momentum transfer incurred in normal or near-normal collisions with the perturbations of a rough surface. To obtain a more

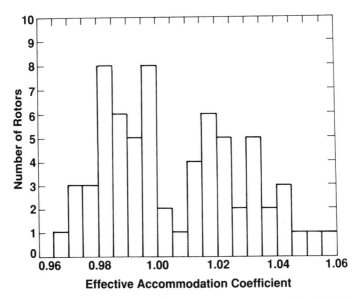

Figure 2.21 Nitrogen effective accommodation coefficients for a sample of 67 "smooth" (as fabricated) bearing balls. All balls were cleaned with solvents and baked under vacuum at 200°C before calibration.

accurate value of σ_{eff} than can be estimated from Figure 2.21, it is necessary to calibrate the gauge against a vacuum standard. For a given ball there can be an additional variation in σ_{eff} of up to about 2% from one gas to another [102].

Published changes that have been observed in effective accommodation coefficients over longer terms are typically 0.5% over a few months, and generally no more than 1 or 2% over 1- or 2-years time [102–104]. Analysis of the data obtained for a group of spinning rotor gauges used in an international comparison of primary vacuum standards [105] shows, after excluding one set of data because of obvious changes in the surface of the balls, that the average change in the calibrations during the 1–4 months between recalibrations against a reference standard was 0.4% for argon and 0.5% for hydrogen. These data are all consistent in predicting that for nonreactive gases, if the ball's surface is not mechanically or chemically altered, the spinning rotor gauge can be expected to be stable to within about 0.3–0.5% over a period of a few months, and to within 1–2% over a 1- or 2-yr period.

These gauges are relatively expensive, slow to operate, cannot be used for pressures much below 10^{-4} Pa, and become nonlinear in the transition region starting around 10^{-1} Pa. Their superior stability, however, is a major advantage where measurements accurate to within a few percent are required, and they serve very well as transfer or calibration standards in the high-vacuum range. With viscosity corrections they can be used for pressures as high as 10 Pa. Their lack of a hot filament or electrical discharge is a significant advantage, and they should be relatively compatible with many reactive gases, although gases that will alter the ball's surface will clearly cause changes in the gauge's sensitivity. Furthermore, they are, to at least first order, immune to thermal transpiration effects (Section 6.3). Efforts in several laboratories are under way to better understand and improve the performance of this gauge.

5.2.4 Partial Pressure Analyzers

In many vacuum systems the total pressure is of less interest than the partial pressures of one or more molecular species. Partial pressure or residual gas analyzers (*PPAs* or *RGAs*) are widely used for this purpose. These are mass spectrometers of relatively small size that are designed to be attached to a vacuum system as an appendage instrument. Most of them can be baked and are constructed to be compatible with UHV systems. They are generally designed for high sensitivity so that they can detect and analyze the residual gases in a vacuum system. They consist of an electron-impact ionizer, in some cases similar to the filament and grid structure of a BA gauge; extraction electrodes to remove ions from the ionizer and accelerate them into a mass filter; and a detector at the far end of the mass filter. Most PPAs have used mass filters of the magnetic sector or quadrupole type. Over the last decade quadrupole PPAs have come to dominate the commercial market. A schematic of a typical quadrupole PPA is shown in Figure 2.22. A combination of DC and radio frequency (rf) voltages in the megahertz range are impressed on the quadrupole rods. The applied rf

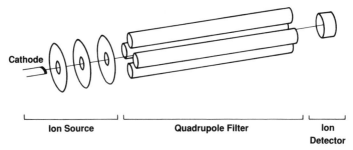

Figure 2.22 Basic schematic of a quadrupole partial pressure analyzer. Ion source designs vary considerably from one instrument to another, and Faraday cups and/or secondary electron multipliers are used for ion detection.

allows ions with the selected charge-to-mass ratio to travel down the axis of the quadrupole structure to the detector; other ions are deflected to the rods or the surrounding structure. The detector may be a Faraday cup, as shown in the figure, or, for increased sensitivity, a secondary electron multiplier (SEM), either of the discrete or continuous dynode types. The SEM may be used as an analog current amplifier, or at very low pressures it can be combined with a fast-rise amplifier and used to detect the arrival of individual ions. The mass ranges of different instruments vary considerably, but ranges of 50–200 charge-to-mass ratio are typical. The pressure range of these instruments is limited at high pressures by nonlinearities in the ion source to total pressures of the order of 10^{-3}–10^{-1} Pa. The low-pressure limit, or minimum detectable partial pressure, is determined by a number of factors, most notably, instrument sensitivity and detector characteristics. Using Faraday cup detection and a typical picoammeter, partial pressures of the order of 10^{-7} Pa can be detected, although this limit can vary from one instrument to another by an order of magnitude or more. State-of-the-art ammeters, with noise currents of the order of 10^{-16} A, can reduce this limit by 2 or 3 decades. A similar reduction, and faster response times, can be achieved using SEMs. Ion counting, using a fast-rise SEM, can further improve the minimum detectable partial pressure by about a decade.

Lichtman [106] recently reviewed the development of these instruments, many of which were originally intended to provide only a qualitative assessment of the relative quantities of different gases remaining in a vacuum system at its base vacuum. This capability can be invaluable in diagnosing and troubleshooting a vacuum system or process and has made PPAs or RGAs a common vacuum system accessory. Modern electronics has made possible decreased cost and improved ease of operation, further increasing PPA use. Many users now desire to use PPAs for quantitative gas analysis, such as might be required for on-line process control. This trend has probably been encouraged by the use, in some instruments, of microprocessors to analyze the measured ion currents and display them as equivalent pressures for different gas species. The high cost of

PPAs, relative to ordinary vacuum gauges, and the impressive graphic or multidigit displays now available lead some users to expect correspondingly accurate results. This expectation is generally misleading. While quantitative results accurate at the few percent level are possible with some PPAs, this level of accuracy requires frequent calibration, which in many cases may be complicated and time-consuming. For other instruments, instabilities, non-linearities, and interactions between gases may cause order of magnitude errors even after the instrument has been calibrated.

Leaving aside the problems of interpreting mass spectrometer cracking patterns (the pattern of signals at lower mass-to-charge ratios generated by multiply charged ions and fragments of the parent molecule generated by electron impact), the difficulties in obtaining quantitative partial pressure measurements from PPAs arise from their complicated nature; they are susceptible to a number of design, manufacturing, and operating variables. Some of these can be manipulated by the designer, or, with some instruments, the user, to enhance one or more characteristics at the expense of others. For example, the sensitivities of different instruments, operating with the same electron emission current and detector configuration, may differ by as much as two orders of magnitude. The high sensitivity of the more sensitive instruments may be achieved, in part, by using space charges to create a high-efficiency ion source. However, space charge effects are very pressure dependent, and they may cause the sensitivity to be correspondingly dependent on pressure. The sensitivity of less sensitive instruments, with less efficient ion sources, may be relatively constant with pressure. The more sensitive instrument may be useful for detecting residual gases in a baked UHV system, but variations in their sensitivities with pressure may make them useless for quantitative analysis of partial pressures. For all instruments, drifts with time of electrical potentials or deposition during operation of conducting or insulating layers on insulators or electrodes can cause significant changes in sensitivity. Since the gain of secondary electron multipliers is very sensitive to the condition of the electrode surfaces, their gain, and the instrument sensitivity, may change significantly with time and use. At very low pressures, however, this instability can be greatly reduced by using ion-counting techniques since the SEM pulses counted are independent of amplitude, except for the small fraction near the detector threshold.

In recent years a number of PPA performance studies have been conducted [91, 107–113]). Depending on the instrument and selection of operating parameters, the sensitivity can vary, in some cases by as much as one or two orders of magnitude, with the ion source parameters, the molecular species, and the pressure. The relative sensitivities, that is, the ratios of the sensitivities of an instrument for different gases to the sensitivity of the same instrument for a reference gas (typically nitrogen or argon), can vary from one instrument to another by an order of magnitude (for this reason the relative sensitivities given in Table 2.4 for ion gauges should not be assumed applicable to PPAs). The choice of filament material can significantly alter the relative sensitivities for different molecular species. For some PPAs the sensitivity will vary significantly with total pressure for pressures as low as 10^{-7} Pa.

The nonlinearity with pressure of some instruments will be strongly affected by ion source parameters: electron emission current, electron accelerating voltage, and, especially, the ion extraction voltage. Note that the ion extraction voltage control is labeled *ion energy* on many PPAs; however, because of space charge effects the actual ion energy may differ significantly from this extraction voltage. Similarly, the electron accelerating voltage control is often labeled *electron energy*. A study of five different commercial instruments [113] found that the nonlinearities could be broadly categorized as "low" pressure, typically occurring during operation with high ion extraction voltage settings, and "high" pressure, typically occurring with low ion extraction voltage settings. The range of possible performances can be appreciated from the data of Figures 2.23 and 2.24 for two different PPAs. The instruments were calibrated with argon from 10^{-6} to 10^{-1} Pa for 27 different combinations of emission current, electron accelerating voltage, and ion extraction voltage. The data in Figures 2.23 and 2.24 were selected from the 27 different calibrations to illustrate the conditions that produced the maximum low-pressure nonlinearity (solid circles), the maximum high-pressure nonlinearity (solid squares), and the best linearity (solid line, no symbol). In both figures the sensitivities for a given set of parameters were normalized to 1 at 10^{-4} Pa. The data in Figure 2.23 were obtained from the most linear instrument, and the nonlinearities evident at high pressures were

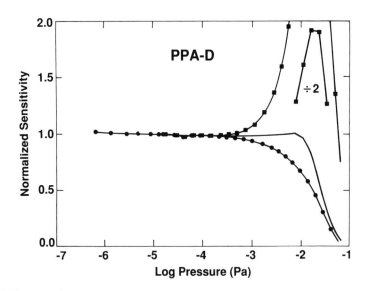

Figure 2.23 Normalized argon sensitivities for a PPA operated with three different combinations of ion source parameters. The electron emission current–electron accelerating voltage–ion extraction voltage combinations (in mA, V, V) were ●, 0.5, 60, 10; —, 1, 100, 10; ■, 2, 100, 3.

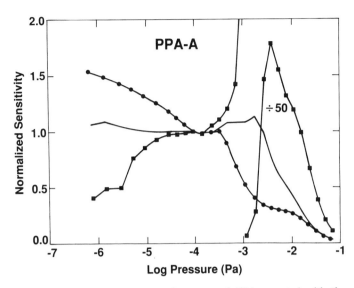

Figure 2.24 Normalized argon sensitivities for a second PPA operated with three different combinations of ion source parameters. The electron emission current–electron accelerating voltage–ion extraction voltage combinations (in mA, V, V) were ●, 0.5, 93, 10; —, 1, 93, 10; ■, 2, 93, 3.

observed only under extreme settings of the operating parameters. No significant nonlinearities could be observed at low pressures. The "best" performance was typical of that observed over a range of ion source parameters. In contrast, the instrument used for Figure 2.24 evidenced significant nonlinearities for all combinations of operating parameters. The solid line was truly the best performance, and large low- and high-pressure nonlinearities were observed for a wide range of operating parameters. Note in particular, that when operated at low ion extraction voltages the sensitivity of this instrument will change by more than two orders of magnitude as a function of pressure.

Clearly, a significant dependence on pressure of the sensitivity complicates the calibration and use of a PPA. A further problem was observed for instruments that showed significant nonlinearities. For these instruments the sensitivity for a specific gas tended to change not just as the partial pressure of that gas is changed, but also varied as the pressures of other gases were changed. Thus, the sensitivity for a trace gas can depend on the background gas pressure, in some cases the changes can be as large as two orders of magnitude, a highly undesirable feature. For all instruments, exposure to active gases, such as, CO_2, O_2, H_2O, caused changes of 10–20% in the sensitivity for all gases. With removal of the active gases the sensitivity returned to original values over periods of days. Over periods of weeks, changes in the sensitivity as large as an order of magnitude have been observed for some PPAs operated with secondary electron multipliers [111]. On the other hand, it has been found [113] that with the use of Faraday cup detectors some PPAs change by less than 10% over 2 months, although others changed by an order of magnitude under the same conditions.

Various combinations of the factors previously noted can combine to cause errors, in some cases as large as several orders of magnitude, in the partial pressures indicated by PPAs. This difficulty should not discourage their use since the qualitative diagnostic information that can be obtained with an RGA generally more than justifies their cost, particularly with UHV systems. However, it does indicate the need to exercise some judgment in the selection of PPAs and is a clear warning that some effort will be required to obtain quantitative measures of partial pressure. If a minimum detectable partial pressure is the primary consideration, the instrument can be selected on the basis of maximum sensitivity alone. But, if the objective is to monitor a trace gas in the presence of a varying background gas, a different selection may be appropriate. In particular, for this application it would be highly desirable to have a constant sensitivity over the operating pressure range. It is possible to calibrate some PPAs and obtain accuracies as good as a few percent in the measured partial pressures. For other instruments this is a fool's errand: some instruments are unpredictable, dependent on too many variables, and are unstable with time and use. Unfortunately, repeated calibrations will be required to distinguish the two categories of instruments. Lacking a primary vacuum standard, PPAs can probably best be calibrated by comparing them, as a function of pressure, with BA ion gauges, operated with reliable controllers. These calibrations should be done for different pure gases of interest, taking into account the different sensitivities of the ion gauge for different gases. For better accuracy, the ion gauge and the PPA can be compared with a molecular drag gauge in the high-vacuum range. Determination of the sensitivity to one gas species as a function of the pressure of other species will require more elaborate calibration apparatus and procedures [113]. If the PPA has both a Faraday cup and a secondary electron multiplier, confidence in the electron multiplier measurements can be improved if it is periodically compared with the Faraday cup in a pressure range where both detectors can be used.

6   GENERAL CONSIDERATIONS

6.1   Adiabatic Heating and Cooling

When pressure is changed, mechanical work is done on, or extracted from, the pressure medium. This work will cause a change in the temperature of the medium and the surroundings, including pressure sensors. This temperature change can cause significant perturbations in the performance of pressure sensors. In the extreme in which the pressure fluid is thermally isolated from its surroundings, so that the process occurs under adiabatic conditions, that is, constant entropy S, the temperature change with pressure can be derived using Maxwell's thermodynamic relations

$$(\partial T/\partial P)_S = T\beta/\rho C_P \tag{13}$$

where β is the volume thermal expansion coefficient, ρ is the fluid density, and C_P is its constant-pressure heat capacity. Similarly, the heat Q generated per volume of fluid V is

$$Q/V = T\beta\Delta P \tag{14}$$

where ΔP is the total pressure change and β is the average thermal expansion coefficient.

In a real apparatus, conditions are, of course, not adiabatic, and the temperature changes will attenuate with a time constant dependent on the thermal behavior of the entire apparatus. However, (13) demonstrates that the upper bound of the temperature change depends on the thermal expansion of the fluid, which is large for liquids and larger still for gases. The initial temperature perturbation of the sensor will depend on its heat capacity relative to that of the pressure medium. In a sealed system, this effect will also cause a relaxation of the initial pressure change as the fluid temperature returns to equilibrium. This relaxation complicates the determination of the magnitude of the sensor perturbation.

An example of this effect can be seen in the operation of temperature-controlled, force-balance quartz spiral gauges, which are available for gas pressures up to 33 MPa (5000 psi). Because of the limited tensile strength of fused quartz, in high-range gauges the pressure is applied to the outside of the quartz spiral and is contained in a surrounding pressure housing. This results in a relatively large volume of pressure fluid; and after a full-scale pressure change, times on the order of $\frac{1}{2}$ h are required before temperature equilibrium is adequate for the perturbation of the pressure reading to be reduced below 0.01%. For mercury manometers, temperature changes of tens of millikelvins can occur for 100 kPa pressure changes. This will cause a significant perturbation for high-accuracy instruments. With liquid (oil) pressure systems, the thermal expansion coefficient is smaller, but the pressure changes can be much larger, so that adiabatic temperature changes may be a significant perturbation for many measurements. With high-pressure systems the thermal situation can also be very complicated as the apparatus often involves thick-walled steel vessels. A pressure increase will heat the oil and the inside of the vessel, but the outside of the vessel will cool because it is under increased tension. In this situation the return to thermal equilibrium may be both long and complicated.

Adiabatic effects can be reduced by minimizing the volume of the pressure fluid and maintaining good thermal contact with a large thermal mass. A thermometer mounted as close as possible to the sensor may help to indicate if the sensor has been significantly affected. Ultimately, accurate measurements may require additional time to allow for a return to thermal equilibrium.

6.2 Hydrostatic Head Corrections

Any pressure fluid in a gravitational field will develop a vertical pressure gradient equal to the density of the fluid times the gravitational acceleration. Thus, a pressure sensor connected to a process at a different elevation will not sense the true process pressure. Similarly, if a pressure gauge is calibrated against a pressure standard that is at a different elevation, the hydrostatic head will contribute an error to the calibration unless an appropriate correction is made. In many cases the consequent error will be trivial; but, in many others it will not be, and it is important, for a given situation, to calculate at least the magnitude of the correction to determine whether it can be ignored.

Since liquids are relatively incompressible, with an approximately constant density, the hydrostatic head correction will be independent of pressure. Thus, for a fluid with a density of 1 g/cm^3 in a gravitational acceleration of 980 cm/s^2, the correction will be -98 Pa for each centimeter increase in height. As an example, for a pressure of 100 MPa (15,000 psi), a 1-m difference in height will require a 98 ppm correction, which will be of consequence for even some routine measurements. At a pressure one decade lower, 10 MPa (1500 psi), for the same conditions the correction will be correspondingly larger, 0.1%.

Since gases are compressible, with a density approximately proportional to pressure, the correction will be a constant fraction of the pressure, per unit height. Thus, for air at 297 K, with a nominal molecular weight MW_{air} of 29, the correction will be -1.15 ppm of the pressure per 1 cm increase in height. The correction for other gases and temperatures will be proportional to the molecular weight and the inverse of the absolute temperature.

For reasons unknown, hydrostatic head corrections seem to be a major cause of confusion and consequent errors. Two rules seem to help: (1) the pressure always decreases as the height increases, and (2) always make corrections based on absolute pressures. Even for differential or gauge pressures the corrections should be independently made to the low and high absolute pressures, and the differential pressure obtained from the difference of the corrected absolute pressures.

An example is shown in Figure 2.25. In this case a pressure transducer is being calibrated in a room temperature environment, 297 K, against a piston gauge with argon, molecular weight MW_{Ar} of 40, at a nominal gauge pressure of 1 MPa. Both the transducer and the piston gauge have an atmospheric reference at a nominal pressure of 100 kPa, so that the nominal absolute high pressure is 1.1 MPa. The pressure port of the transducer is located 50 cm above the reference level of the piston gauge, which is at the bottom of the piston.

The correction to the atmospheric reference pressure will be

$$(-1.15 \text{ ppm/cm}) \times (50 \text{ cm}) \times (10^5 \text{ Pa}) = -5.8 \text{ Pa}.$$

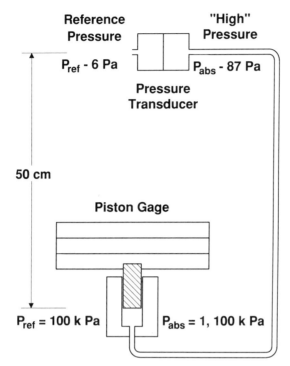

Figure 2.25 Example of hydrostatic head pressure corrections. When making corrections to gauge pressures, independent corrections should be made to both the absolute reference and absolute "high" pressures, and the correction to the gauge pressure is calculated from the difference.

The correction for the nominal high pressure of 1.1 MPa will be

$$(-1.15 \text{ ppm/cm}) \times (MW_{Ar}/MW_{air}) \times (50 \text{ cm}) \times (1.1 \times 10^6 \text{ Pa}) = -87.2 \text{ Pa}.$$

Thus, the corrected gauge pressure at the transducer will be

$$(1{,}100{,}000 - 87.2) - (100{,}000 - 5.8) = 999{,}919 \text{ Pa}.$$

This amounts to a -81 ppm correction to the gauge pressure.

6.3   Thermal Transpiration

In higher pressure gases, where the mean free path between collisions is small compared to dimensions of the apparatus and molecule–molecule collisions dominate, the frequent collisions between gas molecules maintain temperature and molecular speed equilibrium on a local scale. This condition is known as the viscous flow regime and prevails for typical laboratory apparatus at pressure above about 10 Pa. At lower pressures, where molecule–molecule collisions are rare, the speed and temperature of a molecule will be determined primarily by the temperature of the last surface it struck, and may be quite different from other molecules in the immediate vicinity. This condition is known as the molecular flow regime and, for typical laboratory apparatus, prevails for pressures below about 0.1 Pa. If two connected chambers are maintained at different temperatures in the molecular flow regime, the gas molecules can intermix with two different velocity distributions. Thus, as in Figure 2.26, gas molecules in the warmer chamber will pass through the connection between the chambers at a faster rate than those from the cooler chamber, until an equilibrium is established, where the same number of molecules pass each way per second. Using the kinetic theory of gases it can be shown that, if the chambers are connected by an orifice, at equilibrium

$$\rho_1/\rho_2 = (T_2/T_1)^{1/2} \tag{15}$$

where ρ and T are the corresponding densities and absolute temperatures. The pressures can similarly be related by

$$P_1/P_2 = (T_1/T_2)^{1/2} \tag{16}$$

As the pressure is raised from the molecular flow regime through the transition regime into the viscous flow regime the pressures in the two chambers will approach equality. The departure from pressure equality, which is quite real, is generally known as the *thermal transpiration* or thermomolecular effect.

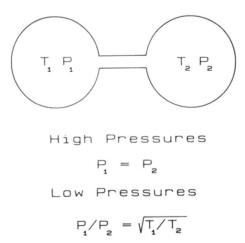

THERMAL TRANSPIRATION

High Pressures

$$P_1 = P_2$$

Low Pressures

$$P_1/P_2 = \sqrt{T_1/T_2}$$

Figure 2.26 Thermal transpiration. As the pressure is lowered to the point that collisions between molecules become infrequent, a pressure difference will develop between the two chambers. At very low pressures gas kinetic theory predicts a limiting difference, $P_1/P_2 = (T_1/T_2)^{1/2}$.

In most practical cases two volumes of gas at different temperatures will communicate by something more complicated than an orifice, and some or all of the molecules will collide with a wall at an intermediate temperature during the passage from one chamber to another. In the simplest approximation, (15) and (16) will still apply. However, there is experimental evidence [114, 115] that for such an apparatus the actual pressure difference in the molecular flow regime will be less than that predicted by (16). This deviation appears to depend on the condition of the surfaces between the two vessels, which presumably affects the efficiency of energy exchange at these surfaces. The situation for pressures above the molecular flow regime is even more difficult. Although it is clear that the pressures gradually approach equilibrium with increasing pressures, theoretical predictions [116, 117] of the pressure differences as a function of pressure through the transition range have met with only limited success.

Theoretical difficulties aside, the thermal transpiration effect must be taken into account when measuring pressures in the transition or molecular flow regimes. If the pressure gauge is at a different temperature than the region where one wants to know the pressure, the gauge will sense a different pressure. This is of particular concern for capacitance diaphragm gauges and instruments employing a hot-cathode.

Capacitance diaphragm gauges, discussed in Section 5.1, are widely used throughout the transition range. Generally, these gauges are controlled at an elevated temperature, typically 35–50°C, to improve thermal and mechanical stability. Generally, the manufacturer's calibration of these gauges does not take thermal transpiration into account, so that as the pressure is reduced through the low-vacuum range into the high-vacuum range, there is an increasing difference between the pressure reading of the gauge and the pressure it is supposed to measure. Since this effect is controlled primarily by mean free paths, it will depend not only on the pressure, but also on the gas molecular weight, temperature of the vacuum chamber, control temperature of the particular gauge, and the dimensions and surface condition of the plumbing between the gauge and the vacuum system. Experimental investigations of this effect [118, 119] show the offset at low pressures to be between 2 and 4%, depending on the gauge. Data obtained over the transition range at the National Physical Laboratory, Teddington, UK [118] were fit as a function of pressure to the equations proposed by Liang [116] and Takaishi and Sensui [117]. The best fit was found for the Takaishi and Sensui equation. Similar data [119] were successfully fit to the Takaishi and Sensui equation for two gauges. Data for two other gauges, however, did not conform to this functional form.

Tubulated hot-cathode ionization gauges operate well above ambient temperature due to the power, typically about 10–20 W, dissipated by the hot filament. The temperature of the gauge structure and envelope is very nonuniform, and it is almost impossible to determine an effective temperature. Whatever the temperature, the thermal transpiration effect will be taken into account if the gauge is calibrated against a vacuum standard. However, if the gauge is subsequently used in a different environment or if its filament power should change so that its effective temperature is different from when it was calibrated, additional errors will be incurred. With reasonable care these can be kept small; a 70°C change will cause an error of only about 10%. If the operating gauge is insulated, however, the temperature can increase by as much as 100–150°C, with consequent errors as large as 20%. Similar effects can occur with PPAs or RGAs.

If the thermal transpiration follows the ideal behavior predicted by (16), it compensates for the temperature dependence predicted by (12) for a molecular drag gauge. Therefore, to first approximation, a molecular drag gauge is not affected by thermal transpiration.

7 FURTHER READINGS

Both the high- and low-pressure extremes are covered by general reviews of the measurement science and technology. The review by Decker and co-workers [120] is older and directed primarily to pressures above the range of this discussion, but it is very comprehensive with an extensive set of references. The review by Liu and co-workers [121], although also oriented toward higher pressures, covers the high- and very high pressure ranges in more detail. The book edited by Peggs on high-pressure measurement [122] includes chapters on both static and dynamic pressures, fixed points, and high-pressure transducers. Pavese and Molinar [123] cover gas-pressure measurements from the low-vacuum to high-pressure ranges. Berman's book [124], as the title suggests, discusses "total," as opposed to partial, pressure measurements from atmospheric pressures to UHV. Leck's book [125] is an update of his earlier standard reference and addresses partial pressure measurements as well as total pressure or vacuum measurements. General readers should be aware that vacuum measurement technology has evolved so recently and has been such an active field that general reviews often include discussions and references to instruments that are no longer available commercially, or may exist only in a few specialized laboratories, if at all. Blake [126] discusses differential pressure measurements, with emphasis on probes and sensors used for flow measurements.

References

1. *Compilation of ASTM Standard Definitions*, 6th ed., American Society for Testing and Materials, Philadelphia, PA, 1986.

2. *Standard Metric Practice Guide*, Publication E380-89a, American Society for Testing and Materials, Philadelphia, PA, 1989.

3. *American National Standard Metric Practice*, ANSI/IEEE Std 268-1982, The Institute of Electrical and Electronic Engineers, New York, 1982.

4. J. R. Roebuck and H. W. Ibser, *Rev. Sci. Instrum.*, **25**, 46 (1954).

5. H. Bauer, J. Gielessen, and J. Jager, *PTB-Mitteilungen*, **87**, 384 (1977).

6. L. A. Guildner and J. Terrien, "Mercury Absolute Manometers," in B. LeNeindre and B. Vodar, Eds., *Experimental Thermodynamics*, Vol. II, Part 1, Butterworths, London, 1975, Chap. 4, pp. 115–132.

7. S. Ruthberg, "Pressure Measurements for the Range 1 kPa to 100 μPa," in B. LeNeindre and B. Vodar, Eds., *Experimental Thermodynamics*, Vol. II, Part 6, Butterworths, London, 1975, Chap. 4, pp. 229–272.

8. G. N. Peggs, *J. Phys. E*, **13**, 1254 (1980).

9. M. C. Wilkinson, *Chem. Rev.*, **72**, 575 (1972).

10. R. H. Orcutt, *J. Vac. Sci. Technol.*, **10**, 506 (1973).

11. A. M. Thomas and J. L. Cross, *J. Vac. Sci. Technol.*, **4**, 1 (1967).

12. J. L. Truffier and P. S. Choumoff, *Jpn. J. Appl. Phys.* **(2)1**, 139 (1974).

13. K. F. Poulter and P. J. Nash, *J. Phys. E.*, **12**, 931 (1979).

14. P. L. M. Heydemann, C. R. Tilford, and R. W. Hyland, *J. Vac. Sci. Technol.*, **14**, 597 (1977).

15. C. R. Tilford, *Metrologia*, **24**, 121 (1987).

16. International Association of Geodesy, *Geodetic Reference System 1967, Special Publication No. 3,* Bureau Central de L'Association Internationale de Geodesie, Paris, 1967.

17. L. A. Guildner, H. F. Stimson, R. E. Edsinger, and R. L. Anderson, *Metrologia*, **6**, 1 (1970).

18. J. Bonhoure and J. Terrien, *Metrologia*, **4**, 59 (1968).

19. R. Kanedo, S. Sudo, and K. Nishibata, *Bull. Natl. Res. Lab. Metrol. Tokyo*, **9**, 24 (1964).

20. S. J. Bennett, P. B. Clapham, J. E. Daborn, and D. I. Simpson, *J. Phys. E*, **8**, 25 (1975).

21. E. R. Harrison, D. J. Hatt, D. B. Prowse, and J. Wilbur-Ham, *Metrologia*, **12**, 115 (1976).

22. Sheng Yi-tang, Han Nui-wen, Guo Chun-shan, Duan Ming-bo, and Xu Ying-zu, "A New Primary Standard Manometer," *Proceedings of the 11th Triennial World Congress of the International Measurementation Confederation (IMEKO): Metrology,* Houston, TX, October, 1988, pp. 265–270.

23. F. A. Gould and T. Vickers, *J. Sci. Instrum.*, **29**, 85 (1952).

24. B. E. Blaisdell, *J. Math. Phys.* (Cambridge, Mass.), **19**, 186 (1940).

25. J. B. Patterson and D. B. Prowse, *Metrologia*, **21**, 107 (1985).

26. A. H. Cook, *Philos. Trans. R. Soc. London A*, **254**, 125 (1961).

27. H. Adametz and M. Wloka, *Metrologia*, **28**, 333 (1991).

28. J. A. Beattie, R. E. Blaisdell, J. Kaye, H. T. Gerry, and C. A. Johnson, *Proc. Am. Acad. Arts Sci.*, **74**, 371 (1941).

29. T. Grindley and J. E. Lind, Jr., *J. Chem. Phys.*, **54**, 3983 (1971).

30. G. A. Mukhachev, V. A. Borodin, and Yu. A. Poskonin, *Russ. J. Phys. Chem.*, **39**, 1080 (1965).

31. D. P. Johnson and P. L. M. Heydemann, *Rev. Sci. Instrum.*, **38**, 1294 (1967).

32. A. E. Eremeev, *Meas. Tech. (USSR)*, **17**, 1004 (1974).

33. K. Nishibata, S. Yamamoto, and R. Kaneda, *Jpn. J. Appl. Phys.*, **11**, 2245 (1980).

34. P. L. M. Heydemann and B. E. Welch, "Piston Gauges," in B. LeNeindre and B. Vodar, Eds., *Experimental Thermodynamics*, Vol. II, Part 3, Butterworths, London, 1975, Chap. 4, pp. 147–202.

35. R. S. Dadson, S. L. Lewis, and G. N. Peggs, *The Pressure Balance: Theory and Practice*, Her Majesty's Stationary Office, London, 1982.

36. C. M. Sutton, *J. Phys. E*, **12**, 466 (1979).

37. C. R. Tilford and R. W. Hyland, "The NBS Ultrasonic Interferometer Manometer and Studies of Gas-Operated Piston Gauges," *Proceedings of the 11th Triennial World Congress of the International Measurementation Confederation (IMEKO): Metrology*, Houston, TX, October, 1988, pp. 277–289.

37a. C. R. Tilford, R. W. Hyland, and S. Yi-tang, "Non-geometric Dependencies of Gas-Operated Piston Gage Effective Areas," in F. Molinar, Ed., *High Pressure Metrology*, Monograph 89/1, Bureau International des Poids et Mesures (BIPM), Sevres, France, 1989, pp. 105–113.

38. A. H. Bass and E. Green, *ISA Trans.*, **11**, 113 (1972).

39. G. N. Peggs, K. W. T. Elliott, and S. Lewis, *Metrologia*, **15**, 77 (1979).

40. B. E. Welch, R. E. Edsinger, V. E. Bean, and C. D. Ehrlich, "Observations of Gas Species and Mode of Operation Effects on Effective Areas of Gas-Operated Piston Gages," in F. Molinar, Ed., *High Pressure Metrology*, Monograph 89/1, Bureau International des Poids et Mesures (BIPM), Sevres, France, 1989, pp. 81–94.

41. K. F. Poulter, *J. Phys. E*, **10**, 112 (1977).

42. J. K. N. Sharma, H. K. Dwivedi, and D. R. Sharma, *J. Vac. Sci. Technol.*, **17**, 820 (1980).

43. C. G. J. Jansen and A. Venema, *Vacuum*, **9**, 219 (1959–1960).

44. H. H. Podgurski and F. N. Davis, *Vacuum*, **10**, 377 (1960).

45. K. W. T. Elliott, D. M. Woodman, and R. S. Dadson, *Vacuum*, **17**, 439 (1967).

46. L. B. Thomas, R. E. Harris, and C. L. Krueger, *Proc. Roy. Soc. London*, **A 397**, 311 (1985).

47. Gunter Messer, *Phys. Bl.*, **33**, 343 (1977).

48. W. Jitschin, J. K. Migwit, and G. Grosse, *Vacuum*, **40**, 293 (1990).

49. M. Bergoglio, A. Calcatelli, L. Marzola and G. Rumiano, *Vacuum*, **38**, 887 (1988).

50. A. Berman, *Vacuum*, **29**, 417 (1979).

51. K. F. Poulter, *Vacuum*, **28**, 135 (1978).

52. C. R. Tilford, S. Dittmann and K. E. McCulloh, *J. Vac. Sci. Technol. A*, **6**, 2853 (1988).

53. M. Hojo, M. Ono, and K. Nakayama, "A Gauge Calibration System for 10^{-2}–10^{-7} Pa Range," in R. Dobrozemsky, F. Rudenauer, F. P. Viehbock, and A. Breth, Eds., *Proceeding of the Seventh International Vacuum Congress and the Third International Conference on Solid Surfaces*, Vol. I, R. Dobrozemsky et al., Vienna, 1977, pp. 117–120.

54. G. Grosse and G. Messer, *Vacuum*, **20**, 373 (1970).

55. K. Poulter, *J. Phys. E*, **7**, 39 (1974).

56. C. R. Tilford, *J. Vac. Sci. Technol. A*, **1**, 152 (1983).

57. J. K. N. Sharma, P. Mohan, and D. R. Sharma, *J. Vac. Sci. Technol. A*, **8**, 941 (1990).

58. K. E. McCulloh, C. R. Tilford, C. D. Ehrlich, and F. G. Long, *J. Vac. Sci. Technol. A*, **5**, 376 (1987).

59. G. N. Peggs, *Vacuum*, **26**, 321 (1976).

60. C. Andeen, J. Fontanella, and D. Schuele, *Rev. Sci. Instrum.*, **42**, 495 (1971).

61. J. H. Colwell, "A Solid-Dielectric Capacitive Pressure Transducer," in K. D. Timmerhaus and M. S. Barber, Eds., *High-Pressure Science and Technology*, Vol. 1, Plenum, New York, 1979, pp. 798–804.

62. G. N. Peggs, *High Temp. High Pressures*, **12**, 1 (1980).

63. G. F. Molinar, L. Bianchi, J. K. N. Sharma, and K. K. Jain, *High Temp. High Pressures*, **18**, 241 (1986).

64. G. F. Molinar, *Physica*, **139, 140B**, 743 (1986).

65. R. W. Hyland and C. R. Tilford, *J. Vac. Sci. Technol. A*, **3**, 1731 (1985).

66. K. F. Poulter, *Le Vide*, **36**, 521 (1981).

67. P. J. Nash and T. J. Thompson, *J. Vac. Sci. Technol. A*, **1**, 172 (1983).

68. G. Grosse and G. Messer, *J. Vac. Sci. Technol. A*, **5**, 2463 (1987).

69. W. Jitschin, *J. Vac. Sci. Technol. A*, **8**, 948 (1990).

70. K. F. Poulter, M. Rodgers, and K. Ashcroft, *J. Vac. Sci. Technol.*, **17**, 638 (1980).

71. R. N. Peacock, N. T. Peacock, and D. S. Hauschulz, *J. Vac. Sci. Technol. A*, **9**, 1977 (1991).

72. G. F. Weston, *Vacuum*, **29**, 277 (1979).

73. P. Nash, *Vacuum*, **37**, 643 (1987).

74. C. R. Tilford, *J. Vac. Sci. Technol. A*, **3**, 546 (1985).

75. S. D. Wood and C. R. Tilford, *J. Vac. Sci. Technol. A*, **3**, 542 (1985).

76. M. Hirata, M. Ono, H. Hojo, and K. Nakayama, *J. Vac. Sci. Technol.*, **20**, 1159 (1982).

77. I. Warshawsky, *J. Vac. Sci. Technol. A*, **3**, 430 (1985).

78. K. F. Poulter and C. M. Sutton, *Vacuum*, **31**, 147 (1981).

79. U. Harten, G. Grosse, and W. Jitschin, *Vacuum*, **38**, 167 (1988).

80. R. N. Peacock and N. T. Peacock, *J. Vac. Sci. Technol. A*, **8**, 3341 (1990).

81. C. R. Tilford, K. E. McCulloh, and H. S. Woong, *J. Vac. Sci. Technol.*, **20**, 1140 (1982).

82. W. J. Lange, *Phys. Today*, **25(8)**, 40 (August, 1972).

83. J. M. Lafferty, *J. Vac. Sci. Technol.*, **9**, 101 (1972).

84. P. A. Redhead, *J. Vac. Sci. Technol. A*, **5**, 3215 (1987).

85. *Proceedings of the Ultrahigh Vacuum Gauging Workshop*, *J. Vac. Sci. Technol. A*, **5**, 3213 (1987).

86. A. R. Filippelli, *J. Vac. Sci. Technol. A*, **5**, 3234 (1987).

87. H. C. Hseuh and C. Lanni, *J. Vac. Sci. Technol. A*, **5**, 3244 (1987).

88. R. Holanda, *J. Vac. Sci. Technol.*, **10**, 1133 (1973).

89. J. R. Young, *J. Vac. Sci. Technol.*, **10**, 212 (1973).

90. K. Nakayama and H. Hojo, *Jpn. J. Appl. Phys.*, Supple. 2, Pt. 1, p. 113 (1974).

91. J. D. Sankey and A. H. Bass, *Vacuum*, **40**, 309 (1990).

92. P. Kleber, *Vacuum*, **25**, 191 (1975).

93. R. A. Haefer, *Vacuum*, **30**, 193 (1980).

94. I. Arakawa, M. Kim, and Y. Tuzi, *J. Vac. Sci. Technol. A*, **2**, 168 (1984).

95. J. K. Fremerey, *Vacuum*, **32**, 685 (1982).

96. J. K. Fremerey, *J. Vac. Sci. Technol. A*, **3**, 1715 (1985).

97. J. Setina, *Vacuum*, **40**, 51 (1990).

98. K. E. McCulloh, S. D. Wood, and C. R. Tilford, *J. Vac. Sci. Technol. A*, **3**, 1738 (1985).

99. B. E. Lindenau, *Vacuum*, **38**, 893 (1988).

100. M. Hirata, H. Isogai, and M. Ono, *J. Vac. Sci. Technol. A*, **4**, 1724 (1986).

101. S.-H. Choi, S. Dittmann, and C. R. Tilford, *J. Vac. Sci. Technol. A*, **8**, 4079 (1990).

102. G. Comsa, J. K. Fremerey, B. Lindenau, G. Messer, and P. Rohl, *J. Vac. Sci. Technol.*, **17**, 642 (1980).

103. K. E. McCulloh, *J. Vac. Sci. Technol. A*, **1**, 168 (1983).

104. S. Dittmann, B. E. Lindenau, and C. R. Tilford, *J. Vac. Sci. Technol. A*, **7**, 3356 (1989).

105. G. Messer, W. Jitschin, L. Rubet, A. Calcatelli, F. J. Redgrave, A. Keprt, F. Weinan, J. K. N. Sharma, S. Dittmann, and M. Ono, *Metrologia*, **26**, 183 (1990).

106. D. Lichtman, *J. Vac. Sci. Technol. A*, **8**, 2810 (1990).

107. F. M. Mao, J. M. Yang, W. E. Austin, and J. H. Leck, *Vacuum*, **37**, 335 (1987).

108. F. M. Mao and J. H. Leck, *Vacuum*, **37**, 669 (1987).

109. R. J. Reid and A. P. James, *Vacuum*, **37**, 339 (1987).

110. A. Calcatelli, M. Bergoglio, and G. Rumiano, *J. Vac. Sci. Technol. A*, **5**, 2464 (1987).

111. W. R. Blanchard, P. J. McCarthy, H. F. Dylla, P. H. LaMarche, and J. E. Simpkins, *J. Vac. Sci. Technol. A*, **4**, 1715 (1986).

112. Abstracts from a *Workshop on Calibration of Residual Gas Analyzers*, *Vacuum*, **35**, 629 (1985).

113. L. Lieszkovszky, A. R. Filippelli, and C. R. Tilford, *J. Vac. Sci. Technol. A*, **8**, 3838 (1990).

114. T. Edmonds and J. P. Hobson, *J. Vac. Sci. Technol.*, **2**, 182 (1965).

115. J. P. Hobson, *J. Vac. Sci. Technol.*, **6**, 257 (1969).

116. S. C. Liang, *J. Phys. Chem.*, **57**, 910 (1953).

117. T. Takaishi and Y. Sensui, *Trans. Faraday Soc.*, **59**, 2503 (1963).

118. K. F. Poulter, M. J. Rodgers, P. J. Nash, T. J. Thompson, and M. P. Perkin, *Vacuum*, **33**, 311 (1983).

119. W. Jitschin and P. Rohl, *J. Vac. Sci. Technol. A*, **5**, 372 (1987).

120. D. L. Decker, W. A. Bassett, L. Merrill, H. T. Hall, and J. D. Barnett, "High-Pressure Calibration: A Critical Review," *J. Phys. Chem. Ref. Data*, **1**, 773 (1972).

121. C. Y. Liu, K. Ishizaki, J. Paauwe, and I. L. Spain, *High Temp. High Pressures*, **5**, 359 (1973).

122. G. N. Peggs, Ed., *High Pressure Measurement Techniques*, Applied Science, New York, 1983.

123. F. Pavese and G. Molinar, *Modern Gas Based Temperature and Pressure Measure-*

ments in K. D. Timmerhaus, A. F. Clark, and C. Rizzuto, Eds., *International Cryogenic Monograph Series*, Plenum, New York (1992).

124. A. Berman, *Total Pressure Measurements in Vacuum Technology*, Academic, Orlando, FL, 1985.

125. J. H. Leck, *Total and Partial Pressure Measurement in Vacuum Systems*, Blackie & Son, Glasgow, 1989.

126. W. K. Blake, "Differential Pressure Measurement," in R. J. Goldstein, Ed., *Fluid Mechanics Measurements*, Hemisphere, Washington, DC, 1983, Chap. 3, pp. 61–97.

Chapter **3**

EXPERIMENTAL METHODS FOR STUDYING DIFFUSION IN GASES, LIQUIDS, AND SOLIDS

Peter J. Dunlop, Kenneth R. Harris, and David J. Young

Physical Methods of Chemistry, Second Edition Volume Six: Determination of Thermodynamic Properties   Edited by Bryant W. Rossiter and Roger C. Baetzold
ISBN 0-471-57087-7   Copyright 1992 by John Wiley & Sons, Inc.

1   INTRODUCTION

Isothermal diffusion is the term used to describe the *macroscopic* process by which *relative* motion takes place between the components of a fluid or solid. Diffusion is an irreversible process that causes transport of material within a system until a final state of equilibrium is established. Motion of a system or part of a system as a whole in the absence of relative movement of components is *not* a diffusion process and is sometimes known as *bulk flow*. It is, of course, possible that a system may undergo both diffusion and bulk flow simultaneously. Should this be the case, a method must be available to separate the two effects if a study of the diffusion process alone is desired. This problem is discussed in some detail in Section 2.1.

Here we describe some of the more important methods that are used frequently by research workers for studying the diffusion process in binary, ternary, and quaternary systems of nonreacting components. The discussion is confined to transport in one phase, that is, in a gas, liquid, or solid phase. Some techniques that were described in detail in the previous edition of this series [1] have been omitted because their use has rarely been reported in the literature. The present chapter supplements other accounts [2–7].

All techniques for studying diffusion aim at determining coefficients necessary to describe the relative motion of the components of the system. These coefficients are proportionality factors that relate the component flows to the various concentration gradients present in the system. The equations relating the component flows to the concentration gradients are known as *flow equations*; and, because they are basic to all experimental methods, they are treated in detail in the following section.

2   GENERAL THEORY

2.1   Flow Equations and Frames of Reference

Isothermal diffusion is an irreversible process in which relative motion occurs between the components of a single phase at a constant temperature T. The diffusion process is, however, a *macroscopic* concept that does not describe the continual relative motion that takes place between molecules when a system is in equilibrium.

From both the experimental and theoretical points of view, flow equations are a satisfactory means of describing the relative motion of components during the diffusion process. These equations are as fundamental to a study of diffusion as Ohm's law (a flow equation) is to a study of electricity. A flow of matter can take place in three dimensions, as may also a flow of electric charge; however, here, except in one or two cases, we choose to restrict the discussion to relative motion of components along the x direction of a Cartesian coordinate system, usually taken to be the vertical direction to ensure gravitational stability of the system. The flow of each component is a vector quantity, which gives the amount (moles or grams) of the component that crosses unit area at right angles to the x coordinate in unit time. Note, however, that vector notation is not used in this chapter. *Experimental* flow equations [8–10] assume each flow to be proportional to concentration gradients; *theoretical* flow equations [10] assume each flow to be proportional to gradients of chemical potential. Since both types of equations describe the same phenomena, they must be related to one another [9], but we discuss only experimental flow equations.

Normally, diffusion measurements are made in a coordinate system that is fixed with respect to the cell [10]. It is often necessary, however, to relate a flow relative to the cell to a flow relative to another frame of reference [10, 11]. Because of complications due to possible volume changes of mixing [9, 10], flows relative to the cell are not always simply related to the diffusion coefficient in Fick's first law [8]. In the following we consider only binary and ternary systems; thus, whenever possible, explicit equations are given. Generalization to systems of more than three components follows quite naturally [12, 13]. When necessary, the subscripts 0, 1, and 2 are used to denote the solvent and the two solutes, respectively.

2.1.1 Binary Systems and Fick's First Law

Because they are so important to the experimentalist, we first consider flows measured with respect to a *volume* frame of reference [9, 10] denoted by a superscript v and defined by

$$\bar{V}_0 J_0^v + \bar{V}_1 J_1^v = 0 \tag{1}$$

where J_0^v and J_1^v are the flows relative to this frame of reference and $\bar{V}_0$ and $\bar{V}_1$ are the component partial molar volumes. Then, according to Fick's first law [8], these flows are given by the expressions

$$J_0^v = -D_0^v \left(\frac{\partial C_0}{\partial x} \right)_t \tag{2a}$$

$$J_1^v = -D_1^v \left(\frac{\partial C_1}{\partial x} \right)_t \tag{2b}$$

where D_0^v and D_1^v are diffusion coefficients for the volume frame of reference, $(\partial C_0/\partial x)_t$ and $(\partial C_1/\partial x)_t$ are concentration gradients at a particular time t, and C_0 and C_1 are concentrations in moles per centimeter cubed. Other concentration scales are introduced from time to time; from this point, all subscripts t are dropped from concentration gradients. For the volume frame of reference, combination of (1) and (2) with the thermodynamic relation

$$C_0 \bar{V}_0 + C_1 \bar{V} = 1 \tag{3}$$

indicates [14, 15] that

$$D \equiv D_0^v = D_1^v \tag{4}$$

and, thus, since a binary system can be described by a single *mutual* diffusion coefficient D, the subscripts and superscripts in (4) can be deleted.

If there is a volume change on mixing during the diffusion process in a cell *closed* at the bottom and open at the top, the liquid in the cell will experience a local bulk velocity, u^{vc}, which Onsager [14] has defined by

$$u^{vc} = \bar{V}_0 J_0^c + \bar{V}_1 J_1^c \tag{5}$$

where the superscript c denotes a flow with respect to the cell. The bulk velocity u^{vc} (the velocity of the volume-fixed frame of reference relative to the cell) is a function of both x and t and differs from zero if the partial molar volumes $\bar{V}_0$ and $\bar{V}_1$ vary in the system during the course of an experiment. Thus, to describe the solute flow relative to the cell, (2b) must be modified [9, 10] to include a flow contribution $C_1 u^{vc}$, which takes into account the additional flow due to the presence of the bulk velocity [a similar term $C_0 u^{vc}$ must be added to (2a)]. Thus, (2b) becomes [9, 10]

$$ J_1^c = -D\left(\frac{\partial C_1}{\partial x}\right) + C_1 u^{vc} \tag{6a} $$

or

$$ J_1^c = J_1^v + C_1 u^{vc} \tag{6b} $$

In almost all experiments used to determine the mutual diffusion coefficients D, the conditions are chosen so that $\bar{V}_0$ and $\bar{V}_1$ are essentially constant. With this restriction, u^{vc} is zero, and the cell and volume frames of reference are identical [10]. Therefore, under these conditions D can be determined from measurements with respect to the cell, and this value suffices to characterize the flows of both the solvent and the solute.

Two other frames of reference are important [10] when considering theoretical flow equations. The first is the *local mass-average* frame designated by a superscript m and defined by

$$ M_0 J_0^m + M_1 J_1^m = 0 \tag{7a} $$

where M_0 and M_1 are molecular weights; and the second is the solvent reference frame designated by a superscript 0 and defined by

$$ J_0^0 = 0 \tag{7b} $$

2.1.2 Ternary Systems

Experiments indicate that Fick's first law is not always sufficient to describe the flows in ternary systems; hence, flow equations of a more general nature must be employed. Onsager [14] proposed an extremely general set of flow equations in which each flow depends on the concentration gradients of all components. For ternary systems his equations have the form

$$ J_i^v = -\sum_{j=0}^{2} D_{ij}^v \frac{\partial C_j}{\partial x} \tag{8} $$

where $i = 0, 1, 2$.

In the absence of volume changes on mixing, however, neither the flows nor the concentration gradients in (8) are independent due to the restrictions

$$\sum_{i=0}^{2} \bar{V}_i J_i^v = 0 \tag{9a}$$

$$\sum_{i=0}^{2} \bar{V}_i \left(\frac{\partial C_i}{\partial x} \right) = 0 \tag{9b}$$

Therefore, Onsager imposed the conditions [14]

$$\sum_{i=0}^{2} \bar{V}_i D_{ik}^v = 0 \tag{10a}$$

where $k = 0, 1, 2$, and

$$\sum_{k=0}^{2} D_{ik}^v C_k = 0 \tag{10b}$$

where $i = 0, 1, 2$, in order to define the nine D_{ik}^v in (8). It should be noted [14] that (10a) and (10b) imply that

$$\sum_{i=0}^{2} \sum_{k=0}^{2} \bar{V}_i D_{ik}^v C_k = 0 \tag{10c}$$

and hence only five of the six restrictions (10a) and (10b) are independent. Thus, Onsager's equations (8) contain four *independent* diffusion coefficients.

A set of experimental flow equations based on those of Onsager was proposed by Baldwin and co-workers [16–18] and tested by many workers [19–23]. (Other references include ternary diffusion in discussions of the diaphragm cell and Gouy interferometer techniques.) These flow relations use four diffusion coefficients to describe the *solute* flows in a ternary system; that is,

$$J_1^v = -D_{11}^v \left(\frac{\partial C_1}{\partial x} \right) - D_{12}^v \left(\frac{\partial C_2}{\partial x} \right) \tag{11a}$$

$$J_2^v = -D_{21}^v \left(\frac{\partial C_1}{\partial x} \right) - D_{22}^v \left(\frac{\partial C_2}{\partial x} \right) \tag{11b}$$

where D_{11}^v and D_{22}^v are the main diffusion coefficients and D_{12}^v and D_{21}^v are cross-term diffusion coefficients [16]. Dole [24] indicated how to compute the nine dependent coefficients in Onsager's equations (8) from the four experimental values in (11). When the experiments are performed so that the volume and cell frames of reference are identical (no volume changes on mixing), the four D_{ij}^v can be measured directly; hence, the superscripts v are omitted in the remainder of this chapter. It should be noted that these four coefficients are not independent [10], since for a ternary system Onsager [14] derived one relation between them (the Onsager reciprocal relation) from the assumption of *microscopic reversibility* [25, 26].

2.1.3 Tracer Diffusion Coefficients

Inspection of (11) indicates that [21, 22, 27, 28]

$$D_{12} \to 0 \quad \text{as} \quad C_1 \to 0 \tag{12a}$$

and

$$D_{21} \to 0 \quad \text{as} \quad C_2 \to 0 \tag{12b}$$

These conditions and (11) are now used to define certain limiting values of the main diffusion coefficients that can be measured experimentally.

1. $D_{12} \to 0$ when $C_1 \to 0$. When component 1 is present in a ternary mixture $(0, 1, 2)$ in vanishingly small (tracer) amounts, then D_{11} becomes the tracer diffusion coefficient D_{T1} of 1 in mixtures of 0 and 2; D_{22} becomes the mutual diffusion coefficient of components 0 and 2; and D_{21} is the *nonzero*, cross-term diffusion coefficient.

2. $D_{21} \to 0$ when $C_2 \to 0$. When component 2 is present in the ternary mixture $(0, 1, 2)$ in vanishingly small amounts, then D_{11} becomes the mutual diffusion coefficient of components 0 and 1; D_{22} becomes the *tracer diffusion coefficient* D_{T2} of 2 in mixtures of 0 and 1; and D_{12} is the *nonzero*, cross-term diffusion coefficient.

It has been shown by two different methods [29–32] that, *when components 2 and 1 become chemically identical yet remain distinguishable (radioactive isotopes)*, a particularly simple relation exists between the three limiting diffusion coefficients necessary to describe the type of ternary systems in cases 1 and 2 above. For case 2 this relation is

$$D = D_{T2} + D_{12} \tag{13}$$

where D is the mutual diffusion coefficient for components 0 and 1 (the limit of D_{11} as $C_2 \to 0$).

For a ternary system consisting of a solvent 0 and two solutes 1 and 2, which are chemically identical although physically distinguishable, Albright and Mills [30] showed that, when C_1 and C_2 are finite, the system is characterized by one coefficient that is named the intradiffusion coefficient $D^\dagger$. This coefficient is given by the relations

$$D^\dagger = D_{11} - D_{12} = D_{22} - D_{21} \tag{14}$$

where $D^\dagger$ is a function of $(C_1 + C_2)$ but independent of the ratio of C_1 to C_2 when their sum remains constant. This equation becomes identical with (13) when $C_2 \rightarrow 0$, since

$$D_{11} \rightarrow D \qquad \text{as} \qquad C_2 \rightarrow 0$$

$$D_{22} \rightarrow D_{T2} \qquad \text{as} \qquad C_2 \rightarrow 0 \tag{15}$$

$$D_{21} \rightarrow 0 \qquad \text{as} \qquad C_2 \rightarrow 0$$

Thus, D_{T2} is the limit of $D^\dagger$ as $C_2 \rightarrow 0$; similarly, D_{T1} is the limit of $D^\dagger$ as $C_1 \rightarrow 0$. At present only values of D_{T2} (or D_{T1}) have been measured; however, when diffusion in systems consisting (say) of two optically active isomers in a solvent are studied, then *intradiffusion* coefficients are necessary to describe the transport in such a system.

As a first step in obtaining another important relation between mutual and tracer diffusion coefficients, it is convenient to relate the mutual diffusion coefficient for 0 and 1 to be the tracer diffusion coefficient for another ternary system. In this system, component 1 becomes the solvent, component 0 becomes one of the solutes, and the second solute is a new component numbered 3 and is a radioactive isotope of component 0, which is present in the system $(0, 1, 3)$ in vanishingly small amounts. When (11) are rewritten with 3 replacing 2, and 0 replacing 1, the required relationship corresponding to (13) is

$$D = D_{T3} + D_{03} \tag{16}$$

where D_{T3} is the tracer diffusion coefficient of component 3 (an isotope of 0) in mixtures of components 0 and 1, and D_{03} is the cross-term diffusion coefficient. The equation of interest relating the mutual diffusion coefficient for the binary system of 0 and 1 to the two tracer diffusion coefficients D_{T2} and D_{T3}, associated with this binary system, but measured in two ternary systems $(0, 1, 2)$ and $(0, 1, 3)$, respectively, is obtained by adding the two equations obtained by multiplying (13) and (16) by N_0 and N_1, respectively, to give

$$D = (N_0 D_{T2} + N_1 D_{T3}) + (N_0 D_{12} + N_1 D_{03}) \tag{17}$$

where N_0 and N_1 are the mole fractions of 0 and 1, respectively. For the almost ideal system benzene–chlorobenzene, experiments [32] indicate that probably

$$D = (N_0 D_{T2} + N_1 D_{T3}) \tag{18a}$$

and

$$N_0 D_{12} + N_1 D_{03} = 0 \tag{18b}$$

The sum in (18b) is not 0 for nonideal systems [32], and thus it appears that the reasons for the departure of this sum from 0 should be investigated.

For the systems $(0, 1, 2)$ and $(0, 1, 3)$ defined previously, the values of $(D_{T2})_{c_0 = 0}$ and $(D_{T3})_{c_1 = 0}$ are the tracer diffusion coefficients of 2 and 3 in 1 and 0, respectively, and are binary diffusion coefficients. The corresponding self-diffusion coefficients D_{S1} and D_{S0} are defined by the relations

$$D_{S1} \equiv \lim_{m_2 \to m_1} (D_{T2})_{C_2 = 0} \tag{19a}$$

$$D_{S0} \equiv \lim_{m_3 \to m_0} (D_{T2})_{C_1 = 0} \tag{19b}$$

where the m_i are the molecular masses, and the limits in (19) indicate that not only the masses, but also the chemical properties of the components become identical. Self-diffusion coefficients cannot be measured directly, except by nuclear magnetic resonance (NMR) and light-scattering techniques, but must be obtained from tracer coefficients measured with isotopes of the species in question. In the case of gases the Chapman–Enskog theory [33] indicates that the diffusion coefficient is inversely proportional to the square root of the reduced mass of the system, so that the self-diffusion coefficient can be calculated from a single experimental result. Recent studies with hydrogen and helium isotopes [34] illustrate the validity of this approach. However, for liquids the inverse square root law is not valid so that a series of measurements with different isotope masses must be employed, and the self-diffusion coefficient obtained by extrapolation. Mills [35] obtained the self-diffusion coefficients of H_2O and D_2O by this procedure.

2.2   The Equations of Continuity and Fick's Second Law

Several experimental techniques for studying diffusion are based directly on Fick's first law as expressed by (2). These methods usually involve measuring the flow of a given solute component under certain convenient conditions and then computing the diffusion coefficient from the concentration changes that take place during transport. Techniques involving Fick's first law are known as *steady-state methods*, and they yield diffusion coefficients that may not be absolute (i.e., it may be necessary to calibrate the apparatus with data obtained by an absolute method). Most other methods are based on a differential equation obtained by combining Fick's first law for each solute component [or (11)] with the corresponding *equation of continuity* [9, 36]. These latter relations are mathematical conditions for the conservation of each component of the system. For a system of three nonreacting components diffusing in one dimension the continuity equations are

$$\left(\frac{\partial C_i}{\partial x}\right)_t = -\left(\frac{\partial J_i}{\partial x}\right)_t \tag{20}$$

where $i = 0, 1, 2$.

2.2.1 Binary Systems

When Fick's law for the second component (2b) is combined with (20), a partial differential equation is obtained:

$$\left(\frac{\partial C_1}{\partial t}\right)_x = \left\{\frac{\partial}{\partial x}\left[D\left(\frac{\partial C_1}{\partial x}\right)\right]\right\}_t \tag{21a}$$

and, if the diffusion coefficient is independent of concentration, this relation becomes

$$\left(\frac{\partial C_1}{\partial t}\right)_x = D\left(\frac{\partial^2 C_1}{\partial x^2}\right)_t \tag{21b}$$

an equation that is known as Fick's second law.

Equations (22) show Fick's second law in three dimensions [37] for binary systems with a constant diffusion coefficient: cartesian coordinates (x, y, z)

$$\left(\frac{\partial C_1}{\partial t}\right)_{x,y,z} = D\left(\frac{\partial^2 C_1}{\partial x^2} + \frac{\partial^2 C_1}{\partial y^2} + \frac{\partial^2 C_1}{\partial z^2}\right)_t \tag{22a}$$

cylindrical coordinates (r, z, θ)

$$\left(\frac{\partial C_1}{\partial t}\right)_{r,z,\theta} = \frac{1}{r}\left\{\frac{\partial}{\partial r}\left[rD\left(\frac{\partial C_1}{\partial r}\right)\right] + \frac{\partial}{\partial \theta}\left[\frac{D}{r}\left(\frac{\partial C_1}{\partial \theta}\right)\right] + \frac{\partial}{\partial z}\left[rD\left(\frac{\partial C_1}{\partial z}\right)\right]\right\}_t \tag{22b}$$

spherical coordinates (r, θ, ϕ)

$$\left(\frac{\partial C_1}{\partial t}\right)_{r,\theta,\phi} = \frac{1}{r^2}\left\{\frac{\partial}{\partial r}\left[Dr^2\left(\frac{\partial C_1}{\partial r}\right)\right] + \frac{1}{\sin\theta}\frac{\partial}{\partial \theta}\left[D\sin\theta\left(\frac{\partial C_1}{\partial \theta}\right)\right]\right.$$
$$\left. + \frac{D}{\sin^2\theta}\left(\frac{\partial^2 C_1}{\partial \phi^2}\right)\right\}_t \tag{22c}$$

In the next sections on experimental techniques, simplified forms of (22) are used to develop methods to obtain binary diffusion coefficients.

2.2.2 Ternary Systems

When the equations of continuity for the two solutes are combined with (11), the relations that are basic for diffusion in all ternary systems are obtained:

$$\left(\frac{\partial C_1}{\partial t}\right)_x = D_{11}\left(\frac{\partial^2 C_1}{\partial x^2}\right)_t + D_{12}\left(\frac{\partial^2 C_2}{\partial x^2}\right)_t \tag{23a}$$

$$\left(\frac{\partial C_2}{\partial t}\right)_x = D_{21}\left(\frac{\partial^2 C_1}{\partial x^2}\right)_t + D_{22}\left(\frac{\partial^2 C_2}{\partial x^2}\right)_t \qquad (23b)$$

Solutions to (23), which assume that the diffusion coefficients are independent of concentration, can be used to compute the four D_{ij} from experimental data.

3   METHODS OF MEASUREMENT—GASES

Some of the older techniques for studying diffusion in gases have not been used recently by active research workers and hence the reader is referred to detailed discussions of those methods in a previous edition of this series [1]. The gas chromatographic and NMR spin–echo methods have been included in Section 4.

It is possible with some techniques to obtain a precision of 0.1% in the measured diffusion coefficient and sometimes the precision and the accuracy are essentially the same, particularly in the vicinity of 300 K. However, at high temperatures where one would expect experimental difficulties to be much greater, the agreement between results reported for the same system varies by as much as 10% at 1200 K (see Table 3.1). In general, experimental conditions were

Table 3.1ᵃ   Smoothed Limiting Diffusion Coefficients $\mathscr{D}_{12}^0$ for He–Ar as a Function of Temperature

T (K)	Hogervorst[b]	Cain–Taylor[c]	Zwakhals	Suetin[c]	Theory[d]
300	0.73		0.86	0.73	0.73
400	1.18	1.22	1.34	1.17	1.19
500	1.72	1.77	1.90	1.69	1.73
600	2.33	2.40	2.53	2.28	2.36
700	3.00	3.12	3.23	2.89	3.06
800	3.75	3.94	4.01		3.85
900	4.56	4.86	4.87		4.70
1000	5.42	5.86	5.83		5.61
1100	6.33	6.94	6.87		6.60
1200	7.33	8.10	7.97		7.66
		±1.3	±1.0	±0.7	

ᵃUnits of $\mathscr{D}_{12}^0$ cm²/s.
ᵇReported as smoothed values from [38].
ᶜSmoothed values with average deviations at the bottom of each column.
ᵈCalculated from Chapman–Cowling theory [33] using the potential function of R. A. Aziz, P. W. Riley, U. Buck, G. Maneke, G. Schleusener, G. Scoles, and U. V. Valbusa, *J. Chem. Phys.*, **71**, 2637 (1979).

chosen so that the cell and volume frames of reference should be almost identical.

The binary diffusion coefficient[†] $\mathscr{D}_{12}$ is given by the Chapman–Enskog theory [33] to the mth approximation by

$$\mathscr{D}_{12} = [D_{12}]_1 f_D^{(m)} \tag{24a}$$

where

$$[D_{12}]_1 = \frac{3}{8}\left(\frac{k^3 T^3}{2\pi\mu_{12}}\right)^{1/2}(P\sigma^2\Omega_{12}^{(1,1)*})^{-1} \tag{24b}$$

and $f_D^{(m)}$ describes the higher terms. In these expressions P, T, μ, k, and σ denote pressure, temperature, reduced mass, Boltzmann's constant, and the position of the zero of the potential energy of interaction, respectively; and the subscripts 1 and 2 denote the interacting molecules. The reduced collision integral $\Omega^{(1,1)*}$ depends on the reduced temperature $T = (kT/\varepsilon)$, where ε is the depth of the potential well.

Because the binary diffusion coefficient depends only on the unlike interactions to the first approximation, it is important that the experimental methods be improved to enable better results to be obtained at high temperatures so that more accurate potential energy functions can be determined.

3.1   Loschmidt-Type Techniques

An excellent method for measuring diffusion coefficients of gas mixtures that was introduced by Loschmidt [39] has been employed by many workers, some using the original [39–46] and others, a modified method of operation [47–52]. Those methods that used an optical system to monitor concentrations continuously in a Loschmidt-type cell were described in detail previously [1]. In these methods relative motion of components occurs along the vertical axis of a long cell of uniform cross section.

The diffusion cells usually consist of two identical symmetrical sections or halves joined together about a common pivot. Each section consists of a long cylinder (or tube of rectangular cross section) closed at one end and attached to a disk in an off-center position at the other. Each disk is covered with a very thin layer of a suitable grease and the two are clamped together with a central rod and a pair of strong springs. O-rings may be installed at suitable positions between the disks to prevent leakage at high pressures. Rotation of the upper disk about the common center enables the upper half-cell to be isolated from, or brought into perfect alignment with, the lower half-cell. Each half-cell can be evacuated or filled with gas using valves attached to vents in the two disks. A typical Loschmidt cell is shown in Figure 3.1 [53].

[†]For gases the symbol $\mathscr{D}$ is used to denote the experimental diffusion coefficient; the symbol D is used to denote theoretical values [see (24)].

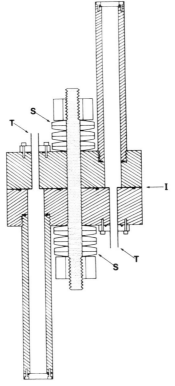

Figure 3.1 Loschmidt cell for measuring diffusion coefficients to 20-atm pressure. The two cell halves rotate about the interface I. Springs S permit the pressure between the plates to be varied; the gases enter through the inlet tubes T. Reprinted with permission from I. R. Shankland and P. J. Dunlop, *Physica*, **100A**, 64 (1980).

To perform a classical experiment the cell is adjusted in a vertical position so that the two halves are isolated, and then each half is filled with a pure gas or a gas mixture so that the pressures are identical. The two halves are then carefully aligned as quickly as possible and diffusion is allowed to proceed for an optimum time and then the halves are isolated while diffusion is still proceeding. The apparatus is then left until complete mixing has been achieved in each half. The concentration of the heavy component C_1 is then determined in each half of the cell. The binary diffusion coefficient can be determined from the two concentrations, the cell length, and the diffusion time.

If we denote the vertical direction of the cell as the x axis and take the bottom of the cell at $x = 0$, then $x = l/2$ corresponds to the center of the cell of total length l. If the cell halves are filled with two binary gas mixtures so that the concentration of the heavy component is C_1^l in the lower half cell and C_1^u in the upper, the boundary conditions for the solution of Fick's second law [8] are

$$C_1 = C_1^l \tag{25a}$$

where $0 < x < l/2$ and $t = 0$ and

$$C_1 = C_1^u$$

where $l/2 < x < l$ and $t = 0$ and, since no flow can take place through the ends of the cell

$$(\partial C_1/\partial x)_t = 0 \tag{25b}$$

where $x = 0$, l at all t.

The solution for these conditions is

$$C_1 = \sum_{n=0}^{\infty} A_n \exp(-n^2\pi^2 \mathscr{D}_{12}t/l^2) \cos(n\pi x/l) \tag{25c}$$

where the coefficients A_n are determined by the initial concentration distribution. When, as is usually the case, pure gases are used in each cell half, (25c) becomes

$$C_1 = \bar{C}_1 + \frac{2\Delta C_1}{\pi} \sum_{n=1}^{\infty} \frac{\sin(n\pi/2)}{n} \cos\left(\frac{n\pi x}{l}\right) \exp\left(\frac{-n^2\pi^2 \mathscr{D}_{12}t}{l^2}\right) \tag{26a}$$

where

$$\bar{C}_1 = (C_1^l + C_1^u)/2 \tag{26b}$$

$$\Delta C = (C_1^l - C_1^u) \tag{26c}$$

and $\mathscr{D}_{12}$ is assumed to be independent of concentration.

Equations (26) can be used to derive expressions for the average concentrations $\langle C_1^l \rangle$ and $\langle C_1^u \rangle$ in the lower and upper halves of the cell, respectively, and from those expressions it is a simple matter to deduce the following relation

$$f \equiv \frac{\langle C_1^l \rangle - \langle C_1^u \rangle}{\langle C_1^l \rangle + \langle C_1^u \rangle} \tag{27a}$$

$$= \frac{8}{\pi^2} \sum_{n=0}^{\infty} \frac{1}{(2n+1)^2} \exp[-(2n+1)^2\pi^2 \mathscr{D}_{12}t/l^2] \tag{27b}$$

Provided the experiment is allowed to proceed for the optimum time, $t_{\text{opt}} \geq l^2/\pi^2 \mathscr{D}_{12}$ [54], it can be shown that the second term in (27b) is less than 0.005% of the first; thus $\mathscr{D}_{12}$ can be determined from

$$f = \frac{8}{\pi^2} \exp(-\pi^2 \mathscr{D}_{12}t/l^2) \tag{27c}$$

by measuring $\langle C_1^l \rangle$, $\langle C_1^u \rangle$, t, and l. Ljunggren [51] and Paul [55] have both shown that if $\mathscr{D}_{12}$ is linearly dependent on concentration, then the measured value of the diffusion coefficient corresponds to the final equilibrium concentration in the cell. Equation (27a) assumes that volume concentrations have been measured. If gas concentrations are determined as mole fractions N_1 and if these mixtures are nonideal, (27a) becomes [53]

$$f = \frac{[\langle x_1^l \rangle z_m^u - \langle x_1^u \rangle z_m^l]}{[\langle x_1^l \rangle z_m^u + \langle x_1^u \rangle z_m^l]} \tag{27d}$$

where the z_m are defined by

$$z_m = 1 + B_m P + \cdots \tag{28}$$

and the B_m are pressure second virial coefficients for the mixtures.

The absolute Loschmidt technique was used by Shankland and Dunlop [53] to measure the pressure dependence of $\mathscr{D}_{12}$ for the systems N_2–Ar, N_2–O_2, Ar–O_2, and Ar–Kr up to 25 atm. These systems were chosen because the pressure change on mixing is quite small in all cases, so that the cell and volume frames [10] can be identified. The final concentrations in each cell half were analyzed with a mass spectrometer calibrated with gas mixtures that were accurately prepared by weighing. The results are estimated to be accurate to $\pm 0.3\%$. A graph of results for the pressure dependence of the diffusion coefficient of the system N_2–O_2 is shown in Figure 3.2 [53].

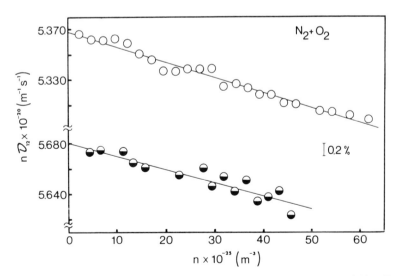

Figure 3.2 Results for the pressure dependence of the diffusion coefficient of N_2—O_2; $\bigcirc$, $T = 300\,\text{K}$; $\ominus$, $T = 323\,\text{K}$; n is the number density. Reprinted with permission from I. R. Shankland and P. J. Dunlop, *Physica*, **100A**, 64 (1980).

3.2 Restricted Diffusion

Restricted diffusion takes place in a cell of constant cross section that is closed at both ends; a gradient is formed and the concentration is monitored until it becomes uniform. The concentration distribution is described by a Fourier series; but, by choosing appropriate positions to measure the variation with time of the concentration, only simple expressions are required to obtain the diffusion coefficients.

3.2.1 Restricted Diffusion—Thermistor Sensors

A modification of the Loschmidt technique was suggested by Onsager to Harned, who used it [56, 57] to study diffusion in aqueous electrolyte solutions. In this method *differences* in concentrations between two fixed points ($l/6$) from either end of the cell are monitored as a function of time. For these conditions the even terms in (26a) cancel and the third term is zero, so that

$$\Delta C_1(t) \equiv C_1(l/6, t) - C_1(5l/6, t) \tag{29a}$$

$$\Delta C_1 = A'_1 \exp(-\pi^2 \mathcal{D}_{12} t/l^2) - A'_5 \exp(-25\pi^2 \mathcal{D}_{12} t/l^2) + \cdots \tag{29b}$$

where $A'_n = \sqrt{3} A_n \ (n \geqslant 1)$.

After a very short time the second term becomes negligible, so that (29b) becomes

$$\Delta C_1(t) = A' \exp(-\pi^2 \mathcal{D}_{12} t/l^2) \tag{29c}$$

For gases Dunlop and co-workers [58–60] positioned sensitive matched thermistors at distances ($l/6$) from each end of a Loschmidt cell and measured the difference between them as a function of time by means of a simple Wheatstone bridge, a digital voltmeter, and a data logger. The difference in resistance was shown to be proportional to the corresponding difference in concentration [61]. Thus, for two perfectly matched thermistors (29c) becomes

$$\Delta R(t) \equiv R(l/6, t) - R(5l/6, t) \tag{30a}$$

$$= B \exp(-\pi^2 \mathcal{D}_{12} t/l^2) \tag{30b}$$

In general, however, perfectly matched thermistors are not available, so (30b) becomes

$$\Delta R(t) - \Delta R(\infty) = B \exp(-\pi^2 \mathcal{D}_{12} t/l^2) \tag{30c}$$

It is possible to obtain $\mathcal{D}_{12}$ in two ways [62]: (1) measure $\Delta R(\infty)$ and least square as a two-parameter problem or (2) nonlinear least square as a three-parameter problem and obtain $\mathcal{D}_{12}$, $\Delta R(\infty)$, and B. For a perfect exponential the value of $\mathcal{D}_{12}$ should be independent of the number of points used, and the value of $\Delta R(\infty)$

Table 3.2   Results for a Diffusion Experiment for
the System He—C_2H_6 at 300 K

N	$\Delta R(t)$	$\Delta R(\infty)_{calcd}$	$\mathscr{D}_{12}$
0	256.450	−0.644	0.5070
10	232.520	−0.632	0.5072
20	210.402	−0.630	0.5072
30	190.356	−0.631	0.5072
40	172.220	−0.631	0.5072
50	155.808	−0.632	0.5072
60	140.986	−0.631	0.5072
70	127.554	−0.630	0.5072
80	115.401	−0.630	0.5072
90	104.429	−0.630	0.5072
100	94.476	−0.629	0.5072
110	85.469	−0.630	0.5072
120	77.338	−0.627	0.5073
130	69.959	−0.625	0.5073
140	63.321	−0.625	0.5073

calculated should be equal to the value measured. Table 3.2 shows results for an experiment in which C_2H_6 diffused in helium at 300 K; N is the number of points omitted from the least-square procedure, and the other quantities are defined by (30c). It was shown [63] that this method and the Loschmidt technique yield results that agree within 0.2%.

In this method the experiment is commenced in exactly the same way as in the classical Loschmidt technique, but diffusion is allowed to proceed until the concentration is uniform throughout the cell; it is not necessary to know the duration of the experiment; only accurate concentration differences and corresponding times are required. Diffusion coefficients were obtained as functions of concentration [64], temperature [65], and pressure [59, 60, 66]. Unfortunately, the range of temperature available to a Loschmidt-type cell is limited by the availability of a suitable grease to lubricate and seal the two moving plates. A suitable diffusion cell [59] is illustrated in Figure 3.3. The concentration dependence of $\mathscr{D}_{12}$ for ethane diffusing in helium was measured in this cell and is shown in Figure 3.4.

3.2.2   Restricted Diffusion—Scintillator Sensors

Codastefano and co-workers [67] used essentially the same technique as that used by Dunlop and co-workers [58–60], but they replaced the thermistors by plastic scintillators to study tracer diffusion of ^{85}Kr in several noble gases as a function of pressure [68, 69]. Their cell is shown in Figure 3.5. The β particles emitted from ^{85}Kr were counted simultaneously at distances ($l/6$) from each end of the cell by two photomultiplier tubes. Corrections were applied for the

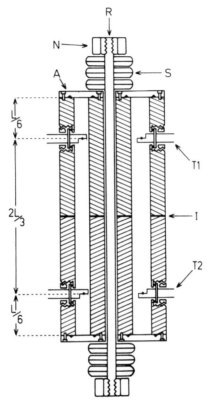

Figure 3.3   Cell in which two experiments can be performed simultaneously; the top half rotates with respect to the bottom half at the interface I. The T_1 and T_2 are thermistors; disk-washers S permit pressures as high as 20 atm to be employed; A is the end plate, N is a tension nut on the stainless steel rod R. Reprinted with permission from G. R. Staker, and P. J. Dunlop, *Chem. Phys. Lett.*, **42**, 419 (1976).

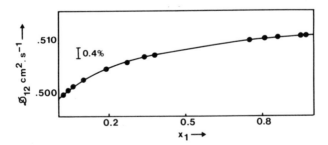

Figure 3.4   Concentration dependence of the diffusion coefficient $\mathscr{D}_{12}$ of He—C_2H_6 at 300 K.

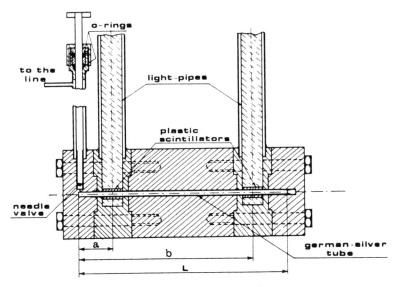

Figure 3.5   Cell for measuring diffusion coefficients using plastic scintillators as detectors situation $(l/6)$ from each end of the cell. Reprinted with permission from P. Codastefano, A. DiRusso, and V. Zanza, *Rev. Sci. Instrum.*, **48**, 1650 (1977).

background counts and the difference in the coefficients of the two tubes. In designing their cell they recognized the real possibility that [85]Kr might be absorbed into the plastic scintillator, so they placed an extremely thin layer of german silver between the diffusion gases and the detectors. The cell used was not of Loschmidt design, but rather similar to one used by Carson and co-workers [58] with a thermistor bridge. In this design the concentration gradient is initiated by carefully adding a second component through a needle valve at one end of the cell.

The solution to Fick's law for the above experimental conditions yields [67] a working relation that is exactly the same as for the thermistor bridge

$$\Delta n(t) \equiv n^l(l/6, t) - n^u(5l/6, t) \tag{31a}$$

$$= B \exp(-\pi^2 \mathcal{D}_{12} t/l^2) \tag{31b}$$

where $\Delta n(t)$ is the difference between the corrected number of counts generated at the lower and upper detectors. We can obtain the diffusion coefficient by using least-square techniques. A graph of data obtained when [85]Kr diffused in xenon at 298 K is shown in Figure 3.6. The average deviation of the experimental data from the straight line is $\pm 2.3\%$; the value of $\mathcal{D}_{12}$ at 1-atm pressure is 0.0747 cm^2/s to be compared with the theoretical value of 0.0750 calculated from the Chapman–Cowling theory [33] and the potential function of Aziz and van Dalen [70].

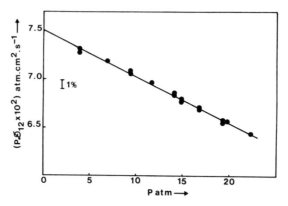

Figure 3.6   Pressure dependence of the diffusion coefficients $\mathscr{D}_{12}$ of a trace of ^{85}Kr in xenon at 298 K [69].

3.2.3   Restricted Diffusion—Mass Spectrometer Sensor

Hogervorst [38, 71] used a quartz cell 130 cm in length and with constant cross section to study diffusion in noble gas mixtures in the temperature range 300–1400 K. By means of a clever cataphoretic technique [72], he was able to form a linear concentration gradient of one noble gas (1–2%) in another and then initiate a diffusion measurement with the initial and boundary conditions

$$(\partial C_1/\partial x) = 0 \tag{32a}$$

for $x = 0$ and l, and

$$C_1(x, 0) = C_1(0, 0) - ax \tag{32b}$$

He solved Fick's second law to give

$$C_1(x, t) = C_1(x, \infty) + \sum_{n=1}^{\infty} \frac{4al}{\pi^2 n^2} \cos \frac{n\pi x}{l} \exp(-n^2\pi^2 \mathscr{D}_{12} t/l^2) \tag{32c}$$

where l is the length of the cell. If x is chosen to be $(l/6)$, the second term in the summation is zero and all further terms become negligible if $\mathscr{D}_{12}t > 200$ cm^2 for $l = 130$ cm. Thus, $\mathscr{D}_{12}$ can be derived from the simple relation

$$\Delta C_1(t) \equiv [C_1(l/6, t) - C_1(l/6, \infty)] = B \exp(-\pi^2 \mathscr{D}_{12} t/l^2) \tag{33}$$

where

$$B = (4al/\pi^2)$$

Hogervorst monitored the concentration of the minor component at position $(l/6)$ with a mass spectrometer as a function of time using a molecular leak. Cell pressures varied from 5 to 20 torr and diffusion times were about 100 s. Diffusion coefficients were obtained with an accuracy of approximately 2%. Table 3.1 lists his smoothed values of $\mathcal{D}_{12}$ (400–1200 K) for argon (ca. 1%) diffusing in helium and compares them with similar results obtained by two independent methods; the differences between corresponding results is somewhat greater than the accuracies claimed for the individual values.

3.3   Two-Bulb Method

In this method two bulbs are connected by a vertical tube of constant cross section and filled with a gas of known pressure. A small amount of a second gas (heavier than the first) is carefully added to the bottom bulb so that diffusion takes place through the connecting tube. The apparatus is designed so that any pressure gradient in the connecting tube is extremely small.

Ney and Armistead [73] appear to be the first workers to use such an apparatus for gases and derived a working equation to enable a diffusion coefficient to be calculated. However, Barnes [74] was first to derive the working equation that has been now accepted as correct; Annis and co-workers [75] obtained the same relation at a later date.

Barnes solved Fick's second law subject to the conditions

$$\left(\frac{\partial C_1^l}{\partial t}\right) = \frac{-\mathcal{D}_{12}A}{V^l}\left(\frac{\partial C_1^T}{\partial x}\right)_{x=0} \tag{34a}$$

$$\left(\frac{\partial C_1^u}{\partial t}\right) = \frac{-\mathcal{D}_{12}A}{V^u}\left(\frac{\partial C_1^T}{\partial x}\right)_{x=l} \tag{34b}$$

where C_1^l, C_1^u, and C_1^T are concentrations of heavy component in the lower and upper bulbs and the connecting tube, respectively; V^l and V^u are the volumes of the two bulbs; and A is the cross section of the tube of length l. These equations describe the rate of change of the concentration of component 1 in each bulb.

In addition Barnes used further boundary conditions

$$C_1^T(0, t) = C_1^l(t) \quad \text{and} \quad C_1^T(l, t) = C_1^u(t) \tag{34c}$$

to obtain the result

$$\Delta C_1(t) \equiv C_1^l(t) - C_1^u(t) = [C_1^l(0) - C_1^u(0)]\exp(-t/\tau) \tag{34d}$$

for the variation with time of the difference in concentration of component 1 in the two-bulb cell, where

$$\tau = \left[\left(1 - \frac{V^T}{6V^l}\right)\frac{\mathcal{D}_{12}A}{l}\left(\frac{1}{V^l} + \frac{1}{V^u}\right)\right]^{-1} \tag{34e}$$

is the relaxation time and V^T is the volume of the connecting tube that is assumed to be small compared to V^u and V^l. Equations (34d) and (34e) only differ from those obtained by Ney and Armistead by the factor $[1-(V^T/6V^l)]$, which appears because Barnes did not assume that the concentration distribution in the correcting tube was in a quasistationary state.

Because the concentration gradient is not confined to the connecting tube a small correction must be applied to its length. Similar corrections were used in theories of sound [76] and electricity [77]. For infinite flanges [76] at the ends of the tube the total length correction was found to be $2 \times 0.82\,r$, where r is the tube radius; whereas, when no flange [76, 78] exists, the correction is $2 \times 0.58\,r$. Wirz [79] investigated the corrections for tubes with intermediate flange widths. When we construct a two-bulb cell, end corrections can be minimized by choosing long tubes of small radius, but not so small as to generate pressure gradients in the tube.

When studying the application of the end corrections outlined above, Arora and co-workers [80] found that, if an accuracy of 0.1% in $\mathscr{D}_{12}$ is required, the end corrections outlined above did not yield results that agreed with those obtained in a Loschmidt-type cell, which is an absolute method. Those authors suggested that two-bulb cells should be calibrated with accurate results obtained with an absolute method.

Equation (34d) has exactly the same form as (30c), (31), and (33) so that we can determine $\mathscr{D}_{12}$ by using two matched thermistors, one in each bulb, and following the difference in their resistances with a simple Wheatstone bridge, an accurate digital voltmeter, and a data logger. Thus, as was the case previously [60], we can obtain $\mathscr{D}_{12}$ by treating the data either as a two- or a three-parameter problem using nonlinear least-square techniques. van Heijningen and co-workers [81, 82] were the first to use thermistors in this way and measured $\mathscr{D}_{12}$ values that were accurate to 1–2%. Since then Dunlop and co-workers [83–86] reported diffusion coefficients for many systems with an accuracy of 0.1–0.2%.

Instead of monitoring the difference in concentration between the bulbs directly Cain and Taylor [87, 88] used a technique that was introduced by Dubro and Weissman [89]. Samples are taken from both bulbs at known times and analyzed with a mass spectrometer that has been calibrated with mixtures of known concentration. Because the pressure in the cell decreases when samples are taken, (34d) becomes

$$\Delta C_1(t_n) = \Delta C_1(0) \exp[\gamma f(t_i, P_i)] \tag{35a}$$

where

$$\gamma = \left(1 - \frac{V^T}{6V^l}\right) \frac{A\mathscr{D}_{12}^{P=1}}{l_{\text{eff}}} \left(\frac{1}{V^l} + \frac{1}{V^u}\right) \tag{35b}$$

and

$$f(t_i, P_i) = \left[t_n/P_n + \sum_{i=1}^{\infty} t_{i-1}(P_i - P_{i-1})/P_i P_{i-1} \right] \tag{35c}$$

where $\mathscr{D}_{12}^{P=1}$ is the diffusion coefficient at 1-atm pressure, and t_i and P_i refer to the ith samples taken simultaneously from the two bulbs.

Cain and Taylor [87, 88] studied several noble gas systems in a cell fabricated from Hastelloy-X steel, and reported data in the temperature range 350–1300 K with an estimated accuracy of $\pm 2\%$. In Table 3.1 their data for He—Ar, corrected to 0 concentration of argon, are compared with the results of Hogervorst who claimed an accuracy of 1–2% for his data.

Trappeniers and Michels [90] used a two-bulb cell to study the pressure dependence of the self-diffusion coefficient of krypton using the β-emitter ^{85}Kr as a tracer. In their experiments the lower bulb and the connecting tube were filled with a krypton–^{85}Kr mixture to a known pressure and a special valve in the upper bulb was used to isolate the lower bulb and tube, while the upper bulb was filled with natural krypton to exactly the same pressure. The valve was constructed so that it could be opened or closed without causing pressure changes within the cell.

The concentration of tracer in the lower bulb was measured in terms of the ionization current caused by the decaying ^{85}Kr between two coaxial electrodes, the outer connected to an accurate constant-voltage supply and the inner to the input circuit of a vibrating-reed electrometer. The diffusion coefficient was obtained from the relation

$$i(t) - i(\infty) = [i(0) - i(\infty)] \exp(-t/\tau) \tag{36}$$

Where $i(0)$ and $i(\infty)$ are the measured ionization currents before the diffusion commences and after complete mixing, respectively, and τ is given by (34e). Some of their data for the pressure dependence of the self-diffusion coefficient of krypton at 298 K are shown in Figure 3.7; the precision is 0.1%.

Suetin and co-workers [91–94] used a two-bulb cell to measure diffusion coefficients for the systems He—Ar, H_2—He, and H_2—Ar for temperatures between 290 and 700 K. Their results were obtained for a mole fraction of 0.5 with an estimated accuracy of 1.5%. Using concentration dependences calculated from the Chapman–Enskog theory and the potential functions reported in the literature, this group's smoothed results for He—Ar were reduced to limiting values $\mathscr{D}_{12}^0$ and are listed in Table 3.1 for comparison with the results of other workers.

3.4   Dual Two–Bulb Method

An excellent method for measuring ratios of diffusion coefficients for a binary system as a function of temperature was introduced and developed by Vugts and co-workers [95]. Their apparatus is shown schematically in Figure 3.8. It

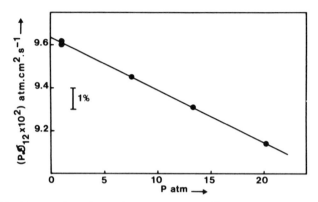

Figure 3.7   Pressure dependence of tracer amounts of ^{85}Kr in normal Kr at 298 K [90].

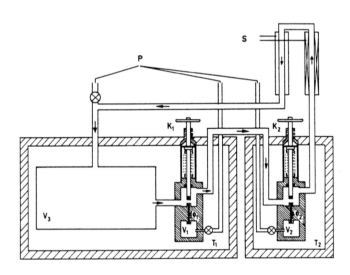

P	Pumps and gas filling system.
S	Thermosyphon.
K_1, K_2	Valves.
V_1	Small reservoir, thermostated at temperature T_1.
V_2	Small reservoir, thermostated at temperature T_2.
V_3	Large reservoir, thermostated at temperature T_1.
ϕ_1, ϕ_2	Diffusion fluxes of the enriched isotope.

Figure 3.8   Dual two-bulb cell for measuring accurate diffusion coefficient ratios. Reprinted with permission from H. F. Vugts, A. J. H. Boerboom, and J. Los, *Physica*, **44**, 219 (1969).

consists essentially of two almost identical two-bulb cells with lower volumes V_1 and V_2 (ca. 50 cm^3) both connected with tubes to a much larger common volume V_3 (ca. 5000 cm^3). The connecting tubes may be closed at their upper ends by two valves K_1 and K_2. The reference cell and V_3 are positioned in a thermostat T_1, and the other cell is positioned in a second thermostat that can be adjusted to be at a lower or higher temperature T_2 than T_1.

To perform an experiment the valves K_1 and K_2 are closed and the volumes V_1 and V_2 are filled to the same pressure with a binary gas mixture, while V_3 is filled with a similar mixture of lower density to a slightly lower pressure. The valves K_1 and K_2 are then opened simultaneously $(t = 0)$ and diffusion is allowed to proceed through both connecting tubes. The gas mixture in V_3 is maintained of uniform concentration throughout by a thermosyphon.

After a suitable time t_1 the valve K_2 is closed and then the valve K_1 is closed at time t_2 so that the final concentrations in V_1 and V_2 are almost identical. The gas mixtures in V_1 and V_2 are then analyzed with a mass spectrometer.

Now if $C_1(0)$ is the initial concentration in V_1 and V_2 at $t = 0$, and $C_3(0)$ is the corresponding concentration in V_3, Fick's first law can be used to describe the transport through both connecting tubes of lengths l_1 and l_2 and cross sections O_1 and O_2 to derive [95] the following expression for the ratio of the diffusion coefficient of the system at two temperatures T_1 and T_2

$$\frac{(\mathscr{D}_{12})_1(T_1)}{(\mathscr{D}_{12})_2(T_2)} = \frac{t_1}{t_2} \frac{O_2}{l_2 V_2} \left(\frac{O_1}{l_1 V_1}\right)^{-1} \Big/ AB \qquad (37a)$$

$$A = \left[1 - \frac{O_1(\mathscr{D}_{12})_1}{l_1 V_3}(t_2 - t_1)\right] \qquad (37b)$$

$$B = 1 + \frac{[C_1(t_2) - C_2(t_1)]\left\{2 - \dfrac{O_1(\mathscr{D}_{12})_1}{l_1 V_3}(t_1 + t_2) - \dfrac{1}{2}\left[\dfrac{O_1(\mathscr{D}_{12})_1}{l_1 V_1}(t_2 - t_1)\right]^2\right\}}{[C_1(t_2) + C_2(t_1) - 2\overline{C(0)}]\left(\dfrac{O_1(\mathscr{D}_{12})_1}{l_1 V_1}\right)t_2} \qquad (37c)$$

The quantity $[O_1(\mathscr{D}_{12})_1/l_1 V_3]$ can be approximated quite accurately by

$$\frac{O_1(\mathscr{D}_{12})_1}{l_1 V_3} = -\frac{V_1}{V_1 + V_3} t_2^{-1} \ln\left\{\left[\frac{C_1(t_2) - C_3(0)}{C_1(0) - C_3(0)} - \frac{V_1}{V_1 + V_3}\right]\frac{V_1 + V_3}{V_3}\right\} \qquad (37d)$$

while the geometry factor

$$\frac{O_2}{l_2 V_2}\left(\frac{O_1}{l_1 V_1}\right)^{-1} \qquad (37e)$$

can be determined from an experiment in which $t_1 = t_2$ and $T_1 = T_2$ so that the diffusion coefficient ratio is unity.

In these equations a subscript on a concentration indicates the volume to which it refers, and $\overline{C(0)}$ is the average concentration in the total volume $(V_1 + V_2 + V_3)$ at all times. The concentration difference $\Delta C(t) = [C_1(t_2) - C_2(t_1)]$ can be monitored with a mass spectrometer fitted with a double inlet system and a double collector so that t_2 can be chosen in (37c) to make $\Delta C(t)$ close to 0 and hence B almost unity. The apparatus described above yields diffusion coefficient ratios that are accurate to 0.1%.

In Table 3.3 absolute values of $\mathscr{D}_{12}^0$, experimental values extrapolated to $x_1 = 0$, for hydrogen–noble gas systems [34, 96] are compared [97] with corresponding values derived for these systems from diffusion-coefficient ratios reported by Wahby and Los [98] who used a reference temperature of 299.16 K. Absolute values of $\mathscr{D}_{12}^0$ at this temperature were taken from the results reported in the left-hand column of the results for each system. Since the $\mathscr{D}_{12}^0$ values in each left-hand column have an estimated accuracy of 0.3%, the agreement between the two sets of data is excellent. Many papers have been published in which the dual-bulb cell method has been used [99–101].

Table 3.3 Comparison of Limiting Diffusion Coefficients $\mathscr{D}_{12}^0$ for H_2–Noble Gas Systems

| | He—H_2 | | Ar—H_2 | |
| | $\mathscr{D}_{12}^0$ | | $\mathscr{D}_{12}^0$ | |
$T(K)$	[34]	R	[96]	R
240			0.556_2	1.002_0
260			0.640_6	1.002_5
280	1.428	1.001_4	0.729_9	1.001_4
300	1.603	1.000_0	0.823_6	0.999_6
320	1.785	0.998_9	0.921_2	0.998_0

| | Kr—H_2 | | Xe—H_2 | |
| | $\mathscr{D}_{12}^0$ | | $\mathscr{D}_{12}^0$ | |
$T(K)$	[96]	R	[96]	R
240	0.486_9	1.003_3	0.418_3	0.997_8
260	0.561_9	1.002_3	0.482_9	1.002_5
280	0.641_1	1.000_8	0.551_2	1.003_1
300	0.723_9	0.999_9	0.622_8	1.001_8
320	0.809_9	0.999_4	0.697_5	0.999_6

[a]The values R in the right-hand column for each system are ratios of results obtained from [98] to the values in the corresponding left-hand column $\mathscr{D}_{12}^0$.

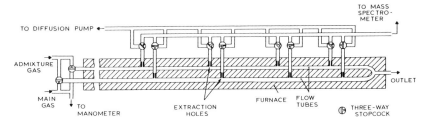

Figure 3.9 Cell for measuring diffusion coefficients by the backdiffusion technique; see text for details. Reprinted with permission from K. W. Reus, C. J. Zwakhals, and J. A. Smit, *Physica*, **100C**, 221 (1980).

3.5   Back-Diffusion Method

Recently, Reus and co-workers [102, 103] revived a technique that was used in 1933 by Harteck and Schmidt [104] to study diffusion in *ortho–para* hydrogen mixtures at 20 K. In this method, the back-diffusion technique, one component (the minor) diffuses upstream against another flowing with a known average velocity $\bar{v}$ and constant pressure P in a tube of constant cross section A.

Figure 3.9 [102] illustrates the primary points of the technique. The main component of the mixture to be studied enters one of the two tubes (either one can be used) and the minor component enters through the other. A needle valve (not shown in the diagram) at the outlet controls the rate at which the gases are removed by an oil diffusion pump. By adjusting the pressures of the two gases and the exit needle valve it is possible to obtain a constant concentration $C(0)$ of the minor component at the right-hand extraction hole and a steady-state concentration gradient from the right to the region of the left-hand hole as the minor component diffuses backward against the flow of the major component.

In the region of the gradient the rate of passage of the minor component through a given cross section due to flow in the tube is

$$\phi_v = \int_A vC \, dA \tag{38}$$

where v is the velocity at a given point in the cross section. The corresponding rate of passage of the minor component through the same cross section due to diffusion is

$$\phi_D = -\int_A \left(-\frac{dC}{dx} \right) \mathcal{D}_{12} \, dA \tag{39}$$

so that in the stationary state

$$\int \left[Cv + \mathcal{D}_{12} \left(\frac{dC}{dx} \right) \right] dA = 0 \tag{40}$$

If one now assumes that the tube has been designed so that C and (dc/dx) are constant over any cross section and that $\mathscr{D}_{12}$ is constant throughout the gradient, (40) becomes

$$C\bar{v} + \mathscr{D}_{12}\left(\frac{dC}{dx}\right) = 0 \tag{41}$$

Thus as $\bar{v}$ and $\mathscr{D}_{12}$ are independent of position, (41) can be written

$$C(x) = C(0)\,\exp(-x\bar{v}/\mathscr{D}_{12}) \tag{42a}$$

or

$$C(x)/C(x') = \exp[-(x-x')\bar{v}/\mathscr{D}_{12}] \tag{42b}$$

Thus, a knowledge of the ratio of the concentrations at two points in the cell, x and x', and $\bar{v}$ permits $\mathscr{D}_{12}$ to be calculated at the pressure of the experiment.

The two tubes of the cell used by Reus and co-workers [102, 103] (see Figure 3.9) were made of fused silica 88 cm in length positioned in a furnace capable of withstanding temperatures up to 1600 K. Each tube had a different diameter so that experimental conditions could be varied; one tube was used for the major component, the other to add the minor. Four pinholes, small enough to ensure molecular flow, were constructed at known positions along each tube so that samples of mixture could be extracted and analyzed with a mass spectrometer to give $C(x)$ and $C(x')$. The expression

$$\bar{v} = G/AP \tag{43}$$

was used to calculate the average gas velocity from the quantity G of the major component lost per unit time from a standard volume A and P. Thus we can determine $(P\mathscr{D}_{12})$ from (42b) using $C(x)$, $C(x')$, and $\bar{v}$.

The authors have given a detailed description of the apparatus construction, of the method used to obtain the diffusion coefficients, and of the corrections that must be applied to the raw experimental data. Unfortunately, their results for noble gas mixtures in the temperature range 300–1300 K are not very accurate. It appears that this technique needs to be modified further if results capable of testing theoretical developments are to be obtained. Table 3.1 gives a comparison of the authors' smoothed results [102] for He—Ar with those of other workers.

4 METHODS OF MEASUREMENT—LIQUIDS

The major techniques for the determination of diffusion coefficients in liquids fall naturally into five groups. The first of these is *free diffusion*, where the solution components are transported across a boundary between two solutions of different composition and the timescale of the experiment is such that there is no change in composition at the ends of the cell. The second is *restricted diffusion*, where such changes are allowed to occur. The third is *steady-state diffusion*, where a time invariant concentration gradient is set up. The fourth is *diffusion in a flowing stream*, where a small volume of one solution is injected into a flowing stream of the other and the concentration distribution or peak broadening is determined after passage through a long, uniform tube. The last is made up of techniques that do not require macroscopic concentration gradients, that is, by light scattering, where use is made of microscopic concentration fluctuations due to the thermal motion of the molecules, and the NMR spin–echo technique, a T_2 relaxation time determination, where a dephasing mechanism additional to the spin–spin interaction is provided by the thermal motion in a magnetic field gradient of the molecules of the component containing the resonant nuclei. This grouping is retained in what follows.

Since 1972, when the previous review in this series [1] appeared, several books have been published on diffusion in liquids, together with many reviews about particular techniques. The latter are cited at the appropriate points in the text. Of the former, one is a general monograph dealing with theory, methods of measurement, and experimental results [6]; a second deals with multicomponent diffusion [5]; and a third, a more general text by the same author as the second, covers diffusion and mass transport with application to practical problems [7].

Table 3.4 makes a comparison between the precision, accuracy, costs of equipment, and time and the range of temperatures and pressures covered by the techniques described in this section.

4.1   Free Diffusion

The 1972 review [1] contained a detailed account of the optical interferometric methods used to obtain binary and ternary diffusion coefficients. The Gouy, Rayleigh, shearing (Bryngdahl), and wedge interferometric techniques were described. This material is not repeated here for reasons of economy and also because these methods had been essentially developed fully by that date, with little furtherance since. This is not to say that the methods are no longer important [105]. On the contrary, the Gouy and Rayleigh diffusiometers provide the most precise absolute techniques yet developed for the measurement of diffusion coefficients ($\pm 0.1\%$ in binary systems), and many of the other techniques depend for calibration of the apparatus on data obtained by these methods. Accordingly, the reader is referred to the earlier review for a detailed account of these methods and to the monograph by Tyrrell and Harris [6], which also includes other, less commonly used, optical free diffusion techniques.

4.1.1   Gouy Interferometry

The Gouy method, like the other optical techniques, has generally relied on photographic recording of interference fringes. The optical system is shown in

Table 3.4 Comparison of the Techniques for Liquid Diffusion Measurements

Method	Precision	Accuracy	Capital Cost	Temperature Range	Pressure Range	Time Required
Optical						
Gouy	High	High	High	Restricted	Atmospheric	1–2 days
Rayleigh	High	High	High	Restricted	Atmospheric	1–2 days
Bryngdahl	High	High	High	Restricted	Atmospheric	hours
Restricted diffusion						
Harned	High	High	Low	Restricted	Atmospheric	days
Capillary	Moderate	Moderate	Low–moderate	Wide	Atmospheric	days
Diaphragm cell	High–moderate	High–moderate	Low–moderate	Moderate	To 400 MPa	days
Taylor dispersion	Moderate	Moderate	Low–high	Moderate	To 70 MPa	Tens of minutes
Light scattering	Moderate	Uncertain	High	Wide	Wide	hours
NMR	Moderate	Moderate	High	Wide	Wide	Tens of minutes

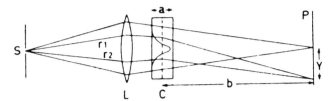

Figure 3.10 Optical system for the single lens Gouy diffusiometer [9]. Rays of light r_1 and r_2 from the source slit S are focused by means of the lens L onto the photographic plate P. Light passing through the cell C where a sharp boundary is formed between two solutions of different concentrations, which are then allowed to mix, gives rise to interference fringes displaced a distance Y from the optic axis. Part a is the cell thickness parallel to the optic axis and b is the optical lever arm. Reprinted with permission from P. J. Dunlop, B. J. Steel, and J. E. Lane, "Experimental Methods for Studying Diffusion in Liquids, Gases and Solids," in A. Weissberger and B. W. Rossiter, Eds., *Physical Methods of Chemistry*, Vol. 1, Part IV, Wiley-Interscience, New York, 1975, p. 205. Copyright © 1975 by John Wiley & Sons, Inc.

Figure 3.10 and a typical set of interference patterns taken at different times during a diffusion experiment is illustrated in Figure 3.11.

Renner and Lyons [106] devised a method of rapidly scanning (3–4s at 60 Hz) the image at the focal plane with a moving slit set in front of a photomultiplier and storing the resultant intensity data in the memory of a PDP12 computer. Fluctuations in the signal, due to the alternating current (ac) mercury lamp used as a light source, were removed by using the computer as a phase-sensitive detector. The positions of the lowest five fringe minima were used to compute mutual diffusion coefficients from the theoretical treatment of Gosting and Onsager [107]. This technique is identical with the normal photographic one. Excellent agreement (within $\pm 0.1\%$) was obtained with standard water–KCl data [108] at a mean concentration of 0.2250 mol/L. The diffusion coefficient was also determined from the *variation* of the intensity of the fringes, testing the Gosting–Onsager analysis directly, and excellent results were again obtained. Figure 3.12 shows a comparison of measured and calculated intensities for the first four fringes from such an experiment.

It was found that the amount of computing time required for a full fit of the intensity distributions was excessive. (This is probably no longer true with modern laboratory computers.) Renner and Lyons [106], therefore, used a second technique with a fixed slit offset from the optic axis of the diffusiometer by about 20 mm. The fringe intensity at this point was recorded continuously and the times when minima and maxima passed the detector were determined. From this information, diffusion coefficients were obtained with no less accuracy than by the former method.

The Gouy technique has been extensively used to determine ternary diffusion coefficients [1, 5, 6]. Kim [12, 13] developed procedures for the calculation of the 3×3 matrix of diffusion coefficients required to describe a quaternary system from free, restricted, and steady-state diffusion experiments. His work suggested that a combination of Gouy and Rayleigh free diffusion experiments should give

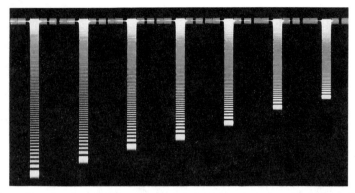

Figure 3.11 Gouy interference fringes photographed at intervals during an experiment in which an aqueous 1% solution mixed with water at 25°C. Rayleigh reference fringes defining the position of the undeviated slit image lie on either side of each Gouy pattern. Each set was photographed in a separate exposure of the plate after removing the lens L shown in Figure 3.10, replacing it with a cylindrical lens, and placing the appropriate masks in front of the cell. Reprinted with permission from P. J. Dunlop, B. J. Steel, and J. E. Lane, "Experimental Methods for Studying Diffusion in Liquids, Gases and Solids," in A. Weissberger and B. W. Rossiter, Eds., *Physical Methods of Chemistry*, Vol. 1, Part IV, Wiley-Interscience, New York, 1975, p. 205. Copyright © 1975 by John Wiley & Sons, Inc.

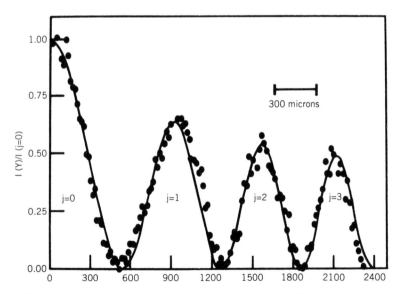

Figure 3.12 Comparison of measured (●) and theoretical (−) intensities for the first four Gouy fringes as a function of displacement from the center of the zeroth fringe. Displacement in microns, from $j = 0$ maximum. (Solid line theoretical. Filled dots experimental.) Reprinted with permission from T. A. Renner and P. A. Lyons, *J. Phys. Chem.*, **78**, 2050 (1974). Copyright © 1974 American Chemical Society.

good results. More recent work by Miller [109] indicates that good results can be had from either optical system separately. However, both procedures remain to be tested by experiment.

4.1.2 Shearing Interferometry

In this method, polarized light is passed through the refractive index gradient provided by the interdiffusing solutions and then through a Savart plate. This plate consists of two uniaxial quartz crystals, with principal axes at 90° to one another cemented together. The plate is oriented such that the plane of polarization of the incident light is at 45° to both halves. The plate produces two rays of equal intensity, in phase, polarized at right angles to one another, and displaced by an amount depending on the thickness of the plate and the refractive indices of these ordinary and extraordinary rays. This displacement, or lateral shear, parallel to the diffusion coordinate, is removed on reunification of the two beams by a second Savart plate. The interference pattern that results is due to the different paths the two rays have traveled. The theory for this version of the method, originally devised by Bryngdahl and Ljunggren [110, 111] for small shear, has been given and tested by Becsey and Bierlein [112] for any shear and with correction for any aberations producing asymmetric fringes.

In the notation of [1], the path difference between the two cofocused rays, each at conjugate values $\pm x$ of the diffusion coordinate at time t, is given by:

$$\Delta P(t, \pm x) = a(\Delta n/2)\left[\mathrm{erf}\left(\frac{x+b/2}{2\sqrt{Dt}}\right) - \mathrm{erf}\left(\frac{x-b/2}{2\sqrt{Dt}}\right)\right] \qquad (44)$$

where b is the shear, a is the width of the cell, Δn is the difference in the refractive index of the solutions, and D is the mutual diffusion coefficient. The area under a fringe resulting from ΔP, A', is obtained by integration of (44):

$$A'(t, \pm x) = a\Delta n\sqrt{Dt}\, \{2[u\,\mathrm{erf}(u) - v\,\mathrm{erf}(v)] + \mathrm{erf}'(u) - \mathrm{erf}'(v)\} \qquad (45)$$

where

$$u, v = \frac{(x \pm b/2)}{2\sqrt{Dt}} \qquad (46)$$

and

$$\mathrm{erf}'(w) = \frac{2}{\sqrt{\pi}}\exp(-w^2) \qquad (47)$$

$w = u, v.$

In practice, it is convenient to measure the area enclosed by the convex portion of the fringe at $\pm x$, and to apply the usual zero-time correction for unavoidable diffusion occurring during boundary formation and sharpening. The nth area for the mth interferogram A_{mn} then becomes

$$A_{mn} = A(t_m, \pm x_{mn})$$
$$= a\Delta nb\{0.5[\text{erf}(u_{mn}) + \text{erf}(v_{mn})]$$
$$+ \sqrt{D(t_m + t_0)}[\text{erf}'(u_{mn}) - \text{erf}'(v_{mn})]/b\} \tag{48}$$

where

$$u_{mn}, v_{mn} = \frac{(x_{mn} \pm b/2)}{2\sqrt{D(t_m + t_0)}} \tag{49}$$

For each interferogram (Figure 3.13), taken at time t_m, A_{mn} is measured for a convenient choice of x_{mn}. Becsey and Bierlein [112] did this by means of a digitizer and simple summation. The parameters a, Δn, b, t_0, and D can be obtained by minimization of the sum of the squares of the area residuals for the measured fringes, or by nonlinear least-squares techniques based on (48). (A good general method for nonlinear least-squares computations, used by two of the authors in other contexts, is that of Wentworth [113]. It has not been tried with this particular problem.)

Becsey and Bierlein [112] obtained a standard error of 0.3% for a mean concentration of 0.00625 mol/L of water–KCl at 25°C, an accuracy comparable

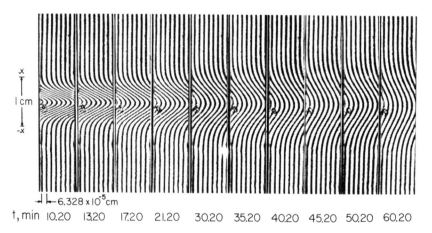

Figure 3.13 Successive Savart plate interferograms obtained for the interdiffusion of 0.0125 mol/L KCl solution into water at 25°C. Reprinted with permission from J. Becsey and J. A. Bierlein, *Rev. Sci. Instrum.*, **49**, 227 (1978). Copyright © 1978 American Physical Society.

with that of Pepela and co-workers [114] for aqueous sucrose and other solutions using the Bryngdahl single Savart plate method described in [1].

4.1.3 Ternary Systems

Tanigaki and co-workers [115, 116] developed a method for determining ternary diffusion coefficients directly from the concentration profiles that are measured in a free diffusion experiment. These profiles were measured by determination of optical absorbances simultaneously at more than one wavelength with a position scanning spectrophotometer. This method is restricted to solutions of components with resolvable absorption bands such as are the property of the nickel, cobalt, and copper salt systems investigated by these workers.

Miller [117], following earlier work by Albright and Sherrill [118], developed a method for the determination of ternary diffusion coefficients from Rayleigh fringe patterns by a nonlinear least-squares fit of the fringe number j as a function of the reduced position

$$y_j = (x_j - x_{J-j})/4\sqrt{t} \tag{50}$$

There are a total of J fringes, and this working definition of y_j, due to Creeth [119], has the effect of canceling certain optical effects and others due to the dependence of the refractive index on concentration. The working equation [116, 118, 120] for the reduced fringe number $f(j)$ is

$$\begin{aligned} f(j) &= (2j - J)/J \\ &= (a + b\alpha_1)\,\text{erf}(s_+ y_j) + (1 - a - b\alpha_1)\,\text{erf}(s_- y_j) \end{aligned} \tag{51}$$

where the parameters a, b, s_+, and s_- are functions of the diffusion coefficients D_{ij} and the refractive index increments R_1 and R_2. The latter are found from the relation

$$J = R_1\Delta C_1 + R_2\Delta C_2 \tag{52}$$

using total fringe numbers from several experiments carried out at the same mean composition, but with different initial ΔC_i. The value of α_1 is obtained from the definition $(R_1\Delta C_1/J)$. The diffusion coefficients are functions of a, b, s_+, s_-, R_1, and R_2 [117, 118]. Corrections are necessary because the initial boundary is not initially sharp and there are minor optical imperfections in the system. Data within a fixed range of $f(j)$ are used for consistency. The method is an iterative one; and as more than the minimum number of experiments are carried out, the *precision* of the resultant D_{ij} is excellent, 0.001–0.006 10^{-5} cm^2/s. Rard and Miller [121] estimated the *accuracy* of this method to be perhaps 10 times less, by comparison of results for the system $H_2O–NaCl–KCl$ from their own Rayleigh and Gouy diffusiometer and from other laboratories.

Miller [109] recently developed a similar nonlinear least-squares analysis for application to Gouy interferometry for ternary and quaternary systems. This is believed to be superior to the procedures of Revzin (ternary) [122] and Kim (quaternary) [12, 13].

4.1.4   Boundary Formation

In methods where very small concentration gradients are used, such as the shearing interferometric techniques, the gravitational stability of the diffusion column, which depends on the difference in density of the diffusing solutions, can be difficult to maintain. Therefore, the better optical systems are isolated carefully from vibration. The boundary is also difficult to form, and special methods are usually required [114, 123].

Tyrrell and co-workers [124, 125] overcame these problems neatly in viscous media by applying a temperature gradient across a film of liquid 1 mm thick in a thermal diffusion cell. Once the resulting concentration gradient is established, thermal equilibrium is quickly restored, and the diffusion process is monitored by shearing interferometry. This was applied successfully to the measurement of diffusion coefficients of about 10^{-7} cm^2/s, well below that normally accessible by Gouy or Rayleigh interferometry.

4.1.5   Static and Dynamic Instabilities in Interdiffusing Multicomponent Systems

In diffusion experiments in multicomponent systems, certain criteria must be met to prevent convective mixing of the two interdiffusing solutions. This can occur when coupled transport results in the development of layers within the boundary with densities greater than that of the lower solution or less than that of the upper. These criteria for *static* stability were laid down by Wendt [126] and by Reinfelds and Gosting [127] for ternary free diffusion and by Kim [128] for quaternary systems undergoing free diffusion or the quasisteady-state diffusion of the diaphragm cell. Since that time, a wide literature has developed on the subject of convective instabilities and disturbances due to *fingering* or *dynamic* instabilities observed in systems that should be gravitationally stable. This seems to occur irrespective of whether the cross diffusion terms are large or small. A theory accounting for this was developed by McDougall [129, 130]. It was extended to the general free diffusion case and its predictions were confirmed by the experiments of Miller and Vitagliano [131]. The criteria for the onset of convective instabilities were also investigated by Comper et al. [132].

4.2   Restricted Diffusion

4.2.1   The Harned Conductance Method

The application of this method to binary systems was described in [1] and was extended by Lyons and Leaist [133–135] to multicomponent systems, following an earlier treatment by Fujita [136].

Diffusion is allowed to occur in a cell of rectangular or cylindrical cross section until the solute concentration changes at both ends of the cell. Usually the gradient is small as the method is applied most often to dilute solutions, and the boundary is formed near the bottom of the cell by shearing or injection of the denser solution below the less dense one from a syringe. The boundary conditions for the solution of Fick's second differential equation are

$$\left(\frac{\partial C}{\partial x}\right)_t = 0 \tag{53}$$

at $x = 0, l$ and the solution for a binary system is a Fourier series

$$C(x, t) = A_0 + \sum_{n=1}^{\infty} A_n \exp\left(-\frac{n^2\pi^2 Dt}{l^2}\right)\cos\left(\frac{n\pi x}{l}\right) \tag{54}$$

where the A_n are constants. By measuring concentration *differences* at $x = l/6$, and $x = 5l/6$, one obtains a rapidly convergent series

$$C\left(\frac{5l}{6}, t\right) - C\left(\frac{l}{6}, t\right) = 2A_1 \exp\left(-\frac{\pi^2 Dt}{l^2}\right)$$

$$+ 2A_5 \exp\left(-\frac{25\pi^2 Dt}{l^2}\right) + \cdots \tag{55}$$

After a short initial period, the second term of (55) becomes negligible compared to the first, and the mutual diffusion coefficient can be obtained from the relation

$$D = -\left(\frac{l}{\pi}\right)^2\frac{d \ln \Delta C(t)}{dt} \tag{56}$$

where $\Delta C(t)$ is the difference in concentration at the two planes of measurement. In practice, small platinum electrode pairs are mounted at these positions, and the electrolyte resistances are measured carefully with an alternating current Jones bridge. To a first approximation, for small gradients, the conductivity difference $\Delta\kappa$ is proportional to $\Delta C^\dagger$, and D is evaluated from

$$D = -\left(\frac{l}{\pi}\right)^2\frac{d \ln \Delta\kappa(t)}{dt} \tag{57}$$

The theoretical treatment of Fujita [136] of the case of ternary diffusion applies to symmetrically sheared cells where the initial boundary is formed at the middle plane of the diffusion cell. This type of boundary formation was found to be unreliable for dilute multicomponent solutions [134], and formation of the

---

†Of course, this method is inapplicable at the concentration of the conductivity maximum.

boundary at the cell bottom is preferred. We outline here Leaist's adaption [135] of the Fujita analysis.

The three component diffusion equations are

$$\left(\frac{\partial C_1}{\partial t}\right)_x = D_{11}\left(\frac{\partial^2 C_1}{\partial x^2}\right)_t + D_{12}\left(\frac{\partial^2 C_2}{\partial x^2}\right)_t \tag{58a}$$

$$\left(\frac{\partial C_2}{\partial t}\right)_x = D_{21}\left(\frac{\partial^2 C_1}{\partial x^2}\right)_t + D_{22}\left(\frac{\partial^2 C_2}{\partial x^2}\right)_t \tag{58b}$$

The boundary condition (53) applies to each component, and the initial conditions for a boundary formed at $x = \varepsilon l$ are

$$C_i = \bar{C}_i + (1-\varepsilon)\Delta C_{io} \qquad 0 < x < \varepsilon l \tag{59a}$$

$$C_i = \bar{C}_i - \varepsilon\Delta C_{io} \qquad \varepsilon l < x < l \tag{59b}$$

where $\bar{C}_i$ is the mean cell concentration and ΔC_{io} the initial concentration difference across the boundary. The solutions to (58) for this set of conditions are [134, 135]

$$\Delta C_1(t) = [\alpha/(D_1 - D_2)]\{[(D_{11} - D_2)\Delta C_{10} + D_{12}\Delta C_{20}]\exp(-\beta D_1 t)$$
$$+ [(D_{22} - D_2)\Delta C_{10} - D_{12}\Delta C_{20}]\exp(-\beta D_2 t)\} \tag{60a}$$

$$\Delta C_2(t) = -[\alpha/(D_1 - D_2)]\{[-D_{21}\Delta C_{10} + (D_{11} - D_1)\Delta C_{20}]\exp(-\beta D_1 t)$$
$$+ [D_{21}\Delta C_{10} + (D_{22} - D_1)\Delta C_{20}]\exp(-\beta D_2 t)\} \tag{60b}$$

where the ΔC_i is again the concentration difference between the measurement planes at $x = l/6$ and $5l/6$, and α and β are the geometric factors

$$\alpha = (2\sqrt{3}/\pi)\sin(\pi\varepsilon/l) \tag{61}$$

$$\beta = \pi^2/l^2 \tag{62}$$

and D_1 and D_2, the eigenvalues of the diffusion coefficient matrix, are given by

$$D_1, D_2 = \tfrac{1}{2}\{(D_{11} + D_{22}) \pm [(D_{22} - D_{11})^2 + 4D_{12}D_{21}]^{1/2}\} \tag{63}$$

The measured, normalized (specific) conductivity difference $\Delta\kappa(t)$ is written in terms of separately determined partial conductivities κ_i as

$$\frac{\Delta\kappa(t)}{\Delta\kappa(0)} = \frac{\kappa_1\Delta C_1(t) + \kappa_2\Delta C_2(t)}{\kappa_1\Delta C_1(0) + \kappa_2\Delta C_2(0)} \tag{64}$$

and using (60) this becomes

$$\frac{\Delta\kappa(t)}{\Delta\kappa(0)} = A_1 \exp(-\beta D_1 t) + (1-A_1)\exp(-\beta D_2 t) \tag{65}$$

where the constant A_1 is given by

$$A_1 = \{(\kappa_1/\kappa_2)D_{12}-D_{11}+D_1+X_1[(\kappa_2/\kappa_1)D_{21}$$
$$-(\kappa_1/\kappa_2)D_{12}+2D_{11}-D_1-D_2]\}/(D_1-D_2) \tag{66}$$

and

$$X_1 = \kappa_1\Delta C_{10}/(\kappa_1\Delta C_{10}+\kappa_2\Delta C_{20}) \tag{67}$$

is that fraction of the initial conductivity difference due to solute 1.

The zeroth and first time moments of the normalized conductivity difference are

$$I_0 = \int_0^\infty [\Delta\kappa(t)/\Delta\kappa(0)]\, dt \tag{68}$$

and

$$I_1 = \int_0^\infty t[\Delta\kappa(t)/\Delta\kappa(0)]\, dt \tag{69}$$

or, on integration,

$$I_0(X_1) = I_0(0) + [I_0(1)-I_0(0)]X_1 \tag{70}$$

and

$$I_1(X_1) = I_1(0) + [I_1(1)-I_1(0)]X_1 \tag{71}$$

where

$$I_0(0) = [D_{11}-(\kappa_1/\kappa_2)D_{12}]/\beta D_1 D_2 \tag{72}$$

$$I_0(1) = [D_{22}-(\kappa_2/\kappa_1)D_{21}]/\beta D_1 D_2 \tag{73}$$

$$I_1(0) = [D_{11}I_0(0)-(\kappa_1/\kappa_2)D_{12}I_0(1)]/\beta D_1 D_2 \tag{74}$$

$$I_1(1) = [D_{22}I_0(1)-(\kappa_2/\kappa_1)D_{21}I_0(0)]/\beta D_1 D_2 \tag{75}$$

These moments are determined by numerical integration of conductivity–time data from two experiments, the first with an initial gradient only in solute one ($\Delta C_{20} = 0$, $X_1 = 1$) and the second with the initial gradient only in the second solute. The D_{ij} can be then obtained by solution of (72)–(75),

$$D_{11} = [I_0(0)I_0(1) - I_1(0)]/F \tag{76}$$

$$D_{12} = (\kappa_2/\kappa_1)[I_0^2(0) - I_1(0)]/F \tag{77}$$

$$D_{21} = (\kappa_1/\kappa_2)[I_1(1) - I_0^2(1)]/F \tag{78}$$

$$D_{22} = [I_1(1) - I_0(0)I_0(1)]/F \tag{79}$$

with

$$F = \beta[I_0(0)I_1(1) - I_0(1)I_1(0)] \tag{80}$$

The necessary condition for omission of higher terms in the Fourier series is that conductivity differences should be used only at times greater than $l^2/50D'$, where D' is the smaller of D_1 and D_2.

The precision of this method is not as good as that of the Gouy or Rayleigh methods, the uncertainty in the diffusion coefficients being of the order of 0.01– 0.05 10^{-5} cm^2/s, that is, an order of magnitude larger.

4.2.2   Open-Ended Capillary Method

A modification of this method for application to dilute electrolyte solutions, which has some similarity to the Harned method, was described by Agar and Lobo [137] and Lobo and Teixeira [138] and is so described at this point.

In the original technique [1], a capillary closed at one end is filled with a solution of a given composition and then immersed in a stirred bath of a different composition, and diffusion is allowed to proceed. After a suitable interval, the capillary is removed, and its contents are extracted for analysis. The solution to the diffusion equation is again a Fourier series; but as all the terms are valid, convergence can be slow. The method is particularly suited to tracer diffusion, but it has also been commonly used with electrolytes and molten salts [6]. At best, the precision is typically $\pm 0.5\%$.

The Agar–Lobo method uses two capillaries, each closed at one end by an unplatinized platinum electrode, placed on the same axis a short distance from one another in a gently stirred bath. The third electrode is a platinum wire placed midway between the capillary openings. The apparatus is shown in Figure 3.14. The bath lies in a thermostat, and initially the solutions in the bath and the capillaries are of the same composition. After thermal equilibrium is reached, the capillary contents are replaced carefully using syringes, with one capillary being filled with solution less concentrated (say 25%) than the bath solution and one correspondingly more concentrated. The ratio of the re- sistances of the capillary solutions measured from the center tap is then followed with time. At the end of the experiment, the capillaries are flushed with bath

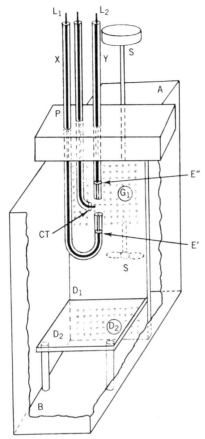

Figure 3.14 Dual open-ended capillary apparatus of Agar and Lobo for mutual diffusion of electrolyte solutions: E', E'', platinum electrodes; CT, center tap; S, G_1, G_2, D_1, D_2, arrangement to stir solution and maintain a steady flow over the capillary ends. Reprinted with permission from J. N. Agar and V. M. M. Lobo, *J. Chem. Soc. Faraday Trans. 1*, **71**, 1659 (1975). Copyright © 1975 Royal Society of Chemistry.

solution and the infinite time resistance ratio is determined. Due to slight differences in the lengths of the capillaries and the small asymmetry in the position of the central electrode, this is not exactly unity.

The theory for this method was given by Agar and Lobo [137]. The experiments typically take 1–2 days as the solution to the diffusion equation is a Fourier series for which convergence is slow. The precision depends very much on a proper choice of the rate of flow of solution across the mouths of the capillaries. With careful attention to detail, however, Lobo and Teixeira [138] obtained a precision of ±0.02–0.05% for dilute aqueous HCl solutions. The absolute accuracy indicated by comparison with results from the Harned conductance method and the complete Onsager–Fuoss theory for electrolyte diffusion [139] is 0.2–0.4%.

The theory of the open-ended capillary method has been extended to ternary systems by Toukubo and Nakanishi [140]. Two experiments are required to give the four diffusion coefficients.

4.2.3 Closed Capillary Method

Passiniemi and co-workers [141, 142] revived the closed capillary technique for tracer diffusion in aqueous electrolytes. The cell, a uniform bore capillary 40 mm long and with an internal diameter of 0.87 mm, is half-filled with the tracer-labeled solution by means of a micropipet and then topped up with unlabeled material. It is closed with a screw in the capillary mounting and inserted in a well in a plastic scintillator for β-emitting tracers (e.g., $^{36}Cl^-$) or in a lead shielded cavity above the scintillator in the case of γ-emitting tracers (e.g., $^{22}Na^+$). The capillary and scintillator lay in an air bath surrounded by a constant temperature water bath. The integrated counting rate was determined over a period of days with corrections for background, the tracer half-life and the dead time of the counting apparatus.

A general theory for a closed diffusion cell of any geometry with a detector enveloping the cell or a short distance away, measuring a time dependent integrated signal has been given [143]. The concentration distribution function, which is a solution to Fick's law, is a complicated series with terms that are the products of trigonometric, exponential, and Bessel functions. The working equation for the radiation detector signal for one-dimensional tracer diffusion is [141]

$$I(t) = I_\infty + \sum_{m=1}^{\infty} I_m \exp(-m^2\pi^2 Dt/l^2) \tag{81}$$

When the initial distribution is such that exactly one-half of the capillary contains labeled material and one-half, unlabeled, the even coefficients are 0. After sufficient time, only the first term of (81) is required, and I_∞ can be eliminated by taking the difference $I(t) - I(t+\Delta t)$, where Δt is a fixed interval.

Passiniemi [142] made a comparison of data for $^{22}Na^+$ and $^{36}Cl^-$ diffusing in aqueous NaCl solutions in the range 10^{-5}–1.5 mol/L at 25°C with results of other workers obtained with the diaphragm cell and the open-ended capillary. The precision is about $\pm 0.3\%$, and accuracy is about $\pm 1\%$.

4.3 Steady-State Diffusion—The Diaphragm Cell

Fick's equations describe the decay of concentration gradients by pure diffusion. Other mixing due to thermal or mechanical disturbances must be eliminated as far as is possible. One way of doing this is to confine the flow to narrow channels, such as uniform bore capillaries [1, 6] mentioned earlier, or those to be found in thixotropic gels [144], Millipore filters [145, 146] and glass [147], or metal [148] sintered frits.

The diaphragm cell design of Stokes [149] is based on the sintered frit and is the standard for this type of work. The reader is again referred to [1] for details of the theory and operational technique for binary, ternary, and tracer diffusion, as well as to the detailed account of Mills and Woolf [148].[†]

The technique has remained a popular one, with many modifications for specialized purposes. In addition, it has been extended to pressures as high as 400 MPa in the studies of Woolf and co-workers [150, 151] and to quaternary diffusion [152].

In the ordinary diaphragm cell measurement of binary (mutual) diffusion, the experiment is started with solutions of different composition in the two cell compartments and these are analyzed after the elapse of a suitable time [1]. The diffusion coefficient is not normally independent of concentration; and as the concentration differences used are quite large, the result of the experiment is an integral diffusion coefficient $\bar{D}$, which is a complicated average over concentration and time of the required differential mutual diffusion coefficient D. Thus,

$$\bar{D} = \frac{1}{t} \int_0^t \bar{D}(t)\, dt$$

$$= \frac{1}{t} \int_0^t \left[\frac{1}{\Delta C(t)} \int_{C_T(t)}^{C_B(t)} D(C)\, dC \right] dt \tag{82}$$

where C_B is the time dependent concentration in the bottom compartment, C_T is that in the top, and ΔC is the difference between these concentrations.

To an approximation sufficient in most cases [153],

$$\bar{D} = \frac{1}{\Delta C} \int_{\bar{C}_T}^{\bar{C}_B} D\, dC \tag{83}$$

where the bars indicate the mean of the initial and final values. The mutual diffusion coefficient can be obtained by fitting integral values to a suitable function of concentration, followed by differentiation. Alternatively, the differentiation can be done numerically. The value of $\bar{D}$ is found from

$$\bar{D} = -\frac{1}{\beta t} \ln \left[\frac{\Delta C(t)}{\Delta C(0)} \right] \tag{84}$$

where β is the cell constant, found by calibration, usually with aqueous KCl solutions.

The effect of the differentiation is to obtain the differential coefficient by

---

[†]Available from the Atomic and Molecular Physics Laboratory, Research School of Physical Sciences, Australian National University, Canberra, ACT 2600, Australia.

addition or subtraction of a correction term to or from the integral coefficient. The accuracy of this procedure depends very much on the strength of the concentration dependence of D. Firth and Tyrell [154] found that, in 3 mol/L aqueous silver nitrate, the correction term amounted to some 25% of the integral coefficient; and, in consequence, the precision of their result was somewhat less than that of the experimental data from which it was derived. This problem does not arise, however, in tracer diffusion experiments where ΔC is very small.

Baird and Frieden [155] attempted to give a more rigorous theory of the diaphragm cell, which avoids the approximations inherent in the use of Gordon's integral diffusion coefficients. Their solution, applicable to the case of equal compartment volumes, and no volume change on mixing, is given in a series form,

$$t = A_0 + A_1 \ln(\Delta C) + A_2(\Delta C)^2 + A_4(\Delta C)^4 + \cdots \tag{85}$$

where ΔC is the concentration difference determined at the end of the experiment and t is time. The coefficient A_1 determines the differential mutual diffusion coefficient:

$$D(\bar{C}) = -\frac{1}{\beta A_1} \tag{86}$$

Use of this equation still requires more than one experiment; at least three are required at each mean composition.

It has long been recognized that if the compartment concentrations can be determined during the experiment, then the differential diffusion coefficient can be obtained directly from a single experiment at a given mean composition. This could be done by use of (83) or by the fitting procedure of Baird and Frieden [155].

One successful design for this purpose is that of Collings and co-workers [156] used for aqueous electrolyte solutions. The composition of the top compartment is determined at intervals during the course of the experiment by drawing 200 μL of this solution into a capillary tube attached to the side of the cell. This tube acts as both conductance cell and syringe, with one of the pair of electrodes acting as the plunger. This electrode was attached to a vernier calliper and a dial gauge, so that the conductivity cell constant could be set to suitable, previously determined values. The concentrations of the capillary tube contents were determined from known conductivity data. The plunger action seemed not to disturb the diffusive flow, and results were obtained consistent to better than 0.5% with the best literature data for aqueous KCl and NaCl solutions.

Reference [156] is also of interest for the design of a simple but effective stopper for diaphragm cells, illustrated in Figure 3.15. It seals tightly, without exerting undue pressure on the joint, and can be flushed easily at the back with a syringe, so that traces of the initial solutions do not contaminate those withdrawn for analysis (when the traditional diaphragm cell method is used).

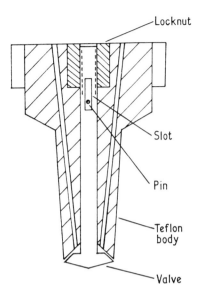

Figure 3.15 Design for diaphragm cell stoppers. Reprinted with permission from A. F. Collings, D. C. Hall, R. Mills, and L. A. Woolf, *J. Phys. E*, **4**, 425 (1971). Copyright © 1971 IOP Publishing Ltd.

4.3.1 High-Pressure Diaphragm Cells

The Canberra school also developed the diaphragm cell for tracer diffusion studies at pressures to 400 MPa [150, 157–159]. The cell is of stainless steel construction, with a platinum or stainless steel sintered diaphragm and with a bellows as the wall of the lower compartment (Figure 3.16). It is contained in a capped beryllium–copper or titanium alloy pressure vessel, and the bellows allows the cell contents to be compressed by an external hand pump without contamination by the hydraulic fluid. The top compartment is partly constructed of removable cylindrical rings, sealed by captured O-rings, so that its volume can be adjusted to match that of the bottom at the operating pressure. The contents are stirred magnetically by double-bladed stirrers, one blade sweeping the diaphragm surface, the other the cell compartment proper. The cell is loaded and unloaded at atmospheric pressure. As compression drives a certain amount of solution from the bottom to the top compartment, the initial condition is the solvent-filled diaphragm [1] and the cell constant is

$$\beta = \frac{A}{l}\left[\frac{1}{V_T} + \frac{1}{V_B}\right]\left[1 - \frac{V_D}{3(V_T + V_B)}\right] \tag{87}$$

where A/l is the effective area to channel length ratio for the diaphragm, obtained by calibration at atmospheric pressure, and V_T, V_B, and V_D are the compartment and diaphragm volumes. The value of V_B is found for a given run from the known V_T, V_D, and the compressibility of the solution, so an equation of

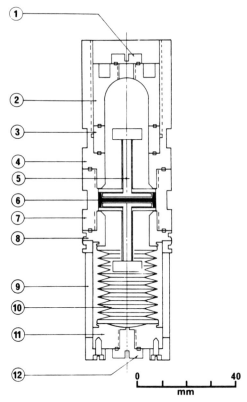

Figure 3.16 High-pressure diaphragm cell: 1, top compartment stopper; 2, top compartment end piece; 3, volume adjustment ring; 4, top compartment body; 5, magnetic stirrer; 6, stainless steel or platinum diaphragm; 7, support ring; 8, 9, and 10, bellows, support and guide; 11 and 12, bellows end piece and bottom stopper (NB the screws shown are released during an experiment; these hold the bellows extended during sampling at its end). Reprinted with permission from L. A. Woolf, *J. Chem. Soc. Faraday Trans.* 1, **71**, 784 (1975). Copyright © 1975 Royal Society of Chemistry.

state is a prerequisite for this work. Decompression at the end of the run drives solution from top to bottom, but a correction for this can be made [6, 157], taking into account the expansion of the liquid in the diaphragm; expansion in the top compartment forcing liquid into the diaphragm; and, for the release of the higher pressures, flow of liquid through the diaphragm. The precision is from ± 2–4%, depending on the pressure and the compressibility. Results for the self-diffusion of benzene have been checked for consistency against NMR spin–echo results of ± 1–2% accuracy [160, 161].

Another type of cell, used to only 4 MPa, was described by Kircher and co-workers [162]. This cell uses flexible membranes, attached to both compartments, for the transmission of pressure, which is more cumbersome than the single bellows apparatus. The range of temperature covered by this apparatus was quite large, from 25 to 180°C; but the systems tested were water–alcohol mixtures, a rather unfortunate choice as there is a large volume of mixing effect to be taken into account. The precision claimed was $\pm 3\%$.

4.3.2 Other Special Purpose Diaphragm Cells

Diaphragm cells are normally used in water or oil containing thermostat baths and, due to the need for manual filling and emptying, are rarely operated above 50°C. Attempts were made to extend this range [163, 164], but the cells that were developed had large attached dead volumes and were difficult to manipulate. Easteal and co-workers [165] designed a high-temperature cell where the bottom stopper is replaced by a sidearm tube (Figure 3.17). This can be closed at its base by an elongated stopper into which narrow capillaries for flushing are inserted. Therefore, the bottom stirrer is incorporated into its compartment during manufacture, and its density must be chosen carefully to match that of the solutions to be studied. The cell is manipulated while mounted on a cradle placed in an air bath maintained at the operating temperature.

The cell is used for tracer diffusion and is filled completely through the top compartment by gentle pressure from a solution reservoir or by suction from the sidearm. After insertion of the radiotracer, top stirrer, and stopper, any bubbles

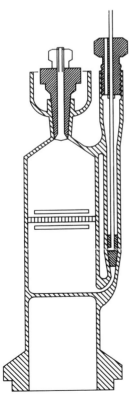

Figure 3.17 Diaphragm cell for high-temperature experiments. Reprinted by courtesy of Dr. W. E. Price, University of Wollongong, NSW, Australia.

remaining in the lower compartment are removed by manipulation of the cell so that they lodge at the capillary opening and are then withdrawn by application of pressure or suction. The bottom stopper is then inserted, the sidearm flushed, the cell placed in its thermostat, and stirring begun. On completion of the run, the top compartment sample is withdrawn in the normal manner [1, 6, 148], the sidearm stopper is loosened, and the solution is withdrawn from the bottom compartment through the diaphragm. This mixture of top, diaphragm, and bottom solutions is removed completely with a syringe and needle and then discarded. The bottom compartment is then sampled by the sidearm tube.

Diaphragm cells have also been designed for operation with air excluded from the system. Ahn and Derlacki [166] used such a cell to study, in a number of solvents, the tracer diffusion of vanadyl acetylacetonate, a substance that is unstable in the presence of dissolved oxygen.

4.3.3   Multicomponent Diffusion

The use of diaphragm cells for the study of ternary diffusion was outlined in [1]. Where the concentration dependence of the D_{ij} is unfavorable, or where only a single physical property of the solutions is determined, the precision of this method is perhaps 10–100 times worse than that of the optical free diffusion methods [119, 121, 167], but the diaphragm cell is simple to set up and its operation is straightforward, provided some care is taken to avoid bubble formation, to ensure smooth and consistent stirring, and to obtain precise analyses of the solutions.

Cullinan and co-workers [152, 168] extended this technique to encompass quaternary systems and gave a procedure for determination of the optimum time of the experiment. A minimum of three experiments at the same mean composition, but with different initial concentration gradients, is required to determine the 3×3 diffusion coefficient matrix. The precision is improved, however, by overdetermination. Rai and Cullinan [168] repeated each experiment on the system acetone–benzene–carbon tetrachloride–n–hexane five times, giving 15 in all. Leist [169] measured the change in each ΔC for the system H_2O–HCl–NaI as a function of time using each of three sets of initial gradients. Values of $\Delta C(t)$ obtained from eight such runs per set were fitted to a quadratic by the method of least squares; interpolated values were obtained at a common value of βt; and the diffusion coefficient matrix was calculated from these.

4.4   The Taylor Dispersion Method

This technique is known by a variety of names, including *chromatographic peak broadening* [170], *perturbation chromatography* [171], or *Taylor dispersion* [172]. *Chromatography* is not, of course, a term apt for a diffusion process, but its use derives from the observation of diffusive peak broadening and splitting in analytical chromatography and the common usage of commercial gas–liquid chromatography (GLC) and high-performance liquid chromatography (HPLC) instrumentation in this type of work.

The method is based on the dispersion of a small volume of fluid of one composition in another of different composition in laminar flow within a straight tube of uniform circular cross section. It was originally applied to gaseous systems independently by Giddings and Seager [173] and Bohemen and Purnell [174]. Although dispersion due to diffusion was first observed by Griffiths in 1911 [175], a mathematical theory was not developed until the 1950s when Taylor [176] and Aris [177] published their treatments. The necessary and sufficient conditions for the measurement of mutual diffusion coefficients were further developed by others [178–182]. Application to tracer diffusion in dense gases at high pressure [170, 171, 183], to mutual diffusion in liquids [184–186], and to self-diffusion using deuterated [187] and radiotracers [188] has followed. The theory and practice for ternary systems has also been examined [189, 190]. Figures 3.18 and 3.19 are schematic diagrams for atmospheric pressure [188] and high pressure (supercritical fluids) [191], respectively.

Recently, Leist has devised methods for the direct evaluation of ternary and

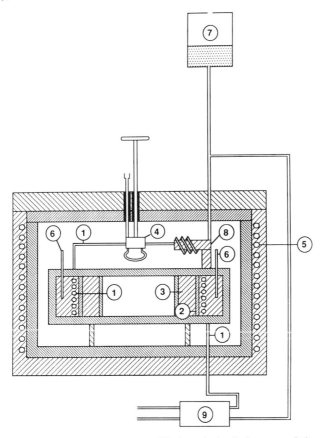

Figure 3.18   Peak-broadening apparatus: 1–3, diffusion tube in air thermostat 5; 4, sample injector; 6, thermometer; 7, solution reservoir; 8, equilibration coil; 9, detector. Reprinted with permission from W. A. Wakeham, *Faraday Symp. Chem. Soc.*, **15**, 145 (1980). Copyright © 1980 Royal Society of Chemistry.

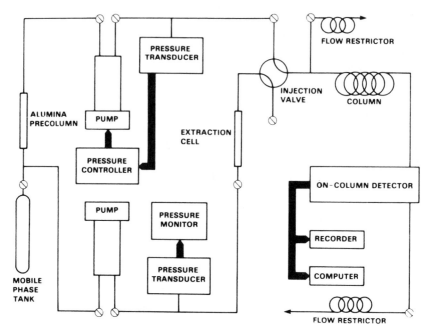

Figure 3.19 Peak-broadening apparatus for supercritical fluids. Reprinted with permission from M. Roth, J. L. Steger, and M. V. Novotny, *J. Phys. Chem.*, **91**, 1645 (1987). Copyright © 1987 American Chemical Society.

quaternary diffusion coefficients from Taylor dispersion methods [410–412]. It has also been demonstrated that least squares fits of the peak concentration profiles are a better means of data analysis than the direct calculation of moments by integration [411, 413].

The early work was summarized by Tyrrell and co-workers [6, 192]. The gaseous diffusion data of the Giddings school [193] seem to be in error as they report that D_{12} increases with increasing mole fraction of the lighter component in some systems, contrary to the predictions of the well-tested Chapman–Enskog theory for dilute gases and to experimental results [194] from the Loschmidt and two-bulb cells described in the previous section of this chapter.

The original theory of Taylor [176] and Aris [177] is outlined as follows. A fluid mixture, solute concentration C', flows through a straight tube of uniform cross section with velocity $u(r)$ in the laboratory reference frame, where r is the radial distance perpendicular to the axis of the tube and a is the radius. If d is r/a, and the mean velocity is $\bar{u}$, one-half $u(0)$, then for laminar flow the velocity profile is parabolic

$$u(r) = 2\bar{u}(1 - d^2) \tag{88}$$

At zero time, a small pulse of fluid of concentration C'', perhaps 20 μL in a tube volume of 10 mL, is injected into the flowing stream. This is dispersed along the axial coordinate x. The flow velocity in the frame moving with the mean velocity $\bar{u}$ is then

$$v(r) = u(r) - \bar{u} = \bar{u}(1 - 2d^2) \tag{89}$$

and the flux in this frame is the sum of diffusive and convective terms

$$J_x = -D_{12}\left(\frac{\partial C}{\partial x}\right) + C\bar{u}(1 - 2d^2) \tag{90}$$

The radial flux is given by

$$J_r = -D_{12}\left(\frac{\partial C}{\partial r}\right)$$

$$= -\frac{D_{12}}{a}\left(\frac{\partial C}{\partial d}\right) \tag{91}$$

For this system, Fick's second law (22) becomes

$$\frac{\partial^2 C}{\partial d^2} + \frac{1}{d}\left(\frac{\partial C}{\partial d}\right) + a^2\left(\frac{\partial^2 C}{\partial x^2}\right) = \frac{a^2}{D_{12}}\left[\left(\frac{\partial C}{\partial t}\right)_x + \bar{u}(1 - 2d^2)\frac{\partial C}{\partial x}\right] \tag{92}$$

The term $\partial^2 C/\partial x^2$ can be neglected provided that [176]

$$l/\bar{u} \gg 2a^2/14.4D_{12} \tag{93}$$

which is the first of several conditions to be imposed on the experiment. Equation (92) then reduces to

$$\frac{\partial^2 C}{\partial d^2} + \frac{1}{d}\left(\frac{\partial C}{\partial d}\right) = \frac{a^2}{D_{12}}\left[\left(\frac{\partial C}{\partial t}\right)_x + (1 - 2d^2)\frac{\partial C}{\partial x}\right] \tag{94}$$

The ideal detector determines a mean concentration at a selected value of z in the mean velocity reference frame

$$\bar{C} = 2\int_0^1 Cd\,d\,d \tag{95}$$

Then (94) simplifies [176, 177] to

$$\frac{\partial \bar{C}}{\partial t} = K\left(\frac{\partial^2 \bar{C}}{\partial x^2}\right) \tag{96}$$

where the dispersion coefficient K is defined by

$$K = D_{12} + a^2\bar{u}^2/48D_{12} \tag{97}$$

For the initial condition of a uniform δ pulse of injected solution at $t = 0$, $z = 0$, and the boundary conditions of an infinitely long tube with

$$\frac{\partial C}{\partial r} = 0 \tag{98}$$

at $r = a$, (96) has the solution

$$\bar{C} = \frac{n}{\pi a_0^2 (4\pi K t)^{1/2}} \exp(-x^2/4Kt) \tag{99}$$

where n is the amount of solute injected in the pulse (mol). This is a Gaussian of variance $(2Kt)$. If the time taken for the moving reference frame to traverse a length l of the tube to the position of the laboratory detector is $\bar{t}$, then x is $\bar{u}(t - \bar{t})$, and

$$\bar{C} = \frac{n}{\pi a_0^2 \bar{u} (4\pi K t/\bar{u}^2)^{1/2}} \exp\left[-\frac{(t - \bar{t})}{2(2Kt/\bar{u}^2)} \right] \tag{100}$$

where the variance σ^2 is $2Kt/\bar{u}^2$.

If it can be assumed that no dispersion occurs in the detector (a limitation in the measurement of small D_{12}), then the variance is

$$\sigma^2 = 2K\bar{t}/\bar{u}^2 \tag{101}$$

or, using (97)

$$\sigma^2 = \frac{2D_{12}\bar{t}^3}{l^2} + \frac{a^2\bar{t}}{24D_{12}} \tag{102}$$

The term $\sigma^2 \bar{u}^2/l$ is sometimes written as H, the plate height of the diffusion column, by chromatographers, and (102) can be rewritten as

$$H = \frac{2D_{12}}{\bar{u}} + \frac{a^2\bar{u}}{24D_{12}} \tag{103}$$

For gaseous systems, where $D \simeq 1$ cm^2/s, only the first terms of (102) and (103) are significant, and for liquids, with $D \simeq 10^{-5}$ cm^2/s, only the second terms are significant. However, many workers prefer to solve (103) for D_{12}, which yields

$$D_{12} = \frac{H \pm \sqrt{H^2 - a^2/3}}{4} \tag{104}$$

the choice of the positive or negative root being dependent on the range of $\bar{u}$

used. Giddings and co-workers [170, 173, 193] determined H from the variance observed with tubes of different lengths

$$H = \frac{(l_2 - l_1)(\sigma_2^2 - \sigma_1^2)}{(\bar{t}_2 - \bar{t}_1)^2} \tag{105}$$

to correct for instrument *dead volume* and possible dispersion by the detector. For liquid systems, where the first term of (103) can be neglected, others [191] calculated H directly from σ^2 and obtained D_{12} from the reciprocal slope of a plot of H against $\bar{u}$.

Wakeham and co-workers [172, 186–188] treated the problem a little differently by using solutions to Fick's law for the spatial moments of the concentration distribution. The amount of solute injected into the flowing stream is written in terms of the δ function

$$n = \pi a^2 \int_{-\infty}^{\infty} \delta(x) \Delta C \, dx \tag{106}$$

where ΔC is $(C'' - C')$. The dispersion equation is written

$$\nabla^2(\Delta C_1) = \frac{1}{D_{12}} \left[\frac{\partial(\Delta C_1)}{\partial t} + 2\bar{u}(1 - 2d^2) \frac{\partial(\Delta C_1)}{\partial x} \right] \tag{107}$$

where ∇^2 is the three-dimensional operator in cylindrical coordinates. Therefore, this equation differs from (92) in retaining angular terms.

Spatial moments of C are defined by

$$C_p(r, \theta, t) = a^{p+1} \int_{-\infty}^{\infty} \xi^p \Delta C_1(\xi, r, \theta, t) d\xi \tag{108a}$$

where

$$\xi = (z - \bar{u}t)/a \tag{108b}$$

cross-sectional average moments by

$$m_p = \frac{1}{\pi a^2} \int_0^{2\pi} \int_0^a r C_p(r, \theta, t) dr \, d\theta \tag{109}$$

and finally the normalized moments by

$$\mu'_p = m_p/m_0 \tag{110}$$

As mass is conserved in the system, m_0 is constant; m_1 is 0, as the mean of the distribution travels with the average velocity of the flow. The normalized second moment obtained by solution of (107) [172, 177] is given in terms of D_{12} by

$$\mu'_2 = 2\left(D_{12} + \frac{\bar{u}^2 a^2}{48D_{12}}\right)t - 128\left(\frac{\bar{u}^2 a^4}{D_{12}}\right)\sum_{n=1}^{\infty} \alpha_{0n}^2[1 - \exp(-\alpha_{0n}^2 D_{12}t/a^2)]$$ (111)

where α_{0n} is the nth zero of the differential of the zeroth-order Bessel function J_0.

This complex expression approaches a normal distribution as $t \to \infty$, and provided

$$\bar{u} > 700D_{12}/a$$ (112)

and

$$D_{12}t/a^2 > 700$$ (113)

it can be reduced to

$$\mu'_2 = \left(\frac{\bar{u}^2 a^2}{24D_{12}}\right)t$$ (114)

In practice, concentrations are measured as a function of time at the detector, not as functions of axial position x. If the first *temporal* moment at $x = l$ is defined as

$$\bar{t} = \int_0^{\infty} t\Delta C(l, t)\, dt \bigg/ \int_0^{\infty} \Delta C(l, t)\, dt$$ (115)

and the variance σ^2 is

$$\sigma^2 = \int_0^{\infty} (t - \bar{t})^2 \Delta C(l, t)\, dt \bigg/ \int_0^{\infty} \Delta C(l, t)\, dt$$ (116)

then the mutual diffusion coefficient, in terms of these quantities, is given by

$$D_{12} = \frac{a^2}{48t}\left[\frac{(1 + 4k^2)^{1/2} + 3}{(1 + 4k^2)^{1/2} + 2k^2 - 1}\right][1 + (1 - \delta_a)^{1/2}]$$ (117)

where $\delta_a = 0.26666\bar{u}^2 a^2/lD_{12}$ and $k = \sigma/\bar{t}$.

There are corrections to be made [172] for the finite length of the detector, that of the injected pulse, and any difference in radius of the diffusion tube and tubing connecting the detector. These corrections, δt and $\delta\sigma^2$, are added to $\bar{t}$ and σ^2:

$$\bar{t} = \bar{t}_{\text{expt}} + \sum_i \delta \bar{t}_i \tag{118a}$$

$$\sigma^2 = \sigma^2_{\text{expt}} + \sum_i \delta \sigma^2_i \tag{118b}$$

The formulas for these corrections are summarized in Table 3.5, which also contains formulas for the several conditions required in the experiment.

Although this method is relatively rapid, the precision is only moderate, $\pm 1\%$ at best.

4.5 Light Scattering

It has long been standard practice to determine the size and shape of particles in solution by means of the intensity and angular dependence of the scattering of light. With the development of lasers, it has become possible to examine the spectrum of the scattered radiation more readily. This contains information about fluctuations in molecular orientations, number density, and other properties. The coupling with number density, that is, concentration, leads to information about diffusion processes.

Laser light scattering is an extensive subject with a wide literature [195–198], but the basic theory and its relationship to the time correlation function formalism for transport processes are outlined in an excellent text [199] and its application to diffusion in liquids is reviewed [6, 192].

The theory was developed originally for independent scatterers dispersed through the fluid [6, 192, 199]. Experimenters have used poly(styrene) latex spheres [200], macromolecules [196], and emulsions [201] as such scatterers in the examination of the Brownian motion of these large particles. Mixtures of spheres of different refractive indexes have even been used to determine tracer diffusion in model colloidal systems [202].

The question whether the light scattering diffusion coefficient is in fact equal to that given by the application of classical diffusion techniques to such systems has caused some confusion and controversy. Yoshida [203] argued that the concentration fluctuations of the scatterers occurred at constant temperature and chemical potential of the solvent and not at constant temperature and pressure, the normal conditions for a mutual diffusion coefficient measurement. Phillies [204] presented a counterargument based on the nonequilibrium thermodynamics descriptions of Kirkwood and Goldberg [10, 205] and the scattering theory of Mountain and Deutch [206] and this has been accepted [207].

This same scattering theory of Mountain and Deutch was applied to binary mixtures of small molecules (nonelectrolyte solutions) by several groups [208–213] as well as to aqueous electrolyte solutions [214]. Here both components scatter light and the result is an effective mutual diffusion coefficient whose exact

Table 3.5 Formulas and Conditions for Taylor Dispersion Experiments

Conditions

Mathematical: $\bar{u} > 700 D_{12}/a$

$$D_{12}\bar{t}/a^2 > 10$$

Neglect of $\partial^2 C/\partial x^2$: $\dfrac{1}{\bar{u}} \gg 2a^2/14.4 D_{12}$

Laminar flow: $2a\rho\bar{u}/\eta < 2000$

Coiling of diffusion tube, radius R: $R/a > 100$

$$a_0 R\bar{u}^2 \rho/\eta D_{12} < 5$$

where ρ is the density, η is the shear viscosity, and the other symbols are defined in the text.

Correction Formulas

Sample introduction, loop volume V: $\delta t = -(V/2\pi a^2 \bar{u})$

$$\delta\sigma^2 = -(\delta t)^2/3$$

Detector, length δl: $\delta t = -\delta l/2\bar{u}$

$$\delta\sigma^2 = -\left(\frac{l}{\bar{u}}\right)\left[\zeta\left(\frac{\delta l}{l}\right) + \frac{1}{12}\left(\frac{\delta l}{l}\right)^2\right]$$

where

$$\zeta = \frac{\bar{u}a_0^2}{48 D_{12} l}$$

Detector, with perfect mixing, volume V: $\delta t = \left(\dfrac{l}{\bar{u}}\right)(3\zeta - V/\pi a^2 l)$

$$\delta\sigma^2 = \left(\frac{l}{\bar{u}}\right)[13\zeta - (V/\pi a^2 l)^2 - 2\zeta(V/\pi a^2 l)]$$

Correction tubing for detector, length l', radius a':

$$\delta t = (l'/\bar{u})(a'/a)^2\{1 + (a/a')^2(l/l')\zeta[1 + (a'/a)^2]\}$$
$$\delta\sigma^2 = (la'/\bar{u}a)^2\{2(l/l')\zeta + (l/l')^2\zeta^2[3(a'/a)^2 + 2]\}$$

relationship to the classical mutual diffusion coefficient is as yet unclear. Pecora and co-workers [210] and Lucas and co-workers [211] seem to have concluded that the two coefficients are the same on the basis that the composition dependence in certain systems is similar to that obtained by classical techniques. Unfortunately, this comparison was made for data sets obtained at different temperatures in the range 19–27°C and not for systems where accurate Gouy or Rayleigh interferometric results are available. The best that can be said is that the light scattering results of different groups are consistent with one another for the systems acetone–CS_2 [210–212], methanol–benzene, n-hexane–benzene, acetone–benzene [211, 212], and benzene–nitromethane [210, 212], within the 1–5% error claimed.

Notwithstanding these objections, this mutual diffusion method and the NMR spin–echo self-diffusion technique share the advantage of not requiring macroscopic concentration gradients. Therefore, it can be used to examine difficult systems such as gas-liquid solutions under pressure [215].

It should be noted that theoretical relationships are available for the example of scatterers of the same size, interacting with the same potential, but with different scattering amplitudes [202]. In this case the structure factor contains both mutual and self-diffusion terms. These have been separated in experiments with mixtures of silica suspensions where the refractive index of the majority of the particles was matched with that of the solvent. Thus, only the unmatched particles scattered light and tracer diffusion coefficients were obtained. Other experiments with uniform solute particles yielded the mutual diffusion coefficient.

The intensity of the light scattered from a binary solution is dependent on fluctuations in the permittivity tensor of the solution, and can be expressed in terms of the Fourier transform of the space–time autocorrelation function for this quantity. The response of the system to deviations from equilibrium was calculated by Mountain and Deutch [206] from linearized hydrodynamic equations to determine the modes by which the system returns to equilibrium and their amplitudes, and from thermodynamic fluctuation theory for temperature, pressure, and concentration fluctuations.

The spectrum consists of a Rayleigh peak centered on the frequency of the incident beam and two Brillouin peaks on either side. For one component systems the central Rayleigh peak can be used to obtain information about self-diffusion [216, 217], or, for symmetric top molecules [217], collective rotational diffusion; and the total spectrum yields information on the frequency dependence of viscoelastic processes [217].

For a binary system, the Rayleigh peak contains the combined effect of temperature and concentration fluctuations, and cross terms are required for a full thermodynamic description. Mountain and Deutch [206] considered only the conditions for which the two effects are separable. The Brillouin lines are related to fluctuations in the pressure of the system, that is, to pressure diffusion (see [10]).

The intensity of the scattered light is a function of the distance l of the sample origin to the point of observation, the change k in the wave vector, and the change ω in the frequency of the scattered light

$$I(l,\, k,\, \omega) = I_0 \left(\frac{Lk_0^4}{32\pi^3 l^2}\right) \sin^2\phi S(k,\, \omega) \tag{119}$$

where ϕ is the angle between the electric vector of the incident light and ω, and the magnitude of k is given in terms of the solution refractive index n and the scattering angle θ by

$$k = 2nk_0 \sin(\theta/2) \tag{120}$$

The structure factor $S(k,\omega)$ is written in terms of the fluctuations of the permittivity ε as

$$S(k,\, \omega) = 2\,\mathrm{Re}\int_0^\infty \int\int \langle \delta\varepsilon(r+r',\,t)\delta\varepsilon(r',0)\rangle\, \exp[i(k\cdot r)-\omega t]\,dr\,dr'\,dt \tag{121}$$

where $\delta\varepsilon(r,t)$ is the fluctuation in the local permittivity at point r at time t; Re indicates the real part of the function.

The fluctuation $\delta\varepsilon$ is expressed in terms of partial differentials in the thermodynamic variables T, p, and C as

$$\delta\varepsilon(r,\, t) = \left(\frac{\partial\varepsilon}{\partial p}\right)_{T,C} \delta p(r,\, t) + \left(\frac{\partial\varepsilon}{\partial T}\right)_{p,C} \delta T(r,\, t) + \left(\frac{\partial\varepsilon}{\partial C}\right)_{T,p} \delta C(r,\, t) \tag{122}$$

The *time* dependence of these fluctuations in T, p, and C is given by the following linearized hydrodynamic equations:

$$\frac{\partial\rho}{\partial t} + \rho_0 \,\mathrm{div}\,v = 0 \tag{123}$$

$$\rho_0(\partial v/\partial t) = -\,\mathrm{grad}\,p + \eta_s\nabla^2 v + \tfrac{1}{3}(\eta_s+\eta_v)\,\mathrm{grad}\,\mathrm{div}\,v \tag{124}$$

$$\frac{\partial C}{\partial t} = D\left(\nabla^2 C + \left(\frac{k_T}{T_0}\right)\nabla^2 T + \left(\frac{k_p}{p_0}\right)\nabla^2 p\right) \tag{125}$$

and

$$\frac{\lambda}{\rho_0}\nabla^2 T = C_p\frac{\partial T}{\partial t} - k_T\left(\frac{\partial\mu}{\partial C}\right)_{p,T}\frac{\partial C}{\partial t} + T_0\left(\frac{\partial S}{\partial p}\right)_{T,C}\frac{\partial p}{\partial t} \tag{126}$$

where subscript 0 indicates the equilibrium value; ρ is the density; v is the velocity in the mass-fixed frame; η_s and η_v are the shear and bulk viscosities, respectively; C_p is the heat capacity at constant pressure; μ is the total chemical potential of the mixture; λ is the thermal conductivity; k_T is the ratio of the thermal diffusion and mutual diffusion coefficients D_T/D; and

$$k_p = -\frac{(p_0/\rho_0^2)(\partial\rho/\partial C)_{p,T}}{(\partial\mu/\partial C)_{p,T}} \tag{127}$$

Mountain and Deutch [206] next select three statistically independent variables in terms of which the probability w of a fluctuation can be expressed by the Boltzmann equation

$$\Delta S = k \ln w = -\frac{1}{2T_0}\left[\frac{C_p}{T_0}(\delta\phi)^2 + \frac{\beta_S}{\rho_0}(\delta p)^2 + \left(\frac{\partial\mu}{\partial C}\right)_{T,p}(\delta C)^2\right] \tag{128}$$

where the new variable

$$\phi = T - \left(\frac{T_0\alpha_T}{C_p\rho_0}\right)p \tag{129}$$

has been introduced; α_T is the isothermal expansion coefficient and β_S is the adiabatic compressibility. They then write (122) in terms of spatial Fourier transforms as

$$\varepsilon(\boldsymbol{k},\,t) = \left(\frac{\partial\varepsilon}{\partial p}\right)_{\phi,c}p(\boldsymbol{k},\,t) + \left(\frac{\partial\varepsilon}{\partial\phi}\right)_{p,C}\phi(\boldsymbol{k},\,t) + \left(\frac{\partial\varepsilon}{\partial C}\right)_{\phi,p}C(\boldsymbol{k},t) \tag{130}$$

and the equations (123)–(126) are also rewritten in terms of the new set of variables. After some lengthy manipulation, an expression for the structure factor $S(\boldsymbol{k},\omega)$ is obtained.

The central Rayleigh peak consists of the superposition of two Lorentzian functions whose amplitudes depend on many parameters due to the thermodynamic coupling between diffusion and heat flow. However, the expression for $S(\boldsymbol{k},\omega)$ can be simplified under certain conditions.

For instance, as the solute concentration C becomes small, k_T tends to zero. Thus, in dilute solutions, the central Rayleigh line should be resolvable into heat conduction and diffusion components. In the favorable case that conduction occurs more rapidly than diffusion, which is usual, that is, the *thermal diffusivity*

$$\chi = \lambda/\rho_0 C_p \tag{131}$$

(not to be confused with the thermal diffusion coefficient) is much greater than the diffusion coefficient, then the observed Rayleigh peak is essentially due to diffusion alone. However, the assumption about k_T remains and the effect of nonzero values has not usually been considered.

With these restrictions the structure factor becomes

$$S(\boldsymbol{k}, \omega) = \left(\frac{\partial\varepsilon}{\partial C}\right)^2_{p,T} \frac{kT_0}{(\partial\mu/\partial C)_{p,T}} \left[\frac{2Dk^2}{(Dk^2)^2 + \omega^2}\right] \qquad (132)$$

and D can be determined by plotting the half-width of the Lorentzian against k^2, k being given by (120) [208].

These half-widths however, are very small (about 600 kHz for $\lambda \simeq 500$ nm, $D \simeq 10^{-5}$ cm^2/s, and a refractive index of 1), and the usual experimental method is that of photon correlation spectroscopy [218].

This method, heterodyne optical mixing, relies on the measurement of the *intensity* autocorrelation function

$$g^{(2)}(\tau) = \langle I(k, 0)I(k, \tau)\rangle/\langle I\rangle^2 \qquad (133)$$

where $I(0)$ is an intensity at a given zero time and $I(\tau)$ is that determined after an interval τ. In practice, the light signal scattered by the sample mixes with stray light reflected from the walls of the sample cell. A schematic diagram of the apparatus of McKeigue and Gulari [213] is shown in Figure 3.20. They used 514.5-nm light from an argon ion laser with low scattering angles of 4–18°, and passed the signal from an ITT FW-130 photomultiplier through an amplifier

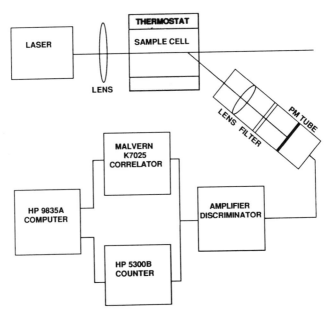

Figure 3.20 Light-scattering apparatus. Reprinted with permission from K. McKeigue and E. Gulari, *J. Phys. Chem.*, **88**, 3472 (1984). Copyright © 1984 American Chemical Society.

and discriminator to a 64-channel Malvern K7025 correlator and a digital counter, both interfaced to a small computer.

The term $g^{(2)}$ is related to the field autocorrelation function $g^{(1)}$, which is the Fourier transform of the structure factor $S(k, \omega)$

$$g^{(1)}(k, \tau) = \langle E^*(k, 0)E(k, \tau)\rangle/\langle E(k, 0)\rangle^2 \qquad (134)$$

where E is the electric field and E^* its conjugate, by the Siegert relationship [218, 219]

$$g^{(2)}(k, \tau) = 1 + |g^{(1)}(k, \tau)|^2 \qquad (135)$$

For a solution of identical, spherical, independent scatterers with independent positions and orientations in a nonscattering solvent [198, 218, 220], $g^{(1)}$ decays exponentially

$$g^{(1)}(k, \tau) = A \exp(-Dk^2\tau) \qquad (136)$$

These conditions, together with that for k_T mentioned earlier, are met by dilute solutions of large particles or macromolecules. It is not obvious that they are met by mixtures of similarly sized molecules, where both components scatter light.

Equations (135) and (136) may be combined to give the working equation

$$g^{(2)}(k, \tau) = 1 + A' \exp(-2Dk^2\tau) \qquad (137)$$

where A' is a time independent constant for a given apparatus and system.

The function $g^{(2)}$ is determined using the digital autocorrelator. Due to the optical mixing mentioned earlier, the number of photoevents occurring at the detector in short, successive time intervals varies. If $N(t_i, \Delta t)$ is the number of photoevents detected in a set interval Δt, centered on t_i, and $N(t_i + \tau, \Delta t)$ is that in the same interval after the elapse of time τ, then the product of these quantities can be stored, and an average of M such products can be obtained. If $\bar{N}$ is the average number of events occurring at the detector in the interval Δt, then $g^{(2)}$ is given by

$$g^{(2)}(\tau) = \frac{1}{M\bar{N}^2} \sum_{i=1}^{M} N(t_i, \Delta t)N(t_i+\tau, \Delta t) \qquad (138)$$

Single-channel clipped correlators have been described for this purpose, but now multichannel devices have been put into practice [211, 212, 221].

Once $g^{(2)}$ is obtained, it can be fitted to (137) as a function of scattering angle and τ by a method of nonlinear least squares. Sometimes the thermal diffusivity term is retained where the decay terms for χ and D are of the same order of magnitude.

Methods for testing the experimental apparatus and optimizing the conditions for its use were given by Lucas and co-workers [211].

One can also observe light scattering from pure liquids [216, 217]. In this case the central Rayleigh peak can be related to the self-diffusion coefficient, and radiotracer and NMR spin–echo data have been utilized to interpret scattering results for liquid argon, CCl_4, $GeCl_4$, and some hydrocarbons. The method has not been used to determine self-diffusion coefficients directly.

4.6 Nuclear Magnetic Resonance Spin–Echo Measurement of Self-Diffusion Coefficients

The NMR spin–echo technique is a rapid and now reasonably accurate method ($\pm 2\%$) for the measurement of self-diffusion coefficients. It can be used to obtain results for gases or liquids over several orders of magnitude (10^{-1}–10^{-8} cm^2/s). Only small samples are required, and hence control of temperature and pressure are relatively straightforward and can be made as exact as necessary. Low-frequency, steady-gradient machines have been used for some time to measure self-diffusion in pure fluids or in mixtures where only one component bears the resonant isotope, but Fourier transform techniques used with high-resolution, high-frequency machines now allow simultaneous determination of as many self-diffusion coefficients as there are resolvable resonances. There are several variants of the method that can be adapted to given experimental requirements.

The method is based on the Hahn two-pulse spin–echo method for the measurement of the NMR T_2 relaxation time [222]. Hahn realized that diffusion due to thermal motion gave rise to more relaxation additional to the normal spin–spin T_2 mechanisms, which is observed when the magnetic field is inhomogeneous. Deliberate degradation of the field by application of a small gradient increases the size of this effect, and measurement of the echo height as a function of the pulse separation and gradient strength allows the determination of the self-diffusion coefficient.

Recent books and reviews on pulsed NMR spectroscopy and instrumentation suitable for diffusion include those by Farrar and Becker [223], Shaw [224, 225], Geiger and Holz [226], Redfield [227], Laszlo [228], Callaghan [229], and Stilbs [230], together with the annual *Specialist Periodical Reports* of the Royal Society of Chemistry [231].

The basic apparatus required for pulsed NMR spectroscopy consists of a high-field magnet, radiofrequency (rf) generator, pulse programmer and transmitter, sample probe, receiver, amplifier, and data recorder. For spin–echo diffusion studies, magnetic field gradient coils, a (constant or pulsed) current source for these, and a thermostat are also required. If Fourier transform techniques are to be used, signal digitization and computer equipment are also necessary.

The magnet should produce a homogeneous field over the sample with long-term stability of 0.1 ppm or better. Higher fields give better signal-to-noise ratios (S/N) as this varies as (frequency)$^{3/2}$ [225]. With weak signals, data accumulation is necessary and field stabilization is then more important.

The pulse programmer and transmitter must produce intense microsecond rf pulses with short rise and fall times. A single coil acting both as transmitter and receiver is used now almost uniformly. The receiver must be able to recover quickly from overloads generated by the transmitter as the pulse separation is as small as 1 ms and the signal width can be much less than this. It is essential that the sample be contained within the most homogeneous region of both the rf field and the magnetic field gradient, a point often overlooked in early work, and this is most easily done by winding them on a common former [232, 233].

Receivers may be diode, phase-sensitive, phase-locked-loop, or quadrature detectors. The older diode receivers do not require as stable a field as the phase-sensitive types, but they do require calibration because of their nonlinear response and are now used infrequently. For low-field work, the phase-locked-loop detector is particularly convenient [234].

Methods used for the production of the magnetic field gradient include the use of opposed Helmholtz coils [235, 236] and variants [237], quadrupole coils [238], and the modified shim coils of high-field magnets [239]. The quadrupole coils require smaller currents than the Helmholtz coils, have a more convenient geometry, and have a lower inductance to gradient ratio, making them ideal for pulsed gradient work. Other coil types for superconducting magnets were discussed by Stilbs [230], and pulsed gradient circuitry, by many workers, including Callaghan and co-workers [239] and, again, Stilbs [230]. The determination of the strength of the gradient produced is discussed further below.

Examples of NMR diffusion apparatus described in the literature that have given excellent results include those of Gerritsma and Trappeniers [236], Jonas [240], Cantor and Jonas [241], Harris and co-workers [232], Callaghan and co-workers [239], and Stilbs and Moseley [242].

The theory of the method was detailed by Tyrrell and Harris [6] following earlier accounts by Hahn [222], Carr and Purcell [243], Muller and Bloom [244], and Gerritsma [245].

Briefly, this is as follows: Consider nuclei with a nonzero magnetic moment in an applied magnetic field. The total magnetization of the sample of nuclei precesses at a frequency, known as the Larmor frequency, due to its interaction with the main field. NMR experiments are made to take place in a frame of reference rotating with this frequency ω. A short rf pulse at this same frequency is applied from a coil normal to the main field for time t_p. This has two circularly polarized components, one of which is in phase with the rotating frame $\boldsymbol{B}_1$ and causes the magnetization to precess about it with angular velocity $-\gamma\boldsymbol{B}_1$ and through an angle

$$\theta = \gamma|\boldsymbol{B}_1|t_p \tag{139}$$

Normally, the restoration of the system to equilibrium occurs by means of natural relaxation processes. These are described by two relaxation times defined by the symmetry of the field T_1, being that for the z component in the main field direction, T_2 being that in the $x-y$ plane. The signal observed due to emission is called the free induction decay (fid).

The Hahn method requires two pulses with lengths sufficient to tip the magnetization vector through 90° and then, after an interval τ, through 180°. During this interval the magnetization loses phase coherence, there being a spread of frequencies in the sample because of the natural slight inhomogeneity of the field, and in consequence some nuclear moments precess more quickly than average, some more slowly. The effect of the second pulse is to reverse the moments about the axis of B_1, which thus gain phase coherence and then lose it. The signal observed is a reversed fid followed by its mirror image, giving the characteristic *spin–echo*, Figure 3.21. The echo height is less than that of the original fid and T_2 can be obtained by its measurement as a function of the pulse spacing τ. If an additional magnetic field gradient of known strength is applied, further diminishment of the echo occurs due to self-diffusion of the molecules bearing the resonant nuclei in this gradient. Gerritsma [245] has given the full solution for the echo height for a cylindrical sample of height h, radius r, with an applied gradient g and pulse angles θ_1 and θ_2.

$$A = A^0 \sin\theta_1 \sin^2(\theta_2/2) \exp\left\{-\frac{t}{T_2} - \frac{2}{3}\gamma^2[\mathbf{g}\cdot\mathbf{g}]D_S(t-\tau)^3\right\}$$

$$\cdot\frac{\sin[\frac{1}{2}\gamma g_x h(t-2\tau)]}{[\frac{1}{2}\gamma g_x h(t-2\tau)]}\cdot\frac{J_1[\frac{1}{2}\gamma(g_y^2+g_z^2)^{1/2}(t-2\tau)]}{[\frac{1}{2}\gamma(g_y^2+g_z^2)^{1/2}(t-2\tau)]} \qquad (140)$$

where J_1 is a first-order Bessel function and g has been expanded in terms of its rectangular components. This equation contains all the information to be obtained from the magnitude, shape, time- and field-dependence of the echo signal.

In the simplest form of the experiment D_S is determined from measurements

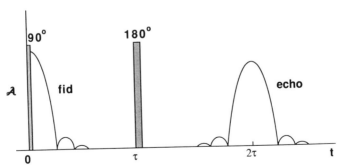

Figure 3.21 Pulse and signal sequence for the Hahn spin–echo experiment: A, signal amplitude; τ, rf pulse spacing; and t, time.

of the echo height as a function of pulse spacing τ in the presence and absence of a known field gradient. The ratio of these heights is from (140)

$$\exp[-\tfrac{2}{3}\gamma^2(\mathbf{g}\cdot\mathbf{g})D_S\tau^3]$$

This is independent of T_2, θ_1, and θ_2, although the peak amplitude is a maximum for the 90–180° pulse sequence.

However, systematic errors may occur if the residual inhomogeneity of the main field is neglected. If $\mathbf{g}$ is the sum of this term $\mathbf{g}_m$ and that applied through the gradient coils $\mathbf{g}_a$, then (117) can be written as

$$A = A'\exp\{-\tfrac{2}{3}\gamma^2 D_S\tau^3[\mathbf{g}_a\cdot\mathbf{g}_a]-\tfrac{4}{3}\gamma^2 D_S\tau^3[\mathbf{g}_a\cdot\mathbf{g}_m]\} \qquad (141)$$

The values for D_S and $\mathbf{g}_m$ can then be obtained from a nonlinear least-squares fit. Combination of these values with those determined by varying τ with $\mathbf{g}$ held at fixed values of both signs leads to quite good reproducibility and precision, depending on the density of resonant nuclei in the sample. For instance, $\pm 1\text{–}2\%$ was obtained with ^{1}H- and ^{19}F-containing substances [232].

A variant of the Hahn two-pulse method is to use a train of 180° pulses after the 90° pulse, at intervals of 2τ [6, 243, 246]. This gives a train of echoes midway between the rf pulses, the heights of which can be recorded in a single experiment. The method is commonly used in the Fourier transform pulsed gradient spin–echo method described next.

4.6.1 Pulsed Gradient and Fourier Transform Methods

The steady gradient spin–echo method outlined above has proved to be a very versatile technique, and the dependence of the echo attenuation on the two parameters g and τ allows a wide range of diffusion coefficients, from 1 to 10^{-6} cm^2/s to be determined in this way. The lower limit is set by the maximum field gradient that can be used, a typical value being 1 T/m. There are two consequences of the use of a large gradient. One is an increase in the NMR line width with a corresponding decrease in the width of the fid and echo; the second is that the rf pulses do not then uniformly excite the spectrum. The broader line width necessitates an increase in the detector bandwidth, thereby decreasing the S/N ratio, and the requirement that the rf field amplitude be greater than the line width may not be met.

The use of pulsed gradients was examined as early as 1955 by Hahn and his colleagues [6], but the first systematic investigation of the use of time-dependent gradients was that of Stejskal and Tanner [247, 248]. This work has been the subject of excellent reviews by Callaghan [229] and Stilbs [230], the latter being particularly thorough in its coverage of the historical development of the topic, apparatus, technique, and applications.

The result for the echo amplitude at $t = 2\tau$, where a gradient pulse of width δ

is applied at time $t = t_1$ after the $90°$ rf pulse and a second is applied after an interval Δ such that it falls between the $180°$ pulse and the echo, is

$$A = A^0 \exp\left(-\frac{2\tau}{T_2} - \gamma^2 D_S \left\{\frac{2}{3}\tau^3 \boldsymbol{g}_m \cdot \boldsymbol{g}_m + \delta^2 \left(\Delta - \frac{\delta}{3}\right) \boldsymbol{g}_a \cdot \boldsymbol{g}_a \right.\right.$$
$$\left.\left. - \delta\left[(t_1^2 + t_2^2) + \delta(t_1 + t_2) + \frac{2\delta^2}{3} - 2\tau^2\right]\boldsymbol{g}_a \cdot \boldsymbol{g}_m\right\}\right) \qquad (142)$$

where

$$t_2 \equiv 2\tau - (t_1 + \Delta + \delta) \qquad (143)$$

(See Figure 3.22.) This equation can be simplified by choosing pulse widths δ and time intervals τ such that $(\delta \boldsymbol{g}_a)$ is much larger than $(\tau \boldsymbol{g}_m)$; then, with neglect of the T_2 term,

$$A = A^0 \exp\{-\delta^2 D_S \delta^2 [\Delta - (\delta/3)]\boldsymbol{g}_a \cdot \boldsymbol{g}_a\} \qquad (144)$$

Thus, the attenuation of the echo is independent of τ. Its width, however, depends on $\boldsymbol{g}_m$, the natural field inhomogeneity of the magnet, not on the applied gradient $\boldsymbol{g}_a$; and thus diffusion coefficients as low as 10^{-8} cm^2/s can be determined with gradients of as much as 100 T/m, but with a satisfactory S/N.

Echo distortion and displacement can take place due to phase errors resulting from the rapid switching of the high currents required. It is common practice to adjust the phase of the $180°$ pulse and the width of the gradient pulses to minimize these effects. Another technique is to manipulate the background gradient $\boldsymbol{g}_m$ [249].

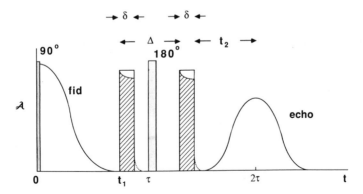

Figure 3.22 Pulse and signal sequence for the Stejskal–Tanner spin–echo experiment: A, signal amplitude; τ, spacing between the rf pulses; t, time; t_1, time interval between the $90°$ rf pulse and the first gradient pulse width δ; Δ, time interval between the gradient pulses; t_2, time interval between the second gradient pulse and the echo maximum.

For nuclei with spin greater than one-half, quadrupole interaction further attenuates the echo. The dependence of the echo height on g is still given by the Stejskal–Tanner equation, but the coefficient A^0 is less than the signal height observed at zero applied gradient. These matters were discussed by Callaghan and co-workers [250] and the system D_2O was examined carefully. Deuterium has a spin of unity, and accurate values for $D_S(D_2O)$ were calculated [35] from radiotracer data for DTO in D_2O and interferometric data for HDO in D_2O. Measurements have also been made on both H and D in deuteriated benzene with excellent results [250].

An important application of the pulsed gradient technique has been in measurement of restricted diffusion, that is, diffusion where molecules may strike barriers during the course of the experiment. The diffusion time Δ is well defined in this case and may be adjusted so that (144) can be employed. Barrier separations can then be determined from the attenuation measured at larger values of this parameter [6, 229, 230].

The techniques discussed above are carried out in the time domain, and the spectral contributions to the fid and echo are unresolved. Spin–spin coupling due to neighboring nuclear spins is a τ-dependent relaxation process that can render impossible diffusion measurements by the normal methods in systems with nonequivalent resonant nuclei. This difficulty can be removed by Fourier transformation of the echo, whereupon a frequency spectrum is obtained and the attenuation of each of the several peaks can be determined. By this means it is then possible to extract the self- or intradiffusion coefficients of more than one species in a mixture, provided that the resonant frequencies can be resolved.

The first application of this method was that of James and MacDonald [251] to the system water–dimethyl sulfoxide (DMSO). The gradient was applied during the period 2τ, except during the rf pulse periods, and the half-echo following that interval transformed to give the frequency spectrum. The true pulsed experiment was carried out by Kida and Uedaira [252] on acetic acid–water and water–methanol, but the interpretation of the results in these systems was not without ambiguity due to proton exchange between the components.

Apparatus and procedures giving a precision of 1–2% were described by several groups [239, 242, 253]. The final spectrum depends on the setting of the phase detector; delay between the time of the echo peak maximum and the start of the signal acquisition; length of the acquisition time, which must not be less than $3T_2$; and filtering. Time averaging is usually required, together with sufficient computing capacity for handling data on- and offline. Computational requirements for Fourier transformation of NMR signals were comprehensively discussed by Dumoulin and Levy [254]. Good temperature control is essential as short-term fluctuations can cause variations in the T_2 relaxation and hence have an indirect, nondiffusive influence on the echo shape [230].

4.6.2   The Calibration of Magnetic Field Gradient Coils

The simplest way to determine the field gradient as a function of the applied current is to use accurate tracer data for calibration purposes. Table 3.6 lists

Table 3.6 Self- or Intradiffusion Coefficients, $D_S/10^{-5}$ cm^2/s, of Water and Benzene

T (°C)	H_2O [35]	D_2O [35]	$H_2{}^{18}O$ [165, 256]	C_6H_6 [257]
1	1.149			
4	1.276			
5	1.313	1.015		
5.29			1.309	
15	1.777		1.751	1.860
25	2.299	1.872	2.275	2.207
				2.219^a
35	2.919			2.605
40			3.186	
45	3.575	2.979		3.010
50			3.960	
55				3.45^b
60				3.70^b
65			5.138	
80			6.517	
90			7.496	

^{a} Reference [258] corrected for a small mass dependence of the tracer values.
^{b} Reference [259].

such data for H_2O, D_2O, $H_2^{18}O$, and benzene. The values for H_2O and D_2O were calculated by Mills [35] by combination of diaphragm cell data for HTO and DTO tracer diffusion coefficients in these liquids with the two limiting mutual diffusion coefficients for H_2O–D_2O, that is, the tracer coefficients of HDO in H_2O and D_2O, obtained from the interferometric results of Longsworth [255]. The latter have an absolute accuracy of $\pm 0.1\%$, and the diaphragm cell tracer measurements have a relative precision of $\pm 0.2\%$, based on calibration with standard conductimetric and interferometric data for the system H_2O–KCl at 25°C, which are also accurate to $\pm 0.1\%$ [1].

A second method is to make use of the shape of the echo. The first zeros of the Bessel function factor in (117) are given by the relation

$$\tfrac{1}{2}\gamma r(t - 2\tau)(g_y^2 + g_z^2)^{1/2} = 3.83171 \tag{145}$$

and g_a for a given gradient producing current can be obtained provided the radius of the sample is accurately known. Murday [260] listed the precautions to be taken if this procedure is to be used. Hrovat and Wade [249] described a method to determine g_a and the angle between g_a and g_m in pulsed gradient experiments that makes use of deliberately mismatched gradient pulses.

Jonas and co-workers [261, 262] used a refinement of this method, fitting the echo shape to obtain the gradient for each echo amplitude measurement. This is used directly in the diffusion expression (140), in its working form, thus including the g_0 contribution. The need for an independent calibration is thus eliminated, which is a distinct advantage when operating in regions of temperature and pressure where calibration data are not available. Jonas and co-workers [261, 262] claim a precision of $\pm 1\%$ in the gradient and an overall accuracy of $\pm 3\%$.

4.6.3   Sample Cells and High-Pressure Studies

It is imperative for high precision in spin–echo studies that the rf and gradient fields be made as uniform as possible over the sample volume, and careful design of the sample cell is necessary. For dense gas studies it is sufficient to have a space filling former for the coil(s) [233, 234] in the pressure vessel, but for liquids some sort of compression cell is required. Cells fitted with bellows [240, 263, 264] and pistons [265] have been extensively used, as well as thick-walled glass capillaries treated with hydrofluoric acid [266]. Nonmagnetic, nonmetallic materials such as Teflon polymer are preferable [263, 264], but even a non-magnetic stainless steel bellows fitted to a glass cell has been used successfully with low-field electromagnets [267].

Pressure vessels have usually been manufactured from Be—Cu alloy for cryogenic to ambient temperatures or titanium IMI 680 alloy [268] for higher temperatures, although titanium alloys have poor thermal conductivity and pressure vessels made from these take much longer to come to thermal equilibrium following a pressure or temperature change. A diamond anvil cell was used at pressures as high as 5200 MPa for relaxation time determinations, and it should be possible to use similar cells for diffusion measurements [269].

An interesting method complementary to the usual techniques has been used by Jonas and co-workers [270] for very viscous materials at high pressures. The self-diffusion coefficient is related [271] to the frequency dependence of the rotating-frame, spin–lattice relaxation time $T_{1\rho}$.

$$D = \left[-\frac{\sqrt{2}\gamma^4\hbar^2\pi n}{20} \frac{d(\omega^{1/2})}{d(1/T_{1\rho})} \right]^{2/3} \tag{146}$$

where n is the spin number density, γ is the gyromagnetic ratio, and ω is γB_1. Two conditions are required for this equation to be valid: the B_1 field must exceed the local field due to dipolar broadening and $(\omega\sigma^2/D)^{1/2} \ll 2$, σ being the molecular diameter. A precision of only $\pm 30\%$ [270] is claimed, but the technique allowed measurement [262] of diffusion coefficients as low as 10^{-14} m^2/s.

The use of superconducting magnets requires special attention to the design of the cell and the rf and field-gradient coils. In some designs, it may be difficult to assure gradient homogeneity over the whole of the sample volume; however, it has been reported that pulsed gradients as high as 1 T/m have been used without quenching of the superconducting magnet [230].

Two groups [272–274] reported the use of high-pressure vessels in superconducting magnets. Both used the steady gradient method. Trappeniers and co-workers [273, 274] designed a Be—Cu pressure vessel for use to 300 MPa. The field gradient was determined from the variation of the Larmor frequency of a jig-mounted capillary sample of D_2O with position in the field, and only a single value of 81 mT/m was used. Under these conditions the precision of the gradient was ±0.5%, and the overall precision at 75 MHz was ±1.5%. Prielmeier and co-workers [272] combined Yamada's design of an HF-etched, thick-walled glass cell [266] with a quadrupole gradient coil placed in the bore of a Varian XL-100 spectrometer. They claim only ±10% inaccuracy, at pressures up to 200 MPa. This group [275] also reported the successful application of the pulsed field-gradient technique with this design in a Bruker MSL 300 spectrometer, giving results with an uncertainty of ±5%.

Braun and Weingärtner [276] also described a probe for a Bruker BH300/89 wide bore magnet for steady gradient experiments, claiming ±2% for the diffusion of alkali metal ions in aqueous solution.

5   METHODS OF MEASUREMENTS—SOLIDS

The study of diffusion in solids is based on the same principles that are used for fluid phases. However, because of the possible anisotropy in the properties of a solid system, a more general form of Fick's first law is needed. The diffusion *coefficient* is a second-order tensor; within rectilinear coordinate axes, it relates the flux vector to the concentration gradient vector for a binary system through the expression

$$
\begin{pmatrix} J_x \\ J_y \\ J_z \end{pmatrix} = - \begin{pmatrix} D_{xx} & D_{xy} & D_{xz} \\ D_{yx} & D_{yy} & D_{yz} \\ D_{zx} & D_{zy} & D_{zz} \end{pmatrix} \begin{pmatrix} \dfrac{\partial C}{\partial x} \\ \dfrac{\partial C}{\partial y} \\ \dfrac{\partial C}{\partial z} \end{pmatrix}
\tag{146}
$$

The terms J_x, J_y, and J_z are the x, y, and z components of the flux of one of the components; and $\partial C/\partial x$, $\partial C/\partial y$, and $\partial C/\partial z$ are the concentration gradients of one of the components in the x, y, and z directions; the D_{ab} are the nine elements of the diffusivity tensor, and their values are dependent on both the orientation and the velocity of the coordinate axes. In a multicomponent system each D_{ij} of the generalized Fick's law [8] is a second-order diffusion tensor with nine elements.

In an isotropic system the diffusivity tensor takes a particularly simple form in which

$$D_{xx} = D_{xy} = D_{zz} = D \tag{147}$$

$$D_{xy} = D_{xx} = D_{yx} = D_{yz} = D_{zx} = D_{zy} = 0 \tag{148}$$

where D is regarded as the diffusion coefficient. Equations (147) and (148) are applicable to fluid systems and to polycrystalline solids. They may not be satisfied for some single crystals, as was demonstrated for zinc [277–279] and for ice [280, 281].

The diffusivity tensor is a function of temperature, composition, and the stress tensor [282, 283]. In this chapter it is assumed that the nine elements σ_{ij} of the stress tensor (in a rectilinear coordinate system) satisfy the requirements

$$\sigma_{xx} = \sigma_{yy} = \sigma_{zz} = -P \tag{149}$$

$$\sigma_{xy} = \sigma_{xz} = \sigma_{yx} = \sigma_{yz} = \sigma_{zy} = 0 \tag{150}$$

where P is often described as the *pressure* of the solid. There are several diffusion studies [284–289] with solid systems in which elaborate precautions are made in an attempt to satisfy the requirements (150) and (151) for every diffusion run, each run having a different value of P. The variation of D with P was similar to the variation of D with pressure in fluid systems. There have been attempts to determine the values of D when (149) and (150) are not satisfied, but the results have been conflicting. It now seems clear from experiments on metals in the plastic region, just below the melting point, that nonzero values for the shearing stresses of (150) have little effect on D in this temperature range [290–294]. This may not be true at lower temperatures, where much higher shearing stresses can be sustained without the occurrence of plastic flow.

In writing (146) it has been assumed that the system is subject to a uniform stress. If there are stress gradients, then (146) will need additional terms to allow for the resultant convective flow [295] and the diffusive flows that arise from the variation of chemical potential with stress [283]. Unfortunately, many solid diffusion couples show strong evidence (pores, necking, and other distortion) of the existence of stress gradients during the diffusion experiment; therefore, they are unsuitable for quantitative analysis. In the remainder of this chapter it is assumed that (147)–(150) are applicable to the systems under discussion.

Further distinguishing features of solid-state diffusion are the slow rates and high activation energies characteristic of the process. This often means that high temperatures must be used for the diffusion experiment. However, it also means that diffusion rates at room temperature are usually negligible, and subsequent analysis of the diffused sample can be carried out at leisure.

Even isotropic, single-phase solids possess microstructural features that can affect their diffusional properties. In a typical polycrystalline solid, diffusion can occur by the (imperfect) solid lattice, along grain and subgrain boundaries, and by dislocations. It is therefore necessary to determine which mechanisms of diffusion are in effect in the temperature regime of interest. Because the activation energies for these different diffusional mechanisms are quite varied, extrapolation from high-temperature data to much lower temperatures is frequently misleading. For this reason, considerable effort has been expended on extending the accessible range of measurement to lower values of D. Further improvements are desirable.

Solid-state diffusion measurements are of two types. In the first, and most widely used, the movement of a system toward homogeneity is observed and compared with the predictions of the diffusion equations. These equations, together with boundary conditions appropriate to the experiment, yield solutions of the general form

$$C_i = C_i(x, t, D) \tag{151}$$

Thus D is evaluated by fitting theoretical curves to concentration profiles, $C_i = C_i(x, t)$. In these experiments, two different mixtures are brought into contact to form a diffusion couple, and diffusion is observed in a direction orthogonal to the original interface. Obviously, the method is appropriate to the measurement of diffusion in the presence of a concentration gradient (i.e., chemical diffusion) but may also be used for self-diffusion studies where the only gradient is that of a radiotracer.

Alternatively, measurements can be carried out on completely homogeneous materials by observing the interaction of the solid's constituent atoms or ion with a transient perturbation, or their subsequent relaxation to a low-energy state. Where translational particle motion provides the means of interaction of relaxation, these processes display a time-dependent that involves particle mobilities. Measurement of this time-dependence leads, therefore, to a determination of D. Techniques of this type can be used to measure self-diffusion. However, because they are not sensitive to position within a sample they cannot reveal gradient effects, and they will be described here only briefly.

5.1   Self-Diffusion Measurements in Homogeneous Materials

5.1.1   Nuclear Magnetic Resonance

By far the most commonly used technique for observing self-diffusion in homogeneous materials is that of NMR. The principles of this method and the different ways in which it can be applied were covered in Section 4.6. Its application to solids was reviewed by Stokes [296].

Measurements of the spin–lattice relaxation times T_1 and T_{1p} have been used to measure diffusion coefficients over a wide range, the latter measurement giving access to D values as low as 10^{-15} cm^2/s [297]. The more direct pulsed field gradient spin-echo technique is limited so far to values of D greater than about 10^{-8} cm^2/s [298] and is therefore not suitable for many solids.

5.1.2   Anelastic Relaxation Measurements

Self-diffusion in solids occurs by the movement of lattice defects. The presence of these defects causes a slight displacement of adjacent lattice species and hence leads to a change in specimen dimensions. Depending on the lattice and defect types, these changes may be anisotropic. The orientations of the defects will be random in an unstressed solid. However, when a directional stress is applied to the solid, some of the defects will be oriented unfavorably. These will relax to a lower energy state by a local migration process, that is, by diffusion. The resulting change in specimen dimension is an anelastic strain and its rate of development is related to defect mobility that can thereby be measured [299–302].

Anelastic behavior is usually measured by observing vibration damping (or *internal friction*) as a function of temperature. For a single, thermally activated process a single internal friction peak is found when the measurement is repeated over a range of temperatures. The peak is measured at several different vibration frequencies. In this way the diffusion coefficient and its activation energy are determined. In fact, a spectrum of relaxation processes (and times) may be encountered. A technique has been proposed [303] for the analysis of internal friction results found in this case.

Internal friction measurements permit observation of very short range events, essentially single defect jumps. The related technique of Gorsky relaxation allows the measurement of much longer range movement. In the Gorsky experiment an initially flat sheet sample is bent around a mandrel of constant curvature to produce a stress gradient across the sheet thickness. Defects then diffuse to an equilibrium distribution under the effect of this gradient. When the stress is removed, the sheet returns almost, but not exactly, to its original slope. As the defects then diffuse back to a uniform distribution, the sheet slowly unwinds. The rate of this final, inelastic relaxation is controlled by the defect diffusion rate. The diffusion distance is about that of the sheet thickness and the technique is therefore limited to highly mobile defects like hydrogen in metals [302].

5.1.3   Quasielastic Neutron Scattering

Self-diffusion measurements by quasielastic neutron scattering have been reported for many solids [304–306]. Quasielastic scattering results from interaction of the neutrons with randomly moving (diffusing) particles as distinct from interaction with periodic vibration of particles about their mean positions, which gives rise to inelastic scattering. Thus, a broadening of the scattering angle results, the extent of the broadening being dependent on the magnitude of D. The technique has several advantages, but it is strictly limited in applicability: The diffusing atoms must have significant neutron cross sections; they must have significantly different coherent and incoherent scattering efficiencies; and they must possess diffusion coefficients greater than about 10^{-7} cm^2/s.

5.1.4   The Mössbauer Effect

Mössbauer spectroscopy can be used to measure self-diffusion [307–309]. The effect is observed by oscillating a Mössbauer source along the radiation direction being observed. A Mössbauer absorber will then absorb at the resonant frequency and also at frequencies shifted by the Doppler effect, which was induced by oscillating the source. If the Mössbauer atoms in the absorber move with a jump time comparable to or less than the nuclear lifetime characteristic of the Mössbauer event, then a similar Doppler effect results. In practice this is observed as a line broadening. There are difficulties with the technique [310] and its application is restricted to Mössbauer active nuclei (principally ^{57}Fe) with relatively high-diffusion coefficients ($> 10^{-11}$ cm^2/s for ^{57}Fe). Nonetheless, crystal orientation effects are easily studied [311] and the correlation factor f can be estimated [312], so the technique continues to be of great interest.

5.2   Diffusion Couple Experiments

A diffusion couple consists of two different mixtures, each initially homogeneous, joined at a planar interface. If the dimensions of the couple and the period of diffusion are such that concentrations do not change at the ends of the system during the period of observation, then the experiment is described as an infinite diffusion couple, which is used to measure chemical diffusion coefficients. Tracer diffusion measurements, on the other hand, involve the formation of a couple consisting of a thin film of isotopically labeled material on the end of a long bar of unlabeled but compositionally identical material. In both cases the experiment involves preparation of the couple (without inducing diffusion), an annealing period at constant temperature in a controlled environment during which diffusion occurs, subsequent analysis of the concentration profiles, and calculation of the diffusion coefficients.

5.2.1   Preparation of Diffusion Couples

The halves of a diffusion couple must first be ground and polished flat at the mating surfaces. They are then homogenized by annealing at the highest possible temperature without inducing melting or another phase transformation. In the case of a compound, the desired stoichiometry is obtained by annealing under an atmosphere of controlled partial pressure of electronegative species. If low partial pressures are controlled by a gas-phase equilibrium such as

$$CO_2(g) = CO(g) + \tfrac{1}{2}O_2(g)$$

problems in attaining a true equilibrium between solid (s) and gas (g) may be encountered [313–318]. In these cases it may be preferable to use a metal–metal oxide couple

$$MO(s) = M(s) + \tfrac{1}{2}O_2(g)$$

to establish the oxygen partial pressure (p_{O_2}), within a closed system. If the homogenization temperature is higher than that of the diffusion experiment, then accurate knowledge of the deviation from stoichiometry as a function of both T and p_{O_2} is required. Data of such quality are usually not available, and homogenization is often better carried out at the same temperature. A second purpose of the homogenization heat treatment is to remove dislocation structures left from the polishing procedure.

The two halves of an infinite couple are usually joined by clamping them together and subjecting them to a brief heat treatment at a temperature much lower than that of the diffusion experiment [319, 320]. It is wise to verify experimentally that only negligible diffusion occurs during this welding procedure. It is, of course, essential that the clamp does not deform the diffusion couple. If it is desired to perform a Kirkendall experiment, as discussed later, markers consisting of small particles of chemically inert material are placed between the couple halves before welding to locate the interface later. The choice of marker is dependent on the subsequently used analytical technique. For example, ZrO_2, which fluoresces in an electron beam, may be a good marker if an electron optical technique is used.

One-half of a diffusion couple may be a vapor phase, such as a volatile metal like zinc [321] or the electronegative constituent of a compound. In either case, it is essential that the vapor-phase reservoir be sufficiently large that the partial pressure of the diffusing species can be maintained constant. In these experiments gas-phase diffusion is, of course, very much faster than in the solid, and the concentration of diffusing species at the solid surface is thus fixed at the same value as in the gas phase. Tracer diffusion experiments for oxygen in oxides, nitrogen in nitrides, and so on are performed in this way [322–324].

A problem sometimes encountered with solid–gas diffusion couples is that of evaporation during the diffusion anneal. Corrections for the resulting surface recession are available [325].

Most tracer diffusion experiments involve the formation of the isotopically labeled part of the diffusion couple as a very thin film on the surface of a slab of homogeneous material. This technique is appropriate for metal diffusion in metals, alloys, and compounds, provided the film of labeled material is of the same chemical composition as the unlabeled portion of the diffusion couple. For metals and alloys, electrolytic and electroless plating can be used [326–329]. Since electroplating can lead to very high defect densities, a preliminary, low-temperature annealing treatment may be necessary. Thin films can also be prepared by chemical or physical vapor deposition. Since uncontrolled specimen heating can occur during these operations, it is necessary to verify experimentally that no significant amount of diffusion results from the pre-paration process.

When metal tracer diffusion in a compound is being studied, a metal film is first deposited by one of the above techniques and then chemically converted to the desired compound. The necessary reaction is carried out at a temperature low enough to prevent significant tracer diffusion or change in stoichiometry of the compound. Because the amount of labeled compound is very much less than that of unlabeled substrate material, its stoichiometry will be adjusted to match that of the substrate during the diffusion anneal. Because the isotope film thickness is very much less than the diffusion distances usually studied, the time taken for this adjustment to occur may be neglected.

An alternative technique is to deposit a liquid salt solution of the isotope on the unlabeled substrate surface. Vacuum or thermal evaporation of the solvent is then followed by chemical conversion to the desired state [330, 331].

5.2.2   The Diffusion Anneal

The requirements of a diffusion anneal experiment are that the temperature be uniform and precisely controlled, the time of the anneal be accurately known, and the composition of the diffusion couple be unaffected by matter exchange with the environment. An inevitable difficulty arises because of the need to raise and lower the temperature at the beginning and end of the annealing period. However, these periods of changing temperature can generally be kept very much shorter than the anneal, and therefore can be neglected. A correction method is available [332] for those situations in which diffusion during the heating or cooling periods cannot be ignored.

Metallic diffusion couples are protected from oxidation by using vacuum or inert gas environments. Compound stoichiometries are preserved by keeping the couple under a gas atmosphere with which it is in equilibrium. This is most easily done by sealing the couple within a small enclosure [333] so that the volume of, for example, gaseous oxygen that is originally within the container is small, as is the amount released from or taken up by the solid oxide (in coming to equilibrium) compared to the very much larger amount remaining in the solid. Alternatively, the partial pressure of oxygen can be controlled by a flowing gas mixture. Where sample vaporization is significant, the use of a closed container will restrict the magnitude of the effect to negligible proportions.

5.2.3   Analysis of Concentration Profiles

At the conclusion of a diffusion anneal a concentration profile is established along the diffusion direction, normal to the couple interface. In tracer diffusion studies the most widely used method of analysis involves serial sectioning of the couple normal to the diffusion axis and subsequent analysis of the sections.

SECTIONING METHODS

Sectioning methods have been reviewed by Rothman and co-workers [332] and are described only briefly here. Because surface diffusion of the tracer may be rapid, it is necessary first to grind off the sample surfaces that are parallel to the diffusion axis. In this way, any contribution to tracer concentrations in the sections due to surface diffusion or vapor transport is eliminated.

Sections are most commonly cut using mechanical means: lathe, grinding machine, or microtome. For metals and alloys a precision lathe is suitable, but accuracy is limited to about 250 μm. Greater precision is obtainable by grinding, the thickness being gauged by weighing or direct micrometer measurement. This method is also suitable for materials that are too brittle or hard for lathe sectioning and thus is appropriate to ceramics and other inorganic compounds. The sample is mounted in a holder with mechanical adjustments for alignment, and its end is ground with an abrasive of the desired coarseness. A common arrangement is for the sample to be mounted to a piston in a cylinder, with the axis of the cylinder kept at right angles to the grinding surface. With a fine diamond abrasive on a flat glass plate, sections of thickness about 1 μm can be obtained [334].

Electrochemical and chemical sectioning methods have been developed for the valve metals, silicon, and the noble metals [332]. Variations of these methods involve direct chemical dissolution of the sample surface; anodic dissolution of the surface; and formation of a thin, uniform surface film of oxide by anodic oxidation followed by chemical dissolution of this oxide. The advantages of these techniques are (1) collection of the removed section is obviously very easy and (2) very thin sections, of about a few nanometers, are possible in principle. These methods depend on the removal of *uniform* sections of material. The development of a method for a particular material therefore centers around the avoidance of a localized mode of attack such as pitting, cracking, grain boundary attack, or selective dissolution.

The sputter sectioning technique is a method of sectioning in which material sputtered with an ion beam from the solid surface is collected for subsequent analysis [335–337]. Low-energy ions are employed and sputter ion currents kept low to maintain low erosion rates and therefore achieve thin sections. Section thicknesses are determined either by an independent calibration and a reliance on uniform sputtering rates or, preferably, by using a quartz crystal film thickness transducer at the section collection location. Since it is necessary to collect a substantial number of sections, a magazine of collectors is required. These are brought sequentially into the collecting position by a mechanism operated from outside the evacuated sputtering chamber [338, 339].

Although very thin sections (ca. <1 nm) can be removed, the true spatial resolution is degraded by the detailed mechanics of sputtering: surface roughening, preferential sputtering, sputter-induced mixing, and so on. The true resolution is probably about 5 nm in metals and somewhat better in oxides.

Subsequent analysis of the sections can be performed to almost any desired level of precision and/or sensitivity using radiotracer counting, neutron activation, or other chemical techniques. In the case of radioactivity measurements, it is sometimes preferable to measure the residual activity of the partially sectioned couple rather than that of the sections themselves. Where difficulty is experienced in collecting the sections, this may be the only viable method.

An alternative approach to the sectioning problem is to erode material continuously from the sample surface by means of ion beam sputtering along the diffusion axis. Simultaneous concentration analyses are performed either by mass spectrometric analysis of the sputtered material or by analysis of the freshly revealed surface. Thus, a continuous sectioning experiment is performed. Clearly, the requirements of such an experiment are that (1) the analytical technique be fast enough relative to the sputtering rate and compositional gradients, (2) the sputtering process not perturb the diffusion profiles, and (3) the sputtering produces a uniform erosion of the surface at a precisely known and controlled rate.

Speed of analysis appears not to be a problem. However, ion sputtering can cause mixing of the specimen atoms in subsurface regions. These effects are minimized by using low-energy sputter ions of high mass and high angles of incidence [340, 341]. The effects are much larger in alloys than in stable compounds like base metal oxides.

The sputtering process produces a crater whose bottom should be flat. This is reasonably true for single-crystal materials, but considerable roughening can occur in polycrystalline substances [342]. Several sputtering rate calibration procedures are available [343].

Sputtering of insulating materials poses a particular problem because of specimen charging. The usual procedure is to first coat the specimen with a thin conductive layer and, in addition, to flood the surface with low-energy electrons during the sputtering operation.

Sectioning methods are rather slow and laborious because of the need for a substantial number of precisely defined and analyzed sections. For this reason most diffusional experiments depend on nonsectioning techniques. The most commonly used of these is electron probe microanalysis (EPMA).

ELECTRON PROBE MICROANALYSIS

Recent reviews of electron probe microanalysis and related techniques are available [344, 345]. The technique of EPMA was first announced by Castaing [346] and has since been developed into a standard instrumental technique. Many accounts of the principles of operation have been given [345, 347–349] together with tabulations of the physical constants required in the interpretation of the data.

A diagram of an electron microprobe is shown in Figure 3.23. A beam of high-energy electrons is focused to a diameter of a few to a few hundred nanometer. This is directed onto the flat sample surface. An imaging system, either optical or of secondary or backscattered electrons, is used to view the sample and locate the probe beam. The high-energy electrons excite surface and subsurface atoms, which then emit X-rays. These are characteristic of the atomic number of their parent atoms and may be identified by wavelength or energy-dispersive spectrometry (ENS). Measurement of X-ray intensities relative to those excited under the same conditions in standards of known composition leads to quantitative elemental analysis.

All microprobes are equipped with wavelength-dispersive spectrometers.

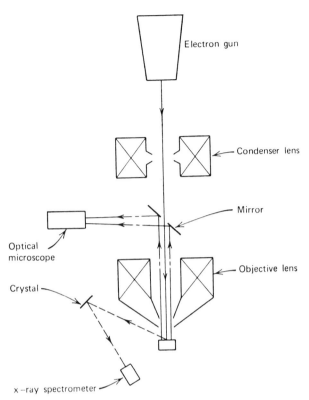

Figure 3.23   Schematic diagram of an electron probe microanalyzer: (——) electron path; (– – –) X-ray path; (———) optical path. Reprinted with permission from P. J. Dunlop, B. J. Steel, and J. E. Lane, "Experimental Methods for Studying Diffusion in Liquids, Gases and Solids," in A. Weissberger and B. W. Rossiter, Eds., *Physical Methods of Chemistry*, Vol. I, Part IV, Wiley-Interscience, New York, 1975. Copyright © 1975 by John Wiley & Sons, Inc.

These permit elemental detection for atomic numbers of 4 or more; good X-ray resolution (equivalent to about 5 eV) permitting high-quality discrimination; and, by providing a good S/N, yield a detectability of about 100 ppm. The diffracting crystals used for X-ray discrimination are appropriate to a limited range of wavelengths, and several different crystals are required. Simultaneous analysis for different elements requires a separate spectrometer for each, and the instrument cost is consequently high.

An alternative means of analyzing X-rays is provided by EDS. In this technique a single detector discriminates among X-ray according to their energies and can simultaneously analyze all incident X-rays. This technique provides signal identification for atomic numbers of 11 or more (although windowless detectors extend the range down to 6), rather poor energy resolution (ca. 150 eV), and a rather low S/N yielding a detectability of around 1000 ppm or more. The energy resolution limitation is serious; whereas wavelength-dispersive spectrometry discriminates between sulfur and molybdenum [350], this is quite impossible with EDS. However, the EDS technique provides rapid multielement analysis and is well suited to preliminary sample examination.

The conversion of X-ray intensity ratios to elemental mass concentrations requires calculation of the magnitude of several physical phenomena occurring within the sample. These are the X-ray excitation efficiency; absorption within the sample of emerging X-ray beams; and fluorescence due to characteristic radiation or, less importantly, the continuous spectrum. These effects are functions of the atomic numbers of the constituent atoms within the sample and their concentrations; methods for their calculation are given in [345, 347–349]. Note, the effects are explicit functions of the *take-off angle*, that is, the angle between the incident electron beam and the path of the emergent X-rays being detected. In the case of wavelength-dispersive spectrometers this angle is defined precisely, and quantitative computation is possible. In the case of energy-dispersive analysis, the detector was originally located adjacent to the sample, subtending a large angle. Consequently, quantitative analysis was of poor quality. This deficiency has been rectified in modern instruments by locating the detector at a position remote from the sample.

The analysis of binary systems has at times been carried out with the aid of calibration curves constructed from homogeneous mixtures. It will be apparent that this procedure is inappropriate to multicomponent systems and that computer calculation is mandatory. Nonetheless, it will be found that standards of composition not too far removed from that of the experimental sample are to be preferred. The solid terminals of a diffusion couple are usually suitable.

Absolute errors are usually less than 1 wt%. As the concentration decreases, the absolute error is reduced but the relative error becomes larger. However, since large numbers of replicate analyses are readily obtainable, the shape of the concentration profiles can be determined exactly. Figure 3.24 shows a concentration profile for an iron–iron, 24.2% nickel diffusion couple annealed at 1288°C as determined by Goldstein and co-workers [351] using three scans of at least 30 points each. Even in the case of inorganic compounds quite good results can be obtained, as shown in Figure 3.25 [352].

Spatial resolution is a function of the electron-sample interactions that lead to electron-scattering within the sample. These are the same processes that determine the X-ray excitation efficiency and are more severe at higher atomic numbers (more scattering) and higher electron beam voltages (longer path length). X-rays will be excited within the volume of solid defined by the electron scattering. A compromise must be reached between the needs for high X-ray intensities and those for small excitation volumes. With modern instruments a beam voltage of 20% greater than the critical excitation voltage of the element in question is probably optimum. A spatial resolution of 2–3 μm is usually attainable.

A diffusion couple being prepared for electron probe analysis is sectioned normal to the original interface, polished flat to avoid false intensity measurements, and cleaned. If the material is not electrically conductive, it must be

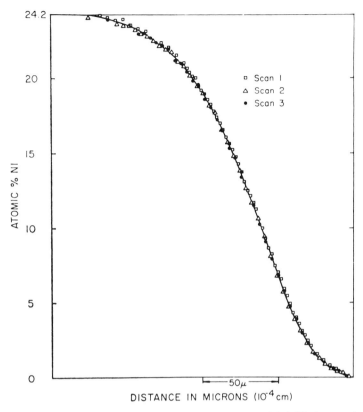

Figure 3.24 Electron microprobe scans across an iron–iron 24.2% nickel diffusion couple annealed at 1561 K [351]. Reprinted with permission from P. J. Dunlop, B. J. Steel, and J. E. Lane, "Experimental Methods for Studying Diffusion in Liquids, Gases and Solids," in A. Weissberger and B. W. Rossiter, Eds., *Physical Methods of Chemistry*, Vol. I, Part IV, Wiley-Interscience, New York, 1975. Copyright © 1975 by John Wiley & Sons, Inc.

coated with a thin film of carbon to avoid specimen charging and beam displacement effects. A diffusion sample is by definition nonhomogeneous. To minimize the effect on the correction calculation procedures of this artifact, the X-rays should always be collected in a direction normal to the diffusion axis. In this case the emergent X-ray traverses a region of the sample having the same composition as the microvolume from which it was emitted. Point counting must be used for accuracy and multiple scans performed. It is highly desirable that all constituents of the sample be analyzed simultaneously. Many diffusional measurements carried out by microprobe analysis have been reported [353], and the method is now routine.

The principal disadvantages of the EPMA technique are its inability to detect isotope differences and its limited spatial resolution. The latter limitation is largely overcome by using an analytical electron microscope.

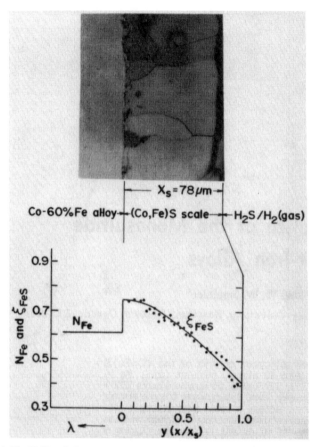

Figure 3.25 A (Co, Fe)S scale formed on Co–60% Fe alloy at 873 K by reaction with a H_2/H_2S gas mixture. Compositional analysis performed by simultaneous EPMA of Co, Fe, and S. Reprinted by permission of the publisher, The Electrochemical Society, Inc., from D. J. Young, T. Narita, and W. W. Smeltzer, *J. Electrochem. Soc.*, **127**, 679 (1980).

ANALYTICAL ELECTRON MICROSCOPY

Analytical electron microscopy is carried out using a scanning transmission electron microscope and analyzing the X-rays produced by interaction between the electron beam and the sample. The sample is prepared as a thin foil (thickness of ca. 100 nm), and it is this feature that leads to greatly improved spatial resolution.

In a bulk sample, as used in EPMA, the extent of electron scattering is independent of the electron beam diameter and depends on the extent to which the electrons penetrate the solid. Thus, as shown in Figure 3.26, a teardrop shaped X-ray excitation volume is formed, and defines the achievable resolution. Clearly, the diameter of this volume can be reduced by shortening the electron path, and the practicable way of doing this is by using a thin foil specimen. The

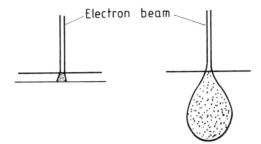

Figure 3.26 Schematic illustration of how electron scattering within a solid increases the X-ray excitation volume. The effect is minimized by using a thin-foil specimen.

spatial resolution is now a function of beam diameter b and film thickness d equal to $(b^2 + d^2)^{1/2}$. Values of 20–40 nm are easily realizable, which is about 100 times better than in EPMA. Because the diffusion distance varies approximately as $(Dt)^{1/2}$, an improvement factor of 10^4 in the minimum value of D measurable results. Despite this dramatic advantage, analytical electron microscopy has not supplanted EPMA, principally because of the difficulties of specimen preparation.

A thin-foil specimen containing the diffusion axis in its plane and encompassing the diffusion distance is required. The specimen preparation procedure is illustrated in Figure 3.27. A 3-mm-diameter cylinder is cut from the couple so as to contain the couple interface. Disks cut from this cylinder are then thinned mechanically and subjected to electrochemical polishing. Final thinning is accomplished by ion beam milling. When dealing with nonmetallic samples, the electrochemical polishing step is replaced by abrasive dimpling [354]. The

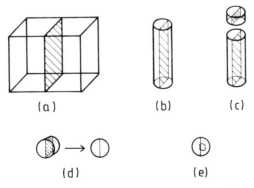

Figure 3.27 Preparation of a diffusion couple, thin-foil sample containing the original couple interface shown here as a shaded plane: (a) original couple, (b) 3-mm rod cut from couple, (c) disks cut from rod, (d) disks thinned mechanically, (e) final thinning to perforation using electrochemical polishing and ion beam milling.

general area of specimen preparation has been reviewed by Goldstein and co-workers [344].

Because beam currents in analytical electron microscopy are so low, wavelength-dispersive spectrometry cannot be employed and EDS is used instead. For EDS analysis of thin foils, X-ray intensities are converted to concentrations using the Cliff–Lorimer [355] ratio equation

$$\frac{C_1}{C_2} = K_{12}\left(\frac{I_1}{I_2}\right) \tag{152}$$

Here the C_i are concentrations, I_i are intensities, and K_{12} is a proportionality constant. This constant is characteristic of the instrument as well as the elements and must be determined by calibrating with a series of homogeneous samples of known composition. An example of the sort of diffusion profile data obtainable is shown in Figure 3.28 [356].

A problem is encountered with samples that are thermal and electrical insulators. Electron beam heating can then lead to specimen damage, including much enhanced diffusion of mobile species. Analysis for these species will show a downward drift with time, and measurement times are of necessity limited.

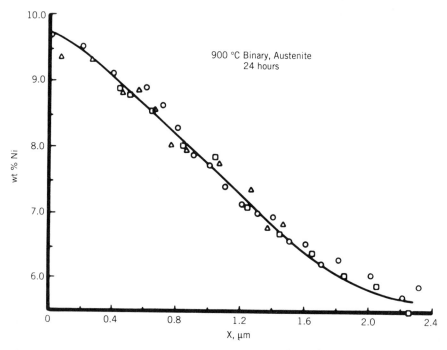

Figure 3.28 Experimental Ni concentration gradient measured by analytical electron microscopy in an Fe–5% Ni versus Fe–10% Ni austenite couple diffused at 1184 K for 24 h. Reprinted with permission from D. C. Dean and J. I. Goldstein, *Metall. Trans. A*, **17A**, 1131 (1986).

As noted earlier, a serious disadvantage of EDS is its insensitivity to light elements. For these elements, therefore, an alternative analytical technique is required. One suited to the electron microscope is electron energy loss spectroscopy (EELS). In this technique, the energy of transmitted electrons is measured, and the energy loss peaks characteristic of the sample elements are defined. Quantification of the results remains difficult, but it has been demonstrated [357] that signal intensity variations due to diffraction, uneven illumination, and varying specimen thickness can be compensated for.

ION BEAM TECHNIQUES

Commercial instrumentation for ion microprobe mass analysis is now available and should prove an extremely useful technique. The diffusion couple is sectioned along the diffusion axis, the resulting surface is polished flat, and the concentration profile is measured in the ion microprobe. A finely focused, high-energy beam of ions is directed at the sample surface where it dislodges the constituent atoms of the sample. The technique is thus destructive. The emergent species are then analyzed by mass spectrometry. This technique has the advantages of very high analytical sensitivities (10^{-9}–10^{-4} atom fraction); the ability to analyze light elements; and, obviously, the capability of isotope analysis. The technique has been described in detail [342, 343, 358, 359].

Spatial resolutions achieved with the ion microprobe are about 1 μm. As this is not significantly better than an electron microprobe; and since calculation of concentrations from signal intensities is more difficult, the choice of this technique is appropriate only for atomic numbers of 8 or less, for isotopic discrimination, or for trace level impurity diffusion. Much improved spatial resolutions are obtained using sputtering to achieve depth profiling.

A sample surface, normal to the diffusion axis, is continuously ion beam sputtered over a relatively large area. The dislodged ions are analyzed by mass spectrometry and this variation of the technique is most commonly known as secondary ion mass spectrometry (SIMS). Because there is no need for a finely focused ion beam, the instrumentation is relatively cheap. Slow sputter rates and high analytical sensitivities yield potentially good spatial resolutions. The use of SIMS in diffusion measurements was described by Macht and Naundorf [342] who estimate a practical lower limit on D of 3×10^{-19} cm^2/s. Application of the technique to oxides was developed by Mitchell and co-workers [360].

Ion beams with energies in the million electronvolts range can also be used to measure compositional variation within a solid without sectioning because high-energy beams can penetrate the solid. The beam is directed at a sample surface that is at right angles to the diffusion axis. Thus the beam penetrates along the axis, sampling the solid over a range of depths. Two types of event can be monitored to gain the desired information. High-energy ions incident on the sample undergo elastic (Rutherford) backscattering when they encounter target nuclei. The scattered ions sustain an energy loss due to target nucleus recoil. The extent of recoil varies with atomic number, thereby leading to element identification. Because the ions lose energy along the path through the solid, depth resolution is possible. Recent applications have been reviewed by Myers

[361]. Spatial resolution and maximum penetration vary with choice of incident ions. With He^+ a resolution of 20 nm and depth of 1 μm are obtained, whereas with H^+ the corresponding figures are 50 nm and 10 μm. Alternatively, it is possible to observe the products of nuclear reactions resulting from inelastic collisions between the incident ions and the target species.

A particularly important example for oxygen diffusion [362, 363] is proton activation of ^{18}O by the reaction $^{18}O(p, \alpha)^{15}N$. A monoenergetic beam of protons is used to bombard an isotopically exchanged solid. The energy spectrum of the α particles corresponds to the range of depths from which they were emitted, and its analysis yields the ^{18}O concentration profile. Spatial resolution is in the range 0.1–1 μm and analysis depths can be several micrometers.

The use of other nuclear reactions in analyzing light element profiles in crystals has been reviewed by Lanford and co-workers [364]. A particularly useful technique is that of resonant nuclear reaction analysis. Here a monoenergetic beam incident on the sample surface is attenuated with increasing penetration depth. When the resonant energy is reached, the nuclear reaction occurs and is detected by its characteristic emission. By varying the incident beam energy one varies the depth at which the reaction is excited and hence obtains a concentration profile.

A novel use of ion beams is in implanting a profile of diffusant species at concentrations that can be greater than thermal solubilities. Thus, easily measurable concentrations are produced, and problems involving phase boundary kinetics associated with the thermal dissolution process are avoided. Observation of the subsequent broadening of the implanted profile leads to an estimate of D, but the results are influenced by the degree of damage induced during implantation [365, 366]. This disadvantage may be overcome by observing the subsequent diffusion of the implanted species through undamaged regions of the host lattice remote from the implanatation zone [366].

AUGER ELECTRON SPECTROSCOPY
A technique that is becoming increasingly widely available, Auger electron spectroscopy (AES), uses an electron beam to excite Auger electrons within surface atoms [367]. The escape depth of these rather low-energy electrons is about 1 nm and high resolution is attainable. Since the technique is one of surface analysis, carefully cleaned surfaces are essential. Ion sputtering is used for this purpose and depth profiling of diffusion gradients in thin solid regions is achieved readily. Auger analysis is performed on the new surface being revealed at the bottom of the sputtered crater. Alternatively, the electron beam can be scanned laterally with a spatial resolution of about 0.1 μm.

5.2.4   Calculation of Diffusion Coefficients

TRACER DIFFUSION COEFFICIENTS
Because the isotopically labeled species used in tracer experiments is chemically identical with the solvent species, and because tracer concentrations are

generally small, it is reasonable to assume that the tracer diffusion coefficient D_{Ti} is independent of concentration and, hence, of position. If the tracer is supplied as a thin film on the solid surface, then the standard solution [368] of the diffusion equation for this boundary condition is

$$C = \frac{M \, \exp(-x^2/4D_{Ti}t)}{(\pi D_{Ti}t)^{1/2}} \tag{153}$$

which can be rearranged as

$$\ln C = \ln \left[\frac{M}{(\pi D_{Ti}t)^{1/2}} \right] - \frac{x^2}{4D_{Ti}t} \tag{154}$$

where M is the amount of the tracer species in the diffusion system at $t = 0$. A plot of $\log C$, as measured in the series of sections, against x^2 is then linear with slope S, if the boundary condition and Fick's law have been satisfied. The diffusion coefficient is then obtained from the slope as

$$D_{Ti} = -\frac{1}{4St} \tag{155}$$

Since a logarithmic plot is used, absolute calibration of radioactivity versus concentration are not required.

If the plot is not linear, the reason for this situation should be examined. The usual explanation of such behavior is that the diffusion process is not random; that is, certain paths such as dislocations or grain boundaries have been favored. Alternative solutions of the diffusion equation for these situations are available and can be tested [369–371].

Where the residual activity in the solid is measured after removal of the sections, an alternative form of (153) is required. If the total initial activity before removal of any sections is $A(0)$, and the activity remaining after removal of a total thickness x is $A(x)$, then integration of (153) over the indicated limits yields

$$\frac{A(x)}{A(0)} = \text{erfc} \left[\frac{x}{(4D_{Ti}t)^{1/2}} \right] \tag{156}$$

Here erfc represents the error function complement $(= 1 - \text{erf})$, tabulated values of which are available [368]. In arriving at this result, it has been assumed that no attenuation in radioactivity results from absorption within the solid. Absorption corrections have been developed [372]. Plotting the inverse function $\text{erfc}^{-1}[A(x)/A(0)]$ against x leads to a straight line whose slope yields an estimate of D_{Ti}. The function erfc^{-1} is defined by $x = \text{erfc}^{-1}(y)$ if $y = \text{erfc}(x)$ and is obtainable from the usual tabulation of $\text{erfc}(x)$.

In a gas–solid diffusion couple, the tracer concentration at the solid's surface data. At high pressures, as the gas becomes denser and the molecular interactions are more frequent and complex, the simple picture of the kinetic theory no longer holds. In fact, one finds that the viscosity of the gas becomes strongly dependent on pressure and its temperature dependence deviates substantially from the low density limit as represented by (20).

A generalized correlation for the viscosity of gases as a function of pressure and temperature can be formulated based on the corresponding states principle. According to this principle a nearly universal response can be obtained when $\mu_r = \mu/\mu_c$ is correlated with $P_r = P/P_c$ and $T_r = T/T_c$, where the subscript c denotes the critical point of the gas and μ_c can be estimated from simple empirical correlations [25]. Although this approach is empirical in origin, it has been fairly successful in describing the pressure and temperature dependence of viscosity for many gases. One such correlation is shown in Figure 5.2 and is seen to cover a wide range of conditions, including the gaseous dilute and dense states as well as the liquid state. Although this correlation is expected to predict the viscosity of a variety of gases to within $\pm 20\%$ [26] and should be used merely for estimation purposes, it clearly illustrates the general effects of temperature and pressure on viscosity. We see, for example, that the temperature dependence for dilute and dense gases is markedly different from that for liquids, thus suggesting that different molecular mechanisms govern momentum transport in the liquid and gaseous states. It is also noted that the kinetic theory [see (20)] can explain satisfactorily only the low-density (low-P) limit but is clearly inadequate at medium and high densities.

Before turning to a discussion of Newtonian liquids, we note that the general techniques for measuring the viscosity of gases are essentially identical to those used for measuring the viscosity of liquids. The main instrumental differences between liquid and gas viscometers arise from the special requirements to contain and move the gas within the viscometer and from the low forces generated by the gas that call for very sensitive stress (pressure, torque) sensors. The high compressibility of gases may also pose some difficulties when the viscosity is measured at high pressures. A general discussion of the design and operation of gas viscometers is given by Chierici and Paratella [27].

2.2 Newtonian Liquids

Molecular models for momentum transport in liquids are not as well developed as those for gases. From a strictly mathematical viewpoint, extension of the kinetic theory to liquids is a formidable task. The main difficulties arise from the fact that liquid molecules are very close-packed compared to gas molecules and their relative motions are governed by complex intermolecular interactions (molecular collisions that govern momentum transport in gases are not an important transport mechanism in liquids!). A general approach for treating the motion of liquids on a microscopic level was taken by Eyring and his co-workers [28]. Their model is based on a general theory of rate processes and although it does not constitute a molecular theory *per se*, it provides some useful insights.

If the diffusion coefficient is essentially constant for all experimental values of x and t, it can be evaluated by any of the standard methods, described in detail by Crank [368]. For a semiinfinite diffusion couple subject to the boundary conditions

$$C = C^0$$

for $x > 0$, $t = 0$ and for $x = +\infty$, $t > 0$, and

$$C = C^1 \tag{160}$$

for $x < 0$, $t = 0$ and for $x = -\infty$, $t > 0$.

The solution to the diffusion equation for fixed D is

$$\frac{C-(C^0+C^1)/2}{C^0-C^1} = \frac{1}{2}\,\mathrm{erf}\left[\frac{x}{(4Dt)^{1/2}}\right] \tag{161}$$

In reality, however, this is very seldom the case, and the following analysis must be used.

Beginning with Fick's second law for a binary system,

$$\frac{\partial C}{\partial t} = \frac{\partial}{\partial x}\left[D\left(\frac{\partial C}{\partial x}\right)\right] \tag{162}$$

with the boundary conditions of (160), and introducing the variable $y = x/\sqrt{t}$, we obtain the ordinary differential equation

$$-\frac{y}{2}\left(\frac{dC}{dy}\right) = \frac{d}{dy}\left[D\left(\frac{dC}{dy}\right)\right] \tag{163}$$

with the conditions

$$C = C^0 \quad (y = +\infty)$$
$$C = C^1 \quad (y = -\infty)$$

This is the analysis of Matano [376] and is based on an earlier result of Boltzmann's research [377].

Equation (163) can be integrated to give

$$\int_{c^0}^{c} \frac{y}{2}\,dc = -D\left(\frac{dC}{dy}\right)\bigg|_c \tag{164}$$

where the fact that $dC/dy = 0$ as $C \to C^0$ has been used. Rearrangement then leads to

$$D = -\frac{1}{2}\left(\frac{dy}{dC}\right)\int_{c^0}^{c} y\,dC \tag{165}$$

with

$$\int_{c^0}^{c^1} x\,dC = 0$$

The second expression defines the origin, known as the Matano interface. This analysis is valid only if the molar volumes of the mixtures are independent of concentration over the range existing in the system. Numerical evaluation of the integral in and of the differential dy/dC is used, with the principal error residing in the latter.

CHEMICAL DIFFUSION COEFFICIENTS IN TERNARY SYSTEMS

In ternary and higher order systems, diffusion equations in the form of

$$J_i = -\sum_{j=1}^{n-1} D_{ij}\left(\frac{\partial C_j}{\partial x}\right) \tag{166}$$

must be used to describe the flux of component i in an n-component system. In general, the off-diagonal coefficients D_{ij} $(i \neq j)$ are nonzero and must be evaluated from the results of the diffusion profile analysis. For a ternary system there are a total of four coefficients to be found.

If the diffusion coefficients can be regarded as constant, the measured concentration profiles must satisfy solutions to the diffusion equations obtained under the assumption of constants D or D_{ij}. The first analysis of this kind for a solid was provided by Kirkaldy [378] in relation to the data of Darken [379]. For the boundary conditions of a ternary or higher order infinite diffusion couple, there are known analytic solutions [380] and trial values for the D_{ij} can be used to predict the concentration profiles. Comparison of these with the experimental curves leads to better estimates of the D_{ij}. An iterative computation based on minimization of the residuals in concentration, $\Sigma[c_i(\text{measured}) - c_i(\text{calculated})]^2$, is employed. The usual first approximation [381] is to assign the D_{ii} values of the self-diffusion coefficients and to set the D_{ij}, $i \neq j = 0$. Schut and Cooper [382] proposed a method for determining the full matrix from a single diffusion couple.

The values for the D_{ij} determined in this way depend on the choice of solvent from among the components. It is therefore essential to specify the solvent when quoting multicomponent diffusion coefficients. A convention is sometimes adopted whereby the solvent component used as reference frame is denoted by a superscript as in D_{ij}^k.

The approximation of concentration-independent values of the D_{ij} is seldom useful over significant ranges of concentration and must be abandoned. Instead, the multicomponent extension of the Matano analysis [378] is used for the infinite couple boundary conditions of (160), the integrals can be written

$$\int_{c_i^0}^{c_i} y \, dC_i = -\sum_{k=1}^{n-1} D_{ik} \frac{dC_k}{dy}\bigg|_c \qquad (167)$$

with the origin given by

$$\int_{c_i^0}^{c_i^1} x \, dC_i = 0 \qquad (168)$$

The value of x given by (168) must be the same for all components if the analysis is to be applicable.

Using (167) and (168), it is possible to evaluate the four D_{ij} for a ternary system from two diffusion experiments, provided the diffusion paths have different terminal compositions but one common composition. Figure 3.29 shows two hypothetical diffusion paths (a) and (b) in a ternary-phase diagram, with a common point A. The D_{ij} that are evaluated refer to the composition at the common point. Therefore, in the most favorable circumstances, a set of K couples can determine all D_{ij} at $(K-1)!$ distinct composition points, although anything approaching this figure is unlikely to be attained in an experimental study. Ziebold and Ogilvie [383] were able to obtain the D_{ij} at 23 compositions of the system copper–silver–gold using 13 different diffusion couples. The difficulty of extension to higher order systems is apparent.

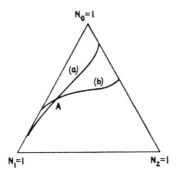

Figure 3.29 Two diffusion paths (a) and (b) in a ternary phase diagram. Each diffusion couple passes through the common composition point A, permitting evaluation of the D_{ij} at this point. Reprinted with permission from P. J. Dunlop, B. J. Steel, and J. E. Lane, "Experimental Methods for Studying Diffusion in Liquids, Gases and Solids," in A. Weissberger and B. W. Rossiter, Eds., *Physical Methods of Chemistry*, Vol. I, Part IV, Wiley-Interscience, New York, 1975. Copyright © 1975 by John Wiley & Sons, Inc.

If one concentration profile has a minimum, a maximum, or both, then it is possible to evaluate one off-diagonal and one on-diagonal coefficient at the concentration C_i^e of each extreme point. If component 1 has the extreme point in the concentration profile, then rearranging (167) gives

$$D_{12}(C_1^e, C_2^e) = -\int_{C_1^0}^{C_1^e} y \, dC_1 \bigg/ \frac{dC_2}{dy}\bigg|_{C_1^e, C_2^e} \tag{169}$$

$$D_{22}(C_1^e, C_2^e) = -\int_{C_2^0}^{C_2^e} y \, dC_2 \bigg/ \frac{\partial C_2}{dy}\bigg|_{C_1^e, C_2^e} \tag{170}$$

All the quantities on the right-hand side of (169) and (170) can be obtained from the two concentration profiles for the single diffusion experiment.

The sensitivity with which off-diagonal coefficients can be measured varies greatly with the experimental design. We consider here a ternary system in which component 3 is the solvent. It has long been recognized [278, 384, 385] as expedient to set the concentration of one component (say C_1) initially constant while providing an initial step change in C_2 as shown in Figure 3.30. Redistribution of component 1 is then due primarily to the diffusional cross effect represented by D_{12}. Conversely, the flow of component 2 is not affected significantly by the small gradients that develop in C_1 and, to a reasonable approximation, we can set $D_{21} = 0$. The consequences were explored analytically by Kirkaldy and co-workers [386, 387]. The available data on ternary metallic systems were summarized and reviewed by Kirkaldy and Young [388].

THE KIRKENDALL EFFECT

In a binary diffusion couple, the intrinsic mobilities of the two components are generally different and lead to a compensating bulk material flow. This effect,

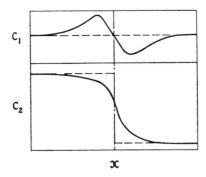

Figure 3.30 Concentration profiles (concentration C versus position x) of components in a ternary diffusion couple. Component 1 has a uniform distribution at $t = 0$ and a maximum and minimum concentration at later times. Reprinted with permission from P. J. Dunlop, B. J. Steel, and J. E. Lane, "Experimental Methods for Studying Diffusion in Liquids, Gases and Solids," in A. Weissberger and B. W. Rossiter, Eds., *Physical Methods of Chemistry*, Vol. I, Part IV, Wiley-Interscience, New York, 1975. Copyright © 1975 by John Wiley & Sons, Inc.

when observed in solids, is known as the Kirkendall effect [326]. The quantitative results are accessible from a comparison between the laboratory frame of reference (within which chemical diffusion measurements are made) and the lattice-fixed reference frame (to which tracer diffusion measurements are relative) [389].

For the system A–B, assuming equal partial molar volumes for the two species, the laboratory frame is defined by

$$J_A = -J_B \tag{171}$$

In the lattice-fixed frame, however

$$J'_A + J'_B = -J_v \tag{172}$$

Where the subscript v denotes vacancies. The two frames are related by

$$J_A = J'_A + C_A v$$
$$J_B = J'_B + C_b v \tag{173}$$

and v is the velocity of the lattice-fixed frame relative to the laboratory frame. It follows then that

$$v = -\frac{J'_A + J'_B}{C_A + C_B} \tag{174}$$

and

$$J_A = N_B J'_A - N_A J'_B \tag{175}$$

where the N_i are mole fractions. Since

$$J'_A = -D_A \nabla C_A, \qquad J'_B = -D_B \nabla C_B \tag{176}$$

where the D_i are designated intrinsic coefficients [390]. Then,

$$v = V_A(D_A - D_B)\nabla C_A \tag{177}$$

for V_A the molar volume of A, and

$$J_A = -\tilde{D}\nabla C_A \tag{178}$$

where

$$\tilde{D} = N_B D_A + N_A D_B \tag{179}$$

It is readily demonstrated [389] that

$$\tilde{D} = [N_{\mathrm{B}} D_{\mathrm{Ti(A)}} + N_{\mathrm{A}} D_{\mathrm{Ti(B)}}] \left[1 + \frac{d \ln \gamma}{d \ln N} \right] \tag{180}$$

where γ is the activity coefficient. Thus, if the thermodynamic factor $d \ln \gamma / d \ln N$ is known, measurement of both $\tilde{D}$ and v yields estimates of the two intrinsic diffusion coefficients.

In infinite diffusion couple measurement, the movement of material is restricted to a rather narrow zone near the couple interface, leaving the ends unaffected. The original interface, therefore, moves relative to the couple ends, and we can observe this movement by using inert markers [391, 392]. Although there may be some small additional corrections to (177) and (180) [393, 394], the result (180) can be viewed as a useful interpolation formula.

The preceding analysis can be extended to ternary systems, but the determination of the six independent intrinsic coefficients by measuring $\tilde{D}$ values and the use of marker experiments is a formidable task [395]. A combination of tracer and chemical diffusion measurements is to be preferred [388, 396–398].

5.3   Relaxation Methods for Chemical Diffusion Coefficients

Diffusion profile measurements are rather time-consuming and/or expensive. Thus, there is a considerable attraction in the idea of measuring some macroscopic property of a sample, such as its mass, while material diffuses into or out of it. The obvious candidate for such a measurement is a solid–gas diffusion couple.

We consider a metal oxide in equilibrium with a given value p_{O_2}. If it is subjected to an abrupt change in p_{O_2}, then the process of reequilibration to a new homogeneous state involves the creation or annihilation of defects at the solid–gas interface and their diffusion into or out of the bulk solid. If the surface process is effectively at equilibrium, the diffusion of the defects, commonly cation defects, controls the rate of reequilibration. The process can be monitored thermogravimetrically [399] or by observing the change in electrical properties of the compound [400, 401] as the level of charged defects changes. For long diffusion times the change in weight Δw in a compound reequilibrating to a new stoichiometry is given by Neuman [402] as

$$\left(1 - \frac{\Delta w}{\Delta w_\infty} \right) = \frac{8}{\pi^2} \exp \left(- \frac{\tilde{D} \pi^2 t}{4 l^2} \right) \tag{181}$$

when the sample is a thin plate of thickness $2l$ and Δw_∞ is the weight change after the new equilibrium state is reached. It has been assumed that $\tilde{D}$ is independent of composition.

A variant on this method has been proposed by Rosenburg [403]. It involves changing the partial pressure of oxidant abruptly during the course of a metal oxidation reaction, thereby perturbing the oxidation rate. A general solution for a sample on which the scale is much thinner than its lateral dimensions was given by Fryt [404]:

$$\frac{\Delta w}{\Delta x_0 \Delta m} = \frac{\tilde{D}t}{x_0^2} + \frac{2}{\pi^2} \sum_{n=1}^{\infty} \frac{1}{n^2} \left[1 - \exp\left(-\frac{n^2 \pi^2 \tilde{D}t}{x_0^2} \right) \right]$$
(182)

where x_0 is the oxide scale thickness at the time of change in p_{O_2}, Δx_0 is the subsequently observed change, and Δm is the molar point defect concentration difference across the scale thickness. Again, it was assumed that $\tilde{D} \neq f(m)$. Several applications of this technique were reviewed by Mrowec [405]. Fryt and co-workers [406] demonstrated that the results obtained by the two methods described by (181) and (182) are closely similar for $Fe_{1-\delta}S$. The situation where the surface process contributes to rate control was analyzed [407] and the appropriately modified form of (182) was developed. It is clear that these methods are limited in applicability to situations in which only one lattice species is mobile and, hence, have value only for binary diffusion studies.

Yet another method for measuring chemical diffusion in an oxide has been proposed by Kofstad [408, 409]. In this method a metal is oxidized to form a uniform scale and then exposed to a high vacuum. Under conditions when the oxide is stable and nonvolatile but the metal has a relatively high-vapor pressure, outward diffusion of metal through the oxide leads to metal vaporization and sample weight loss. Measurement of the rate of weight loss then leads to an estimate for D. The method depends for its success on preservation of both oxide stoichiometry and contact between metal and oxide.

References

1. P. J. Dunlop, B. J. Steel, and J. E. Lane, "Experimental Methods for Studying Diffusion in Liquids, Gases, and Solids," in A. Weissberger and B. W. Rossiter, Eds., *Physical Methods of Chemistry*, Vol. 1, Part IV, Wiley-Interscience, New York, 1972, p. 205.

2. W. Jost, *Diffusion in Solids, Liquids, and Gases*, Academic, New York, 1960.

3. J. Crank and G. S. Park, Eds., *Diffusion in Polymers*, Academic, London and New York, 1968.

4. E. A. Mason and T. R. Marrero, "The Diffusion of Atoms and Molecules," in D. R. Bates and I. Esterman, Eds., *Advances in Atomic and Molecular Physics*, Vol. VI, Academic, New York, 1970, p. 155.

5. E. L. Cussler, *Multicomponent Diffusion*, Elsevier, Amsterdam, 1976.

6. H. J. V. Tyrrell and K. R. Harris, *Diffusion in Liquids*, Butterworths, Borough Green, Kent, 1984.

7. E. L. Cussler, *Diffusion: Mass Transfer in Fluid Systems*, University Press, Cambridge, 1984.

8. A. Fick, *Ann. Phys. Chem.*, **94**, 59 (1855).

9. L. J. Gosting, "Measurements and Interpretation of Diffusion Coefficients of Proteins," in M. L. Anson, K. Bailey, and J. T. Edsall, Eds., *Advances in Protein Chemistry*, Vol. XI, Academic, New York, 1956, p. 429.

10. J. G. Kirkwood, R. L. Baldwin, P. J. Dunlop, L. J. Gosting, and G. Kegeles, *J. Chem. Phys.*, **33**, 1505 (1960).

11. D. G. Miller, V. Vitagliano, and R. Sartario, *J. Phys. Chem.*, **90**, 1509 (1986).

12. H. Kim, *J. Phys. Chem.*, **70**, 562 (1966).

13. H. Kim, *J. Phys. Chem.*, **73**, 1716 (1969).

14. L. Onsager, *Ann. NY Acad. Sci.*, **46**, 241 (1945).

15. G. S. Hartley and J. Crank, *Trans. Faraday Soc.*, **45**, 801 (1949).

16. R. L. Baldwin, P. J. Dunlop, and L. J. Gosting, *J. Am. Chem. Soc.*, **77**, 5235 (1955).

17. L. S. Darken and R. W. Gurry, *Physical Chemistry of Metals*, McGraw-Hill, New York, 1953, p. 438.

18. O. Lamm, *J. Phys. Chem.*, **61**, 948 (1957) and J. S. Kirkaldy, *Can. J. Phys.*, **35**, 435 (1957).

19. P. J. Dunlop and L. J. Gosting, *J. Am. Chem. Soc.*, **77**, 5238 (1955).

20. H. Fujita and L. J. Gosting, *J. Am. Chem. Soc.*, **78**, 1099 (1956).

21. I. J. O'Donnell and L. J. Gosting, "The Concentration Dependence of the Four Diffusion Coefficients of the System NaCl–KCl–H_2O at 25°C," in W. J. Hamer, Ed., *The Structure of Electrolyte Solutions*, Wiley, New York, Chapman & Hall, London, 1959, p. 160.

22. L. A. Woolf, D. G. Miller, and L. J. Gosting, *J. Am. Chem. Soc.*, **84**, 317 (1962).

23. R. P. Wendt, *J. Phys. Chem.*, **66**, 1279 (1962).

24. M. Dole, *J. Chem. Phys.*, **25**, 1082 (1956).

25. L. Onsager, *Phys. Rev.*, **37**, 405 (1931).

26. L. Onsager, *Phys. Rev.*, **38**, 2265 (1931).

27. P. J. Dunlop, *J. Phys. Chem.*, **61**, 994 (1957).

28. L.-O. Sundelöf, *Ark. Kemi*, **20**, 369 (1963).

29. P. J. Dunlop, *J. Phys. Chem.*, **69**, 1693 (1965).

30. J. G. Albright and R. Mills, *J. Phys. Chem.*, **69**, 3120 (1965).

31. P. F. Curran, A. E. Taylor, and A. K. Solomon, *Biophys. J.*, **7**, 879 (1967).

32. K. R. Harris, C. K. N. Pua, and P. J. Dunlop, *J. Phys. Chem.*, **74**, 3518 (1970).

33. E. A. Mason, *J. Chem. Phys.*, **27**, 76 (1957).

34. P. J. Dunlop, C. M. Bignell, W. L. Taylor, and B. A. Meyer, *J. Chem. Phys.*, **87**, 3591 (1987).

35. R. Mills, *J. Phys. Chem.*, **77**, 685 (1973).

36. S. R. deGroot and P. Mazur, *Non-Equilibrium Thermodynamics*, North-Holland, Amsterdam, 1962, p. 168.

37. J. Crank, *The Mathematics of Diffusion*, Oxford University Press, London, 1956, pp. 148, 237.

38. W. Hogervorst, *Physica*, **51**, 59 (1971).

39. J. Loschmidt, *Akad. Wiss. Wien*, **61**, 367 (1870).

40. L. E. Boardman and N. E. Wild, *Proc. R. Soc. London*, **A162**, 511 (1937).

41. F. T. Wall and G. A. Kidder, *J. Phys. Chem.*, **50**, 235 (1946).

42. R. A. Strehlow, *J. Chem. Phys.*, **21**, 2101 (1953).

43. J. N. Holsen and M. R. Strunk, *Ind. Eng. Chem. Fundam.*, **3**, 143 (1964).

44. I. Amdur, J. W. Irvine, E. A. Mason, and J. Ross, *J. Chem. Phys.*, **20**, 436 (1952).

45. I. Amdur and T. F. Schatzke, *J. Chem. Phys.*, **38**, 188 (1963).

46. I. Amdur and A. P. Malinauskas, *J. Chem. Phys.*, **42**, 3355 (1965).

47. C. A. Boyd, N. Stein, V. Steingrimsson, and W. F. Rumpel, *J. Chem. Phys.*, **19**, 548 (1951).

48. R. E. Bunde, *University of Wisconsin Naval Research Laboratory Report CM-850*, University of Wisconsin, Madison, Wisconsin, 1955.

49. P. E. Suetin and B. A. Ivakin, *Zh. Tekh. Fiz.*, **34**, 1155 (1964); *Sov. Phys. Tech. Phys.*, **9**, 866 (1964).

50. P. E. Suetin, G. T. Shchegolev, and R. A. Klestov, *Zh. Tekh. Fiz.*, **29**, 1058 (1959); *Sov. Phys. Tech. Phys.*, **4**, 964 (1959).

51. S. Ljunggren, *Ark. Kemi*, **24**, 1 (1965).

52. M. Manner, "Diffusion in Gases," doctoral dissertation, University of Wisconsin, 1967; University Microfilms Inc., Ann Arbor, MI, 1969.

53. I. R. Shankland and P. J. Dunlop, *Physica*, **100A**, 64 (1980).

54. D. P. Shoemaker and C. W. Garland, *Experiments in Physical Chemistry*, 2nd ed., McGraw-Hill, New York, 1967, p. 96.

55. R. Paul, *Phys. Fluids*, **3**, 905 (1960).

56. H. S. Harned and D. M. French, *Ann. NY Acad. Sci.*, **46**, 267 (1945).

57. H. S. Harned and R. L. Nuttall, *J. Am. Chem. Soc.*, **69**, 736 (1947).

58. P. J. Carson, P. J. Dunlop, and T. N. Bell, *J. Chem. Phys.*, **56**, 5310 (1972).

59. G. R. Staker and P. J. Dunlop, *Chem. Phys. Lett.*, **42**, 419 (1976).

60. T. N. Bell, I. T. Shankland, and P. J. Dunlop, *Chem. Phys. Lett.*, **45**, 445 (1977).

61. M. A. Yabsley and P. J. Dunlop, *J. Phys. E*, **8**, 834 (1975).

62. G. R. Staker, P. J. Dunlop, K. R. Harris, and T. N. Bell, *Chem. Phys. Lett.*, **32**, 561 (1975).

63. I. R. Shankland and P. J. Dunlop, *Chem. Phys. Lett.*, **39**, 557 (1976).

64. M. A. Yabsley and P. J. Dunlop, *Physica*, **85A**, 160 (1976).

65. R. D. Trengove, H. L. Robjohns, and P. J. Dunlop, *Physica*, **128A**, 486 (1984).

66. P. S. Arora and P. J. Dunlop, *J. Chem. Phys.*, **71**, 2430 (1979).

67. P. Codastefano, A. DiRusso, and V. Zanza, *Rev. Sci. Instrum.*, **48**, 1650 (1977).

68. P. Codastefano, M. A. Ricci, and V. Zanza, *Physica*, **92A**, 315 (1978).

69. P. Codastefano, D. Rocca, and V. Zanza, *Physica*, **96A**, 454 (1979).

70. R. A. Aziz and A. van Dalen, *J. Chem. Phys.*, **78**, 2402 (1983).

71. W. Hogervorst and J. Freudenthal, *Physica*, **37**, 97 (1967).

72. J. Freudenthal, *Physica*, **36**, 354, 365 (1967).

73. E. P. Ney and F. C. Armistead, *Phys. Rev.*, **71**, 14 (1946).

74. C. Barnes, *Physics*, **5**, 4 (1934).

75. B. K. Annis, A. E. Humphreys, and E. A. Mason, *Phys. Fluids*, **12**, 78 (1969).

76. J. W. S. Rayleigh, *The Theory of Sound II*, Dover, New York, 1945, pp. 203, 491.
77. J. C. Maxwell, *Electricity and Magnetism*, Vol. I, Oxford University Press, London, 1981, p. 434.
78. L. V. King, *Philos. Mag.*, **21**, 128 (1936).
79. P. Wirz, *Helv. Phys. Acta*, **20**, 3 (1947).
80. P. S. Arora, I. R. Shankland, T. N. Bell, M. A. Yabsley, and P. J. Dunlop, *Rev. Sci. Instrum.*, **48**, 673 (1977).
81. R. J. J. van Heijningen, A. Feberwee, A. van Oosten, and J. J. M. Beenakker, *Physica*, **32**, 1649 (1966).
82. R. J. J. van Heijningen, J. P. Harpe, and J. J. M. Beenakker, *Physica*, **38**, 1 (1968).
83. P. S. Arora, H. L. Robjohns, and P. J. Dunlop, *Physica*, **95A**, 561 (1979).
84. R. D. Trengove and P. J. Dunlop, *Physica*, **115A**, 339 (1982).
85. R. D. Trengove, K. R. Harris, H. L. Robjohns, and P. J. Dunlop, *Physica*, **131A**, 506 (1985).
86. P. J. Dunlop and C. M. Bignell, *Physica*, **145A**, 584 (1987).
87. D. Cain and W. L. Taylor, *J. Chem. Phys.*, **78**, 6220 (1983).
88. D. Cain and W. L. Taylor, *J. Chem. Phys.*, **71**, 3601 (1979).
89. G. A. Dubro and S. Weissman, *Phys. Fluids*, **13**, 2682, 2689 (1970).
90. N. J. Trappeniers and J. P. J. Michels, *Chem. Phys. Lett.*, **18**, 1 (1973).
91. P. E. Suetin, B. A. Kalinin, and A. E. Loiko, *Zh. Tekh. Fiz.*, **40**, 1735 (1970); *Sov. Phys. Tech. Phys.*, **15**, 1349 (1970).
92. A. E. Loiko, B. A. Ivakin, and P. E. Suetin, *Zh. Tekh. Fiz.*, **43**, 416 (1973); *Sov. Phys. Tech. Phys.*, **18**, 266 (1973).
93. A. E. Loiko, B. A. Ivakin, and P. E. Seutin, *Zh. Tekh. Fiz.*, **44**, 682 (1973); *Sov. Phys. Tech. Phys.*, **19**, 434 (1974).
94. A. E. Loiko, B. A. Ivakin, and P. E. Seutin, *Zh. Tekh. Fiz.*, **47**, 873 (1977); *Sov. Phys. Tech. Phys.*, **22**, 522 (1977).
95. H. F. Vugts, A. J. H. Boerboom, and J. Los, *Physica*, **44**, 219 (1969).
96. P. J. Dunlop, H. L. Robjohns, and C. M. Bignell, *J. Chem. Phys.*, **86**, 2922 (1987).
97. P. J. Dunlop, *Physica*, **145A**, 597 (1987).
98. A. S. Wahby and J. Los, *Physica*, **128C**, 243 (1985).
99. H. F. Vugts, A. J. H. Boerboom, and J. Los, *Physica*, **51**, 311 (1971).
100. A. S. M. Wahby, A. J. H. Boerboom, and J. Los, *Physica*, **75**, 500 (1974).
101. A. S. M. Wahby, M. Abdul-Rahman, and J. Los, *Physica*, **112A**, 214 (1982).
102. K. W. Reus, C. J. Zwakhals, and J. A. Smit, *Physica*, **100C**, 221 (1980).
103. C. J. Zwakhals and K. W. Reus, *Physica*, **100C**, 231 (1980).
104. P. Harteck and H. W. Schmidt, *Z. Phys. Chem. (Leipzig)*, **821**, 447 (1933).
105. D. G. Miller, A. W. Ting, J. A. Rard, and L. B. Eppstein, *Geochim. Cosmochim. Acta*, **50**, 2397 (1986), and other work by this group.
106. T. A. Renner and P. A. Lyons, *J. Phys. Chem.*, **78**, 2050 (1974).
107. L. J. Gosting and L. Onsager, *J. Am. Chem. Soc.*, **74**, 6066 (1952).
108. L. J. Gosting, *J. Am. Chem. Soc.*, **72**, 4418 (1950).
109. D. G. Miller, *J. Phys. Chem.*, **92**, 4222 (1988).

110. O. Bryngdahl and S. Ljunggren, *J. Phys. Chem.*, **64**, 1264 (1960).

111. O. Bryngdahl, *J. Opt. Soc. Am.*, **53**, 571 (1963).

112. J. Becsey and J. A. Bierlein, *Rev. Sci. Instrum.*, **49**, 227 (1978).

113. W. E. Wentworth, *J. Chem. Educ.*, **42**, 96, 162 (1965).

114. C. N. Pepela, B. J. Steel, and P. J. Dunlop, *J. Am. Chem. Soc.*, **92**, 6743 (1970).

115. M. Tanigaki, K. Kondo, M. Harada, and W. Eguchi, *J. Phys. Chem.*, **87**, 586 (1983).

116. W. Eguchi, M. Harada, M. Adachi, and M. Tanagaki, *J. Chem. Eng. Jpn.*, **17**, 472 (1984).

117. D. G. Miller, *J. Solution. Chem.*, **10**, 831 (1981).

118. J. G. Albright and B. C. Sherrill, *J. Solution Chem.*, **8**, 201 (1979).

119. J. M. Creeth, *J. Am. Chem. Soc.*, **77**, 6428 (1955).

120. H. Fujita and L. J. Gosting, *J. Am. Chem. Soc.*, **78**, 1099 (1956).

121. J. A. Rard and D. G. Miller, *J. Phys. Chem.*, **91**, 4614 (1987).

122. A. Revzin, *J. Phys. Chem.*, **76**, 3419 (1972).

123. G. R. Staker and P. J. Dunlop, *J. Chem. Eng. Data*, **18**, 61 (1973).

124. M. Mitchell and H. J. V. Tyrrell, *J. Chem. Soc. Faraday 2*, **68**, 385 (1972).

125. C. Skipp and H. J. V. Tyrrell, *J. Chem. Soc. Faraday 1*, **71**, 1744 (1975).

126. R. P. Wendt, *J. Phys. Chem.*, **66**, 1740 (1962).

127. G. Reinfelds and L. J. Gosting, *J. Phys. Chem.*, **68**, 2464 (1964).

128. H. Kim, *J. Phys. Chem.*, **74**, 4577 (1970).

129. T. J. McDougall and J. S. Turner, *Nature (London)*, **299**, 812 (1982).

130. T. J. McDougall, *J. Fluid Mech.*, **126**, 379 (1983).

131. D. G. Miller and V. Vitagliano, *J. Phys. Chem.*, **90**, 1706 (1986).

132. W. D. Comper, R. P. W. Williams, G. J. Checkley, and B. N. Preston, *J. Phys. Chem.*, **91**, 993 (1987).

133. D. G. Leaist and P. A. Lyons, *Aust. J. Chem.*, **33**, 1869 (1980).

134. D. G. Leaist and P. A. Lyons, *J. Phys. Chem.*, **86**, 564 (1982).

135. D. G. Leaist, *Can. J. Chem.*, **63**, 2933 (1985).

136. H. Fujita, *J. Phys. Chem.*, **63**, 242 (1959).

137. J. N. Agar and V. M. M. Lobo, *J. Chem. Soc. Faraday Trans. 1*, **71**, 1659 (1975).

138. V. M. M. Lobo and M. H. S. F. Teixeira, *Electrochim. Acta*, **24**, 565 (1979).

139. L. Onsager and R. M. Fuoss, *J. Phys. Chem.*, **36**, 2689 (1932).

140. K. Toukubo and K. Nakanishi, *J. Phys. Chem.*, **78**, 2281 (1974).

141. P. Passiniemi, S. Liukkonen, and Z. Noszticzius, *J. Chem. Soc. Faraday Trans. 1*, **73**, 1834 (1977).

142. P. Passiniemi, *J. Solution Chem.*, **12**, 801 (1983).

143. Z. Noszticzius, S. Liukkonen, P. Passiniemi, and J. Rastas, *J. Chem. Soc. Faraday Trans. 1*, **73**, 2537, 2836 (1972).

144. R. Freer and J. N. Sherwood, *J. Chem. Soc. Faraday Trans. 1*, **76**, 1021 (1980).

145. T. Hashitani and R. Tamamushi, *Trans. Faraday Soc.*, **63**, 369 (1967).

146. K. Tanaka, T. Hashitani, and R. Tamamushi, *Trans. Faraday Soc.*, **66**, 74 (1970).

147. J. N. Northrup and M. L. Anson, *J. Gen. Physiol.*, **12**, 543 (1928).

148. R. Mills and L. A. Woolf, *The Diaphragm Cell*, ANU Press, Canberra, Australia, 1968, p. 40.

149. R. H. Stokes, *J. Am. Chem. Soc.*, **72**, 763 (1950).

150. A. F. Collings, D. C. Hall, M. A. McCool, and L. A. Woolf, *J. Phys. E*, **4**, 1019 (1971).

151. A. J. Easteal and L. A. Woolf, *J. Phys. Chem.*, **85**, 1066 (1985).

152. G. Kosanovich and H. T. Cullinan, Jr., *Can. J. Chem. Eng.*, **49**, 753 (1971).

153. A. R. Gordon, *Ann. NY Acad. Sci.*, **46**, 285 (1945).

154. J. G. Firth and H. J. V. Tyrrell, *J. Chem. Soc.*, 1658 (1962).

155. J. K. Baird and R. W. Frieden, *J. Phys. Chem.*, **91**, 3920 (1987).

156. A. F. Collings, D. C. Hall, R. Mills, and L. A. Woolf, *J. Phys. E*, **4**, 425 (1971).

157. M. A. McCool and L. A. Woolf, *High Temp. High Pressures*, **4**, 85 (1972).

158. L. A. Woolf, *J. Chem. Soc. Faraday Trans. 1*, **71**, 784 (1975).

159. L. A. Woolf, *J. Chem. Soc. Faraday Trans. 1*, **78**, 583 (1982).

160. K. R. Harris, *Physica*, **94A**, 448 (1978).

161. R. L. Hurle and L. A. Woolf, *J. Chem. Soc. Faraday Trans. 1*, **78**, 2233 (1982).

162. K. Kircher, A. Schaber, and E. Obermeier, *Proceedings of the 8th Symposium on Thermophysical Properties*, Am. Soc. Mech. Engrs., New York, 1982, p. 297.

163. S. A. Sanni and H. P. Hutchinson, *J. Phys. E*, **1**, 1101 (1968).

164. W. F. Calus and M. T. Tyn, *J. Phys. E*, **7**, 561 (1974).

165. A. J. Easteal, W. E. Price and L. A. Woolf, *J. Chem. Soc. Faraday Trans. 1*, **85**, 1091 (1989).

166. M.-K. Ahn and Z. J. Derlacki, *J. Phys. Chem.*, **82**, 1930 (1978).

167. E. L. Cussler and P. J. Dunlop, *J. Phys. Chem.*, **70**, 1880 (1966).

168. G. P. Rai and H. T. Cullinan, Jr., *J. Chem. Eng. Data*, **18**, 213 (1973).

169. D. G. Leaist, *J. Chem. Soc. Faraday Trans. 1*, **83**, 829 (1987).

170. Z. Balenovic, M. N. Myers, and J. C. Giddings, *J. Chem. Phys.*, **52**, 915 (1970).

171. A. T.-C. Hu and R. Kobayashi, *J. Chem. Eng. Data*, **15**, 328 (1970).

172. A. Alizadeh, C. A. Nieto de Castro, and W. A. Wakeham, *Inter. J. Thermophys.*, **1**, 243 (1980).

173. J. C. Giddings and S. L. Seager, *J. Chem. Phys.*, **33**, 1579 (1960).

174. J. Bohemen and H. J. Purnell, *J. Chem. Soc.*, 360 (1961).

175. A. Griffiths, *Proc. Phys. Soc. London*, **23**, 190 (1911).

176. G. Taylor, *Proc. R. Soc. London*, **A219**, 186 (1953); **A223**, 446, 473 (1954).

177. R. Aris, *Proc. R. Soc. London*, **A235**, 67 (1956).

178. O. Levenspiel and W. K. Smith, *Chem. Eng. Sci.*, **6**, 227 (1957).

179. E. van Andel, H. Kramers, and A. de Voogdt, *Chem. Eng. Sci.*, **19**, 77 (1964).

180. M. E. Erdogan and P. C. Chatwin, *J. Fluid Mech.*, **29**, 465 (1967).

181. D. J. McConalogue, *Proc. R. Soc. London*, **A315**, 99 (1970).

182. R. J. Nunge, R. S. Lin, and W. N. Gill, *J. Fluid Mech.*, **51**, 363 (1972).

183. I. Swaid and G. M. Schneider, *Ber. Bunsenges. Phys. Chem.*, **83**, 969 (1979).

184. A. C. Ouano, *Ind. Eng. Chem. Fundam.*, **11**, 268 (1972).

185. E. Grushka and E. J. Kitka, *J. Phys. Chem.*, **78**, 2297 (1974).

186. K. C. Pratt and W. A. Wakeham, *Proc. R. Soc. London*, **A336**, 393 (1974); **A342**, 401 (1975).

187. K. C. Pratt and W. A. Wakeham, *J. Chem. Soc. Faraday Trans. 2*, **73**, 997 (1977).

188. W. A. Wakeham, *Faraday Symp. Chem. Soc.*, **15**, 145 (1980).

189. A. A. Usmanova, M. N. Vostretsov, A. Sh. Bikbulatov, and S. G. D'yakov, *Teplofiz. Svoista Veshchestv Mater.*, **17**, 122 (1982); *Chem. Abstra.*, **99**, 11242a (1983).

190. W. E. Price, *J. Chem. Soc. Faraday Trans. 1*, **84**, 2431 (1988).

191. M. Roth, J. L. Steger, and M. V. Novotny, *J. Phys. Chem.*, **91**, 1645 (1987).

192. H. J. V. Tyrrell and P. J. Watkiss, *Ann. Rep. Chem. Soc. A*, 35 (1976).

193. J. C. Giddings and S. L. Seager, *Ind. Eng. Chem. Fundam.*, **1**, 277 (1962).

194. P. S. Arora, H. L. Robjohns, I. R. Shankland, and P. J. Dunlop, *Chem. Phys. Lett.*, **59**, 478 (1978).

195. P. A. Fleury and J. P. Boon, "Laser Light Scattering in Fluid Systems," in I. Prigogine and S. A. Rice, Eds., *Advances in Chemical Physics*, Vol. XXIV, Wiley, New York, 1973, p. 1.

196. B. Chu, *Laser Light Scattering*, Academic, New York, 1974.

197. B. J. Berne and R. Pecora, *Dynamic Light Scattering*, Wiley, New York, 1976.

198. R. Pecora, *Dynamic Light Scattering and Velocimetry: Applications of Photon Correlation Spectroscopy*, Plenum, New York, 1982.

199. D. A. McQuarrie, *Statistical Mechanics*, Harper and Row, New York, 1976.

200. H. Z. Cummins, N. Knable, and Y. Yeh, *Phys. Rev. Lett.*, **12**, 150 (1964).

201. R. Finsy, A. Devriese, and H. Lekkerkerker, *J. Chem. Soc. Faraday Trans. 2*, **76**, 767 (1980).

202. M. M. Kops-Werkhoven, C. Pathmamanoharan, A. Vrij, and H. M. Fijnaut, *J. Chem. Phys.*, **77**, 5913 (1982).

203. N. Yoshida, *J. Chem. Phys.*, **83**, 1307 (1985).

204. G. D. J. Phillies, *J. Chem. Phys.*, **84**, 5972 (1986).

205. J. G. Kirkwood and R. J. Goldberg, *J. Chem. Phys.*, **18**, 54 (1950).

206. R. D. Mountain and J. M. Deutch, *J. Chem. Phys.*, **50**, 1103 (1969).

207. N. Yoshida, *J. Chem. Phys.*, **84**, 5973 (1986).

208. M. Dubois, P. Berge, and C. Laj, *Chem. Phys. Lett.*, **6**, 227 (1970).

209. M. J. Cardamone and D. P. Eastman, *J. Chem. Phys.*, **59**, 553 (1973).

210. K. J. Czworniak, H. C. Andersen, and R. Pecora, *Chem. Phys.*, **11**, 451 (1975).

211. W. Krahn, G. Schweiger, and K. Lucas, *J. Phys. Chem.*, **87**, 4515 (1983).

212. E. Gulari, R. J. Brown, and C. J. Pings, *Am. Inst. Chem. Eng. J.*, **19**, 1196 (1973).

213. K. McKeigue and E. Gulari, *J. Phys. Chem.*, **88**, 3472 (1984).

214. J. H. R. Clark, G. J. Hills, C. J. Oliver, and J. M. Vaughan, *J. Chem. Phys.*, **61**, 2810 (1974).

215. H. Saad and E. Gulari, *J. Phys. Chem.*, **88**, 136 (1984).

216. P. A. Maddern, *Mol. Phys.*, **36**, 365 (1978).

217. G. D. Patterson and P. J. Carroll, *J. Phys. Chem.*, **89**, 1344 (1985).

218. E. Jakeman, "Photon Correlation," H. Z. Cummins, "Light Beating Spectroscopy," and C. J. Oliver," Correlation Techniques," in H. Z. Cummins and E. R. Pike, Eds., *Photon Correlation and Light Beating Spectroscopy*, Plenum, New York, 1973, pp. 75, 225, 151.

219. A. J. F. Siegert, *Mass. Inst. Technol. Radiat. Lab. Rep.*, 465 (1943).

220. P. N. Pusey, "Macromolecular Diffusion," in H. Z. Cummins and E. R. Pike, Eds., *Photon Correlation and Light Beating Spectroscopy*, Plenum, New York, 1973, p. 387.

221. T. N. Bender and R. Pecora, *J. Phys. Chem.*, **90**, 1700 (1986).

222. E. Hahn, *Phys. Rev.*, **80**, 580 (1950).

223. T. C. Farrar and E. D. Becker, *Pulse and Fourier Transform NMR*, Academic, New York, 1971.

224. D. Shaw, *J. Phys. E* **7**, 689 (1974).

225. D. Shaw, *Fourier Transform NMR Spectroscopy*, Elsevier, Amsterdam (1976).

226. A. Geiger and M. Holz, *J. Phys. E* **13**, 697 (1980).

227. A. G. Redfield, "How to Build a Fourier Transform NMR Spectrometer for Biochemical Applications," in M. M. Pinter, Ed., "Introductory Essays," in P. Diehl, Ed., *NMR 13: Basic Principles and Progress*, Springer-Verlag, Berlin, 1976, p. 137.

228. P. Laszlo, Ed., *NMR of Newly Accessible Nuclei*, Vol. 1, *Chemical and Biochemical Applications*, Academic, New York, 1983.

229. P. Callaghan, *Aust. J. Phys.*, **37**, 359 (1984).

230. P. Stilbs, *Prog. Nucl. Magn. Reson. Spectrosc.*, **19**, 1 (1986).

231. H. Weingärtner, *R. Soc. Chem. Specialist Periodical Rep.*, **14**, 111 (1985).

232. K. R. Harris, R. Mills, P. J. Back, and D. S. Webster, *J. Magn. Reson.*, **29**, 473 (1978).

233. K. R. Harris, *Physica*, **93A**, 593 (1978).

234. B. Arends, "A Spin-Echo Investigation of Self-Diffusion in Liquid Ethylene," doctoral dissertation, Van der Waals Laboratorium, University of Amsterdam, The Netherlands, 1979.

235. W. Berger and H. J. Butterweck, *Arch. Electrotech. (Berlin)*, **42**, 216 (1956).

236. C. J. Gerritsma and N. J. Trappeniers, *Physica*, **51**, 365 (1971).

237. J. S. Blicharski and W. T. Sobol, *J. Magn. Reson.*, **46**, 1 (1982).

238. D. S. Webster, "A Study of Self-Diffusion in Biological Systems Using Pulsed Magnetic Field Gradient-Nuclear Magnetic Resonance Techniques," doctoral dissertation, University of New South Wales, Sydney, 1971; D. S. Webster and K. H. Marsden, *Rev. Sci. Instrum.*, **45**, 1232 (1974).

239. P. T. Callaghan, C. M. Trotter, and K. W. Jolley, *J. Magn. Reson.*, **32**, 247 (1980).

240. J. Jonas, *Rev. Sci. Instrum.*, **45**, 1232 (1974).

241. D. M. Cantor and J. Jonas, *J. Magn. Reson.*, **28**, 157 (1977).

242. P. Stilbs and M. E. Moseley, *Chem. Scr.*, **15**, 176 (1980).

243. H. Y. Carr and E. M. Purcell, *Phys. Rev.*, **94**, 630 (1954).

244. B. Muller and M. Bloom, *Can. J. Phys.*, **38**, 1318 (1960).

245. C. J. Gerritsma, "Spin-Roosterrelaxatic van Protonen in Methaan: R.F.—

Pulsspectrometer voor het Meten van Relaxatietijden en Zeldiffusie coëfficiënten," doctoral dissertation, Van der Waals Laboratorium, University of Amsterdam, The Netherlands, 1969.

246. S. Meiboom and D. Gill, *Rev. Sci. Instrum.*, **29**, 688 (1958).

247. E. O. Stejskal and J. E. Tanner, *J. Chem. Phys.*, **42**, 288 (1965).

248. E. O. Stejskal, *J. Chem. Phys.*, **43**, 3597 (1965).

249. M. I. Hrovat and C. G. Wade, *J. Magn. Reson.*, **44**, 62 (1981).

250. P. T. Callaghan, M. A. Le Gros, and D. M. Pinder, *J. Chem. Phys.*, **79**, 6372 (1983).

251. T. L. James and G. C. MacDonald, *J. Magn. Reson.*, **11**, 58 (1973).

252. J. Kida and H. Uedaira, *J. Magn. Reson.*, **27**, 253 (1977).

253. T. Cosgrove, J. S. Littler, and K. Stewart, *J. Magn. Reson.*, **38**, 207 (1980).

254. C. L. Dumoulin and G. C. Levy, "Computational Considerations," in P. Laszlo, Ed., *NMR of Newly Accessible Nuclei*, Vol. 1, *Chemical and Biochemical Applications*, Academic, New York, 1983, Chap. 3.

255. L. G. Longsworth, *J. Phys. Chem.*, **58**, 770 (1954); **64**, 1914 (1960).

256. A. J. Easteal, A. V. J. Edge, and L. A. Woolf, *J. Phys. Chem.*, **89**, 1064 (1985).

257. A. F. Collings and R. Mills, *Trans. Faraday Soc.*, **66**, 2761 (1970).

258. S. J. Thornton and P. J. Dunlop, *J. Phys. Chem.*, **78**, 846 (1974).

259. A. F. Collings and L. A. Woolf, *J. Chem. Soc. Faraday Trans. 1*, **71**, 2296 (1975).

260. J. S. Murday, *J. Magn. Reson.*, **10**, 111 (1973).

261. D. M. Lamb, P. J. Grandinetti, and J. Jonas, *J. Magn. Reson.*, **72**, 532 (1987).

262. N. A. Walker, D. M. Lamb, S. T. Adamy, J. Jonas, and M. P. Dare-Edwards, *J. Phys. Chem.*, **92**, 3675 (1988).

263. K. R. Harris and L. A. Woolf, *J. Chem. Soc. Faraday Trans. 1*, **76**, 377 (1980).

264. A. J. Easteal and L. A. Woolf, *J. Phys. Chem.*, **90**, 2441 (1986).

265. J. Jonas, T. E. Bull, and C. A. Eckert, *Rev. Sci. Instrum.*, **41**, 1240 (1970).

266. H. Yamada, *Rev. Sci. Instrum.*, **45**, 640 (1974).

267. K. R. Harris, *J. Chem. Soc. Faraday Trans. 1*, **78**, 2265 (1982).

268. J. G. Powles and M. C. Gough, *Mol. Phys.*, **16**, 349 (1969).

269. S.-H. Lee, K. Luszcynski, R. E. Norberg, and M. S. Conradi, *Rev. Sci. Instrum.*, **58**, 415 (1987).

270. N. A. Walker, D. M. Lamb, J. Jonas, and M. P. Dare-Edwards, *J. Magn. Reson.*, **74**, 580 (1987).

271. L. J. Burnett and J. F. Harmon, *J. Chem. Phys.*, **57**, 1293 (1972).

272. F. X. Prielmeier, E. W. Lang, and H.-D. Lüdemann, *Mol. Phys.*, **52**, 1105 (1984).

273. P. W. E. Peereboom, H. Luigjes, K. O. Prins, and N. J. Trappeniers, *Physica*, **139** and **140B**, 134 (1986).

274. P. W. E. Peereboom, K. O. Prins, and N. J. Trappeniers, *Rev. Sci. Instrum.*, **59**, 1182 (1988).

275. E. W. Lang, F. X. Prielmeier, H. Radkowitsch, and H.-D. Lüdemann, *Ber. Bunsenges. Phys. Chem.*, **91**, 1017, 1025 (1987).

276. B. M. Braun and H. Weingärtner, *J. Phys. Chem.*, **92**, 1342 (1988).

277. G. A. Shirn, E. S. Wajda, and H. B. Huntington, *Acta Metall.*, **1**, 513 (1953).

278. A. P. Batra and H. B. Huntington, *Phys. Rev.*, **154**, 569 (1967).

279. C. J. Santoro, *Phys. Rev.*, **179**, 593 (1969).

280. K. Itagaki, *J. Phys. Soc. Jpn.*, **22**, 427 (1967).

281. R. O. Ramseier, *J. Appl. Phys.*, **38**, 2553 (1967).

282. J. W. Gibbs, *Collected Works*, Vol. 1, Dover, New York, 1961, pp. 184–186.

283. H. B. Callen, *Thermodynamics*, Wiley, New York, 1960, Chap. 13.

284. N. H. Nachtrieb, J. A. Weil, E. Catalono, and A. W. Lawson, *J. Chem. Phys.*, **20**, 1189 (1952).

285. E. D. Albrecht and C. T. Tomizuka, *J. Appl. Phys.*, **35**, 3560 (1964).

286. H. R. Curtin, D. L. Decker, and H. B. Vanfleet, *Phys. Rev.*, **139**, 1552 (1965).

287. A. Ascoli, B. Bollani, G. Guarini, and D. Kustudic, *Phys. Rev.*, **141**, 732 (1966).

288. J. I. Goldstein, R. E. Hanneman, and R. E. Ogilvie, *Trans. Metall. Soc. AIME*, **233**, 812 (1965).

289. R. E. Hanneman, R. E. Ogilvie, and H. C. Gatos, *Trans. Metall. Soc. AIME*, **233**, 691 (1965).

290. R. W. Balluffi and A. L. Ruoff, *J. Appl. Phys.*, **34**, 1634 (1963).

291. A. L. Ruoff and R. W. Balluffi, *J. Appl. Phys.*, **34**, 1848 (1963).

292. A. L. Ruoff and R. W. Balluffi, *J. Appl. Phys.*, **34**, 2862 (1963).

293. A. F. Brown and D. A. Blackburn, *Acta Metall.*, **11**, 1017 (1963).

294. A. C. Damask, G. J. Dienes, H. Harman, and L. E. Katz, *Philos. Mag.*, **20**, 67 (1969).

295. S. R. de Groot and P. Mazur, *Non-equilibrium Thermodynamics*, North-Holland, Amsterdam, 1962, p. 14.

296. H. T. Stokes, "Study of Diffusion in Solids by Pulsed Nuclear Magnetic Resonance," in G. E. Murch, H. K. Birnbaum, and J. R. Cost, Eds., *Nontraditional Methods in Diffusion*, Metallurgical Society of AIME, New York, 1984.

297. R. Messer and F. Noack, *Appl. Phys.*, **6**, 79 (1975).

298. S. C. Chen, J. C. Tarczon, W. P. Halperin, and J. O. Brittain, *J. Phys. Chem. Solids*, **46** 895 (1985).

299. C. Zener, *Elasticity and Anelasticity of Solids*, Chicago University Press, Chicago, IL, 1948.

300. A. S. Nowick and B. S. Berry, *Anelastic Relaxation in Crystalline Solids*, Academic, New York, 1972.

301. A. S. Nowick, in W. P. Mason and R. N. Thurston, Eds., *Physical Acoustics*, New York, 1977, pp. 1–28.

302. B. S. Berry and W. C. Pritchet, *Phys. Rev. B*, **24**, 2299 (1981).

303. J. R. Cost, "Analysis of Anelastic Relaxations Controlled by a Spectrum of Relaxation Times," in G. E. Murch, H. K. Burnbaum, and J. R. Cost, Eds., *Nontraditional Methods in Diffusion*, Metallurgical Society of AIME, New York, 1984.

304. T. Springer, *Quasielastic Neutron Scattering for the Investigation of Diffusive Motions in Solids and Liquids*, Springer Tracts in Modern Physics, Springer, Berlin, 1972.

305. H. Zabel, "Quasi-Elastic Neutron Scattering: A Powerful Tool for Investigating Diffusion in Solids," in G. E. Murch, H. K. Birnbaum, and J. R. Cost, Eds.,

Nontraditional Methods in Diffusion, Metallurgical Society of AIME, New York, 1984.

306. G. Lucazeau, J. R. Gavarri, and A. J. Dianoux, *J. Phys. Chem. Solids*, **48**, 57 (1987).

307. K. A. Singwi and A. Sjolander, *Phys. Rev.*, **120**, 1093 (1960).

308. S. L. Ruby, J. C. Love, P. A. Flinn, and B. J. Zabransky, *Appl. Phys. Lett.*, **27**, 320 (1975).

309. Th. Wichert, G. Grubel, and M. Deicher, *International Seminar on Solute-Defect Interaction*, Canadian Institute of Metals, Kingston, Canada, August, 1985.

310. J. G. Mallen, "Mössbauer Diffusion Studies," in G. E. Murch, H. K. Birnbaum, and J. R. Cost, Eds., *Nontraditional Methods in Diffusion*, Metallurgical Society of AIME, New York, 1984.

311. S. Mantl, W. Petry, K. Schroeder, and G. Vogl, *Phys. Rev B.*, **27**, 5315 (1983).

312. D. Wolf, *Appl. Phys. Lett.*, **30**, 617 (1977).

313. F. Pettit, R. Yinger, and J. B. Wagner, *Acta Metall.*, **8**, 617 (1960).

314. F. Pettit and J. B. Wagner, *Acta Metall.*, **12**, 35 (1964).

315. H. J. Grabke, *Ber. Bunsenges, Phys. Chem.*, **69**, 48 (1965).

316. C. Wagner, *Ber. Bunsenges, Phys. Chem.*, **70**, 775 (1966).

317. A. Gala and H. J. Grabke, *Arch. Eisenhuettenwes.*, **43**, 463 (1972).

318. J. P. Orchard and D. J. Young, *J. Electrochem. Soc.*, **133**, 1734 (1986).

319. N. L. Peterson and R. E. Ogilvie, *Trans. TMS-AIME*, **227**, 1083 (1963).

320. A. D. Romig, Jr., *J. Appl. Phys.*, **54**, 3172 (1983).

321. M. Hillert and G. R. Purdy, *Acta Metall.*, **26**, 333 (1978).

322. Y. Oishi and W. D. Kingery, *J. Chem. Phys.*, **33**, 480 (1960).

323. W. K. Chen and R. A. Jackson, *J. Phys. Chem. Solids*, **30**, 1309 (1969).

324. J. Desmaison and W. W. Smeltzer, *J. Electrochem. Soc.*, **122**, 354 (1975).

325. J. L. Routbort and S. J. Rothman, *Defects and Diffusion Data*, **40**, 1 (1985).

326. A. Smigelskas and E. Kirkendall, *Trans. AIME*, **171**, 130 (1947).

327. G. A. Shirn, E. S. Wajda, and H. B. Huntington, *Acta Metall.*, **1**, 513 (1953).

328. N. L. Peterson and R. E. Ogilvie, *Trans. TMS-AIME*, **227**, 1083 (1963).

329. C. J. Santoro, *Phys. Rev.*, **179**, 539 (1969).

330. H. H. Woodbury and R. B. Hall, *Phys. Rev.*, **157**, 641 (1967).

331. N. L. Peterson and W. K. Chen, *J. Phys. Chem. Solids*, **43**, 29 (1982).

332. S. J. Rothman, in G. E. Murch and A. S. Nowick, *Diffusion in Crystalline Solids*, Academic, Orlando, FL, 1985.

333. J. L. Routbort and S. J. Rothman, *J. Phys. Chem. Solids*, **47**, 993 (1986).

334. G. W. Weber, W. R. Bitler, and V. S. Stubican, *J. Phys. Chem. Solids*, **41**, 1335 (1980).

335. K. Maier and W. Schule, *EURATOM Rep.*, 5234d (1974).

336. G. Gupta, *Thin Solid Films*, **25**, 231 (1975).

337. A. Atkinson and R. I. Taylor, *Thin Solid Films*, **46**, 291 (1977).

338. A. Atkinson, M. L. O'Dwyer, and R. I. Taylor, *J. Mater. Sci.*, **18**, 2371 (1983).

339. J. Harvath, F. Dyment, and H. Mehrer, *J. Nucl. Mater.*, **126**, 206 (1984).

340. H. H. Andersen, *Appl. Phys.*, **18**, 131 (1979).

341. P. Williams, *Appl. Phys. Lett.*, **36**, 758 (1980).

342. M.-P. Macht and V. Naundorf, *J. Appl. Phys.*, **53**, 7551 (1982).

343. W. T. Peturkey, "Diffusion Analysis Using Secondary Ion Mass Spectroscopy," in G. E. Murch, H. K. Birnbaum, and J. R. Cost, Eds., *Nontraditional Methods in Diffusion*, Metallurgical Society of AIME, New York, 1984.

344. J. I. Goldstein, M. R. Notis, and A. D. Romig, Jr., *AIME Symposium on Atomic Transport in Concentrated Alloys and Intermetallic Compounds*, Detroit, September, 1984, Metallurgical Society of AIME, Warrendale, PA, 1985.

345. A. D. Romig, Jr., D. L. Humphreys, J. I. Goldstein, and M. R. Notis, in J. Armstrong, Ed., *Microbeam Analysis*, San Francisco Press, San Francisco, CA, 1985.

346. R. Castaing, "Application des sonoles electronique a une method d'analyse ponctuelle chimique et crystallographique," doctoral dissertation, University of Paris, France, 1951.

347. R. Theisen, *Quantitative Electron Microprobe Analysis*, Springer-Verlag, Berlin, 1965.

348. J. I. Goldstein and H. Yakowitz, *Practical Scanning Microscopy: Electron and Ion Microprobe Analysis*, Plenum, New York, 1975.

349. B. M. Siegel and D. R. Bearman, *Physical Aspects of Electron Microscopy and Microbeam Analysis*, Wiley, New York, 1975.

350. D. J. Young, W. W. Smeltzer, and J. S. Kirkaldy, *Metall. Trans.*, **6A**, 1205 (1975).

351. J. I. Goldstein, R. E. Hannerman, and R. E. Ogilvie, *Trans. Metall. Soc. AIME*, **233**, 812 (1965).

352. D. J. Young, T. Narita, and W. W. Smeltzer, *J. Electrochem. Soc.*, **127**, 679 (1980).

353. J. S. Kirkaldy and D. J. Young, *Diffusion in the Condensed State*, Institute of Metals, London, 1988.

354. Zoung Young-An, *Electron Microscopy and Analysis 1983, Institute of Physics Conference Series No. 68*, Institute of Physics, London, 1984.

355. G. Cliff and G. W. Lorimer, *J. Microsc.*, **103**, 203 (1975).

356. D. C. Dean and J. I. Goldstein, *Metall. Trans. A*, **17A**, 1131 (1986).

357. D. F. Mitchell and M. J. Graham, *J. Electrochem. Soc.*, **133**, 2433 (1986).

358. C. A. Evans, Jr., *Thin Solid Films*, **19**, 11 (1973).

359. K. Wittmaack, *Nucl. Instrum. Methods*, **168**, 343 (1980).

360. D. F. Mitchell, R. J. Hussey, and M. J. Graham, *Trans. Jpn. Inst. Met. Suppl.*, **24**, 121–125 (1983).

361. S. M. Myers, "Ion Beam Analysis and Ion Implantation in the Study of Diffusion," in G. E. Murch, H. K. Birnbaum, and J. R. Cost, Eds., *Nontraditional Methods in Diffusion*, Metallurgical Society of AIME, New York, 1984.

362. H. Yinnon and A. R. Cooper, *Phys. Chem. Glasses*, **21**, 204 (1980).

363. K. P. R. Reddy, S. M. Oh, L. D. Major, and A. R. Cooper, *J. Geophys. Res.*, **85**, 322 (1980).

364. W. A. Lanford, R. Benenson, C. Burman, and L. Wielunaki, "Nuclear Reaction Analysis for Diffusion Studies," in G. E. Murch, H. K. Birnbaum, and J. R. Cost, Eds., *Nontraditional Methods in Diffusion*, Metallurgical Society of AIME, New York, 1984.

365. J. P. Biersack and D. Fink, "Proceedings of the 8th Symposium on Fusion Technology," *EURATOM Rep.* 5182e, 1974.

366. W. Moller, M. Hufschmidt, and Th. Pfeiffer, *Nucl. Instrum. Methods*, **149**, 73 (1978).

367. S. Hofman and J. Erlewein, *Scr. Metall.*, **10**, 857 (1976).

368. J. Crank, *The Mathematics of Diffusion*, Oxford University Press, Oxford, 1957.

369. T. Suguoka, *J. Phys. Soc. Jpn.*, **19**, 839 (1964).

370. A. Atkinson and R. I. Taylor, *Philos. Mag.*, **A43**, 979 (1981).

371. A. D. Le Claire, *Br. J. Appl. Phys.*, **14**, 351 (1963).

372. P. L. Gruzin, *Izv. Akad. Nauk SSSR Otd. Tekh. Nauk*, **3**, 353 (1953).

373. P. C. Carman and R. A. W. Haul, *Proc. R. Soc. London*, **A222**, 109 (1954).

374. J. E. Reynolds, B. Averbach, and M. Cohen, *Acta Metall.*, **5**, 29 (1957).

375. T. O. Ziebold, "Ternary Diffusion in Cu—Ag—Au Alloys," doctoral dissertation, Massachusetts Institute of Technology, Cambridge, MS, 1965.

376. C. Matano, *Jpn. Phys.*, **8**, 109 (1933).

377. L. Boltzmann, *Ann. Phys.*, **53**, 96 (1894).

378. J. S. Kirkaldy, *Can. J. Phys.*, **35**, 435 (1957).

379. L. S. Darken, *Trans. AIME*, **180**, 430 (1949).

380. J. S. Kirkaldy and D. J. Young, *Diffusion in the Condensed State*, Institute of Metals, London, 1988, Chap. 5.

381. E. L. Cussler, *Multicomponent Diffusion*, Elsevier Scientific Publishing Company, Amsterdam, 1976, p. 72 ff.

382. R. J. Schut and A. R. Cooper, *Acta Metall.*, **30**, 1957 (1982).

383. T. O. Ziebold and R. E. Ogilvie, *Trans. AIME*, **239**, 942 (1967).

384. V. K. Chandhok, J. P. Hirth, and E. J. Dulis, *Trans. AIME*, **224**, 858 (1962).

385. J. S. Kirkaldy, G. R. Mason, and W. J. Slater, *Trans. Can. Inst. Min. Metall.*, **64**, 53 (1961).

386. J. S. Kirkaldy, Zia-Ul-Haq, and L. C. Brown, *Trans. Am. Soc. Met.*, **56**, 834 (1963).

387. J. S. Kirkaldy, R. J. Brigham, and D. H. Weichert, *Acta Metall.*, **13**, 907 (1965).

388. J. Kirkaldy and D. J. Young, *Diffusion in the Condensed State*, Institute of Metals, London, 1988, Chap. 7.

389. L. S. Darken, *Trans. AIME*, **175**, 184 (1948).

390. D. E. Meyer, *The Kinetic Theory of Gases*, translated by R. E. Baynes, Longmans Green, London, 1899, pp. 248, 255.

391. L. C. Correa da Silva and R. Mehl, *Trans. AIME*, **191**, 155 (1951).

392. J. E. Reynolds, B. L. Averbach, and M. Cohen, *Acta Metall.*, **5**, 29 (1957).

393. J. R. Manning, *Diffusion Kinetics for Atoms in Crystals*, Van Nostrand, Princeton, NJ, 1968, p. 36 ff., p. 83 ff.

394. J. S. Kirkaldy and D. J. Young, *Diffusion in the Condensed State*, Institute of Metals, London, 1988, Chap. 6.

395. J. S. Kirkaldy and J. E. Lane, *Can. J. Phys.*, **44**, 2059 (1966).

396. J. Philibert and A. G. Guy, *Compt. Rend.*, **227**, 2281 (1963).

397. T. O. Ziebold and A. R. Cooper, *Acta Metall.*, **13**, 465 (1965).

398. H. Schonert, *Z. Phys. Chem. Frankfurt am Main*, **119**, 53 (1980).

399. R. L. Levin and J. B. Wagner, Jr., *Trans. AIME*, **233**, 159 (1965).

400. J. B. Price and J. B. Wagner, Jr., *Z. Phys. Chem. Frankfurt am Main*, **49**, 257 (1966).

401. K. Weiss, *Ber. Bunsenges. Phys. Chem.*, **73**, 338 (1969).

402. A. B. Neuman, *Trans. Am. Inst. Chem. Eng.*, **27**, 203 (1931).

403. A. J. Rosenburg, *J. Electrochem. Soc.*, **107**, 795 (1960).

404. E. M. Fryt, *Oxid. Met.*, **12**, 139 (1978).

405. S. Mrowec, in Z. A. Foroulis and W. W. Smeltzer, Eds., *Metal-Slag-Gas Reactions and Processes*, The Electrochemical Society Inc., Princeton, NJ, 1975, p. 414.

406. E. M. Fryt, W. W. Smeltzer, and J. S. Kirkaldy, *J. Electrochem. Soc.*, **126**, 673 (1979).

407. F. Gesmundo, F. Viani, and V. Dovi, *Oxid. Met.*, **23**, 141 (1985).

408. P. Kofstad and K. P. Lillerud, *Oxid. Met.*, **17**, 177 (1982).

409. P. Kofstad, *J. Phys. Chem. Solids*, **44**, 129 (1983).

410. D. G. Leaist, *J. Phys. Chem.*, **94**, 5180 (1990).

411. D. G. Leaist, *J. Chem. Soc. Faraday Trans.*, **87**, 597 (1991).

412. D. G. Leaist, *Bur. Bunsenges. Phys. Chem.*, **95**, 117 (1991).

413. K. R. Harris, *J. Solution Chem.*, **20**, 595 (1991).

Chapter **4**

DETERMINATION OF SOLUBILITY

Dorothy K. Wyatt and Lee T. Grady

Physical Methods of Chemistry, Second Edition Volume Six: Determination of Thermodynamic Properties   Edited by Bryant W. Rossiter and Roger C. Baetzold
ISBN 0-471-57087-7   Copyright 1992 by John Wiley & Sons, Inc.

1   INTRODUCTION TO SOLUBILITY

The solubility of compounds is a fundamental property with a variety of applications: appropriate solvents furnish media for reactions; differential solubility relations underlie several methods for the isolation, purification, and determination of substances; solubility analysis reveals the purity of substances; comparison of the solutions one compound forms with others provides information about its molecular structure and the nature and extent of intermolecular forces; solubility tests are one of the bases of systematic qualitative analysis. It is fortunate that so useful a property as solubility is so often easy to measure with adequate precision.

This chapter is intended to serve as a working guide for making determinations of solubility and interpretating the results. It does not discuss the analytical methods used to measure final concentrations of solutions since these do not differ, per se, from other analyses of solutions.

1.1   Definition of Solubility

Solubility can be defined as the capacity of two or more substances to form spontaneously, one with the other, without chemical reaction, a homogeneous molecular, or colloidal dispersion. First, note that the state of the dispersion is not limited; it can be gaseous, liquid, crystalline, mesomorphic, or amorphous. A swollen polymer, such as rubber in benzene, is as much a solution as is alcohol in water. A colloidal dispersion of a water-insoluble dye in an aqueous soap solution may be considered a category of solution. It is required that the dispersion be formed spontaneously, so that the formation of the solution is accompanied by a decrease in Gibbs free energy of the system. This change is reversible with concomitant precipitation of crystallization. Free energy changes due to chemical reaction, as in the "solution" of base metals in acid, are ruled out.

The capacity of any system of substances to form a solution has definite limits and is defined by the Gibbs phase rule:

$$F = C + 2 - P \qquad (1)$$

where F is the number of degrees of freedom (temperature, pressure, compositions) in a system of C components with P phases, which is not under the influence of gravitational, electrical, or magnetic forces. At constant temperature and pressure this simplifies to $F^0 = C - P$, where $F^0 = F - 2$. Thus for two components and two phases (solid and liquid, two liquids, or two solids) under the pressure of their own vapor and at constant temperature, $F^0 = 0$, or in other words, composition is not free to vary. If one of the phases consists solely of one component (i.e., is a pure substance), the quantitative measure of solubility is a single number, namely, the amount of the substance (solute) that is contained in saturated solution in a unit amount of the other component (solvent). A saturated solution is one in which no more solute will dissolve at a given temperature. Therefore, whenever F^0 is zero a definite, reproducible solubility equilibrium can be reached.

The partial molal free energy or chemical potential μ_i of any component i in solution at saturation must equal the partial molal Gibbs free energy or chemical potential of the same component in the undissolved phase $\mu_{i,s}$

$$\mu_i = \mu_{i,s} \qquad (2)$$

More solute will solubilize when $\mu_{i,s} > \mu_i$. Precipitation or recrystallization will take place when $\mu_i > \mu_{i,s}$. At equilibrium $\mu_i = \mu_{i,s}$. In the presence of excess solute, the relationship is invariant. A binary system containing a solute dissolved in a liquid not at saturation (equilibrium) is completely defined by the composition of the system alone ($F^0 = 1$).

The three phases: gas, solid, and liquid, give rise to nine classes of binary solutions. Three classes of solution are predominant in the interest of organic chemists and are discussed in further detail, solids in liquids, liquids in liquids, and gases in liquids.

1.2   Expression of Solubility

In a saturated solution ($F^0 = 0$) the expression of solubility is the set of numbers giving the composition of each of the phases at the given temperature. Complete representation of the solubility relations requires determination of the phase diagram, which gives the number, composition, and relative amounts of each phase present at any temperature in a system containing the components in any specified proportion.

In a binary system containing solute dissolved in solvent, the chemical potential can be defined as

$$\mu_i = \mu_{i,s} - RT \ln a_{i,s} \tag{3}$$

where μ_i and $\mu_{i,s}$ are the chemical potentials of a component i in solution and in the solid phase; $a_{i,s}$ is the activity of this component. The terms R and T are the gas constant and temperature, °K, respectively.

The activity, $a_{i,s}$ is the concentration of molar fraction, or molality, or volume fraction of component i in solution, which has been multiplied by the empirical factor termed by G. N. Lewis as the activity coefficient. The activity coefficient is unity when the solution obeys van't Hoff's or Raoult's law. This coefficient is thus a measure of the difference in chemical potential of an actual component in solution and the chemical potential anticipated on the basis of either of these laws.

Interconversion of these expressions is simple if the respective densities are known. Solubility in grams per 100 mL of solvent or solution is equal to solubility in grams per 100 g of solvent or solution multiplied by the density of the solvent or the solution, respectively. Solubilities of solid or liquid compounds normally are expressed as the weight of solute per weight or volume of solvent or solution (see Table 4.1 [1]).

Other terms such as molarity, molality, normality, and mole fraction are used commonly as solubility expressions. Solutions by weight or volume commonly are expressed by percentage. In any event, the expression of solubility includes temperature and, where applicable, partial pressures of the system.

Solubilities of gases find some rather different expressions. Mole fraction $\bar{X}$ is used commonly, additionally in the form $-R \ln X_2$ as by Hildebrand [2]. The classical expression for the solubility of a gas in a liquid is the Bunsen coefficient α [3]. The Bunsen coefficient or absorption coefficient is defined as the volume of

Table 4.1 Solubility of Various Compounds Expressed in Different Units[a,b]

			Solubility			
Compound	Solvent	Temperature (°C)	g/100 g Solvent	g/100 mL Solvent	g/100 g Solution	g/100 mL Solution
Tartaric acid	50% by wt Ethanol-water	25	88.6	80.9	47.0	55.7
o-Nitrophenol	Ethanol	30.2	60.6	47.3	41.0	44.5
Sodium chloride	Water	20	35.8	35.8	26.4	31.6

[a]Reference [1].
[b]Reprinted with permission from W. J. Mader and L. T. Grady, "Determination of Solubility," in A. Weissberger and B. W. Rossiter, Eds., *Physical Methods of Chemistry*, Vol. 1, Part V, Wiley, New York, 1971. Copyright © 1971 by John Wiley & Sons, Inc.

gas V_0, reduced to $0°C$ and 1 atm, dissolved by unit volume of solvent V_p, at the temperature of the experiment under a partial pressure p of the gas of 1 atm.

$$\alpha = \frac{V_0}{V_p} \tag{4}$$

Bunsen used the ideal gas law to reduce the gas volume to standard conditions. Since this law is not exact, the coefficients found by different methods, namely, physical and chemical, can be expected to differ. Many of the past workers have not controlled the total pressure carefully, or have neglected solvent partial pressure. Some have corrected results at other partial pressures to 1 atm using Henry's law, which describes a direct, linear relation between solubility and partial pressure: $P_2/X_2 = K_2 = P_2^0/X_2^0$.

If gas solubility is calculated according to the Bunsen coefficient, except that the amount of the solvent is 1 g, the result is known as the Kuenen coefficient. Markham and Kobe [4] express the volume of gas, reduced to standard conditions, dissolved by the quantity of solution containing 1 g of solvent; this is designated by S and is proportional to gas molality. If the solubility is calculated as grams of gas dissolved per 100 mL of solvent at the temperature of the experiment and a partial gas pressure of 760 mmHg, the result is known as the *Raoult absorption coefficient*. The Ostwald coefficient of solubility β is defined as the volume of gas, measured under the temperature and pressure at which the gas dissolves, taken up by unit volume of the liquid. Thus, the Ostwald and Bunsen coefficients are related by $\beta = \alpha(T/273)$.

Mixed solids also can require characterization. The solubility index is used to express the solubility of some foods. It is the volume of solvent required to dissolve a specified weight of a food preparation.

1.3   Measurements of Solubility

Generally, the methods used in the determination of solubility are classified as *synthetic* or *analytic*, the terms applied originally by Alexejew [5] to the type of determination. Synthetic refers to those methods applied to an arbitrary system of solute and solvent in which the temperature or the pressure or both are varied until the solute just dissolves. An analytical method refers to a method in which the composition of the solution phase is determined by analysis in a system containing excess solute at a given temperature and pressure.

Methods for determining solubility can be classified on the basis of the phase rule. Thus, Hill [6] would group all the existing methods of solubility determination according to a constant factor, that is, thermostatic, plethostatic (constant composition), and barostatic. Williams [7] called those methods in which sampling is not required *isosystic*.

In general, the analytical methods consist of obtaining a saturated solution at equilibrium and analyzing the resulting solution by some suitable physical or chemical method. Equilibration can be obtained by intimately mixing the solute and solvent, percolation of the solvent through the solute, and convection of the solvent through the solute. When solvent and solute are mixed intimately, separation of the phases can be accomplished by decantation, possibly after centrifugation, or filtration. Alternatively, undissolved solute is measured in some procedures.

This chapter does not describe the safe laboratory practices required for the handling of volatile or toxic materials as these practices are presumed to be in general application.

1.4   Factors Affecting Solubility Determinations

The solubility of a substance in a solvent at equilibrium is a function of temperature, pressure, and purity of solute and solvent. Equilibration, in turn, is dependent on many factors.

From a practical standpoint, one cannot generally calculate with confidence the solubility of a solid in a liquid, and even a rough experimental measurement is probably more accurate than calculated values. Moreover, calculation requires data not likely to be available for compounds whose solubility, at least in common solvents, has not been determined. Hildebrand [2] found that the most significant parameter in calculating solubilities of nonelectrolytes in normal liquids is the energy of vaporization of the solute at its boiling point, this being most indicative of entropy. The natural solubility of an ideal solute also can be related to the heat of fusion and the melting point of the solute. If the melting point of the solute is high, the natural solubility is small; if the melting point is low, the natural solubility is large; if the solute is liquid, the natural solubility is infinite. Abnormally high solubilities (negative partial pressure curves) can be explained by supposing that some kind of combination occurs between the components. Abnormally low solubilities (positive partial pressure curves) can be shown to be given by two components that are approaching the temperature at which they will separate into layers.

1.4.1   Temperature

The solubility diagram records the change in equilibrium concentration with temperature. In general, the solubility of a solid increases with increasing temperature (Figure 4.1, curve D). However, there are available examples that illustrate little or no change (curves B and C) and a decrease (curve E) with increase of temperature. The solubility curve may exhibit a maximum (curve A) or a minimum (curve B). According to the phase rule, $F = C - P$ only when temperature and pressure of the system are constant.

As temperature has such a great effect on solubility, a good thermostatic system and an accurate temperature-measuring device are prerequisites for precise solubility determinations.

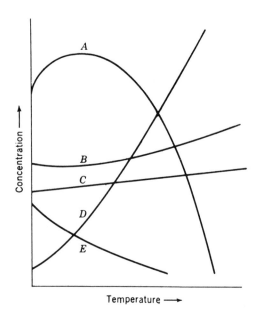

Figure 4.1 Temperature-concentration curves. Reprinted with permission from W. J. Mader and L. T. Grady, "Determination of Solubility," in A. Weissberger and B. W. Rossiter, Eds., *Physical Methods of Chemistry*, Vol. 1, Part V, Wiley, New York, 1971. Copyright © 1971 by John Wiley & Sons, Inc.

1.4.2   Purity

Solubility measurements are of little value unless the nature and amount of impurities in both solute and solvent are considered, an impurity being any substance other than the solute and solvent of the system under consideration. The importance that an impurity will assume in a solubility measurement is dependent on the amount present, the number of impurities present, the component that is being determined, and the affinity or reactivity that the impurity has for the solute or solvent. There is considerable leeway for the exercise of the experimenter's judgment as to how rigorously purification of the solute and solvent must be carried out. The stability of the solute in the solvent must always be determined prior to solubility determination.

The following examples illustrate the considerations governing minimum purity acceptable.

1.   In measuring the solubility of anhydrous sodium stearate in alcohol, both solute and solvent must be very dry, since in the presence of even small amounts of water the equilibrium phase is not anhydrous sodium stearate but a stoichiometric hydrate [8]. If 1 g of anhydrous sodium stearate is equilibrated with 1000 g of ethyl alcohol, the latter must contain less than

about 0.07% moisture if the equilibrium solid is not to be entirely converted to $C_{17}H_{35}COONa \cdot \frac{1}{2}H_2O$, the lowest reported hydrate.

2. The solubility of succinic acid in absolute acetone [9] at 40°C is 7 g/100 g of solvent, whereas for acetone containing 5% water it is 13 g/100 g of solvent. For acetone adulterated with 5% carbon tetrachloride or methanol, ethanol, or 2-propanol, the solubility of succinic acid is 6.2, 9.3, 8.0, and 7.5 g/100 g, respectively. From these figures one can readily calculate (assuming a linear mixture rule) how pure the acetone must be if the measured solubility is to be correct within any given precision. The differences just cited are not extraordinarily large, although the effect of chemically similar impurities, as toluene in benzene, will be expected to be small. Thus the solubility of m-nitroaniline in benzene is diminished by only 0.6% for a contamination of benzene with 5% toluene.

3. Similarly the solubility of naphthalene in benzene, toluene, and ethyl-benzene exhibits the same mole fraction singly or in binary solvent mixtures [10]. However, other normal solvents give different equilibrium solubilities, and prediction of solubility behavior in ternary mixtures can only be made empirically.

4. Surfactants have also been found to solubilize small organic molecules [11].

Conversely, the effect of solute impurities on solubility measurements is the basis for the determination of the purity of the solute, as discussed in later sections.

1.4.3 Equilibration

The third factor and the one that often causes the most difficulty is the attainment of equilibrium. The chemical and physical nature of the solute and solvent have a great effect; consequently, the equilibration time varies from system to system. For example, the absorption of solutes by polymers is characterized by long equilibration periods even in films.

A reliable method used to determine whether an equilibrium condition has been obtained is one in which equilibration is approached from both directions (i.e., undersaturation and supersaturation). When using this method, heat one of two identical samples to a temperature well above the equilibration temperature so that the equilibrium solubility is exceeded. Then both these two samples are placed in a thermostatic bath to equilibrate. If identical analytical results on aliquots of these two samples are obtained after a period of time, adequate equilibration has been reached.

The attainment of equilibrium within a system also can be revealed by periodic sampling; when the analytical values so obtained establish a plateau, adequate equilibration has been reached. One must be careful to interpret sample data only where the elapsed times are meaningful; in other words, analyses of samples taken a few minutes apart cannot define the equilibrium condition of slowly dissolving solutes.

Equilibration may not be readily obtainable when dealing with macro-molecules such as proteins. These incorporate numerous functional groups capable of retaining ions and water molecules that are in solution. Typically, inorganic reagents and buffers are used to solubilize proteins. To establish equilibrium and/or ascertain its attainment, it may be necessary to wash the protein preparation with successive portions of fresh solvent until a constant solubility is obtained [12].

To decrease the time necessary to obtain an equilibrated system, some means of continual agitation is usually necessary. A rotating wheel that has clamps attached to it is the best agitation apparatus. This rotating device is placed in a constant temperature bath in such a manner that the sample containers clamped on the wheel are always completely submerged. A less reliable, but somewhat faster method uses rapid vibration equipment and in this case best results are obtained by attaching the sample containers directly to the vibrator. Equilibration can be hastened in some cases by, for example, using a column of solid solute through which the solvent is allowed to percolate at a predetermined rate [13].

Chugaev and Khlopin [14] suggested that a solution be equilibrated at the boiling point of the saturated solution, thus eliminating the need for a constant temperature bath. The desired temperature in such a system is obtained by varying the pressure. Thorough mixing is ensured by vigorous boiling. The apparatus used is simply a wide-mouthed flask, equipped with a sensitive thermometer and coupled to a vacuum pump, in which the solution is boiled in the presence of excess solute. In the upper part of the flask is suspended a weighing tube that can be filled with the saturated solution by a syphon reaching to the bottom of the flask.

According to the theory of Noyes and Whitney [15–18] solubility or dissolution is a diffusion process (see Section 10 on dissolution for a more thorough treatment). A thin layer of solvent surrounding the solute crystal first becomes saturated, and material is transferred from this layer to the solvent by diffusion. The higher the viscosity of the solvent, the slower is the diffusion; the higher the temperature, the faster is the diffusion. The higher the solubility, the faster equilibrium is established, since diffusion velocity is proportional to the concentration gradient. The particle size and surface area of a solute will affect the rate of equilibration, and the mathematical relationship has an area term. The solubility one obtains for a solute in a solvent at equilibrium is independent of particle size (colloidal suspensions excepted).

The state of hydration of solids also contributes to the rate at which equilibration is obtained. For some drugs, Higuchi [19] reported that the hydrate and anhydrous form dissolve at different rates.

1.4.4   Proteins

Proteins (and other macromolecules such as polymers) acquire a unique configuration in solution that affects their solubility. This structure is stabilized by covalent and noncovalent forces of varying intensity such as covalent disulfide linkages that are formed between amino acids and redox active —SH groups. The frequency varies with the protein that contains the amino acids. Temperature also affects protein solubility. Proteins are stable over a limited temperature range and are typically denatured above 60°C.

Electrostatic interactions result from the inherent positive or negative charge of the proteins and are dependent on solution pH. At isoelectric pH, constant temperature, and ionic strength, the net negative and positive charges are equal and protein solubility is at a minimum. At either side of the isoelectric pH, protein molecules are polarizable and solubility in polar solvents is facilitated. Similarly, in acidic or alkaline solvents solubility is enhanced at either side of the isoelectric pH. Electrostatic bonds, however, are disrupted at the extremes of acidic or alkaline pH.

Hydrogen bonds are formed with proton donor and acceptor solvents. These tend to stabilize the protein structure. Intra- and intermolecular hydrogen bonds are also formed with other protein molecules. Hydrogen bonds are due to the presence of amido and carbonyl groups in the peptide bond configuration.

Hydrophobic bonds also affect protein structure and solubility. These are typically formed between organic species that perturb the structure of water molecules around them. Bond formation is due to unfavorable entropic changes leading to a more organized water structure as well as attraction of these organic species for one another. These bonds can also be formed in organic solvents but are usually much weaker than in water if formed at all.

2   APPLICATIONS OF SOLUBILITY IN CHEMISTRY

The uses of solubility are many and varied and it is not the intention of the authors to consider all of them. However, a number of examples will be considered to manifest what can be done with quantitative or semiquantitative knowledge of this fundamental physical property, solubility.

2.1   Isolation and Identification of Organic Compounds

Differential solubility relationships have long constituted the major means of separating, and thereby distinguishing, one organic compound from another. Differential solubilities of solutes, or partitioning, between two liquid phases are outstanding in significance and form the basis of extraction, countercurrent distribution, and partition chromatography techniques. Similarly, gas chromatography depends on partition behavior. These important techniques for isolation and identification are discussed elsewhere in this series.

2.1.1   Recrystallization

Differences in the solubility of two compounds in a given solvent as a function of temperature can be used to effect a more or less complete separation in which the less soluble compound is produced readily in pure form. This kind of

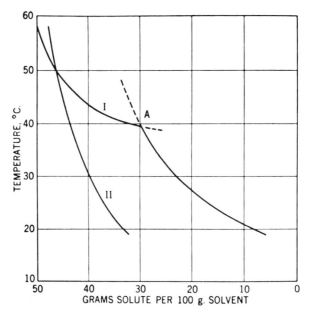

Figure 4.2 Typical solubility curves. Reprinted with permission from W. J. Mader and L. T. Grady, "Determination of Solubility," in A. Weissberger and B. W. Rossiter, Eds., *Physical Methods of Chemistry*, Vol. 1, Part V, Wiley, New York, 1971. Copyright © 1971 by John Wiley & Sons, Inc.

separation is made evident by Figure 4.2, which shows two hypothetical solubility curves. Suppose, for example, that it desired to separate compound I from compound II, which is contained in I to the extent of about 20%. A sample of 50 g of the mixture, containing 40 g of I and 10 g of II, will dissolve completely in 100 g of solvent when heated to 45°C. Upon cooling to room temperature (20°C), II will remain entirely in solution, while 31 g of I will crystallize out, since its solubility at 20°C is only 9 g/100 g of solvent. Thus, pure I is obtained in a 77.5% yield by one recrystallization. In this elementary example, complete independence of the two curves has been assumed, the possibility of coprecipitation is neglected, and the problem of washing the crystals of pure I free from the solution containing a considerable amount of II is ignored. For details on fractional crystallization, see Tipson [18]. The solubility curve for compound I is not smooth; the inflection at point A indicates either that there are two allotropic forms of I between which a transition occurs at 39.5°C or that I forms one or more solvates with this solvent. It is necessary to ascertain whether the solid formed at room temperature is pure compound I.

2.1.2   Salting-Out

The solubility of solutes can be decreased by changes in the solvent. The separation of many biochemical preparations, particularly proteins and protein complexes, is carried out by the addition of successive amounts of some salt, for example, ammonium sulfate or potassium citrate, to a buffered solution of a protein mixture in aqueous solution. This technique also indicates the number of components present in a complex mixture. Standard fractionation procedures for plasma proteins are described in biochemistry texts. Some proteins, for example, globulins, can be solubilized by the addition of a low concentration of salts in aqueous solution (salting-in).

Solutes of low polarity show reduced solubility in polar solvents as the ionic strength increases. Gordon and Thorne [20] studied the activity coefficient of naphthalene in aqueous electrolyte solutions composed of mixtures of two salts and, from Setschenow plots, determined that inorganic ion salting-out effects were additive; however, organic salts were not additive with the inorganic, tending towards greater naphthalene solubility.

Organic solvents are used frequently to suppress the solubility of slightly soluble salts. The selection of the organic solvent and its most effective concentration has usually been determined by trial and error. The common ion effect can be employed simultaneously. Jentoft and Robinson [21] presented an objective method that is based upon mathematical derivation and involves the graphical analysis of solubility data to determine the ratio of organic solvent to water for the most complete separation; tangents are drawn to the plotted solubility curve; the point giving the lowest tangent intercept represents the most efficient solvent composition.

2.1.3   Identification Schemes

The solubility of a pure compound at any given temperature is a characteristic of the compound just as are other physical properties such as the melting point or the boiling point. Reeve and Adams [22] suggest that when conventional melting points and mixed melting points fail to prove the identity of two materials or to serve as criterion of purity, owing to decomposition, a procedure based on the temperature of complete solution often can be used. When thermal analysis of binary mixtures by melting point curve is not feasible, the substitution of a similar type of curve based on the temperature of complete solution as determined by a solubility procedure should be attempted. In this section we later describe a convenient method of determining the temperature at which a solute will dissolve in a fixed amount of solvent and demonstrate how this temperature, designated as the *solubility temperature*, can be used in the characterization of compounds and in the analysis of mixtures.

The qualitative solubility behavior of organic compounds toward a selected group of solvents is the basis of one of four general procedures used for the systematic identification of unknown compounds [23]. The solvents employed for such general classifications are usually water and ether, as representatives of polar and nonpolar solvents; dilute solutions of acid and base; and, finally, more concentrated acids. More elaborate systems based on the use of a sequence of organic solvents are employed only in special cases, for example, in the partition pairs and equipment worked out by Beroza and Bowman [24] for pesticides and residues. In general, the solubility relations of substances are too specific to permit a division of compounds into mutually exclusive small classes.

High polymers [25] show a behavior toward solvents quite different from that of substances with smaller molecules. Some, notably chain polymers, swell continuously in certain solvents until finally a homogeneous solution is formed. Conversely, the solvent is said to function as a plasticizer. There are no saturation equilibria and no numerical expressions of solubility. Rather, at a given temperature, liquids can be classed informally as solvents or nonsolvents for the particular polymer. For some solvents, a critical temperature exists above which *solubility* of the polymer is unlimited and below which it is negligible. Other polymers, notably space and not polymers, swell to a limited extent in various solvents, but no molecular dispersion forms in the equilibrated solvent. The behavior of a given polymer toward various standard solvents is a way to identify unknown samples.

The amounts of nonsolvent required to precipitate a polymer from its solution in a miscible solvent, at varying polymer concentrations, can be used to estimate the molecular weight of the polymers [26]. The separation of a given polymer into fractions of varying molecular weight by the progressive addition of a nonsolvent, or precipitant, to a solution of the polymer has become of very great physical importance. The separation is based on the generalization that, for chains of like structure, those of high molecular weight are precipitated by smaller amounts of nonsolvent, although examples are known of the reverse relation for relatively short chains where the contribution of the terminal groups to solubility is not negligible [27].

2.2   Purity Determination for Compounds

2.2.1   Phase Solubility Analysis

Phase solubility analysis is the application of precise solubility measurements to the determination of the purity of a substance. This method is applicable to all species of molecules. A complete knowledge of the nature of the impurities is not required. The solubility method has been established on the sound theoretical principle of Gibbs' phase rule. Temperature and pressure are held constant so that $C = P + F^0$, and the degrees of freedom are expressed only in composition. For a pure solid in solution, one phase is present and one degree of freedom is possible as the concentration varies from zero to a saturated solution.

Mader [28, 29] published a comprehensive treatment and methodology of phase solubility analysis. Consequently, we will not treat this subject extensively here but merely include the following illustrative example. An aqueous solution of pure DL-isoleucine in water will give the solubility curve *ABS* in Figure 4.3, while a mixture of 85% DL-isoleucine and 15% L(+)-glutamic acid in aqueous solution will yield the solubility curve *ADEF*. From this solubility curve *ADEF*, the following information can be extracted:

1.   The solute is composed of two components.

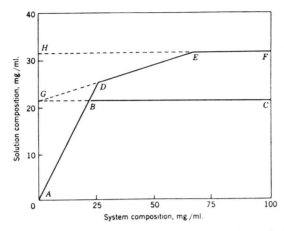

Figure 4.3 Examples of solubility curves obtained in phase-solubility analysis. Reprinted with permission from W. J. Mader and L. T. Grady, "Determination of Solubility," in A. Weissberger and B. W. Rossiter, Eds., *Physical Methods of Chemistry*, Vol. 1, Part V, Wiley, New York, 1971. Copyright © 1971 by John Wiley & Sons, Inc.

2. The principle component of the solute is present to the extent of 85% (100%—slope *GDE*).
3. The impurity is present to the extent of 15% (slope *GDE*).
4. The solubility of the principle component is 21.2 mg/mL of water (point *G*).
5. The solubility of the impurity is 10.3 mg/mL of water (point *H*—point *G*).

However, as differing polymorphic forms of the same compound may have different solubilities in a specific solvent, care must be exercised in the identification of a second slope as an impurity where multiple polymorphic forms are known to occur. A mixture of solids can form solid solutions also yielding misleading results. In addition, in purity determinations of compound salts, such as the hydrochloride, slope can be altered such that purity results are greater than 100%. This is a result of common-ion suppression of solubility. In such cases, the presence of an impurity can be masked by this salt effect.

The presence of differing polymorphic forms can be substantiated by other techniques such as thermal analysis. Individual samples used in the determination of the phase solubility plot can be tested by other methods such as thin-layer chromatography (TLC) or, depending on solvent, gas or liquid chromatography to verify the presence of an additional component. Analysis in multiple solvents can be helpful in ascertaining the presence of a solid solution.

The solid phase present in equilibrium with the solution formed during phase solubility analysis may not be identical to the original compound as it may be enriched in insoluble or slightly soluble impurities. Concentrations of soluble impurities will be decreased. In addition, as determined by MacDonald and North [13], the solid phase in equilibrium during phase solubility analysis under some high-pressure conditions is a mixture of the original anhydrous and a hydrate form for some slightly soluble salts such as calcium carbonate and strontium sulfate.

One should also not overlook the possibility of decomposition in the solubility solvent during the phase solubility analysis. Where the potential for such decomposition exists, a portion of the material in solution can be held at analytical conditions simulating phase solubility analysis and then tested by TLC or other means against freshly prepared sample for the formation of or increase in existing impurities.

Tarpley and Yudis [30] discovered that the classical methods of purity determination, namely, freezing and melting curves, were not suitable for steroids, the steroids being both heat labile and prone to undergo polymorphic modifications. They were, however, able to apply the phase-solubility method of purity determination to these systems with great success. The purity of a slightly soluble salt can be determined by use of another solubility procedure in which the conductance of a saturated solution, in the presence of a solid solute, is determined. Then the supernatant liquid is poured off and fresh solvent is added to the solute. The conductance of this second saturated solution should be the same as the first if the solid is pure. This procedure is of no value if the impurities in question are nonconducting, however.

Phase solubility analysis has also recently been applied to purity analysis of organophosphate pesticides [31] and fungicides [32] by Lesser and Massil. It has significant pharmaceutical applications as well [24, 33–35].

An additional technique based on phase solubility analysis is called *swish purification* [36]. A phase equilibrium is established for the purpose of either purifying sample or concentrating impurities. In Figure 4.4 the solution is unsaturated at system concentrations below X_1 (line DA). The system is saturated with respect to the main component at X_1, and subsequent points at

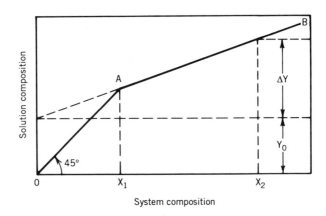

Figure 4.4   Solubility analysis diagram. Reprinted from G. B. Smith and G. V. Downing, Jr., *The Practice of Phase-Solubility Analysis*, 21st Annual Conference on Pharmaceutical Analysis, University of Wisconsin Extension-Pharmacy, Madison, WI, 1981.

system concentrations greater than X_1 produce line AB. Extrapolation of line AB to the y axis yields the solubility of the main component (Y_0).

In swish purification, equilibration is accomplished at system concentration X_2. The enrichment of impurities in solution can be calculated using the assumption that the solid phase at equilibrium contains only the main component. In this case

$$m_1 = \Delta Y/X_2$$
$$m_2 = \Delta Y/(Y_0 + \Delta Y)$$

(5)

where m_1 and m_2 equal the fraction of impurity present in the original sample and in the solute at equilibrium, respectively. The ratio m_2/m_1 is the impurity enrichment factor, that is, the impurity content of the solute is m_2/m_1 times that of the original sample.

$$m_2/m_1 = X_2/(Y_0 = \Delta Y)$$

(6)

The enrichment factor is approximately equal to the ratio of system composition (X_2) to Y_0 for minor impurities. This same factor applies to each minor impurity present in the sample.

Thorough agitation is required although temperature control is optional in this application. Equilibration of the solid phase is desirable but not essential in this application. Purification of the solid phase must be substantiated by other techniques such as TLC. The solution that is enriched in impurities can be used in other analytical techniques to substantiate or better identify impurities determined using these techniques.

2.2.2 Solubility Temperature

Solubility temperature supplies an alternate procedure [28, 29] for purity determinations. The temperature at which the last crystal of solute dissolves in a given volume of solvent, at constant pressure, is plotted against added levels of impurities. A disadvantage of this method is that the kinds of impurities should not vary and that the solubility temperature of pure compound be known. Where a pure sample is not available, phase solubility analyses of samples having several levels of impurities can be used to construct the calibration curve. The main advantage of solubility temperature, and a significant one, is relative speed, which may be used to advantage in routine process control after initial definition by phase solubility analysis. Satterfield and Haulard [37] describe the magnitude of overshoot of the true solubility temperature caused by various rates of temperature increase and recommend 3 h of agitation at the solubility temperature. Heric [38] used their [37] data to develop a rapid dynamic method based on extrapolation of data from varying heating rates, say 1–10°C/min, to zero rate, finding the temperature so obtained agreed with the value obtained statically.

2.3   Study of Intermolecular Forces

The mutual solubility of two substances has long been used as a qualitative measure of the extent of the interaction between their molecules, varying from simple departure from the laws of ideal solution (Raoult's law and Henry's law) to a genuine formation of a compound or stoichiometric complex between the solvent and the solute.

The question of the existence of solution complexes between aromatic hydrocarbons and aluminum halides has, because of its importance to a knowledge of the mechanism of Friedel–Crafts reactions, attracted the attention of several workers. The use of solubility data in this investigation serves as an excellent example of the application of solubility data to physical chemical problems. The existence of a chemical solvent–solute interaction, in a strictly binary aluminum chloride–hydrocarbon system was not established readily. This occurred because the low solubility of aluminum chloride in hydrocarbon solutions containing neither hydrogen chloride nor moisture made it impractical to study a possible complex formation by vapor-pressure lowering or the formation of univariant two-component systems. Such a system can be studied, however, by a solubility method, because complex formation in solution is reflected in the temperature dependence of solubility. Fairbrother and co-workers [39] determined the solubility of aluminum chloride in several hydrocarbons under rigorously anhydrous conditions, from 20 to 70°C. Their solubility results gave evidence of weak complex formation at room temperature in aromatic hydrocarbons. This complex formation was found to become greater as the electron-donating character of the hydrocarbon increased and less as the temperature increased.

Another example that illustrates the use of solubility data is the study of Gill and co-workers [40], in which diketopiperazine was used as a model compound in an investigation of hydrogen-bond interactions between the peptide bond and urea. They determined the solubility of diketopiperazine in aqueous urea solutions as a function of temperature and urea concentrations. The enthalpy of solution of diketopiperazine was estimated and consequently provided thermodynamic information on the interaction effects of diketopiperazine and urea in aqueous solutions.

The effects of additional —CH_2 groups on solubility of a long chain or branched compound is of interest in learning the steric and inductive effects imparted by the —CH_2 group. Bell [41] accumulated solubility data on hydrocarbons. Fragment solubility constants were determined and tabulated by Warita and co-workers [42] for several functional groups including ketone, alcohols, aldehydes, CN, ethers, esters, and NO_2, as predictors of aqueous solubility of drugs.

The parameters promoting solubility and the solubility trends for known carbohydrate polymers were investigated by Glass [43].

Solubility measurements can be used to determine the stability constant and the stoichiometric ratio of a complex or chelate. Higuchi and Lach [44] used a

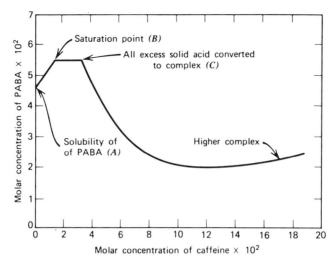

Figure 4.5   Phase diagram of PABA–caffeine system in water at 30°C. Reprinted with permission from T. Higuchi and J. L. Lach, *J APhA Sci. Ed.*, **43**, 349 (1954) and with permission of the copyright owner, the American Pharmaceutical Association.

solubility method to investigate the complexation of *p*-aminobenzoic acid (PABA) by caffeine (Figure 4.5). Obviously, the concentration of PABA at the initial point of the curve, that is, at the vertical axis intersection, is the solubility of PABA in water. At low caffeine concentration a straight line relationship is obtained between the concentration of PABA and caffeine in solution. This increase in solubility of PABA is due to complex formation and continues until the solubility limit of the complex is reached. The saturation point of the solution with respect to the complex is noted by a sharp break in the curve, point *B*. On further addition of caffeine the complex continues to form but precipitates out of the already saturated system. The total concentration of PABA in solution is constant in this region as shown by the flat plateau and is independent of the amount of caffeine, because any acid precipitated in the form of the complex is replaced by further dissolution of the excess solid PABA. The next break in the curve, point *C*, is noted when there is no more excess solid PABA; that is, all the PABA has either passed into solution or been converted to the complex. Further addition of caffeine results in depletion of PABA in solution. Although the solid PABA is exhausted and the solution is no longer saturated, some of the PABA remains uncomplexed in solution, and it combines further with caffeine to form higher complexes, as shown by the curve at the right of the diagram. The stoichiometric ratio of the components of the PABA–caffeine complex, which was formed in the plateau region, can be calculated from the phase diagram. Once the stoichiometry has been established the stability constant *K* also can be calculated.

3   DETERMINATION OF THE SOLUBILITY OF SOLIDS IN LIQUIDS

Methods of determining the solubility of solids in solvents vary in the degree of accuracy and precision obtainable and in the type of systems to which they are best suited. The first few methods considered here are methods that are applicable to common systems that have no unusual restraints for which special treatment must be given. After these methods of general applicability are described, some specialized methods are treated.

3.1   General Solubility Method

Reilly and Rae [45] described a method of general applicability with medium accuracy that requires a method of analysis for the solute. Weigh out, crudely, four samples of the solid of a size that will not dissolve completely at the given temperature in the volume of solvent contemplated. The solid should consist partly of fine and coarse crystals. Place the sample in each of four flasks or cylinders, and pour in the determined volume of solvent so that the flasks are about two-thirds full. Stopper tightly. Heat two of the samples well above the final temperature of measurement to approach equilibrium from the side of supersaturation, as most substances have a positive temperature coefficient for solubility. Place all four flasks on a suitable agitation device in such a way that they are submerged completely in the thermostat. After equilibrium seems likely to have been reached, discontinue agitation and set the flasks upright in the thermostat to facilitate settling of the excess solid to the bottom of the cylinder. If the flask has a long neck and the bath temperature is not very different from room temperature, the stopper may be allowed to project above the bath to facilitate removal of the saturated solution. Weigh samples of the solution that had a known volume at the temperature of the experiment. This procedure provides data for interconversion of units. The removal of a sample from the flasks is usually done with a pipet to which there is attached a filtering device of some sort. The filtering device may be made from sintered glass, cotton, or filter paper held to the pipet by rubber bands. Analyze the solution for solute content. All four results should agree within the precision called for in the determination, and there must be no significant difference between the pair originally supersaturated and the pair originally undersaturated.

To ensure that a change in the solid phase has not occurred during equilibration, obtain a sample of the solid by filtering the residue and examine a crystal under a hand magnifier or a low-power microscope. If any doubt exists about its identity with the original, free a sample from the excess solution by rapid washing on a filter with fresh solvent or a more volatile solvent miscible with the first, and analyze the residue for solute content or take a melting point of it alone and in admixture with the original.

3.2   Rapid Methods

Some of the more popular rapid methods are described below. These procedures do not ordinarily require special equipment or manipulations. One usually sacrifices accuracy and precision in gaining time.

Ward [46] suggested the following method. A test tube containing 3 mL of solvent is immersed in a beaker of water at 10–20°C above the temperature at which the solubility is to be determined. Solid is added in divided portions until a portion remains undissolved. Then the test tube is transferred to another beaker of water at a temperature at which the data is required and held for 10 min with occasional shaking. Push a small thimble made of folded filter paper, similar to a miniature Soxhlet thimble, down into the liquid and allow filtration to proceed from the exterior to the interior. Pipet a definite volume of the clear solution to a tared container. Evaporate the solvent and weigh the residue. Individual results can be obtained within 1 h.

A rapid solubility approximation was presented by Pastac and Lecrivain [47]. Weigh out twice the amount of A, 2A g of sample, where A represents more solute than can be dissolved in 10 mL of solvent. Then this 2A g of sample is divided into portions weighing 0.5 of A, 0.25 of A, 0.125 of A, and so on. Each portion is placed separately into a test tube containing 10 mL of solvent. The tubes are heated to 70°C and then cooled to the desired temperature. The last tube that shows undissolved sample is noted.

Another method that was proposed for the rapid estimation of solubility is based on the rate of solution [48]. This method is particularly useful in measuring the slight solubility of some electrolytes in concentration solutions of others, the solubility in systems that tend toward supersaturation, and the solubility of some electrolytes in supersaturated unstable solutions of other electrolytes. An analytical method is not required. This method is based on the relation:

$$V = K(C - C_2) = \frac{\rho^{2/3}(P_0^{1/3} - P^{1/3})}{2\tau} \tag{7}$$

where V is the rate of solution at concentration C_2, which is close to the true solubility, C; ρ is the specific gravity of the solid solute; P_0 is the initial weight of the dissolving crystals; P is the final weight of the dissolving crystals; τ is the time; and K is the coefficient of solution rate. Values of V are determined experimentally, and $C_2 = C$ (when $V = 0$) is determined graphically or by the equation. This equation is for cubic, crystalline solutes and factors other than "2" are required for other crystal geometries.

The residue–volume method of solubility determination is a rapid method with reasonable precision and can be used with pure substances, compound mixtures, and substances containing insoluble impurities [49]. Some of the advantages and disadvantages of this method are illustrated in Table 4.2. The

Table 4.2   Residue–Volume Method of Solubility Determination[a]

Advantages	Disadvantages
Accuracy sufficient for most commercial applications	Method not applicable to solutes of density lower than solvent nor effectively to solutes having very slow rates of solution
Quickly gives preliminary view of solubility limits in many systems	
Solubility found from number of individual determinations that operate to check on one another and to define precision of determination	The particle size of residue must be fine enough to give a compact, reproducible residue in capillary and must be coarse enough to be readily precipitated
No analysis needed	
Gives good results even when insoluble impurity present	The time and force of centrifuging must be constant
Difficulties with supersaturation not likely to be met	Temperature control is difficult since centrifuging causes a small rise in temperature of the solutions

[a]Reprinted with permission from W. J. Mader and L. T. Grady, "Determination of Solubility," in A. Weissberger and B. W. Rossiter, Eds., *Physical Methods of Chemistry*, Vol. 1, Part V, Wiley, New York, 1971. Copyright © 1971 by John Wiley & Sons, Inc.

principle of this method rests on the fact that when the solubility limit of a solute in a solvent has been reached, further additions of solute result in a proportional increase in the amount of undissolved residue. The volume of these residues can be measured for a series of solute–solvent ratios. These volumes are plotted as a straight line and extrapolated to zero residue volume as the solubility limit. When insoluble impurities are present it will be found that the plotted points may be corrected by two straight lines. Below the solubility limit, the points will yield a line of very low slope corresponding mainly to traces of impurities. Above the solubility limit a line of very steep slope will be obtained. The intersection of these two lines is taken as the solubility limit.

Several pesticide solubilities have been determined by Fuerer and Geiger [50], who used a rapid method in which the solute particles were finely ground to micron size. This is added to solvent at a concentration well above the predicted solubility. The suspension obtained is diluted in a stepwise manner until a loss of turbidity is observed. While no special apparatus is required and the method is applicable over a wide range of solubility, it does require a pure sample.

A rapid method for the determination of solubility of solids in viscous liquids was developed by Grant and Abougela [51] that measures the temperature at which a stirred solute–solvent mixture of predetermined composition just forms a homogeneous solution.

In this way, the equilibrium solubility temperature can be determined within 1°. From the solubility temperature data thus obtained changes in free energy, enthalpy, and entropy for the solution process can be calculated readily, and physiochemical interactions between a solid drug and a molten excipient can be studied at regions of composition in which other thermal methods of analysis may not be suitable.

3.3   Apparatus for Solubility of Solids in Liquids

One of the most efficient solubility devices was originally designed by Campbell [52] and further modified [53] as is shown in Figure 4.6. In using this apparatus, bottle A is charged with the solvent and excess solute and is rotated until equilibrium is reached. The glass jacket then is inverted while it is still in the thermostatic bath. The saturated solution is permitted to filter through the tube containing glass wool from *A* to *B*, the air being displaced simultaneously from *B* to *A* through the capillary. The filtration thus occurs at exactly the same temperature as that at which the saturated solution is prepared.

Lombardo [54] devised a two-chamber automatic-filtration cell (Figure 4.7) by using readily available glassware, and it is operated by pulses of pressure

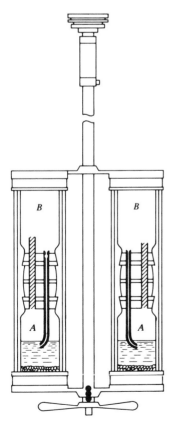

Figure 4.6   Campbell solubility apparatus. Reprinted with permission from A. N. Campbell, *J. Chem. Soc. (London)*, 179 (1930).

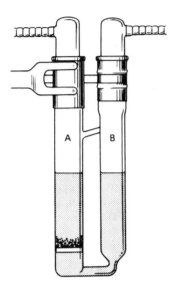

Figure 4.7 Lombardo apparatus. Reprinted with permission from J. B. Lombardo, *J. Chem. Educ.*, **44**, 600 (1967).

delivered from a plastic bellows. The whole unit can be submerged in a bath; therefore, the apparatus can be operated at otherwise inconvenient temperatures. Solute and solvent are placed in side *A*, which was prepared from a filter tube. The bellows is connected by tubing to side *A* and operated at about two impulse cycles per minute.

3.4   Determination in Volatile Solvents

The procedure of Reilly and Rae [45], previously described, can be used for solvents of moderate volatility, in other words, 50–100 mmHg at the temperature of the thermostat, if time is allowed for the air space in the flasks to become saturated with the vapor before stoppering. When the vapor pressure of the solvent is high, that is, 1 atm or more, special equipment is required.

One such piece of equipment is a modified solubility tube that has been used to determine the solubility of solids in volatile liquids by a synthetic method [55]. In this procedure known weights of the solute and solvent are contained in a sealed glass apparatus. The amount of solvent admitted to the tube containing the solute can be varied at will without opening the apparatus. The solvent is kept in the graduated reservoir tube *D* (Figure 4.8). The solvent is distilled from *D* to *A*, the solubility vessel containing the solid-wider investigation, through trap *C*. Trap *C* is closed during determinations by melting a pellet of silver iodide–silver chloride mixture above the constriction in the U tube and *C*, and allowing it to run into the constriction. This method permits the solubility of nonvolatile solids in volatile solvents to be measured with an accuracy of greater than 1% over a wide temperature range.

Butter [56] also designed a system (Figure 4.9) for determination of solids in

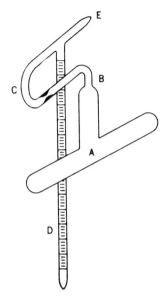

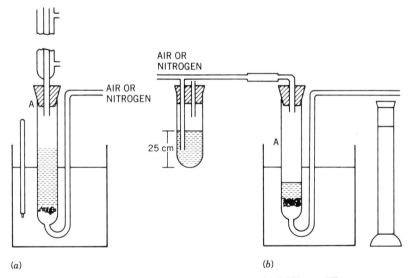

Figure 4.8 Apparatus for determination of solubility of solids in volatile liquids according to Menzies. Reprinted with permission from A. W. Menzies, *J. Am. Chem. Soc.*, **58**, 934 (1936). Copyright © 1936 American Chemical Society.

Figure 4.9 (*a*) Equilibrium apparatus for the determination of solubility at different temperatures. Tube *A* contains excess solute and solvent; the condenser prevents loss of solvent at high temperatures. Mixing is achieved by bubbling air or nitrogen. (*b*) Filtration apparatus for determination of solubility at different temperatures. Filtration of solution from tube *A* is achieved by applying pressure; *B* is merely a tube filled with mercury acting as a pressure-relief bubbler to avoid excessive buildup of pressure. Reprinted with permission from S. A. Butter, *J. Chem. Educ.*, **51(1)**, 70 (1974).

volatile solvents. The apparatus shown in Figure 4.9 is also very useful in determining solubility at different temperatures. The excess solute is first mixed with a known quantity of solvent. Nitrogen is typically bubbled through the apparatus to promote mixing. The condensor prevents the loss of solvent. The condensor is replaced with a stopper fitted with a tube when equilibration is achieved. A slight positive pressure is applied through this tube so that the solution from tube *A* is filtered through the tip of the other tube. The filtered solution is then collected and dried to constant weight. The undissolved solute can be dried and weighed also. The amount of solute dissolved can be determined by subtracting the weight from the amount added.

3.5   Slightly Soluble Solutes in Liquids

Gross [57] used the interferometer to determine the solubility of slightly soluble liquids in water. Adams [58] earlier stated that the interferometer is well adapted to determine a single varying component in a transparent mixture, whether that component be solute or solvent, electrolyte or nonelectrolyte.

Mitchell [59] extended the use of the interferometer method by developing a procedure that could be used in the determination of the solubility of any sparingly soluble solute, including liquids, in any solvent. A double cell was constructed through which liquid could be circulated, and readings were made with pure solvent circulating through one side of the cell and the solution through the other. The equilibrium-saturated solution was prepared and interference compensation data for serial dilutions of this solution were plotted against the percentage of saturation. In all but one case, that being benzene, a linear relationship was obtained. In view of this one noted exception, it is always necessary as a preliminary to a solubility determination by this method to investigate the relationship between compensator readings and the percentage of saturation. A change in solute concentration of 2 ppm will be detected by this method.

Strongly absorbing substances are measured easily by ultraviolet (UV) spectrophotometry, so this analytical method is very good for determining the solubility of slightly soluble substances. A good example of this technique is the determination of aromatic hydrocarbons in water by Bohon and Claussen [60]. Fluorescence and phosphorescence methods also may find application in the determination of slightly soluble solutes, particularly in the presence of other absorbing solutes. Solubility parameters have also been determined for polymers at infinite dilution by fluorescence [61].

Gas chromatography (GC) has been used to determine the solubility properties in polymers and biological media [62]. A head space method has been used to determine solubilities of hydrocarbons in aqueous solution [63]. High-performance liquid chromatography (HPLC) has also been used in the determination of solubility parameters [64, 65]. Reverse-phase TLC has been used to determine solubility of polynuclear aromatics and chlorinated biphenyls in aqueous solution [66].

Weyl [67] devised an apparatus for the conductometric determination of the solubility of slightly soluble electrolytes in a closed system. The unique internal pumping system consists of a column of mercury that changes position when the apparatus is gyrated mechanically. This process of equilibration is monitored continuously by measuring the conductance in a cell in the line of the liquid flow. Figure 4.10 is a perspective drawing of the solubility apparatus, which consists of two conductance cells, a bulb filled with the electrolyte under investigation, and a pump for circulating the solution. The entire apparatus is made from glass and can be filled completely with the solution under

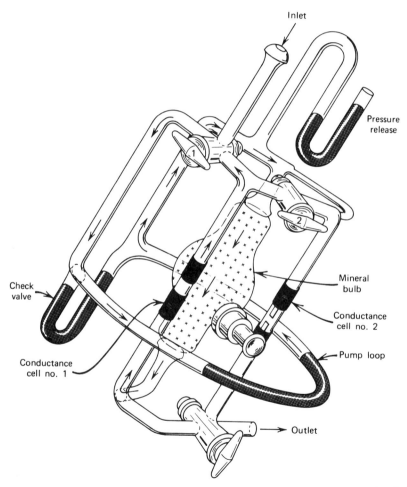

Figure 4.10 Weyl solubility apparatus for determination of slightly soluble solids. Reprinted with permission from P. K. Weyl, *Rev. Sci. Instrum.*, **28**, 722 (1957) and from W. J. Mader and L. T. Grady, "Determination of Solubility," in A. Weissberger and B. W. Rossiter, Eds., *Physical Methods of Chemistry*, Vol. 1, Part V, Wiley, New York, 1971. Copyright © 1971 by John Wiley & Sons, Inc.

investigation. The mineral bulb is terminated by two coarse glass frits to retain the material under investigation. Filling and emptying is accomplished through a side port, normally closed by a glass stopper. The mounting for this apparatus is so constructed that the entire apparatus can be immersed in a thermostatic bath. By replacing the conductance cell by a suitable counter, this apparatus can be used to measure the solubility of radioactive solids.

Methods have been developed for determining the solubility of slightly soluble materials by use of radioactive indicators [68, 69]. In the procedure proposed by Jordan [68] the tracer is added to a saturated solution of the substance at room temperature. The mixture is warmed to the temperature of the solubility determination. At first, the radioisotope is evenly distributed in solution but on cooling it is distributed both in solution and in the solid. The activity in the solid and liquid phases is determined and the solubility is calculated by the ratio. Batra [70] also used the tracer method to determine the solubility of some steroid hormones. Lo and co-workers [71] used tracer methods for the determination of organic solvents in water.

Aqueous solubilities of fatty acids and alcohols have been determined by a film-balance technique. Robb [72] measured interfacial tensions and related these to concentration. He found that the logarithm of the solubility varied linearly with chain length, with some deviation due to fatty acid ionization.

3.6 Apparatus for High-Temperature Solubility

Marshall [73] obtained the solubility data in the dilute region for the system uranium trioxide–sulfuric acid–water at elevated temperatures and, in so doing, devised a method for sampling an equilibrated solution at elevated temperatures. In this method the pressure bomb, in which is found the solution plus solid, contains a length of thin-walled capillary tubing attached at one end to a pressure valve equipped with a sampling tip (Figure 4.11). Part of the capillary passes through a wet ice bath so that the solution is cooled rapidly during the sampling process, thereby eliminating a separate cooling or isolation chamber and facilitating the procurement of many samples per run. With this direct sampling technique there is neither a distillation problem nor a need for correction for loss of solvent or other components to the vapor phase at the equilibration temperature, as the liquid phase alone is sampled at the elevated temperature and can be analyzed for all components. However, with systems having a positive temperature coefficient of solubility, the precipitation characteristics and solubility range of the individual system would have to be considered and evaluated. The solubility relations for the system that Marshall investigated fell within the most sensitive range of pH variation with mole ratio, and thus the solubility of the uranium trioxide was determined by comparing the pH of the isolated sample at 25°C with control pH data. For the application of this method to other systems the sensitivity of pH in the particular solubility range must be considered. There are, however, possible modifications of the analytical method.

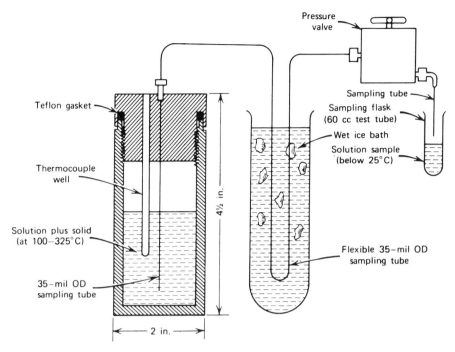

Figure 4.11 Solubility apparatus for determinations at elevated temperatures. Reprinted with permission from W. J. Marshall, *Anal. Chem.*, **27**, 1923 (1955). Copyright © 1955 American Chemical Society.

To determine the solubility of salts at high temperatures, a method suggested by d'Ans [74] is used, which involves the determination of the temperature at which a known amount of salt just dissolves. This determination is made in small tubes placed in an air thermostat and observed through a glass window.

In the apparatus devised by Breusov and co-workers [75], there are two vessels connected by two tubes. The bottom tube is equipped with a fritted-glass filter. An upper tube equalizes the pressure between the two vessels. The entire apparatus is immersed in a bath at the desired temperature. Alkyl bromides have been determined using this apparatus.

Susarla and co-workers [76] determined salts using a new apparatus at temperatures of 30–200°C. Potassium chromate, ammonium sulfate, and sodium sulfate were determined at temperatures of 10–90°C using an apparatus designed by Butter [56]. Menzies [55] used an apparatus for the determination of salts in water at different temperatures.

3.7   Apparatus for Low-Temperature Solubility

Solubility determination at low temperature and freezing point determination can be accomplished using a special apparatus (Figure 4.12) designed to avoid condensation of solvent vapors [77]. The apparatus consists of a constant

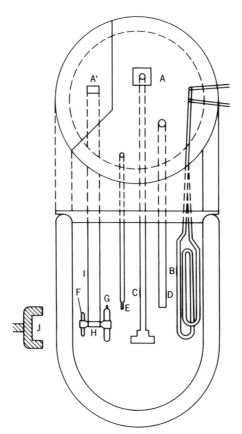

Figure 4.12 Special apparatus for determination of solubility at low temperatures. Solute and solvent are placed in sample tube *A*, which is rotated by the action of a magnetic motor *J* on magnetic stirrer *F*. Coolant is circulated through the cooling coils *B*; *C* and *D* are stirrer and thermoregulator, respectively; and *A* and *A'* are parts of lid. Reprinted with permission from J. A. Harris, A. V. Bailey, and E. L. Skau, *J. Am. Oil Chem. Soc.*, **45(9)**, 639 (1968).

temperature bath in a large unsilvered Dewar flask. The lip *A* of the flask lid is partly recessed on the flask top. The cooling coils, *B* a bimetallic thermoregulator *D*, and a stirring mechanism *C* are attached to the area of the lid *A*, which is permanently attached to the flask. A rubber gasket between the flask and the cover lip section of the lid ensures consistent temperature.

The sample assembly is attached to the other part of lid *A'*. The lid *A* can be removed from the apparatus using the lifting handle. A magnetic stirring bar *F* and a clip *G* for holding the sample tube make up the sampling assembly. The stirring bar and the sample tube are connected by a horizontal shaft covered by a 0.75-in.-long Teflon sleeve *H*, which is mounted on the stainless steel support *I* attached to *A'*. A horseshoe magnet *J*, which is located outside the flask, is rotated by a motor; thus, the magnetic stirring bar F and the shaft H are rotated along the axis H. The sample tube then turns end to end at a desired rate, facilitating proper mixing. A focusing microscope light is placed behind the apparatus. This shines through the sample tube as an aid to visualizing the contents of the sample tube G during analysis.

3.8   Solubility of Mixtures

Phase solubility analysis and solubility temperature, as outlined in Section 2 constitute the primary tools for solubility analysis of mixtures.

Where only the more soluble fraction of a mixture is of interest, and time is limiting, a method is available that has been applied to soluble fractions of wax [78]. In this method the total amount of the more soluble fraction of a mixture is calculated from the amounts of material dissolved in each of several successive extractions. An accurately weighed sample of the mixture is extracted successively with equal volumes of solvent to obtain a series of fractions. Following removal of the solvent by evaporation, each of the fractions is weighed. The total weight extracted w is then plotted against the number of extractions n in the form n/w versus n, so that the reciprocal slope of the graph gives the amount of soluble fraction.

The kinetic method can be used to determine the solubility of one compound in a solution of another [79]. In this method, the first solid is extracted using solutions of varying concentrations of the second solid. The rate of dissolution of the first solid is calculated using Fick's second law of diffusion. The solubility of the solid is estimated from it using this formula

$$\delta c/\delta t = D(\delta^2 c/\delta^2 x) \qquad (8)$$

where c is the concentration, x is the distance to which the boundary has moved at time t, and D is the diffusion coefficient.

The kinetic method has also been used to determine the solubility of oxygen [80] and other gases [81].

3.9   Solubility in Liquified Gases

There have been several methods published over the years for the determination of solids in liquid ammonia, two of which are discussed here. Schenk and Tulhoff [82] determined solubilities of salts as a function of temperature by condensing liquid ammonia on known solute samples until all dissolved; the amount of ammonia delivered was measured by loss of pressure in an ammonia storage vessel. Such a procedure would be generally applicable to solutes and gases.

The 25°C solubility data for several inorganic halide salts in liquid ammonia and liquid sulfur dioxide have been determined from weight measurements or chemical analysis obtained during the course of a process based on effecting solution in a glass filter tube, separating the saturated solution from excess solid by centrifugation, and subsequently removing the solvent [83]. The exact procedure of these authors follows. A weighed sample of solid, approximately 0.1 g, is introduced into one end of a glass filter tube (fritted-glass disk at midpoint). This end of the tube is sealed and the tube and contents are weighed. The tube is flushed out with anhydrous gas, after which the open end of the tube is attached to the source of the gas. The end of the tube containing the sample is immersed in a dry ice–acetone bath ($-75°C$) and a suitable quantity of solvent, 0.8–1.0 g, is condensed on the solid sample. The open end of the tube is sealed off under conditions that permit the determination of the weight of gas removed in making the seal. The sealed tube is allowed to warm to room temperature and then weighed and agitated in a thermostat ($25 \pm 1°C$) for 48 h. The tube is removed and centrifuged, thus effecting separation of the saturated solution and the excess undissolved solid. The end containing the saturated solution is cooled to $-75°C$ and the other end is drawn out to a fine capillary through which solvent is allowed to escape. Thereafter the tube is maintained at 10^{-2} mmHg for 1–2 h. The weight of solute in the saturated solution is determined by removing the portion of the tube that contains the sample of the solid that was dissolved, weighing it before and after removal of the solid (recommended procedure), or by obtaining a similar determination of the weight of the excess solid and the weight of the dissolved portion by difference. The weight of the solvent can be measured by a determination of the weight of the tube assembly before and after introduction of solvent and application of the correction for the weight of the glass removed.

4   DETERMINATION OF THE SOLUBILITY OF LIQUIDS IN LIQUIDS

4.1   General Methods

There are several methods, both analytical and synthetic, that have broad applicability to solubility determinations in liquid–liquid systems.

When an analytical procedure is used there are essentially two operations, namely, securing equilibrium between the two liquid phases during a sufficient equilibration time, with the customary precaution of approach to the final state from both sides (i.e., higher and lower temperature) and obtaining samples of each layer at the temperature of equilibration.

The following general analytical method is suggested. Measure out the desired volumes of the two liquids into two suitable containers that can be stoppered tightly. If either liquid has an appreciable vapor pressure, allow the vessels to stand uncovered for several minutes before stoppering, or reduce the pressure inside after stoppering by means of a suitable pump connected through the stopper and a stopcock. Place one vessel in a beaker of cold water and agitate thoroughly. Place the second in hot water and agitate. Then submerge the two in a thermostat at the desired temperature and rock mechanically for a suitable time interval. With adequate mixing, if the liquids are not too viscous, equilibrium should be reached in a few minutes. However, if the liquids tend to emulsify, the only permissible agitation is a gentle rotation, which renews the liquid near the surface without rupturing the meniscus separating the two liquids. In such a case, several hours may be required to establish equilibrium, depending on the ratio of interfacial area to volumes of the phases; and, in this case long, narrow tubes are to be preferred. To withdraw samples, allow the vessels to come to rest in an upright position with their stoppers barely projecting above the surface of the bath. If distillation occurs because of the temperature gradient thus established along the neck of the flask, leave the vessels completely submerged but so arranged that they can be raised without disturbing the liquid layers. The samples desired can be taken from the top layer simply by pipet. A sample of the bottom layer can also be secured by maintaining a slight air pressure on the pipet to prevent contamination as it is lowered through the top layer into the bottom. Alternatively, a graduated pipet with the tip sealed by a thin-blown glass membrane can be used to sample the lower layer, the membrane being broken against the bottom of the vessel. Weigh known volumes of the two layers to obtain the data needed for expressing the results in any units desired. Use any convenient procedure to determine the composition of the layers. If the necessary calibrations have been made, physical methods such as measurements of refraction or density may be adequate. When separation occurs readily into two liquid layers, each of moderate volume, the authors prefer the method described earlier.

Synthetic methods, however, are also generally useful. In one such synthetic method, the "cloud-point" method, the temperature of incipient separation into two phases is determined as the isotropic, single-phase liquid solution is cooled. Minute droplets of the second phase begin to form throughout the formerly homogeneous system and give it a cloudy appearance. The temperatures of appearance of the cloud on cooling, and of its disappearance or heating, are usually the same within experimental error, $\pm 0.1°C$. However, instances of real discrepancies of up to $3°C$ have been reported [84]. A single experiment gives no information about the composition of the second phase. To determine the composition of both phases at any given temperature, a series of experiments must be performed so that the complete solubility curve can be constructed.

Sometimes no readily visible cloud is formed. If a dye can be found whose color differs in the two phases, a minute amount dissolved in the solution may change color sharply when the second phase separates out. Klobbie [85] used this effect successfully in studying the system diethyl ether–water. To increase the visibility of the cloud one can place the sample in a heavy-walled tube held in a horizontal position. A wire network is viewed through the tube, which acts as a lens. Distortion or disappearance of the image shows when a second phase begins to form. In the special cases in which the separating phase is liquid crystalline rather than liquid, observation through crossed polaroid plates renders its appearance conspicuous.

The thermostatic method of Hill [6] is a procedure of wide application in liquid–liquid systems. This involves the phase-rule principle that two-component systems consisting of two liquid phases under their vapor pressure are univariant, that is, the fixing of one condition determines the system in all other conditions. Thus, if the temperature is fixed, the composition of the liquid phase as well as the vapor pressures will be fixed at equilibrium. The two liquids are mixed in two different ratios by weight in two separate experiments at the same temperature, using the flasks shown in Figure 4.13. The m and m' represent the weights of the first component used in the two experiments; x represents the first component's concentration in grams per milliliter at equilibrium in the upper phase in both experiments, since by the phase rule the concentration at saturation cannot vary. Similarly, y represents the concentration in the lower phase in both experiments. If the measured volumes of the upper phase are a and a' and the measured volumes of the lower phase are b and b', then $ax + by = m$, and $a'x + b'y = m'$.

Solving the equations for x and y will give the concentration of the first component in each phase. If the equations are again solved, substituting the weights n and n' of the second component in place of m and m', then the second component's concentration in each phase becomes known. By adding together the weight of each component present in 1 mL of a given phase, the weight per milliliter or density is obtained and the percentage of composition by weight thus follows.

Evans [86] suggested the use of an oil centrifuge bottle instead of Hill's original apparatus. He looked into the question of what volume ratio gives maximum accuracy in this method and determined graphically that a and b should be as small as possible.

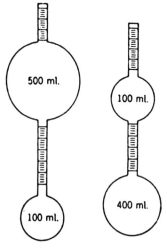

Figure 4.13 Solubility flasks used in the Hill solubility method. Reprinted with permission from W. J. Mader and L. T. Grady, "Determination of Solubility," in A. Weissberger and B. W. Rossiter, Eds., *Physical Methods of Chemistry*, Vol. 1, Part V, Wiley, New York, 1971. Copyright © 1971 by John Wiley & Sons, Inc.

4.2   Rapid Methods

Herz [87] described an approximate procedure for the solubility of one liquid in another at room temperature. One liquid is added to the other dropwise from a buret, with continuous agitation. The first excess gives the whole liquid in the flask a cloudy appearance, which can be amplified by the modifications mentioned earlier. This procedure also is applicable to ternary liquid systems. A refinement of the preceding method is reported by Sobotka and Kahn [88], who used minute, jagged crystals insoluble in the first liquid as indicator particles. The slightest excess of the second liquid rapidly dissolved the indicator particles, converting them from jagged crystals into rounded droplets of a different color. They determined the water solubility of the ethyl esters of the homologous series from propionic to capric acid and from malonic to sebacic acid using Sudan IV as the dye. An accuracy of $\pm 0.001\%$ or 0.01 mL in 1 L was obtained.

4.3   Slightly Soluble Liquids in Liquids

Analytical difficulties have often characterized determinations of the slightly soluble liquids. Physical methods often are inapplicable. Gas chromatography is most useful, particularly where the slightly soluble compound is the more volatile. Some specific applications of other methods are listed here. Methods of choice in the determination of water in organic solvents are GC, Karl Fischer titrimetry [89], and infrared spectroscopy.

Gross [57] used an interferometer to determine the solubility of slightly soluble liquids in water. The interferometer scale is calibrated by measuring synthetic solutions. The estimated errors of the solubility values obtained in this manner are $1-2\%$, which represents an error of only 0.01% in the total composition of the solution when the solubility of the solute is only 1%.

Hayashi and Sasaki [90] determined the solubility of sparingly soluble organic liquids in water by turbidity tritrations using Tween 80. There is a linear relationship between the measured turbidity and the solute concentration in water.

Hilder [91] determined the solubility of water in rapeseed, coconut, and palm–coconut oil between 60 and 100°C using the solute isopiestic method. At a constant temperature in an enclosed space if a sample is exposed to the vapors of a solute in the presence of a reference substance containing a known amount of the solute, the amount of solute dissolved by the sample is determined by the vapor pressure of the solute and can be calculated by knowing the amount of solute in the reference substance and the vapor pressure of the solute at that temperature; thus, solubilities of water in fats and oils at constant humidity can be determined.

The apparatus consists of two conical flasks connected by a T joint, with the oil sample in one conical flask and the reference substance, for example, aqueous sodium hydroxide solution, in the other. Water determinations can be performed to ascertain the water content of the oil samples.

5   MICROMETHODS FOR SOLUBILITY DETERMINATION

Although many solubility methods can be modified to determine the solubility of microsamples, very few are designed specifically for such use. Yet, it is quite important that one have available solubility determination method that is generally applicable when only a limited amount of sample is available.

Such a general procedure was devised by Nash [92]. In this procedure one determines the solubility through the measurement of the vapor pressure lowering in a saturated solution. The accuracy of the measurements depend on the value of P_0 (the vapor pressure of pure solvent at the temperature of the measurement) and on the intrinsic solubility of the solute. However, an accuracy of at least 5% can generally be maintained when dealing with moderately soluble substances that obey Raoult's law. Only enough sample to saturate approximately 0.1 mL of solvent is required, and no weighings are involved. However, Raoult's law must apply, and molecular weights must be known or determined. The apparatus that is required for this submerged bulblet method is rather elaborate. Figure 4.14 is a diagram of the equipment used in this method of solubility determination. However, if microscale determinations are required in a laboratory with any frequency at all, it might be advisable to construct this apparatus. This same apparatus can be used for vapor pressure, decomposition pressure, phase study, purity of solute or solvent, and molecular weight as a function of concentration measurements.

The procedure used in determining solubility follows. A quantity of solute estimated to be sufficient to saturate 0.1–0.2 mL of the solvent at the maximum temperature of the equipment is introduced into the bulblet L, and 0.2–0.3 mL of the solvent is added. The vapor jacket H is charged with the same solvent, and the apparatus is assembled as shown in Figure 4.14. The pressure in the jacket is set at an appropriate low value. The air is swept from the bulblet; and after a pause to establish equilibrium, the pressure in the jacket P_0 is read from barometer A, and the value of Δp is taken from manometer B or C. The temperature of the experiment is fixed by the value of P_0; and if X is taken as the mole fraction of the solute, then from Raoult's law:

$$\frac{1}{X} = \frac{P_0}{\Delta p} \tag{9}$$

If the molecular weight of the solvent and solute is known, the mole fraction can be converted readily to a solubility value.

The attainment of equilibrium is checked by successive approaches from under- and oversaturation. The assumption of Raoult's law can be checked with the same sample by a determination of the apparent molecular weight, even if association or dissociation occurs, as a function of concentration. If the results are independent of concentration, then that value (which may, of course, be only an apparent molecular weight of a substance either associated or dissociated)

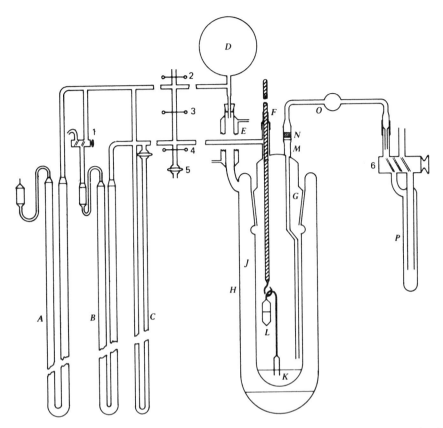

Figure 4.14 Submerged-bulblet apparatus for solubility determination of microsamples. Reprinted with permission from L. K. Nash, *Anal. Chem.*, **21**, 1405 (1949). Copyright © 1949 American Chemical Society.

can be used in converting the mole fractions to gram solubilities. If the apparent molecular weight varies with concentration, this method is not valid.

Accurate measurements of the solubilities of various compounds of the synthetic elements have been obtained on a sample as small as a few hundredths of a microgram [93]. Such measurements are facilitated by radiometric assay of the solution in equilibrium with the solid phase. For example, in the case of ^{229}Pu, as little as 0.01 μg can be determined with an error of less than 2%. Solubility measurement of a nonradioactive material usually requires several micrograms of sample, as precise chemical analysis is quite difficult with smaller samples. In this procedure, the solid and solvent are placed in a glass capillary, which is sealed and attached to the periphery of a notched wheel that is rotated by means of a motor. This rotation produces a continuous mixing action within a specially designed liquid thermostat. After equilibration, the solid phase is centrifuged to the bottom of the tube and the sample is withdrawn. The sample is then spread on a thin plate and dried for radiometric or chemical assay.

6   DETERMINATION OF THE SOLUBILITY OF GASES IN LIQUIDS

Determination of gas solubilities in liquids can be categorized by the type of measurement made. Physical methods are based on the gas law and measurements commonly are manometric within constant volume, isothermal systems. Physical methods can be classed further as saturation or extraction methods. A saturation method measures the amount of gas required to saturate the previously degassed solvent. An extraction method measures the gas extracted or liberated from a saturated solution of that gas in the solvent. Several systems of general usefulness are considered here.

The generally applied methods are physical ones, mostly of the saturation type. Precision varies according to the apparatus used and the specific gas–liquid pairs. Accuracy depends heavily on the approach to ideality or the validity of correction for nonideality. In all physical methods, the liquid must be gas-free at the start and this is a major source of error if not carried out scrupulously. This condition is usually obtained by boiling, followed by vacuum cooling [95] or vacuum sublimation [94].

Chemical methods are those that measure a property of the gas molecule other than gas, law relations, whether that measurement be spectrophotometric, radiometric, or by any other "physical" determination or by chemical reaction. As such methods are specific to the gas–liquid pair, extensive treatment is not required in this chapter.

6.1   Rapid Methods

Two generally applicable methods are available that attain equilibrium rapidly.

6.1.2   Dymond and Hildebrand Apparatus

Dymond and Hildebrand [95] described a glass apparatus for the accurate and rapid determination of gas solubility in liquids (Figure 4.15). Bulb A contains a known volume of liquid and opens into bulb B into which a known amount of gas is introduced. The solvent is pumped into the upper bulb by a sealed magnetic pump in the side arm between bulbs A and B. The liquid then runs down the walls of bulb B and in this way fresh interface is exposed continually to the gas. Equilibrium is determined when consecutive manometric readings on the undissolved gas remain constant, typically 1–3 h. Measurements made by reequilibrating at several temperatures permit calculation of the entropy of solution. Equilibrium for nonviscous solvents was obtained in 1–3 h, and accuracy was reported as better than 1%.

To operate, the whole apparatus is evacuated; mercury from reservoir S is

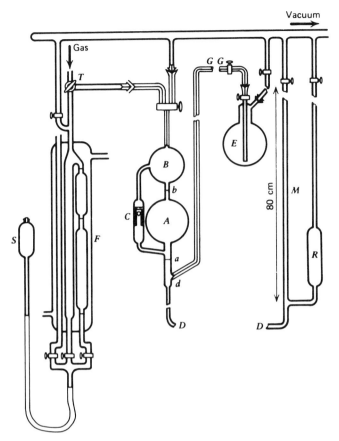

Figure 4.15 Apparatus for determining solubility of gases. Reprinted with permission from J. Dymond and J. H. Hildebrand, *Ind. Eng. Chem. Fundam.*, **6**, 130 (1967). Copyright © 1967 American Chemical Society.

admitted to the gas buret until the meniscus is level with the lower mark. Purified gas is admitted to the buret by the three-way stopcock *T* until the pressure is about 1 atm. The mercury height in the central column is adjusted until the pressure equals 1 atm. Mercury from reservoir *R* is admitted to *D* until the meniscus is just below *d*. Purified solvent in *E* is degassed scrupulously by a combination of pumping on the frozen solvent and boiling away a portion of it under vacuum. The degassed solvent is then impelled under its own vapor pressure by *G* into *A* until the bulb is nearly full. In the case of a liquid of insufficient vapor pressure, helium at a pressure of about 50 mmHg is used to force the liquid over. After the liquid has attained the temperature of the bath, the mercury level in *D* is raised to cut off the capillary sidearm and to bring the level of the liquid to *b*. The distance from the mercury meniscus to *a* is measured and the volume of solvent is calculated exactly.

Gas from the buret is admitted slowly to B while the amount of mercury from reservoir R is increased, so as to keep the mercury meniscus just below a. The gas is shut off as the pressure in B approaches 1 atm, as shown on the manometer M. The mercury height in F is adjusted until the pressure of the gas remaining is 1 atm and the amount of gas introduced into B is calculated from the difference between the initial and final buret readings. The motor and attached eccentric operating the magnet are turned on and slugs of liquid at the rate of about 1 per second are pumped into B, where they dissolve gas as they run down the inside of the bulb. Under these conditions, no bubbles of gas are carried into A. The pressure in B becomes constant in from 1 to 3 h, depending on the system studied. More gas is added, if necessary, to raise the equilibrium pressure to at least 300 mmHg. The mercury level in D is then adjusted to bring the liquid meniscus to b, and the pumping is continued for a short time. The number of moles of gas undissolved in B is calculated from the observed pressure less the vapor pressure of the solvent and the pressure of the head of liquid. More gas is added and equilibrated as a check on the attainment of equilibrium and, in the case of a very soluble gas, to test the applicability of Henry's law. To obtain values for heats or entropies of solution, solubilities at a series of temperatures are measured just by changing the temperature and reequilibrating the system.

This apparatus has been used to study the solubility of fluorocarbon gases in cyclohexane [96], as well as inert gases and lower alkanes [95]. The solubility of inert gases was interpreted by Miller and Hildebrand [97].

6.1.2   Cukor and Prausnitz Apparatus

The apparatus designed by Cukor and Prausnitz [98] is similar to that of Dymond and Hildebrand [95]. It is especially useful in the determination of gases at elevated temperatures. Solubility measurements must be made at progressively decreasing temperatures. As shown in Figure 4.16 bulbs A and B, solvent flask E, and sidearm C are identical to those in Figure 4.15. Gas buret J is water jacketed. There are predetermined volumes between mark b and the stopcock at D, and between mark b and the stopcock at G. There are three 40-mL and four 10-mL chambers. At 10 mL calibrated buret K is used to adjust the gas pressure inside the gas buret with the atmospheric pressure.

The special pressure measuring device (Figure 4.17), used at elevated temperatures, is designed to eliminate errors due to condensation of solvent vapor, which would normally occur at a temperature lower than the thermostatic temperature.

The whole apparatus is evacuated during a solubility determination after zeroing of the photocells in a null meter. Mercury from reservoir R fills the gas buret until the menisci in both tubes are level with the calibrated buret. Taps S and T are opened to allow solute gas to enter until the manometer reads about 1 atm. The amount of gas entering the buret is calculated by adjusting calibrated buret K to read exactly 1 atm. The solvent is kept in flask E, where it is repeatedly degassed by freeze–thaw cycles in liquid nitrogen.

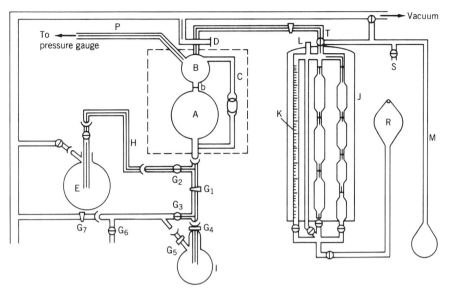

Figure 4.16 Apparatus for determination of gas solubility at elevated temperatures. This apparatus is similar to the one designed by Dymond and Hildebrand (Figure 4.14). For determination of solubility at elevated temperatures, the assembly containing bulbs *A* and *B* and plunger *C* can be jacketed by a constant-temperature bath. Reprinted with permission from P. M. Cukor and J. M. Prausnitz, *Ind. Eng. Chem. Fundam.*, **10(4)**, 638 (1971). Copyright © 1971 American Chemical Society.

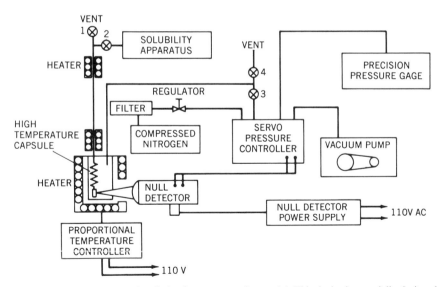

Figure 4.17 Pressure-measuring device for apparatus in part (*a*). This device is especially designed for determination of pressure of a gas in solution which is at a much higher temperature than the device. Reprinted with permission from P. M. Cukor and J. M. Prausnitz, *Ind. Eng. Chem. Fundam.*, **10(4)**, 638 (1971). Copyright © 1971 American Chemical Society.

The degassed solvent passes through sidearm H into bulb A. The vapor pressure of very volatile solvents can be enough to drive solvent over otherwise, it may be necessary to heat the flask. Enough solvent is introduced into bulb A so that it overshoots mark b. Stopcock G_2 is closed tightly and the solvent is allowed to attain the temperature of the bath. Stopcock G_3 is then closed and G_1 and G_4 are opened to allow the solvent to collect in flask I until the solvent meniscus in bulb A is level with mark b. Stopcocks G_1, G_4, and G_5 are closed tightly, G_3 and G_6 are opened, and the flask I is removed and weighed accurately. The thermostatically controlled bath is then raised to the highest temperature at which the solubility measurements are to be made, flask I is attached, the tubing between G_7 and G_1 is evacuated, and the solvent in bulb A is adjusted to mark b. Flask I is removed and weighed. Solvent from flask I is added to bulb A when cooling. Therefore, the exact amount of solvent in bulb A is always known.

Bulb B is filled with solute gas by adjusting the height of the mercury in reservoir R until the gas pressure in bulb B reaches 1 atm. The height of mercury in buret K is adjusted again to bring the pressure of the gas remaining in the buret to 1 atm. The exact amount of gas in bulb B is calculated using the difference between initial and final readings in buret K. The solvent from bulb A is mixed with solute in bulb B at a recommended rate of approximately 5 mL/s to avoid formation of gas bubbles in bulb A. The pressure in bulb B is monitored until a constant value is obtained. Equilibration should be rapid. It is checked by adding excess gas.

6.1.3   Loprest Apparatus

Loprest [99] developed an apparatus (Figure 4.18) for a physical saturation method employing manometric measurements to calculate solute distribution in the system.

Prior to carrying out a determination, the volume of flask H is calibrated from mark h up to and including the bore of stopcock 2, about 100 mL; and the volume of flask I, whose weight is already known, is calibrated to mark i and to the mercury surface in U tube, J, about 120 mL. Initially flask I is empty. With stopcock 1 open, the level of mercury in H is set at the mark h using leveling bulb B. That stopcock is closed and the system is evacuated through 2, 4, 5, and 9. Stopcocks 5 and 9 are closed and solute gas is introduced from P, a gas cylinder, through 8 until atmospheric pressure is reached and the gas is allowed to bubble into the atmosphere through Q, a mercury bubbler. Stopcock 5 is then opened slowly so that the bubbling continues through Q. The valve of tank P and 8 are then closed. Stopcock 9 is then opened slightly and some solute gas is pumped out of the system until the pressure falls to some value below atmospheric where the gas-measuring operation is to be carried out. The introduction of solute gas into the system is carried out in this way to avoid possible contamination of the gas, which may occur if the valve on the gas tank P were opened to the vacuum manifold. The temperature in the room is noted and the temperature of the air

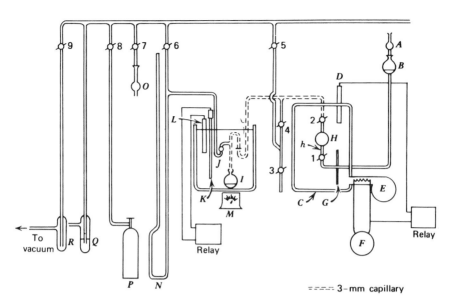

Figure 4.18 Apparatus for the rapid determination of gas in liquid solubility. Reprinted with permission from F. J. Loprest, *J. Phys. Chem.*, **61**, 1128 (1957). Copyright © 1957 American Chemical Society.

thermostat C is set only a few degrees higher. (Temperature control is obtained in this Lucite box C by means of a thermoregulator D and a hair dryer E, with its heating element connected to an electronic relay through a Variac F.) This reduces the amount of heat transfer and permits accurate control of the temperature of the airbath ($\pm 0.05°C$). After sufficient time is allowed for thermal equilibrium to occur, the pressure of the gas is measured on the manometer N using a cathetometer. Stopcock 2 is then closed rapidly and the number of moles of gas in H can then be calculated.

The system, exclusive of H, is brought to atmospheric pressure with air or nitrogen by opening 7. Flask I is removed and an appropriate amount of solvent and the magnetic stirring bar are added. The solvent is degassed in the following manner. First it is frozen by placing a low-temperature bath around flask I. The system is then evacuated through stopcocks 4 and 5. The solvent is allowed to thaw and a portion of it is "boiled off." The boiling-off operation can be eliminated if solvent must be conserved by performing a series of freeze–evacuate–thaw cycles.

With the constant temperature bath K in place, the vapor pressure of the solvent is measured. This is compared with the literature value or, if no literature value is available, the degassing cycle is repeated until a constant vapor pressure is obtained. The pressure measurement is made after closing stopcocks 4 and 5 and opening 6 and 7, thus allowing air or nitrogen to build up the pressure to the point where the levels of the mercury in both arms of J (a mercury-filled U tube containing fritted-glass disks that do not permit the passage of mercury) are equal. The pressure is then read on the manometer N with a cathetometer. If the surfaces in J are not level, the difference is determined with the cathetometer, and appropriate corrections to the pressure reading and the volume of I are made. With stopcock 4 closed, 1 and 2 are opened and mercury is allowed into H and into the capillary lines to the mark i. The solvent is stirred magnetically, and the system is allowed to come to equilibrium. The pressure is again measured with sufficient time allowed between the leveling of the surface in J and the final pressure measurement with N to ensure complete equilibration.

The number of moles of solute gas remaining in the gas phase is calculated and the amount dissolved is determined by difference. The temperature of bath K is either raised or lowered, and the system again is equilibrated. It is, therefore, possible to obtain solubility data over a wide temperature range with a single charging of flask I. The weight of the solvent is determined at the conclusion of the experiment.

Equation (10) is used to obtain the solubility data is

$$n = \frac{76}{WR}\left[\left(\frac{P_1 \pm V_1}{(P_2 - P_v)T_1}\right) - \left(\frac{V_2 - V_0}{T_2}\right)\right] \tag{10}$$

where n is the solubility in moles of gas per gram of solvent at T_2 K, and at a partial pressure of gas of 1 atm; W is the weight of solvent in grams; R is the gas constant in centimeters, milliliters per moles kelvin; P_1 is the pressure of initial quantity of gas in H at T_1 in centimeters; P_2 is the total pressure at equilibrium in the solubility vessel I at T_2 in centimeters; p_v is vapor pressure of the solvent at T_2 in centimeters; V_1 is the volume of H in milliliters; V_2 is the volume of I in milliliters; V_0 is the volume of solvent at T_2 in milliliters; T_1 is the temperature of air bath C and gas in flask H in kelvin; T_2 is the temperature of bath K and contents of I at which equilibration is carried out, in kelvin.

This equation assumes that the ideal gas laws are obeyed, that the vapor pressure of the solvent in the saturated solution is the same as the pure solvent, and that Henry's law is obeyed up to a pressure of 1 atm. These assumptions produce deviations well within experimental error under the conditions employed in the experiments. It is believed that a precision of $\pm 0.5\%$ can be obtained easily with this apparatus when n is about 10^{-5}. The precision is poorer for solubilities of a lower order of magnitude.

The Loprest method has several advantages over other apparatus previously used. A liquid bath whose temperature is controlled easily at a constant value is used and the need for elaborate air thermostats is eliminated. Solvent does not come in contact with mercury surfaces. The necessity of reading a gas buret with the attendant error also is eliminated. In this apparatus the liquid is degassed easily. The vapor pressure of the solvent can be determined. The trend to equilibrium from above and below saturation can be followed easily, and equilibrium is attained rapidly (often in less than 20 min). The solubility can be measured at various partial pressures; however, if Henry's law is assumed to be applicable, the solubility can be determined on a single sample of solvent and an entire gas–liquid system can be characterized in several hours.

6.1.4 Other Methods

Armitage and co-workers [100] designed a continuous dilution apparatus (Figure 4.19), which is useful for measurement of solubilities of a gas at any temperature and over a range of binary mixture compositions. It is necessary to account for the partial vapor pressure of the solvent. The method is unsuitable for solvents that react with mercury.

The apparatus consists of cell C connected to pump P equipped with a metal-in-glass plunger. This plunger is driven by a magnet-driven electric motor forcing the liquid in cell C over D. The solvent burets B_1 and B_2 are solvent burets connected to cell C by stopcocks T_7 and T_9, which dip into mercury reservoirs. The levels of mercury are controlled by connecting either stopcocks T_4 and T_{12} to nitrogen or T_5 and T_{11} to vacuum. Bulbs G_1 and G_2 are used to degas the solvent.

Prior to a solubility determination, the solvent is degassed as follows. Bulb

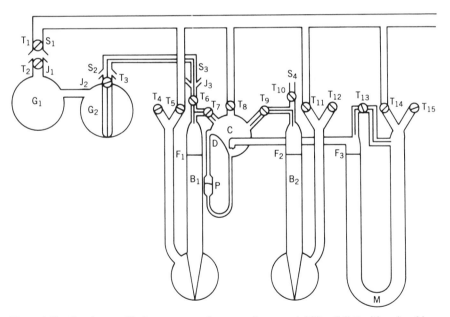

Figure 4.19 Continuous dilution apparatus for measuring gas solubility. Cell C with a glass blown cheek D contains the solute gas. Burettes B_1 and B_2 contain the solvent, which is sucked into cell C by system evacuation. Pump P is used to force solvent over D, which allows a fresh, thin film of solvent to be exposed to gas continuously. Reprinted with permission from D. A. Armitage, R. G. Linford, and D. G. T. Thornhill, *Ind. Eng. Chem. Fundam.*, **17(4)**, 362 (1978). Copyright © 1978 American Chemical Society.

G_1 is filled with about 250 mL of solvent; degassing is preceeded by boiling under nitrogen and pumping onto the frozen solid. Bulb G_1 is warmed and G_2 is cooled with liquid nitrogen to facilitate sublimation of the solvent; T_1 and T_2 are opened to remove any air present in G_1. The sublimation processes is repeated until the pressure gauge E does not show any rise in pressure. Buret B_1 or B_2 is evacuated together with a section of the tubing as far as T_3. Bulb G_2 is warmed; tap T_7 or T_9 is closed; and T_3 is opened to fill the buret with solvent. During this process nitrogen at a pressure slightly above the vapor pressure of the liquid is applied to T_4 or T_{12}.

Cell C, pump P, and U tube M are evacuated, and C is filled with the solute gas; T_{13} and T_{15} are closed; M is evacuated again, and the pressure of gas introduced is determined. Nitrogen pressure is applied at T_4 and T_{12}, forcing degassed solvent into cell C. The exact volume of solvent is determined by the difference between the initial and the final mercury levels in the buret.

The solvent in cell C is allowed to trickle onto the glass-blown cheek D using pump P, presenting a fresh thin film of solvent to the gas. This is done until a constant pressure is obtained.

In the semiwet mode, cell C, pump P, and the left arm of tube M can be evacuated and filled with fresh degassed solvent, then a volume of gas can be added to cell C through the precision-bore right arm of M, T_{13}, and the capillary tube W. This latter tube prevents diffusion of solvent vapors out of cell C while introducing the solute gas.

An apparatus (Figure 4.20) designed by Dean and co-workers [101] has been used for the determination of slightly soluble gases. Bulb B consists of stopcock T, sidearm A, and magnetic stirring bar F. The volumes enclosed between mark b and stopcock T and by the precision bore tubing C are predetermined. The solution is stirred by F to facilitate thorough mixing of the gas and solvent. Tube C is the high-pressure arm, and tube M is the low-pressure arm of a manometer. Mercury reservoir R is used to adjust the level of mercury in the solution vessel within tube C. Vessel D is used as a solvent storage vessel. A suitable cathetometer is used to make pressure measurements and a regular thermostatically controlled bath can be used to regulate the temperature of the apparatus. The solvent is degassed repeatedly by vacuum distillation and condensation during the solubility measurements. For toxicological studies, the solvent used is often water or an aqueous mixture that requires extra care to ensure complete degassing. The degassed solvent is stored in vessel D; and T_1, T_2, and T_3 are closed.

The solute gas is transferred from the gas buret to the solution vessel. The moles of gas used is determined by measuring the pressure of the solute gas that entered the solution vessel. Calibrated solvent vessel D is completely filled and allowed to equilibrate to the temperature of the bath. The mercury level in the solution vessel is lowered below point W. Stopcock T_3 is opened, mercury from reservoir R is allowed to enter the solvent vessel, and the mercury level in tube C reaches above point W. Proper mixing is achieved by stirring. Solubility

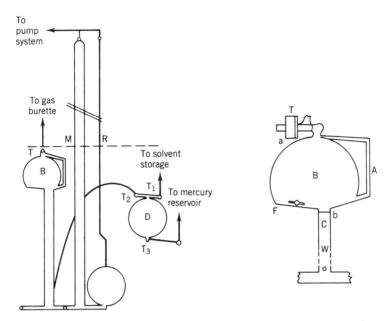

Figure 4.20 Apparatus for determination of solubility of slightly soluble gases. Vessel D contains degassed solvent. Calibrated vessel B contains solute gas. Solvent from D is forced into B by allowing mercury to fill bulb D. Reprinted with permission from C. R. S. Dean, A. Finch, and P. J. Gardner, *J. Chem. Soc., Dalton Trans.*, **23**, 2722 (1973).

determinations at other temperatures can be carried out by adjusting the manometer reservoir arm. The equilibrium time is 3 h.

In the Benson and co-workers [102] method the composition of the liquid phase and the vapor phase in equilibrium is determined by classical *PVT* measurements. The apparatus consists of a degassing device; an equilibrator; and an extractor, the Töpler pump, for transferring dry gas to the manometric system, thermostats and thermometers. Numerous modifications to the system exist [103].

A typical system consists of a 3-L suction flask equipped with Teflon stopcocks, an O-ring joint for adding solvent, a condensor to prevent solvent loss, and a magnetic stirring bar. The solvent is vacuum degassed for approximately 1 h. After attaining equilibrium, accurately known volumes of the vapor phase in the vapor-phase sample bulb and liquid solution in the liquid-phase sample bulb are isolated. The gas in the vapor-phase sample bulb is dried and pumped into the manometric system where amount is determined. The dissolved gas in the liquid-phase sample bulb is extracted, dried, and transferred to the manometric system where it is measured. The amounts of dried gas in the vapor and liquid phases are determined using these volumes and temperature and pressure measurements.

Gas chromatography has been used extensively in solubility determination. It has been found to be a useful method for the determination of very soluble gases [103]. It is unsuitable for light gases that are not very soluble in most solvents that, thus, elute rapidly.

It is beyond the scope of this chapter to list all methods currently in use for the determination of gases in liquids. Comprehensive reviews of methods used for determination of solubility of gases in liquids have been published by Wilhelm [103] and others [104]. Procedures for determining gas solubility in polymer melts have also been addressed [105, 106].

6.2   Classical Saturation Apparatus

In general, the kinds of gas saturation apparatus in use today are modifications of the apparatus and technique used by Henry in 1803, Bunsen in 1855, and Ostwald in 1890. However, the Ostwald method and its various modifications have largely displaced the others.

6.2.1   Markham–Kobe Modification, Ostwald

The Markham and Kobe [4, 107] modification of the Ostwald apparatus and technique will be used to indicate the general nature of the Ostwald method. In this method a measured volume of gas is brought in contact with a measured quantity of gas-free liquid, equilibrium is established by agitation; the volume of gas remaining is measured; and the change in volume gives the amount dissolved by the liquid. Thus, the preceeding methods also are generally related.

The Markham and Kobe apparatus (Figure 4.21) differed from others in the method of providing a gas–liquid interface and in the provision for agitating the absorption flask. The buret A is connected at the bottom by a T tube to the mercury leveling bulb B and the manometer tube C open at the top. The cock D is between the buret and the manometer tube. At the top, the buret is connected by the ground-glass joint E to the T tube F. One branch of this T tube connects to the vacuum through the stopcock G. The other ends in a straight tube at H. The absorption flask consists of two bulbs J and K, one having the volume of the other. They are connected at the bottom through the three-way cock L and at the top through the three-way cock M. A capillary tube leads from M to O.

During a run the absorption flask is in a thermostat with a water level nearly up to N. The buret and manometer tube C are jacketed in a large glass tube through which water flows from the thermostat. The tube from E to H is capillary, so only a very small volume of gas is outside the thermostat. A framework of metal supports the absorption flask. It connects to a motor that oscillates the flask through an arc of about 10°, with O–N as an axis, at 160 oscillations per minute. The buret and the bulbs of the absorption flask are calibrated by filling them with mercury.

These steps are used to determine the solubility of a gas in a liquid. Boil the

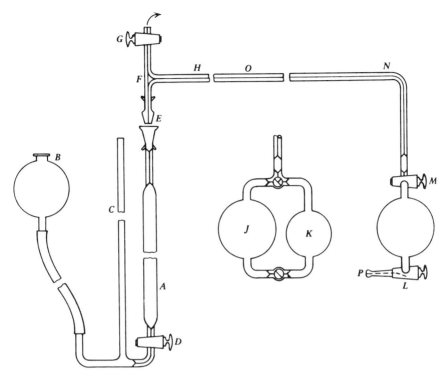

Figure 4.21   Markham and Kobe modification of Ostwald apparatus. Reprinted with permission from A. F. Markham and K. A. Kobe, *Chem. Rev.*, **28**, 519 (1941). Copyright © 1941 American Chemical Society.

solvent under reflux to free it from dissolved gas, seal the condenser, and cool the solvent under vacuum to the temperature of the thermostat. The liquid is then withdrawn from the reflux flask by means of a withdrawal tube that dips to the bottom of the flask and is connected to the absorption flask at *P*. Cocks *L* and *M* are so arranged that there is a passage through one bulb from *P* to *O*. Suction is applied at *O* and the bulb filled with the solvent. After one bulb is filled with liquid, the cocks are turned to give free passage through the other bulb from *P* to *O*, thereby isolating the full bulb. Several drops are then drawn into the empty bulb and the refluxing flask and vacuum disconnected. The buret is filled with mercury by raising the leveling bulb and the cock *D* is closed. The absorption flask is put into the thermostat, connected to the gas supply at *P*, with *O* and *H* connected by tubing. The gas from the cylinder goes first through a saturator filled with the solvent under study at the thermostat temperature. The several drops of solvent previously drawn into the bulb ensure saturation. With the gas and suction connected at *P* and *G*, respectively, the bulb is alternately evacuated and filled four times with gas by manipulation of the cocks *L* and *G*. Then cock *D* is opened, the leveling bulb is lowered, and the buret is filled with gas. Cock *L* is shut and the gas supply is disconnected. The leveling bulb is adjusted to such a level that the partial pressure of the gas is exactly 760 mmHg, allowance being made for the density of the mercury, capillary effects, barometric pressure, and the vapor pressure of the solution. The buret level at this time is recorded. Cock L is opened between the two bulbs, and cock M is opened between the bulbs and to the buret. The entire apparatus is then shaken. The leveling bulb is raised as solution proceeds to maintain pressure relationships. After there is no perceptible change in the buret levels, shaking is stopped. The leveling bulb is adjusted as before, and the final buret reading is recorded.

This apparatus gives a precision of 0.2% and is simple in construction. The possible sources of error in using this method have been considered by Markham and Kobe. According to their findings, supersaturation is not a problem. A small error may reflect solvent expansion due to the dissolved solute gas, which thereby affects the observed volume of gas.

6.2.2   Cook and Hansen

It should be noted in the preceding procedure that the gas was saturated with liquid vapor before the buret was filled. Other investigators have kept the gas in the buret dry. If the gas in the buret is saturated, the vapor pressure of the solvent is of little consequence. If the gas is dry, however, the vapor pressure must be known accurately, since all gas coming into the free space above the liquid in the absorption vessel picks up vapor, increasing its volume to an extent determined by the vapor pressure. On the other hand, if the gas in the buret is saturated, any part of the apparatus that is not in the thermostat may collect condensed solvent if the thermostat is above room temperature. The capillary between the buret and the absorption vessel is usually out of the thermostat. Drops of liquid in this capillary would make the pressure adjustment in the buret uncertain.

If the gas in the buret is dry, the temperature of the whole apparatus can be changed and thus a range of temperatures can be covered with one filling, which is a valuable feature where the thermodynamics of solution are of interest. The Ostwald-type apparatus involves a mercury surface contact with the gas and sometimes with the solvent as well. This is a serious drawback when dealing with a system that reacts with or is soluble in mercury.

Cook and Hansen [108] gave an excellent discussion of the problems involved in saturation methods of solubility determinations. It struck these authors that deviations in gas solubility data of greater than 1%, common in the literature, are not in accord with the accuracy obtainable in the physical measurements involved, for surely an accuracy of 0.1% in the measurement of a pressure, a volume, or a mass can be obtained with minimal care. They reasoned that if the purity of the materials is satisfactory and the measurements of temperature, volume, and pressure are sufficiently accurate, the discrepancies in the solubility values must result from other sources of error. These errors can be one or more of the following: failure to reach equilibrium, failure to degas the solvent completely, failure to ascertain the true amount of gas dissolved, and failure to ensure that the transfer of gas from a primary container to the apparatus does not involve air contamination. Cook and Hansen proceeded to design an apparatus that would eliminate these errors, and this apparatus is shown in Figure 4.22.

6.3 Classical Extraction Apparatus

6.3.1 Van Slyke

One of the best and the most popular types of apparatus for solubility determination by the extraction method is that of Van Slyke and Neill [109, 110]. This manometric apparatus is based on the principle of extraction of the gas from the liquid and subsequent measurement of the pressure of the liberated gas. The apparatus is shown in Figure 4.23. The 50-mL short pipet A has several graduations on it and α corresponds to 2 mL. The pipet is connected to the manometer and to the mercury leveling bulb. The sample of a solution of the gas is introduced through stopcock b by a special pipet in such a way that the solution does not come in contact with the air. The gas solution is evacuated by lowering the leveling bulb, and the pipet is shaken for 2–3 min to assist in liberating the gas. The liberated gas is compressed into the volume α and the pressure read on the manometer.

A later modification of the Van Slyke–Neill apparatus eliminates the transfer step [109]. Modified Van Slyke designs have been used extensively in biochemical studies. Orcutt and Seevers [111] determined blood levels of the anesthetic gases, cyclopropane, ethylene, nitrous oxygen, and carbon dioxide.

The general equation for calculating the total gas content of a solution from the amount of gas extracted in an evacuated chamber of definite volume is

$$V^0 = \frac{V_t(P_b - P_v)}{760(1 + 0.000367t)}\left[1 + \alpha'\left(\frac{S}{A-S}\right)\right] \tag{11}$$

where V^0 is the volume of gas, measured at $0°C$, 1 atm in the solution analyzed; V_t is the volume of gas at t; $P_b - P_v$ is the barometric pressure in millimeters of Hg corrected for the vapor pressure of the liquid at the temperature of the experiment; t is the temperature in degree Celsius; α' is the Ostwald distribution coefficient of the gas between gas and liquid phases; that is, $\alpha' = (t/273)$; A is the volume of extraction chamber; and S is the volume of solution in extraction chamber.

The factor in (11) that corrects for the gas that remains unextracted is $[1 + S\alpha'/(A - S)]$.

6.3.2 Baldwin–Daniel Apparatus for Viscous Liquids

Baldwin and Daniel [112] designed an extraction apparatus to measure the solubility of gases in liquids, especially viscous liquids, which consists essentially of three operations: deaeration of the liquid, saturation of the liquid with the test

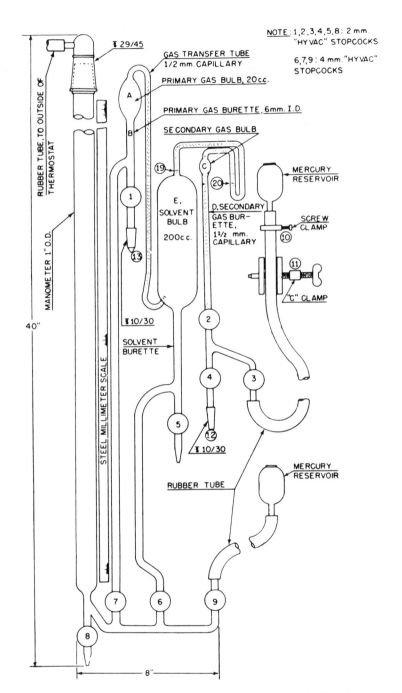

Figure 4.22 Cook and Hansen apparatus for determination of the solubility of a gas in liquid. Reprinted with permission from M. W. Cook and D. N. Hansen, *J. Chem. Phys.*, **28**, 370 (1957) and from W. J. Mader and L. T. Grady, "Determination of Solubility," in A. Weissberger and B. W. Rossiter, Eds., *Physical Methods of Chemistry*, Vol. 1, Part V, Wiley, New York, 1971. Copyright © 1971 by John Wiley & Sons, Inc.

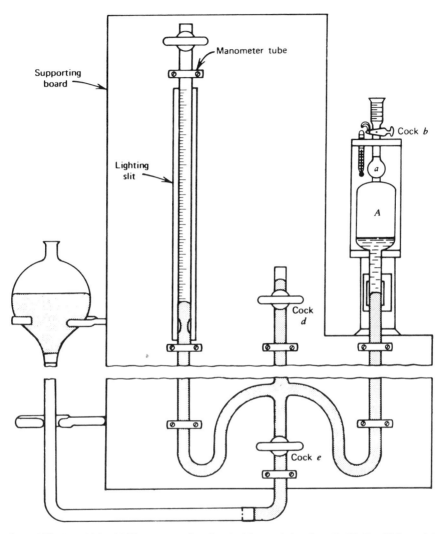

Figure 4.23 Van Slyke–Neill apparatus. Reprinted with permission from D. W. Van Slyke and J. M. Neill, *J. Biol. Chem.*, **61**, 523 (1924).

gas at atmospheric pressure, and determination of the amount of gas liberated under vacuum from a known volume of the saturated liquid.

The apparatus used for the first two steps is shown in Figure 4.24. The solvent is placed in funnel F and allowed to drip slowly into vessel A, which is evacuated. It might be expected that this procedure would remove all dissolved gas, but tests showed that 2–3% of the original gas still remained in solution. Hence, the gas under examination is passed into liquid through tap T_1 for 2–3 h.

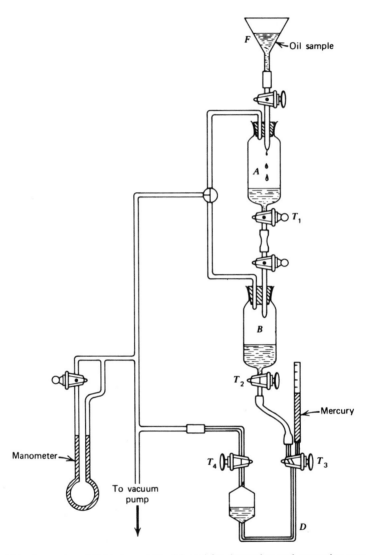

Figure 4.24 Apparatus of Baldwin and Daniel used for deaeration and saturation procedure in their solubility method. Reprinted from *J. Appl. Chem.*, **2**, 161 (1952) and with permission from W. J. Mader and L. T. Grady, "Determination of Solubility," in A. Weissberger and B. W. Rossiter, Eds., *Physical Methods of Chemistry*, Vol. 1, Part V, Wiley, New York, 1971. Copyright © 1971 by John Wiley & Sons, Inc.

The gas supply is then stopped, and the liquid is allowed to drip into vessel *B*, which is evacuated continuously. The liquid finally is free of the dissolved gas initially present (usually air) and contains only a small fraction of the gas under examination.

For the saturation and storage of the liquid solvent before test, a specially designed displacement buret D is used. The evacuated liquid from B is introduced into D through taps T_2 and T_3, tap T_4 of the buret being open to the vacuum pump. When the displacement buret is full, taps T_3 and T_4 are closed, and the buret is disconnected from the apparatus and placed in the thermostat. Taps T_3 and T_4 are then opened and the saturating gas bubbled into the liquid through tap T_2 until saturation is completed. When no gas bubbles can be detected in the bulk of the liquid, the buret tube above T_3 is filled with mercury to displace the gas from the buret. When all the gas has been expelled, tap T_4 is closed, and T_3 is left open so that there is a head of some 100 mmHg in the buret tube, which suffices to keep the gas in solution before test begins.

The apparatus used for determining the amount of gas is shown in Figure 4.25. It consists essentially of a vessel V with a sidearm, taps, and so on, for connection to the displacement buret D; and it is connected by a ground-glass joint at the top to a Töpler pump P, manometer M, and calibrated vessel S. The whole apparatus is initially evacuated, and the gas released from the liquid is

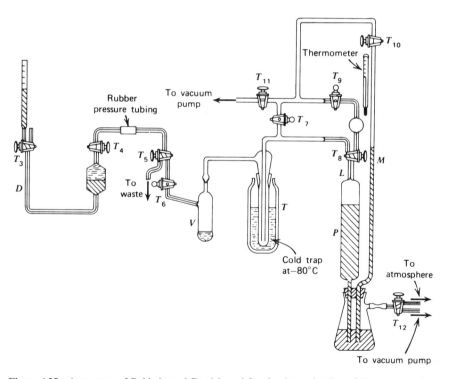

Figure 4.25 Apparatus of Baldwin and Daniel used for the determination of the amount of gas liberated from the liquid in their solubility method. Reprinted from *J. Appl. Chem.*, **2**, 161 (1952) and with permission from W. J. Mader and L. T. Grady, "Determination of Solubility," in A. Weissberger and B. W. Rossiter, Eds., *Physical Methods of Chemistry*, Vol. 1, Part V, Wiley, New York, 1971. Copyright © 1971 by John Wiley & Sons, Inc.

transferred by the Töpler pump to the calibrated vessel S, the pressure it exerts in this volume being measured by the manometer M.

Tap T_5 is closed; the apparatus is evacuated through T_{11} by a high-vacuum pump; and T_7 is then closed to prevent loss of gas to the vacuum pump. The displacement buret is then connected to the solubility apparatus by a short length of pressure tubing. The space between T_4 and T_5 is filled with liquid by displacing, with mercury, a little of the liquid to waste through T_5. The evacuated space between T_5 and T_6 is then filled. The mercury reading in the buret tube is noted. Tap T_6 is opened and liquid is allowed to run into vessel V, which is immersed in bath at 150°C to assist gas solution. During this time, the Töpler pump is connected to the vessel V through T_8 and the mercury is maintained at the bottom of its stroke. When about 10 mL of liquid has been introduced into V, T_6 is closed, and the change in mercury level in the buret tube is noted. In this manner, the volume of oil introduced can be measured to ± 0.02 mL. Tap T_8 is reversed, and the collected gas is forced into S; T_8 is again reversed so as to connect to vessel V; and the mercury is drawn to the bottom of the Töpler pump, the process being repeated until all liberated gas has been collected.

To measure the final amount of gas collected, the Töpler pump is brought to a fixed mark on the neck of the pump at L by adjustment of the pressure at the base of the pump. The readings of the manometer M and the thermometer are then noted. To obtain the zero of the manometer, the volume S and the Töpler pump are evacuated through T_9 and the manometer reading is again noted when the mercury level in the Töpler pump is again brought to L. The difference in the two manometer readings gives the pressure of the gas, and the volume of the gas is known from the calibration of S and the additional volume between T_8 and the mark L.

The volume of gas collected is corrected to standard conditions by the gas laws. This method has an accuracy of at least $\pm 1\%$.

6.4 Special Applications

The following techniques serve to illustrate the range of solubility problems encountered with gases.

Durrill and Griskey [113] modified the pressure vessels of Newitt and Weale [114] and Lundberg and co-workers [115], which allow determination of both the solubility and diffusivity of simple gases in molten or thermally softened polymers. In these, the unusual observation was made that Henry's law held up to 20 atm.

Bott and Schulz [116] studied the solubility of gaseous chlorine in brine solutions as a function of pH in order to improve the operation of electrolytic cells. The study was complicated by reaction of chlorine with water to yield hydrochloric and hypochlorous acids and with chloride ion to form a complex, which reactions in turn were influenced by pH and salt concentration. The complexity of problems illustrates the need to understand the overall chemistry of any system for which solubility data are desired.

The physical solubility of gases at 1 atm in fused silica was investigated by Doremus [117]. He found that, in marked contrast to most gas solubility situations, gas molecular solubilities in fused silica were relatively insensitive to temperature, pressure, and molecular size. These observations are consistent with an inert matrix with some free volume, and it is this free volume alone that is available for gas solutions.

7   DETERMINATION OF DISSOLUTION RATE

Dissolution-rate studies hold the attention of organic and, especially, pharmaceutical chemists. Such rate studies are important for the understanding and the design of experiments in partition, adsorption, and dialysis involving the transfer of solids to solutions. Dissolution-rate data can guide the pharmaceutical scientist in designing the solid oral dosage forms that represent the majority of drugs on the market. Orally administered drugs must pass through the gastrointestinal barrier, and to accomplish this most drugs first must be in solution. For some drugs it is desirable that absorption be rapid and for others that absorption should be delayed by modification of rate of dissolution.

Dissolution rate can be expressed by the equation advanced by Noyes and Whitney [15, 16]: $dC/dt = k(C_s - C)$, where C is the concentration of solute in the solvent at time t, C_s is the solubility of the solute in the solvent and k is a constant with the dimension t^{-1}. They developed the relationship from studies on benzoic acid and lead chloride of essentially constant surface area prepared by depositing molten solute on a glass core and noted that rate studies previously were lacking because of the constant surface problem. Subsequently [17], the surface area of the solute S was incorporated in equation form. The modern equation is

$$dW/dt = k_1 S(C_s - C) \tag{12}$$

where dW is the mass of solute entering solution, in which case the first-order rate constant k_1 has the dimensions mass/(area)(time).

Dissolution rate is viewed thus as a simple transfer process involving a concentration gradient between a thin film of solvent surrounding the solute, and saturated by it, and the bulk of the solution. Broadly, the dissolution rate is directly proportional to the equilibrium solubility. The dissolution rate constant k can be interpreted in terms of diffusivity and, therefore, solution viscosity. Similarly, as k depends on the thickness of saturated film, it can be viewed in terms of speed of convection or turbulence of the solvent in the vicinity of undissolved solute.

Vibration of the system has direct effect on k and thus requires strict control.

The thermodynamics of the transfer of a solute molecule across the interface of the solid and liquid phases is not a component of k, as this transfer is a feature of the equilibrium solubility C_s.

Dissolution methods can be classified into two general categories—intrinsic and apparent methods. Intrinsic methods determine the rate in terms of mass/(area)(time) [milligrams per square centimeter hour] under conditions of known or controlled surface area. Apparent methods measure the total mass dissolved per unit of time. Dissolution methods are further classified as *sink* or *nonsink* methods. A broad interpretation of a sink method is one in which the solute is removed from the media as dissolution proceeds or where the solute available cannot establish more than a fraction of the concentration at saturation.

As there are almost as many methods for determining dissolution as there are workers, we have taken the liberty of selecting examples of only some of the methods now in general use.

7.1   Determination of Intrinsic Dissolution Rate

7.1.1   Fine Particle Method

Edmundson and Lees [118] developed a procedure for dissolution rates of fine particles in which the changing solute particle-size distribution is followed in a stirred solvent by means of a particle-size counter. The dissolution rate is related to loss of particle diameter per unit of time. They obtained dissolution rate data on crystalline hydrocortisone acetate by this method. The major disadvantages of this method are the requirement for an expensive particle-size distribution counter, poor precision, and difficult calibration and control of stirring dynamics.

7.1.2   Rotating Disk Method

Levy and Sahli [119] proposed a procedure wherein the solute is compressed by a hydraulic press into a plane-faced disk that is placed in a holder that in turn mounts at the end of a stirrer shaft. The shaft extends into a 500-mL round-bottom flask (three-neck) containing 200 mL of dissolution medium. The shaft is rotated at the desired speed and samples are withdrawn through the other necks. Observed dissolution rates were highly sensitive to variations in rotation speed, so Levy and Tanski [120] developed a power unit from commercial pump motors and speedometer cable that allowed constancy of rotation over a 3–400-rpm range. Another modification [121] featured a compression die that was used directly as the disk holder, eliminating several manipulations and ensuring that a single planar surface was exposed to the medium.

Control and constancy of rotation rates, and absence of shaft wobble are critical to this method. The method is usable only where direct compression of the solute is feasible and where the compressed disk does not disintegrate in the presence of dissolution medium.

7.1.3 Hanging Pellet Method

Another compressed-disk procedure [122] involves mounting the disk on a strip of aluminum with wax so as to expose just one surface of the disk. The strip is then suspended in the dissolution medium from the arm of a suitable balance. The loss of weight of the disk is observed with time. This is a static method and depends, therefore, on solvent convection and viscosity.

7.2 Determination of Apparent Dissolution Rate

General methods are available to determine apparent or total dissolution rate, mass per unit time. Methods for apparent rates need not control surface area.

7.2.1 Pharmacopeial Method

The rotating basket was first proposed by Pernarowski and co-workers [123] and was adopted with modifications by the United States Pharmacopeia (USP) [124] and the National Formulary [125] in 1970 as an official method for determining the apparent dissolution rate of solid oral-dosage forms (Figure 4.26). The dosage form is placed in the stainless-steel, 40-mesh screen basket, which clamps onto the end of the stainless steel shaft. This shaft is rotated by suitable means, such as a variable-speed stirrer at a carefully controlled speed between 50 and 150 rpm. The basket is immersed in 900 mL of the selected dissolution medium, which is maintained at 37°C by locating the specified 1.1-L round-bottomed flask in a constant temperature bath. Samples are removed at periodic intervals and analyzed by a specified procedure. Standard calibration procedures are available. The control of vibration from any source is critical to the methods [126].

A second stirring element was adopted in 1978, which is a rotating paddle (Figure 4.27). The two stirring elements of these official USP tests are interchangeable, allowing a common, extended apparatus configuration and computerized autoanalysis.

Water is the preferred medium because it least augments the essential processes of dissolution of drugs from tablets and capsules. Several pharmaceutical formulation and manufacturing mistakes are well established as resulting in diminished bioavailability of medications. All such mistakes result in diminished amounts of dissolution as measured by the pharmacopeial tests [127].

The pharmacopeial assembly is commercially available internationally and its inherent versatility permits a much broader application than just for standardization of pharmaceuticals.

7.2.2 Beaker Method

This method, widely employed in pharmaceutical research, was introduced by Parrott and co-workers [128] in a study of benzoic acid spheres. Levy and Hayes [129] modified this for drugs in tablet form and generally the procedure is as follows: 250 mL of dissolution medium at 37°C is placed in a 400-mL beaker

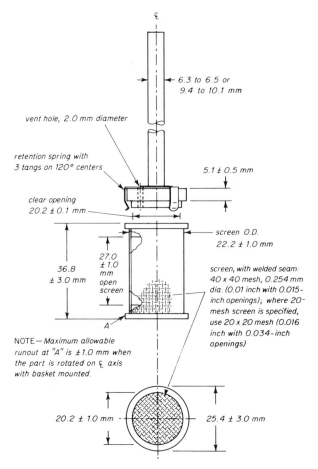

Figure 4.26 Basket-stirring element. Copied from *United States Pharmacopeia Twenty Second Revision* © 1989, The United States Pharmacopeial Convention. Permission granted.

and a 5-cm, three-bladed polyethylene stirrer is immersed at the exact center of the beaker to a depth of 2.7 cm. The stirrer usually is rotated at about 60 rpm. The tablet is dropped down the side of the beaker to the bottom where it disintegrates into granules covering a small (but uncontrollable) area. Samples are withdrawn using a sintered-glass immersion filter.

A variety of modifications of this basic beaker method have been reported. The advantages of the method are simplicity and ready availability of components. The disadvantages are poor reproducibility due to variable geometry between solute and stirrer, and restriction to solutes more dense than the medium (capsules, e.g., float). Further modifications have included supernatant, immiscible layers [130] to supply a partition effect. Immiscible layers and adsorbents such as charcoal have been used to maintain *sink* conditions.

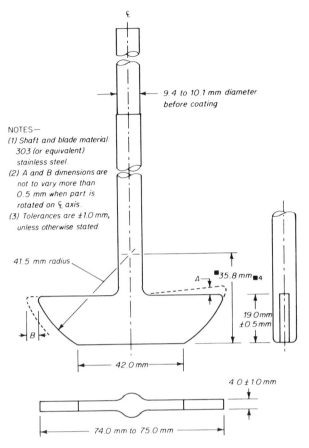

Figure 4.27 Paddle-stirring element. Copied from *United States Pharmacopeia Twenty Second Revision* © 1989, The United States Pharmacopeial Convention. Permission granted.

7.2.3 Solvometer

The apparatus developed by Klein [131] features a boat or pan suspended from a float, which in turn is suspended from the arm of a balance or some form of graduated scale. The pan and float are submerged in the medium and loss of weight of solute is observed. Widespread disintegration of solute or carrying-off of particles by convection is a serious problem and places limits on the agitation that can be given to the medium. The principle and equipment is simple and this method would be particularly convenient for moderately dense solids, polymer, or mineral samples.

7.2.4 Adhesive Tape Method for Particles

In this procedure, particles of solute are pressed onto $\frac{3}{4}$-in.-wide pressure-sensitive tape and the tape placed in a framelike holder. The beaker containing dissolution medium is fitted with runners into which the frame containing the solute tape is slipped and held firmly. A variable speed stirrer also projects into the medium. The method is suitable to multiparticulate systems and was used originally [132] to determine dissolution rates of benzoic acid and salicylamide.

References

1. A. Seidell, *Solubilities of Inorganic and Organic Compounds*, Vol. 2, 3rd ed., Van Nostrand, Princeton, NJ, 1941; A. Seidell and W. F. Linke, *Supplement*, 1952; A. Seidell, Vol. 2, 4th ed., American Chemical Society, 1965.

2. J. H. Hildebrand, *Proc. Natl. Acad. Sci.*, **57**, 542 (1967).

3. R. W. Bunsen, *Philos. Mag.*, **9**, 116, 181 (1855).

4. A. E. Markham and K. A. Kobe, *J. Am. Chem. Soc.*, **63**, 449 (1941).

5. W. Alexejew, *J. Prakt. Chem.*, **25**, 518 (1882).

6. A. E. Hill, *J. Am. Chem. Soc.*, **45**, 1143 (1923).

7. V. C. Williams, *J. Phys. Colloid Chem.*, **52**, 1263 (1948).

8. M. J. Buerger, L. B. Smith, A. de Boetteville, Jr., and F. V. Ryer, *Proc. Natl. Acad. Sci.*, **28**, 526 (1942).

9. W. D. Bancroft and F. J. C. Butler, *J. Phys. Chem.*, **36**, 2515 (1932).

10. E. L. Heric and C. D. Posey, *J. Chem. Eng. Data*, **9**, 35 (1964).

11. T. Fumikatsu, *Bull. Chem. Soc. Jpn.*, **43(3)**, 934 (1970).

12. J. A. V. Butler, *J. Gen. Physiol.*, **24**, 189 (1940).

13. A. W. MacDonald and N. A. North, *Can. J. Chem.*, **52(18)**, 3181 (1974).

14. L. A. Chugaev and V. G. Khlopin, *J. Russ. Phys. Chem. Soc.*, **46**, 1659 (1914).

15. A. A. Noyes and W. R. Whitney, *Z. Phys. Chem. (Leipzig)*, **23**, 689 (1897).

16. A. A. Noyes and W. R. Whitney, *J. Am. Chem. Soc.*, **19**, 930 (1897).

17. L. Bruner and St. Tolloczko, *Z. Phys. Chem. (Leipzig)*, **35**, 283 (1990); F. Brunner, *Z. Phys. Chem. (Leipzig)*, **47**, 56 (1904).

18. R. S. Tipson, "Crystallization and ReCrystallization," in A. Weissberger, *Technique of Organic Chemistry*, Vol. 3, Part 1, 2nd ed., Interscience, New York, 1956.

19. T. Higuchi, *Proc. Am. Assoc. Coll. Pharm. Teachers' Seminar*, **13**, 119 (1961).

20. J. E. Gordon and R. L. Thorne, *J. Phys. Chem.*, **71**, 4390 (1967).

21. R. E. Jentoft and R. J. Robinson, *Anal. Chem.*, **26**, 1156 (1954).

22. W. Reeve and R. Adams, *Anal. Chem.*, **22**, 755 (1950).

23. R. L. Shriner, R. C. Fuson, and D. Y. Curtin, *Systematic Identification of Organic Compounds* (Laboratory Manual), 5th ed., Wiley, New York, 1964, p. 67 ff.

24. M. Beroza and M. C. Bowman, *Anal. Chem.*, **38**, 837 (1966).

25. K. H. Meyer, *Natural and Synthetic High Polymers*, Interscience, New York, 1942, pp. 9, 25, 565.

26. G. V. Schulz and B. Jirgensons, *Z. Phys. Chem.* (*Leipzig*), **B46**, 105, 137 (1940).

27. D. R. Morey and J. W. Tamblyn, *J. Phys. Colloid Chem.*, **51**, 721 (1947).

28. W. J. Mader, "Phase Solubility Analysis," in *Organic Analysis*, Vol. 2, Interscience, New York, 1954, p. 253.

29. W. J. Mader, *Crit. Rev. Anal. Chem.*, **1**, 193 (1970).

30. W. Tarpley and M. Yudis, *Anal. Chem.*, **25**, 121 (1953).

31. S. E. Massil and J. H. Lesser, *Anal. Lett.*, **17 (A11)**, 1307 (1984).

32. J. H. Lesser and S. E. Massil, *J. Assoc. Off. Anal. Chem.*, **70(4)**, 638 (1987).

33. L. T. Grady, D. K. Wyatt, B. L. Goldman, and V. J. Jackson, *Pharmacop. Forum*, **7**, 1431 (1981).

34. L. T. Grady, D. K. Wyatt, S. E. Hays, R. H. King, W. J. Mader, H. R. Klein, and R. O. Zimmerer, Jr., *J. Pharm. Sci.*, **62**, 456 (1973).

35. Zhi-Jihong and Yang-Lahu, *Yaowu Fenxizazhi*, **7(4)**, 252 (1987).

36. G. B. Smith and G. V. Downing, *Anal. Chem.*, **51**, 2290 (1979).

37. R. G. Satterfield and M. Haulard, *J. Chem. Eng. Data*, **10**, 397 (1965).

38. E. L. Heric, *J. Chem. Eng. Data*, **12**, 71 (1967).

39. F. Fairbrother, N. Scott, and H. Prophet, *J. Chem. Soc.* (*London*), 1164 (1956).

40. S. J. Gill, J. Hutson, J. R. Clopten, and M. Downing, *J. Phys. Chem.*, **65**, 1432 (1961).

41. G. H. Bell, *Chem. Phys. Lipids*, **10**, 1 (1973).

42. K. Wakita, M. Yoshimoto, S. Mikamoto, and H. Watanabe, *Chem. Pharm. Bull.*, **34**, 4663 (1986).

43. J. E. Glass, *Adv. Chem. Ser.*, **213**, 3 (1986).

44. T. Higuchi and J. L. Lach, *J. Am. Pharm. Assoc., Sci. Ed.*, **43**, 349 (1954).

45. J. Reilly and W. N. Rae, *Physico-Chemical Methods*, Vol. 1, 3rd ed., Van Nostrand, Princeton, NJ, 1939, p. 589.

46. T. J. Ward, *Analyst*, **44**, 137 (1919).

47. I. A. Pastac and R. Lecrivain, *Chim. Anal.*, **30**, 28 (1948).

48. A. B. Zdanovskii, *Zh. Neorg. Khim.*, **1**, 1279 (1956) in Israel Program for Scientific Translations, *J. Inorg. Chem.*, **1(6)**, 164 (1956).

49. T. H. Vaughn and E. G. Nutting, Jr., *Ind. Eng. Chem., Anal. Ed.*, **14**, 454 (1942).

50. R. Fuerer and M. Geiger, *Pestic. Sci.*, **8(4)**, 337 (1977).

51. D. J. W. Grant and I. K. A. Abougela, *Int. J. Pharm.*, **16(1)**, 11 (1983).

52. A. N. Campbell, *J. Chem. Soc.* (*London*), 179 (1930).

53. G. Aravamudan and K. R. Krishnaswami, *Current Sci.* (*India*), **25**, 287 (1956).

54. J. B. Lombardo, *J. Chem. Educ.*, **44**, 600 (1967).

55. A. W. C. Menzies, *J. Am. Chem. Soc.*, **58**, 934 (1936).

56. S. A. Butter, *J. Chem. Educ.*, **51(1)**, 70 (1974).

57. P. Gross, *J. Am. Chem. Soc.*, **51**, 2362 (1929).

58. L. H. Adams, *J. Am. Chem. Soc.*, **37**, 1181 (1915).

59. S. Mitchell, *J. Chem. Soc.* (*London*), 1333 (1926).

60. R. L. Bohon and W. F. Claussen, *J. Am. Chem. Soc.*, **73**, 1571 (1951).

61. Li Xiaobai, M. A. Winnik, and J. E. Guillet, *Macromolecules*, **16**, 992 (1983).

62. M. H. Abraham, G. J. Buist, P. L. Greillier, and R. A. McGill, *J. Chromatogr.*, **409**, 15 (1987).

63. S. P. Wasik, F. P. Schwartz, and S. G. Tewari, *J. Res. Natl. Bur. Stand. (U.S.)*, **89**, 273 (1984).

64. F. M. Yamamoto, S. Rukushika, and H. Hatano, *J. Chromatogr.*, **408**, 21 (1987).

65. H. A. Cooper and R. T. Hurtubise, *J. Chromatogr.*, **328**, 81 (1985).

66. W. A. Bruggeman, J. Van Der Steen, and O. Hutzinger, *J. Chromatogr.*, **238**, 335 (1982).

67. P. K. Weyl, *Rev. Sci. Instrum.*, **28**, 722 (1957).

68. P. Jordan, *Z. Phys. Chem. Neue Folge*, **9**, 187 (1956).

69. N. B. Mikheev, *Int. J. Appl. Radiat. Isot.*, **5**, 32 (1959).

70. S. Batra, *J. Pharm. Pharmacol. Commun.*, **27**, 777 (1975).

71. J. M. Lo, C. L. Tseng, and J. Y. Yang, *Anal. Chem.*, **58**, 1596 (1986).

72. I. D. Robb, *Aust. J. Chem.*, **19**, 2281 (1966).

73. W. L. Marshall, *Anal. Chem.*, **27**, 1923 (1955).

74. J. d'Ans, *J. Chem. Appar.*, **28**, 197 (1941).

75. O. N. Breusov, N. I. Kashina, and T. V. Revzina, *Prom. Khim. Reakt. Osobo. Chist. Veshchestv.*, **9**, 103 (1967).

76. V. R. K. S. Susarla, A. Eber, and E. U. Franck, *Proc. India Acad. Sci. Chem. Sci.*, **99**, 195 (1987).

77. J. A. Harris, A. V. Bailey, and E. L. Skau, *J. Am. Oil Chem. Soc.*, **45(9)**, 639 (1968).

78. W. B. Bunger, *J. Am. Oil Chem. Soc.*, **36**, 466 (1959).

79. V. Y. Abramov and L. P. Narushevich, *Zh. Fiz. Khim.*, **51(9)**, 2350 (1977).

80. G. Reynafarje, L. E. Costa, and A. L. Lehninger, *Anal. Biochem.*, **145**, 406 (1985).

81. *Huadong Huagong Xueyuan Xuebao*, **97** (1982).

82. W. Schenk and H. Tulhoff, *Ber. Bunsenges., Phys. Chem.*, **71**, 206 (1967).

83. G. W. Watt, W. A. Jenkins, and C. V. Robertson, *Anal. Chem.*, **22**, 330 (1950).

84. H. S. Davis, *J. Am. Chem. Soc.*, **38**, 1166 (1916).

85. H. A. Klobbie, *Z. Phys. Chem.*, **24**, 615 (1897).

86. T. W. Evans, *Ind. Eng. Chem., Anal. Ed.*, **8**, 206 (1936).

87. W. Herz, *Ber.*, **31**, 2669 (1898).

88. H. Sobotka and J. Kahn, *J. Am. Chem. Soc.*, **53**, 2935 (1931).

89. M. M. Acker and H. A. Frediani, Jr., *Ind. Eng. Chem., Anal. Ed.*, **17**, 793 (1945).

90. M. Hayashi and T. Sasaki, *Bull. Chem. Soc. Jpn.*, **29**, 857 (1956).

91. M. H. Hilder, *J. Am. Oil Chem. Soc.*, **45(10)**, 703 (1968).

92. L. K. Nash, *Anal. Chem.*, **21**, 1405 (1949).

93. B. B. Cunningham, *Nucleonics*, **5(5)**, 62 (1949).

94. J. H. Hibben, *J. Res. Natl. Bur. Std.*, **3**, 97 (1929).

95. J. Dymond and J. H. Hildebrand, *Ind. Eng. Chem. Fundam.*, **6**, 130 (1967).

96. K. W. Miller, *J. Phys. Chem.*, **72**, 2248 (1968).

97. K. W. Miller and J. H. Hildebrand, *J. Am. Chem. Soc.*, **90**, 3001 (1968).

98. P. M. Cukor and J. M. Prausnitz, *Ind. Eng. Chem. Fundam.*, **10(4)**, 638 (1971).

99. F. J. Loprest, *J. Phys. Chem.*, **61**, 1128 (1957).

100. D. A. Armitage, R. G. Linford, and D. G. T. Thornhill, *Ind. Eng. Chem. Fundam.*, **17(4)**, 362 (1978).

101. C. R. S. Dean, A. Finch, and P. J. Gardner, *J. Chem. Soc. Dalton Trans.*, **23**, 2722 (1973).

102. B. B. Benson, D. Krause, Jr., and M. A. Peterson, *J. Sol. Chem.*, **8**, 655 (1979).

103. E. Wilhelm, *CRC Crit. Rev. Anal. Chem.*, **16**, 129 (1985).

104. B. I. Morsi, *Nato ASI Ser. E.*, **72**, 53 (1983).

105. J. Kolmacka and P. Smilek, *Plasty Kauc.*, **23**, 13 (1986).

106. J. Kolmacka and P. Smilek, *Plasty Kauc.*, **22**, 364 (1985).

107. A. E. Markham and K. A. Kobe, *Chem. Rev.*, **28**, 519 (1941).

108. M. W. Cook and D. N. Hansen, *Rev. Sci. Instrum.*, **28**, 370 (1957).

109. D. D. Van Slyke, *J. Biol. Chem.*, **130**, 545 (1939).

110. D. D. Van Slyke and J. M. Neill, *J. Biol. Chem.*, **61**, 523 (1924).

111. F. S. Orcutt and M. H. Seevers, *J. Biol. Chem.*, **117**, 501, 509 (1937).

112. R. R. Baldwin and S. G. Daniel, *J. Appl. Chem. (London)*, **2**, 161 (1952).

113. P. L. Durrill and R. G. Griskey, *Am. Inst. Chem. Eng. J.*, **12**, 1147 (1966).

114. D. M. Newitt and K. E. Weale, *J. Chem. Soc. (London)*, 1541 (1948).

115. J. L. Lundberg, M. B. Wilk, and M. J. Huyett, *Ind. Eng. Chem. Fundam.*, **2(1)**, 37 (1963).

116. T. R. Bott and S. Schulz, *J. Appl. Chem. (London)*, **17**, 356 (1967).

117. R. H. Doremus, *J. Am. Ceram. Soc.*, **49**, 461 (1966).

118. I. C. Edmundson and K. A. Lees, *J. Pharm. Pharmacol.*, **17**, 193 (1965).

119. G. Levy and B. A. Sahli, *J. Pharm. Sci.*, **51**, 58 (1962).

120. G. Levy and W. Tanski, Jr., *J. Pharm. Sci.*, **53**, 679 (1964).

121. J. H. Wood, J. E. Syarto, and H. Letterman, *J. Pharm. Sci.*, **54**, 1668 (1965).

122. E. Nelson, *J. Pharm. Sci.*, **47**, 297 (1958).

123. M. Pernarowski, W. Woo, and R. O. Searl, *J. Pharm. Sci.*, **57**, 1419 (1968).

124. *The Pharmacopeia of the United States of America*, 18th Revision, U.S. Pharmacopeial Convention, General Tests, Rockville, MD, 1970. Twenty-Second Revision, 1989.

125. *The National Formulary*, 13th ed., American Pharmaceutical Association, Washington, DC, 1970.

126. K. D. Thakker, *Pharmacop. Forum*, **6**, 177 (1980).

127. L. T. Grady, *Pharm. Ind.*, **45**, 640 (1983); *Am. J. Hosp. Pharm.*, **39**, 1546 (1982).

128. E. L. Parrott, D. E. Wurster, and T. Higuchi, *J. Am. Pharm. Assoc. Sci. Ed.*, **44**, 270 (1955).

129. G. Levy and B. A. Hayes, *New Engl. J. Med.*, **262**, 1053(1960).

130. B. Gibaldi and S. Feldman, *J. Pharm. Sci.*, **56**, 1238 (1967).

131. L. Klein, *Bull. Biolistes. Pharmaciens.*, **273** (1932); as reprinted by G. H. Elliott, *Pharm. J.*, **131**, 514 (1933).

132. A. H. Goldberg, M. Gibaldi, J. L. Kanig, and J. Shanker, *J. Pharm. Sci.*, **54**, 1722 (1965).

Chapter **5**

VISCOSITY AND ITS MEASUREMENT

Jehuda Greener

Physical Methods of Chemistry, Second Edition Volume Six: Determination of Thermodynamic Properties   Edited by Bryant W. Rossiter and Roger C. Baetzold
ISBN 0-471-57087-7   Copyright 1992 by John Wiley & Sons, Inc.

1   INTRODUCTION

1.1   Scope

Viscosity is an important transport property of matter, expressing its resistance to irreversible deformation (flow). More specifically, this property reflects the extent to which relative motion of adjacent liquid layers is retarded and it can be generally regarded as a measure of internal friction of the liquid. Since it governs the dynamics and motion of fluids (gases, liquids, and viscoelastic solids), precise knowledge of this property is essential in a wide range of fluid transport problems in chemistry, physics, medicine, and engineering. Also, owing to its close association with molecular structure and composition, viscosity is an important characterization tool used extensively in the study of gases, "simple" liquids, and macromolecular fluids. This property is one of several rheological characteristics defining the dynamic response of materials to the myriad of possible deformations. Viscosity represents the dissipative (irreversible) portion of the dynamic response, whereas properties such as Young's modulus represent the elastic or recoverable component of the response. The methodology of measuring viscosity—*viscometry*—is a branch of a broader discipline, *rheometry*, which deals with the general techniques and instrumentation for measuring the gamut of rheological properties of materials.

In this chapter we confine the discussion to several important classes of viscometric techniques with emphasis on recent advances and the state of the art. To cover a sufficiently broad ground the treatment is necessarily selective and emphasis is placed on the underlying concepts. For more exhaustive discussion of the topics covered the reader should consult the pertinent references as well as several excellent monographs on the subjects of rheometry and viscometry [1–6]. This chapter deals primarily with the shear viscosity coefficient (henceforth referred to simply as *viscosity*) with some, albeit limited, discussion of the dynamic and extensional viscosities. Other rheological properties are mentioned only in passing in the context of the viscometric techniques being discussed. The terminology used is consistent with the official Society of Rheology nomenclature [7] and the units follow the standard Systéme International d'Unités System (SI). A glossary of terms commonly used in viscometric studies is given in Table 5.1 together with common notation and units.

As a prelude to the discussion of the various viscometric techniques, an overview of basic kinematic and dynamic concepts is presented, followed by a cursory review of some phenomenological aspects of viscosity (Section 2). The discussion of the various viscometric techniques is organized according to the main classes of flow employed in the measurement of viscosity. These include Poiseuille-type systems (Section 3), rotational and Couette systems (Section 4), and falling body or Stokesian systems (Section 5). In Poiseuille-type viscometers the liquid is driven through a conduit with a known geometry by an externally imposed pressure gradient, whereas in Couette and rotational viscometers flow is induced by drag forces imparted by the motion of one or more surfaces of the system within which the fluid is confined. In the Stokesian viscometers the liquid bulk is generally quiescent and flow is localized in the vicinity of an object (usually a sphere) that moves within the liquid mass. The motion of the object is typically induced by gravity or buoyancy. Dynamic techniques, discussed in Section 6, are used mainly for studying linear viscoelastic response of materials, which goes beyond the scope of this chapter. Nonetheless, these techniques are reviewed briefly since they can be used, through simple analogies, to obtain steady-state viscosity data. Dynamic techniques are particularly useful in the study of relatively rigid samples that cannot be readily subjected to steady-state flow.

1.2 Kinematics and Flow Classification

The motion and deformation of continuous media is commonly described by a physically based mathematical formalism known as *kinematics*. Kinematic quantities can be derived from a detailed mapping of fluid particles over time or from knowledge of the velocity field $u(x)$. These quantities represent the deformation or deformation rate experienced by fluid elements over the duration of the motion and must be expressed in tensor form to conform to the principle of *material objectivity* [8]. Generally, two classes of kinematic tensors are employed: strain tensors and strain-rate tensors. The former express the extent

Table 5.1   Nomenclature Frequently Used in Viscometric Studies

Symbol	Name	Definition	Units (SI)
1	Direction of flow		
2	Direction of velocity change		
3	Neutral direction		
u_i	Velocity	ith Component of velocity vector	m/s
$\dot{\gamma}$	Shear rate	Equation (5)[a]	1/s
τ_{12}	Shear stress	Force per unit area acting in direction 1 on surface $x_2 = $ constant	Pa
τ_{ii}	Normal stress (deviatoric)	Force per unit area acting in direction i normal to surface $x_i = $ constant	Pa
N_1	First normal stress difference	$\tau_{11} - \tau_{22}$	Pa
N_2	Second normal stress difference	$\tau_{22} - \tau_{33}$	Pa
μ	Viscosity (Newtonian)	$\tau_{12}/\dot{\gamma}$	Pa·s
η	Viscosity (non-Newtonian)	—"—	Pa·s
η_0	Zero-shear viscosity	$\lim_{\dot{\gamma} \to 0} (\tau_{12}/\dot{\gamma})$	Pa·s
η_∞	Infinite-shear viscosity	$\lim_{\dot{\gamma} \to \infty} (\tau_{12}/\dot{\gamma})$	Pa·s
v^b	Kinematic viscosity	η/ρ	m²/s
η_s, μ_s	Solvent viscosity		Pa·s
η_r	Relative viscosity[c]	η/η_s	
η_{sp}	Specific viscosity	$\eta_r - 1$	
η_{red}	Reduced viscosity[d]	η_{sp}/c	m³/kg
$[\eta]$	Intrinsic viscosity[e]	$\lim_{c \to 0} (\eta_{sp}/c)$	m³/kg
η_{inh}	Inherent viscosity[f]	$\ln \eta_r/c$	m³/kg
η^*	Complex dynamic (shear) viscosity	Equation (175)	Pa·s
η'	Dynamic (shear) viscosity	$\mathrm{Re}\{\eta^*\}$	Pa·s
ω	Oscillation (vibration) frequency		rad/s
Ψ_1	Primary normal stress coefficient	$N_1/\dot{\gamma}^2$	Pa·s²
Ψ_2	Secondary normal stress coefficient	$N_2/\dot{\gamma}^2$	Pa·s²

[a] For simple shear flows in a Cartesian coordinate frame $\dot{\gamma} = du_1/dx_2$.
[b] $\rho = $ Density of liquid.
[c] The term *viscosity ratio* is recommended by IUPAC.
[d] The term *viscosity number* is recommended by IUPAC; $c = $ concentration.
[e] The term *limiting viscosity number* is recommended by IUPAC.
[f] The term *logarithmic viscosity number* is recommended by IUPAC.

of deformation undergone by a fluid element at a given instant and can be formulated in several alternative forms. These tensors are of little use in the discussion of viscometry and viscosity and are not considered further in this chapter. For a comprehensive discussion of these tensors and other kinematic concepts the reader should consult the books by Lodge [9], Bird and co-workers [5], Coleman and co-workers [10], and Middleman [2]. More pertinent to this discussion are the strain-rate tensors, which express the instantaneous rate at which a fluid element is being deformed. The simplest rate tensor is the velocity gradient $\mathbf{Vu}$. This quantity, however, is not a pure measure of the rate of deformation since it represents also the rigid body motions of the liquid. By simple decomposition of the velocity gradient into symmetric and antisymmetric tensors the two types of motion represented in $\mathbf{Vu}$ can be decoupled. We can write in general

$$\mathbf{Vu} = \tfrac{1}{2}(\dot{\gamma} + \omega) \tag{1}$$

where

$$\dot{\gamma} \equiv \mathbf{Vu} + \mathbf{Vu}^{\mathrm{T}} \tag{2}$$

and

$$\omega \equiv \mathbf{Vu} - \mathbf{Vu}^{\mathrm{T}} \tag{3}$$

The term $\mathbf{Vu}^{\mathrm{T}}$ is the transpose of $\mathbf{Vu}$; $\dot{\gamma}$, the strain-rate tensor, expresses the rate at which a fluid element deforms in a flow field $\mathbf{u}(\mathbf{x}, t)$, whereas ω, the vorticity, represents the rigid body rotation imparted to the element during flow. Since ω does not concern the deformation of the liquid, it is of no rheological (viscometric) significance. The strain rate tensor $\dot{\gamma}$, on the other hand, is an important kinematic quantity often used in flow classification schemes and in the formulation of rheological constitutive equations.

Despite the large diversity of flows, most flow systems can be grouped into one of three general classes: (1) shear, (2) extensional (shear-free), and (3) mixed flows. Shear flows are most commonly used in viscometric experiments and are also the most extensively studied of the general flow classes. These flows (motions) can be visualized as a "sheaf of inextensible material surfaces in relative sliding motion" [11]. A more rigorous definition of shear flows can be given in terms of $\dot{\gamma}$ as

$$\dot{\gamma} = \dot{\gamma} \begin{pmatrix} 0 & 1 & 0 \\ 1 & 0 & 0 \\ 0 & 0 & 0 \end{pmatrix} \tag{4}$$

where $\dot{\gamma}$, the *shear rate*, is a characteristic scalar function of time and space having units of 1/time (typically 1/sec). This parameter can be defined in terms of the second invariant of $\dot{\gamma}$, II_γ, as

$$\dot{\gamma} = \sqrt{1/2\mathrm{II}_\gamma} \tag{5}$$

where $\mathrm{II}_\gamma = \dot{\gamma} : \dot{\gamma}$ (: is the scalar product for second rank tensors; see [5] for a more complete definition and an overview of tensor algebra).

Shear flows can be divided into several major subclasses depending mainly on the functional form of $\dot{\gamma}$ [10–12]. The flow is said to be viscometric if $\dot{\gamma} \neq \dot{\gamma}(t)$; that is, the shear rate may vary spatially, but it is independent of time (steady). If $\dot{\gamma} \neq \dot{\gamma}(\mathbf{x})$, the flow is homogeneous; that is, the shear flow may be unsteady, but the shear rate is uniform throughout the flow space. The flow is unidirectional if the streamlines (trajectories of fluid particles) coincide with the shearing surfaces (planes of constant $\dot{\gamma}$). Finally, if the flow field is independent of the rheological (viscometric) characteristics of the fluid and is strictly a function of the imposed kinematic boundary conditions, it is said to be *controllable*. Examples of various shear flow types are given in Table 5.2. In most standard viscometric tests the flow is, by implication, viscometric; that is, it is steady but not necessarily homogeneous. Indeed, one of the major tasks in interpreting viscometric data is to specify the operative (apparent) shear rate for the flow at hand. It is thus advantageous, though not necessary, to employ flows that are both homogeneous and controllable.

Figure 5.1 depicts a simple case of shear flow—a steady plane Couette flow—in which a fluid is sheared between two parallel surfaces, one stationary and the other moving at a constant speed U. In accordance with our convention (see Table 5.1) coordinate 1 is the flow direction, coordinate 2 is the direction of velocity change, and coordinate 3 is the neutral axis. If the fluid adheres fully to both surfaces (*no slip*) and the flow is relatively slow (see Section 1.3), then the velocity field is

$$\mathbf{u} = \left(\frac{U}{H} x_2, 0, 0\right) \tag{6}$$

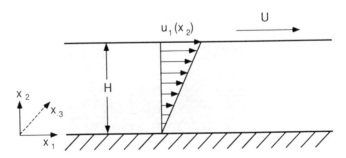

Figure 5.1 Plane Couette flow.

and

$$\dot{\gamma} = \frac{U}{H} \begin{pmatrix} 0 & 1 & 0 \\ 1 & 0 & 0 \\ 0 & 0 & 0 \end{pmatrix} \tag{7}$$

where

$$\dot{\gamma} = \frac{du_1}{dx_2} = \frac{U}{H} \tag{8}$$

and the physical meaning of $\dot{\gamma}$ becomes apparent. It must be stressed, however, that although the shear rate and the velocity gradient are closely related, they are not necessarily synonymous; A strict definition of the shear rate is given in (5). In terms of the classification of shear flows given above, the flow in Figure 5.1 is viscometric, homogeneous, and controllable. The space between the plates can be viewed as a series of shearing planes, or lamina, that slip past each other at a relative velocity $(U/H)dx_2$. In this flow momentum is transferred through the liquid lamina in the direction of velocity change (coordinate 2). As we will see in Section 1.3 the rate at which momentum *diffuses* through the liquid is expressed by its viscosity.

Extensional or shear-free flows represent a drastically different mode of fluid motion. While in shear flows a fluid element is sheared between two planes that traverse at different speeds, in extensional flows a fluid filament (or surface) is stretched between two points that move at varying speeds. This qualitative difference is of little consequence in the case of low molecular weight fluids (gases or liquids), but for complex liquids such as polymer melts and solutions, it may lead to fundamental differences in the dynamic response of the system. Since in extensional flows shearing deformations are absent, one can write

$$\dot{\gamma} = \begin{pmatrix} \dot{\gamma}_{11} & 0 & 0 \\ 0 & \dot{\gamma}_{22} & 0 \\ 0 & 0 & \dot{\gamma}_{33} \end{pmatrix} \tag{9}$$

which constitutes a formal definition of extensional flows. Also, if the fluid is incompressible, conservation of mass dictates

$$\sum_i \dot{\gamma}_{ii} = 0. \tag{10}$$

Extensional flows are sometimes called *irrotational flows* since $\omega = 0$ [cf. (3)] for all flows represented by (9). Because of their experimental and analytical complexity and since our understanding of these flows is still incomplete [13], as it stands now, rheological measurements in extensional flows are of limited use

Table 5.2 Simple Shear Flows[a]

Flow	Coordinates			Comments
	1	2	3	
Poiseuille flow in a tube	z	r	θ	Inhomogeneous, uncontrollable
Poiseuille flow in a slit die	x	y	z	Inhomogeneous, uncontrollable
Cylindrical Couette flow	θ	r	z	Homogeneous and controllable for narrow gaps only

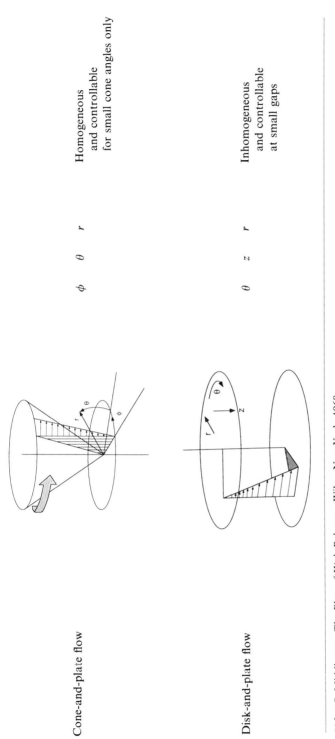

Cone-and-plate flow

$\phi \qquad \theta \qquad r$

Homogeneous
and controllable
for small cone angles only

Disk-and-plate flow

$\theta \qquad z \qquad r$

Inhomogeneous
and controllable
at small gaps

[a]After S. Middleman, *The Flow of High Polymers*, Wiley, New York, 1968.

to the nonspecialist and are, therefore, not discussed further in this chapter. For a comprehensive discussion of extensional flows see [13–15].

Mixed flows are some combination of shear and extensional flows; that is, $\dot{\gamma}$ for these flows contains both diagonal (extensional) and off-diagonal (shear) components. Although mixed deformations are common in many industrial flows, for example, lubrication systems, and flow through converging dies, these flows are unattractive in the viscometric sense since they are not readily amenable to analysis. A simple classification scheme for mixed flows was proposed by Tanner [16] based on the relative importance of the extensional components in $\dot{\gamma}$. Generally, mixed flows can be used in viscometric studies provided they are sufficiently "weak"; that is, the relative contribution of the extensional kinematics to the flow dynamics is small.

1.3   Dynamics of Purely Viscous Liquids

The forces acting within a continuous medium that is subjected to flow can be described by the equations of motion, which are derived from Newton's second law. These equations can be written in the form [17]

$$\rho\left(\frac{D\mathbf{u}}{Dt}\right) = -\nabla P - [\nabla \cdot \tau] + \mathbf{f} \tag{11}$$

where ρ is the density, D/Dt is the substantial or Stokes derivative, P is the pressure, and $\mathbf{f}$ is a body force (e.g., a gravitational force per unit volume ρg). The deviatoric stress τ is the part of the stress tensor that is associated with the rheological (*nonhydrostatic*) response of the material. Equation (11) states that a unit mass of the fluid will accelerate in response to the net sum of forces—pressure, viscous, and body tractions—acting upon it. However, for this equation to be usable, it must be accompanied by an equation specifying the relationship between τ and the velocity field or the motion of the fluid. This is formally expressed by the constitutive law (rheological equation-of-state), which essentially couples the rheological character of the liquid with the dynamics of the flow system. In its most general form for the so-called *simple incompressible liquid* the constitutive law can be written as [18]

$$\tau = \overset{\infty}{\underset{s=0}{\hbar}}\, F(s) \tag{12}$$

where $F(s)$ is a kinematic tensor representing the displacement gradient at time s and $\hbar$ is some functional of F that runs over the deformation history of the fluid—from the distant past, $s = \infty$, to the present, $s = 0$. Equation (12) simply states that the stress in any liquid element depends on the entire deformation history experienced by that element. Naturally, (12) can assume many, indeed infinite, alternative forms that will satisfy the general criteria of a simple liquid, but only a few will conform to reality. Indeed, one of the most challenging tasks facing the discipline of rheology is to identify the equation within this spectrum that is most consistent with observation. As of this writing, no single equation applies universally for all flow situations and all fluids, although some are remarkably successful in describing the response of many liquids under a wide range of flow conditions [19, 20]. For the purpose of this discussion suffice it to consider a specialized class of constitutive equations—the generalized Newtonian fluid (GNF)—that is well suited to describe the viscous response of most liquids. The GNF can be written as

$$\tau = -\eta\dot{\gamma} \tag{13}$$

where η is the shear viscosity function or simply the viscosity having units of stress × time; η is an even scalar function of shear rate, and it also depends on temperature and pressure. (The negative sign implies that the material under shearing motion is in a state of tension.) The dependence of η on $\dot{\gamma}$ is a rheological property of the fluid, whereas the effects of temperature and pressure on η express its thermodynamic character. When η is independent of $\dot{\gamma}$, (13) reduces to Newton's law of viscosity or Newton's hypothesis. For shear flows this law reads

$$\tau_{12} = -\mu\dot{\gamma} \tag{14}$$

where μ is the Newtonian or constant viscosity of the liquid. Equation (14) is a special case of (13) and it states that the shear stress imparted by the liquid in response to an externally imposed shearing motion is proportional to the shear rate, with μ being the proportionality constant. The viscosity is a fundamental transport property of matter expressing the inherent rate (diffusivity) at which momentum is transferred in direction 2 (direction of velocity change) by liquid lamina moving in direction 1 (flow direction). The value of μ can vary over many decades depending on the physical state and molecular structure of the material. Typically, the viscosity of gases ranges from 10^{-6} to 10^{-4} Pa·s, the viscosity of low molecular weight liquids is in the range 10^{-4}–1 Pa·s, and that of concentrated polymer solutions and melts is in the range 0.1–10^5 Pa·s. The viscosity of amorphous materials close to their glass transition temperature may be as high as 10^{12} Pa·s.

Newton's law was considered to be universally valid for over two centuries and only with the advent of synthetic polymers, at the turn of this century, was it realized that it is not obeyed by a substantial number of liquids. Liquids that follow Newton's hypothesis, which includes the majority of gases, low molecular weight liquids, and dilute polymer solutions, are called *Newtonian*. Those that do not follow (14), either because μ depends on $\dot{\gamma}$ or because the liquid exhibits elastic character (normal stresses and transient behavior) in shear flows, are *non-Newtonian*. In a strict sense the GNF model, (13), is not valid because it cannot account for elastic and transient phenomena usually observed in conjunction with shear rate dependence of viscosity [19]. Indeed, η is only one of three viscometric functions that define the rheological response of simple liquids in simple shear flows [see (12)]. In viscometric flows these liquids will give rise to shear stresses (*viscous response*) as well as to normal stresses (*elastic response*). Accordingly, one can define two rheological functions in addition to η: the primary normal stress coefficient

$$\Psi_1 \equiv \frac{N_1}{\dot{\gamma}^2} = \frac{\tau_{11} - \tau_{22}}{\dot{\gamma}^2} \tag{15}$$

and the secondary normal stress coefficient

$$\Psi_2 \equiv \frac{N_2}{\dot{\gamma}^2} = \frac{\tau_{22} - \tau_{33}}{\dot{\gamma}^2} \tag{16}$$

where N_1 and N_2 are the first and second normal stress differences. These characteristic material functions are not predicted by (13) and in this sense the GNF is a "purely viscous" model. Despite this deficiency the GNF model can be quite useful in analyzing a wide range of flow problems where the viscous response of the liquid is dominant. The function η, which provides a rheological fingerprint of the fluid, may assume many different forms to be discussed in the next section. Indeed, the main object of viscometric tests is to evaluate this function over the widest possible range of shear rates, or at least over a range that corresponds to a given flow process under study.

For the special case of Newtonian and incompressible liquids, (11) reduces to the familiar Navier–Stokes equation [17]

$$\rho \left(\frac{D\mathbf{u}}{Dt} \right) = -\nabla P + \mu \nabla^2 \mathbf{u} + \mathbf{f} \tag{17}$$

which is a fundamental tool of fluid dynamics. Dimensional analysis of this equation yields an important flow characteristic, the Reynolds number (Re), which expresses the relative magnitude of inertia-to-viscous forces in a given flow. The Reynolds number is generally defined as [21]

$$\text{Re} = \frac{HU}{\nu} \tag{18}$$

where H and U are characteristic dimension and speed, respectively, and ν is the kinematic viscosity ($= \mu/\rho$) of the liquid. If the liquid is non-Newtonian, ν is given by η/ρ, where η is evaluated at some nominal shear rate, characteristic of the flow process under study. The Reynolds number not only expresses the relative importance of inertia forces, but it also serves to indicate whether the flow will be stable (laminar). A substantial body of data [22] suggests that when Re exceeds some critical value (2100 for flow in a tube) the flow becomes "chaotic" or turbulent and practically useless for viscometric tests. Thus, turbulence must be avoided for a viscometric test to be successful. As will be shown in Section 4.6, turbulence is but one of several inertia-driven flow instabilities to watch for in viscometric experiments; even under stable flow conditions large inertia forces (high Re) may give rise to various secondary flows that are unaccounted for in simple viscometric analyses, and it is thus necessary, in some cases, to strive for Re → 0 (the *creeping flow* limit) to allow realistic and accurate interpretation of data.

2 PHENOMENOLOGY AND MOLECULAR ORIGINS OF VISCOSITY

2.1 Gases

The viscosity of gases, especially at low pressures, has been studied extensively over the past century and is relatively well understood in terms of molecular parameters and mechanisms. In the classical kinetic theory of gases gas molecules are represented as noninteracting rigid spheres undergoing random thermal fluctuations (Brownian motion). When such an ensemble is subjected to a shearing deformation, momentum transfer between neighboring lamina is assumed to arise strictly from intermolecular collisions. The viscosity, or the rate of momentum transfer, for this idealized system is given by [23]

$$\mu = \frac{5}{16\pi^{3/2}d^2} \sqrt{\frac{MkT}{N_A}} \tag{19}$$

where d is the collision diameter of the molecule (not necessarily its geometric diameter), M is the molecular weight, k is Boltzmann's constant, T is the absolute temperature, and N_A is Avogadro's number. Thus, according to (19), μ is independent of pressure and has a square-root dependence on temperature. For many gases, however, the temperature dependence of μ is far stronger than that predicted by (19), which is attributed to intermolecular forces ignored in the original treatment of the kinetic theory. The Chapman–Enskog extension of the kinetic theory [24] accounts for such forces by considering pairwise interactions using an arbitrary potential function. When the Chapman–Enskog theory is used in conjunction with the Lennard-Jones potential, the following expression for the viscosity is obtained

$$\mu = \frac{A}{\sigma^2\Omega(T)} \sqrt{MT} \tag{20}$$

where A is a numerical constant, σ is a parameter (*collision diameter*) of the Lennard-Jones potential function, and Ω is a slowly varying function of T. Thus, here, too, the viscosity is independent of pressure, but the temperature dependence of μ is stronger than in (17), which is more in line with experimental data. At high pressures, as the gas becomes denser and the molecular interactions are more frequent and complex, the simple picture of the kinetic theory no longer holds. In fact, one finds that the viscosity of the gas becomes strongly dependent on pressure and its temperature dependence deviates substantially from the low density limit as represented by (20).

A generalized correlation for the viscosity of gases as a function of pressure and temperature can be formulated based on the corresponding states principle. According to this principle a nearly universal response can be obtained when $\mu_r = \mu/\mu_c$ is correlated with $P_r = P/P_c$ and $T_r = T/T_c$, where the subscript c denotes the critical point of the gas and μ_c can be estimated from simple empirical correlations [25]. Although this approach is empirical in origin, it has been fairly successful in describing the pressure and temperature dependence of viscosity for many gases. One such correlation is shown in Figure 5.2 and is seen to cover a wide range of conditions, including the gaseous dilute and dense states as well as the liquid state. Although this correlation is expected to predict the viscosity of a variety of gases to within $\pm 20\%$ [26] and should be used merely for estimation purposes, it clearly illustrates the general effects of temperature and pressure on viscosity. We see, for example, that the temperature dependence for dilute and dense gases is markedly different from that for liquids, thus suggesting that different molecular mechanisms govern momentum transport in the liquid and gaseous states. It is also noted that the kinetic theory [see (20)] can explain satisfactorily only the low-density (low-P) limit but is clearly inadequate at medium and high densities.

Before turning to a discussion of Newtonian liquids, we note that the general techniques for measuring the viscosity of gases are essentially identical to those used for measuring the viscosity of liquids. The main instrumental differences between liquid and gas viscometers arise from the special requirements to contain and move the gas within the viscometer and from the low forces generated by the gas that call for very sensitive stress (pressure, torque) sensors. The high compressibility of gases may also pose some difficulties when the viscosity is measured at high pressures. A general discussion of the design and operation of gas viscometers is given by Chierici and Paratella [27].

2.2   Newtonian Liquids

Molecular models for momentum transport in liquids are not as well developed as those for gases. From a strictly mathematical viewpoint, extension of the kinetic theory to liquids is a formidable task. The main difficulties arise from the fact that liquid molecules are very close-packed compared to gas molecules and their relative motions are governed by complex intermolecular interactions (molecular collisions that govern momentum transport in gases are not an important transport mechanism in liquids!). A general approach for treating the motion of liquids on a microscopic level was taken by Eyring and his co-workers [28]. Their model is based on a general theory of rate processes and although it does not constitute a molecular theory *per se*, it provides some useful insights.

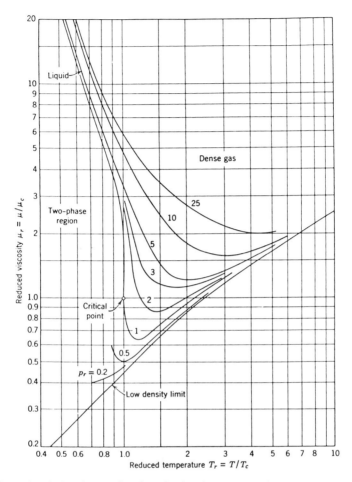

Figure 5.2   Reduced viscosity as a function of reduced temperature for several values of reduced pressure for gases and liquids. Reprinted from R. B. Bird, W. E. Stewart, and E. N. Lightfoot, *Transport Phenomena*, Wiley, New York, 1960, p. 16 and O. A. Uyehara and K. M. Watson, *Natl. Pet. News*, Tech. Sect. 36, 764 (Oct. 4, 1944).

Eyring and co-workers [28] picture the liquid continuum as being an ensemble (lattice) of randomly distributed cages and holes. A cage is a site occupied by a molecule or a liquid *particle* and a hole is an empty site. At rest, a dynamic equilibrium exists between the populations of cages and holes that involves randomly oriented jumps from cages to holes. To escape from its cage to a neighboring hole, a molecule must surpass an energy barrier ΔG_0^*, which is the molar free energy for flow activation. Unlike the kinetic theory of gases, Eyring's treatment does not invoke Newton's law, and, in fact, it yields in general a non-Newtonian viscosity even for low molecular weight liquids. In the limit of low shear stress, however, the viscosity becomes independent of stress and is given by

$$\mu \cong \frac{N_A h}{\tilde{V}} \exp\left(\frac{\Delta G_0^*}{RT}\right) \tag{21}$$

where N_A is Avogadro's number, $\tilde{V}$ is the molar volume, and h is Planck's constant. The temperature dependence predicted by this equation is followed closely for many nonpolar liquids. The activation energy for flow ΔG_0^* is a fundamental property of the liquid expressing its "cohesiveness," that is, its ability to accommodate random rearrangements in the lattice due to thermal fluctuations, and it can be estimated independently from the boiling point [29]. Equation (21) suggests that temperature exerts a strong effect on viscosity and should be controlled tightly in viscometric tests.

One of the key assumptions in the Eyring model is that the total volume fraction of holes in the lattice is independent of temperature and pressure; that is, the liquid is incompressible, having a constant molar volume $\tilde{V}$. In general, however, this may not be the case, particularly for glass-forming liquids near their glass transition temperature T_g, for example, melts of amorphous polymers. This was recognized by Doolittle and Doolittle [30] who proposed a simple phenomenological equation relating viscosity to free volume

$$\mu = A \exp\left[B\left(\frac{1}{f} - 1\right)\right] \tag{22}$$

where f is the fractional free volume in the liquid (equivalent to the volume fraction of holes in Eyring's model) and A and B are constants. When viewed in the context of Eyring's model, this equation merely states that the energy barrier for jumps is lowered if f is increased. Consequently, if f is temperature dependent, the temperature can be viewed as having two independent effects: (1) it helps overcome an existing barrier by thermally induced fluctuations, and (2) it may reduce the height of the energy barrier by increasing the fraction of holes in the lattice; that is, $\Delta G_0^* = \Delta G_0^*(T)$. Thus, from knowledge of the dependence of f on temperature and pressure one can modify Eyring's theory, (21), to account for changes in free volume as well as to estimate the effects of pressure and temperature on viscosity for the general case where $\Delta G^* = \Delta G^*(T, P)$. Since such effects are particularly important in polymeric fluids, we reserve a more detailed consideration of this point to the discussion of non-Newtonian liquids (Section 2.3).

Finally, we note that the effects of temperature and pressure can also be estimated to a reasonable degree of accuracy from the corresponding states principle using, for example, the chart in Figure 5.2. This chart suggests that for low molecular weight Newtonian liquids the viscosity is a slowly increasing function of pressure with a negligible effect for $P_r < 1$. For a more detailed discussion on the molecular theory of liquids the reader should consult the books by Hirschfelder and co-workers [23], Frenkel [31], Kirkwood [32], and Bondi [33]. A large compilation of viscosity data for low molecular weight liquids is given in a book by Touloukian and co-workers [34].

2.3   Non-Newtonian Liquids

It was noted in Section 1.3 that Newton's law of viscosity is not obeyed by many liquids. In fact, many technologically important liquids such as polymer solutions and melts, suspensions, latexes, and pastes, exhibit distinct non-Newtonian character in shear flow. An important, although not the only, manifestation of non-Newtonian behavior is the shear rate dependence of viscosity, which can be expressed by some empirical equation of the form $\eta = \eta(\dot{\gamma})$. The exact form of $\eta(\dot{\gamma})$, sometimes referred to as the *flow curve*, defines, at least in part, the rheological character of the liquid and it can provide some insights into its molecular structure (see Section 2.4). Phenomenologically, many functional forms of $\eta(\dot{\gamma})$, some more common than others, have been observed. Some typical flow curves are shown schematically in Figure 5.3a. The most common response is represented by curve I. This response is typical of polymeric liquids, both solutions and melts, and various heterogeneous liquids (see Section 2.5). Generally, curve I can be divided into five distinct regimes: (1) the zero-shear regime or the Newtonian plateau ($\dot{\gamma} < \dot{\gamma}_I$), (2) the low-shear transition regime ($\dot{\gamma}_I < \dot{\gamma} < \dot{\gamma}_{II}$), (3) the power law regime ($\dot{\gamma}_{II} < \dot{\gamma} < \dot{\gamma}_{III}$), (4) the high-shear transition regime ($\dot{\gamma}_{III} < \dot{\gamma} < \dot{\gamma}_{IV}$), and (5) the infinite-shear regime or the second Newtonian plateau ($\dot{\gamma} > \dot{\gamma}_{IV}$). (We note that the term Newtonian is used here somewhat loosely since the response of the liquid in this regime appears to be Newtonian but may, in fact, be non-Newtonian.) The viscosity in the zero-shear Newtonian regime (η_0—the zero-shear viscosity) is the most sensitive to variations in molecular structure, topology (molecular weight, molecular weight distribution, branching, etc.), composition, and temperature. In some cases involving melts and solutions of high molecular weight polymers, the Newtonian regime is limited to very low shear rates and is sometimes unattainable by conventional viscometric techniques. In the power law regime the viscosity can be expressed by a simple empirical formula of the form

$$\eta = K\dot{\gamma}^{n-1} \tag{23}$$

where K is the consistency index and n is the power law index. Equation (23) is sometimes called the *Ostwald–de Waele* model [35]. The temperature dependence of viscosity is incorporated into K while n represents the shear sensitivity of the viscosity function and is generally independent of temperature. The material is Newtonian when $n = 1$ for which K becomes μ; it is shear thinning (pseudoplastic) when $0 < n < 1$, and it is shear thickening (dilatant) when $n > 1$. Polymeric liquids are typically shear thinning with n ranging from 0.15 to 0.7. The second Newtonian plateau (characterized by the infinite-shear viscosity η_∞) is hardly ever observed in conventional liquid systems, and the

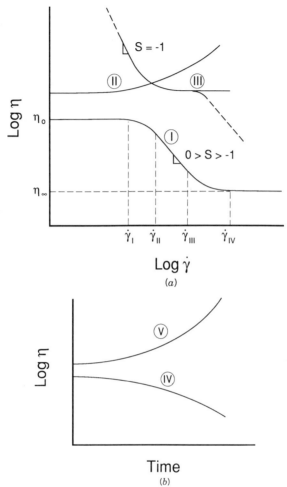

Figure 5.3 (a) Schematic flow (viscosity vs. shear rate) curves for various liquids: (I) typical shear thinning (pseudoplastic) liquid, (II) shear thickening (dilatant) liquid, and (III) "plastic" liquid. (b) Schematic viscosity-time curves for time-dependent liquids: (IV) thixotropic liquid and (V) rheopectic liquid.

information concerning this regime is relatively scant, mainly due to experimental difficulties in attaining $\dot{\gamma}_{IV}$. Studies of the behavior of liquids in the second Newtonian plateau have been reported by Merril and co-workers [36] and by others [37, 38].

Although many empiricisms have been proposed for expressing curve I mathematically, a commonly used form is the four-parameter model of Carreau [39]

$$\frac{\eta - \eta_\infty}{\eta_0 - \eta_\infty} = [1 + (\lambda\dot{\gamma})^2]^{(n-1)/2} \qquad (24)$$

where λ is some characteristic time constant that can be estimated roughly as $\lambda \sim 1/\dot{\gamma}_{\mathrm{II}}$ and n is the power law index of (23). Since in most cases $\eta_\infty \ll \eta$, (24) can be considered practically a three-parameter model. Another useful empirical model is the Cross–Carreau–Yasuda equation [40–42]

$$\frac{\eta - \eta_\infty}{\eta_0 - \eta_\infty} = \left[1 + \left(\frac{\eta_0 \dot{\gamma}}{\tau^*} \right)^b \right]^{(n-1)/b} \tag{25}$$

where b is a parameter controlling the shape of the viscosity function in the transition regimes (Figure 5.3) and $\tau^*(=\eta_0/\lambda)$ is a characteristic stress or modulus. Equation (25) provides generally a better description of the response in the transition regimes than does (24), but it requires the specification of one additional parameter (b). Both τ^* and b are material-specific, temperature-independent parameters; τ^* represents the critical shear stress for the onset of shear thinning and is generally a constant for a particular polymer class, while b is closely related to the polydispersity of the polymer species [42]. Other, less commonly used viscosity equations are the Ellis model [43], the Powell–Eyring model [44], and the truncated power law model [45]. The latter is a computationally useful empiricism that can be effectively employed in various numerical schemes for non-Newtonian flow calculations. Shear thickening behavior as represented by curve II in Figure 5.3a is very rare and is observed predominantly in the case of suspensions and slurries that contain deflocculating agents [46] (see Section 2.5). An apparent shear thickening has also been observed for some polymer solutions with species that are capable of inter-molecular association through various physical interactions (ionic, dipolar, hydrogen bonding, etc.) [47]. Curves IV and V in Figure 5.3b are characteristic of complex liquids, that is, liquids undergoing structural changes, physical or chemical, during the viscometric test. If the viscosity increases with time under steady-state conditions, the liquid is *thixotropic*, and if it drops, the liquid is considered to be *rheopectic*. Such behavior is marked by distinct hysteresis in the shear stress–shear rate response curve and is generally observed in heterogeneous liquids that contain flocs or aggregates and in liquids with chemically reactive species. (Hysteresis may arise strictly from transient viscoelastic effects, which should not be confused with thixotropy or rheopexy [48]!)

The last group of non-Newtonian liquids is that of "plastic" liquids. These liquids are characterized by a threshold or yield stress τ_y, which must be exceeded for flow to commence, and are represented schematically by curve III in Figure 5.3a. If the applied stress is lower than the yield stress, the material responds in a rigid, solidlike manner, while for stresses higher than τ_y the material is a purely viscous or viscoelastic liquid. Examples of plastic liquids are highly loaded suspensions of rigid particles, various pastes, and blood. Three well-known empiricisms for viscoplastic behavior are the Bingham model [49] (one of the earliest non-Newtonian models formulated)

$$\eta = \begin{cases} \mu_0 + \tau_y/\dot{\gamma} & \text{for } \tau_{12} > \tau_y \\ \infty & \text{for } \tau_{12} \leqslant \tau_y \end{cases} \tag{26}$$

the Herschel–Bulkley (HB) model [50]

$$\eta = \begin{cases} K\dot{\gamma}^{n-1} + \tau_y/\dot{\gamma} & \text{for } \tau_{12} > \tau_y \\ \infty & \text{for } \tau_{12} \leqslant \tau_y \end{cases} \tag{27}$$

and the Casson model [51]

$$\eta = \begin{cases} \mu_0 + \tau_y/\dot{\gamma} + 2\sqrt{\tau_y \mu_0/\dot{\gamma}} & \text{for } \tau_{12} > \tau_y \\ \infty & \text{for } \tau_{12} \leqslant \tau_y \end{cases} \tag{28}$$

In all cases τ_y is the yield stress and μ_0 for the Bingham and the Casson models is the Newtonian viscosity at sufficiently large stresses ($\mu_0\dot{\gamma} \gg \tau_y$). The HB model approaches the power-law model when $K\dot{\gamma}^n \gg \tau_y$, with n and K being the usual power law parameters, see (23). Although the general concept of viscoplastic behavior has been criticized on fundamental grounds [52, 53], (26)–(28) have proven nonetheless useful in describing the apparent response of liquids that exhibit yield stress. Various techniques for measuring the yield stress are described by Yoshimura and co-workers [54].

Although the viscosity of non-Newtonian liquids is generally less temperature-sensitive than that of Newtonian liquids, temperature is a key variable in viscometric tests that must be closely monitored and controlled. The effect of temperature on the viscosity function of a typical polymeric liquid is illustrated in Figure 5.4, clearly showing a weaker dependence in the power law regime than in the Newtonian regime. To examine the temperature effect more closely we rewrite (24) with a temperature term added and with η_∞ set to zero

$$\eta = \frac{\eta_0 a_T}{[1 + (\lambda a_T \dot{\gamma})^2]^{(1-n)/2}} \tag{29}$$

where a_T is the time-temperature shift factor, a function that effectively expresses the temperature dependence of the zero-shear viscosity. This factor is defined as [55]

$$a_T = \frac{\eta_0(T) T_r \rho_r}{\eta_0(T_r) T \rho} \tag{30}$$

where T_r, an arbitrary reference temperature, is usually taken as T_g, the glass-transition point, and ρ_r is the density at that temperature. We now recall the semiempirical Doolittle equation, (22), which relates η to the fractional free volume f. For glass-forming liquids (most synthetic polymers) f is a linear function of temperature

$$f = f_g + \alpha_f(T - T_g) \tag{31}$$

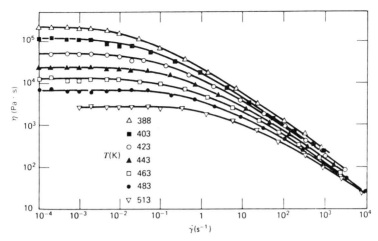

Figure 5.4   Flow curves of a low-density poly(ethylene) melt at different temperatures. Reprinted with permission from J. Meissner, *Kunststoffe*, **61**, 576 (1971), Carl Hanser Verlag, Munich.

where f_g is the fractional free volume at T_g and α_f is the thermal expansion coefficient of the free volume above T_g. When this equation is combined with (22), the following expression for a_T is obtained

$$\log a_T = \frac{-c_1(T - T_g)}{c_2 + (T - T_g)} \tag{32}$$

where c_1 and c_2 are "nearly" universal constants. This expression, known as the Williams–Landel–Ferry (WLF) equation [55], was found to be very effective in describing the response of most molten polymers in the range $T_g < T \lesssim T_g + 100°C$. At higher temperatures the temperature dependence of a_T reverts to an Arrhenius form predicted by Eyring's model [see (21)]

$$\log a_T = \frac{\Delta E}{R}\left(\frac{1}{T} - \frac{1}{T_r}\right) \tag{33}$$

where ΔE, the activation energy for viscous flow, is closely related to the free energy of activation in (21). Equation (33) is equivalent to the empirical equation proposed by Andrade [56] and is sometimes called the *Andrade equation*. The transition from a WLF to an Arrhenius response is illustrated in Figure 5.5 for a poly(methyl methacrylate) melt with a T_g of 100°C. The activation energy for flow in the Arrhenius regime is governed by barriers to relative translation of molecules and is generally observed to increase with the extent of intermolecular interactions (e.g., hydrogen bonding) and the presence of steric barriers (e.g., bulky side groups) [57]. The value of ΔE usually ranges from 5 to 60 kJ/mol for low molecular weight liquids and from 15 to 150 kJ/mol for polymeric liquids. A

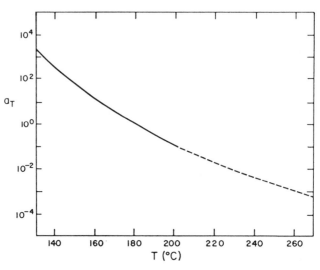

Figure 5.5 Time–temperature shift factor for a poly(methyl methacrylate) melt: $T_r = 180°C$ and $T_g = 100°C$; solid curve: WLF equation and dashed curve: Arrhenius equation. Reprinted with permission from G. H. Pearson, "Rheology," in M. D. Baijal, Ed., *Plastics Polymer Science and Technology*, Wiley, New York, 1982, p. 519. Copyright © 1982 by John Wiley & Sons, Inc.

simple correlation and a graphical procedure for estimating the activation energy of polymer melts from molecular structure is given by van Krevelen [58]. An equivalent form of the WLF Equation is the Vogel–Fulcher equation [59, 60], which expresses the viscosity as a function of temperature in a generalized form

$$\log \mu = A' + B'/(T - T_\infty) \tag{34}$$

where A', B', and T_∞ are characteristic constants closely related to c_1, c_2, and T_g in the WLF equation. It is simple to show that when $T \gg T_\infty$, (34) reverts to an Arrhenius form. Angell [61] classifies liquids as *strong* if they follow Arrhenius activation (constant activation energy) and *fragile* if they depart from the Arrhenius form, that is, if their activation energy is dependent on temperature.

   Inspection of (29) suggests that the temperature dependence of viscosity in the power-law regime $(a_T\dot{\gamma}\lambda \gg 1)$ is given by a_T^n $(n < 1)$ compared to a_T for the Newtonian case, that is, one should expect a lower temperature sensitivity for shear thinning liquids, in line with observation (see Figure 5.4). Another important consequence of (29) is that one can reduce flow curves for several temperatures into a single *master curve* making use of the time-temperature superposition principle [55]. This is readily accomplished by plotting either $\eta(T)/[\eta_0(T_0)a_T]$ versus $\lambda a_T\dot{\gamma}$ or simply $\eta(T)/\eta_0(T)$ versus $\eta_0(T)\dot{\gamma}$ [cf. (24) and (25)]. Otherwise, the data reduction can be done by shifting the curves horizontally and vertically to some reference curve (temperature) until a visually good overlap is obtained. The shifting technique can be used to extend the effective range of a flow curve by repeating the test at a series of temperatures. Generally, higher temperatures should be used when higher shear rates are sought and vice versa. The shifting procedure can be extended to other independent variables such as concentration, molecular weight (cf. Figure 5.8), and pressure.

By generalizing the expression for f to include the effect of pressure, the pressure dependence of viscosity can be estimated in much the same way. Generally, one can write [55]

$$f = f_g + \alpha_f(T - T_r) - \beta_f(P - P_r) \tag{35}$$

where P_r is a reference pressure (usually ambient) and β_f, the compressibility of the free volume, can be estimated from

$$\beta_f \cong \alpha_f \, dT_g/dP$$
$$\approx \Delta\alpha \, dT_g/dP \tag{36}$$

with $\Delta\alpha$ being the difference in thermal expansion between the material in the liquid and glassy states. Similarly, in analogy to the temperature effect [cf. (30)], one can define a time-pressure shift factor

$$a_P \cong \frac{\eta_0(P)}{\eta_0(P_r)} \frac{\rho(P_r)}{\rho(P)} \tag{37}$$

that can be incorporated into (29) together with a_T to account for the effect of pressure. Because of scarcity of data, expressions for a_P are not as well developed as are those for a_T, although the experimental studies of Westover [62] and Crowson and co-workers [62a] suggest that

$$\log a_P = \Phi(P - P_r) \tag{38}$$

should be a reasonable approximation. Here, Φ is an empirical constant of the order of 10^{-8} Pa^{-1}. Other forms of a_P have been proposed by Moonan and Tschoegl [63] and Utracki [64].

2.4   Structural and Topological Effects

In this discussion we focus on the effects of concentration, molecular weight, and its distribution on the viscosity of concentrated polymer solutions and melts. Some consideration of dilute polymer solutions is given in Section 3.5 where intrinsic viscosity techniques are discussed. When the zero-shear viscosity of polymer melts is plotted against molecular weight (M_w, the weight-average molecular weight in the case of polydisperse systems), a most striking relationship is observed. Below some critical molecular weight M_c the dependence of η_0

on M_w is linear for all polymer systems; but once M_c is surpassed, a strong, 3.4th power relationship is observed

$$\eta_0 = K'M_w^{3.4} \tag{39}$$

The sharp change in slope on the $\log \eta_0 - \log M$ curve is attributed to the onset of entanglement coupling and M_c is usually taken as [65]

$$M_c \cong 2M_e \tag{40}$$

where M_e is the average molecular weight between entanglements. This quantity varies in the range 2,000–60,000 for synthetic polymers and is loosely related to the stiffness (persistence length) of the polymer chain. Equation (39) is observed for many polymer systems and it underlies the great sensitivity of η_0 to fluctuations in M_w once M_c is exceeded. Examples of the M_w dependence of η_0 for various polymers are shown in Figure 5.6. A similar dependence is observed for concentrated polymer solutions ($c \gtrsim 0.25$ g/mL), except that here M_w is replaced with the product cM_w and $(cM_w)_c$ now represents a critical composition for entanglement coupling or a change in slope in the $\log \eta_0 - \log(cM_w)$ curve. This implies that the characteristic molecular weight for entanglement formation in concentrated polymer solutions is proportional to $1/c$ although recent data suggest some deviation from this rule [66]. The product $(cM_w)_c$ depends on the stiffness of the polymer chain and, to a lesser extent, on the nature of the diluent and the polymer–diluent interaction [67, 68]. Generally, the nature of the solvent and the polymer–solvent interaction play an important role only in the dilute solution regime. As the volume fraction of the polymer exceeds approximately 0.15–0.20, the effect of solvent quality on viscosity becomes secondary. Figure 5.7 shows the combined effects of c and M on the viscosity of poly(vinyl acetate) in diethyl phthalate (a good solvent) and in cetyl alcohol (a poor solvent). The value of $(cM)_c$ and the shape of the η versus (cM) curve are clearly independent of concentration and the nature of the diluent over the range tested.

The general effects of M_w and c on η_0 below M_c are properly described by the molecular bead-spring models of Rouse [69] and Zimm [70] in which the polymer molecule is represented as a chain of beads connected in series by massless springs. The plausible role of molecular entanglements in the viscous dissipation process for $M_w > M_c$ has been recognized for some time [71–73]. Bueche [72] contends that entanglements involve two distinct dissipation mechanisms; the first relates to the tendency of molecules to drag along entangled partners and the second corresponds to additional translational motion between entanglements. Graessley [73], in an early phenomenological theory, employed the entanglement concept to explain shear thinning effects in concentrated polymer solutions and melts. While the role of entanglements is generally well accepted, theoretical resolution of the 3.4th power law in the $\eta_0 - M_w$ relationship [cf. (39)] is not fully settled. The tube model of Doi and

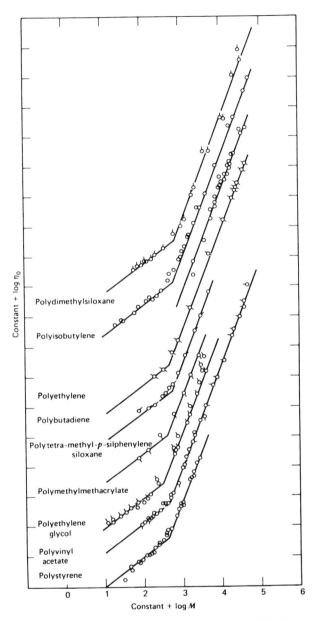

Figure 5.6 Zero-shear viscosity versus weight-average molecular weight for several polymers. Reprinted with permission from G. C. Berry and T. G. Fox, *Adv. Polym. Sci.*, **5**, 261 (1968).

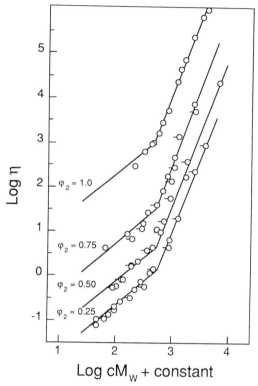

Figure 5.7 Zero-shear viscosity versus cM_w for solutions of poly(vinyl acetate) in diethyl phthalate (○) and cetyl alcohol (—○) at 396 K. Reprinted with permission from G. C. Berry and T. G. Fox, *Adv. Polym. Sci.*, **5**, 261 (1968).

Edwards [74, 75] and its various offshoots are probably the ones that come closest to resolving the strong molecular weight dependence of η_0. According to the tube model an entangled chain can rearrange its conformations by moving along its own contour (reptating) through a "virtual tube" formed by neighboring molecules. The terminal (longest) relaxation time of this reptation process can be computed and expressed as viscosity.

Generally, the effects of M_w and c on the onset of shear thinning (γ_1 in Figure 5.3) are well represented by the models of Bueche [72] and Graessley [73] in terms of the Rouse relaxation time λ_R; that is, the reduced viscosity can be generally expressed by

$$\frac{\eta}{\eta_0} = f(\lambda_R \dot{\gamma}) \tag{41}$$

where

$$\lambda_R = \frac{12(\eta_0 - \eta_s)M}{\pi^2 cRT}$$

and η_s is the solvent viscosity [cf. (24)]. A striking illustration of the effectiveness of (41) is shown in Figure 5.8 for solutions of an homologous series of narrow distribution poly(styrene) in n-butylbenzene. If we define the critical shear rate for the onset of shear thinning $\dot{\gamma}_0$ as the shear rate where η/η_0 falls to 0.8, then, based on experimental data, $\dot{\gamma}_0\lambda_R \doteq 1-4$, depending on the quantity cM_w [76]. A qualitative representation of the effect of shear rate on the cM dependence of viscosity is shown in Figure 5.9. Consistent with (29) and (41), the sensitivity of η to variations in M and/or c is much lower in the shear thinning regime compared to the Newtonian regime. The effects of molecular weight distribution on the viscosity function are not as yet fully understood although extensions of the Graessley and Bueche theories [73, 77] as well as recent modifications of the tube model for polydisperse systems [78] are generally consistent with experimental data. The effect of polydispersity on the viscosity function can be illustrated by using the modified theory of Graessley, as shown in Figure 5.10. In this figure theoretical viscosity curves for a polymer with the Schultz–Zimm distribution are plotted for several values of the distribution parameter z [when $z = \infty$ the polymer is monodisperse, and when $z = 0$ the polymer has a "most probable" distribution $(M_w/M_n) = 2$]. Thus, it appears that for a polymer system with a narrow distribution the transition from Newtonian to shear thinning behavior is sharper and the onset of shear thinning occurs at a higher shear rate compared to a polydisperse system with the same M_w. For a more complete discussion of the effects of molecular topology and composition on the viscosity of polymer solutions and melts, the reader should consult the reviews by Berry and Fox [67], Carpenter and Westerman [68], Kumar [79], and Graessley [65, 75].

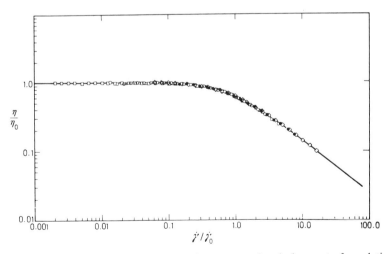

Figure 5.8 A master curve of reduced viscosity versus reduced shear rate for solutions of poly(styrene) in n-butylbenzene at various concentrations, molecular weights, and temperatures. Reprinted with permission from W. W. Graessley, R. L. Hazelton, and L. R. Lindeman, *Trans. Soc. Rheol.*, **11**, 267 (1967).

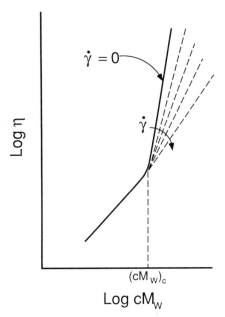

Figure 5.9   Log viscosity versus log(cM): A qualitative illustration of the effect of shear rate.

2.5   Heterogeneous Liquids

Many technologically and biologically important liquids are heterogeneous; that is, they contain two or more distinct components or "phases." Examples of such liquids are blood, milk, paint, ink, photographic "emulsions" (this is a misnomer!; see below), and molten polymer composites. The major liquid

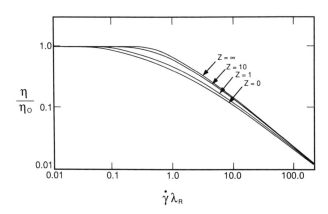

Figure 5.10   Effect of polydispersity on reduced flow curve. Calculated from Graessley's theory for polymers with a Schultz–Zimm distribution. Reprinted with permission from W. W. Graessley, *J. Chem. Phys.*, **47**, 1942 (1967).

component in such a system is commonly referred to as the *continuous* phase and the minor component(s) is called the *discrete* or *dispersed* phase. Heterogeneous liquids (dispersions) can be generally categorized in terms of the physical state of the discrete phase. If the discrete phase consists of solid particles, then the liquid is a *suspension*. If it is an immiscible liquid dispersed in the continuous phase, then the heterogeneous liquid is an *emulsion*; and when the discrete phase consists of gas bubbles, the liquid is a *foam*. The viscous response of a dispersion is expected to depend on a large number of factors including the state of the discrete phase, the shape and size of the dispersed particles, the concentration of particles, the rheological properties of the continuous and discrete phases, and the type and extent of interparticle interactions. The complex interplay among these factors makes the analysis of dispersion hydrodynamics a nontrivial and computationally intensive endeavor (see, e.g., [80, 81]), but with reasonable assumptions and approximations one can develop relatively simple working relations for the viscosity of dispersions.

The classical study of Einstein [82] on the viscosity of a dilute suspension of spheres constitutes the first attempt to derive an expression for the viscosity of a dispersion. From evaluation of the excess energy dissipation in shear flow due to the addition of a spherical particle to a Newtonian liquid, Einstein derived the following expression for the viscosity of the suspension

$$\eta = \eta_s(1 + \tfrac{5}{2}\phi) \tag{42}$$

where η_s is the viscosity of the continuous phase and ϕ is the volume fraction of spheres in the liquid. Although this simple result is valid only for very dilute suspensions of uniform spheres ($\phi < 0.01$), it serves as a fundamental formula in dispersion rheology, which can be extended, empirically or theoretically, to include concentration, inertia, particle deformability and nonsphericity, and other effects not considered in the original Einstein treatment. For more concentrated suspensions the empirical Thomas equation [83] is commonly used

$$\eta_r \equiv \frac{\eta}{\eta_s} = 1 + 2.5\phi + 10.05\phi^2 + 2.73 \cdot 10^{-3} \exp(16.6\phi) \tag{43}$$

which is typically valid up to a concentration of $\phi = 0.55$. Another, somewhat simpler but very successful formula is the equation of Kitano and co-workers [84, 85], based on an earlier suggestion by Maron and Pierce [86], which is valid for suspensions of particles of various shapes and sizes

$$\eta_r = \left[1 - \frac{\phi}{\phi_{max}} \right]^{-2}. \tag{44}$$

ϕ_{max} is a parameter expressing the maximum packing concentration, which strongly depends on the size and shape of particles in the discrete phase.

Kitano and co-workers [85] find that $\phi_{max} \cong 0.68$ is a reasonable value for suspensions of smooth spheres while for fibers ϕ_{max} drops with increase in the aspect ratio of the fiber ($\phi_{max} = 0.18$ for aspect ratio of 27). Equation (44) can be generalized to the form [87]

$$\eta_r = [1 - k\phi]^{-a} \tag{45}$$

where k and a are adjustable parameters. Special cases of this formula have been suggested in the literature for various types of suspensions [88–90].

Extension of Einstein's equation to dilute suspensions of spheroids was carried out by Kuhn and Kuhn [91] and independently by Simha [92]. For prolate spheroids, for example, Kuhn and Kuhn derive the following expression for the intrinsic viscosity [see (95)] of the dispersion

$$[\eta] \equiv \lim_{\phi \to 0} \frac{\eta - \eta_s}{\eta_s \phi} = \begin{cases} 2.5 + 0.4075(r-1)^{1.508} & 1 < r < 15 \\ 1.6 + \dfrac{r^2}{5}\left[\dfrac{1}{3(\ln 2r - 1.5)} + \dfrac{1}{\ln 2r - 0.5}\right] & r > 15 \end{cases} \tag{46}$$

where r is the axis ratio (= major axis/minor axis) of the spheroid. Einstein's result ($[\eta] = 2.5$) is recovered when $r = 1$ (a sphere). The rheological response of dispersions containing spheroidal and other nonspherical particles is quite complex because of the flow-induced orientation of the particles in the flow field, which may lead to intrinsic shear thinning and elastic effects [93]. Inertia forces can also lead to deviations from Einstein's result. Lin and co-workers [94] have shown that for a dilute suspension of spheres

$$\eta_r = 1 + (2.5 + 1.34\,\mathrm{Re}_p^{3/2})\phi \tag{47}$$

where Re_p, the particle Reynolds number, is given by

$$\mathrm{Re}_p = \frac{\rho_s r_p U}{\eta_s} \tag{48}$$

r_p is the radius of the sphere, U is the average speed of the liquid, and ρ_s is the density of the continuous phase. Thus, the Einstein relation will underpredict the viscosity when inertia forces are relatively high.

A qualitative difference between suspensions and emulsions is that the particles in an emulsion are deformable; that is the shape of the particles is intimately related to the global flow field. Since the particle in an emulsion has a finite viscosity η_p, it is reasonable to expect the viscosity of the dispersion to depend on the viscosity ratio $\eta_{pr} = \eta_p/\eta_s$. This has first been shown in the classical study of Taylor [95] and later by Frankel and Acrivos [96] who derive a simple expression for the viscosity of a dilute emulsion of spheres

$$\eta_r = 1 + \left(\frac{1 + 2.5\eta_{pr}}{1 + \eta_{pr}} \right) \phi \qquad (49)$$

The Einstein limit is recovered when $\eta_{pr} \to \infty$ (corresponding to the rigid sphere case). At the other extreme, when $\eta_{pr} \to 0$ (a foam), the intrinsic viscosity drops from 2.5 to 1.0; but, paradoxically, the viscosity of the foam is still higher than the viscosity of the pure solvent. Extension of (49) to the nondilute case was carried out by Choi and Schowalter [97] who show that

$$\eta_r = 1 + \phi \left(\frac{1 + 2.5\eta_{pr}}{1 + \eta_{pr}} \right) \left[1 + 2.5\phi \left(\frac{1 + 2.5\eta_{pr}}{1 + \eta_{pr}} \right) + O(\phi^{5/3}) \right] \qquad (50)$$

It was also found that a dispersion of deformable spheres exhibits normal stresses and other viscoelastic effects that depend on the value of η_{pr} and the interfacial tension between the dispersed and continuous liquids.

We finally note that viscometric testing of heterogeneous liquids is fraught with difficulties arising mainly from the tendency of the dispersed particles to settle, migrate, coalesce, flocculate, or deflocculate during a standard shearing test. One or any combination of these phenomena may give rise to various anomalous effects including thixotropy, shear-thickening [98], and "plastic" (yield stress) response, as noted in 2.3, which must be carefully evaluated and checked for reproducibility. It is especially important to ascertain that the measured viscosity (or any other rheological property) is not dependent on the viscometer configuration and dimensions which is usually indicative of some experimental artifact. For a more complete discussion of the rheology of heterogeneous liquids the reader should consult [99–103].

3 POISEUILLE-TYPE VISCOMETERS

3.1 Poiseuille Flows

Poiseuille flows, named after the French physicist J. L. Poiseuille, are simple shear flows [cf. (4)] induced by a pressure gradient acting along the flow trajectory of the liquid. Two examples of such flows, to be discussed below, are flow through a tube and flow in a slit die (see Figure 5.11 and Table 5.2). Other Poiseuille flows, for example, pressure-driven flow in the annular space between two concentric cylinders and radial flow between parallel disks, can also be used in viscometric tests but are less popular than tube and slit flows and are not considered further in this chapter. Since Poiseuille flows are neither homogeneous nor controllable (see Table 5.2), they may not seem to be well suited for viscometric tests. Yet, because of their operational and geometric simplicity,

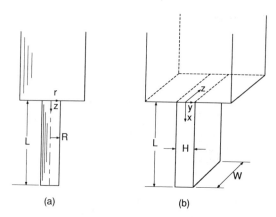

Figure 5.11   Flow geometries for Poiseuille-type viscometers: (a) capillary viscometer and (b) slit-die viscometer.

these flows have played, historically, a major role in the viscometric characterization of liquids and are used to this day in a wide variety of viscometric
systems.

In 1840 Poiseuille published the results of studies on the flow of liquids
through fine bore capillary tubes [104]. His data were shown to follow the
empirical equation

$$Q = CPR^4 \tag{51}$$

where Q is the total volume of liquid discharged per unit time, P is the pressure
driving the liquid through the capillary, R is the bore radius, and C is some
empirical constant that was shown to depend on the test liquid and the
temperature. A similar result was found by one of Poiseuille's contemporaries,
G. Hagen. Using Newton's law of viscosity [see (14)] Hagen [105] arrived at a
more explicit expression for the flow rate

$$Q = \frac{\pi R^4 \Delta P}{8 \mu L} \tag{52}$$

where ΔP is the pressure drop along the capillary, L is the capillary length, and μ
is the viscosity. This result, known as the Hagen–Poiseuille (HP) law, forms the
basis for most capillary viscometric techniques. If gravity forces are relatively
large, the pressure drop in (52) must be corrected by

$$\Delta \bar{P} = \Delta P + \rho g_1 h \tag{53}$$

where g_1 is the component of gravitational acceleration operating in the
direction of flow, and h is the liquid head at the entry point to the capillary. The

HP law can be derived from the Navier–Stokes equations [see (17)] subject to the following assumptions and conditions: (1) The flow is laminar and inertia-less; that is, $Re \rightarrow 0$; (2) the flow is isothermal; (3) the fluid is incompressible ($\rho = \text{const}$); (4) the flow is fully developed; that is, the capillary is sufficiently long that effects of velocity profile rearrangement at the entrance and exit regions are negligible; (5) the fluid is Newtonian [$\mu \neq f(Q)$]; (6) there is no slip of liquid layers at the tube wall; $u(R) = 0$; and (7) the flow is steady; that is, the pressure and flow rate are not time-dependent. Naturally, any deviation from these restrictions will lead to errors and inconsistencies in the determination of μ based on (52). A close examination of these assumptions and discussion of various sources of error are given in Section 3.4. Solution of the Navier–Stokes and continuity equations leads to several important results in addition to the HP law. From inspection of the dynamic equations it can be shown that the only operative stress component in this flow, τ_{12}, is a linear function of radial position (see Figure 5.11a and Table 5.2)

$$\tau_{12} = \tau_w \left(\frac{r}{R} \right) \tag{54}$$

where

$$\tau_w = \frac{\Delta P R}{2L} \tag{55}$$

is the shear stress at the tube wall ($r = R$). Equation (55) can also be derived from a simple force balance on the capillary boundaries without invoking the dynamic equations. The velocity profile for this case can be represented by a parabola with a maximum at the centerline

$$u_z = 2U \left[1 - \left(\frac{r}{R} \right)^2 \right] \tag{56}$$

where

$$U \equiv \frac{Q}{\pi R^2} \tag{57}$$

is the average velocity in the capillary. Since Poiseuille flows are not con-trollable, the HP law and the related results are limited to Newtonian liquids and are of no use if the rheological character of the test liquid is not known a priori. To derive expressions for the more general case of a purely viscous, non-Newtonian liquid it is necessary to solve the complete dynamic equations for a specific viscous model subject to the conditions and assumptions listed above (except for the constant viscosity assumption). Inspection of the dynamic equations reveals that (54) and (55) are still valid regardless of the constitutive model and are, in fact, general model-invariant results for capillary flow. For the special case of a power-law liquid [see (23)] it is easy to show that the velocity profile becomes nonparabolic

$$u_z = \left(\frac{3n+1}{n+1}\right) U \left[1 - \left(\frac{r}{R}\right)^{(n+1)/n}\right] \tag{58}$$

where U is still given by (57). Thus, as $n \to 0$ the profile flattens and departs increasingly from a parabola. From a simple statement of mass conservation one can derive an analogous expression to the HP law for the power-law liquid

$$Q = \frac{n\pi R^3}{3n+1} \left(\frac{\Delta P R}{2KL}\right)^{1/n} \tag{59}$$

It follows from (58) that the operative (nominal) shear rate at the capillary wall is

$$\dot\gamma_w = -\left.\frac{\partial u_z}{\partial r}\right|_R = \frac{4Q}{\pi R^3}\left(\frac{3n+1}{4n}\right) \tag{60}$$

and the corresponding apparent viscosity is given by

$$\eta = \frac{\tau_w}{Q/\pi R^3}\left(\frac{n}{3n+1}\right) \tag{61}$$

Thus, the shear rate and the shear stress are maximum at the capillary wall ($r = R$) while the viscosity itself is minimum at this position if the liquid is shear thinning.

For the Bingham liquid [see (26)], as well as for other liquids exhibiting yield stress, flow is possible only if $\tau_w > \tau_y$, where τ_w is given by (55). If this condition is met, then the flow rate-pressure relation is given by the Buckingham–Reiner equation [106]

$$Q = \frac{\pi \Delta P R^4}{8\mu_0 L}\left[1 - \frac{4}{3}\frac{\tau_y}{\tau_w} + \frac{1}{3}\left(\frac{\tau_y}{\tau_w}\right)^4\right] \tag{62}$$

Thus, the discharge rate of a Bingham fluid from a capillary is generally smaller than that for a corresponding Newtonian liquid ($\tau_y = 0$). The velocity profile for this liquid consists of two regimes: (1) a plug flow regime for $r < r_0\ (= 2\tau_y L/\Delta P)$, where

$$u_z = \frac{\Delta P R^2}{4\mu_0 L}\left(1 - \frac{r_0}{R}\right)^2 \tag{63}$$

and (2) a shearing flow regime for $r > r_0$, where

$$u_z = \frac{\Delta P R^2}{4\mu_0 L}\left[1 - \left(\frac{r}{R}\right)^2\right] - \frac{\tau_y R}{\mu_0}\left[1 - \left(\frac{r}{R}\right)\right] \tag{64}$$

and

$$\dot{\gamma}_w = \left(\frac{4Q}{\pi R^3}\right)\frac{\left(1 - \dfrac{r_0}{R}\right)}{\left[1 - \dfrac{4}{3}\left(\dfrac{r_0}{R}\right) + \dfrac{1}{3}\left(\dfrac{r_0}{R}\right)^4\right]} = \frac{\tau_w - \tau_y}{\mu_0} \tag{65}$$

Since the expression for $\dot{\gamma}_w$, and hence for $\eta(\dot{\gamma}_w)$, is model specific, it may seem that some prior knowledge of the rheological character of the test liquid is needed to select the proper equation for analyzing viscometric data obtained with a capillary device. This, however, may not be necessary. Based on a simple mass balance combined with (54), it can be shown that

$$\dot{\gamma}_w = \frac{Q}{\pi R^3}\left[3 + \frac{d\,\ln(Q/\pi R^3)}{d\,\ln \tau_w}\right] \tag{66}$$

for any arbitrary viscosity model. This expression known as the *Weissenberg–Rabinowitsch* (WR) equation is based on an early derivation by Rabinowitsch [107], and is a general model-invariant result for capillary flow. The departure from unity of the derivative on the right-hand side of (66) expresses the departure of the fluid from Newtonian behavior. Once $\dot{\gamma}_w$ and τ_w are known, the viscosity is obtained from

$$\eta(\dot{\gamma}_w) = \frac{\tau_w}{\dot{\gamma}_w} \tag{67}$$

which defines the *apparent* viscosity of the liquid at $\dot{\gamma}_w$. Thus, to evaluate the viscosity of a liquid with an unknown rheological character it is necessary first to determine the slope S of a $\ln Q - \ln \Delta P$ curve for any $Q - \Delta P$ pair. Then, with S known, $\dot{\gamma}_w$ and η are readily obtained from (66) and (67). The slope S is constant for a Newtonian $(=1)$ or a power-law $(=1/n)$ liquid, but it varies with Q for liquids in the transition regime (see Figure 5.3).

Another Poiseuille flow system commonly used in viscometric studies is the slit die (see Figure 5.11b and Table 5.2). The analysis of flow in slits parallels that for capillary tubes. In deriving the relevant equations for this geometry we use the same conditions and assumptions as in the previous case, in addition to assuming that the aspect ratio of the flow channel, W/H, is relatively large. This latter assumption implies that the flow in the slit is effectively one dimensional (rectilinear) with negligible effect of the side walls on the flow dynamics. Here, as for tube flow, the shear stress is a linear function of the gapwise position y

$$\tau_{12} = \tau_w \frac{2y}{H} \tag{68}$$

where

$$\tau_w = \frac{\Delta P H}{2L} \tag{69}$$

is the nominal (wall) shear stress. For the case of a power law fluid the velocity profile in the slit, at positions sufficiently removed from the side walls, takes the form

$$u_x = \left(\frac{1+2n}{1+n}\right) U \left[1 - \left|\frac{2y}{H}\right|^{1 + 1/n}\right] \tag{70}$$

where

$$U \equiv \frac{Q}{WH} \tag{71}$$

As for tube flow, the profile is parabolic in the Newtonian case, $n = 1$, and it becomes progressively flatter as $n \to 0$. The corresponding equations for the flow rate and the nominal shear rate are

$$Q = \frac{nWH^2}{2(1+2n)} \left(\frac{\tau_w}{K}\right)^{1/n} \tag{72}$$

and

$$\dot{\gamma}_w = -\left.\frac{\partial u_x}{\partial y}\right|_{H/2} = \frac{6U}{H}\left(\frac{2n+1}{3n}\right) \tag{73}$$

An expression similar to the Buckingham–Reiner equation (62) can be derived for slit flow for the Bingham model (26) with similar considerations regarding the general shape of the velocity profiles [108].

If the Weissenberg–Rabinowitsch scheme is extended to the slit die geometry, an equation similar in form to the WR equation (66) is obtained for the nominal shear rate

$$\dot{\gamma}_w = \frac{2Q}{WH^2}\left[2 + \frac{d\ln(6Q/WH^2)}{d\ln \tau_w}\right] \tag{74}$$

As before, since this expression is not restricted to a particular viscosity model, it can be used to reduce viscometric data for fluids with an unknown rheological character and construct a complete flow curve.

3.2   Capillary Viscometers

Many types of capillary viscometers have been used over the years to study the rheological properties of liquids and gases. Details of the design and operation of many of these devices can be found in the books by van Wazer and co-workers [1], Walters [3], and Dealy [109] and in reviews by McKie and Brandts [110] and Carpenter and Westerman [68]. The following discussion is thus confined to only the major operational and design considerations in state-of-the-art capillary viscometers and only selected examples of commercial systems are given.

Based on (55) and (66) a viscometric run with a capillary device should entail imposition of either a constant pressure gradient and measurement of the discharge rate Q or vice versa. These two options define, in effect, two broad classes of capillary viscometers: (1) stress-controlled and (2) rate-controlled devices. In the former the shear stress (pressure) is imposed either pneumatically, for example, by compressed air or nitrogen, or by a hydrostatic head, and the resultant discharge rate is measured. In the second category, a driving piston moving at a controlled speed (flow rate) is forcing the liquid from a large reservoir through a known capillary bore into the atmosphere. The force needed to maintain a steady motion of the piston is measured by a pressure-sensing device, typically a load cell or a pressure transducer. The stress-controlled viscometers are used primarily for low viscosity liquids that require low driving pressures. In the case of dilute polymer solutions or liquids having viscosities of the order of $1\,\mathrm{mPa\cdot s}$, gravity drainage viscometers are most commonly used. The rate-controlled viscometers are typically used for characterizing viscous liquids (e.g., concentrated polymer solutions and melts) or gases.

The most common gravity drainage capillary viscometers are made of glass. Some typical glass viscometers are shown in Figure 5.12. The design and operation of such viscometers are based on a modified version of the HP law (52): If the pressure forcing the liquid through the capillary is a constant hydrostatic head $\rho g \Delta h$, where Δh is the height difference between the liquid surfaces at the entry and exit points, then (52) takes the form

$$\mu = \frac{\pi R^4 \rho g \Delta h\, t}{8LV} = At \tag{75}$$

and the nominal shear stress becomes

$$\tau_w = \frac{\rho g \Delta h R}{2L} \tag{76}$$

where V is the total volume discharged at time t. Thus, if the discharge volume and height difference are fixed, the factor A can be considered a viscometer constant and the determination of viscosity requires only measurement of the flow time t between two marked positions. The Ostwald U-tube system (see Figure 5.12a) is the simplest of this class of viscometers. This viscometer is operated by filling a known volume of liquid into the wide arm and drawing it with some suction device to the upper fiducial mark above the bottom bulb. The liquid is then allowed to drain through the capillary and the time for the upper surface to reach the bottom mark is recorded and used to determine the viscosity. The viscometer constant is obtained by measuring the flow time of a known reference liquid. Two major drawbacks of this design are (1) The liquid volume should be measured precisely to obtain a constant head at every run, and (2) the viscometer should be positioned vertically to ensure that the operative gravity force is consistent from run to run. If the viscometer is tilted by an angle α from the vertical, then the effective gravitational acceleration in the direction of flow is

$$g_1 = g \cos \alpha \tag{77}$$

leading to an error in the calculation of viscosity. The problem of vertical mounting is somewhat mitigated in the design of the Canon–Fenske viscometer (Figure 5.12b). By bending both arms in the viscometer, slight deviation from vertical mounting has a lesser effect on the effective gravity force. The Ubbelhode viscometer (Figure 5.12c) is designed such that a precise volume of

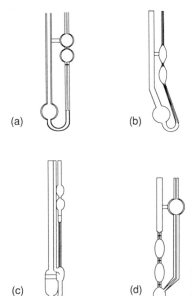

(a)

(b)

(c)

(d)

Figure 5.12   Typical gravity-drainage glass viscometers: (a) Ostwald U-tube viscometer, (b) Cannon–Fenske viscometer, (c) Ubbelhode (suspended arm) viscometer, and (d) variable shear stress viscometer.

liquid is not necessary to maintain a constant head because the effective lower liquid surface is always open to the atmosphere at the junction to the middle arm. Thus, regardless of the amount of liquid in the large reservoir arm, the head always corresponds to the height difference between the top surface and the junction point of the middle arm. This arrangement, often referred to as the *suspended level system*, is very convenient for running intrinsic viscosity measurements (see Section 3.5) since one can prepare solutions at different concentrations by successive dilutions directly in the viscometer. If proper care is taken in maintaining constant temperature, for example, by a tightly controlled thermostated bath, and the various sources of error in this system are properly accounted for (see Section 3.4), the Ubbelhode viscometer can be used to obtain highly precise values of viscosity [111]. Precise timing of flow in glass viscometers can be automatically performed by means of an electrooptical sensor.

Since the nominal shear stress is always constant in most gravity drainage viscometers [see (76)], it is clear that non-Newtonian effects cannot be evaluated independently and may introduce large errors if (75) is used. A modified version of the Ubbelhode design with a variable head (or shear stress), Figure 5.12*d*, allows one to assess the shear sensitivity of the liquid by repeating the measurement for several heads. Generally, when $\mu = \mu(\Delta h)$, the liquid may be considered non-Newtonian. Another option is to apply a variable positive pressure to the capillary arm of the viscometer, superposed on the hydrostatic head, and thereby vary the shear stress continuously over a convenient range. Some guidelines for selecting a viscometer constant in gravity drainage systems are given by McKie and Brandts [110].

Generally, because of the limited range of shear stress (or shear rate) in glass viscometers these devices are not well suited for measuring flow curves [$\eta = \eta(\dot{\gamma})$] for non-Newtonian liquids. The shear rate-controlled viscometers are much more versatile in this regard. Two such systems are shown in Figures 5.13 and 5.14. Both viscometers are used primarily for measuring viscosity of polymer melts and highly viscous liquids and they can operate typically at shear rates of 10–20,000 s^{-1}. The principal difference between these systems is in the way the pressure is measured. In the Mertz–Colwell Instron Capillary rheometer (MCR) the pressure is recorded by a compression load cell that is attached directly to the driving piston, whereas in the Göttfert Rheograph 2001 rheometer the pressure is measured by a transducer positioned just upstream from the capillary entrance. The latter arrangement ensures that pressure losses in the barrel and errors due to the compressibility of the molten material do not bias the effective pressure drop in the capillary. In both cases a typical run consists of driving the piston through the barrel at a certain range of speeds (flow rates) and measuring the corresponding loads (pressures). These values are then used to establish a flow curve following the procedure outlined in Section 3.1. Although most conventional capillary rheometers are limited to shear rates of < 20,000 s^{-1}, these devices can be modified, through special design of the drive and pressure-sensing units, to attain extremely high shear rates. One such

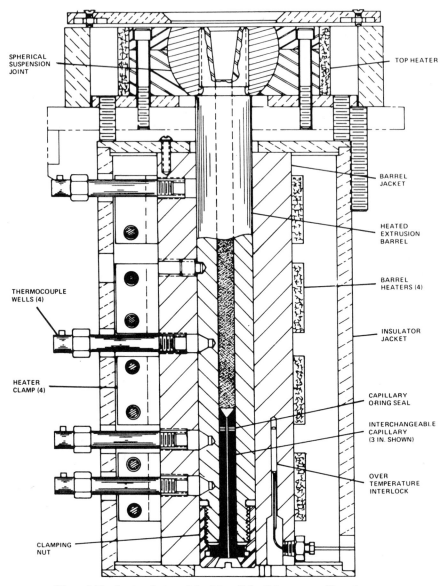

SPHERICAL
SUSPENSION
JOINT

TOP HEATER

BARREL
JACKET

HEATED
EXTRUSION
BARREL

BARREL
HEATERS (4)

THERMOCOUPLE
WELLS (4)

INSULATOR
JACKET

HEATER
CLAMP (4)

CAPILLARY
ORING SEAL

INTERCHANGEABLE
CAPILLARY
(3 IN. SHOWN)

OVER
TEMPERATURE
INTERLOCK

CLAMPING
NUT

Figure 5.13 Sectional view of the Merz Colwell Instron capillary rheometer.

system, described by Takahashi and co-workers [112], was reported to approach shear rates of $10^7 \, \text{s}^{-1}$ by using an ultrafast hydraulically driven injection unit.

A simplified capillary viscometer commonly used to rank the viscous response of polymer melts is the *melt indexer* [109]. This device is an indexer rather than a viscometer insofar as it does not measure the viscosity per se but

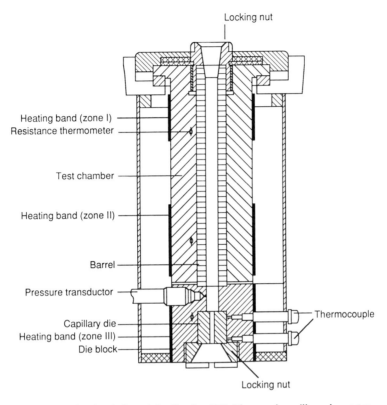

Locking nut

Heating band (zone I)
Resistance thermometer

Test chamber

Heating band (zone II)

Barrel

Pressure transductor

Capillary die
Heating band (zone III)
Die block

Thermocouple

Locking nut

Figure 5.14 Sectional view of the Gottfert 2001 Rheograph capillary rheometer.

rather a quantity—the melt (flow) index (MFI), which is closely related to the viscosity of the melt. The principle of operation of the melt indexer is similar to that of a controlled-stress capillary viscometer. The polymer is heated to a standard temperature in a round barrel and then forced through a short capillary tube at the bottom of the barrel by a standard weight placed on a round plunger in the barrel. The MFI is defined as the weight of polymer (in grams) issuing from the capillary over a 10 min interval. Since the conditions of the MFI test (temperature, capillary dimensions, and driving weight), as prescribed by ASTM Standard Test Method D1238, vary for different polymer classes, the MFI is useful only for ranking polymer melts within a particular class. Indeed, the limitations of the melt indexer as a rheological characterization tool are quite obvious, yet it is an inexpensive, simple, and robust instrument with potential utility in quality assurance and process control applications. Shenoy and Saini [112a] have proposed a general scheme for converting MFI data to complete flow curves based on the assumption that the viscosity functions of polymers from the same generic class have a similar shape (cf. (25)).

3.3   Slit-Die Viscometers

The major design and operational considerations for slit-die viscometers are similar to those for capillary viscometers with the exception that flow in slit dies is never truly one-dimensional. For the flow to approach one-dimensional (plane) Poiseuille flow closely it is necessary to employ slits with large aspect (width-to-thickness) ratios. Typically, aspect ratios of 10–20 are considered sufficient for achieving an effective one-dimensionality [113]. Despite the historical role and widespread use of capillary devices, slit-die viscometers have some unique and attractive features and they offer several tangible advantages over comparable capillary systems: (1) Slit-die systems provide considerable geometric flexibility. By allowing a variable channel depth, which can be readily incorporated into the rheometer design, one can cover an extremely wide range of shear rates with a single geometrical unit. (2) Pressures can be measured directly along the flow path in the die by means of flush-mounted transducers; thus, the problems associated with end effects can be circumvented (see Section 3.4.1). (3) Pressure measurements in the slit can be used to extract information on the elastic as well as on the viscous properties of the test liquid (see below). Because of these attributes, slit-die rheometers have received much attention in recent years, particularly with the introduction of designs by Lodge (the Stressmeter) [114–116], Han (the Seiscor/Han rheometer) [117], and Laun [118].

Developed by A. S. Lodge, the Stressmeter is a novel device that generates simultaneously data on the shear viscosity and the first normal stress difference N_1. The basic unit consists of a slit die with three pressure ports as shown in Figure 5.15. One transducer is flush-mounted on one surface; a second is mounted in a slot (hole) directly opposite the first; and the third is mounted in an identical slot some distance downstream from the second. From the pressure difference $(P_2 - P_3)$ the wall shear stress (or η) can be calculated [see (69)], while from the pressure difference $(P_1 - P_2)$ the first normal stress difference N_1 [cf. (15)] can be estimated based on the hole-pressure method [119]. The hole-pressure effect and the procedure for estimating normal stresses are discussed briefly in Section 3.4.5. Because of the typically small values of the pressure difference $(P_1 - P_2)$, the transducers in the Stressmeter must be sufficiently sensitive to detect small pressure variations. Also, by properly adjusting the slit height, this system can attain shear rates as high as $10^6 \, \mathrm{s}^{-1}$—an important feature when evaluating the rheological response of lubricating oils, various high-speed coating liquids, and other fluids that are processed at high rates of deformation. The possibility of viscous heating—a common impediment in high shear viscometry (Section 3.4.7)—can be evaluated by measuring simultaneously the temperatures in the inlet and outlet streams. Various designs of the Stressmeter were used for testing low viscosity liquids as well as highly viscous polymer melts [120].

The Seiscor/Han rheometer (SHR) is a similar device designed primarily for studying molten polymers. The melt is forced through the slit by an extruder or

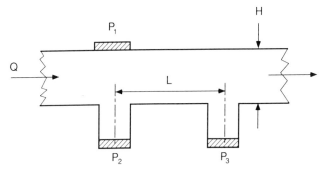

Figure 5.15 Schematic diagram of the central section of the die in Lodge's Stressmeter rheometer; P_1 is a flush-mounted pressure transducer and P_2 and P_3 are hole-mounted pressure transducers.

other melt-metering pump capable of delivering melt streams at controlled rates. The wall shear stress and the viscosity are measured exactly as in the Stressmeter, but here evaluation of the normal stress difference is based on the exit pressure method proposed by Han [121]. According to this method the exit pressure (pressure at the exit point) should be finite for a viscoelastic liquid and its magnitude should be related directly to N_1, the first normal stress difference. As of this writing the theoretical validity of Han's method awaits final resolution. Some studies are, in fact, critical of the exit pressure technique as a quantitative method for the determination of N_1 [122].

Both the Mertz–Colwell Instron Capillary rheometer and the Göttfert Rheograph system, discussed in the previous section (Figures 5.13 and 5.14), can be converted into slit-die rheometers by replacing the capillary dies with compatible slit dies. In the former, the pressure is measured by a compression load cell attached to the driving piston, as with the capillary dies, while in the latter provisions exist for attaching several transducers along the slit. We finally note that for identical flow rates, the nominal shear rates in capillaries are higher than in slits with $H = D$. Based on (60) and (73), the ratio of the nominal shear rates for these devices for a power law liquid is

$$\frac{\dot{\gamma}_{w,cap}}{\dot{\gamma}_{w,slit}} \approx \frac{W}{R} \tag{78}$$

Since $W \gg H \ (= 2R)$, the shear rates in the capillary will be greater than in an equivalent slit die. Likewise, for the same length and flow rate, the pressures generated in a capillary die will be higher than in an equivalent slit die

$$\frac{\Delta P_{cap}}{\Delta P_{slit}} \approx \left(\frac{W}{R}\right)^n \tag{79}$$

Thus, the lower apparent shear rates in the slit [cf. (78)] will correspond to lower driving pressures.

3.4   Sources of Error

The analysis of viscometric data obtained with slit or capillary devices may be subject to systematic error if one or more of the assumptions used to derive the working equations for these systems is violated. It must be emphasized that these are "errors" only in the sense that the simplified equations used to describe the flow in the corresponding viscometers are invalid. Otherwise, the data may reflect a perfectly legitimate physical situation that is not captured by the simplified analysis. In the next sections we discuss briefly the main sources of error in Poiseuille-type viscometers and how to remedy them.

3.4.1   End (Couette) Effects

One of the assumptions used in deriving the flow equations for capillaries and slits is that the flow is fully developed; that is, the velocity profile is fully established throughout the entire length of the device and the measured pressure drop represents only this portion of the flow system. In practice, the measured pressure contains contributions from flow in the reservoir just upstream from the device, frictional losses (in the case of plunger-driven systems) and losses due to velocity profile rearrangement at the entrance and exit regions. In many devices the fluid undergoes an abrupt kinematic change as it enters into the controlled flow space from a large reservoir (see Figure 5.11), and losses at the entrance region (Couette effects) are often significant. The sudden contraction at the entry point may require a distance of several diameters (or channel heights), depending on the Reynolds number and the rheological nature of the liquid, before a fully developed flow is established. Rigorous evaluation of this distance—the entrance length L_e—may be quite involved, but it can be estimated fairly accurately for Newtonian liquids from the following correlations [123]

$$\frac{L_e}{D} = 0.59 + 0.056 \, \text{Re} \tag{80}$$

for capillaries, and

$$\frac{L_e}{H} = 0.63 + 0.044 \, \text{Re} \tag{81}$$

for slit dies. For highly elastic liquids L_e is expected to be considerably higher than that given by (80) and (81) [123, 124]. The nature of the velocity profile at the exit region is less clear, although it has been speculated that some distance upstream from the end (L_{ex}) it will deviate from its fully developed form mainly

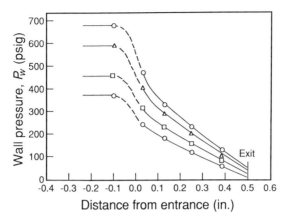

Figure 5.16 Pressure profiles for flow of high-density poly(ethylene) melt from a reservoir into a capillary of $L=0.5$ in. and $D=0.125$ in. Nominal shear rates are (from top): 790, 616, 313, and $160\,\mathrm{s^{-1}}$. Reprinted with permission from C. D. Han, M. Charles, and W. Philippoff, *Trans. Soc. Rheol.*, **14**, 393 (1970).

due to elastic phenomena [121, 122]. End effects are typically manifested by nonlinearities in the pressure profile near the end regions, as illustrated in Figure 5.16 for a poly(ethylene) melt flowing into a capillary under various nominal shear rates; only in the section with a linear pressure gradient a fully developed flow may be assumed. To correct for end effects that are inherently present in any Poiseuille-type system it is not necessary to have a detailed knowledge of the hydrodynamics at the entrance and exit regions. According to one correction scheme, proposed by Bagley [125], the tube is artificially extended by a section of length $L_B = eR$, which represents the excess pressure dissipation corresponding to the end regions. Thus, the effective wall shear stress for the extended tube is

$$\tau_w = \frac{\Delta P\,R}{2(L + L_B)} \tag{82}$$

Since this quantity must be constant for a fixed $Q/\pi R^3$ (an apparent shear rate), plotting L/R versus ΔP should yield a straight line with an intercept of $-e$ and a slope of $1/2\tau_w$. This procedure, generally referred to as the Bagley method, involves repeating measurements of ΔP for flow through capillaries with various values of L/R and extracting e from a plot of L/R versus ΔP for a given $Q/\pi R^3$. In the case of a slit die L_B is extracted from the intercept of an equivalent "Bagley plot" in which L/H is plotted versus ΔP for a constant Q/WH^2. An alternative scheme that is somewhat easier to execute experimentally involves separate measurements with two capillaries having identical radii but different lengths (L_S and L_L) [126]. If both L_S and L_L are larger than $L_e + L_{ex}$, then the pressure gradient over the fully developed flow section is given by $(\Delta P_L - \Delta P_S)/(L_L - L_S)$, where ΔP_L and ΔP_S are the pressure drops for the two capillaries measured at a fixed value of Q. We can thus write

$$\tau_w(Q) = \frac{\Delta P_L - \Delta P_S}{L_L - L_S}\left(\frac{R}{2}\right) \tag{83}$$

and equivalently for a slit die

$$\tau_w(Q) = \frac{\Delta P_L - \Delta P_S}{L_L - L_S}\left(\frac{H}{2}\right) \tag{84}$$

Here both H and W must be identical for slits L and S. Calculation of the nominal shear rate for the fully developed flow section in capillaries or slits follows the regular procedure outlined in Section 3.1. It should be noted that the Bagley method and the two-capillary method are fundamentally identical.

3.4.2   Kinetic Energy

For relatively thin liquids ($\mu \lesssim 1 \, \text{mPa} \cdot \text{s}$) the driving pressure is expended not only on overcoming the viscous resistance of the liquid in the capillary, but also on imparting a finite kinetic energy to the liquid jet issuing from the capillary. Thus, to use (52) or its non-Newtonian analogues, we must subtract the kinetic energy contribution from the applied pressure. A general expression for the effective (corrected) pressure is given by [127]

$$\Delta P_{eff} = \Delta P - \chi\left(\frac{\rho Q^2}{\pi^2 R^4}\right) \tag{85}$$

where the constant χ depends on the rheology of the liquid and the precise geometry of the capillary ends. For a power law fluid and an abrupt expansion at the end of the capillary [128]

$$\chi = [3(3n+1)^2]/[(4n+2)(5n+3)] \tag{86}$$

For a Newtonian liquid, $\chi = 1$ for an abrupt expansion and $\chi = 0.32$ (for $Re < 50$) for a trumpet-shaped capillary end [68, 127]. In any case, inspection of (85) reveals that the kinetic energy contribution would be significant only when the dimensionless number $\rho Q/8\pi L\mu$ is large.

3.4.3   Surface Tension

Surface tension may obscure viscometric measurements especially in gravity drainage viscometers when low-viscosity liquids are tested. If the interfacial forces at the upper and lower meniscii of the liquid column are not balanced, the overall hydrostatic head will contain a contribution from the net surface tension force. The rise of the liquid due to this force is [110]

$$\Delta h_\sigma = \frac{2\sigma}{\rho g}\left(\frac{\cos\theta_t}{r_t} - \frac{\cos\theta_b}{r_b}\right) \tag{87}$$

where σ is the surface tension, r_t and r_b are the radii of curvature, and θ_t and θ_b are the contact angles at the top and bottom meniscii. Strictly speaking, (87) should include dynamic, rather than static, contact angles since the liquid column is in motion during the test, but this effect is likely to be small under typical testing conditions since flow speeds in gravity-drainage viscometers are usually quite low. For aqueous liquids θ is nearly zero when a glass capillary is used and (87) is considerably simplified. Generally, the value of Δh_σ should be subtracted from the nominal head Δh to obtain the operative head that effectively drives the liquid through the capillary. Clearly, if $(\cos\theta_t)/r_t = (\cos\theta_b)/r_b$, the surface tension effect cancels out regardless of the magnitude of σ. In the Ubbelhode viscometer, for example (see Figure 5.12), the suspended level arm is designed such that surface tension effects are minimized. This is achieved by proper dimensioning of the juncture of the arm to the main column. Generally, surface tension effects are negligible when $\sigma R^3/LQ\mu \ll 1$; that is, errors due to surface tension should be considered only when the liquid viscosity and the discharge rate are low, and the capillary bore is relatively large.

3.4.4 Variable Head

One of the main drawbacks of gravity-drainage viscometers is the nonconstancy of the driving hydrostatic head; since the head varies during the run, (76) cannot be used strictly. This, however, can be fairly easily rectified by using an effective value for Δh [68]

$$\Delta h_{eff} = \frac{\Delta h_i - \Delta h_e}{\ln(\Delta h_i/\Delta h_e)} \tag{88}$$

where Δh_i and Δh_e are the actual values of the head at the beginning and end of the run. Equation (88), known as the Meissner equation [110], applies only to a cylindrical bulb. Similar expressions can be derived, based on simple geometric arguments, for other bulb shapes, for example, a sphere, a cone, and so on [68]. Either expression can be incorporated directly into the viscometer constant in (75), which can be used explicitly, with Δh replaced by Δh_{eff}. Generally, one can reduce variations in head by using large bulbs with closely spaced fiducial marks.

3.4.5 The Hole-Pressure Error

Direct measurement of pressure drop along the flow path of the liquid eliminates the need to correct for end effects provided the pressure ports (two or more) are located within the fully developed flow regime in the channel. If, however, the pressure sensors are recessed in a slot off the channel walls, the reading is likely to be in error. This was discussed earlier in connection with the Lodge Stressmeter system [114], which utilizes the so-called hole-pressure effect to measure the elastic viscometric functions (see Section 3.3). The hole-pressure error depends on the shape of the slot and the rheological character of the liquid, and its magnitude can be estimated from the HPBL equations [114]

$$\left(\frac{d \ln P_h}{d \ln \tau_w}\right) P_h = \begin{cases} \dfrac{N_1 - N_2}{3} & \text{for circular holes} \\[2mm] \dfrac{N_1}{2} & \text{for transverse slots} \\[2mm] -N_2 & \text{for parallel slots} \end{cases} \tag{89}$$

where N_1 and N_2 are the first and second normal stress differences [cf. (15) and (16)] and P_h is the hole-pressure error ($= P_1 - P_2$ in Figure 5.15), that is, the difference in pressure between the value recorded by a flush-mounted transducer and the value measured by a transducer recessed in a hole. Although these equations should be regarded at this time as a semiempirical conjecture, a growing body of data seems to support their validity over a wide range of conditions. Since typically $|N_2| \cong 0.1 N_1$, it is expected, based on (89), that circular and transverse slots will give rise to larger error than will parallel slots. It was found that inertial effects may also contribute to the hole-pressure error [129], which can be estimated from

$$P_h = \begin{cases} -0.033\tau_w \, \text{Re} & \text{for transverse slots} \\ -0.024\tau_w \, \text{Re} & \text{for circular holes} \end{cases} \tag{90}$$

Thus, inertial hole-pressure errors are expected to be opposite in sign to errors caused by elastic effects. It is finally noted that since the depth of the slot does not appear in (89, 90), the hole-pressure error may be equally severe for shallow as for deep slots.

3.4.6 Spurt Fracture: The 'No-Slip' Assumption

Various low Reynolds number instabilities in capillary flow, usually manifested by a visible distortion in the issuing jet, have been attributed to viscoelastic effects and are referred to collectively as *melt fracture* [130, 131]. Generally, elastic instabilities may originate in the barrel just upstream from the capillary (the inlet zone), the flow zone in the capillary, or the end zone. Instabilities at the inlet zone are manifested by an abrupt change in the flow pattern above some critical shear stress and a distinct distortion in the extruded filament. Such instabilities are typical of branched polymers, for example, low-density poly(ethylene) (LDPE), which tend to form distinct recirculating patterns at the entry zone in flows through abrupt contractions. It has been noted, however, that the effect of such instabilities can be minimized by increasing the L/D ratio of the capillary [132]. Also, these instabilities are of no direct concern in viscometric tests since they do not affect the flow curve significantly. Indeed, it is often difficult to detect the onset of such instabilities from inspection of the flow curve. Instabilities that originate at the capillary land, sometimes called *spurt fracture*, are of more immediate concern since they have a decisive effect on the flow curve [133]; above some critical nominal shear stress the flow curve becomes discontinuous and practically independent of shear stress. Spurt fracture has been observed primarily for linear polymers and is usually attributed to slip of the polymer melt at the capillary walls [134]. The breakdown of the no-slip assumption introduces an error in the sense of (55) and should be clearly avoided. Based on data for many liquids, both melts and solutions, Lyngaae-Jörgensen and Marcher [135] proposed an empirical correlation for the critical shear stress and shear rate for the onset of spurt fracture as a function of the critical molecular weight for entanglement formation M_c and temperature. Their criterion is consistent with previous suggestions by Vinogradov [136] and Vlachopoulos and Alam [137]. Spurt fracture may be observed also in slit dies, although the body of data for these systems is limited.

3.4.7 Viscous Heating

Given the strong dependence of viscosity on temperature [e.g., (21)], proper control of temperature is critical in any viscometric test. Strictly speaking, however, viscometric flow, being inherently a dissipative process, is nonisothermal to some extent regardless of the quality of the temperature control device used to maintain constant temperature. The energy dissipated during flow is manifested by a temperature rise that may be substantial if the liquid is relatively viscous and/or the corresponding shear rates are high. This effect, commonly referred to as *viscous* or *shear heating*, must be considered carefully in any viscometric test, but especially when polymer melts are evaluated in Poiseuille viscometers under high shear rates. Viscous heating may affect viscometric data in two ways: (1) The effective temperature during the test may be higher than the preset temperature thus yielding lower apparent viscosities and (2) the velocity profile in a capillary or slit die may be altered thus rendering (66) and (74) invalid. Viscous heating in Poiseuille-type viscometers is often difficult to detect because it is manifested as a rheological shear-thinning effect; in the case of Newtonian liquids viscous heating can lead to a pseudo shear-thinning response, while for non-Newtonian liquids the extent of shear thinning can be artificially enhanced. For the purpose of our discussion here suffice it to say that viscous heating in capillaries and slits can be characterized by a dimensionless parameter, the Nahmé–Griffith number, which expresses the ratio of heat generated by viscous heating and heat removed by conduction. For a power-law liquid this parameter is defined by [138]

$$N_{NG} = \begin{cases} \dfrac{K_0 R^2}{k\,\Delta T_{Rheol}} \left(\dfrac{1+3n}{n} \dfrac{U}{R} \right)^{1+n} & \text{for capillaries} \\[3mm] \dfrac{K_0 H^2}{4k\,\Delta T_{Rheol}} \left(\dfrac{1+2n}{n} \dfrac{2U}{H} \right)^{1+n} & \text{for slits} \end{cases} \tag{91}$$

where k is the thermal conductivity, K_0 is the value of the consistency index at the initial temperature T_0, and

$$\Delta T_{\text{Rheol}} \equiv -\frac{1}{(d \ln \eta/dT)_{\dot{\gamma}}} \qquad (92)$$

expresses the temperature sensitivity of the viscosity function. Generally, the value of N_{NG}, which is closely associated with the extent of temperature rise due to viscous heating, must be minimized to assure uniform temperature during the test. From inspection of (91) we conclude that the most significant parameters influencing viscous heating are the flow rate Q (through U), the consistency index K (or the viscosity in the case of a Newtonian liquid) and the geometry of the conduit, that is, R or H. A general procedure for estimating temperature rise in capillaries for power-law liquids is presented by Middleman [139] based on the analytical study of Bird [140]. For a more complete discussion of the viscous heating problem the reader is referred to articles by Pearson [138], Sukanek and Laurence [141], Cox and Macosko [142], and Winter [143].

3.5   Intrinsic Viscosity

Intrinsic viscosity (IV) is a physical quantity (not a viscosity !), extractable from the viscosity of a dilute polymer solution, that carries some information about the size and shape of a single macromolecule in solution (or a particle in an heterogeneous liquid; see Section 2.5). Because of its close association with certain molecular and topological parameters IV has been used extensively to characterize and study various polymer systems and heterogeneous liquids. We chose to discuss this topic in this section since IV is measured predominantly by Poiseuille-type viscometers and particularly by the Ubbelhode gravity-drainage viscometer discussed in Section 3.3 (cf. Figure 5.12). Measurement of IV by rotational viscometers, although uncommon, is possible if the torque transducer is capable of providing precise measurements for dilute polymer solutions. Some of the terminology and notation used in dilute solution work is listed in Table 5.1.

In the limit of exceedingly low concentration c the viscosity of polymer solutions can be represented by a power series expansion of the form

$$\eta = \eta_s[1 + a_1c + a_2c^2 + \cdots] \qquad (93)$$

where η_s is the solvent viscosity. This polynomial can be rearranged conveniently to

$$\frac{\eta - \eta_s}{c\eta_s} \equiv \frac{\eta_{sp}}{c} = a_1 + a_2c + a_3c^2 + \cdots \equiv [\eta] + k_1[\eta]^2c + \cdots \qquad (94)$$

where η_{sp} is the specific viscosity and $[\eta]$ is the intrinsic viscosity. Thus, based on

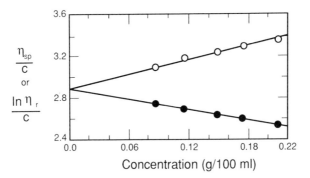

Figure 5.17 Reduced viscosity (○) and inherent viscosity (●) versus concentration for solutions of poly(styrene) ($M_w = 1.40 \times 10^6$) in CCl_4 at 30°C. Reprinted with permission from D. K. Carpenter and L. Westerman, "Viscometric Methods for Studying Molecular Weight and Molecular Weight Distribution," in P. E. Slade, Ed., *Polymer Molecular Weights*, Part II, Marcel Dekker, Inc., New York, 1975, Chap. 7.

(94), $[\eta]$ is defined by

$$[\eta] = \lim_{c \to 0} \left(\frac{\eta_{sp}}{c} \right)$$ (95)

This quantity expresses the fractional increase in the viscosity of a unit volume of solvent due to addition of polymer in the limit of vanishingly low concentrations, and it has units of inverse concentration. At these concentration levels the polymer molecules are presumed to be sufficiently far apart that they do not pervade each other's space. Equation (94) is known as the Huggins equation and k_1 is the Huggins constant. This parameter is related to the "goodness" of the solvent for the particular polymer and it decreases with increase in solvent power. Typically, k_1 varies in the range 0.3–0.8 [144]. A similar equation derived by Kraemer [145] can also be used to determine $[\eta]$

$$\frac{\ln \eta_r}{c} = [\eta] + k_1'[\eta]^2 c + \cdots$$ (96)

where η_r is the *relative* viscosity ($= \eta_{sp} + 1$) and $\ln \eta_r/c$ is the *inherent* viscosity. This equation is related directly to (94) through

$$k_1 - k_1' = 1/2$$ (97)

Thus, following (94) and (96) we can evaluate $[\eta]$ by plotting η_{sp}/c and/or $\ln \eta_r/c$ versus c and extrapolating to zero concentration. It is customary to plot both equations and take $[\eta]$ as the point of intersection of both curves at $c = 0$ (see Figure 5.17). If the curves do not intersect at $c = 0$, $[\eta]$ can be taken as the average of both intercepts. Equations (94) and (96) can be used as long as the corresponding curves are linear. This is usually the case when $c[\eta] < 1.0$, corresponding to the *dilute* solution limit. At higher concentrations, when a slight curvature is observed, it may be necessary to resort to a technique proposed by Maron and Reznik [146] who extend Huggins' analysis by carrying the polynomial expansion to the second order in c

$$\frac{\eta_{sp}}{c} = [\eta] + k_1[\eta]^2 c + k_2[\eta]^3 c^2 \tag{98}$$

and

$$\frac{\ln \eta_r}{c} = [\eta] + k_1'[\eta]^2 c + k_2'[\eta]^3 c^2 \tag{99}$$

where the second-order coefficients are interrelated by

$$k_2 - k_2' = k_1 - 1/3 \tag{100}$$

and (97) still holds. Equations (98) and (99) can now be combined to give

$$\frac{\eta_{sp} - \ln \eta_r}{c^2} = \frac{[\eta]^2}{2} + (k_1 - 1/3)[\eta]^3 c \tag{101}$$

which can be used to extract both $[\eta]$ and k_1. Maron and Reznik have shown that use of (89) yields better values of $[\eta]$ and k_1 compared to the linear equations. If the curvature is more severe, it may be necessary to use Martin's equation [147]

$$\log\left(\frac{\eta_{sp}}{c}\right) = \log[\eta] + k_1''[\eta]c \tag{102}$$

in place of (101). This equation tends to remove the curvature and may be easier to extrapolate than (101), but its validity at infinite dilutions is not firmly established. Equation (102) was generally noted to provide a good description of the effect of concentration on viscosity for polymer solutions in the *semidilute* regime where $1 < c[\eta] < 10$. Another equation that can be used when curvature effects are appreciable is the Schulz–Blaschke equation [148]

$$\frac{\eta_{sp}}{c} = [\eta] + k^*[\eta]\eta_{sp} \tag{103}$$

which is effectively a second-order equation since η_{sp} is quadratic in c. This equation can be rearranged into a more convenient form proposed by Heller [149]

$$\frac{c}{\eta_{sp}} = \frac{1}{[\eta]} - k^*c \qquad (104)$$

Generally, the dilute solutions used to determine IV can be considered Newtonian under typical testing conditions. This, however, may not be the case for rodlike, stiff molecules such as deoxyribonucleic acid (DNA) or helical polypeptides and for very large ($M_w > 10^6$), flexible polymer chains. Since most structural correlations involving IV [cf. (105)] are restricted to the zero-shear Newtonian regime, it is important to ascertain that the solution is indeed Newtonian under the conditions of the experiment and, in case it is not, to extend the measurements to sufficiently low shear rates. Non-Newtonian behavior is manifested by the dependence of the relative (or specific) viscosity on the nominal shear stress. The effect of shear stress can be evaluated by varying the hydrostatic head in a gravity-drainage viscometer or by conducting the measurements in a low torque rotational viscometer under a wide range of shear rates (see Section 4). Zimm and co-workers [150, 151] constructed a specialized viscometer designed to measure the viscosities and IV values of aqueous DNA solutions under extremely low levels of stress (<0.03 Pa).

When determining the IV of polyelectrolytes, close attention must be given to electrostatic effects that may obscure the data at low concentrations. At low ionic strength aqueous polyelectrolyte solutions exhibit an upward turn at low c when η_{sp}/c is plotted versus c. This anomaly is caused by the expansion of the molecular coil as a result of intramolecular repulsion of charged sites in each isolated chain, which may lead to erroneous interpretation of IV data. This effect can be minimized by conducting the measurements under high ionic strength conditions whereby the charged sites are "screened" by the addition of appropriate counterions.

We noted earlier that the intrinsic viscosity is closely associated with the shape and dimensions of an isolated macromolecule in solution. This is best expressed by the Mark–Houwink (MH) equation [152]

$$[\eta] = KM_v^a \qquad (105)$$

where K and a are empirical constants specific to the polymer–solvent system and M_v is the viscosity–average molecular weight defined by

$$M_v = \left(\frac{\sum_i n_i M_i^{a+1}}{\sum_i n_i M_i} \right)^{1/a} \qquad (106)$$

where n_i is the number of species i having a molecular weight M_i. Although the

IV technique is not a primary method for determining molecular weight, it has been used extensively because of its relative ease and accessibility. Extensive compilation of K and a can be found in the literature for many polymer–solvent systems [153]. The parameter a is intimately related to the conformation and shape of the macromolecule in solution. For random-coil, flexible chain polymers $0.5 < a < 0.8$ and $\bar{M}_v$ is intermediate between the number- and weight-average molecular weights, being usually closer to the weight-average molecular weight.

The theoretical foundation of the MH equation is firmly established. Based on an early work by Kirkwood and Riseman [154], Flory and Fox [155] have shown that

$$[\eta] = \bar{K}M^{1/2}\alpha^3 \tag{107}$$

where $\bar{K} = \Phi_0[\langle R_g^2 \rangle_\theta M]^{3/2}$, Φ_0 is a nearly universal constant for sufficiently large chains [156], $\langle R_g^2 \rangle^{1/2}$ is the root-mean-square radius of gyration of the polymer chain, and α is an expansion factor

$$\alpha = \left[\frac{\langle R_g^2 \rangle}{\langle R_g^2 \rangle_\theta}\right]^{1/2} \tag{108}$$

which represents the fractional increase in the radius of gyration of the polymer molecule relative to its value at the theta temperature θ, that is, when the chain is collapsed to its unperturbed random-coil dimensions. Using thermodynamic arguments, Flory [157] has shown that α depends on molecular weight through

$$\alpha^5 - \alpha^3 = A\left(1 - \frac{\theta}{T}\right)M^{1/2} \tag{109}$$

where A is a characteristic constant. Since the quantity $\langle R_g^2 \rangle_\theta M$ in (107) is also a constant (based on the definition of $\langle R_g^2 \rangle_\theta$), it can be shown through (109) that the MH parameter a should vary in the range 0.5–0.8 as the temperature varies from $T = \theta$ to $T \gg \theta$, consistent with observation. Since IV is strictly a measure of the hydrodynamic volume of the polymer molecule, one can use the IV technique to infer the chain structure and topology. For example, it is common to define a branching index for a branched polymer system as [158]

$$g = \left(\frac{[\eta]_\theta^b}{[\eta]_\theta^l}\right)^{2/3} \tag{110}$$

where the superscripts refer to branched and linear systems having the same molecular weights. Similar parameters can be defined for other topological structures, such as star, comb, and ring molecules [68]. The IV technique can also be used to study the molecular structure of rigid biopolymers [159] and the shape of particles suspended in a viscous liquid (see Section 2.5).

4   COUETTE-TYPE ROTATIONAL VISCOMETERS

4.1   Generalized Couette Flows

Most modern viscometers are based on hydrodynamic systems commonly referred to as *generalized Couette flows*. These flows are simple shear flows, see (4), induced by the motion of the boundary surfaces confining the liquid; that is, the liquid is driven strictly by *drag* forces. In a typical setup a thin liquid layer is confined between two rigid curvilinear surfaces, one of which is set in motion while the other is stationary, and from knowledge of the speed of the boundary surface and the force needed to maintain that speed the viscosity of the liquid can be determined. Generalized Couette flows are widely used in a variety of commercial instruments and are practiced primarily in a rotational mode. Three common flow configurations used in rotational viscometers are (see Figure 5.18): (1) tangential flow between coaxial cylinders (cylindrical Couette flow), (2) flow between a cone and a plate, and (3) flow between parallel disks (torsional flow). Viscometers employing these flows offer several tangible advantages over typical Poiseuille-type devices:

1.   Under some conditions (e.g., narrow gaps in the cylindrical Couette system or small cone angles in the cone-and-plate system) the flow is controllable and homogeneous, thus making the data reduction relatively simple and direct.

2.   Through simultaneous measurements of normal tractions and torque, the cone-and-plate and disk-and-plate systems can be used to obtain the normal stress coefficients and the viscosity function in a single run.

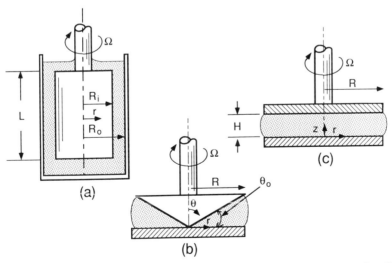

Figure 5.18   Schematic diagrams of Couette-type rotational viscometers: (*a*) concentric cylinder (Couette) viscometer, (*b*) cone-and-plate viscometer, and (*c*) disk-and-plate viscometer.

3. Rotational viscometers require typically small amounts of liquid that can be confined in a closed, readily thermostated system.

4. The generalized Couette viscometers can be designed to cover an extremely wide range of shear rates. Conventional systems are operative from very low to intermediate shear rates ($10^{-3} \lesssim \dot{\gamma} \lesssim 10^3 \text{ s}^{-1}$), but by placing the two members of the viscometer a very small distance apart extreme values of shear rate ($\gtrsim 10^5 \text{ s}^{-1}$) can be attained [160–162].

5. Most commercial rotational rheometers offer outstanding geometric versatility; that is, the cone-and-plate, parallel-disk, and concentric cylinder systems can be used interchangeably with the same drive and torque sensing units. These options are featured in several state-of-the-art commercial rheometers; for example, Rheometrics' Mechanical Spectrometer, Sangamo's Rheogoniometer, Instron's Rotary Rheometer [109], and the CarriMed Controlled Stress Rheometer. In these systems a wide range of geometries is available and tests can be run in steady-state (a standard viscometric test), transient, and dynamic (oscillatory) modes.

Couette-type viscometers can also be operated in a translation (linear) mode with either conicylindrical (axial motion) or parallel plate geometries [163, 163a], but such configurations are less common and are not well suited for steady-state measurements. Since the steady-state viscosity is the focus of this chapter, translational Couette systems are not considered further in this section but are briefly discussed in Section 6.2 (cf. Figure 5.26). Common rotational viscometers are rate-controlled; that is, the rotational speed (related to shear rate) of the rotating member is set while the torque (shear stress) needed to maintain that speed is measured by an appropriate torque transducer. In some systems (e.g., the CarriMed Controlled Stress Rheometer and the Rheometrics Stress Rheometer) the force (torque) is set and the resultant rotational speed is measured by a motion detection sensor. The latter class of rheometers is particularly useful in the study of liquids exhibiting yield stress [cf. (26)–(28)].

4.2   The Cylindrical Couette Viscometer

A typical cylindrical Couette (CC) viscometer is shown schematically in Figure 5.18a (see also Table 5.2). In this system the liquid is confined in the annular space between two coaxial cylinders and is set in motion by the rotation of either the inner or the outer cylinder. The inner cylinder is sometimes referred to as the *bob* and the outer cylinder is the *cup*. In a conventional setup the bob rotates at a constant rotational speed Ω while the cup is stationary. A transducer attached to the bob records the torque needed to balance the viscous forces acting on its surface as it rotates at the specified speed. In other configurations the cup rotates and the torque required to hold the bob in place is measured. The latter configuration can be found, for example, in the Contraves Low Shear 30 system.

In either case the measured torque M_i is used to calculate the shear stress at the wall of the inner cylinder

$$\tau_i = M_i/2\pi R_i^2 L \tag{111}$$

where the subscript i refers to the inner cylindrical surface. Of course, to evaluate the viscosity one needs an expression for the shear rate at $r = R_i$. Such an expression can be derived from a series solution to the integral equation of the CC system subject to the proper kinematic boundary conditions. If inertial terms and end effects are neglected [$Re \rightarrow 0$, $L/(R_o - R_i) \gg 1$], the nominal shear rate at the inner cylindrical surface is given by [164]

$$\dot{\gamma}_i = -r\frac{du_\theta}{dr}\bigg|_{R_i} = 2\Omega \sum_{k=0}^{\infty} s^{2km} \cdot m(s^{2k}\tau_i) \tag{112}$$

where

$$m \equiv \frac{d\log\Omega}{d\log M_i}$$

and

$$s \equiv \frac{R_i}{R_o}$$

This series converges slowly when the gap is small so that (112) is useful only for large-gap systems. A more useful form for large gaps was derived by Yang and Krieger [165] based on an earlier derivation by Krieger [166]:

$$\dot{\gamma}_i = \frac{2\Omega m}{(1-s^{2m})}\left[1 + m^{-2}\frac{d(m)}{d\ln\tau_i}f_1(m\ln s) + m^{-3}\frac{d^2(m)}{d(\ln\tau_i)^2}f_2(m\ln s) + \cdots\right] \tag{113}$$

where f_1 and f_2 are specified analytical functions. Yang and Krieger noted that this series is rapidly convergent and, neglecting higher order terms, would lead to errors no greater than 1%. A general numerical scheme for evaluating shear rates in the CC viscometer is described by MacSporran [167].

We can now examine several special cases. If the fluid is Newtonian ($m = 1$), (113) reduces to

$$\dot{\gamma}_i = \frac{2\Omega}{(1-s^2)} \tag{114}$$

and combining with (111) gives an expression for the viscosity of Newtonian liquids

$$\mu = \frac{M_i(1 - s^2)}{4\pi L \Omega R_i^2} \tag{115}$$

known as the *Margules equation*. For power law liquids $m \ (= 1/n)$ is a constant and (113) gives

$$\dot{\gamma}_i = \frac{2\Omega}{n(1 - s^{2/n})} \tag{116}$$

which is simply the leading term in the Krieger equation. It is thus clear that large errors can be incurred in determining $\dot{\gamma}_i$ if non-Newtonian effects or large gaps are not properly accounted for. For a power-law liquid the relative error in using (114) can be estimated from

$$\varepsilon \equiv \frac{\dot{\gamma}_{i,PL} - \dot{\gamma}_{i,N}}{\dot{\gamma}_{i,PL}} = 1 - \frac{n(1 - s^{2/n})}{1 - s^2} \tag{117}$$

This error will be larger the larger the gap and the smaller the value of n. If the annular gap is sufficiently small ($s \sim 1$), it is easy to show, using (114), that the shear rate across the gap is uniform and is given by

$$\dot{\gamma} = \frac{\Omega s}{1 - s} \tag{118}$$

Since this result is independent of the rheology of the liquid, in this limiting case only, cylindrical Couette flow may be considered homogeneous and controllable. Also, this case corresponds to infinite curvature and is indistinguishable from the equivalent plane Couette flow. Equation (118) is practically valid when $s > 0.99$ [164]. Another interesting limiting case is when the bob is immersed in a "sea of liquid"; that is, the gap is infinitely large. For this case $s = 0$ and (112) reduces to

$$\dot{\gamma}_i = 2\Omega m \tag{119}$$

which is a good approximation for $s < 0.1$ [164]. An attempt to extend the Yang–Krieger analysis to liquids that exhibit yield stress was made by Darby [168] who noted that large errors in the evaluation of shear rate may be incurred if the liquid exhibits plug flow with a sharp stress discontinuity within the annular gap.

Based on an early design by Barber [169], Porter and co-workers [160, 170] have constructed a modified CC system for measurements at very high shear rates. By using concentric cylinders with extremely narrow annular gaps ($<1 \mu$m) they were able to attain shear rates approaching $10^6 \, \text{s}^{-1}$. Concentricity at such gaps was maintained by making the inner (rotating) cylinder self-centering through attachment to a flexible drive. It was also claimed that by confining the liquid to very narrow gaps the temperature uniformity was improved and inertia-driven instabilities (cf. 4.3.2) were effectively eliminated.

4.3   Sources of Error in the Cylindrical Couette Viscometer

4.3.1   End Effects

The calculation of shear rate as given by (112) assumes that the flow in the annular gap is tangential; that is, fluid particles have circular trajectories coaxial with the cylinders, and the fluid velocity does not vary in the axial direction. Also, the use of (111) for calculating the shear stress assumes that the measured torque is only imparted on the cylindrical surface of the inner cylinder by the idealized tangential flow in the annular space. These assumptions break down if the bob is relatively short and/or the bottom of the bob is in close proximity to the bottom of the cup. The contribution of the end regions can be minimized by using a long bob, generally, such effects cannot be easily eliminated. To alleviate these problems it is customary to construct a bob with a conical bottom (see Figure 5.19a) such that the flow at the bottom of the cup is effectively a cone-and-plate flow (cf. Section 4.4) having a uniform shear rate. To maintain a constant shear rate throughout the fluid the cone angle must be [cf. (118) and (124)]

$$\theta_0 = \frac{1-s}{s} \tag{120}$$

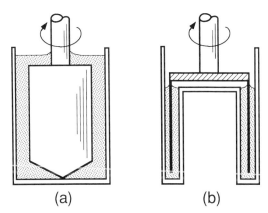

(a)   (b)

Figure 5.19   Improved designs of a concentric cylinder viscometer: (*a*) bob with a conical bottom and (*b*) double Couette viscometer.

assuming a narrow gap ($s \rightarrow 1$), and the shear stress is given by

$$\tau_i = \frac{3M_i}{2\pi R_i^2 (3L + R_i)} \tag{121}$$

Another approach for reducing losses at the bottom portion of the bob is to use an annular cup together with a bell-shaped bob or a bob with a recessed bottom as shown in Figure 5.19b. In this arrangement, known as the double Couette system, shear stress is imparted on both sides of the bob wall so that the surface area in (111) becomes $4\pi R_i L$. If the bob wall is sufficiently thin, end effects due to stress dissipation at the bottom of the cup can be safely neglected. At the upper portion of the bob the liquid may climb up the shaft and thereby alter the effective area of the shearing surfaces. This *rod climbing* phenomenon, known as the *Weissenberg effect*, would be important only for viscoelastic liquids at very high shear rates.

4.3.2   Inertial Effects

Two inertia-driven disturbances in cylindrical Couette viscometers are turbulence and a secondary flow known as the *Taylor instability* [171]. As was noted in Section 1.3 turbulence onsets at a critical Reynolds number (Re) and leads to a virtual breakdown of the viscometric nature of the flow in the annular gap. But even at lower values of Re the viscometric flow will give way to a secondary flow in the form of ordered toroidal cells that circulate perpendicularly to the main tangential flow. These so-called Taylor vortices appear when [172]

$$T_a \equiv \frac{\rho^2 \Omega^2 (R_o - R_i)^3 R_i}{\mu^2} = 1700 \tag{122}$$

which is applicable only for Newtonian liquids and narrow gaps. The critical value of T_a, the Taylor number, is likely to be intimately related to the rheological nature of the liquid [172, 173] although the precise formulation of this condition for non-Newtonian liquids is not well established. Equation (122) applies only in the case of a rotating bob and a stationary cup. For the case of a rotating cup and a stationary bob the flow in the CC system has been shown to be stable up to the onset of turbulence at [174]

$$Re \equiv \frac{\rho \Omega (R_o - R_i) R_i}{\mu} \cong 50,000. \tag{123}$$

In general, the various inertial effects will give rise to an increase in torque compared to the idealized, inertialess flow, because of the increase in energy dissipation in the system.

4.4   The Cone-and-Plate Viscometer

The cone-and-plate (CP) viscometer is a very common and a most ingenious device whose development is credited to K. Weissenberg. A schematic setup of a CP system is shown in Figure 5.18b (see also Table 5.2). The test fluid is placed in the gap between a stationary plate and a shallow cone rotating at a rotational speed Ω, wherein it is confined by a surface tension force acting at the rim (at $r = R$). If the cone angle θ_0 is sufficiently small, a uniform shear rate is established in the gap between the two members and is given by

$$\dot{\gamma} = -\frac{1}{r}\left(\frac{du_\phi}{d\theta}\right) = \frac{\Omega}{\theta_0} \tag{124}$$

Thus, the shear rate (and the shear stress) is constant and dependent only on the cone angle and the rotational speed. This result is derived from a simple kinematic argument; namely, the flow at any radial position can be regarded as a plane Couette flow between two parallel surfaces separated by a distance $r\theta_0$. It is assumed, in addition, that the flow is inertialess, the surface tension force at the rim is negligible, and the fluid–air interface is spherical in shape with a radius equal to R. The last two assumptions simply imply that edge effects are neglected and the CP flow extends all the way to R. The shear stress is also independent of position and is simply related to the torque by

$$\tau_{12} = \frac{3M}{2\pi R^3} \tag{125}$$

where M is the torque required to rotate the cone at the specified speed. Thus, for relatively small cone angles, the CP flow is essentially homogeneous and controllable and herein lies its primary advantage; the flow curve $[\eta(\dot{\gamma})]$ is obtained directly without the need to manipulate the raw data as with the capillary, slit die, and the wide-gap cylindrical Couette viscometers. Another important advantage of the CP viscometer is its ability to provide information on the elasticity (normal stress coefficients) of the liquid through simultaneous measurement of the total normal thrust F and the radial pressure distribution $P(r)$. The first and second normal stress differences, see (15) and (16), can be obtained from [5]

$$N_1 = \frac{2F}{\pi R^2} \tag{126}$$

and

$$N_1 + 2N_2 = -\frac{\partial P}{\partial \ln r} \tag{127}$$

where the pressure profile $P(r)$ must be measured by transducers flush-mounted at the surface of the plate (see Section 3.4.5). Also, the CP device typically requires small volumes of liquid and is relatively easy to load and maintain. The main drawback of the CP system is its limited shear rate range compared with the Poiseuille, cylindrical Couette and the disk-and-plate viscometers. This is mainly due to secondary flows and various irregularities at the rim that appear at relatively low shear rates (see Section 4.5). Under normal operating conditions $\dot{\gamma} < 1000 \, s^{-1}$ for relatively thin liquids and $\dot{\gamma} < 100 \, s^{-1}$ for molten polymers.

Since the homogeneity of flow is an important feature of the CP device, the validity of (124) was evaluated by several investigators. Adams and Lodge [175] found that the error in shear rate due to nonuniformity of the shear field should not exceed 2% for a cone angle of 10° and 0.32% for a cone angle of 4°. Paddon and Walters [176] computed somewhat larger errors for shear thinning liquids, but the error is still well within the experimental uncertainty for this system, especially for $\theta_0 < 4°$. We finally note that two variations of the basic design in Figure 5.18b are often used in practice. To prevent direct contact and friction between the cone and the plate it is customary to truncate the cone tip slightly and maintain a gap equal to the truncated height between the rotating cone and the stationary plate. Also, the CP device is sometimes operated in a "sea-of-liquid" mode to eliminate errors due to solvent evaporation and asphericity of the interface at $r = R$. For the case of aqueous and some organic solutions, solvent evaporation can be reduced by spraying a thin coat of silicone oil around the rim.

4.5   The Disk-and-Plate and Rotating Disk Viscometers

The configuration of the disk-and-plate (DP) viscometer is similar to that of the CP viscometer except that the cone is replaced by a flat parallel disk (see Figure 5.18c and Table 5.2). The gap between the disks is set at some known value H, where $H \ll R$. As in the CP viscometer, the flow field in this system can be deduced from simple kinematic statements without resorting to the equations of motion. When $H \ll R$, it is reasonable to assume that at any radial position the flow is tangential and equivalent to a plane Couette flow between parallel plates with the velocity of the upper plate being equal to Ωr. In this so-called *torsional* flow the velocity field is given by

$$u_\theta = \frac{\Omega r z}{H}$$ (128)

and the shear rate is

$$\dot{\gamma} = \frac{\partial u_\theta}{\partial z} = \frac{\Omega r}{H}$$ (129)

Thus, torsional flow is in general controllable; but unlike the CP flow, it is clearly inhomogeneous since $\dot{\gamma} = \dot{\gamma}(r)$. To evaluate the shear stress it is necessary to resort to a scheme similar to that used by Weissenberg and Rabinowitsch for capillary flow (see Section 3.1). Integration by parts of an equation for the torque yields the following expression for the shear stress at the rim (*nominal* shear stress) [5]

$$\tau_R = \frac{M}{2\pi R^3}\left[3 + \frac{d \ln(M/2\pi R^3)}{d \ln \dot{\gamma}_R}\right] \tag{130}$$

and the corresponding (*nominal*) shear rate is given by,

$$\dot{\gamma}_R = \frac{\Omega_R}{H} \tag{131}$$

which is also the maximum shear rate in this system [cf. (129)]. For a power-law liquid [cf. (23)] the derivative on the right-hand side of (130) is simply n, but for a liquid in the transition zone (cf. Figure 5.3) it should be evaluated explicitly for each shear rate.

The inhomogeneity of the torsional flow field puts the DP device at some disadvantage relative to the CP system. However, because of the ready adjustment of the gap over a wide range, the DP system offers a considerably greater operational flexibility than the CP device. By varying the gap setting in the DP system a wide range of shear rates can be probed in a single run without changing the fixture or even the test sample. Indeed, it has been shown that the DP device can be used to measure viscosities at very high shear rates simply by maintaining a small uniform gap between the disks. Connelly and Greener [161] demonstrated that by setting the gap at approximately 50 μm not only the operational shear rates are very high (approaching 10^5 s^{-1}), but also some of the disturbances that typically impair rheological tests with rotational devices (Section 4.6) are practically eliminated. Of course, operation at such small gaps requires close attention to the parallelism and alignment of the disks and careful calibration with Newtonian liquids. A modified version of the DP system, which permits operation with extremely small gaps, is the torsional balance rheometer developed by Binding and Walters [162]. In this device the lower plate is rotating while the upper plate is free to float on the test liquid under a fixed normal load that is balanced by the elastic normal thrust generated by the liquid. This arrangement eliminates the problem of misalignment and allows viscometric measurements at shear rates $> 10^5$ s^{-1}.

As in the CP system, the DP viscometer can be used to obtain information on the normal stress coefficients through measurement of the normal thrust F and the radial pressure distribution in the gap $P(r)$. The normal stress differences N_1 and N_2 [cf. (15) and (16)] can be determined from the following relations [5]

$$(N_1 - N_2)|_R = \frac{M}{\pi R^2}\left[2 + \frac{d \ln(M/\pi R^2)}{d \ln \dot{\gamma}_R}\right] \tag{132}$$

and

$$(N_1 + N_2)|_R = \frac{dP(0)}{d \ln \dot{\gamma}_R} \tag{133}$$

Thus, the procedure for obtaining the normal stress differences, N_1 in particular, with the DP system is less direct than with the CP device, but these relations can be used nonetheless for extracting information on the elastic response of the liquid in steady-state shear flow.

An interesting extension of the DP viscometer is the case when the rotating disk is fully immersed in a vat of the test liquid and is far removed from the walls of the vessel containing the liquid. This configuration—the rotating disk system—is featured in the Brookfield viscometer and is very simple to operate and maintain. If the disk is relatively thin and is wetted on both sides by the liquid, the viscosity can be obtained from the measured torque and rotational speed, based on the classical Newtonian analysis of Jeffrey [177]

$$\mu = 0.09375M/(R^3\Omega) \tag{134}$$

This assumes that the rod holding the disk is sufficiently thin so that it does not contribute to the measured torque. Although the corresponding flow is clearly inhomogeneous and nonviscometric, one can define average values for the wall shear stress and wall shear rate [178] as

$$\tau_{ave} = 0.75M/(\pi R^3) \quad \text{and} \quad \dot{\gamma}_{ave} = 8\Omega/\pi \tag{135}$$

which is valid only for the Newtonian case. Wein [179] has extended Jeffrey's analysis to the power-law case with the result

$$\tau_{ave} = 1.273f(n)K\Omega^n \quad \text{and} \quad \dot{\gamma}_{ave} = g(n)\Omega \tag{136}$$

where $g(n) = 2.546/n$ and $f(n) = 0.785[g(n)]^n$ and K and n are the usual power-law parameters. Somewhat different results for $g(n)$ and $f(n)$ were obtained by Kale and co-workers [180]. An attempt to employ the rotating disk (Brookfield) viscometer to establish a complete flow curve was made by Williams [178], who developed an elaborate numerical scheme for determining τ_{ave} and $\dot{\gamma}_{ave}$ for fluids having an arbitrary viscosity function. Williams reports that viscosity data generated with this procedure were in good agreement with data obtained by more conventional viscometric techniques. The applicability of the rotating disk viscometer to highly elastic liquids and high shear rates should be called into question, however, because of the appearance of a distinct secondary flow that is strongly influenced by liquid elasticity [181].

4.6   Sources of Error in the Cone-and-Plate and Disk-and-Plate Viscometers

4.6.1   End Effects

The assumptions that the simple shear flow in the CP and DP systems extends all the way to the rim and that the liquid–air interface at the rim assumes a particular shape (spherical in the CP system and cylindrical in the DP system) are often violated. The corresponding error, however, can be critically assessed by conducting experiments in the "sea-of-liquid" configuration where the outside periphery of the flow system is flooded. Based on data by Kaye and co-workers [182], Ginn and Metzner [183], and theoretical calculations by Griffiths and Walters [184], Walters [185] concludes that errors in torque measurement arising from the dynamics of the interface are very small ($<2\%$) for sufficiently small gaps ($H/R < 0.1$ for DP and $\theta_0 < 4°$ for CP). Surface tension forces, however, can introduce significant error when normal forces are measured for mildly elastic liquids [186] but will have limited effect on the apparent shear stress. *Shear fracture* is another source of error closely associated with the dynamics at the interface. This effect is manifested by a rapid fall in stress above some critical shear rate coupled with a gross distortion of the interface at the rim; the interface "caves in" toward the center of the apparatus and the fluid film in the gap appears to have ruptured (*fracture*). Fracture is presumably caused by viscoelastic radial *hoop* stresses, which force the liquid towards the center. Hutton [186] explained this effect in terms of a critical elastic strain energy and showed that fracture occurs when

$$N_1 > A\sigma/H \tag{137}$$

where σ is the surface tension and A is an unspecified constant. More recently, Tanner and Keentok [187] have shown, based on fracture mechanics argu-ments, that the condition for fracture is

$$|N_2| > 2\sigma/3a \tag{138}$$

where a is the size of the fracture ($a \sim H$). This result is consistent with (137) since, in general, $N_1 \propto N_2$. In any case, it is clear that the fracture effect should be of concern only for viscoelastic liquids and it can be circumvented by operating at very small gaps.

4.6.2   Inertial Effects

As with the cylindrical Couette viscometer (Section 4.3), inertial forces in conjunction with viscoelastic effects are expected to induce secondary flows in the CP and DP systems before the onset of turbulence. These flows, usually manifested as toroidal vortices, alter the idealized viscometric character of the flow and may lead to large errors in the measurement of viscosity and the normal stress coefficients. Studies on the nature of these flows by Cheng [188], Giesekus [189], Savins and Metzner [190], and Walters and co-workers [191], among others [192], indicate that the onset of secondary flow is somehow related to an inertial parameter $N_1 = \Omega a^2 \rho / \eta$ (equivalent to Re) where a is a characteristic gap thickness. However, because of the great sensitivity of this phenomenon to the rheological nature of the liquid, it is not possible to express these results in terms of a critical value of N_1. Radial migration is another disturbance associated with inertial effects. When the centrifugal forces acting on the liquid during flow exceed a certain critical value, the liquid will be thrown abruptly out of the gap. For radial migration to occur the centrifugal stress must exceed the surface tension stress that confines the liquid to the gap [187], that is,

$$\frac{3}{20} \rho \Omega^2 R^2 > \frac{\sigma}{H} \tag{139}$$

where σ is the interfacial tension of the liquid. This condition can be cast in terms of a critical shear rate for the onset of radial migration, which, for the CP system, is given by

$$\dot{\gamma}_c = 2.58 \sqrt{\sigma / \rho R^3 \theta_0^3} \tag{140}$$

A similar expression can be derived for the DP system. Thus, by operating with small gaps (or small cone angles) the range of the instrument can be extended to higher shear rates. Equation (140) is strictly valid for inelastic liquids. As was noted earlier, viscoelastic liquids will develop radial *hoop* stresses acting in the opposite direction (inward), which may lead to shear fracture before the onset of radial migration. Generally, we expect the operational range for inelastic liquids to be dictated by the radial migration problem, while that for viscoelastic liquids, by the shear fracture effect, but in both cases the limiting shear rate can be increased by decreasing the gap (or the cone angle) between the two members, cf. [161].

4.6.3   Viscous Heating

In trying to assess the effect of viscous heating on the CP instrument we turn to the studies of Bird and Turian [193, 194] and Turian [195]. For a power-law liquid Turian derives the following expressions for the maximum temperature rise and the corresponding maximum drop in torque for a CP system with isothermal boundaries

$$\Delta T_{max} \cong \Delta T_{Rheol} \left[\frac{N_{NG}}{4} - \frac{N_{NG}^2}{24n} \left(\frac{n}{4} - \frac{1}{16} \right) \right] \tag{141}$$

$$\Delta M_{max} \cong M_0 \frac{N_{NG}}{20n} \tag{142}$$

where T_0 and M_0 are the initial temperature and the torque measured at that temperature and N_{NG}, the Nahmè–Griffith number for the CP system, is given by

$$N_{NG} = K_0 \left(\frac{\Omega}{\theta_0}\right)^{n-1} \frac{R^2\Omega^2}{k\Delta T_{Rheol}} \tag{143}$$

where K_0 and n are parameters of the power-law model evaluated at the initial temperature, k is the thermal conductivity, and ΔT_{Rheol} is defined in (92). It is important to note that (141) and (142) are derived from a perturbation analysis for the effect of temperature on viscosity and they represent only first-order approximations. Nonetheless, these equations can be used to estimate viscous heating effects in the CP system and they suggest that temperature rise can be minimized by using devices with small diameters and shallow cones. Although a similar analysis for the DP system is not available at present, we expect ΔT_{max} and ΔM_{max} to have a similar dependence on N_{NG}, which is given by

$$N_{NG} = K_0 \left(\frac{\Omega R}{H}\right)^{n-1} \frac{R^2\Omega^2}{k\,\Delta T_{Rheol}} \tag{144}$$

It is interesting to point out that for the same nominal shear rate the DP system is likely to generate a lower temperature rise than a CP system with the same radius and gap opening because its space-averaged shear rate

$$\bar{\dot{\gamma}} = \frac{\Omega}{RH}\int_0^R r\,dr = \frac{\Omega R}{2H} \tag{145}$$

is lower by a factor of 2 compared to the nominal shear rate in the CP system. We close by reiterating that viscous heating effects may have a profound effect on viscometric measurements due to the strong dependence of viscosity on temperature [cf. (21)], and are of particular concern when the test fluids are relatively viscous and the operational shear rates are high. For a more exhaustive discussion of the viscous heating problem the reader should consult [138, 142, and 143].

5 STOKESIAN AND FALLING BODY VISCOMETERS

5.1 Stokes Flow

The viscometric systems discussed in this section are among the simplest both from a design and an operation standpoint. This coupled with the ability to attain very low shear rates ($< 10^{-3}\,s^{-1}$) is the main attraction of Stokesian viscometers. (As noted earlier, the importance of obtaining rheological information at very low shear rates stems from the great sensitivity of the zero-shear properties of the liquid to changes in molecular structure and composition.) The term *Stokesian viscometer* is used here somewhat loosely; of the systems described in this section only the falling ball viscometer (FBV) is based on Stokes' law and is thus the only true Stokesian viscometer. The other systems, namely, the rolling ball and the falling cylinder viscometers, are considered under this general category only because they are similar operationally to the FBV. The flows in these systems, however, are quite dissimilar from the idealized Stokes' flow and, consequently, the design considerations and data reduction schemes for these viscometers are different from those for the FBV.

The operational principle of the FBV derives from a classical problem in fluid dynamics that was originally solved by Stokes about a century ago [196, 197]. The problem considered by Stokes involves the unidirectional flow of an incompressible liquid past a rigid, stationary sphere of radius R_s and density ρ_s. The fluid is unbounded and it flows at a uniform speed u_∞ far away from the sphere. The problem at hand is to evaluate the force exerted on the sphere by the flowing liquid. This force is obtained by integrating over the surface area of the sphere expressions for the pressure and shear stress acting at the surface. For the special case of a creeping flow (see below) of a Newtonian liquid the classical analysis yields the following expression for the total force

$$F = \tfrac{4}{3}\pi R_s^3 \rho_1 g + 2\pi\mu R_s u_\infty + 4\pi\mu R_s u_\infty \tag{146}$$

where ρ_1 is the density of the liquid. The first term on the right-hand side of (146) represents the stationary contribution of the force due to buoyancy; the second term, known as the *form drag*, represents the drag force induced by pressure; and the third term, the *friction drag*, represents the drag force due to shear stress at the surface of the sphere. The combined form and friction drag terms are the kinetic force

$$F_k = 6\mu R_s u_\infty \tag{147}$$

that is, the force imparted by the motion of the liquid. Equation (146), also known as *Stokes' law*, is valid only in the limit of creeping flow conditions; namely,

$$\mathrm{Re} \equiv \frac{2R_s u_\infty \rho_1}{\mu} \lesssim 0.1 \tag{148}$$

whereby the flow is said to be *Stokesian*. It is customary to express (146) in terms of the Fanning friction factor $f = f(\mathrm{Re})$, where

$$f = F_k/(\pi R_s^2 \tfrac{1}{2}\rho_1 u_\infty^2) \tag{149}$$

Rearrangement of (149) yields the well-known result for Stokes flow

$$f = \frac{24}{Re} \tag{150}$$

which is valid, again, for Re < 0.1. Equation (150) can be generalized for higher values of Re by writing

$$f = \frac{24}{Re} \chi(Re) \tag{151}$$

where χ is some empirical function of Re. In fact, for $2 < Re < 500$, $\chi = 0.77\,Re^{0.4}$ [198] and the actual force exerted on the sphere is larger than the force F_k predicted by Stokes' law.

The velocity distribution for Stokes flow is given by

$$u_r = u_\infty \left[1 - \frac{3}{2}\left(\frac{R_s}{r}\right) + \frac{1}{2}\left(\frac{R_s}{r}\right)^3 \right] \cos\theta \tag{152}$$

$$u_\theta = u_\infty \left[1 - \frac{3}{4}\left(\frac{R_s}{r}\right) - \frac{1}{4}\left(\frac{R_s}{r}\right)^3 \right] \sin\theta \tag{153}$$

where r is the radial coordinate originating at the center of the sphere and θ is the angular coordinate defining the angle relative to the flow direction. Inspection of these equations reveals that this flow is nonviscometric and that interpretation of rheological data for non-Newtonian liquids is not straightforward. Deviations from Stokes' law due to non-Newtonian effects and finite dimensions of the containing vessel are considered briefly in Section 5.2.

5.2   The Falling Ball Viscometer

Due to its simple design and operation, the FBV has been widely used for measuring the viscosity of Newtonian liquids. However, because of the nonviscometric nature of Stokes' flow and the general lack of exact solutions of the Stokes' problem for non-Newtonian liquids, the utility of the FBV in establishing a complete flow curve $[\eta(\dot\gamma)]$ was, until recently, quite limited. The recent work of Cho and co-workers [199] is the first successful attempt to extend the range of the FBV to intermediate shear rates. In the FBV a rigid sphere of known density is dropped into a large cylinder of radius R_c and length L filled with the test liquid wherein it moves down, under the influence of gravity, at a constant terminal velocity u_∞. Even though the flow in this system is the reverse of the classical Stokes flow; that is, the liquid bulk is quiescent while the sphere is in motion, it is obvious that the two flows are physically indistinguishable. Force balance on the sphere [cf. (146)] gives

$$\tfrac{4}{3}\pi R_s^3 \rho_s g = \tfrac{4}{3}\pi R_s^3 \rho_1 g + 6\pi\mu u_\infty R_s \tag{154}$$

which can be rearranged to

$$\mu = \tfrac{2}{9}R_s^2 \frac{(\rho_s - \rho_1)g}{u_\infty} \tag{155}$$

Thus, determination of μ requires measurement of only one parameter, u_∞. All other parameters are generally known. This is usually done by timing the motion of the sphere between two points along the cylinder. The timing zone must be sufficiently removed from the top and bottom ends of the liquid column to avoid errors due to end effects. Tanner [200] has shown that these effects are limited to only one cylinder radius from either end, while Sutterby [201] claimed that these effects can be safely neglected if the measurement is done at the middle third of the liquid column for sufficiently large cylinders ($R_s/R_c < 0.125$ and $L/R_c > 4$). When the liquid is transparent, timing can be done either visually, using a stopwatch, or by an appropriate electrooptical sensor. For opaque liquids and metal spheres one can use, for example, electromagnetic sensors [202]. It is also important to properly scale the radius of the cylinder to that of the sphere. Stokes' law is practically valid for $R_s/R_c \lesssim 0.01$ thus requiring very small spheres or alternately very large cylinders. In most practical instances, however, it is necessary to correct Stokes' law to account for finite cylinder dimensions. The correction of Faxén [203] is the most widely used

$$\frac{\mu}{\mu_s} = 1 - 2.104 \frac{R_s}{R_c} + 2.09 \left(\frac{R_s}{R_c}\right)^3 - 0.95 \left(\frac{R_s}{R_c}\right)^5 + \cdots \tag{156}$$

where μ_s is the Stokes viscosity given by (155) and μ is the actual viscosity of the liquid. The first-order term in (156), known as the *Ladenburg correction*, is valid for $R_s/R_c < 0.06$. The complete Faxén correction can be used up to $R_s/R_c = 0.32$. Since the wall effect is coupled with inertia effects, the Faxén correction is valid only in the limit of small Re ($\ll 0.1$). The effect of inertia can likewise be expressed using the Oseen–Goldstein correction [204]

$$\frac{\mu}{\mu_s} = 1 + \frac{3}{16}\mathrm{Re} - \frac{19}{1280}\mathrm{Re}^2 + \frac{71}{20,480}\mathrm{Re}^3 + \cdots \tag{157}$$

where the first-order term is obtained from the classical analysis of Oseen, while the higher order terms were derived by Goldstein. This correction can be used up to Re = 2.0. For higher values of Re the empirical correlation for the friction factor [see (151)] is a reasonable approximation. More complete discussions of the combined effects of the cylinder wall and inertia can be found in the studies by Turian [205] and Sutterby [201]. Inspection of (156) and (157) reveals that Stokes' law, (155), will always overpredict the viscosity of the liquid if finite cylinder effects are present and it will underpredict if inertia effects are important.

Because of the nonviscometric nature of Stokes' flow and the difficulty of accounting for non-Newtonian effects, studies of non-Newtonian liquids with the FBV have met with great difficulties [206]. To date, most of the theoretical studies of Stokes' flow for viscoelastic liquids have employed unrealistic constitutive equations and, consequently, a theoretical resolution of this problem has not been achieved. For the purpose of estimation only one can define an average nominal shear rate for the flow in the immediate vicinity of the sphere by

$$\dot{\gamma}_{av} \equiv \frac{\tau_{av}}{\mu} = \frac{2R_s g(\rho_s - \rho_l)}{9\mu} \tag{158}$$

where τ_{av} is the shear stress at the surface of the sphere averaged over its entire area. Thus, the shear rate can be varied by varying the diameter and/or the density of the sphere. Also, the shear rate is liquid dependent (!) and is generally higher the lower the viscosity.

Because it is difficult to specify an effective shear rate for non-Newtonian liquids, measurements with the FBV have been limited to the zero-shear Newtonian regime, which can often be reached only by extrapolation. Some of the commonly used extrapolation techniques were reviewed by Gottlieb [207] and Chhabra and Uhlherr [208]. Gottlieb observes that best results are obtained when $\ln(\eta)$ is plotted versus τ_{av} and the data are extrapolated to zero stress. Chhabra and Uhlherr, on the other hand, obtain the most satisfactory results when the method of Caswell [209] is used—plotting $1/\eta$ versus $(\eta u_\infty/R_s)^2$ and extrapolating to zero. However, as Chhabra and Uhlherr point out, none of the extrapolation techniques are completely satisfactory; if the data lie within the shear-thinning regime, these techniques are likely to overestimate, to some extent, the true value of η_0.

The analytical approach of Cho and co-workers [199] constitutes the first systematic attempt to extend the FBV technique to the non-Newtonian flow regime. Based on the variational analysis of Wasserman and Slattery [210] for a power-law liquid, Cho and co-workers derive the following expressions for the average shear stress

$$\tau_{av}^u = (0.2827 + 0.8744n + 0.4562n^2$$
$$- 0.7486n^3)\tfrac{2}{9}gR_s(\rho_s - \rho_l) \tag{159a}$$

$$\tau_{av}^l = (0.6388 + 0.6418n - 0.4344n^2$$
$$+ 0.156n^3)\tfrac{2}{9}gR_s(\rho_s - \rho_l) \tag{159b}$$

where n is the power-law index and the superscripts u and l correspond, respectively, to upper and lower bound approximations obtained from the variational analysis. The corresponding expressions for the average shear rate are

$$\dot{\gamma}_{av}^{u} = \left(-1.731 + 41.28n - 116.0n^2 + 123.9n^3 - 46.72n^4 \right) \frac{u_{\infty}}{R_s} \qquad (160a)$$

$$\dot{\gamma}_{av}^{l} = \left(-2.482 + 54.35n - 160.1n^2 + 178.2n^3 - 69.04n^4 \right) \frac{u_{\infty}}{R_s} \qquad (160b)$$

and the apparent viscosity is simply

$$\eta = \frac{\tau_{av}}{\dot{\gamma}_{av}} \qquad (161)$$

where $\dot{\gamma}_{av}$ and τ_{av} have the same general meaning as those defined in (158) for Newtonian liquids. If the power-law index does not change appreciably over a small range of $\dot{\gamma}$ then

$$n = \frac{d \ln \tau_{av}}{d \ln \dot{\gamma}_{av}} = \frac{d \ln[R_s(\rho_s - \rho_l)]}{d \ln[u_{\infty}/R_s]} \qquad (162)$$

Since all the parameters in (162) are measurable, n can be obtained directly from falling ball experiments. This value can then be inserted in (159) and (160) and a complete flow curve can be constructed. Results for several poly(acrylamide) solutions, shown in Figure 5.20, compare data obtained by conventional cone-and-plate and capillary instruments with falling ball data obtained by the scheme outlined above. It is noted that the data from all instruments agree closely and that the difference between the upper and lower bound solutions is rather small. Cho and co-workers [199] observe that the measurement of terminal velocity for highly elastic liquids can be subject to considerable error if the time interval used to evaluate u_{∞} is too short. This is due to the long diffusion (*relaxation*) times needed for the fluid to rearrange itself in the wake of the falling sphere. Time intervals in the range of 10–100 min (depending on the elasticity of the liquid) may be needed to minimize this effect. This, of course, severely curtails the practical utility of the FBV technique and limits it to only mildly elastic liquids. Chhabra and co-workers [206] have also observed a significant drop in the drag coefficient for creeping flow around a sphere due to elasticity. Although the flow in the FBV is not viscometric, Hassager and Bisgaard [211] used numerical simulation of the flow of a viscoelastic fluid around a sphere to obtain not only the zero-shear viscosity but also the first normal stress difference N_1 from measurements in a standard FBV experiment.

A device similar to the FBV is the bubble rise viscometer. Here, spherical gas

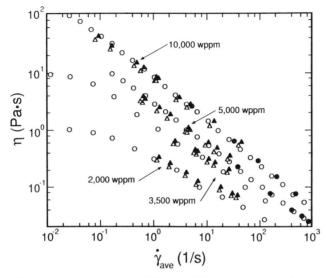

Figure 5.20 Flow curves of aqueous poly(acrylamide) solutions at various concentrations. Comparison of data obtained by three different techniques: (○) cone-and-plate viscometer, (●) capillary viscometer, (△, ▲) falling ball viscometer. Reprinted with permission from Y. I. Cho, J. P. Hartnett, and W. Y. Lee, *J. Non-Newtonian Fluid Mech.*, **15**, 61 (1984).

bubbles generated at the bottom of a liquid column move upward at a terminal velocity u_∞ driven by buoyancy forces. If the bubbles move sufficiently slowly so that inertia effects are negligible and the spherical shape of the bubbles is undistorted by the shearing forces, then (155) is still valid, but in this case ρ_s can be neglected and the radius of the bubbles is controlled by the interfacial tension of the gas–liquid system. Because of difficulties to operate within the creeping flow regime and the restricted range of R_s, the use of this technique has been fairly limited. The bubble rise viscometer is, nonetheless, a simple and viable device.

An interesting extension of the traditional falling ball viscometer was proposed and implemented by Sobczak and co-workers [212–214] who have shown that the measurement of viscosity can be more accurate and convenient if the sphere is driven through the liquid medium by a magnetic, rather than a gravitational, force. In this so-called magnetoviscometer the terminal speed of the (metal) sphere can be more precisely controlled and measured and it can be varied over a wider range.

5.3   The Rolling Ball and Falling Cylinder Viscometers

The viscometers described below are not Stokesian because the flow fields generated in these devices do not fall under the general category of Stokes flow. However, these viscometers are very similar operationally to the falling ball viscometer and are very simple to use and maintain.

The rolling ball viscometer (RBV) is a particularly simple device. A typical setup consists of a cylinder of radius R_c containing a sphere whose radius R_s is slightly smaller than the radius of the cylinder. The cylinder is filled with the test fluid and then it is inclined at an angle β with respect to the horizontal, thereby causing the sphere to roll down the inclined cylinder under the influence of gravity (Figure 5.21). The translational velocity of the sphere, or the time for the sphere to traverse between two marked positions along the cylinder, is used to calculate the viscosity. A general expression for the viscosity of a Newtonian liquid was derived by Lewis [215]

$$\mu = 0.267 \frac{R_c^2(\rho_s - \rho_1)g \, \sin \beta}{U} \left(\frac{R_c - R_s}{R_c} \right)^{5/2} \tag{163}$$

where ρ_s and ρ_1 are the densities of the sphere and the fluid, respectively, and U is the translational speed of the sphere. Thus, for a particular cylinder–sphere combination one can define a viscometer constant and extract the viscosity directly from the rolling time. Equation (163) is valid only for small gaps between the sphere and the cylinder, or $R_s/R_c \to 1.0$, but this result can be considered adequate for $R_s/R_c > 0.95$. Extension of Lewis' analysis to the power-law case, carried out by Bird and Turian [216] can be used to derive an expression analogous to (163),

$$\eta = \frac{n(2R_c)^2(\rho_s - \rho_1)g \, \sin \beta}{3\pi U J_n(2n+1)} \left(\frac{R_c - R_s}{R_c} \right)^{2n+1/2} \tag{164}$$

where n is the power-law index and the constants J_n are defined by

$$J_n = 2 \int_0^\infty \frac{d\xi}{[I_n(\xi^2)]^n} \tag{165}$$

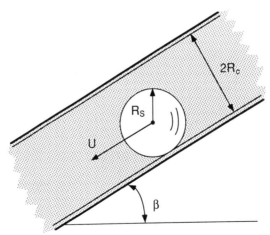

Figure 5.21 Schematic diagram of the rolling ball viscometer.

where

$$I_n(\alpha) = \int_{-\pi}^{\pi} (\cos^2 \tfrac{1}{2}\theta + \alpha)^{2+(1/n)} d\theta \tag{166}$$

Extensive tabulation of J_n versus n was prepared by Šesták and Ambros [217], who tested (164) experimentally. Although the corresponding flow is not truly viscometric, one can define an operative shear rate for the RBV system and, together with expression for the apparent viscosity (164), construct a flow curve for the power-law liquid. Some design and operational considerations for the RBV system are discussed by Van Wazer and co-workers [218].

The falling cylinder viscometer (FCV) is also non-Stokesian, but it is similar operationally to the FBV and RBV. In this system a long cylindrical shaft or a short cylindrical slug of radius R_i is allowed to fall in a concentric tube filled with the test fluid. The motion of the cylinder is induced by the gravity of its own weight or by an added "dead weight," or, in the case of relatively viscous fluids, it can be driven by an external, constant speed motor. From knowledge of the speed of the moving cylinder (fall time) and the applied force one can estimate the corresponding shear rate and viscosity. Since the containing tube is sealed at the bottom of the viscometer, the motion of the falling cylinder must induce flow reversal, that is, a velocity profile with both "positive" and "negative" regimes. In general, unlike the RBV and FBV, the flow in the annular space between the moving cylinder and the tube can be considered viscometric if the speed U is constant and the cylinder is concentric and relatively long. The problem of the falling cylindrical slug was solved by Lohrenz and co-workers [219] and later analyzed by Bird and co-workers [220]. The viscosity for Newtonian liquids in the falling slug system is given by

$$\mu = \frac{(\rho_s - \rho_l)gR_i^2}{2U} \left[\ln \left(\frac{1}{s} \right) - \left(\frac{1-s^2}{1+s^2} \right) \right] \tag{167}$$

where s is the ratio of the radii of the falling cylinder and the tube, and ρ_s and ρ_l are the densities of the slug and the liquid, respectively. The corresponding (nominal) shear rate at the surface of the moving cylinder is

$$\dot\gamma_i \equiv - \frac{du_z}{dr} \bigg|_{R_i} = \frac{U}{R_i} \left[\frac{1-s^2}{(1-s^2)-(1+s^2)\ln(1/s)} \right] \tag{168}$$

Thus, as for the other viscometers discussed in this section, one can extract the necessary viscometric information from knowledge of the fall time ($\propto 1/U$) and a given geometric (viscometer) constant. Equations (167) and (168) can be extended to the power-law case and other non-Newtonian models, but generally the cylindrical slug system is operable at relatively low-shear rates. The shear rate range can be extended by using a system with a long shaft driven externally at high speeds. Rudin and co-workers [221–223] studied this version of the FCV by using a system adapted from the Mertz–Colwell Instron rheometer (see Figures 5.13 and 5.22). They report data at shear rates greater than $10^3 \, s^{-1}$ that agree favorably with data obtained by conventional techniques.

6   DYNAMIC VISCOMETRY

6.1   Oscillatory Shear Measurements

Dynamic techniques have become increasingly popular in the study of the rheological behavior of liquids and are widely employed in a host of commercial

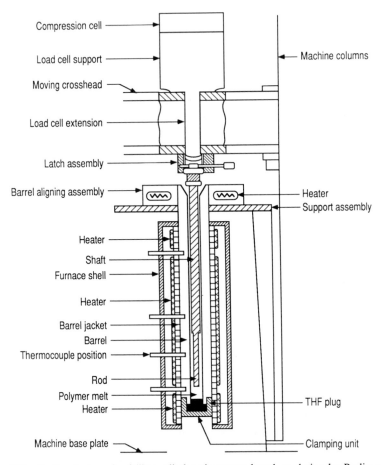

Figure 5.22 Sectional view of a falling cylinder viscometer based on design by Rudin and co-workers [223]. Reprinted with permission from K. K. Chee and A. Rudin, *Rheol. Acta*, **16**, 635 (1977).

rheometers. Improvements in electronic circuitry and electromechanical actuation have made these techniques relatively easy to execute and analyze. The
main attraction of dynamic techniques as vis-a-vis steady-state viscometry lies in
their potential to probe simultaneously the viscous and elastic response of
liquids over a wide range of frequencies [analogous to shear rates; cf. (178)] and
the ability to test a wide variety of materials ranging from dilute polymer
solutions and thin Newtonian liquids to rigid viscoelastic solids. The latter
makes these techniques especially attractive in the study of reactive polymer
systems and molecular network formation, which typically involve substantial
changes in the rheological properties of the material over the course of the test
[224].

Dynamic shear experiments are commonly performed in a forced-oscillation
(or forced-vibration) mode in which a sinusoidal shear strain wave

$$\gamma = \gamma^0 \sin \omega t \tag{169}$$

is imposed on the liquid. γ^0 is the strain amplitude and ω is the radian frequency
of the shear wave. The corresponding *shear rate* wave is

$$\dot{\gamma} = \dot{\gamma}^0 \cos \omega t \tag{170}$$

where $\dot{\gamma}^0 = \omega \gamma^0$. The dynamic response of the liquid can be expressed in general
by

$$\tau = \tau^0 \sin(\omega t + \delta) \tag{171}$$

where δ is a phase angle expressing the lag of the shear stress τ behind the
imposed strain. If the liquid is purely viscous, $\delta = 90°$; that is, the stress is in
phase with the strain rate. Otherwise, if the liquid is viscoelastic, $\delta < 90°$.
Equation (171) can be decomposed into two components, one in phase with the
strain and the other in phase with the strain rate,

$$\tau = \gamma^0(G' \sin \omega t + G'' \cos \omega t) \tag{172}$$

where G' and G'' are the dynamic shear moduli of the liquid; G', the storage
(shear) modulus, expresses the elastic character of the liquid, whereas G'', the loss
(shear) modulus, represents the viscous (dissipative) portion of its response. Both
are characteristic functions of frequency and serve as rheological fingerprints of
the material. Simple manipulation of (171) and (172) gives

$$G' = (\tau^0/\gamma^0) \cos \delta \tag{173a}$$

$$G'' = (\tau^0/\gamma^0) \sin \delta \tag{173b}$$

If the stress is expressed as a complex variable, then one can define a complex shear modulus as

$$G^* \equiv \tau^*/\gamma^* = G' + iG'' \qquad (174)$$

where $i = \sqrt{-1}$ and only the real components are physically meaningful. Similarly, we can define a complex viscosity by

$$\eta^* \equiv \tau^*/\dot{\gamma}^* = \eta' - i\eta'' \qquad (175)$$

where η', the dynamic viscosity, represents the component of the dynamic response in phase with the rate of strain, whereas η'' is the component in phase with the strain. It follows that

$$\eta' = G''/\omega \qquad (176a)$$

$$\eta'' = G'/\omega \qquad (176b)$$

and in the limit of small frequencies it can be shown that η' approaches η_0, the zero-shear viscosity

$$\lim_{\omega \to 0} \eta' = \eta_0 \qquad (177)$$

Although G^* and η^* are interchangeable, it is common to use η^* to describe the viscoelastic behavior of liquids as opposed to solids. Equation (173) suggests that to extract the viscoelastic functions in a forced-oscillation experiment it is necessary to determine the stress amplitude τ^0 and the phase angle δ. However, as is shown in the following sections, other less direct methods can be used to evaluate the viscoelastic properties. Since usually the linear viscoelastic properties are sought, linearity must be checked routinely in dynamic oscillatory experiments by ensuring that the strain amplitude is sufficiently small such that the measured material functions, for example, η' and η'', are independent of strain (γ^0).

It has already been noted that one can obtain the zero-shear viscosity, a steady-state property, from a dynamic shear experiment [cf. (177)]. A more general analogy between dynamic and steady-state rheological properties of liquids is given by the empirical Cox–Merz (CM) rules [225]:

$$\eta(\dot{\gamma}) = |\eta^*(\omega)|_{\omega = \dot{\gamma}} = \sqrt{\eta'^2 + \eta''^2} \qquad (178)$$

and

$$\frac{d\tau(\dot{\gamma})}{d\dot{\gamma}} = \eta'(\omega)|_{\omega = \dot{\gamma}} \qquad (179)$$

Although the existence of such analogy is not immediately obvious given the distinct kinematic and dynamic differences between oscillatory and steady-shear experiments, it has been shown to work effectively for a wide variety of polymeric liquids, both solutions and melts. Some exceptions are filled polymers [226], immiscible polymer blends [227], and highly nonlinear systems such as high molecular weight poly(styrenes) and concentrated solutions of poly(acrylamide) [228]. Despite these exceptions, the CM rules can be used to estimate the steady-shear viscosity function (the *flow curve*) from small-strain (linear) oscillatory shear data. The theoretical basis for the CM rules is discussed by Bird and co-workers [229], Vinogradov and Malkin [230], and Booj and co-workers [231]. The effectiveness of the CM rule is illustrated in Figure 5.23 where steady-shear and dynamic viscosity data for a solution of a narrow distribution poly(styrene) in 1-chloronaphthalene are compared. The results show excellent overlap between the steady and complex viscosities, in line with predictions of the CM rule.

Although the techniques for obtaining linear viscoelastic properties of liquids are widely varied, they can be grouped into three broad classes: (1) damped oscillations, (2) *gap-loading* techniques, and (3) *surface-loading* and wave propagation techniques. In damped (free) oscillation experiments one of the surfaces of the apparatus in contact with the liquid undergoes damped oscillations from which the properties of the liquid are inferred. However, since interpretation of data in such an experiment is not straightforward, it is rarely used in practice and it is not discussed further in this chapter. The other two

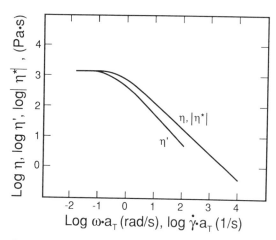

Figure 5.23   Comparison of steady-state viscosity (vs. shear rate) and complex viscosity (vs. radian frequency) for 0.15 g/mL of solution of narrow distribution poly(styrene) in 1-chloronaphthalene. Reprinted with permission from K. Yasuda, R. C. Armstrong and R. E. Cohen, *Rheol. Acta*, **20**, 163 (1981).

classes involve forced oscillations. In gap-loading techniques the imposed shear wave is uniform across the gap between the driving and reflecting surfaces; that is, the stress and the strain do not experience periodic *spatial* variations—only *temporal* variations. In surface-loading techniques the driving surface and the fluid response are oblivious to the surface across the gap; that is, the wave is fully damped before reaching the opposite (reflecting) surface. In wave-propagation techniques the wave is only partially damped across the gap and the properties of the fluid are extracted from its attenuation. The principal difference between gap loading and surface loading techniques lies in the relative magnitude of inertia forces in the liquid. In gap-loading experiments inertial effects are negligible and the instantaneous shear field is identical to the steady-shear field for the particular test geometry. By contrast, in surface-loading and wave-propagation techniques inertia forces are dominant. Inertial effects in sinusoidal shear flows are represented by the dimensionless parameter [232]

$$N_I \cong \frac{\omega \rho H^2}{|\eta^*|} \simeq \frac{\omega \rho H^2}{\eta} \tag{180}$$

which expresses the ratio of inertia-to-viscous forces (a dynamic Reynolds number!). Here H is the gap within which the fluid is confined and ρ is the fluid density. Thus, gap-loading techniques are limited to systems where $N_I \ll 1$; that is, systems involving low frequencies, small gaps, and high viscosities, while the surface loading limit is reached, in systems where $N_I \gg 1$. A detailed discussion of the criteria for using gap-loading and surface-loading techniques is given by Schrag [232]. The general principles and some illustrative examples of both types of dynamic experiments are presented in the next sections. For a more complete discussion of dynamic viscometry the reader should consult the books by Ferry [233], Walters [234], and Malkin and co-workers [235] and a review article by Harrison and Barlow [236].

6.2 Gap-Loading Techniques

With proper modification of the drives and the analysis circuitry, gap-loading dynamic measurements can be performed with conventional rotational viscometers. This feature is exploited in several commercial rotational devices; for example, the Rheometrics System Four, the Sangamo Weissenberg Rheogoniometer, and the CarriMed Controlled Stress Rheometer, where identical fixtures are used for both dynamic and steady-state tests.

In the dynamic tests, forced oscillations with a frequency ω and an angular amplitude θ_1^0 are imposed on one member (the *driving* member) and the response of the other member (the *driven* member), namely, its angular amplitude θ_2^0 or torque M and the phase lag of *these variables* behind the imposed strain wave, are recorded. Expressions relating these parameters to the dynamic functions η' and η'' can be derived from the equation of motion for the driven member combined with a linear–viscoelastic constitutive equation for the liquid. If the motion of the driving member is expressed by

$$\theta_1 = \theta_1^0 \exp(i\omega t) \tag{181}$$

then the motion of the driven member can be described in general by

$$\theta_2 = \theta_2^0 \exp[i(\omega t + \alpha)] \tag{182}$$

where θ_2^0 and α are unknown. These parameters can be obtained from the equation of motion for the driven member

$$I\ddot{\theta}_2 + K\theta_2 + b\eta*(\dot{\theta}_2 - \dot{\theta}_1) = 0 \tag{183}$$

where I is the moment of inertia of the driven member, K is the torsion constant of the wire to which it is attached ($M = K\theta_2$, where M is the torque), and b is a geometric form factor specific to the test configuration. Form factors for several common test geometries are listed in Table 5.3. Conveniently, solution of (181)–(183) can be expressed by the following equations [235]

$$\eta' = \frac{(K - I\omega^2)v \sin \alpha}{b\omega(v^2 - 2v \cos \alpha + 1)} \tag{184a}$$

$$\eta'' = \frac{(K - I\omega^2)(v \cos \alpha - v^2)}{b\omega(v^2 - 2v \cos \alpha + 1)} \tag{184b}$$

where v is the amplitude ratio θ_2^0/θ_1^0. For the special case when the torsion wire is very stiff, such that $K/I\omega^2 \gg 1$ and $v \ll 1$, (184) reduces to

$$\eta' = Kv \sin \frac{\alpha}{b\omega} \tag{185a}$$

$$\eta'' = \frac{Kv \cos \alpha}{b\omega} \tag{185b}$$

and only for this limiting case the phase angle α is equal to the phase lag δ (171).

Of the torsional configurations listed in Table 5.3 the parallel disk (torsion) system is usually the preferred configuration, mainly because of the ability to change the gap between the plates and thus adjust the strain without changing the test specimen or the fixtures. Also, inertial effects (in the liquid) can be accommodated very simply in this configuration compared to other rotational geometries. Indeed, as noted by Walters [234], "... the privileged position of the cone-and-plate geometry (in steady-state rheological measurements) does not carry over to oscillatory flow." Inspection of (184) reveals that each rotational system possesses a characteristic frequency

$$\omega_0 = \sqrt{K/I} \tag{186}$$

Table 5.3 Shape Factors for Geometries Used in Dynamic Shear Experiments

Flow Geometry	Shape Factor, b
Parallel plates[a] (Figure 5.1)	A/H
Coaxial cylinders (annular pumping)[b] (Figure 5.18)	$2\pi L/[\ln q - (q^2-1)/(q^2+1)]$
Coaxial cylinders (torsion) (Figure 5.18)	$4\pi LR_o^2/(q^2-1)$
Cone and plate (torsion) (Figure 5.18)	$2\pi R^3/3\theta_0$
Disk and plate (torsion) (Figure 5.18)	$\pi R^4/2H$
Coaxial cylinders (axial motion)[c] (Figure 5.18)	$2\pi L/\ln q$

[a] A is surface area of plate.
[b] Inner cylinder moves axially, bottom of outer cylinder is closed; $q = R_o/R_i$.
[c] Segel–Pochettino configuration: inner cylinder moves axially and outer cylinder is open-ended.

known as the natural frequency for which $\eta' = \eta'' = 0$; that is, the system response at this frequency is oblivious to the properties of the liquid. The natural frequency should, therefore, be avoided in oscillatory tests. It is, in fact, not recommended to operate in close proximity to ω_0 where large variations in the measured properties can be readily misinterpreted and lead to significant error.

An interesting variant of the torsional systems discussed above was developed by Morrison and co-workers [237] and later improved by Duiser [238], den Otter [239], and te Nijenhuis and van Donselaar [240]. In this system a cylindrical bob suspended between two wires is oscillating inside a coaxial cylindrical cup as shown in Figure 5.24. From measurement of the ratio of the amplitudes of the driving shaft and the bob v and the phase lag between them α the components of the complex viscosity can be obtained through

$$\eta' = \frac{1}{b\omega} K_1 v \sin \alpha \tag{187a}$$

$$\eta'' = \frac{1}{b\omega}[(K_1 v \cos \alpha - 1) - K_2 + I\omega^2] \tag{187b}$$

where K_1 and K_2 are the torsion constants of the upper and lower wires, I is the moment of inertia of the bob, and b is the form factor for a coaxial cylinder configuration in torsion (Table 5.3). This rheometer can be operated without the lower wire, in which case $K_2 = 0$. The improved version of te Nijenhuis and van Donselaar [240] is reported to have excellent resolution of phase angles over the entire range 0–90°. This is achieved by a high-precision optical train consisting of a laser light source and a carefully aligned dual mirror system capable of resolving angular deflections as low as 25" (see Figure 5.24). The reported frequency and modulus ranges for this system are, respectively, $3 \times 10^{-5} - 600$ rad/s and $1 \times 10^{-2} - 3 \times 10^5$ N/m^2. With such a wide operational range this rheometer is particularly well suited for studying network formation

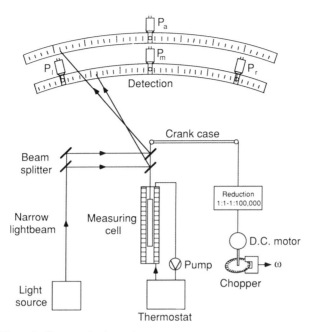

Figure 5.24 Schematic diagram of a dynamic coaxial cylinder viscometer based on design by K. te Nijenhuis and R. van Donselaar. Reprinted with permission from *Rheol. Acta*, **24**, 47 (1985).

and gelation phenomena in polymer systems and other processes involving large changes in rheological properties [241].

When the driven member is fixed in place, it is customary to measure both the force and the displacement on the driving member. Analysis of this configuration is often done in terms of the complex mechanical impedance

$$Z_M^* \equiv \frac{f^*}{u} = R_M + iX_M \tag{188}$$

where, by analogy to alternating current (ac) electrical circuits, R_M and X_M are the mechanical resistance and reactance, respectively; f^* is the complex force with a real component in phase with the velocity u and an imaginary component that is 90° out of phase. The force and the velocity are two independent quantities that can be either measured directly by appropriate displacement and force transducers or expressed in terms of equivalent electrical outputs and measured as such. The quantities R_M and X_M are related to the force and displacement amplitudes f_0 and X_0 by [233]

$$R_M = \frac{f_0 \sin \alpha}{\omega X_0} \tag{189}$$

and

$$X_M = -\frac{f_0 \cos \alpha}{\omega X_0} \tag{190}$$

where α is the phase angle between the force and the displacement. These quantities are related to the loss and storage moduli by

$$R_M = \frac{bG''}{\omega} \tag{191}$$

and

$$X_M = -\frac{bG'}{\omega} - \frac{K'}{\omega} + m\omega \tag{192}$$

where b again is a geometric form factor (Table 5.3), K' is the stiffness of the driving member (a known quantity for a given system), and m is its mass. The parameter K' is sometimes expressed as the elastance S_M^0, where

$$S_M^0 = \frac{K'}{b} \tag{193}$$

In the limit of low frequencies and a soft driving element $(K' \to 0)$ one gets

$$\eta'' = \frac{f_0 \cos \alpha}{\omega b X_0} \tag{194a}$$

and

$$\eta' = \frac{f_0 \sin \alpha}{\omega b X_0} \tag{194b}$$

where $\alpha = \delta$, the phase angle between the stress and the strain.

An example of an instrument based on this general approach is the Birnboim apparatus [242] shown schematically in Figure 5.25. (The system shown is a modified version designed by Massa and Schrag [243].) In this system a cylindrical rod is driven axially into the liquid sample by an electromagnetic drive with a frequency range of 0.01–500 Hz. The liquid is held in a coaxial cylindrical cup with a nominal volume of approximately 1 cm³. The displacement is measured by a capacitance transducer and the force is obtained from the magnitude of the alternating current passing through the driving coil. The apparatus can operate either in an annular pumping mode (shown in Figure 5.25) or in a coaxial open-ended mode based on the Segel–Pochettino con-

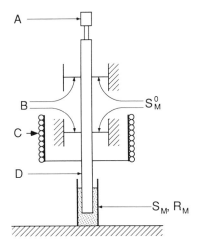

A

B —— S_M^0

C

D

S_M, R_M

Figure 5.25 Schematic diagram of the moving element in the Birnboim apparatus: (A) moving element of capacitor, (B) suspension springs, (C) drive coil assembly, and (D) driving rod. Reprinted with permission from J. D. Ferry, *Viscoelastic Properties of Polymers*, 3rd ed., Wiley, New York, 1980, p. 111. Copyright © 1980 by John Wiley & Sons, Inc.

figuration (see Table 5.3), where the change in configuration requires only substitution of the proper form factor in (194).

Conceptually simpler but less common dynamic viscometers are those based on linear actuation. A typical system based on a design by Miles [244] is shown in Figure 5.26. In this system the liquid is confined by surface tension to the narrow gap between two parallel plates. One plate is driven by a linear actuator that generates forced periodic displacements with an amplitude X_1^0. The second plate (the driven member) is attached to a spring with a constant K and it oscillates in its plane with an amplitude X_2^0 and a phase lag α behind the driving plate. As before, expressions for the dynamic functions can be derived from the equation of motion for the driven plate and assuming a linear–viscoelastic response for the liquid. The equation of motion for the driven plate takes the form

$$m\ddot{X}_2 + KX_2 + b\eta^*(\dot{X}_2 - \dot{X}_1) = 0 \tag{195}$$

where m is the mass of the driven plate and b is a geometric form factor (Table 5.3). Form factors for only the two most common *linear* systems, the parallel plate system and the Segel–Pochettino configuration discussed earlier, are listed in Table 5.3. Solution of (195) yields

$$\eta' = \frac{(K - m\omega^2)v \sin \alpha}{\omega b(v^2 - 2v \cos \alpha + 1)} \tag{196a}$$

$$\eta'' = \frac{(K - m\omega^2)(v \cos \alpha - v^2)}{\omega b(v^2 - 2v \cos \alpha + 1)} \tag{196b}$$

where v is the amplitude ratio X_2^0/X_1^0. Note the similarity of these equations to (184) with the only difference being the substitution of I for m. Also, as before, the

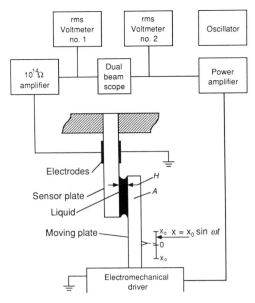

Figure 5.26   Linearly driven dynamic rheometer. Schematic diagram of the simple shear vibrator of Miles. The displacement is specified at the moving plate and the force is measured at the sensor plate. Reprinted with permission from J. D. Ferry, *Viscoelastic Properties of Polymers*, 3rd ed., Wiley, New York, 1980, p. 109. Copyright © 1980 by John Wiley & Sons, Inc.

natural frequency for the system $\omega_0 = \sqrt{K/m}$ and its vicinity should be strictly avoided. For the special case of a very stiff spring ($K \gg m\omega^2$, $v \ll 1$), (196) reduces to, see (185),

$$\eta' = \frac{Kv \sin \alpha}{\omega b} \tag{197a}$$

$$\eta'' = \frac{Kv \cos \alpha}{\omega b} \tag{197b}$$

and, again, only in this limiting case $\alpha = \delta$ [cf. (171)].

A unique device for measuring the dynamic properties of liquids is the eccentric rotating disk (ERD) system, sometimes called the Maxwell Orthogonal Rheometer after one of its originators. In this system the liquid is confined in the gap between two parallel disks of radius R whose centers are displaced a small distance a from one another (see Figure 5.27). The upper disk is driven at a constant angular velocity Ω, and the lower disk is assumed to follow at the same velocity. Although the flow field generated in this system is steady in the laboratory (Eulerian) frame, the fluid experiences periodic shear deformations; that is, the flow oscillates in the Lagrangian frame. Gent [245] appears to be the first to recognize the possibility of using the ERD system for measuring the

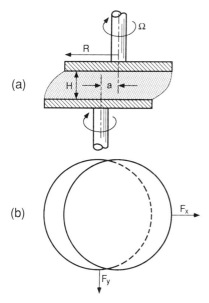

(a)

(b)

Figure 5.27 Schematic diagram of the eccentric rotating disk (ERD) system: (*a*) side view and (*b*) top view.

linear viscoelastic properties of liquids. He proposed the following expressions for the components of the complex viscosity

$$\eta' = F_x \left(\frac{H}{\pi R^2 a} \right) \tag{198a}$$

and

$$\eta'' = F_y \left(\frac{H}{\pi R^2 a} \right) \tag{198b}$$

where F_x and F_y are orthogonal forces acting in the plane of the disks as shown in Figure 5.27. These relations were later tested thoroughly and rationalized by Maxwell and Chartoff [246], Blyler and Kurtz [247], and Macosko and Davis [248], among others. Macosko and Davis conducted a particularly thorough evaluation of the Gent relations and they point out the need to correct for instrument compliance when testing relatively stiff (viscous) materials. They outline a general scheme for correcting for instrument compliance in the ERD system. The compliance problem was later reexamined by Gottlieb and Macosko [249] for the ERD system as well as for a conventional forced-oscillation apparatus. The ERD system has been popularized in the 1970s through a commercial rheometer, the Rheometrics Mechanical Spectrometer, which was the first commercial apparatus to allow eccentric alignment of the disks and measurement of lateral forces. In general, the ERD method is not practical for use with polymer melts because of the large compliance error associated with such liquids.

We close this section by reiterating the need to operate always under low strains. The expressions for η' and η'' presented in this section, including the Gent relations [(198)], are strictly valid only within the linear–viscoelastic range, that is, when the material functions are independent of the strain amplitude. In conventional oscillatory devices the strain amplitude is θ_1^0/H, whereas in the ERD system the strain amplitude is a/H.

6.3 Surface-Loading and Wave-Propagation Techniques

When studying the rheological properties of low-viscosity liquids, it is especially important to conduct measurements at very high frequencies, often in the kilohertz and megahertz ranges, to probe the elastic and non-Newtonian characteristics of the liquid. Unfortunately, most of the commercial dynamic viscometers that operate predominantly in the gap-loading mode are invariably limited to frequencies below ca. 100 Hz. As noted earlier, the gap-loading techniques are limited to relatively low frequencies, narrow gaps, and high viscosities [cf. (180)] corresponding to a shear wavelength that exceeds the thickness of the sample. If the sample is relatively thick, a wave generated on one surface will be attenuated as it propagates toward the opposite (reflecting) surface. For a sufficiently large gap the wave decays completely before reaching the opposite surface and no reflection occurs. This is referred to as the *surface-loading limit*. In this limit the only relevant geometric parameter is the area of the driving surface in contact with the liquid.

Since in this case all the measurements of force and displacement must be conducted on the driving surface, as in the Birnboim apparatus (Figure 5.25), the concept of mechanical impedance is again invoked. This quantity is defined as the ratio of the complex force to velocity [see (188)] and R_M and X_M, the mechanical resistance and reactance, have the same meaning as before. In departure from the gap-loading case, however, the dynamic moduli depend on both components of Z_M^* and are given by [233]

$$\eta' = 2R_M X_M/\omega\rho A^2 \tag{199a}$$

$$\eta'' = (R_M^2 - X_M^2)/\omega\rho A^2 \tag{199b}$$

where ρ is the density of the liquid, A is the surface area of contact, and R_M and X_M represent the dynamic contributions from the liquid itself; that is, after subtraction of the inertia, elastance, and viscous (frictance) losses in the driving member. Although (199) is strictly valid only for plane waves, it can also be used for curved surfaces, provided the attenuation of the waves is much smaller than the radius of curvature of the surface (the waves die out very close to the surface!). For a purely viscous Newtonian liquid (199) reduces to

$$\eta_0 = 2R_M^2/\omega\rho A^2 \tag{200}$$

which can serve as a check on the validity of (199) and used to calibrate the instrument.

Clearly, most of the dynamic systems described in the previous section can operate in principle in the surface-loading mode, provided the oscillation drives can generate sufficiently high frequencies and the measurement of force and displacement can be conducted on the driving member. The apparatus of Birnboim [242] and its variants (cf. Figure 5.25) is particularly adaptable to surface-loading tests since it allows simultaneous measurement of force and displacement on the driving surface.

The multiple-lumped resonator developed by Schrag and Johnson [250] based on an earlier proposal by Birnboim was designed specifically for surface loading measurements. The resonator (Figure 5.28) consists of five cylinders connected by torsion rods of different diameters. Torsional oscillations can be generated at one of five possible resonance frequencies of the five-cylinder assembly without a change in the apparatus. The oscillations are generated by an electromagnetic drive with a frequency range of 0.1–8 kHz. The torque is derived from the current in the driving coil, and a high-precision optical system is used to measure the displacement. The instrument can resolve stresses as low as 0.1 Pa and is suitable for relatively thin liquids ($\eta < 0.05$ Pa·s). A more versatile technique, pioneered by Mason [251] and later improved by Nomura and co-workers [252], involves immersion of an oscillating cylindrical quartz crystal in the test liquid. From the resistance of the crystal and its resonant

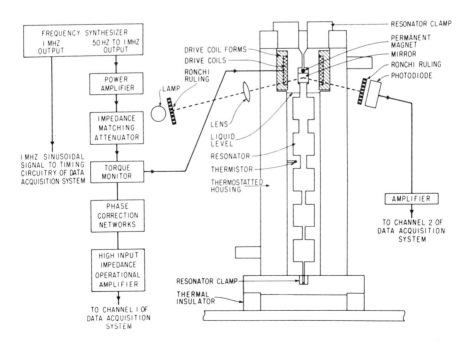

Figure 5.28 Schematic diagram of the multiple-lumped resonator. Reprinted with permission from J. L. Schrag and R. M. Johnson, *Rev. Sci. Instrum.*, **42**, 224 (1971).

frequency one can determine R_M and X_M and use this information to obtain the dynamic moduli via (199). Because of the strong damping in this system the generated shear waves are assumed to be planar despite the cylindrical configuration of the crystal. The crystal in this setup is driven piezoelectrically in the torsion mode and it can generate frequencies up to 100 kHz. (The frequency can be varied by changing the crystal.) Yet higher frequencies (>1 MHz) are reported for instruments employing piezoelectric crystals in conjunction with various ultrasonic methods. These techniques are described in detail in reviews by Barlow and co-workers [236, 253] and McSkimin [254] and are not considered further here because of their specialized nature.

If the damping of the shear wave is not too severe, it is possible to obtain the viscoelastic properties of the liquid from the characteristics of the exponentially decaying wave train in the liquid medium (see Figure 5.29). Experimentally, the plane wave can be generated by a vertically vibrating flat plate immersed in the test liquid. The attenuation φ and wavelength λ of the decaying wave can be detected, for example, by using flow birefringence techniques whereby the shear strain in the liquid can be related to the observed birefringence based on the stress-optical law [255]. This latter approach is limited, of course, to transparent and highly birefringent liquid media. Once the wavelength and attenuation are known, the components of the complex viscosity can be obtained from [256]

$$\eta' = \frac{4\pi\rho\lambda^3\omega\varphi}{[4\pi^2 + (\lambda\varphi)^2]^2} \tag{201a}$$

$$\eta'' = \frac{\rho\lambda^2\omega[4\pi^2 - (\lambda\varphi)^2]}{[4\pi^2 + (\lambda\varphi)^2]^2} \tag{201b}$$

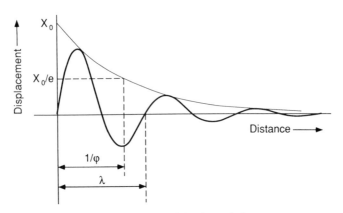

Figure 5.29 Exponentially damped shear wave.

where ω is the vibration frequency of the plate. From inspection of these equations it is easy to show that for small damping ($\lambda\varphi/2\pi \ll 1$)

$$\omega\lambda \approx 2\pi\sqrt{G'/\rho} \tag{202}$$

and

$$\tan \delta = \lambda\varphi/\pi \tag{203}$$

In this limit, however, reflections from the walls of the vessel containing the liquid may obscure the results and lead to large errors. (In an extreme case the gap-loading limit may be reached!) On the other hand, if the damping is high ($\lambda\varphi > 3$), large errors may result from difficulties in measuring the wavelength. The wave-propagation technique has been used at frequencies of 0.004–5 kHz.

7   OVERVIEW

It is hoped that this review has exposed the reader to the wide range of options available for performing viscometric evaluations on various liquids. The choice of a particular viscometric technique should be dictated by the viscosity range of the test liquid, the temperature and shear-rate range of the flow process under evaluation, the special requirements for confining the liquid in the viscometer and the ease of use and robustness of the viscometric instrument. Generally, the controlled-rate Poiseuille-type viscometers (Sections 3.2 and 3.3) are most suitable for the characterization of relatively viscous liquids (10–10^6 Pa·s) at moderate-to-high shear-rates ($10 - 3 \times 10^4$ 1/s), while the Couette-type rotational viscometers (Sections 4.2, 4.4, and 4.5) are often used with low-to-moderate viscosity liquids (10^{-3}–10^4 Pa·s) at low-to-moderate shear rates ($10^{-2} - 500$ 1/s). Gravity-drainage and various specialized rotational viscometers are typically used with low viscosity liquids ($10^{-4} - 1$ Pa·s) at relatively low shear-rates. If extreme shear rates are sought various specialized viscometers must be used. For example, the Lodge stressmeter (Section 3.3) and a modified version of the cylindrical Couette viscometer (Section 4.2) can be used to attain very high shear rates ($> 10^5$ 1/s), while the falling ball viscometer (Section 5.2) and especially designed rotational viscometers (Sections 4.2, 4.4, and 4.5) can be used to generate data at extremely low shear rates ($< 10^{-2}$ 1/s). Since most viscometers are torque- or stress-limited, especially at high shear rates, the actual shear-rate range of the instrument is dictated by the viscosity of the liquid,

$$\dot{\gamma}_{\lim} = \frac{\tau_{\lim}}{\eta} \tag{204}$$

where $\dot{\gamma}_{\text{lim}}$ and τ_{lim} are the extreme values of shear rate and shear stress attainable by the instrument. Clearly, if the instrument is rate limited its limiting stress will also be dictated by the viscosity of the liquid according to (204). The dynamic techniques discussed in Section 6 can be used to obtain viscometric information over a wide range of shear rates (frequencies) and viscosities provided that the Cox–Merz rule (178) is applicable.

Recent advances in instrumentation and computer technology have made the execution of viscometric tests with state-of-the-art viscometers a relatively straightforward and easily automated task. However, correct interpretation of viscometric data still requires careful attention to various experimental details. Generally, three classes of experimental artifacts ("errors") may hamper the accuracy and reliability of viscometric evaluations: (1) instrumental artifacts, (2) material artifacts, and (3) hydrodynamic artifacts. Instrumental artifacts usually relate to inadequate temperature control, insensitive or over-loaded stress transducers, geometric misalignment within the viscometer, and so on. Examples of material artifacts are inhomogeneities in the test liquid due to improper mixing, molecular degradation, solvent evaporation, shear-induced phase-separation, particle agglomeration and flocculation. The latter problems are of particular concern in the study of heterogeneous liquids (see Section 2.5). Hydrodynamic artifacts usually involve flow instabilities, secondary flows, end effects, transient effects due to fluid elasticity, and other hydrodynamic phenomena not accounted for by the working equations used to extract the viscosity from the raw data (cf. Sections 3.4, 4.3, and 4.6). Any one or a combination of these artifacts can lead to significant error and a corresponding misinterpretation of the rheological data. The most direct way to confirm the validity of viscometric data is to repeat the test on two (or more) devices with an overlapping shear-rate range but, preferably, with different operating modes such as Poiseuille-type and Couette-type viscometers (see, for example, [38]). Agreement between the two techniques confirms that the measured viscosity is a true, instrument-invariant material function; otherwise, the experimental error must be diagnosed and corrected.

On-line measurements of viscosity and other rheological properties for use in control and monitoring of various chemical and polymer processes have gained considerable popularity in recent years [257, 258]. The great sensitivity of viscosity to composition and molecular structure (see Section 2), and recent improvements in instrumentation [259] have made such measurements very attractive for process control applications. In general, on-line rheometry can be conducted in two distinct modes: "in-line" and "side-stream." In the former the measurement is conducted directly on the liquid along the process stream, while in the latter the viscometric measurement is performed on a small liquid stream that is bled off the main process stream and then either recycled back or discarded. Various viscometric techniques have been proposed for use in on-line applications including capillary, slit die, and several types of rotational (torsional) rheometers [259], but the area of on-line rheometry is fertile ground for further development and innovation.

ACKNOWLEDGMENTS

The author gratefully acknowledges Dr. D. J. Massa and Dr. R. H. Colby of Eastman Kodak Company and Professor S. Middleman of the University of California at San Diego for reviewing this manuscript and providing helpful comments.

References

1. J. R. Van Wazer, J. W. Lyons, K. Y. Kim, and R. E. Colwell, *Viscosity and Flow Measurements*, Interscience, New York, 1963.

2. S. Middleman, *The Flow of High Polymers*, Wiley-Interscience, New York, 1968.

3. K. Walters, *Rheometry*, Chapman & Hall, London, 1975.

4. A. Ya. Malkin, A. A. Askadsky, V. V. Kovriga, and A. E. Chalykh, *Experimental Methods of Polymer Physics, Part 4*, Prentice-Hall, Englewood Cliffs, NJ, 1983.

5. R. B. Bird, R. C. Armstrong, and O. Hassager, *Dynamics of Polymeric Liquids*, 2nd ed., Vol. 1, *Fluid Mechanics*, Wiley, New York, 1987.

6. J. M. Dealy and K. F. Wissbrun, *Melt Rheology and Its Role in Plastics Processing*, Van Nostrand-Reinhold, New York, 1990.

7. J. M. Dealy, *J. Rheol.*, **28**, 181 (1984).

8. B. D. Coleman, H. Markovitz, and W. Noll, *Viscometric Flows of Non-Newtonian Fluids*, Springer-Verlag, New York, 1966, pp. 14–16.

9. A. S. Lodge, *Body Tensor Fields With Applications to Polymer Rheology*, Academic, New York, 1974.

10. B. D. Coleman, H. Markovitz, and W. Noll, *Viscometric Flows of Non-Newtonian Fluids*, Springer, Berlin, 1966.

11. A. C. Pipkin and R. I. Tanner, in S. Nemat-Nasser, Ed., *Mechanics Today*, Vol. 1, Pergamon, New York, 1972, pp. 262–321.

12. R. B. Bird, R. C. Armstrong, and O. Hassager, *Dynamics of Polymeric Liquids*, 2nd ed., Vol. 1, *Fluid Mechanics*, Wiley, New York, 1987, p. 153.

13. J. Meissner, *Annu. Rev. Fluid Mech.*, **17**, 45 (1985).

14. J. M. Dealy, *Polym. Eng. Sci.*, **11**, 433 (1971).

15. C. J. S. Petrie, *Elongational Flows*, Pitman, London, 1979.

16. R. I. Tanner, *AIChE J.*, **22**, 910 (1976).

17. R. B. Bird, W. E. Stewart, and E. N. Lightfoot, *Transport Phenomena*, Wiley, New York, 1960, p. 79.

18. W. Noll, *Arch. Ration. Mech. Anal.*, **2**, 197 (1958).

19. R. B. Bird, R. C. Armstrong, and O. Hassager, *Dynamics of Polymeric Liquids*, 2nd ed., Vol. 1, *Fluid Mechanics*, Wiley, New York, 1987, Chaps. 7 and 8.

20. R. Larson, *Constitutive Equations for Polymer Melts and Solutions*, Butterworth, Boston, MA, 1987.

21. R. B. Bird, W. E. Stewart, and E. N. Lightfoot, *Transport Phenomena*, Wiley, New York, 1960, pp. 76–81.

22. R. B. Bird, W. E. Stewart, and E. N. Lightfoot, *Transport Phenomena*, Wiley, New York, 1960, pp. 183–190.

23. J. O. Hirschfelder, C. F. Curtiss, and R. B. Bird, *Molecular Theory of Gases and Liquids*, Wiley, New York, 1954, Chaps. 1, 8, and 9.

24. S. Chapman and T. G. Cowling, *Mathematical Theory of Non-Uniform Gases*, 2nd ed., Cambridge University Press, Cambridge, UK, 1951.

25. R. B. Bird, W. E. Stewart, and E. N. Lightfoot, *Transport Phenomena*, Wiley, New York, 1960, pp. 15–19.

26. F. M. White, *Viscous Fluid Flow*, McGraw-Hill, New York, 1974, p. 26.

27. G. L. Chierici and A. Paratella, *AIChE J.*, **15**, 786 (1969).

28. H. Eyring, *J. Chem. Phys.*, **4**, 283 (1936); S. Glasstone, K. J. Laidler, and H. Eyring, *Theory of Rate Processes*, McGraw-Hill, New York, 1951, pp. 477–551.

29. J. F. Kincaid, H. Eyring, and A. E. Stearn, *Chem. Rev.*, **28**, 301 (1941).

30. A. K. Doolittle and D. B. Doolittle, *J. Appl. Phys.*, **28**, 901 (1957).

31. J. Frenkel, *Kinetic Theory of Liquids*, Oxford University Press, London, 1946.

32. J. G. Kirkwood, *Documents on Modern Physics*, Gordon & Breach, New York, 1967.

33. A. Bondi, *Physical Properties of Molecular Crystals, Liquids and Glasses*, Wiley, New York, 1968, Chap. 12.

34. Y. S. Touloukian, C. Y. Ho, R. W. Powell, P. G. Klemens, and P. E. Liley, *Thermophysical Properties of Matter*, Vol. 11, Plenum, New York, 1977.

35. R. B. Bird, R. C. Armstrong, and O. Hassager, *Dynamics of Polymeric Liquids*, 2nd ed., Vol. 1, *Fluid Mechanics*, Wiley, New York, 1987, p. 172.

36. E. W. Merrill, H. S. Mickley, and A. Ram, *Trans. Soc. Rheol.*, **6**, 119 (1962).

37. A. Ram, *Chem. Eng. Commun.*, **30**, 285 (1984).

38. P. J. Hamersma, J. Ellenberger, and J. M. H. Fortuin, *Chem. Eng. Sci.*, **38**, 819 (1983).

39. P. J. Carreau, doctoral dissertation, University of Wisconsin, Madison, WI, 1968.

40. M. M. Cross, *Rheol. Acta*, **18**, 609 (1979).

41. K. Yasuda, R. C. Armstrong, and R. E. Cohen, *Rheol. Acta*, **20**, 163 (1981).

42. C. A. Hieber and H. H. Chiang, *Rheol. Acta*, **28**, 321 (1989).

43. S. Matsushisa and R. B. Bird, *AIChE J*, **11**, 588 (1965).

44. R. E. Powell and H. Eyring, *Nature* (*London*), **154**, 427 (1944).

45. R. B. Bird, R. C. Armstrong, and O. Hassager, *Dynamics of Polymeric Liquids*, 2nd ed., Vol. 1, *Fluid Mechanics*, Wiley, New York, 1987, p. 228.

46. A. B. Metzner and M. Whitlock, *Trans. Soc. Rheol.*, **2**, 239 (1958).

47. M. J. Ballard, R. Buscall, and F. A. Waite, *Polymer*, **29**, 1287 (1988).

48. J. Greener and R. W. Connelly, *J. Rheol.*, **30**, 285 (1986).

49. E. C. Bingham, *Fluidity and Plasticity*, McGraw-Hill, New York, 1922, pp. 215–218.

50. W. H. Herschel and R. Bulkley, *Proc. Am. Soc. Test. Mater.*, **26**, 621 (1926).

51. N. Casson, "A Flow Equation for Pigment-Oil Suspensions of the Printing Ink Type," in C. C. Mill, Ed. *Rheology of Disperse Systems*, Pergamon, New York, 1959, p. 84.

52. H. A. Barnes and K. Walters, *Rheol. Acta*, **24**, 323 (1985).

53. T. S. Stephens, H. H. Winter, and M. Gottlieb, *Rheol. Acta*, **27**, 263 (1988).

54. A. S. Yoshimura, R. K. Prud'homme, H. M. Princen, and A. D. Kiss, *J. Rheol.*, **31**, 699 (1987).

55. J. D. Ferry, *Viscoelastic Properties of Polymers*, 3rd ed., Wiley, New York, 1980, Chap. 11.

56. E. N. Andrade, *Nature (London)*, **125**, 309, 582 (1930).

57. G. V. Vinogradov and A. Ya. Malkin, *Rheology of Polymers*, Springer-Verlag, New York, 1980, pp. 116–121.

58. D. W. van Krevelen, *Properties of Polymers*, Elsevier, Amsterdam, 1976, pp. 339–346.

59. H. Vogel, *Phys. Z.*, **22**, 645 (1921).

60. G. S. Fulcher, *J. Am. Chem. Soc.*, **8**, 339 (1925).

61. C. A. Angell, *J. Phys. Chem. Solids*, **49**, 863 (1988).

62. R. F. Westover, *SPE Trans.*, **July**, 222, 1962.

62a. R. J. Crowson, A. J. Scott, and D. W. Saunders, *Polym. Eng. Sci.*, **21**, 748 (1981).

63. W. K. Moonan and N. W. Tschoegl, *Int. J. Polym. Mater.*, **10**, 199 (1984).

64. L. A. Utracki, *J. Rheol.*, **30**, 829 (1986).

65. W. W. Graessley, "The Entanglement Concept in Polymer Rheology," in *Advances in Polymer Science*, **16**, Springer-Verlag, Berlin, 1974.

66. Y. Takahashi, Y. Isono, I. Noda, and M. Nagasawa, *Macromolecules*, **18**, 1002 (1985).

67. G. C. Berry and T. G. Fox, *Adv. Polym. Sci.*, **5**, 261 (1968).

68. D. K. Carpenter and L. Westerman, "Viscometric Methods for Studying Molecular Weight and Molecular Weight Distribution," in P. E. Slade, Ed., *Polymer Molecular Weights*, Part II, Dekker, New York, 1975, Chap. 7.

69. P. E. Rouse, Jr., *J. Chem. Phys.*, **21**, 1272 (1953).

70. B. H. Zimm, *J. Chem. Phys.*, **24**, 269 (1956).

71. J. D. Ferry, *Viscoelastic Properties of Polymers*, 3rd ed., Wiley, New York, 1980, pp. 247–252.

72. F. Bueche, *J. Chem. Phys.*, **48**, 4781 (1968).

73. W. W. Graessley, *J. Chem. Phys.*, **47**, 1942 (1967).

74. M. Doi and S. F. Edwards, *J. Chem. Soc. Faraday Trans. 2*, **74**, 1789 and 1818 (1978); and **75**, 38 (1979). See also M. Doi and S. F. Edwards, *The Theory of Polymer Dynamics*, Clarendon, Oxford, 1986.

75. W. W. Graessley, "Entangled Linear, Branched and Network Polymer Systems–Molecular Theories," in *Advances in Polymer Science*, **47**, Springer-Verlag, Berlin, 1982, pp. 68–117.

76. W. W. Graessley, "The Entanglement Concept in Polymer Rheology," in *Advances in Polymer Science*, **16**, Springer-Verlag, Berlin, 1974, pp. 134–138.

77. S. Middleman, *J. Appl. Polym. Sci.*, **11**, 470 (1967).

78. M. Rubinstein and R. H. Colby, *J. Chem. Phys.*, **21**, 5291 (1988).

79. N. G. Kumar, *J. Polym. Sci. Macromol. Rev.*, **15**, 255 (1980).

80. J. Happel and H. Brenner, *Low Reynolds Number Hydrodynamics*, Prentice-Hall, Englewood Cliffs, NJ, 1965.

81. S. Kim and S. J. Karrila, *Microhydrodynamics: Principles and Selected Applica-*

tions, Butterworth, Boston, MA, 1991.

82. A. Einstein, *Ann. Phys.*, **19**, 289 (1906); **34**, 591 (1911).

83. D. G. Thomas, *J. Colloid Sci.*, **20**, 267 (1965).

84. T. Kataoka, T. Kitano, M. Sasahara, and K. Nishijina, *Rheol. Acta*, **17**, 149 (1978).

85. T. Kitano, T. Kataoka and T. Shirota, *Rheol. Acta*, **20**, 207 (1981).

86. S. H. Maron and P. E. Pierce, *J. Colloid Sci.*, **11**, 80 (1956).

87. G. Pei-Yun, in P. H. T. Uhlherr, Ed., *Proceedings of the Xth International Congress on Rheology*, Vol. 1, Sydney, 1988, p. 374.

88. D. Quemada, *Rheol. Acta*, **17**, 632 (1978).

89. C. R. Wildemuth and M. C. Williams, *Rheol. Acta*, **23**, 627 (1984); **24**, 75 (1985).

90. R. F. Landel, B. G. Moser, and A. J. Bauman, in E. H. Lee, Ed., *Proceedings of the 4th International Congress on Rheology*, Part 2, Wiley, New York, 1965, p. 663.

91. W. Kuhn and H. Kuhn, *Helv. Chim. Acta*, **28**, 97 (1945).

92. R. Simha, *J. Phys. Chem.*, **44**, 25 (1940); *J. Chem. Phys.*, **13**, 188 (1945).

93. E. J. Hinch and L. G. Leal, *J. Fluid Mech.*, **52**, 683 (1972).

94. C. J. Lin, J. H. Perry, and W. R. Schowalter, *J. Fluid Mech.*, **44**, 1 (1970).

95. G. I. Taylor, *Proc. R. Soc. London Ser. A*, **138**, 41 (1932).

96. N. A. Frankel and A. Acrivos, *J. Fluid Mech.*, **56**, 401 (1970).

97. S. J. Choi and W. R. Schowalter, *Phys. Fluids*, **18**, 420 (1975).

98. R. L. Hoffman, *Trans. Soc. Rheol.*, **16**, 155 (1977).

99. C. D. Han, *Multiphase Flow in Polymer Processing*, Academic, New York, 1976.

100. W. B. Russel, *J. Rheol.*, **24**, 287 (1980).

101. A. B. Metzner, *J. Rheol.*, **29**, 739 (1985).

102. M. R. Kamal and A. Mutel, *J. Polym. Eng.*, **5**, 293 (1985).

103. W. B. Russel, D. A. Saville, and W. R. Schowalter, *Colloidal Dispersions*, Cambridge University Press, Cambridge, 1989, Chap. 14.

104. J. L. Poiseuille, *Compt. Rend.*, **1**, 554 (1836); **11**, 961 (1840).

105. G. Hagen, *Ann. Phys. Chem.*, **46**, 423 (1839).

106. M. Reiner, "Phenomenological Macrorheology," in F. R. Eirich, Ed., *Rheology*, Vol. 1, Academic, New York, 1956, Chap. 2, p. 45.

107. B. Rabinowitsch, *Z. Phys. Chem. Abt. A*, **145**, 1 (1929).

108. R. B. Bird, R. C. Armstrong, and O. Hassager, *Dynamics of Polymeric Liquids*, 2nd ed., Vol. 1, *Fluid Mechanics*, Wiley, New York, 1987, p. 229.

109. J. M. Dealy, *Rheometers for Molten Plastics*, Van Nostrand Reinhold, New York, 1980.

110. J. E. McKie and J. F. Brandts, "High Precision Capillary Viscometry," in C. H. W. Hirs and S. N. Timasheff, Eds., *Methods in Enzymology*, Vol. XXVI, *Enzyme Structure*, Part C, Academic, New York, 1972, p. 257.

111. H. Bauer and G. Meerlander, *Rheol. Acta*, **23**, 514 (1984).

112. H. Takahashi, T. Matsuoka, and T. Kurauchi, *J. Appl. Polym. Sci.*, **30**, 4669 (1985).

112a. A. V. Shenoy and D. R. Saini, *J. Appl. Polym. Sci.*, **29**, 1581 (1984).

113. J. L. S. Wales, *The Application of Flow Birefringence to Rheological Studies of Polymer Melts*, Delft University Press, Rotterdam, 1976, p. 74.

114. A. S. Lodge, *Chem. Eng. Commun.*, **32**, 1 (1985).

115. A. S. Lodge, *J. Rheol.*, **33**, 821 (1989).

116. A. S. Lodge, Paper No. 872043, Society of Automotive Engineers, Int. Fuels and Lubricants Meeting, Toronto, Nov., 1987.

117. J. M. Dealy and K. F. Wissbrun, *Melt Rheology and Its Role in Plastics Processing*, Van Nostrand Reinhold, New York, 1990, p. 313.

118. H. M. Laun, *Rheol. Acta*, **22**, 171 (1983).

119. K. Higashitani and W. G. Pritchard, *Trans. Soc. Rheol.*, **16**, 687 (1972).

120. A. S. Lodge and L. de Vargas, *Rheol. Acta*, **22**, 151 (1983).

121. C. D. Han, *Trans. Soc. Rheol.*, **18**, 163 (1974).

122. D. V. Boger and M. M. Denn, *J. Non-Newtonian Fluid Mech.*, **6**, 163 (1980).

123. B. Atkinson, M. P. Brockelbank, C. C. H. Card, and J. M. Smith, *AIChE J.*, **15**, 548 (1969).

124. L. Choplin and P. J. Carreau, *J. Non-Newt. Fluid Mech.*, **9** (1981).

125. E. B. Bagley, *J. Appl. Phys.*, **28**, 624 (1957).

126. S. Middleman, *The Flow of High Polymers*, Wiley, New York, 1968, pp. 16–19.

127. S. Oka, "Principles of Rheometry," in F. R. Eirich, Ed., *Rheology*, Vol. 3, Academic, New York, 1960, Chap. 2.

128. A. B. Metzner, "Non-Newtonian Technology," in T. B. Drew and J. B. Hoppes, Jr., Eds., *Advances in Chemical Engineering*, Vol. 1, Academic, New York, 1956, p. 113.

129. G. A. Alvarez, A. S. Lodge, and H. J. Cantow, *Rheol. Acta*, **24**, 368 (1985).

130. C. J. S. Petrie and M. M. Denn, *AIChE J.*, **22**, 209 (1976).

131. W. Gleissle, *Rheol. Acta*, **21**, 484 (1982).

132. S. Middleman, *Fundamentals of Polymer Processing*, McGraw-Hill, New York, 1977, p. 476.

133. G. V. Vinogradov, N. I. Insarova, B. Boiko and E. K. Borisenkova, *Polym. Eng. Sci.*, **12**, 323 (1972).

134. A. V. Ramamurthy, *J. Rheol.*, **30**, 337 (1980).

135. J. Lyngaae-Jörgensen and B. Marcher, *Chem. Eng. Commun.*, **32**, 117 (1985).

136. G. V. Vinogradov, *Polymer*, **18**, 1275 (1977).

137. J. Vlachopoulos and M. Alam, *Polym. Eng. Sci.*, **12**, 184 (1972).

138. J. R. A. Pearson, *Polym. Eng. Sci.*, **18**, 222 (1978).

139. S. Middleman, *The Flow of High Polymers*, Wiley, New York, 1968, pp. 30–35.

140. R. B. Bird, *SPE J.*, **11**, 35 (1955).

141. P. C. Sikanek and R. L. Laurence, *AIChE J.*, **20**, 474 (1974).

142. H. W. Cox and C. W. Macosko, *AIChE J.*, **20**, 785 (1974).

143. H. H. Winter, *Adv. Heat Transfer*, **13**, 205 (1977).

144. N. Sutterlin, "Concentration Dependence of Dilute Polymer Solutions," in J. Brandrup and E. H. Immergut, Eds., *Polymer Handbook*, 2nd ed., Interscience, New York, 1975, p. IV-135.

145. E. O. Kraemer, *Ind. Eng. Chem.*, **30**, 1200 (1938).

146. S. H. Maron and R. B. Reznik, *J. Polym. Sci. Part A-2*, **7**, 309 (1969).

147. L. Utracki and R. Simha, *J. Polym. Sci. Part A*, **1**, 1089 (1963).

148. G. V. Schulz and F. Blaschke, *J. Prakt. Chem.*, **158**, 130 (1941).

149. W. Heller, *J. Colloid Sci.*, **9**, 547 (1954).

150. B. H. Zimm and D. M. Crothers, *Proc. Natl. Acad. Sci.*, **48**, 905 (1962).

151. L. C. Klotz and B. H. Zimm, *Macromol.*, **5**, 471 (1972).

152. P. S. Flory, *Principles of Polymer Chemistry*, Cornell University Press, Ithaca, NY, 1953, p. 24.

153. M. Kurata, Y. Tsunashima, M. Iwama, and K. Kamada, "Viscosity-Molecular Weight Relationships and Unperturbed Dimensions of Linear Chain Molecules," in J. Brandrup and E. H. Immergut, Eds., *Polymer Handbook*, 2nd ed., Interscience, New York, 1975, Chap. IV.1.

154. J. G. Kirkwood and J. Riseman, *J. Chem. Phys.*, **16**, 565 (1948).

155. P. J. Flory and T. G. Fox, Jr., *J. Am. Chem. Soc.*, **73**, 1904 (1951).

156. K. Osaki, *Macromolecules*, **5**, 141 (1972).

157. P. J. Flory, *Principles of Polymer Chemistry*, Cornell University Press, Ithaca, NY, 1953, p. 600.

158. B. H. Zimm and W. H. Stockmayer, *J. Chem. Phys.*, **17**, 230 (1949).

159. S. I. Abdel-Khalik and R. B. Bird, *Biopolymers*, **14**, 1915 (1975).

160. W.-M. Kulicke and R. S. Porter, *J. Polym. Sci. Polym. Phys. Ed.*, **19**, 1173 (1981).

161. R. W. Connelly and J. Greener, *J. Rheol.*, **29**, 209 (1985).

162. D. M. Binding and K. Walters, *J. Non-Newt. Fluid Mech.*, **1**, 277 (1976).

163. T. Y. Liu, D. W. Mead, D. S. Soong, and M. C. Williams, *Rheol. Acta*, **22**, 81 (1983).

163a. A. J. Giacomin and J. M. Dealy, *J. Rheol.*, **32**, 711 (1986).

164. S. Middleman, *The Flow of High Polymers*, Interscience, New York, 1968, pp. 19–25.

165. T. M. T. Yang and I. M. Krieger, *J. Rheol.*, **22**, 413 (1978).

166. I. M. Krieger, *Trans. Soc. Rheol.*, **12**, 5 (1968).

167. W. C. MacSporran, *J. Rheol.*, **33**, 745 (1989).

168. R. Darby, *J. Rheol.*, **29**, 369 (1985).

169. E. M. Barber, J. R. Muenger, and F. J. Villforth, *Anal. Chem.*, **27**, 425 (1955).

170. L. A. Manrique, Jr. and R. S. Porter, *Rheol. Acta*, **14**, 926 (1975).

171. E. W. Merrill, H. S. Mickley, and A. Ram, *J. Fluid Mech.*, **13**, 86 (1962).

172. M. M. Denn and J. J. Roisum, *AIChE J.*, **15**, 454 (1969).

173. R. Haas and K. Bühler, *Rheol. Acta*, **28**, 402 (1989).

174. J. R. Van Wazer, J. W. Lyons, K. Y. Kim, and R. E. Colwell, *Viscosity and Flow Measurements*, Interscience, New York, 1963, p. 86.

175. N. Adams and A. S. Lodge, *Philos. Trans. R. Soc. London Ser. A*, **256**, 149 (1969).

176. D. J. Paddon and K. Walters, *Rheol. Acta*, **18**, 565 (1979).

177. G. B. Jeffrey, *Proc. London Math. Soc.*, **14**, 327 (1915).

178. R. W. Williams, *Rheol. Acta*, **18**, 345 (1979).

179. O. J. Wein, *J. Non-Newt. Fluid Mech.*, **1**, 357 (1976).

180. D. D. Kale, R. A. Mashelkar, and J. Ulbrecht, *Rheol. Acta*, **14**, 631 (1975).

181. C. T. Hill, *Trans. Soc. Rheol.*, **16**, 213 (1972).

182. A. Kaye, A. S. Lodge, and D. G. Vale, *Rheol. Acta*, **7**, 368 (1968).

183. R. F. Ginn and A. B. Metzner, *Trans. Soc. Rheol.*, **13**, 429 (1969).

184. D. F. Griffith and K. Walters, *J. Fluid Mech.*, **42**, 379 (1970).

185. K. Walters, *Rheometry*, Chapman & Hall, London, 1975, p. 60.

186. J. F. Hutton, *Rheol. Acta*, **8**, 54 (1969).

187. R. I. Tanner and M. Keentok, *J. Rheol.*, **27**, 47 (1983).

188. D. C.-H. Cheng, *Chem. Eng. Sci.*, **23**, 895 (1968).

189. H. Giesekus, *Rheol. Acta*, **6**, 339 (1967).

190. J. G. Savins and A. B. Metzner, *Rheol. Acta*, **9**, 365 (1970).

191. K. Walters, *Rheometry*, Chapman & Hall, London, 1975, pp. 61–66.

192. R. M. Turian, *Chem. Eng. Sci.*, **11**, 361 (1972).

193. R. B. Bird and R. M. Turian, *Chem. Eng. Sci.*, **17**, 331 (1962).

194. R. M. Turian and R. B. Bird, *Chem. Eng. Sci.*, **18**, 689 (1963).

195. R. M. Turian, *Chem. Eng. Sci.*, **20**, 771 (1965).

196. L. M. Milne-Thomson, *Theoretical Hydrodynamics*, 3rd ed., Macmillan, New York, 1955, pp. 555–557.

197. G. K. Batchelor, *An Introduction to Fluid Dynamics*, Cambridge University Press, London, 1967, pp. 229–244.

198. R. B. Bird, W. E. Stewart, and E. N. Lightfoot, *Transport Phenomena*, Wiley, New York, 1960, pp. 190–196.

199. Y. I. Cho, J. P. Hartnett, and W. Y. Lee, *J. Non-Newtonian Fluid Mech.*, **15**, 61 (1984).

200. R. I. Tanner, *J. Fluid Mech.*, **17**, 161 (1963).

201. J. L. Sutterby, *Trans. Soc. Rheol.*, **17**, 559 and 575 (1973).

202. J. A. Klein, masters dissertation, Department of Chemical Engineering, Massachusetts Institute of Technology, Boston, MA, 1980.

203. O. H. Faxén, *Ark. Math. Astron. Fyz.*, **17**, 1 (1922).

204. S. Goldstein, *Proc. R. Soc. London Ser. A*, **123**, 225 (1927).

205. R. M. Turian, *AIChE J.*, **13**, 999 (1967).

206. R. P. Chhabra, P. H. T. Uhlherr, and D. V. Boger, *J. Non-Newtonian Fluid Mech.*, **6**, 187 (1980).

207. M. Gottlieb, *J. Non-Newtonian Fluid Mech.*, **6**, 97 (1979).

208. R. P. Chhabra and P. H. T. Uhlherr, *Rheol. Acta*, **18**, 593 (1979).

209. B. Caswell, *Chem. Eng. Sci.*, **25**, 1167 (1970).

210. M. L. Wasserman and J. C. Slattery, *AIChE J.*, **10**, 383 (1964).

211. O. Hassager and C. Bisgaard, *J. Non-Newt. Fluid Mech.*, **12**, 153 (1983).

212. R. Sobczak, *Rheol. Acta*, **25**, 175 (1986).

213. W. Hermann and R. Sobczak, *Monatsh. Chem.*, **117**, 753 (1986).

214. M. Gahleitner and R. Sobczak, *Rheol. Acta*, **26**, 371 (1987).

215. H. W. Lewis, *Anal. Chem.*, **25**, 507 (1953).

216. R. B. Bird and R. M. Turian, *Ind. Eng. Chem. Fundam.*, **3**, 87 (1964).

217. J. Šesták and F. Ambros, *Rheol. Acta*, **12**, 70 (1973).

218. J. R. Van Wazer, J. W. Lyons, K. Y. Kim, and R. E. Colwell, *Viscosity and Flow Measurements*, Interscience, New York, 1963, pp. 276–281.

219. J. Lohrenz, G. W. Swift, and F. Kurata, *AIChE J.*, **6**, 547 (1960).

220. E. Ashare, R. B. Bird, and J. A. Lescarboura, *AIChE J.*, **11**, 910 (1965).

221. K. K. Chee and A. Rudin, *Can. J. Chem. Eng.*, **48**, 362 (1970).

222. K. K. Chee, K. Sato, and A. Rudin, *J. Appl. Polym. Sci.*, **20**, 1467 (1976).

223. K. K. Chee and A. Rudin, *Rheol. Acta*, **16**, 635 (1977).

224. C. W. Macosko, *Br. Polym. J.*, **17**, 239 (1985).

225. W. P. Cox and E. H. Merz, *J. Polym. Sci.*, **28**, 619 (1958).

226. T. Kitano, T. Nishimura, T. Kataoka, and T. Sakai, *Rheol. Acta*, **19**, 671 (1980).

227. J. M. Dealy and K. F. Wissbrun, *Melt Rheology and Its Role in Plastics Processing*, Van Nostrand Reinhold, New York, 1990, p. 584.

228. W.-M. Kulicke and R. S. Porter, *Rheol. Acta*, **19**, 601 (1980).

229. R. B. Bird, R. C. Armstrong, and O. Hassager, *Dynamics of Polymeric Liquids*, Vol. 1, *Fluid Mechanics*, Wiley, New York, 1977, p. 379.

230. G. V. Vinogradov and A. Ya. Malkin, *Rheology of Polymers*, Mir, Moscow, 1980, Chap. 3.4.

231. H. C. Booj, P. Leblans, J. Palmen, and G. Tiemersma-Thoone, *J. Polym. Sci. Polym. Phys. Ed.*, **21**, 1703 (1983).

232. J. L. Schrag, *Trans. Soc. Rheol.*, **21**, 399 (1977).

233. J. D. Ferry, *Viscoelastic Properties of Polymers*, 3rd ed., Wiley, New York, 1980, Chaps. 5 and 6.

234. K. Walters, *Rheometry*, Chapman & Hall, London, 1975, Chap. 6.

235. A. Ya. Malkin, A. A. Askadsky, V. V. Kovriga, and A. E. Chalykh, *Experimental Methods of Polymer Physics*, Part 3, Prentice-Hall, Englewood Cliffs, NJ, 1983.

236. G. Harrison and A. J. Barlow, "Dynamic Viscosity Measurement," in P. D. Edmonds, Ed., *Methods of Experimental Physics*, Vol. 19, *Ultrasonics*, Academic, New York, 1981, p. 137.

237. T. E. Morrison, L. J. Zapas, and T. W. de Witt, *Rev. Sci. Instrum.*, **26**, 357 (1955).

238. J. A. Duiser, doctoral dissertation, University of Leiden, Leiden, South Holland, 1965.

239. J. L. den Otter, *Rheol. Acta*, **8**, 355 (1969).

240. K. te Nijenhuis and R. van Donselaar, *Rheol. Acta*, **24**, 47 (1985).

241. K. te Nijenhuis, *Colloid Polym. Sci.*, **259**, 522 (1981).

242. M. H. Birnboim and J. D. Ferry, *J. Appl. Phys.*, **32**, 2305 (1961).

243. D. J. Massa and J. L. Schrag, *J. Polym. Sci. Part A-2*, **10**, 71 (1972).

244. D. O. Miles, *J. Appl. Phys.*, **33**, 1422 (1962).

245. A. N. Gent, *Br. J. Appl. Phys.*, **11**, 165 (1960).

246. B. Maxwell and R. P. Chartoff, *Trans. Soc. Rheol.*, **9**, 41 (1965).

247. L. L. Blyler, Jr. and S. J. Kurtz, *J. Appl. Phys.*, **11**, 127 (1967).

248. C. W. Macosko and W. M. Davis, *Rheol. Acta*, **13**, 814 (1974).

249. M. Gottlieb and C. W. Macosko, *Rheol. Acta*, **21**, 90 (1982).

250. J. L. Schrag and R. M. Johnson, *Rev. Sci. Instrum.*, **42**, 224 (1971).

251. W. P. Mason, *Trans. ASME*, **69**, 359 (1947).

252. H. Nomura, T. Konaka, H. Shimizu, S. Kato, and Y. Miyahara, *Rep. Prog. Polym. Phys. Jpn.*, **20**, 127 (1977).

253. A. J. Barlow, G. Harrison, J. Richter, H. Seguin, and J. Lamb, *Lab. Pract.*, **10**, 786 (1961).

254. H. J. McSkimin, "Ultrasonic Methods for Measuring the Mechanical Properties of Liquids and Solids," in W. P. Mason, Ed., *Physical Acoustics*, Vol. 1A, Academic, New York, 1964, Chap. 4.

255. H. Janeschitz-Kriegl, *Polymer Melt Rheology and Flow Birefringence*, Springer-Verlag, New York, 1983, pp. 60–66.

256. F. T. Adler, W. M. Sawyer, and J. D. Ferry, *J. Appl. Phys.*, **20**, 1036 (1949).

257. J. M. Dealy, Proceedings of the 49th Ann. Tech. Conf. of the Society of Plastics Engineers, May 1991, p. 2296.

258. A. Göttfert, Proceedings of the 49th Ann. Tech. Conf. of the Society of Plastics Engineers, May 1991, p. 2299.

259. J. M. Dealy and K. F. Wissbrun, *Melt Rheology and Its Role in Plastics Processing*, Van Nostrand Reinhold, New York, 1990, Chapter 12.

Chapter **6**

TEMPERATURE MEASUREMENT WITH APPLICATION TO PHASE EQUILIBRIA STUDIES

J. Bevan Ott and J. Rex Goates

Physical Methods of Chemistry, Second Edition Volume Six: Determination of Thermodynamic Properties   Edited by Bryant W. Rossiter and Roger C. Baetzold
ISBN 0-471-57087-7   Copyright 1992 by John Wiley & Sons, Inc.

1   INTRODUCTION

Since almost all physical properties are temperature dependent, an understanding of temperature and its measurement [1, 2] is fundamental to scientific experimentation.[†] In this chapter we describe temperature and the methods of temperature measurement, with special application to the measurement of the temperature of equilibrium phase changes. Melting and boiling processes and solid-state transitions are emphasized. Such topics were addressed in the third edition of Volume I in this series [3].

2   TEMPERATURE

In a qualitative sense, temperature[‡] is the potential that determines the direction of heat flow. Like other potentials, temperature is an intensive quantity; that is, it does not depend on quantity of matter, and temperatures, therefore, are not additive. If two objects at different temperatures are in thermal contact, the net heat flow is from the object at the higher temperature to the one at the lower temperature. Thermal equilibrium exists between two objects when the net exchange of heat from one to the other is zero. An extension of this principle, known as the *zeroth law of thermodynamics*, states that two systems in thermal equilibrium with a third are in thermal equilibrium with each other. In other words, if T_1, T_2, and T_3 are the temperatures of three systems with $T_1 = T_3$ and $T_2 = T_3$, then $T_1 = T_2$. This conclusion seems almost trivial, but it is necessary to make since it serves as the basis for all temperature measurements. Thermometers, the devices used to measure temperature, measure only their own temperatures. We are justified in saying that a system is at the same temperature as the thermometer only if thermal equilibrium is established. Failure to establish thermal equilibrium between the thermometer and the system is a common source of error in temperature measurement.

3   TEMPERATURE SCALES

A quantitative description of temperature requires the definition of a temperature scale. The temperature scales of fundamental importance to a scientist are the absolute or ideal gas scale ($°A$) and the thermodynamic or Kelvin scale (K). As we shall see, these two scales are identical, and temperature can be expressed in degrees absolute and kelvin interchangeably. Other temperature scales such as Celsius or centigrade ($°C$) and Fahrenheit ($°F$) are now defined in terms of the absolute or Kelvin scale. Note that we do *not* write degrees kelvin or $°K$.

[†]Reference [1] contains a comprehensive collection of papers describing temperature measurements, and [2] is a second general reference covering this subject.
[‡]References [4–6] are three recent books that describe temperature and its measurement in detail.

Temperature on the thermodynamic scale is expressed as kelvin and temperature changes as kelvins, but not as degrees kelvin.

3.1   Ideal Gas Temperature Scale

The absolute or ideal gas temperature scale is based on the pressure, volume, temperature (p, V, T) relationships for an ideal gas as given by

$$pV = nRT \tag{1}$$

where n and R are the number of moles of gas and the gas constant, respectively. An ideal gas thermometer [7–11] compares the pV product of the gas at two temperatures.

$$T_2 = T_1 \frac{(pV)_2}{(pV)_1} \tag{2}$$

Thus, T_2 can be calculated from T_1 by measuring the pV product of the ideal gas at the two temperatures.

The size of the absolute temperature degree is defined by dividing the temperature interval from absolute zero (the temperature where the volume of the ideal gas becomes zero) to the triple point (tp) of water into exactly $273.16°$A. With this definition, the absolute temperature is obtained by comparing pV at T with $(pV)_{tp}$, the pV product at the triple point. For the ideal gas

$$T = 273.16 \frac{(pV)_T}{(pV)_{tp}} \tag{3}$$

Since it is not possible to construct an ideal gas thermometer, (3) must be modified to apply to a real gas. For real gases, the pV product can be extrapolated to zero pressure (where all gases behave ideally) and (3) can be written in the form

$$T = 273.16 \lim_{p \to 0} \frac{(pV)_T}{(pV)_{tp}} \tag{4}$$

In practice, however, (4) is usually not applied directly. Rather than extrapolate experimental pV values to zero pressure, an equation of state such as a virial equation[†] is used to correct the pV product of the gas to ideal behavior. Helium is the usual choice of gas for the gas thermometer, although other gases, especially H_2, have also been used.

Gas thermometers are usually constructed to measure the pV product of the gas by holding the volume constant and measuring p as the temperature is

---

[†]A discussion of the virial equation and gas imperfection is found in references such as [12–16].

changed, although sometimes the reverse procedure is used; that is, p is held constant and V is allowed to change with T. Figure 6.1 [17] illustrates a very simple constant volume gas thermometer. The temperature of the gas in the constant volume bulb is followed by measuring the pressure with the manometer. In practice, accurate gas thermometers are considerably more complicated. Figure 6.2 is a schematic diagram of a high-temperature constant volume gas thermometer. The gas bulb is the temperature measuring device. The rest of the system is for pressure measurement, gas handling, and gas purification. Although gas thermometers provide the ultimate reference for temperature measurement, they are quite impractical for routine use. Later we describe other, more convenient thermometers for which the calibration is referred back to gas thermometer measurements.

3.2   Thermodynamic Temperature Scale

The thermodynamic or Kelvin temperature is defined from thermodynamic equations relating temperature to other thermodynamic quantities. Several

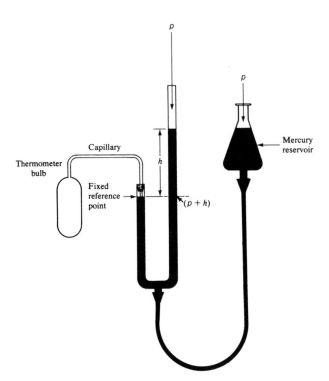

Figure 6.1   Schematic diagram of a simplified gas thermometer. The height h plus the external pressure p is the pressure of the gas in the thermometer bulb. Applications of (4) gives the temperature T in the bulb. Reprinted with permission from P. A. Rock, *Chemical Thermodynamics*, University Science Books, Mill Valley, California, 1983, p. 19.

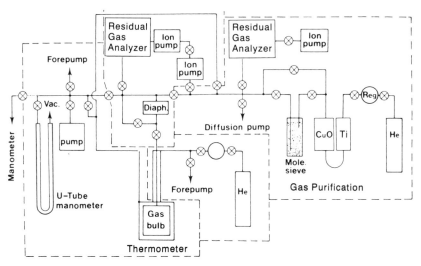

Figure 6.2   Schematic diagram of a constant volume gas thermometer, including the gas handling, purification, and pressure measurement system. The dashed lines indicate different parts of the apparatus that can be isolated for repair or clean up. Adapted from T. J. Quinn, *Temperature*, Academic, New York, 1983, p. 78.

mathematical statements can be used for this purpose [18–21]. A useful one is to express T as the integrating factor that converts the reversible flow of an infinitesimal amount of heat δq in or out of a system to an exact differential (the differential of the state function entropy S)

$$dS = \delta q_{rev}/T \tag{5}$$

An alternative, but equivalent, thermodynamic expression is obtained by defining the temperature of a heat reservoir as being proportional to the heat that flows out of the reservoir and into a Carnot engine. With these or other similar relationships, it can be shown that thermodynamic temperature is proportional to the ideal gas temperature. The Kelvin temperature is then defined with this arbitrary proportionality constant set equal to unity; thus, the absolute and Kelvin temperatures become the same. Temperature is usually expressed in kelvin (K) rather than in degrees absolute ($^\circ$A), even though the experimental measurement of T is traced back to measurements with a gas thermometer.[†]

[†]For a discussion of thermodynamic temperature, see [18–20]. Reference [18] is especially good at relating the thermodynamic and ideal gas temperature scales, and a summary of the different ways to measure thermodynamic temperature is found in [21].

3.3   Other Temperature Scales

The Celsius temperature scale (°C) sets the size of the temperature increment to be the same as with the Kelvin scale, but the Celsius scale is displaced in temperature so that the normal (1 atm) melting point of water saturated with air is at zero. The melting point of water under these conditions is very nearly 0.01 K lower than the triple point, so that 0°C = 273.15 K. In the United States, the Celsius temperature is also referred to as the centigrade temperature.

The size of the temperature increment on the Fahrenheit scale is only five-ninths as large as the centigrade degree, and the scale is displaced so that the normal melting point of ice is 32°F. The Rankine scale (°R) has the same size degree as the Fahrenheit scale, but it is displaced to make 0°R the absolute zero of temperature (-459.67°F). In summary, these temperature scales are related as follows:

$$T(\text{K}) = T(^\circ\text{A}) = 273.15 + t(^\circ\text{C}) = 273.15 + \tfrac{5}{9}[t(^\circ\text{F}) - 32]$$

$$= 273.15 + \tfrac{5}{9}[t(^\circ\text{R}) - 491.67] \tag{6}$$

4   THE INTERNATIONAL PRACTICAL TEMPERATURE SCALE OF 1968

Since the direct measurement of temperature with a gas thermometer is difficult, an *International Temperature Scale*, designated as ITS or IPTS, was defined to provide a way for relatively easy measurement of temperature with highly reproducible thermometers that closely approximate the thermodynamic scale. The *International Practical Temperature Scale of 1968* (IPTS-68), which is an extension of earlier scales, the *International Temperature Scale of 1927* (ITS-27) and the *International Temperature Scale of 1948* (ITS-48), was proposed by the Consultative Committee on Thermometry (CCT), adopted in 1968 by the International Committee of Weights and Measures (CIPM), and amended in 1975 by the same committee at the 15th General Conference of Weights and Measures.[†] As an addition to IPTS-68, there exists a provisional low-temperature scale, *The 1976 Provisional 0.5–30 K Temperature Scale, EPT-76* [27].

In this chapter, we use T as the symbol for absolute or Kelvin temperature and t for centigrade or Celsius temperature. On the IPTS-68 scale, temperatures below the ice point are given in kelvin, expressed as T_{68}; temperatures above the ice point are expressed in centigrade or Celsius temperature.[‡]

Although IPTS-68 was superceded by the *International Temperature Scale of*

[†]Several editions of the *International Practical Temperature Scale* have appeared since the first scale was adopted in 1927. The references that trace the history of IPTS are [22–26].

[‡]References [28–32] offer discussion about the status and future development of the international temperature scale.

1990 (ITS-90), most temperature measurements reported in the literature for the past 22 years are expressed in terms of IPTS-68. The result is a transition period during which it will be helpful to be acquainted with both scales. Accordingly, we discuss the basis for IPTS-68 as well as for ITS-90.

4.1 Fixed Points

The IPTS-68 scale is based on the measurement, with a gas thermometer by as accurate means as possible, of a series of fixed or reference temperatures. These fixed points, which cover the range 13.81–1338 K, are summarized in Table 6.1.

4.2 Temperature Interpolation

The IPTS-68 scale starts at 13.81 K and specifies the type of thermometer to be used in each of several temperature intervals. The thermometer used in the temperature range from 13.81 K to 630.74°C is a platinum resistance thermometer. The experimental temperature determination involves the measurement of electrical resistance of the thermometer, which must be strainfree, annealed, and of high-purity platinum. Its resistance ratio $W(T_{68})$, defined by

$$W(T_{68}) = \frac{R(T_{68})}{R(273.15 \text{ K})} \tag{7}$$

Table 6.1 Fixed Points for the International Practical Temperature Scale, IPTS-68[a]

Equilibrium State	T_{68} (K)	t_{68} (°C)
Triple point of equilibrium hydrogen[b]	13.81	−259.34
Boiling point of hydrogen at $\frac{25}{76}$ atm	17.042	−256.108
Boiling point of hydrogen at 1 atm	20.28	−252.87
Boiling point of neon at 1 atm	27.102	−246.048
Triple point of oxygen	54.361	−218.789
Triple point of argon[c]	83.798	−189.352
Boiling point of oxygen at 1 atm	90.188	−182.962
Triple point of water	273.16	0.01
Boiling point of water at 1 atm	373.15	100.00
Freezing point of tin[c]	505.1181	231.9681
Freezing point of zinc	692.73	419.58
Freezing point of silver	1235.08	961.93
Freezing point of gold	1337.58	1064.43

[a]For additional information, see [25–26, 31–32].
[b]Triple point is the temperature and pressure at which solid, liquid, and vapor coexist.
[c]The triple point of argon and the freezing point of tin are sometimes used as alternatives to the boiling points of oxygen and water, respectively.

where $R(T_{68})$ is the resistance at T_{68} and $R(273.15 \text{ K})$ is the resistance at the ice point, must not be less than 1.39250 at 373.15 K.

The standard thermometer used from 630.74 to 1064.43°C is a (platinum 10% rhodium)/(platinum) thermocouple. The experimental measurement of temperature involves measuring the voltage (electromotive force, emf) of this thermocouple. The platinum wire must be of purity such that $W(373.15 \text{ K})$ is not less than 1.3920. The platinum–rhodium wire should contain nominally 10% rhodium and 90% platinum by mass.

Above 1064.43°C, IPTS-68 is defined in terms of the Planck radiation law and is measured with an optical pyrometer. General discussions of temperature measurement in this range are presented in [33–35].

Interpolation between the fixed points (given in Table 6.1) for the various thermometers described above is accomplished by using a mathematical function; the type of function used depends on the temperature range.

4.2.1   Temperature Interval 13.81–273.15 K

The thermometer specified for the 13.81–273.15 K interval is a platinum resistance thermometer. Its resistance ratio is defined as

$$W(T_{68}) = W_{\text{CCT-68}}(T_{68}) + \Delta W_i(T_{68}) \tag{8}$$

where $W_{\text{CCT-68}}(T_{68})$ is a standard resistance ratio given by the mathematical function

$$T_{68} = \sum_{j=0}^{20} a_j \left[\frac{\ln W_{\text{CCT}-68}(T_{68}) + 3.28}{3.28} \right]^j \tag{9}$$

The coefficients a_j of this reference function are given in Table 6.2.

Equation (9) was reformulated in 1976 from the original version for IPTS-68. Values of T_{68} derived from the two versions agree within $\pm 10 \ \mu\text{K}$ over the entire temperature range.

All IPTS-68 platinum thermometers are based on the same $W_{\text{CCT-68}}(T_{68})$ function. Individual thermometers are calibrated to give the value for $\Delta W_i(T_{68})$ that applies to that specific thermometer. The form of the $\Delta W_i(T_{68})$ function depends on the temperature interval as follows:

Temperature Interval (K)	$\Delta W_i(T_{68})$	Equation Number
13.81–20.28	$\Delta W_1(T_{68}) = A_1 + B_1 T_{68} + C_1 T_{68}^2 + D_1 T_{68}^3$	(10)
20.28–54.361	$\Delta W_2(T_{68}) = A_2 + B_2 T_{68} + C_2 T_{68}^2 + D_2 T_{68}^3$	(11)
54.361–90.188	$\Delta W_3(T_{68}) = A_3 + B_3 T_{68} + C_3 T_{68}^2$	(12)
90.188–273.15	$\Delta W_4(T_{68}) = b_4(T_{68} - 273.15 \text{ K})$ $+ e_4(T_{68} - 273.15 \text{ K})^3 \times (T_{68} - 373.15 \text{ K})$	(13)

Table 6.2 Coefficients a_j of the Reference Function for
Platinum Resistance Thermometers for the Range 13.81–
273.15 K, IPTS-68[a]

j	a_j	j	a_j
0	38.59276	11	524.64944
1	43.44837	12	− 319.79981
2	39.10887	13	− 787.60686
3	38.69352	14	179.54782
4	32.56883	15	700.42832
5	24.70158	16	29.48666
6	53.03828	17	− 335.24378
7	77.35767	18	− 77.25660
8	− 95.75103	19	66.76292
9	− 223.52892	20	24.44911
10	239.50285		

[a]Adapted from [31, 32].

In each temperature interval, the constants are determined by calibration at the fixed points (Table 6.1) appropriate for that temperature range; or, more often, by comparison with another calibrated thermometer that can trace its calibration to these fixed points. A second constraint on the constants is that they must join smoothly where the intervals intersect. The National Institute of Standards and Technology and several commercial companies will, for a reasonable price, calibrate a thermometer in the various temperature intervals and provide values for the calibration constants.

4.2.2   Temperature Interval 0–630.74°C

The thermometer specified for the range 0–630.74°C is again the platinum resistance thermometer. The temperature in centigrade (t_{68}) is defined by

$$t_{68} = t' + 0.045 \left(\frac{t'}{100}\right)\left(\frac{t'}{100} - 1\right)\left(\frac{t'}{419.58} - 1\right)\left(\frac{t'}{630.74} - 1\right) \qquad (14)$$

where t' is given by

$$t' = \left\{ \frac{1}{\alpha}[W(t') - 1] + \delta\left(\frac{t'}{100}\right)\left(\frac{t'}{100} - 1\right) \right\} \qquad (15)$$

with $W(t') = R(t')/R(0°C)$. The constants $R(0°C)$, α, and δ are determined by calibration at the triple point of water, the boiling point of water (or freezing point of tin at 231.968°C, which is sometimes used as an alternative to the boiling point of water), and the freezing point of zinc (see Table 6.1). Again, the calibration is often done against a thermometer whose calibration is traceable to a primary standard.

4.2.3 Temperature Interval 630.74–1064.43°C

The thermometer specified for the 630.74–1064.43°C range is a (platinum–10% rhodium)/(platinum) thermocouple, with its electromotive force (emf), $E(t_{68})$, expressed by the function

$$E(t_{68}) = a + bt_{68} + ct_{68}^2 \qquad (16)$$

The constants a, b, and c are calculated from the values of $E(t_{68})$ at 630.74 $\pm$ 0.2°C as determined by a platinum resistance thermometer and at the freezing points of silver and of gold.

We should note that thermocouple wire used for IPTS-68 must be chosen carefully to have very specific temperature–emf characteristics. Details can be found in the literature [36].

4.2.4 Temperature Range Above 1064.43°C

In the temperature range above 1064.43°C, the optical pyrometer is the temperature measuring device for IPTS-68 [33–35]. With this instrument one compares the spectral concentrations of radiance $L_\lambda(T_{68})$ and $L_\lambda[T_{68}(\mathrm{Au})]$ of a blackbody radiator at a wavelength λ at T_{68} and $T_{68}(\mathrm{Au})$, the melting temperature of gold. These spectral intensities are related by the equation

$$\frac{L_\lambda(T_{68})}{L_\lambda[T_{68}(\mathrm{Au})]} = \frac{\exp\left[\dfrac{c_2}{\lambda T_{68}(\mathrm{Au})}\right] - 1}{\exp\left(\dfrac{c_2}{\lambda T_{68}}\right) - 1} \qquad (17)$$

with $c_2 = 0.014388$ mK. The temperature T_{68} is calculated from (17). It is not necessary, in practice, to specify λ as part of the definition, since the wavelength dependence on T_{68} is negligible.

5   THE 1976 PROVISIONAL 0.5–30 K TEMPERATURE SCALE

The IPTS-68 scale is not applicable below 13.81 K, and yet there is considerable interest in temperature measurement below this value. Various thermometers have been used in this low-temperature range. They measure the temperature effect on such physical properties as the magnetic susceptibility of a paramagnetic salt, the vapor pressure of liquid ^{4}He and ^{3}He, the speed of sound, and electrical resistance; the temperature scale is defined in terms of the physical property measured. To provide some uniformity in temperature measurement at low temperatures, the International Committee of Weights and Measures, approved in 1976 a 0.5–30 K scale called *The 1976 Provisional 0.5–30 K Temperature Scale*, EPT-76 [37]. The scale, which may be thought of as an extension of IPTS-68, was provisional pending the revision and downward extension of IPTS-68 with *The International Temperature Scale of 1990*, ITS-90 [38, 39]. The EPT-76 scale can be realized by using a thermodynamic inter-polating thermometer, such as a gas or a magnetic susceptibility thermometer, calibrated at one or more of the 11 specified reference points given in Table 6.3. It is also permissible to adjust other low-temperature scales, including the helium vapor pressure or any laboratory scale for which tabulated differences from EPT-76 have been measured.

6   THE INTERNATIONAL TEMPERATURE SCALE OF 1990

Very accurate temperature measurements made since the adoption of IPTS-68 have revealed small but significant discrepancies between this scale and the thermodynamic scale of temperature [41–54]. In addition, IPTS-68 has the limitation that it does not extend below 13.81 K (although the provisional scale EPT-76 does extend the measurement to 0.5 K) and the notable deficiency that Pt–Rh thermocouples (used to define the range 630.74–1064.43°C) are not sufficiently reproducible for this purpose. It also has been customary to revise and update the scale approximately every 20 years (ITS-27, ITS-48, and IPTS-68). Accordingly, the International Consultative Committee on Thermometry was charged with formulating a revised international scale (ITS-90) that was adopted January 1, 1990. The ITS-90 scale is designed to give temperatures based on it that do not differ from the Kelvin Thermodynamic Scale by more

Table 6.3   Reference Temperatures T_{76} of EPT-76[a]

Reference Point	T_{76} (K)
Cadmium superconducting transition point	0.519
Zinc superconducting transition point	0.851
Aluminum superconducting transition point	1.1796
Indium superconducting transition point	3.4145
^{4}He normal boiling point	4.2221
Lead superconducting transition point	7.1999
e-H_2 triple point[b]	13.8044
e-H_2 boiling point at $\frac{25}{76}$ standard atmosphere[b]	17.0373
e-H_2 normal boiling point[b]	20.2734
Neon triple point	24.5591
Neon normal boiling point	27.102

[a]Adapted from [37, 38, 40].
[b]Ortho–para equilibrium hydrogen.

Table 6.4 Fixed Points for the International Temperature Scale, ITS-90[a]

Equilibrium State	$T_{90}{}^b$ (K)
Triple point of equilibrium[c] H_2	13.8033
Triple point of Ne	24.5561
Triple point of O_2	54.3584
Triple point of Ar	83.8058
Triple point of Hg	234.3156
Triple point of H_2O	273.16[d]
Melting point of Ga	302.9146
Freezing point of In	429.7485
Freezing point of Sn	505.078
Freezing point of Zn	692.677
Freezing point of Al	933.473
Freezing point of Ag	1234.93
Freezing point of Au	1337.33
Freezing point of Cu	1357.77

[a]For additional information, see [55–59].
[b]T_{90} is the ITS-90 temperature in kelvin.
[c]Equilibrium distribution of ortho and para states.
[d]Exact value.

than the uncertainties of the thermodynamic temperatures on the date of adoption of ITS-90, to extend the temperature range to cover the EPT-76 provisional scale in the low-temperature region, and to replace the thermocouple measurements of IPTS-68 with platinum resistance thermometry. The result is a scale that has better agreement with thermodynamic temperatures and much better continuity, reproducibility, and accuracy than all previous international scales.

As in IPTS-68, temperatures on ITS-90 are defined in terms of fixed points, interpolating instruments, and equations that relate the measured property of the instrument to temperature. The report on ITS-90 of the Consultative Committee on Thermometry is published in *Metrologia* and in the *Journal of Research of the National Institute of Standards and Technology* [55, 56]. The description that follows is extracted from those publications.[†] Two additional documents by CCT further describe ITS-90: *Supplementary Information for the ITS-90* and *Techniques for Approximating the ITS-90* [58, 59].

6.1   Fixed Points

The fixed reference points for ITS-90 (temperatures at specified equilibrium states) are given in Table 6.4. These reference points are selected to calibrate thermometers over different temperature ranges, as we describe later.

[†]Reference [57] is an excellent summary of ITS-90 in the *Journal of Chemical Thermodynamics*.

6.2   Choice of Thermometer

The type of thermometer used to interpolate between the reference points depends on the temperature interval. The specification of the type of thermometer to be used is part of the definition of ITS-90. The choices are listed next.

6.2.1   Temperature Interval 0.65–5.0 K

The range 0.65–5.0 K is defined by vapor pressure–temperature relations of He: 0.65–3.2 K, He^3 vapor pressure thermometer; 1.25–2.1768 K (λ point); and 2.1768–5.0 K, He^4 vapor pressure thermometer. The form of the vapor pressure–temperature relation is

$$T_{90} = A_0 + \sum_{i=1}^{9} A_i \{[\ln p - B]/C\}^i \tag{18}$$

where T_{90} is the ITS-90 temperature in kelvin and p is the vapor pressure in pascals. The values of the coefficients A_1 to A_9 and constants A_0, B, and C are given in Table 6.5.

6.2.2   Temperature Interval 3.0–24.5561 K

The range between 3.0 and 24.5561 K is defined in terms of ^3He or ^4He constant volume gas thermometers (CVGT), calibrated at the triple points of Ne and H_2, and at a temperature between 3.0 and 5.0 K that has been obtained from vapor pressure–temperature relations for He.

Table 6.5   Values of the Coefficients and Constants for (18) Used in Defining the ITS-90 Temperature Scale

Coefficient or Constant	^3He 0.65–3.2 K	^4He 1.25–2.1768 K	^4He 2.1768–5.0 K
A_0	1.053447	1.392408	3.146631
A_1	0.980106	0.527153	1.357655
A_2	0.676380	0.166756	0.413923
A_3	0.372692	0.050988	0.091159
A_4	0.151656	0.026514	0.016349
A_5	−0.002263	0.001975	0.001826
A_6	0.006596	−0.017976	−0.004325
A_7	0.088966	0.005409	−0.004973
A_8	−0.004770	0.013259	0
A_9	−0.054943	0	0
B	7.3	5.6	10.3
C	4.3	2.9	1.9

In the temperature range between 4.2 and 24.5561 K (triple point of neon), T_{90} is defined by the equation

$$T_{90} = a + bp + cp^2 \tag{19}$$

where p is the pressure in the constant volume gas thermometer and a, b, and c are coefficients to be determined by calibration at the three specified temperatures.

When the ^{4}He CVGT is used between 3.0 and 4.2 K (or the He^3 CVGT between 3.0 and 24.5561 K), gas imperfection must be taken into account and (19) becomes

$$T_{90} = \frac{a + bp + cp^2}{1 + B_x(T_{90})N/V}$$

where p, a, b, and c are the same as in (19), $B_x(T_{90})$ is the second virial coefficient for ^{3}He $[B_3(T_{90})]$ or ^{4}He $[B_4(T_{90})]$, and N/V is the gas density (moles per cubic meter) in the CVGT bulb. The values of $B_x(T_{90})$ at any given temperature are calculated from equations specified in the official ITS-90 document [56–59] and also in the *National Institute of Standards and Technology's* (NIST) Technical Note 1265.

6.2.3   Temperature Interval 13.8033–1234.93 K

The thermometer used in the large temperature interval between 13.8033 and 1234.93 K is a platinum resistance thermometer that has specified characteristics and is calibrated at fixed reference points. Temperatures are expressed in terms of $W(T_{90})$, which, as seen in (20), is the ratio of the resistance $R(T_{90})$ of the thermometer at temperature T_{90} and the resistance at the triple point of water $R(273.16\ \text{K})$.

$$W(T_{90}) = R(T_{90})/R(273.16\ \text{K}) \tag{20}$$

The pure platinum strainfree coil of the resistance thermometer must meet one of the following specifications: $W(302.9146\ \text{K}) \geqslant 1.11807$ or $W(234.3156\ \text{K}) \leqslant 0.844235$.

If the resistance thermometer is to be used over the entire range 13.8033–1234.93 K, it must also meet the requirement that $W(1234.93\ \text{K}) \geqslant 4.2844$.

Temperatures T_{90} are calculated from the equation

$$W(T_{90}) = W_r(T_{90}) + \Delta W(T_{90}) \tag{21}$$

in which $W(T_{90})$ is the resistance ratio observed and $W_r(T_{90})$ is the value calculated from a reference function. The deviation of the observed value obtained with a given platinum resistance thermometer and the reference function value is $\Delta W(T_{90})$ and is called the *deviation function.*

There are two reference functions, one for the range 13.8033–273.16 K and the other for the range 273.15–1234.93 K. The reference function $W_r(T_{90})$ for temperatures in the range 13.8033–273.16 K is

$$\ln[W_r(T_{90})] = A_0 + \sum_{i=1}^{12} A_i\{[\ln(T_{90}/273.16)+1.5]/1.5\}^i \tag{22}$$

and for the range 273.15–1234.93 K it is

$$W_r(T_{90}) = C_0 + \sum_{i=1}^{9} C_i \left(\frac{T_{90}-754.15)}{481}\right)^i \tag{23}$$

Inverses of (22) and (23) that are explicit in temperature can also be used. The specified inverse of (22), equivalent to within ± 0.0001 K, is

$$T_{90}/273.16 = B_0 + \sum_{i=1}^{15} B_i \left(\frac{[W_r(T_{90})]^{1/6}-0.65}{0.35}\right)^i \tag{24}$$

The specified inverse of (23), equivalent to within ± 0.00013 K, is

$$T_{90} - 273.15 = D_0 + \sum_{i=1}^{9} D_i \left(\frac{W_r(T_{90})-2.64}{1.64}\right)^i \tag{25}$$

The values for the constants A_0, B_0, C_0, and D_0 and the coefficients A_i, B_i, C_i, and D_i are given in Table 6.6.

6.3   The Deviation Function

The deviation function $\Delta W(T_{90})$ is obtained as a function of T_{90} for various temperature intervals by calibration of the platinum resistance thermometer at specified fixed points. The form of the $\Delta W(T_{90})$ function is dependent on the temperature range in which the thermometer is being calibrated. For example, in the temperature subrange from 234.3156 to 302.9146 K, the form of the deviation function is

$$\Delta W_5(T_{90}) = a_5[W(T_{90})-1] + b_5[W(T_{90})-1]^2 \tag{26}$$

The coefficients a_5 and b_5 are obtained by calibrating the thermometer at the triple points of mercury (234.3156 K) and water (273.16 K) and the melting point of gallium (302.9146 K).

Table 6.7 is a tabulation of subranges in the temperature region 13.8033–1234.93 K, together with the form of deviation equation that applies to each,

Table 6.6   Values of the Coefficients A_i, B_i, C_i, and D_i and of the Constants A_0, B_0, C_0, and D_0 in the Reference Functions (22) and (23) and in the Inverse Functions Given by (24) and (25)[a,b]

Constant or Coefficient	Value	Constant or Coefficient	Value
A_0	−2.13534729	B_{12}	−0.029201193
A_1	3.18324720	B_{13}	−0.091173542
A_2	−1.80143597	B_{14}	0.001317696
A_3	0.71727204	B_{15}	0.026025526
A_4	0.50344027		
A_5	−0.61899395		
A_6	−0.05332322	C_0	2.78157254
A_7	0.28021362	C_1	1.64650916
A_8	0.10715224	C_2	−0.13714390
A_9	−0.29302865	C_3	−0.00649767
A_{10}	0.04459872	C_4	−0.00234444
A_{11}	0.11868632	C_5	0.00511868
A_{12}	−0.05248134	C_6	0.00187982
		C_7	−0.00204472
		C_8	−0.00046122
B_0	0.183324722	C_9	0.00045724
B_1	0.240975303		
B_2	0.209108771		
B_3	0.190439972	D_0	439.932854
B_4	0.142648498	D_1	472.418020
B_5	0.077993465	D_2	37.684494
B_6	0.012475611	D_3	7.472018
B_7	−0.032267127	D_4	2.920828
B_8	−0.075291522	D_5	0.005184
B_9	−0.056470670	D_6	−0.963864
B_{10}	0.076201285	D_7	−0.188732
B_{11}	0.123893204	D_8	0.191203
		D_9	0.049025

[a]Adapted from [55–56].
[b]These functions, coefficients, and constants are part of the definition of ITS-90.

and the calibration points from which the coefficients in the deviation equation are to be obtained.

In summary, to obtain T_{90} for a platinum resistance thermometer, one selects the range of interest, calibrates at the fixed points specified for those ranges, and uses the appropriate function to calculate $\Delta W(T_{90})$ to be used in (21). Companies are available that will perform these calibrations and provide tables of $W(T_{90})$ versus T_{90}, which can be interpolated to give T_{90} for a measured $W(T_{90})$.

Table 6.7 Temperature Subranges, Deviation Functions, and Calibration Points Over the Temperature Range Covered by Platinum Resistance Thermometry for ITS-90[a]

Temperature Subrange (K)	Deviation Function $\Delta W(T_{90}) = W(T_{90}) - W_r(T_{90}) =$	Calibration Points[b] to Determine Coefficients in the Deviation Function
13.8033–273.16	$a_1[W(T_{90})-1]+b_1[W(T_{90})-1]^2+\sum_{i=1}^{5} c_i[\ln W(T_{90})]^{i+2}$	Triplepoint (tp) of H_2, Ne, O_2, Ar, Hg, and two more[c]
24.5561–273.16	$a_2[W(T_{90})-1]+b_2[W(T_{90})-1]^2+\sum_{i=1}^{3} c_i[\ln W(T_{90})]^i$	tp of H_2, Ne, O_2, Ar, and Hg
54.3584–273.16	$a_3[W(T_{90})-1]+b_3[W(T_{90})-1]^2+c_1[\ln W(T_{90})]^2$	tp of O_2, Ar, and Hg
83.8058–273.16	$a_4[W(T_{90})-1]+b_4[W(T_{90})-1]\ln W(T_{90})$	tp of Ar and Hg
273.15–1234.93	$a_6[W(T_{90})-1]+b_6[W(T_{90})-1]^2+c_6[W(T_{90})-1]^3$ $+d[W(T_{90})-W(933.473\,\text{K})]^2$	Freezing point (fp) of Sn, Zn, Al, and Ag
273.15–933.473	$a_7[W(T_{90})-1]+b_7[W(T_{90})-1]^2+c_7[W(T_{90})-1]^3$	fp of Sn, Zn, and Al
273.15–692.677	$a_8[W(T_{90})-1]+b_8[W(T_{90})-1]^2$	fp of Sn and Zn
273.15–505.078	$a_9[W(T_{90})-1]+b_9[W(T_{90})-1]^2$	fp of In and Sn
273.15–429.7485	$a_{10}[W(T_{90})-1]$	fp of In
273.15–302.9146	$a_{11}[W(T_{90})-1]$	Melting point (mp) of Ga
234.3156–302.9146	$a_5[W(T_{90})-1]+b_5[W(T_{90})-1]^2$	tp of Hg; mp of Ga

[a]For additional information, see [55–59].
[b]In addition to the fixed points listed, calibration at the triple point (tp) of H_2O is required.
[c]Two additional points near 17.0 and 20.3 K are required. These may be determined by using either the constant volume gas thermometer or by vapor pressure measurements of H_2.

6.4   Measurement of Temperatures Above 1234.93 K

At temperatures above the melting point of silver (1234.93 K), radiation thermometry is used. The equation that applies is

$$\frac{L_\lambda(T_{90})}{L_\lambda[T_{90}(X)]} = \frac{\exp[c_2/\lambda T_{90}(X)] - 1}{\exp(c_2/\lambda T_{90}) - 1} \tag{27}$$

in which $L_\lambda(T_{90})$ is the spectral concentration of the radiance of a blackbody at wavelength λ at T_{90} and $L_\lambda[T_{90}(X)]$ is the same at $T_{90}(X)$, in which X refers to the freezing point of silver, copper, or gold on ITS-90. The constant c_2 has the value $0.014388 \, \text{m} \cdot \text{K}$ (the same as for IPTS-68).

6.5   Correction of Existing Data to ITS-90

The descriptions of ITS-90 in [55–56] include a table (labeled 6.8) of differences between T_{90} and T_{76} at 1 K intervals between 5 and 26 K; T_{90} and T_{68} at 1 K intervals from 4 to 100 K, and at 10 K intervals between 100 and 200 K; t_{90} and t_{68} at 10°C intervals from -190 to 1000°C, and t_{90} and t_{68} in 100°C intervals from 1000 to 3900°C. These data can be used to update IPTS-68 temperatures to ITS-90.

Figure 6.3 is a graphical representation [55, 56] of the differences

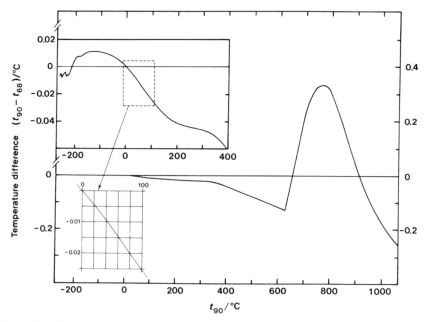

Figure 6.3 Differences between t_{90} and t_{68} as a function of t_{90}°C. Reprinted with permission from H. Preston-Thomas, *Metrologia*, **27**, 4 (1990).

$(T_{90} - T_{68})$ K or $(t_{90} - t_{68})°$C as a function of $t_{90}°$C. The corrections in the region from about 500–1000°C are seen to be substantial, varying all the way from about $+0.35$ to $-0.20°$C.

From Figure 6.3 we can see that differences between t_{90} and t_{68} in the temperature range from -100 to 200°C vary almost linearly with temperature. Using the cubic equation (28), we have been able to express the differences in Table 6.8 in this temperature range to within an uncertainty of 1×10^{-3} K:

$$(t_{90} - t_{68})°\text{C} = -2.32 \times 10^{-4}t - 6.78 \times 10^{-7}t^2 + 4.24 \times 10^{-9}t^3 \qquad (28)$$

with

$$-100°\text{C} \leqslant t \leqslant 200°\text{C}$$

where t is either the t_{90} or t_{68} temperature in centigrade. Equation (28) can be used in this temperature range to convert a temperature from IPTS-68 to its equivalent ITS-90 value.

7 THERMOMETERS: GENERAL

In theory, any physical property that changes with temperature can be used to construct a thermometer, which, with proper calibration, can be used to measure temperature. In addition to the gas thermometers, resistance thermometers, thermocouples, and optical pyrometers referred to earlier, thermometers have been constructed that rely on the thermal expansion of a liquid such as mercury, indium, alcohol, or toluene; magnetic susceptibility of a paramagnetic salt; refractive index; dielectric constant of a gas; speed of sound in a gas (acoustic thermometers); the frequency of oscillation of a quartz crystal; and one that relies on the measurement of thermal noise in an electrical resistor.

Of the above (and this list is by no means all inclusive), the resistance thermometer, thermocouple, and liquid-in-glass (usually mercury) thermometer are the most practical and commonly encountered. We restrict our discussion to these three and refer the reader to the literature [60–65] for descriptions of others.

8 RESISTANCE THERMOMETERS

The advent of the microchip has pushed resistance thermometers to the forefront as temperature measuring devices. The range and variety of resistance thermometers[†] available are remarkable. For a few dollars, small hand-held digital resistance thermometers can be purchased that display temperatures with

---

[†]References [66–74] describe the characteristics of a variety of resistance thermometers.

Table 6.8 Differences Between $(T_{90}$ and $T_{68})$, $(t_{90}$ and $t_{68})$, and $(T_{90}$ and $T_{76})^a$

$(T_{90} - T_{76})$ (mK)

T_{90} (K)	0	1	2	3	4	5	6	7	8	9
0						-0.1	-0.2	-0.3	-0.4	-0.5
10	-0.6	-0.7	-0.8	-1.0	-1.1	-1.3	-1.4	-1.6	-1.8	-2.0
20	-2.2	-2.5	-2.7	-3.0	-3.2	-3.5	-3.8	-4.1		

$(T_{90} - T_{68})$ (K)

T_{90} (K)	0	1	2	3	4	5	6	7	8	9
10	-0.009	-0.008	-0.007	-0.007	-0.006	-0.003	-0.004	-0.006	-0.008	-0.009
20	-0.006	-0.007	-0.008	-0.008	-0.006	-0.005	-0.004	-0.004	-0.005	-0.006
30	-0.006	-0.006	-0.006	-0.006	-0.008	-0.007	-0.007	-0.007	-0.006	-0.006
40	-0.006	-0.005	-0.005	-0.004	-0.006	-0.007	-0.007	-0.007	-0.006	-0.006
50	0.003	0.003	0.004	0.004	-0.003	-0.002	-0.001	0.000	0.001	0.002
60	0.007	0.007	0.007	0.007	0.005	0.005	0.006	0.006	0.007	0.007
70	0.008	0.008	0.008	0.008	0.007	0.008	0.008	0.008	0.008	0.008
80	0.008	0.008	0.008	0.008	0.008	0.008	0.008	0.008	0.008	0.008
90					0.008	0.008	0.008	0.009	0.009	0.009

T_{90} (K)	0	10	20	30	40	50	60	70	80	90
100	0.009	0.011	0.013	0.014	0.014	0.014	0.014	0.013	0.012	0.012
200	0.011	0.010	0.009	0.008	0.007	0.005	0.003	0.001		

$(t_{90} - t_{68})(°C)$

t_{90} (°C)	0	−10	−20	−30	−40	−50	−60	−70	−80	−90
−100	0.013	0.013	0.014	0.014	0.014	0.013	0.012	0.010	0.008	0.008
0	0.000	0.002	0.004	0.006	0.008	0.009	0.010	0.011	0.012	0.012

t_{90} (°C)	0	10	20	30	40	50	60	70	80	90
0	0.000	−0.002	−0.005	−0.007	−0.010	−0.013	−0.016	−0.018	−0.021	−0.024
100	−0.026	−0.028	−0.030	−0.032	−0.034	−0.036	−0.037	−0.038	−0.039	−0.039
200	−0.040	−0.040	−0.040	−0.040	−0.040	−0.040	−0.040	−0.039	−0.039	−0.039
300	−0.039	−0.039	−0.039	−0.040	−0.040	−0.041	−0.042	−0.043	−0.045	−0.046
400	−0.048	−0.051	−0.053	−0.056	−0.059	−0.062	−0.065	−0.068	−0.072	−0.075
500	−0.079	−0.083	−0.087	−0.090	−0.094	−0.098	−0.101	−0.105	−0.108	−0.112
600	−0.115	−0.118	−0.122	−0.125	−0.08	−0.03	0.02	0.06	0.11	0.16
700	0.20	0.24	0.28	0.31	0.33	0.35	0.36	0.36	0.36	0.35
800	0.34	0.32	0.29	0.25	0.22	0.18	0.14	0.10	0.06	0.03
900	−0.01	−0.03	−0.06	−0.08	−0.10	−0.12	−0.14	−0.16	−0.17	−0.18
1000	−0.19	−0.20	−0.21	−0.22	−0.23	−0.24	−0.25	−0.25	−0.26	−0.26

t_{90} (°C)	0	100	200	300	400	500	600	700	800	900
1000		−0.26	−0.30	−0.35	−0.39	−0.44	−0.49	−0.54	−0.60	−0.66
2000	−0.72	−0.79	−0.85	−0.93	−1.00	−1.07	−1.15	−1.24	−1.32	−1.41
3000	−1.50	−1.59	−1.69	−1.78	−1.89	−1.99	−2.10	−2.21	−2.32	−2.43

"Reprinted with permission from H. Preston-Thomas, "The International Scale of 1990 (ITS-90)," *Metrologia*, **27**, 9 (1990).

a precision of 0.1°C (although not to that accuracy). Several probes often can be selected; either Fahrenheit or Celsius temperatures can be read; and, in addition, the instrument sometimes will display the time of day from an internal clock.

At the other extreme are precision resistance thermometers that meet IPTS-68 and ITS-90 specifications and sell for thousands of dollars. These thermometers can be coupled with high-precision and stable resistance bridges to measure temperatures accurately.

Intermediate between these extremes are resistance thermometers of all types and designs that, depending on the need, will measure and monitor temperature or temperature differences accurately over a wide range of temperatures.

In general, resistance thermometers can be classified into two types depending on their temperature versus resistance characteristics. In metallic thermometers, the resistance R decreases with decreasing temperature (dR/dT is positive), with the resistance becoming zero at 0 K. The R versus T relationship is not linear, but it does not deviate greatly from this behavior, allowing these types of thermometers to be used conveniently over wide temperature ranges.

Semiconductors are also used as resistance thermometers. For these substances, dR/dT is negative. Furthermore, the R versus T relationship is quite nonlinear, often being close to exponential in form. These thermometers can usually be used over only relatively narrow temperature ranges because dR/dT is large. On the other hand, this large dR/dT makes these thermometers particularly useful for measuring small temperature differences.

8.1   Platinum Resistance Thermometer

The platinum resistance thermometer is by far the most common of the metallic type, although nickel and copper are also sometimes used. Platinum thermometers have been built with many different designs and construction. Some are delicate and others are rugged, depending on the application. The ITS-90 thermometer is perhaps the most delicate since great care is taken to ensure strainfree construction. It consists of a platinum coil of 0.07-mm wire of length

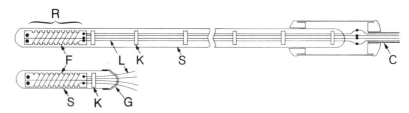

Figure 6.4   Design of a long-stem platinum resistance thermometer (upper) for use in the temperature range from 80 to 900 K and a capsule platinum resistance thermometer (lower) for use in the temperature range from 10 to 400 K: *C*, cable to the resistance meter; *F*, support; *G*, glass cap on the capsule thermometer through which the hermetically sealed leads pass; *K*, support for leads; *L*, leads (usually platinum); *R*, resistance coil; *S*, sheath, usually metal for the capsule type and metal or glass for the long-stemmed type.

to give a resistance of approximately $25\,\Omega$ at $0°C$. It comes in two designs depending on the temperature range (Figure 6.4).

A capsule design is used below liquid nitrogen temperatures (80 K) since in this temperature range, one needs to be able to enclose the entire thermometer within radiation shields and to equilibrate the lead wires with the shield. The platinum coil, which usually contains about 60 cm of wire, is supported on a mica cross or in a pair of twisted glass tubes. This coil is placed inside a platinum sheath joined to a glass seal through which the lead wires (usually platinum) pass. Before sealing, the tube is filled with helium gas at a room temperature pressure of about 30 kPa.

All precision resistance thermometers have a pair of leads sealed to each end of the platinum wire resistor. One lead of each pair carries the electrical current through the coil; the other two leads are for measuring the potential drop across the coil, so that the electrical resistance R can be obtained from

$$R = \frac{E}{i} \qquad (29)$$

where E is the potential drop with a current i flowing (see Figure 6.5). The IPTS-68 scale specifies that a direct current of 1 mA should flow through the

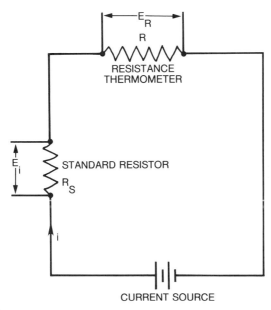

Figure 6.5 Electrical circuit for measuring the resistance of a resistance thermometer. The electrical current through the thermometer is obtained by measuring the voltage E_i across a standard resistor R_S and applying (29) to calculate i. We can calculate the resistance R of the resistance thermometer from (29) by using i and the measured voltage drop across the thermometer E_R.

thermometer, although modern resistance bridges often employ alternating current or pulsed currents for measurements of high accuracy.

The above arrangement results in a four-lead measurement of resistance. Occasionally, a three-lead measurement is used in which the thermometer is made part of a bridge circuit with a single lead on one end of the resistance wire carrying the current and also serving as a potential lead. The three-lead measurement is not recommended for accurate temperature measurement since the assumption must be made that all three leads have the same resistance. It is sometimes used in industrial applications where temperature monitoring does not require high accuracy.

The capsule design can be used at temperatures up to approximately 400 K or 100°C, but above that temperature it is not practical because of a leakage resistance between the leads in the glass seal. Long-stem thermometers (refer again to Figure 6.4) are used to the high-temperature limit of ITS-90, although they can be used at temperatures as low as 80 K. In these thermometers, the platinum coil is wound on a mica cross insulator or supported in silica or aluminum tubes. The thermometer is enclosed in a long-stem glass or metal sheath. The leads, which run from the resistance coil to the top of the sheath, are insulated from one another by mica, silica, or sapphire disks, or by silica or sapphire tubes. These thermometers are usually filled with dry air and annealed before sealing.

The high-precision thermometers described above are not meant for day to day temperature monitoring. Thermometers designed to withstand normal industrial use are less accurate but more rugged. They are manufactured commercially in a variety of sizes and shapes. In these thermometers, the platinum resistance element, which may be a fine platinum wire or a platinum film, is firmly supported, often by being imbedded in glass or ceramic. The reproducibility, although not as good as for the high-precision thermometers, is still better than for thermocouples. For this reason, resistance thermometers are replacing thermocouples in a wide range of applications.

8.2   Thermistor

The thermistor [75–78] is the most commonly encountered semiconductor resistance thermometer. Thermistors are made by sintering mixtures of transition metal oxides into a bead or disk implanted with metallic leads. The composition of the thermistor depends on the temperature range and the resistance characteristics desired. The most stable thermistors used from room temperature up to 250°C are made from mixed oxides of manganese and nickel or manganese, nickel, and cobalt. Their room temperature resistance is usually 5 kΩ, although units with resistance as low as 1 kΩ are available.

Lanthanide oxides are used for thermistors in the temperature range from 300 to 700°C. At still higher temperatures, zirconium oxide doped with small quantities of rare earth oxides is used. At very low temperatures, thermistors are usually made from nonstoichiometric iron oxides.

The advantages of thermistors lie in the large variety of sizes and shapes that can be manufactured and in their large temperature coefficients, and hence, high sensitivity. Beads as small as 0.07 mm in diameter mounted on leads of diameter 0.01 mm are available commercially, which allows for good thermal contact and temperature monitoring in very small spaces. These small beads can also be completely covered with glass to protect them from chemical reaction.

The R versus T relationship for thermistors is of the exponential form that can be approximated by

$$R = R_0 \exp\left\{-B\left(\frac{1}{T} - \frac{1}{T_0}\right)\right\} \tag{30}$$

where B is a constant and R_0 is the resistance at a reference temperature T_0. The constant B, which can be adjusted during sintering, is related to the temperature coefficient α by

$$\alpha = \frac{1}{R}\frac{dR}{dT} = -\frac{B}{T^2} \tag{31}$$

A more accurate representation of the temperature–resistance relationship for a thermistor is the fitting equation

$$T = 1/(a + b \ln R + c \ln^2 R + d \ln^3 R) \tag{32}$$

where a, b, c, and d are constants. This equation is often used in reporting the accurate calibration of a thermistor.

Thermistors have some disadvantages. They are not as stable as metallic resistance thermometers (this is especially true of the disk type of thermistor). To obtain the maximum stability from a room temperature thermistor, it should be aged by holding it at the maximum temperature of intended operation for at least 3 months. Another disadvantage is that the exponential nature of the temperature dependence of resistance makes for a relatively narrow temperature range over which the thermistor is useful. If the temperature becomes too high, the sensitivity becomes too small [see (31)]. On the other hand, if the temperature becomes too low, the resistance becomes too large for convenient measurement [see (30)].

8.3 Measurement of Very Low Temperatures With Resistance Thermometers

Resistance thermometers for use at very low temperatures (<20 K) present special problems [79]. The resistance of pure metals such as platinum becomes very small in this region and the sensitivity is not good. Thermistors can be used, but the large change of resistance with temperature restricts the use of a particular thermistor to a relatively narrow temperature range. Other semicon-

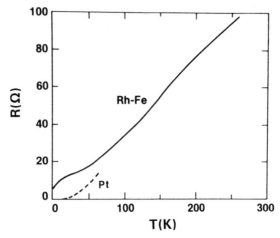

Figure 6.6   Electrical resistance R at temperature T of a $(Rh + 0.5\% \text{ Fe})$ resistance thermometer (solid line). The much higher resistance at low temperature when compared to pure Pt (broken line) makes this thermometer a useful temperature measuring device at low temperatures. Reprinted with permission from J. F. Schooley, *Thermometry*, CRC Press, Boca Raton, FL, 1986, p. 205. Copyright © 1986 CRC Press, Inc., Boca Raton, FL.

ductors have been used. Carbon resistors were the first employed. They are still used, but they lack stability, and they have been replaced in many applications by more stable germanium thermometers [80, 81]. Even the germanium thermometers have been shown to have some instability, and they have temperature–resistance problems associated with being a semiconductor.

A metallic resistance thermometer that shows promise for low-temperature application consists of an alloy wire of rhodium containing 0.5% iron [82, 83]. This thermometer is constructed similar to the low-temperature capsule platinum resistance thermometer (see Figure 6.4). The excellent resistance characteristics of this thermometer as compared to platinum can be seen in Figure 6.6 [84]. It may develop into a secondary temperature interpolating thermometer at low temperatures.

9   THERMOCOUPLES

When dissimilar metals, known as *thermoelectric elements*, are connected in an electrical circuit such as in Figure 6.7, an electrical voltage (emf) is generated between the electrical leads. The size of the emf depends on the type of thermoelectric elements that are joined and is a function of the temperature difference $T - T_{ref}$ or $t - t_{ref}$. For example, Figure 6.8 [85] shows how the emf E varies with t (with t_{ref} at 0°C) for several of the more common thermocouples.[†]

[†]References [86–90] give detailed discussions of thermocouples, their construction, and application.

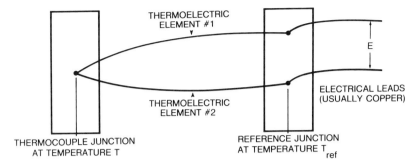

Figure 6.7   Electrical circuit for a thermocouple. The voltage E is a measure of the temperature T. The reference junction is usually an ice bath so that $T_{ref} = 273.15$ K.

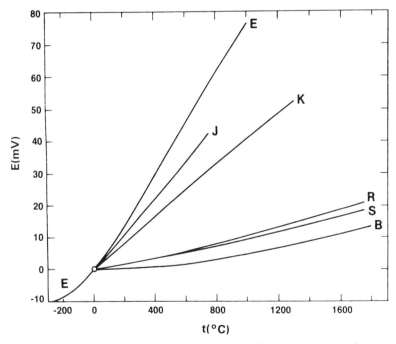

Figure 6.8   The voltage E versus temperature t relationships for the six common thermocouples with Instrument Society of America (ISA) letter designations B, E, J, K, R, and S. A discussion of these thermocouples is given in the text. Reprinted with permission from J. F. Schooley, *Thermometry*, CRC Press, Boca Raton, FL, 1986, p.174. Copyright © 1986 CRC Press, Inc., Boca Raton, FL.

9.1   Advantages of Thermocouples

Thermocouples are popular temperature measuring devices because of their simplicity of construction and low cost. A wide variety of alloy wires specially formulated for use in thermocouples is available commercially in several wire sizes and types of insulation. The wire is packaged individually or in pairs; hence, to make a thermocouple one merely joins the wires on one end and adds leads to the other end. Connectors can be obtained that are made from the same kinds of alloys so that extensions, terminal boards, and switches can be added to the thermocouple circuits. A wide variety of very inexpensive voltmeters are available for measuring E accurately. The meter must have a high input impedance so that no appreciable electrical current flows through the thermocouple during measurement. With the advent of the microprocessor, thermocouple meters can be found that continuously monitor, display, and record temperature directly, and that serve as a sensing element for temperature control devices. Thermocouples are especially useful for high-temperature measurements (above the temperature where resistance thermometers cease to be practical). For example, the nuclear industry relies heavily on thermocouples to monitor the high temperatures present in nuclear reactors.

9.2   Disadvantages of Thermocouples

Compared to resistance thermometers, thermocouples have several shortcomings and are being replaced by resistance thermometers in many applications. One disadvantage of the thermocouple is the inconvenience of needing a reference junction. An ice bath is usually used for this purpose, so that $t_{ref} = 0°C$. Each time a temperature measurement is made, an ice bath must be prepared carefully if accurate measurements are to be made. Some microprocessor-controlled thermocouple temperature measuring devices, however, have built-in compensators that use the ambient temperature as a reference. The ambient temperature is measured and corrections are made to E to account for the different reference temperature.

Another disadvantage of the thermocouple is that it is difficult to make highly accurate temperature measurements with it. In theory, E should depend only on the temperature difference $t - t_{ref}$. However, inhomogeneity and imperfections in the thermoelectric elements cause E to depend on the nature of the temperature gradient between the temperature and reference junctions. Instability and drift are also a problem; annealing can cause changes in the reading. Also, some thermocouples are sensitive to their environment. Some can be used successfully only in an oxidizing atmosphere, while others are useful only when the possibility of oxidation is excluded. With carefully selected thermocouple wire and extensive calibration, thermocouples can be obtained that will measure T to ± 0.05 K. However, for most applications uncertainties range from ± 1 to ± 10 K, depending on the nature of the thermocouple and the temperature range. Thermocouples find their major application when this degree of uncertainty is not a concern.

9.3   Types of Thermocouples

In theory, an essentially infinite number of different thermocouples can be constructed by combination of different thermoelectric elements; and, indeed, many have been made [86–90]. However, in the temperature range from 20 to 2000 K nearly all thermocouple temperature measurements are made with the seven different alloy combinations summarized in Table 6.9.

The thermocouples in Table 6.9 are those for which internationally accepted reference tables of E versus T exist. Accurate temperature measurements with thermocouples require individual calibration, but reference tables can be used to give T from a measured E if high accuracy is not required. The most extensive tabulation of E versus T is found in [91, 92] and other supplements. Other usually more limited tables can be obtained from more readily available sources such as the *Handbook of Chemistry and Physics*. Equivalent tables can also sometimes be obtained from the thermocouple manufacturer.

As an alternative to E versus T tables, E can be expressed as a polynomial function of T. As an example, the type S thermocouple in the temperature range from 630.74 to 1064.43°C (the range in which this thermocouple is the IPTS-68 reference) E is given by the equation

$$E = \sum_{j=0}^{2} a_j T^j \tag{33}$$

with $a_0 = 2.982$, $a_1 = 8.237$, and $a_2 = 1.645$. This equation, as with the tables, does not give an accurate value for T but can be used where high accuracy is not required.

Similar expressions have been used for other temperature ranges and other thermocouples, although these equations often involve higher order polynomials than the above. Values for the coefficients are given in [91, 92] as well as in other sources, such as Quinn [93].

The seven thermocouples summarized in Table 6.9 are listed in terms of a letter designation, a common or trade name, and an elemental composition. The letter designation, which is used in an expanded form for many more thermocouples than the seven given in Table 6.9, was originally introduced by the Instrument Society of America, but is now used worldwide as a shorthand way of referring to a particular thermocouple. Also included in Table 6.9 is the temperature range over which E versus T is tabulated in [91, 92], which is an indication of the temperature region in which the thermocouple can be used. However, care must be taken. Type T (copper–constantan) thermocouples, for example, can be used to 400°C only in a reducing atmosphere. In an oxidizing atmosphere, the limit is 150°C. Table 6.9 also gives the Seebeck coefficient defined as dE/dT. It is a measure of the sensitivity of the thermocouple. In general, the larger the sensitivity, the more easily the thermocouple can be used to make temperature measurements. The Seebeck coefficients given in the table are the values at 25°C and the sensitivity does change with temperature. Figure

Table 6.9 Comparison of Thermocouples[a]

Type	Common Name	Elemental Composition (%)	Temperature Range (°C)	Seebeck Coefficient (μV/K at 298.15 K)
B	30-6	(70.40 Pt + 29.60 Rh)/(93.88 Pt + 6.12 Rh)	0–1820	0.046[b]
E	Chromel–constantan	(90 Ni + 10 Cu)/(45 Ni + 55 Cu)	−270–1000	60.928
J	Iron[c]–constantan[d]	(99.5 Fe)/(45 Ni + 55 Cu)	−180–1200	51.741
K	Chromel–alumel	(90 Ni + 10 Cr)/(95 Ni + 2 Al + 2 Mn + 1 Si)	−270–1372	40.498
R	Platinum–13% Rhodium	(87 Pt + 13 Rh)/(100 Pt)	−50–1768	5.942
S	Platinum–10% Rhodium	(90 Pt + 10 Rh)/(100 Pt)	−50–1768	5.985
T	Copper–constantan	(100 Cu)/(45 Ni + 55 Cu)	−270–400	40.671

[a]In representing thermocouples, a diagonal line designates the separation of the two thermoelements and the elemental composition of each of the thermoelements is enclosed in parentheses. The temperature range given in the table is the one for which E versus T is tabulated in [91, 92]. The Seebeck coefficient S, defined as dE/dT, gives the sensitivity. It varies significantly with temperature. See [91, 92] or its equivalent for values of S at temperatures other than 25°C.

[b]Seebeck coefficient of the type B thermocouple is negative at 0°C and becomes zero at 22°C. It increases significantly at higher temperatures where this thermocouple is most often used ($S = 9.112 \, \mu V/K$ at 1000°C and 11.476 $\mu V/K$ at 1800°C).

[c]Thermoelectric iron used in these thermocouples contains small amounts of C, Mn, S, P, Si, Ni, Cu, and Cr.

[d]The constantan in type J thermocouples is slightly different from the constantan in type E and T.

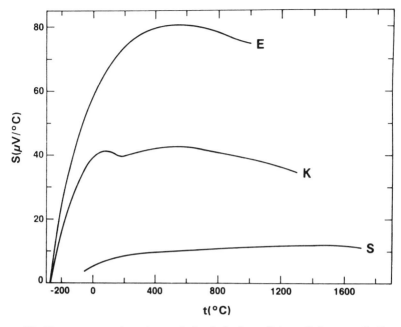

Figure 6.9 Temperature t dependence of the Seebeck coefficients S for type E, K, and S thermocouples. Reprinted with permission from J. F. Schooley, *Thermometry*, CRC Press, Boca Raton, FL, 1986, p. 174. Copyright © 1986 CRC Press, Inc., Boca Raton, FL. The data were taken from [91, 92].

6.9 shows how this coefficient changes with temperature for the type E, K, and S thermocouples.

9.4   Choosing a Thermocouple

Selecting the thermocouple to use for a particular application depends on many factors. Sensitivity is of concern since increased sensitivity allows for easier and more precise measurements. Stability is also of concern. Some thermocouples drift when held at high temperatures or when cycled in temperature. This may be due to oxidation, annealing, or changes in elemental composition of the thermoelectric alloys.

Cost can also be a factor. Thermocouples made from base metals are certainly less expensive than those made from platinum and rhodium. Some manufacturers fabricate thermocouples with various sizes of wire encapsulated in sheaths made of stainless steel, inconel, platinum–rhodium, iridium–rhodium, or tantalum, and insulated with alumina, magnesium oxide, or beryllia. Of the thermocouples given in Table 6.9, some find wider application than others. Type T is a commonly used thermocouple in the low to room temperature range (20–300 K), although type K is sometimes used in this region. Type K is probably the most widely used thermocouple in the ambient to moderately high-temperature region (0–1200°C), although type S is used at the upper end of this region if high stability and reproducibility are more important than high sensitivity. It is because of this high stability that type S is used in IPTS-68.

In addition to the seven thermocouples described above, there are others that are commonly used in special temperature regions. A (gold + 0.07 atomic% iron) or (gold + 0.02 atomic% iron) alloy coupled with copper, silver, or chromel is used at very low temperatures (down to 1 K) [94]. Its sensitivity stays high in this region, while the sensitivity of the more common thermocouples goes to zero. At very high temperatures (up to 3000°C), tungsten–rhenium alloys are used. The alloy combinations usually employed have the following compositions:[†] (W + 26% Fe)/(W), (W + 3% Re)/(W + 25% Re), and (W + 5% Re)/(W + 26% Re). These thermocouples find application in the nuclear industry where very high temperatures must be monitored [95].

9.5   NICROSIL–NISIL Thermocouple

The NICROSIL–NISIL thermocouple was developed for use in the 0–1200°C range. It is a more stable thermocouple than type K, which is commonly used in this region. The compositions of these alloys are: Nicrosil, 84.4% Ni + 14.2% Cr + 1.4% Si and Nisil, 95.5% Ni + 4.4% Si + 0.1% Mg. Tests have shown that thermocouples made with these alloys can be used to higher temperatures, are more resistant to oxidation, and are more stable than type K. A discussion of the development of the Nicrosil–Nisil thermocouple together with E versus T reference tables are found in the literature [96–98].

9.6   Thermopiles

Thermocouples are not good for measuring small temperature differences because of their low sensitivity. For example, measuring a temperature difference of 0.001 K using a thermocouple with a Seebeck coefficient of 50 (a quite sensitive thermocouple) requires a voltage measurement of 0.05 μV, and such small voltages are difficult to measure.

To improve this sensitivity, thermocouples can be combined in series as shown in Figure 6.10. In such an arrangement, the thermocouple junctions must be electrically insulated from one another, but when this is achieved, the voltage effect is cumulative. Thus, a thermopile with 100 junctions will have a sensitivity 100 times that of a single thermocouple.

Thermopiles with as many as 1000 thermocouple junctions are used in the construction of some calorimeters to measure and control the temperature difference between the calorimeter and its surroundings. An advantage of a thermopile in this kind of application is that it gives an average temperature difference over the surfaces for which the temperatures are to be compared. A

[†]In representing thermocouples, a diagonal line is used to separate the two thermoelectric elements with the composition of each element enclosed in parentheses.

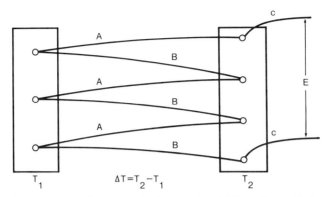

Figure 6.10 Arrangement of thermoelectric elements A and B in a thermopile for measurement from the voltage E of small temperature differences, $\Delta T = T_2 - T_1$. The leads are designated as c.

disadvantage is that with a thermopile there is extensive thermal contact between the calorimeter and its surroundings. In addition, thermopiles are tedious to construct, and they are fragile. Care must be taken to prevent junctions from shorting out or breaking apart.

9.7   Reference Junctions

The reference junction for a thermocouple is almost always $0°C$, obtained from an ice and liquid water mixture, although sometimes this temperature is maintained with another device such as a Peltier cooler. Standard thermocouple tables almost always give E versus T with $0°C$ (or $32°F$) as the reference junction. Temperature-compensated thermocouples are sometimes employed in which the reference junction is maintained at a temperature different from $0°C$, and the measured emf is corrected electronically to the value that would be obtained if the reference junction were $0°C$, so that the standard thermocouple tables can be used.

A mixture of ice and liquid water is the simplest method of obtaining a $0°C$ reference, but care must be taken. Temperature measurements with a thermocouple cannot, of course, be more accurate than the reliability of the ice junction. Distilled water should be used in making the ice bath, and it is best to use ice made from distilled water, although the effect of dissolved impurities usually found in tap water rarely changes the temperature by more than 0.05 K. It has been our experience that an ice bath that differs from $0°C$ by less than $0.01°C$ can be obtained by using distilled water and ice cubes made from tap water (as contrasted to ice flakes obtained in many ice machines). Impurities migrate to the surface of the ice cube during freezing and can be removed by rinsing the cubes with distilled water before crushing them.

For accurate temperature measurement, at least 15 cm of the thermocouple should be immersed in the ice bath. The way to obtain an ice bath that will provide for this immersion and stay at $0°C$ for a long time is to pack a 1-L wide-

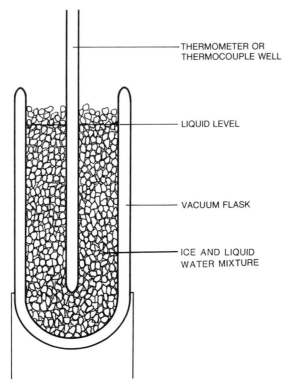

Figure 6.11 Design of an ice bath used as a thermocouple reference junction or for checking the calibration of thermometers.

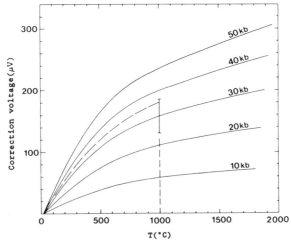

Figure 6.12 Effect of pressure (in kilobars) on the emf of a type S thermocouple. The dashed lines indicate the upper limits of experimentally explored regions. Reprinted with permission from I. C. Getting, and G. C. Kennedy, *J. Appl. Phys.*, **41**, 4552 (1970).

necked vacuum flask with crushed ice and then add distilled water. Adding the water melts some of the ice. More ice and water are added, and the ice is compacted until it fills the flask completely. The water level is adjusted until it is approximately 2 cm below the top of the flask (see Figure 6.11). It is important that the ice reaches to the bottom of the flask. When ice floats on top of the water, temperature gradients arise, with the water warmest at the bottom. This occurs because the density of liquid water near 0°C increases with increasing temperature (maximum density is at 4°C) and warmer water sinks to the bottom. The ice bath must be replaced when enough ice melts that the remaining ice is floating on the liquid water, usually every few hours.

The reference junctions are placed in a closed-end tube located in the center of the flask and extending almost to the bottom. The tube is usually filled to a height of approximately 10 cm with oil. Several thermocouple junctions can be placed in the same tube, but care must be taken that the junctions do not touch one another and short out.

9.8 Pressure Effects on Thermocouples

The emf of a thermocouple depends on pressure, although the effect is not large and needs to be considered only for the most accurate measurements or at high pressures. Experimental measurements of the pressure effect under a variety of temperature and pressure conditions have been reported in the literature [99]. Figure 6.12 shows the results of Getting and Kennedy [100] for a type S thermocouple.

10 MERCURY-IN-GLASS THERMOMETERS

For simplicity of measurement, it is difficult to improve on the mercury thermometer, the oldest of the three thermometers we describe. This thermometer is constructed by attaching a glass bulb to a long capillary tube and filling the bulb with mercury. The glass bulb is brought into thermal contact with the object whose temperature is to be measured. As the temperature changes, the mercury level in the attached capillary moves up and down according to the expansion or contraction of the mercury in the bulb. Markings etched on the capillary tube can be compared to the mercury level to give the temperature (see Figure 6.13) [101].

10.1 Advantages and Disadvantages of the Mercury Thermometer

The advantages of mercury thermometers are simplicity of operation and reproducibility of measurement. There are many disadvantages to their use, however, and these often outweigh the advantages. Since the advent of the microprocessor, other temperature measuring devices, mainly the resistance thermometer and thermocouple, have replaced the mercury thermometer in many applications.

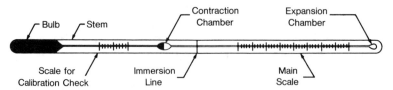

Figure 6.13 Design of a liquid-in-glass (usually mercury) thermometer. Inexpensive models may not include all of the features shown in the figure. Reprinted with permission from J. F. Schooley, *Thermometry*, CRC Press, Boca Raton, FL, 1986, p. 167. Copyright © 1986 CRC Press, Inc., Boca Raton, FL.

One of the most serious disadvantages of the mercury thermometer is the large area of thermal contact required and the slow thermal response. One must wait for the mercury in the bulb to come to temperature equilibrium with the object whose temperature is to be measured. This can take appreciable time and can cause serious errors if the temperature of the object is changing. Another disadvantage is that temperature monitoring requires individual visual measurement. It is not feasible to attach the thermometer to a recording device for continuous temperature monitoring.

Other defects are that the operating range of the thermometer is limited to the liquid range of mercury (-25–$360°C$), and that a large bulb and a small diameter capillary (needed for sensitivity) limit the temperature range that can be covered.

The Beckman mercury thermometer is constructed with an extra large bulb and a fine capillary and is designed for measuring temperature change accurately, but it does not allow for absolute temperature measurement. A "U" tube and mercury reservoir attached to the top of the capillary in this type of thermometer allows one to add to, or remove mercury from, the bulb and hence, to change the temperature range over which measurements can be made. Beckman thermometers, however, are cumbersome, and they have largely been supplanted by other types of sensitive thermometers.

10.2   Sources of Error With Mercury Thermometers

Two significant sources of error when mercury-in-glass thermometers are used for accurate temperature measurement are change in bulb volume and differences in temperature between the bulb and the stem. The bulb of a typical mercury-in-glass thermometer contains mercury equivalent to approximately $5000°C$ of capillary scale length. Hence, small changes in bulb volume can have a relatively large effect on the temperature reading. Thermometers should be annealed thoroughly before use; otherwise, gradual and irreversible changes in bulb volume can occur. Temporary changes in bulb volume due to thermal expansion of the glass occur when the thermometer is heated and then cooled. This can cause errors in the temperature reading as large as $0.01°C$ for each $10°C$ the bulb is heated, and it may take as long as several days for the bulb to readjust.

To minimize errors resulting from changes in bulb volume, the thermometer should have a reference mark or calibration check on its scale corresponding to some readily obtained reference point such as the ice point (refer again to Figure 6.13). It is usually a good assumption that any error observed at this reference point can be applied as a correction to the measurement in the temperature range where the thermometer is calibrated.

The temperature reading of a sensitive mercury thermometer is also affected by atmospheric or hydrostatic pressure. Busse [102] has shown that the pressure coefficient for thermometers having bulb diameters of 5–7 mm is about 0.1°C/atm. Pressure effects resulting from changing the position of a sensitive mercury-in-glass thermometer from horizontal to vertical can also be significant.

The most accurate mercury-in-glass thermometers are usually calibrated to read correctly when both the bulb and the stem are exposed to the temperature to be measured such as in a constant temperature bath. These thermometers are known as *total immersion thermometers.*

In many circumstances, it is not possible to immerse the thermometer totally during a temperature measurement, in which case a "stem correction" should be made that corrects for the error introduced when the mercury in the capillary is at a different temperature from that in the bulb. The stem correction is computed from the equation

$$\Delta t = kn(t_{obs} - t_s)/(1 - kn) \tag{34}$$

where Δt is the correction in degrees, k is the coefficient of expansion of mercury, n is the number of degrees of exposed stem, t_{obs} is the apparent temperature of the bulb or object whose temperature is being measured, and t_s is the temperature of the exposed stem. In most cases $k = 0.00016 \, \text{K}^{-1}$ or $°\text{C}^{-1}$. (In very precise work, correction must also be made for the expansion of the glass.) The stem temperature t_s is usually obtained by placing a second thermometer near the stem of the first. In (34), kn is usually small compared to unity, so that in many cases, the simpler expression

$$\Delta t = kn(t_{obs} - t_s) \tag{35}$$

can be used.

An alternative to making a stem correction is to calibrate the thermometer under the same immersion conditions as that in which it will be used. Often a mark is placed on the stem of the thermometer, indicating this immersion line (refer again to Figure 6.13).[†]

Mercury is by far the most common liquid in glass thermometers; but in special instances, other liquids are also used. Liquid gallium-in-glass thermometers, for example, can be used to measure higher temperatures than mercury-

---

[†]For additional analysis and discussion of mercury thermometers, see [102–105].

in-glass thermometers because of the much higher boiling point of Ga. On the other end, alcohol or toluene can be used in liquid-in-glass thermometers to measure temperatures that are lower than the freezing point of mercury.

11   THERMOMETER CALIBRATIONS

The ITS-90 fixed points given in Table 6.4 can be used to check the calibration of a thermometer, but this measurement requires rather elaborate equipment for some of the fixed points. Measurement of the triple point of water, which serves as the anchor point for any temperature scale, is not difficult. It is made in a triple point cell, whose construction is shown schematically in Figure 6.14.

The cell is filled with high-purity water, evacuated to remove the air, and sealed so that only pure water is present. The cell is placed in an insulated jacket to cut down on heat leak to the surroundings. The triple point is realized by

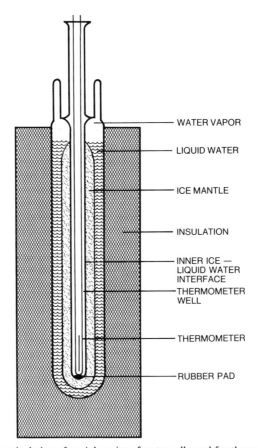

Figure 6.14   Schematic design of a triple point of water cell, used for thermometer calibration.

Table 6.10   Fixed Points for Checking the Temperature Calibration of Thermometers in the Temperature Range from -50 to $50°C$

Description	Reaction[a]	T (K)	t (°C)
Mercury melting point	$Hg(s) = Hg(l)$	234.314	-38.836
Ice point	$H_2O(s) = H_2O(l)$	273.15	0.00
Sodium sulfate decahydrate peritectic	$Na_2SO_4 \cdot 10H_2O(s) = Na_2SO_4(s) +$ saturated solution	305.534	32.384

[a]The letter s designates the solid state; l, the liquid state.

putting liquid nitrogen or dry ice down the thermometer well to freeze a layer of ice around the tube. Alcohol, or some other low-melting liquid, is then poured down the well to melt some of the ice and form a thin layer of liquid water around the well. Thus liquid, solid, and vapor are in equilibrium at the tube surface. The thermometer to be calibrated is placed in the well where it will remain within 0.001 of 273.160 K for at least several hours [106].

Three fixed points that are very easy to obtain without elaborate apparatus or specially purified chemicals can be used to check the calibration of thermometers in the temperature range from -50 to $50°C$, a range where many temperature measurements are made. All three are atmospheric pressure points and are reproducible to better than $0.01°C$. They are summarized in Table 6.10.

The preparation of an ice bath to realize the ice point is the same as described earlier in the preparation of an ice reference junction for thermocouples (see Figure 6.11). It is obtained easily and is probably the first calibration test that should be made to check a thermometer [107].

The mercury melting point is obtained by filling an approximately 25×150-mm test tube to a depth of 10 cm with reagent grade or triple distilled mercury. The test tube is cooled in liquid nitrogen or dry ice until a layer of solid mercury forms on the outer surface of the tube, leaving a hole in the center of the solid that is filled with liquid mercury and is large enough to hold the thermometer (see Figure 6.15). The test tube is placed in a cold vacuum flask and the thermometer is immersed in the liquid mercury. This very simple apparatus will hold at a temperature of 234.31 K for 2–3 h.

The sodium sulfate decahydrate peritectic point is also obtained easily. Reagent grade $Na_2SO_4 \cdot 10H_2O(s)$ is heated in a beaker until approximately one-half of the $Na_2SO_4 \cdot 10H_2O(s)$ decomposes to $Na_2SO_4(s)$ according to the chemical reaction

$$Na_2SO_4 \cdot 10H_2O(s) = Na_2SO_4(s) + \text{saturated solution} \tag{36}$$

With a little practice, it is not difficult to distinguish between $Na_2SO_4 \cdot 10H_2O(s)$ and $Na_2SO_4(s)$ to estimate when approximately one-half is decomposed.

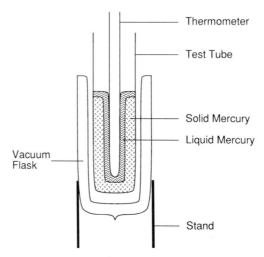

Figure 6.15 Design of an apparatus for checking the calibration of a thermometer at the freezing point of mercury.

The solid–liquid slurry that results is poured into a vacuum flask to a depth of 10–15 cm. The thermometer is pushed down into the mixture in which the temperature will hold at 305.53 K for several hours.

12   ACCURATE TEMPERATURE MEASUREMENT

Two different aspects must be considered when we describe the accuracy of a temperature measurement. The first is the accuracy with which we know the absolute or Kelvin temperature, and the second is the reproducibility with which we can make a measurement. This second aspect is especially important in calibrating a resistance thermometer or thermocouple in terms of the International Temperature Scale.

The triple point temperature of water is the best example from both perspectives. It is the defining point on the Kelvin scale, where it is assigned a value of exactly 273.16 K. At the same time, this temperature can be realized very accurately in an apparatus such as the triple point cell described earlier (see Figure 6.14). With care, one can obtain this temperature within the cell to within 10^{-4} K. The measurement of the triple point of water is impressive. An accuracy of 10^{-4} K represents an error of less than one-half of 1 ppm. Not many other fundamental measurements in science can be made as accurately, and yet as easily, as this one can.

Furukawa and Bigge [108] compared several water triple point cells and showed that they can be used to reproduce the triple point temperature of a platinum resistance thermometer (IPTS-68 or ITS-90) to within 10^{-5} K. The realization of the triple point temperature with this accuracy requires a very careful technique. Care must be taken to obtain ultrapure and gas-free water, since any impurities, including dissolved air, can lower the freezing point significantly. Corrections must be made for the heating effect of the 1-mA current that flows through the thermometer during the resistance measurement. The hydrostatic pressure of the water changes the temperature along the thermometer well $(-7 \times 10^{-6} \text{ K/cm})$, and corrections must be made for this effect. Even the naturally occurring differences in deuterium concentration in water can have a detectable effect.

It is because the triple point cell gives such accurate results that the triple point temperature of water, rather than the normal freezing point, is chosen as the anchor point for all Kelvin temperature measurements on IPTS-68 and ITS-90. Errors compound as one moves away from this temperature. Thus, if one uses a platinum resistance thermometer that is calibrated at the ice point to an accuracy of 0.001 K, the uncertainty in the freezing point of gold (1337.33 K) will be 0.005 K since the resistance of the thermometer has increased by a factor of approximately 5 in going from the ice point to the gold point.

Other fixed points can also be reproduced with high precision. Examples are the triple point temperatures of gallium [109], indium [110], cadmium [111, 112], and aluminum [113, 114] and the melting point of tin [115, 116]. The IPTS-68 and ITS-90 fixed points are, in general, reproducible to the precision expressed in the significant figures given in Tables 6.1 and 6.4.

13   CHOOSING A THERMOMETER

The usual factors to be considered in selecting a thermometer for a particular application are temperature range, stability, sensitivity, convenience, availability, and cost. Table 6.11 summarizes the characteristics of several of the most commonly used general laboratory thermometers. For information about the highly specialized very low and very high temperature ranges, one should consult the original literature. Several references are given in the body of this chapter.

14   MEASUREMENT OF TEMPERATURE WHEN PHASES ARE IN EQUILIBRIUM

The determination of the temperature at which phases are in equilibrium is a measurement of importance in the thermodynamic characterization of a substance.[†] The freezing point or melting point is the temperature at which solid crystals of a substance are in equilibrium with the liquid. If this equilibrium is approached by cooling the liquid, the freezing point is obtained. The melting point is obtained by heating the solid until liquid is present. The boiling point or

---

[†]References [117–121] give detailed discussions of phase equilibria.

Table 6.11 Summary of the Characteristics of Several Commonly Used Laboratory Thermometers[a]

Thermometer	Useful Temperature Range	Advantages	Disadvantages
Mercury-in-glass	−25–350°C	Easy to use, readily available, and inexpensive	A narrow temperature range, slow temperature response, low sensitivity; requires a relatively large area for thermal contact
Copper–constantan thermocouple[b,c]	20–400 K	Simple construction, low cost, and high sensitivity in the low-to-room temperature range	Relatively low stability; requires a reference junction
Chromel–alumel thermocouple[c]	−200–1200°C	Simple construction, low cost, and high sensitivity in the ambient-to-moderately high temperature range	Relatively low stability; requires a reference junction
(Platinum + 10% rhodium)/(platinum) thermocouple[c]	−50–1750°C	High stability and high accuracy (IPTS-68 thermometer for the range 630.74–1064.43°C)	Relatively low sensitivity; requires a reference junction; more expensive to purchase and fabricate than the two previously mentioned thermocouples
Thermistors (semiconductor resistance thermometer)[d]	[e]	Available in a variety of sizes and shapes (beads as small as 0.07 mm in diameter are available commercially);	Relatively low stability and narrow temperature range; requires aging and frequent calibration

494

| Platinum resistance thermometer[f] | 10–900 K (80–900 K for the long-stem version, and 10–400 K for the capsule type; new high-temperature designs extend the upper limit to 1400 K) | high sensitivity, and especially useful for measuring temperature differences; are convenient to use, especially where a small thermal contact is required | High stability and high accuracy (IPTS-68 thermometer for the range from 13.81 K to $630.74°C$) | More expensive, less convenient (requires a relatively large area for thermal contact); slower response than thermocouples or thermistors |

[a]Mercury-in-glass thermometers and some thermocouples and thermistors are available from general scientific supply houses. A wide selection of thermocouples can be found in supply houses that specialize in temperatures, rather than an absolute value of temperature measurements (usually by means of thermocouples and thermistors). The information given is often helpful, especially to someone interested in industrial applications. Platinum resistance thermometers, resistance bridges, and digital voltmeters are available from supply houses that specialize in electrical instruments.

[b]Use is limited to reducing environment at temperatures above 150°C.

[c]One can measure the voltage drop across the thermocouple junctions with high accuracy and convenience by use of a modern digital voltmeter.

[d]Thermistors are most effectively used where differences in temperatures, rather than an absolute value of temperature are of primary concern.

[e]Depending on the composition, one can use thermistors over a temperature range from very low to high. The most stable thermistors used from room temperature to 250°C are made from mixed oxides of manganese and nickel or from manganese, nickel, and cobalt. In the range of 300–700°C, lanthanide oxides are commonly used. At still higher temperatures, zirconium oxide doped with small quantities of lanthanum oxides is practical. At very low temperatures, nonstoichiometric iron oxides may be used.

[f]When used in conjunction with a recording resistance bridge, a platinum resistance thermometer is an ideal instrument for obtaining the time–temperature cooling and heating curves needed in phase equilibria studies.

condensation point results when liquid and vapor are in equilibrium. The boiling point is obtained when a liquid is heated until boiling occurs. The condensation point is obtained by cooling the vapor until droplets of liquid are formed.

Many solids can exist in two or more crystalline forms. The state at which two solid phases are in equilibrium is the *solid-state transition point*. As an example, when rhombic sulfur is heated at atmospheric pressure, it changes crystal structure to a monoclinic form. The two solid phases are in equilibrium at 368.6 K and 1-atm pressure.

An interesting group of compounds are those that form plastic crystals.[†] Each has a solid phase transition at temperatures not far below the melting point. Plastic crystal formation usually has a very pronounced effect on the melting point of the substance, and we discuss it further in Section 18.4.

Another interesting group of substances are those that form liquid crystals. In this case the solid melts to form a "liquidlike" phase that retains order in two dimensions but breaks down in the third dimension. Heating these liquid crystals causes them to melt to a more conventional liquid at a higher temperature. The liquid crystals are classified into different types depending on the nature of the two-dimensional order. Detailed discussions of these interesting compounds can be found in the literature [128–132]. The properties of liquid crystals are summarized in [128–130] and detailed discussions of liquid crystals are written in [131, 132].

15   PHASE DIAGRAM FOR A PURE SUBSTANCE

Temperature (T), pressure (p), and composition (usually mole fraction, x) are the variables that are used to specify phase equilibria. The relationships between these variables are generally summarized through a phase diagram. Figure 6.16 is an example of a simple phase diagram relating p and T for a pure substance ($x = 1$).

Solid and vapor are in equilibrium along line AB. Similarly, solid and liquid are in equilibrium along BD, and liquid and vapor are in equilibrium along BC. If the solid is heated along line $abcd$ at an external pressure p_{ext}, the solid melts at point b, with T_m as the melting point, and boils at point c, with T_b as the boiling point. If the external pressure p_{ext} is 1 atm, T_m and T_b are referred to as the *normal* melting and boiling points, respectively. At point B, called the triple point, all three phases are in equilibrium. It is an invariant point; that is, as long as all three phases are present, the temperature and pressure cannot change. It is this

---

[†]For a review of plastic crystals, see Timmermans [122]. This paper was presented in a symposium entitled "Plastic Crystals and Rotation in the Solid State" held at Oxford, England, in April, 1960. The papers from this symposium were published as part of Vol. 18 of the *Journal of Physics and Chemistry of Solids*. Other papers of thermodynamic interest were presented by Staveley [123], Aston [124], Westrum, Jr. [125], and Weinstock [126]. A subsequent review of plastic crystals is given by Dunning [127].

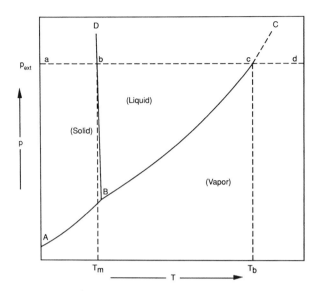

Figure 6.16 Typical pressure p versus temperature T phase diagram for a pure substance. The external pressure is p_{ext}. (Solid + liquid) equilibrium occurs along line BD; (solid + vapor) equilibrium, along AB; and (vapor + liquid) equilibrium, along BC. Point B is a triple point and C is the critical point. Points T_m and T_b are the normal melting and boiling points if $p_{ext} = 1$ atm. The critical pressure is normally very much higher than 1 atm.

temperature invariance that allows the triple point of water to be chosen as the reference for the absolute temperature scale. Point C is the critical point. It is the end of the (vapor + liquid) equilibrium line and is usually at a much higher T and p than the normal melting and boiling points. Liquid does not exist at temperatures greater than the critical temperature.

Line BD is almost vertical, illustrating that pressure has a small effect on the melting temperature unless the pressure becomes large. Thus, changes in atmospheric pressure usually have a negligible effect on the melting point of a substance measured in the atmosphere. For example, the melting temperature for ice changes by 0.073 K/MPa or 0.0074 K/atm around the normal melting point. The small effect of p on T_m causes the triple point temperature and normal melting point to be very nearly the same. For water, the two temperatures differ by 0.01 K (273.16 vs. 273.15 K), and only part of this difference is due to the pressure effect on the melting temperature. Dissolved air, which is present in the water when a melting point is obtained but is not present during a triple point measurement, also contributes to the lowering of the melting point [107].

The quantitative relationship between melting temperature T and pressure p is given by the Clapeyron equation, which is derived from thermodynamic principles:

$$\frac{dp}{dT} = \frac{\Delta_{fus}S}{\Delta_{fus}V} = \frac{\Delta_{fus}H}{T\Delta_{fus}V} \tag{37}$$

where dp/dT is the slope of line BD in Figure 6.16; and $\Delta_{fus}V$, $\Delta_{fus}S$, and $\Delta_{fus}H$ are the volume, entropy, and enthalpy changes, respectively, during melting. For the melting process, $\Delta_{fus}H$ is much larger than $T\Delta_{fus}V$, making dp/dT large, and, accordingly, the slope of line BD steep. From Figure 6.16, we also see that the melting line BD is very nearly a straight line, which requires that $\Delta_{fus}S/\Delta_{fus}V$ be nearly constant. By assuming that this ratio is constant, one can integrate (37) to relate the melting temperature T to the pressure p and the normal melting point T_m:

$$T = T_m + \frac{\Delta_{fus}V}{\Delta_{fus}S}\,(p-1\text{ atm}) \tag{38}$$

Equation (38) is a good approximation for (solid + liquid) equilibrium if the pressure change does not become large. As stated earlier, changes in atmospheric pressure have a negligible effect.

An equation analogous to (37) applies to the boiling process, in which case

$$\frac{dp}{dT} = \frac{\Delta_{vap}H}{T\Delta_{vap}V} \tag{39}$$

where dp/dT is the change in vapor pressure with temperature, and $\Delta_{vap}H$ and $\Delta_{vap}V$ are the changes in enthalpy and volume, respectively, for vaporization. Since $\Delta_{vap}H$ and $T\Delta_{vap}V$ are more nearly the same size than are the corresponding terms in the (solid + liquid) case, line BC (unlike line BD) does not have a steep slope, and the boiling point is sensitive to pressure. Thus, changes in atmospheric pressure can significantly change the boiling temperature. By making some simplifying assumptions, (39) can be integrated to give the Clausius–Clapeyron equation, which can be used to calculate the boiling temperature T at pressure p (in atmospheres) if one knows the normal boiling temperature T_b

$$\frac{1}{T} = \frac{1}{T_b} - \frac{R}{\Delta_{vap}H}\ln p \tag{40}$$

Equation (40) can be used with experimental $\Delta_{vap}H$ values to give a good approximation for T provided p does not differ greatly from 1 atm.[†] Other

---

[†]A summary of literature sources for obtaining enthalpy changes for phase transformations is given by Tamir and co-workers [133]. Heat capacity, entropy, and enthalpy differences for phase changes are summarized by Domalski and co-workers [134]. Reference [135] is a compilation of vapor pressures and enthalpies of vaporization, and [136] summarizes the vapor pressures of 2000 substances and includes graphs of vapor pressure plotted against temperature, and the Antoine constants. For a discussion of vapor pressure measurements on pure substance, see [137, 138]. Vapor pressures and other physical properties of substances that are gaseous at room temperature are summarized in [139].

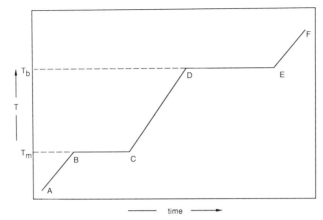

Figure 6.17 Time-versus-temperature warming curve when a substance is heated at a constant rate along line *abcd* in Figure 6.16. The changes that occur along line *ABCDEF* are described in the text.

procedures are described in the literature for predicting more accurately and under more extreme conditions the effect of pressure on boiling temperature [140]. Equation (40) can also be used to calculate vapor pressure p at temperature T, but other more accurate expressions are usually used. The Antoine equation is one of the most commonly encountered semiempirical equations:

$$\log p = A - \frac{B}{(t + C)} \qquad (41)$$

which relates vapor pressure p (usually in torr) to the centigrade temperature t. The Antoine constants A, B, and C are tabulated in many reference sources [136, 141–147].[†] The Antoine equation does a reasonably good job of representing vapor pressure, temperature data, but one must be careful when trying to extend the range of the equation, as it cannot be extrapolated very well.

16   TIME-VERSUS-TEMPERATURE WARMING CURVES

Figure 6.17 is an example of the graph of time-versus-temperature when a solid substance is heated at 1 atm (101.325 kPa) pressure and at a constant rate along the line *abcd* in Figure 6.16. The temperature increases along line *AB* until the melting temperature T_m is reached at *B*. Along *BC*, the solid is melting and the temperature stays constant at T_m. Once the solid is melted, the temperature

---

[†]Compilation of the thermodynamic and physical properties of many pure substances are presented in [141–147].

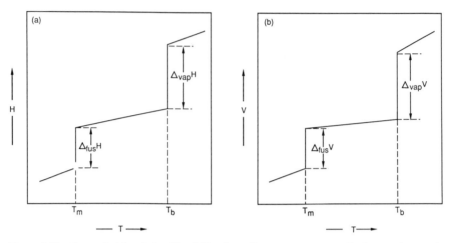

Figure 6.18 Change in (*a*) enthalpy *H* and (*b*) volume *V* versus temperature *T* when a substance is heated along line *abcd* in Figure 6.16. Points T_m and T_b are the melting and boiling temperatures, respectively.

increases again as the liquid is heated from *C* to *D*. At *D*, the liquid boils and the temperature stays at the boiling temperature T_b until all the liquid has evaporated (line *DE*). Further heating increases the temperature of the vapor along line *EF*.

The temperature halts in Figure 6.17 at T_m and T_b can be understood by considering the enthalpy and volume changes that we obtain along line *abcd* in Figure 6.16. Figure 6.18*a* summarizes what happens to the enthalpy as the temperature is increased. When the temperature T_m is reached, the heat added is used to increase the enthalpy *H* as $\Delta_{fus}H$ and the temperature does not change. The process is repeated at T_b with the heat going into the substance as $\Delta_{vap}H$.

Figure 6.18*b* illustrates what happens to the volume during the same heating process with $\Delta_{fus}V$ and $\Delta_{vap}V$ equal to the volume changes on melting and vaporization, respectively. Phase changes that have nonzero enthalpy and volume changes are known as *first-order phase transitions*. There are, however, examples of phase transitions where ΔH and ΔV are zero for the phase changes. Examples are the change of a metal from a normal to a superconducting state and the change of liquid helium from a normal to a superfluid. These changes are known as *second-order phase transitions*. They are not common, but are usually very interesting. The techniques we describe in this chapter apply only to measuring first-order phase changes.[†]

[†]For a general discussion of freezing and melting points methods, see Pistorius [148] and Doucet [149].

17   TIME-VERSUS-TEMPERATURE COOLING CURVES AND SUPERCOOLING EFFECTS

Figure 6.19 illustrates the temperature changes that occur when a substance is cooled along line *dcba* in Figure 6.16. The vapor is cooled along line *AB* until it condenses to a liquid at point *B*. The small dip in the curve at *B* is a nonequilibrium condition caused by supersaturation. This effect is usually small, so that the dip at *B* is usually shallow. Once condensation starts, the temperature increases to T_b and remains at that value until all the vapor is condensed (point *C*). The liquid is cooling along *CD*, with the solid crystallizing at *D*. The dip at *D* below T_m is caused by supercooling of the liquid. Depending on the nature of the liquid, supercooling can be anywhere from a small to a large effect. Once crystallization starts, the temperature increases and remains at T_m until solidification is complete (point *E*). The solid is then cooled along line *EF*.

For some substances, supercooling can be a serious problem. In some instances, it can be so severe that crystallization never occurs. When this happens, the liquid usually becomes more and more viscous until it sets up as a glass, which can be thought of as a liquid with a very high viscosity.

It is of value to be able to predict when extensive supercooling will occur. Symmetry plays an important role, with symmetrical molecules supercooling the least. Thus, benzene, cyclohexane, and naphthalene supercool very little. On

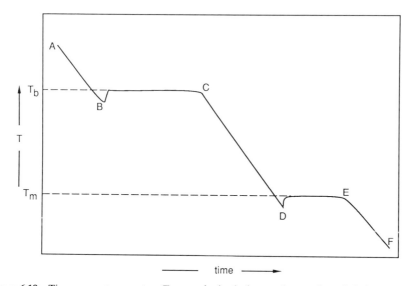

Figure 6.19   Time-versus-temperature T curve obtained when a substance is cooled along line *dcba* in Figure 6.16. Again, T_m and T_b are the melting and boiling temperatures, respectively. The changes that occur along line *ABCDEF* are described in the text; supersaturation of the vapor occurs at *B* and supercooling of the liquid occurs at *D*.

the other hand, supercooling in toluene (methylbenzene), *o*-xylene (1,2-dimethyl-benzene), and *m*-xylene (1,3-dimethylbenzene) is pronounced. As expected, *p*-xylene (1,4-dimethylbenzene) shows much less supercooling than do the asymmetric xylenes.

Hydrogen bonding can also greatly increase the supercooling. Thus, water, which is not a highly asymmetrical molecule, supercools sometimes as much as 20°C. Alcohols supercool even more; and the more hydroxyl groups that are present, the more the supercooling. Glycerin, for example, supercools so badly that special means are required to cause it to crystallize from solution. Purity also affects supercooling. In general, very high purity samples tend to supercool more than those with at least small amounts of impurities.

Various techniques can be employed to minimize supercooling. Stirring (especially vigorous stirring) of a supercooled liquid often serves to initiate crystallization. Scraping the side of the container or adding small particles of an inert material (available in commercial preparations) also may help. One of the best ways to start crystallization is to add a small crystal of the substance to the supercooled liquid. This crystal serves as a nucleus for further crystallization. Supercooling is often much less the second time a substance is frozen, if the melt from the first time is not allowed to warm more than a few kelvins above the melting temperature.

Finally, some substances (usually very viscous liquids or very asymmetrical molecules) are difficult to keep in phase equilibrium even when both solid and liquid are present. Too rapid cooling causes the temperature to decrease below the equilibrium melting point. Solid phase transitions are also often sluggish and hard to keep in equilibrium. In both cases, slow cooling is required. Calorimetric techniques are ideally suited to this type of procedure, and they are described more completely in Chapter 7.

18   EFFECTS OF MOLECULAR SIZE AND SHAPE ON THE MELTING AND BOILING TEMPERATURE

In general, increasing molecular size increases the melting point and the boiling point of a substance. Molecular attractions resulting from polarity also have an effect, especially on the boiling point; and molecular symmetry has an effect, especially on the melting point.

18.1   Effect of Polarity

The effect of the extreme polarity of the hydrogen bond on boiling point is illustrated well by comparing the boiling points of the series HF, HCl, HBr, and HI; H_2O, H_2S, H_2Se, and H_2Te; and NH_3, PH_3, AsH_3, and BiH_3 (see Figure 6.20). In these series, the first member (HF, H_2O, or NH_3) strongly hydrogen bonds, while the other members of the series do not. The three that do, have boiling points that are exceptionally high compared to the other members of the

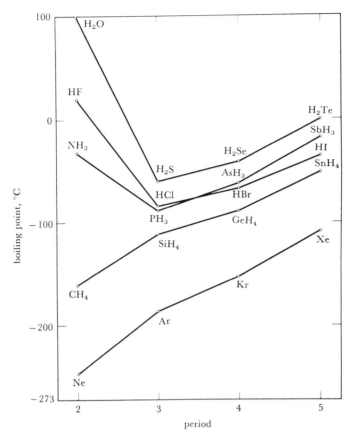

Figure 6.20   Comparison of the boiling points of the hydrides from Groups IV, V, VI, and VII (of the different periods from the Periodic Table of the Elements) and the corresponding noble gases, illustrating the effect of hydrogen bonding in NH_3, H_2O, and HF on this property.

series. The melting points follow the same trend, although the effect is not as pronounced. As another example, the boiling temperatures T_b are compared below for four organic compounds having the same, or very nearly the same, molecular weight; approximately the same shape; but widely differing polarities and extent of hydrogen bonding.

Substance	2-Methylpropane	Acetone	2-Propanol	Acetic acid
Structural formula	C | C—C—C	O || C—C—C	C | C—C—OH	O || C—C—OH
Boiling temperature (K)	272.6	329.4	355.5	391.1

Again, the boiling temperature is seen to increase with increasing polarity, especially when hydrogen bonding is involved.

18.2   Effect of Symmetry

Figure 6.21 summarizes the normal melting and boiling points of the homologous series C_nH_{2n+2}. Note the general increase in melting and boiling points with increasing molecular size. Superimposed upon the general trend is a symmetry effect on the melting point. The hydrocarbons with odd numbers of carbon atoms, being less symmetrical than those with an even number of carbon atoms, have a lower freezing temperature than is predicted simply because of an increase in molecular size. A similar effect is not seen for the boiling points.

An even more dramatic example of the effect of symmetry on melting point can be seen in the following series of alkyl benzenes:

Substance	Benzene	Toluene	o-Xylene	m-Xylene	p-Xylene
Melting temperature (K)	278.68	178.16	247.97	225.28	286.41

The alkyl group in toluene (methylbenzene) destroys the symmetry present in

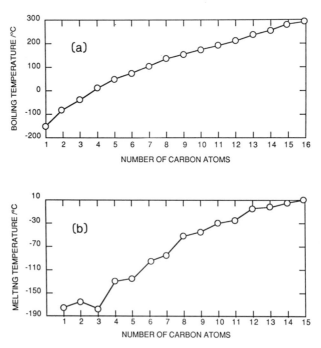

Figure 6.21   Comparison of (a) the normal boiling points and (b) the normal melting points of the homologous series C_nH_{2n+2}.

benzene and lowers the freezing point by over 100 K. The compounds *o*-xylene (1,2-dimethylbenzene) and *m*-xylene (1,3-dimethylbenzene) also lack symmetry and have relatively low melting temperatures; *p*-xylene (1,4-dimethylbenzene), on the other hand, has a high degree of symmetry and a correspondingly high melting point.

18.3   Prediction of Melting Points in Homologous Series

Melting temperature can be related to molecular weight in a homologous series containing the same functional group but with successive members of the series differing by a CH_2 unit. It has been suggested that the melting points in such a series tend to approach a *convergence temperature* of about 390 K as the molecular weight increases [150–152]. Generally, the lower molecular weight members of the series melt below 390 K and increase with molecular weight. Examples are the *n*-alkanes, fatty acids, nitriles, ketones, and the odd-numbered members of the dicarboxylic acid series. Sometimes the melting temperatures start above 390 K and decrease with increasing chain length. Examples are the even-numbered members of the dicarboxylic acid series.

Austin [153] suggested that the melting temperature T_m (in kelvin) or t_m (in °C) for the higher members of a homologous series are related to molecular weight M (g/mol) by the equation

$$\log M = A + BT_m = A_0 + Bt_m \tag{42}$$

where A and B are constants that are characteristic of each homologous series. The relationship in (42) in terms of t_m is due to Skau and Arthur [154]. Figure 6.22 illustrates the conformity of a number of homologous series to the straight-line relationship of (42). Table 6.12 gives values for A, B, and A_0 for several homologous series. These constants can be used with (42) to predict T_m or t_m for unknown members of the series.

There are many exceptions to the general relationship expressed by (42). We have already described the alternation of T_m in the *n*-alkanes for odd and even numbers of carbon atoms, although the alternation does dampen out at high n values. Other series such as fatty acids, glycols, and alkyl malonic acids show similar behavior. Other exceptions are substances containing an acidic and a basic radical, substances containing two amide (or analogous) groups, and a few other isolated series [152].

18.4   Globular Molecules or Plastic Crystals

Timmermans [122] described a group of chemical compounds he refers to as *globular molecules*. They are characterized by a very nearly spherical or globular shape. Examples are methane, tetrachloromethane, 1,1,1-trichloroethane, 2-methyl-2-propanol, cyclohexane, adamantane, and camphor. Globular molecules have solid-state phase transitions, usually at a temperature not far below the melting point, with a plastic or amorphous solid (called a *plastic*

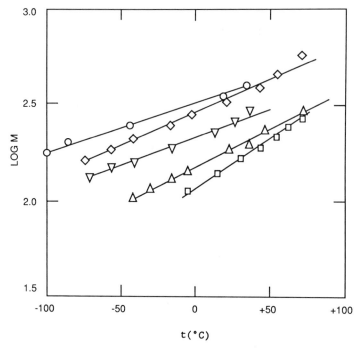

Figure 6.22 Relation between melting points and log *M* for homologous series (*M* is the molecular weight in grams per mole): ○, *n*-alkyl iodides; ◇, acid anhydrides; ▽, methyl esters of *n*-fatty acids; △, *n*-aliphatic alcohols, □, *n*-fatty acids (*n* = even number). Reprinted with permission from E. L. Skau and J. C. Arthur, Jr., "Determination of Melting and Freezing Temperatures," in A. Weissberger and B. W. Rossiter, Eds., *Physical Methods of Chemistry*, Part V, *Determination of Thermodynamic and Surface Properties*, Wiley-Interscience, New York, 1971, Chap. 3, p. 128.

crystal) present in the temperature range between the transition temperature and the melting point.

Plastic crystals have several interesting and unusual properties. They are soft, tacky, and easily deformed. Their triple point vapor pressure is high (some such as C_2Cl_6 sublime at atmospheric pressure), and they have exceptionally low enthalpies of fusion and unusually high melting points. Timmermans [122] lists over 100 compounds that form plastic crystals.

19   EFFECT OF COMPOSITION ON MELTING AND BOILING TEMPERATURES

The discussion so far has been limited to the pressure–temperature relationships during phase equilibria for a pure substance. Composition also has an effect which will now be considered. Referring to Figure 6.23 will help in understanding what happens when composition (*x*) becomes a variable. Figure 6.23 is a

Table 6.12 A, B, and A_0 Constants in (41) and (42) Relating Melting Point (K or °C) and Molecular Weight (g/mol) for Homologous Series[a]

Series Description	A	B	A_0	Minimum Number of C atoms in the Chain for Which Formula Applies
n-Alkanes	1.299	0.0038	2.337	15
n-Alkenes-1	1.240	0.0040	2.333	10
n-Alkynes-1	1.307	0.0036	2.290	7
n-Alkadiynes-1,$(n-1)$	1.252	0.0036	2.235	11
1-Phenyl-n-alkanes	1.563	0.0031	2.410	3
1-Iodo-n-alkanes	1.839	0.0024	2.495	4
n-Alkanols-1	1.092	0.0040	2.185	6
n-Alkanals	1.180	0.0039	2.245	4
n-Alkanones-2	1.120	0.0039	2.185	7
Normal alkanoic acids				
Odd	0.859	0.0047	2.143	5
Even	0.622	0.0053	2.070	6
Methyl esters of n-alkanoic acids	1.485	0.0031	2.332	8
Ethyl esters of n-alkanoic acids	1.535	0.0031	2.382	5
n-Alkanoic acid anhydrides	1.500	0.0035	2.456	4
n-Alkanoic acid chlorides	1.471	0.0034	2.400	6
n-Alkane nitriles	1.302	0.0035	2.258	5

[a]Adapted from [154].

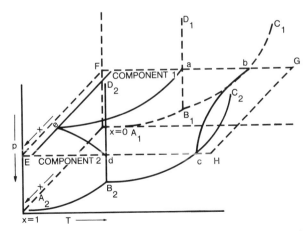

Figure 6.23 Three-dimensional (T, p, x) phase diagram showing the effect of mole fraction x and pressure p on the boiling and melting temperatures T of a binary mixture. The critical points of components 1 and 2 are at C_1 and C_2, respectively, while e is the eutectic point. The other letters are used in the text to describe the diagram.

three-dimensional plot with the x (composition) axis perpendicular to the plane of the paper. The (p, T) plane at $x = 0$ is behind and parallel to the plane of the paper, which is a (p, T) plane at $x = 1$. Both planes contain a p versus T phase diagram similar to Figure 6.16. The first is for pure component 1 and the second is for pure component 2. Lines B_1D_1 and B_1C_1 are the melting and liquid vapor pressure lines, respectively, for component 1; and B_2D_2 and B_2C_2 are the corresponding lines for component 2. The plane $EFGH$ is parallel to the (x, T) plane and intersects the p axes at points E and F. If p_{ext} is 1 atm (101.325 kPa), then the temperature corresponding to point a is the normal melting point of component 1, and point b gives the normal boiling point of the same substance. Similarly, points d and c give the normal melting and boiling points, respectively, of component 2.

Line bc is in plane $EFGH$. It represents the boiling temperature as a function of x at 1 atm pressure as predicted for an ideal solution of mixtures of components 1 and 2 (the ideal solution is described fully in Section 20). Depending on the relative boiling points at b and c, line bc may either increase or decrease as x goes from 0 to 1.

In simple systems such as the one depicted here, adding a second component to a substance lowers the melting temperature. Line ae (in plane $EFGH$) illustrates what happens to the normal melting temperature of component 1 as component 2 is added. Line de (also in plane $EFGH$) illustrates the decrease in the normal melting temperature of component 2 when component 1 is added. Again, these are the predictions for an ideal solution. The two melting lines meet at point e, which is known as the *eutectic point*. It is the lowest temperature at which liquid can be present in this binary mixture. The eutectic is described more fully in Sections 27 and 30.

Pressure affects the boiling and melting temperatures of mixtures. This effect can be represented by drawing planes parallel to $EFGH$ in Figure 6.23 that intersect the vertical axis at the various pressures. Lines B_1D_1 and B_2D_2 are almost vertical, showing that pressure has only a small effect on melting temperature. Boiling temperatures, on the other hand, are affected significantly by pressure since lines B_1C_1 and B_2C_2 are not vertical, nor close to being vertical.

The binary (solid + liquid) (melting) and (vapor + liquid) (boiling) phase diagrams shown in Figure 6.23 are among the simplest to be encountered. We now consider these and more complicated phase diagrams in more detail.[†]

20 BINARY (VAPOR + LIQUID) PHASE EQUILIBRIUM: VAPOR PRESSURES OF IDEAL SOLUTIONS

In a liquid mixture, the two components each exert an equilibrium pressure in the vapor phase. These vapor pressures, p_1 and p_2, are related to the mole fractions $(1-x)$ of component 1 and x of component 2, in the liquid phase.

[†]Some general references on phase equilibria are [155–165].

Raoult's law as given in (43) and (44) is one such relationship.[†]

$$p_1 = (1-x)p_1^*$$ (43)

$$p_2 = xp_2^*$$ (44)

In these equations p_1^* and p_2^* are the vapor pressures of the pure components. By definition, an ideal solution is one in which both components obey Raoult's law over the entire composition range and at all pressures and temperatures.

The total vapor pressure above the solution is given by $p = p_1 + p_2$. Combining this relationship with (43) and (44) gives an equation relating the total vapor pressure p above a liquid solution to the mole fraction x in the liquid phase

$$p = p_1^* + (p_2^* - p_1^*)x$$ (45)

The vapor phase has a different composition than the liquid phase. The total vapor pressure can also be related to the mole fraction y of component 2 in the vapor phase by combining the equation that defines the partial pressure in a mixture of gases

$$p_2 = yp$$ (46)

with (44) and (45) to give

$$p = \frac{p_1^* p_2^*}{p_2^* + (p_1^* - p_2^*)y}$$ (47)

Figure 6.24 is a graph of (45) and (47), indicating how the total vapor pressure above an ideal solution is related to the liquid and vapor phase compositions. In this figure, the upper curve is a graph of p versus x and the lower curve relates p to y. Figure 6.24 is also an example of a (vapor + liquid) phase diagram. If a piston is used to exert a pressure on the mixture, only liquid exists if the pressure is greater than the upper line (liquid line), and only vapor is present if the pressure is less than the lower line (vapor line). Liquid and vapor are in equilibrium only at pressures and compositions corresponding to the region between the two curves. The compositions and relative amounts of the phases present in this two-phase region can be determined from horizontal lines known as *tie lines*. Consider a mixture in Figure 6.24 with pressure and overall composition given by point c. The horizontal dotted line drawn through point c, which intersects the liquid line at point a and the vapor line at point b, is a tie line. The liquid phase composition x is given by point a and the vapor phase

[†]Equations (43)–(47) apply rigorously only when the vapor phase is an ideal gas. Otherwise, fugacities should be used.

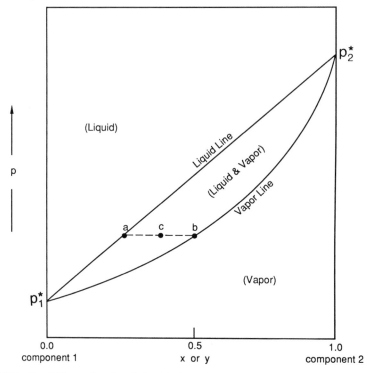

Figure 6.24 Graph illustrating the relationship for an ideal solution between the mole fractions x in the liquid phase and y in the vapor phase and the total vapor pressure p; p_1^* and p_2^* are the vapor pressures of pure components 1 and 2, respectively.

composition y, by point b. Furthermore, the relative amounts of vapor and liquid are given by the lengths $\overline{ac}$ and $\overline{bc}$

$$\frac{\overline{ac}}{\overline{bc}} = \frac{\text{total moles of vapor}}{\text{total moles of liquid}} \tag{48}$$

Equation (48) is known as the *lever rule*.

21 VAPOR PRESSURES OF NONIDEAL SOLUTIONS

Most liquid mixtures do not obey Raoult's law over the entire range of composition, and, hence, are not ideal solutions. For example, Figure 6.25 [166] is a graph of the vapor pressure versus liquid phase mole fraction for (a) (acetone + carbon disulfide) and (b) (acetone + chloroform). In describing phase diagrams, we use the notation $(1-x)A+(x)B$ with x as the mole fraction of component 2(B) and $(1-x)$ is then the mole fraction of component 1(A). Thus,

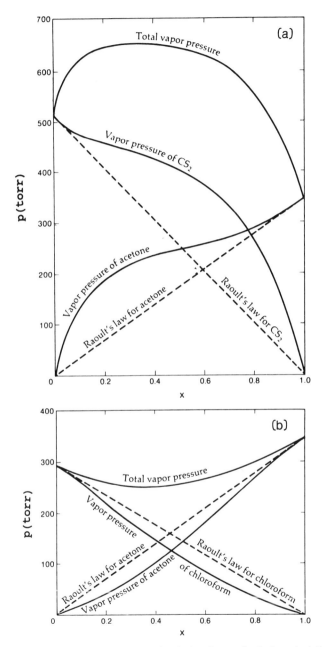

Figure 6.25 Vapor pressure p as a function of mole fraction x of solutions deviating from ideal behavior: (a) $x(CH_3)_2CO + (1-x)CS_2$, an example of a system with positive deviations from Raoult's law, and (b) $x(CH_3)_2CO + (1-x)CHCl_3$, an example of a system with negative deviations from Raoult's law. From J. R. Goates and J. B. Ott, *Chemical Thermodynamics, An Introduction*, Harcourt Brace Jovanovich, New York, 1971, pp. 112–113, copyright © 1971 by Harcourt Brace Jovanovich, Inc. Reprinted by permission of the publisher.

the compositions of the mixtures in Figure 6.25 are described as $x(CH_3)_2CO + (1-x)CS_2$; and $x(CH_3)_2CO + (1-x)CHCl_3$. The broken lines in Figure 6.25 are the ideal predictions (43) and (44), and the solid lines represent the experimental values. The (acetone + carbon disulfide) system is an example of one with positive deviations from Raoult's law; that is, the vapor pressures are greater than they would be if the solution were ideal. On the other hand (acetone + chloroform) shows negative deviations from ideal behavior, the measured vapor pressures being less than the ideal prediction.

Systems can also be formed where positive deviations occur over a portion of the concentration range and negative deviations at other concentrations. Reference [167] describes the different possible types of (vapor + liquid) phase diagrams that can occur. In all cases the solutions show a common characteristic. The solvent (the component present in highest concentration) approaches Raoult's law in the limit of low concentration of solute, and, hence, nearly pure solvent. This can be seen in Figure 6.25 where the vapor pressure curve for acetone approaches the same value and slope as the Raoult's law prediction as the mole fraction of acetone approaches unity. On the other end, where mole fraction of acetone approaches zero, the vapor pressure of the carbon disulfide or chloroform approaches Raoult's law behavior. This is an important conclusion since equations derived to explain phase behavior that is based on the ideal solution can be applied to the solvent in dilute solution even though Raoult's law does not apply over the entire composition range.

In dilute solution, the solute (component present in lowest concentration) does not obey Raoult's law. This is shown in Figure 6.26 [166] for dilute solutions of acetone in carbon disulfide or chloroform, in which we see that the vapor pressure of the acetone is linearly related to mole fraction in the limit of dilute solution, but the slope is not the same as for Raoult's law. This straight line relationship is known as Henry's law, and is represented mathematically by

$$p_2 = k_x x \tag{49}$$

where p_2 and x are the vapor pressure and mole fraction, respectively, of solute; and k_x is the Henry's law constant. Henry's law can be expressed in the alternative forms

$$p_2 = k_m m \tag{50}$$

$$p_2 = k_c c \tag{51}$$

where m and c are molality and molarity, respectively. The three Henry's law constants, k_x, k_m, and k_c, are related by

$$k_x = \frac{1000 k_m}{M_1} = \frac{1000 d_0 k_c}{M_1} \tag{52}$$

where M_1 and d_0 are the molecular weight (g/mol) and density (g/cm³),

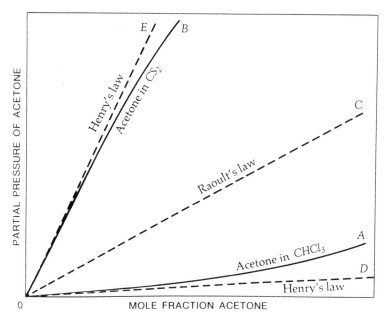

Figure 6.26 Representation of Henry's law and Raoult's law for acetone as a solute in dilute solution: A, vapor pressure of acetone in $CHCl_3$; B, vapor pressure of acetone in CS_2; C, Raoult's law prediction of the vapor pressure of acetone in $CHCl_3$ and in CS_2; D, Henry's law prediction of the vapor pressure of acetone in $CHCl_3$; E, Henry's law prediction of the vapor pressure of acetone in CS_2. The vapor pressures of the acetone deviate widely from Raoult's law, but approximates Henry's law in dilute solution. From J. R. Goates and J. B. Ott, *Chemical Thermodynamics, An Introduction*, Harcourt Brace Jovanovich, New York, 1971, pp. 112–113, copyright © 1971 by Harcourt Brace Jovanovich, Inc. Reprinted by permission of the publisher.

respectively, of the pure solvent. It can be shown from thermodynamic principles that when the solvent obeys Raoult's law, the solute must obey Henry's law.

22   BOILING POINTS OF AN IDEAL SOLUTION

The relationship (for an ideal solution) between vapor pressures at constant temperature and boiling temperatures at constant pressure is shown in Figure 6.27 [168]. In this three-dimensional graph, mole fraction is the horizontal axis, pressure is the vertical axis, and temperature is an axis perpendicular to the plane of the paper. The front surface is the same as in Figure 6.24 and relates total vapor pressure to mole fraction. The side planes at $x = 0$ and $x = 1$ are the vapor pressure against temperature curves for pure component 1 and component 2, respectively, and are the same as line BC in Figure 6.16. The convex and concave surfaces within the figure relate p to x (liquid phase mole fraction) and y

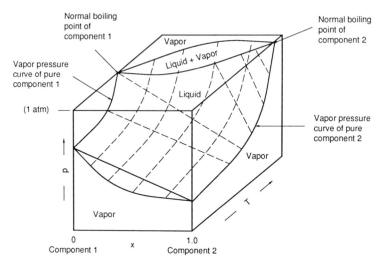

Figure 6.27   Three-dimensional pressure p, temperature T, and mole fraction x (vapor+liquid) phase diagram for a binary system obeying Raoult's law, illustrating the relationship between vapor pressure at constant temperature and boiling temperature at constant pressure. Reprinted with permission from J. Laidler and J. H. Meiser, *Physical Chemistry*, Benjamin/Cummings, Menlo Park, CA, 1982, p. 221.

(gas phase mole fraction) at different temperatures. The solid figure formed from the volume between these surfaces represents the region of (vapor + liquid) equilibrium at the various p, T, and x (or y) combinations. The normal boiling point as a function of composition is obtained from the intersection of the plane with p constant at 1 atm (101.325 kPa) and the solid region where (vapor + liquid) equilibrium can occur. This plane at constant p is the top plane in Figure 6.27 and is shown again in Figure 6.28.[†]

Figure 6.28 is a temperature-composition phase diagram for an ideal solution with pressure set at 1 atm. The lower curve (liquid line) gives the normal boiling point as a function of the liquid phase composition x, while the upper curve (vapor line) expresses the boiling temperature in terms of y, the vapor phase composition. Similar phase diagrams are obtained at other pressures, with the temperatures of the two phase region increasing as the pressure is raised. Figure 6.28 can be used to predict the boiling behavior for an ideal solution. Thus, if a liquid solution with mole fraction and temperature given by point a is heated, boiling starts at a temperature corresponding to point b; and the first vapor to boil off has the composition y, given by point e. With continued boiling, the liquid changes composition along the liquid line, the vapor changes composition along the vapor line; and the boiling temperature increases so that at point c, the liquid has a composition given by point f, and the vapor has a composition given by point g. The relative number of moles of vapor to moles of liquid present are

---

[†]See reference [169] for a discussion of the determination of boiling and condensation temperatures.

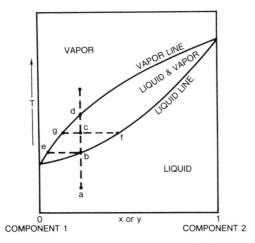

Figure 6.28 Change of boiling temperature T with vapor mole fraction y or liquid mole fraction x for an ideal solution.

given by the ratio of the lengths $\overline{cf}/\overline{gc}$. At point d the vapor has the same composition as the starting liquid. The temperature corresponding to point d gives the highest boiling point for this mixture; above this temperature, only vapor is present.

23 BOILING POINTS OF A NONIDEAL SOLUTION

Most solutions are not ideal, and variations of Figure 6.28 occur. If the components in the mixture are similar, the variations are not severe. However, significant differences in size and polarity between the components can give rise to azeotropy and (liquid + liquid) phase separation or immiscibility. Binary liquid mixtures are classified according to whether they exist in one (homo) or two (hetero) liquid phases and by whether the boiling point-composition diagram of the binary mixture may (azeotropic) or may not (zeotropic) possess maxima, minima, or both. For the purpose of classifying the type of (vapor + liquid) phase equilibria, the mixture fits into one of four categories: (1) homozeotropic, (2) homoazeotropic, (3) heterozeotropic, and (4) heteroazeotropic.

23.1 Homozeotropy

The (x, p) or (x, T) phase diagrams shown for the ideal solution in Figures 6.24 and 6.28 are examples of a homozeotropic system. The liquid line in Figure 6.24 (for the ideal solution) is the straight line shown in Figure 6.29a [169]. Also shown in Figure 6.29a are positive or negative deviations from the linear

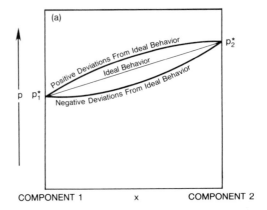

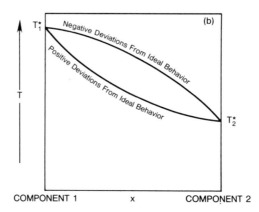

Figure 6.29   (a) Vapor pressure p versus mole fraction x curves and (b) boiling temperature T versus mole fraction x curves of homozeotropic mixtures showing positive or negative deviations from Raoult's law (ideal solution behavior); p_1^*, T_1^* and p_2^*, T_2^* are the vapor pressure or boiling temperature for components 1 and 2, respectively. Each diagram shows more than one curve, with each curve depicting the behavior of a different system. Reprinted with permission from J. R. Anderson, "Determination of Boiling and Condensation Temperatures," in A. Weissberger and B. W. Rossiter, Eds., *Physical Methods of Chemistry*, Part V, *Determination of Thermodynamic and Surface Properties*, Wiley-Interscience, 1971, Chap. 4, p. 207.

behavior. All are examples of homozeotropic behavior; only one liquid phase is present, and the liquid line has neither a maximum nor minimum.

The (x, T) diagram of an ideal system is naturally concave downward (refer again to Figure 6.28). However, its downward concavity can be augmented, nullified, or reversed by deviations from ideality, as shown in Figure 6.29b. Again, the system is homozeotropic unless maxima or minima are present in the (x, T) curve.

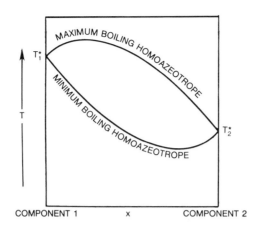

Figure 6.30   Boiling temperature T versus mole fraction x curves of minimum boiling (lower curve) and maximum boiling (upper curve) binary homoazeotropic mixtures. Points T_1^* and T_2^* are the boiling temperatures of pure components 1 and 2, respectively. Reprinted with permission from J. R. Anderson, "Determination of Boiling and Condensation Temperatures," in A. Weissberger and B. W. Rossiter, Eds., *Physical Methods of Chemistry*, Part V, *Determination of Thermodynamic and Surface Properties*, Wiley-Interscience, 1971, Chap. 4, p. 208.

23.2   Homoazeotropy

Homoazeotropy (often described simply as *azeotropy*) occurs when the components are miscible, but maxima or minima occur in the (x, T) curve as shown in Figure 6.30 [169]. These systems are usually referred to as *minimum boiling azeotropes* or *maximum boiling azeotropes*. Occasionally, a system can be found that contains both a minimum and a maximum boiling azeotrope, but this behavior is unusual. An example is the benzene + hexafluorobenzene system [170] shown in Figure 6.31. Azeotropy is a common occurrence in liquid mixtures. Reference books are available in the literature that tabulate the systems known to possess this behavior [171].

23.3   Heterozeotropy

The (x, T) diagram of binary heterozeotropic mixtures contains three regions (Figure 6.32), with the central region consisting of a straight line parallel to the composition axis. This is the region where the mixture separates into two liquid phases. Heterozeotropy is rarely encountered since systems that do not azeotrope are usually miscible.

23.4   Heteroazeotropy

Figure 6.33 illustrates a minimum boiling heteroazeotropic system. (Only minimum boiling heteroazeotropy is known.) The straight line, which is parallel to the x axis, is called the *heteroazeotropic line*. Point b is known as the

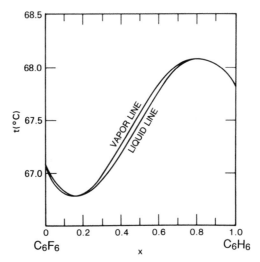

Figure 6.31   Mole fraction x versus temperature t equilibrium diagram for $xC_6H_6 + (1-x)C_6F_6$ at a pressure of 5000 torr (0.664 MPa). This system is an interesting example of one with both a minimum and a maximum boiling homoazeotrope. Reprinted with permission from W. J. Gaw and S. L. Swinton, "Occurrence of a Double Azeotrope in the Binary System Hexafluorobenzene + Benzene," *Nature (London)*, **212**, 284 (1966).

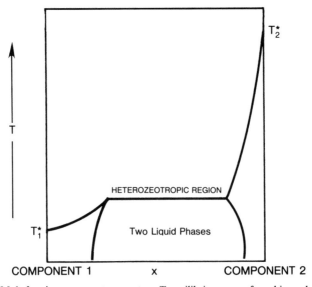

Figure 6.32   Mole fraction x versus temperature T equilibrium curve for a binary heterozeotropic mixture. Points T_1^* and T_2^* are the boiling temperatures of pure components 1 and 2, respectively.

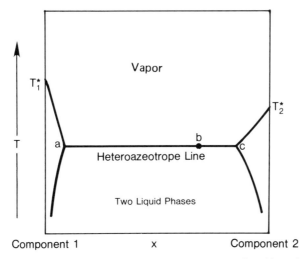

Figure 6.33   Mole fraction x versus temperature T equilibrium curve for a binary heteroazeotropic mixture. Point b is the heteroazeotropic point. Points a and c give the compositions of the two liquid phases in equilibrium at the minimum boiling temperature. Points T_1^* and T_2^* are the boiling temperatures of pure components 1 and 2, respectively.

heteroazeotropic point. It represents the composition of the vapor in equilibrium with the two liquids at the minimum boiling temperature. We will more fully understand the significance of this point and the (x, T) relationships for the different types of systems as we now add the vapor composition lines to Figures 6.29–6.33, and describe the processes that occur as the liquids vaporize or distill in each of the four types of systems.

24   DISTILLATION

The (vapor + liquid) equilibrium properties can be used to separate a system into mixtures enriched in one or the other of the components. This process is known as a *distillation*,[†] and the results of the distillation depend on which of the four systems is present. A simple distillation apparatus (known as an ebulliometer), is shown in Figure 6.34 [180]. A liquid in the flask (pot) is boiled to create vapor. This vapor moves up the flask, down the sidearm, and into the condenser where it is converted to liquid that runs down the tube and collects in the receiver. This liquid is the distillate.

The temperature of the distillation can be followed with the thermometer at the top of the flask. For a pure substance, the distillation temperature should be the same at the beginning and end of the distillation, and should equal the boiling point of the substance at a pressure equal to the atmospheric pressure.

[†]References [172–179] give detailed discussions of distillation procedures.

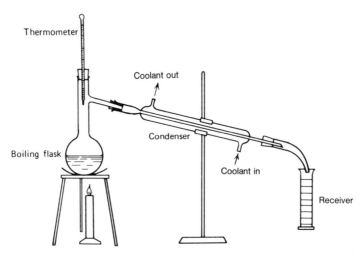

Figure 6.34 Simple distillation apparatus. Reprinted with permission from J. R. Anderson, "Determination of Boiling and Condensation Temperatures," in A. Weissberger and B. W. Rossiter, Eds., *Physical Methods of Chemistry*, Part V, *Determination of Thermodynamic and Surface Properties*, Wiley-Interscience, 1971, Chap. 4, p. 219.

For a mixture, the temperature will change during the distillation and the shape of the distillation curve (temperature vs. amount distilled) is related to the nature and composition of the sample distilled.

24.1 Homozeotropic Distillation

Figure 6.35 illustrates how the temperature and composition change during a homozeotropic distillation. If liquid at point p_1 is heated, boiling starts at point l_1, and the first vapor produced has the composition v_1. With continued boiling, the temperature increases because the liquid left in the pot (the residue) changes composition along the liquid line, becoming more concentrated in the less volatile component 1. At the same time, the distillate follows along the vapor line. Thus, at point p_2, the temperature has increased to T_1 and the compositions of the residue and distillate are r_1 and d_1, respectively. Further separation or purification of the sample can be achieved by collecting the condensed distillate d_1, putting it in a separate ebuillometer, and repeating the distillation process. This condensed distillate will start to boil at T_3 with l_2 and v_2 as the initial compositions of the liquid and vapor, respectively. Continued boiling will result in a residue with composition r_2 and distillate with composition d_2 at a boiling temperature of T_2. The previous procedure can be repeated with the distillate collected, condensed, and revaporized as shown in Figure 6.35 until eventually almost pure component 2 (the more volatile component) is present in the distillate and pure component 1 (the less volatile component) is separated as the residual liquid.

This "batch type" operation would obviously be a very time-consuming

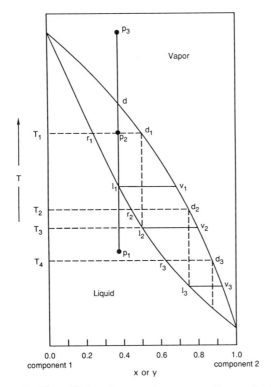

Figure 6.35 (Vapor + liquid) equilibrium diagram of temperature T versus liquid mole fraction x or vapor mole fraction y for a homozeotropic system. The text describes the procedure to be followed during a distillation.

process. Fortunately, the sequence of steps can be combined in an apparatus known as a fractionating column. An example is shown in Figure 6.36. It differs from the simple ebuilliometer in that the vapor produced during the boiling moves up a column containing beads, helices, or some other loosely packed material with a large surface area. The vapor condenses and revaporizes as it moves up the column to the top where part of it is diverted back onto the column and part is collected as distillate. The ratio of condensed vapor returned to the column to that collected is known as the *reflux ratio*.

Condensation and evaporation are occurring all along the fractionating column, and each of these steps is similar to a step in the simple distillation described earlier. The result is an efficient fractionation; and the larger the boiling point separation between the two components, the longer the fractionation column, and the higher the reflux ratio, the more efficient is the separation.

24.2 Homoazeotropic Distillation

Azeotropy occurs when the (x, T) curve has a maximum or a minimum. The phase diagrams are shown in Figures 6.37 and 6.38. The result of the distillation

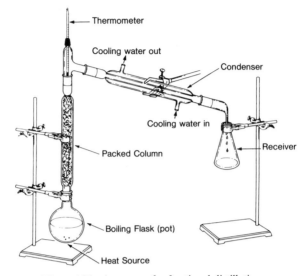

Figure 6.36   Apparatus for fractional distillation.

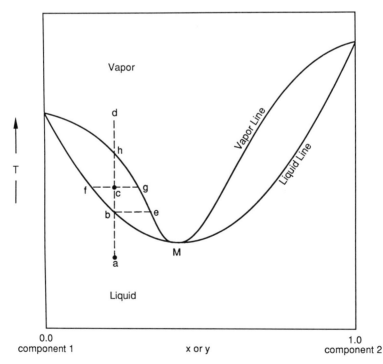

Figure 6.37   (Vapor + liquid) equilibrium diagram of temperature T versus liquid mole fraction x or vapor mole fraction y for a system with a minimum boiling azeotrope of temperature and mole fraction given by M.

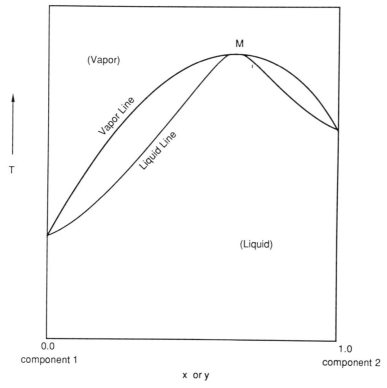

Figure 6.38   (Vapor + liquid) equilibrium diagram of temperature T versus liquid mole fraction x or vapor mole fraction y for a system with a maximum boiling azeotrope of temperature and mole fraction given by M.

of a minimum boiling azeotrope can be predicted by referring to Figure 6.37. If a liquid at (x, T) given by point a is heated, boiling begins at a temperature corresponding to b and the composition of the first vapor is given by e. With continued boiling, the temperature increases, the composition of the residue moves up the liquid curve, and the composition of the distillate moves up the vapor curve, so that at c, the compositions of the residue and distillate are given by f and g, respectively.

If a fractional distillation is performed, the composition of the residue works its way up the curve until essentially pure component 1 is left in the pot, boiling at the boiling temperature of pure component 1. At the same time, the distillate becomes richer in component 2 until point M is reached. The composition at this point is known as the *azeotropic composition* and it represents the maximum enrichment of distillate with component 2, since at point M, the liquid and vapor have the same composition.

A similar process occurs during a fractional distillation of a solution richer in component 2 than point M except that pure component 2 is left as the residue.

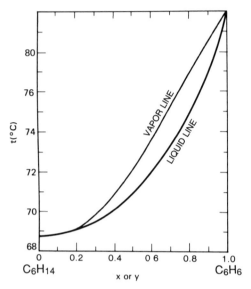

Figure 6.39   (Vapor + liquid) equilibrium diagram of temperature t versus liquid mole fraction x or vapor mole fraction y for the homozeotropic system $xC_6H_6 + (1-x)C_6H_{14}$ [181]. The vapor and liquid lines coalesce at low mole fraction preventing complete separation of the components by distillation.

Once again, the distillate has a final composition given by point M. If a solution with the azeotropic composition (point M) is distilled, the distillate has the same composition as the residue. The distillation appears much like the distillation of a pure substance in that no changes in temperature or composition occur.

The fractional distillation of a maximum boiling azeotrope (Figure 6.38) produces an opposite effect from the distillation of a minimum boiling

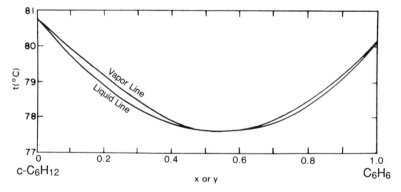

Figure 6.40   (Vapor + liquid) equilibrium diagram of temperature t versus liquid mole fraction x or vapor mole fraction y for $xC_6H_6 + (1-x)c\text{-}C_6H_{12}$ [181]

azeotrope; that is, the residue ends up with composition M and the distillate ends up as essentially pure component 1 or 2, depending on whether the starting mixture has a composition less than or greater than the azeotropic composition (point M), respectively.

Azeotropy prevents separation into pure components and, hence, is a problem when one tries to separate mixtures. The azeotropic composition does, however, change with pressure. As the pressure is decreased or increased, there is a gradual change in one direction or the other of the azeotropic composition until the system becomes zeotropic. Another technique we can use to destroy azeotropy is to add a third component; for example, the minimum boiling azeotrope formed between water and ethanol can be broken by adding benzene to the mixture. That process is used industrially to prepare nearly water-free ethanol, although the product is contaminated by small amounts of benzene.

24.3   Separation Problems Without Azeotropy

It is important to remember that zeotropy (the absence of azeotropy) does not guarantee good separation. The (vapor + liquid) equilibrium diagrams of homozeotropic liquid mixtures may be more concave upward or downward than the ideal solution prediction depending on the direction and extent of the deviation from Raoult's law. If the system is abnormally concave downward as in Figure 6.39 for (benzene + n-hexane) [181], the vapor and liquid curves may coalesce over a rather wide range of compositions in the temperature range near the boiling point of the pure low-boiling component. This makes it difficult to obtain pure distillate. If the abnormal deviations are concave upwards, the curves may coalesce over a rather wide range of compositions near the boiling temperature of the higher boiling component, making it difficult to obtain a pure residue. In some systems, the separation between the liquid and vapor lines may be extremely narrow at either end and/or in intermediate ranges of concentration. In this case, distillation is not a good option for separation.

Concentration regions where the vapor and liquid lines coalesce are not limited to homozeotropes. Figure 6.40 shows the (vapor + liquid) diagram for (benzene + cyclohexane) [181]. A minimum boiling azeotrope is present in which the vapor and liquid lines are so close that it is difficult to find the azeotropic composition (minimum in the curve) without very accurate temperature measurements.

24.4   Heterozeotropic Distillation

The (liquid + vapor) diagram of a heterozeotropic system expanded from Figure 6.32 to include the vapor line, is shown in Figure 6.41 [180]. As mentioned earlier, this type of system is rare. In heterozeotropy as in homozeotropy, all mixtures boil at a temperature between the boiling points of the pure components. A single liquid phase is present in the composition regions below point a and above point b. Distillation behavior in the composition regions below c and above b is similar to that of the homozeotropic systems. If the

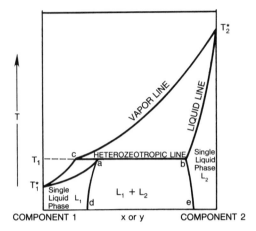

Figure 6.41 (Vapor + liquid) equilibrium diagram of temperature T versus liquid mole fraction x or vapor mole fraction y for a heterozeotropic system. The liquid phases in equilibrium are represented by L_1 and L_2; and T_1^* and T_2^* are the boiling points of pure components 1 and 2, respectively. Reprinted with permission from J. R. Anderson, "Determination of Boiling and Condensation Temperatures," in A. Weissberger and B. W. Rossiter, Eds., *Physical Methods of Chemistry*, Part V, *Determination of Thermodynamic and Surface Properties*, Wiley-Interscience, 1971, Chap. 4, p. 213.

overall composition is between a and c, the boiling temperature starts at a value less than T_1, increases to T_1 (where the liquid separates into two phases), and stays at that temperature until all of the liquid phase represented by a is used up.

Two liquid phases are present in the composition region between a and b at temperatures less than T_1, with compositions of the two phases given by lines ad and be. For these mixtures, the boiling temperature starts at T_1 and stays at that value until all of the phase represented by a is used up. If the distillate collected has a composition between a and b, it will separate into two liquid phases.

24.5 Heteroazeotropic Distillation

Heteroazeotropic behavior is common since water often forms heteroazeotropes with organic liquids. The removal of the water is accomplished easily in appropriately designed apparatus. A typical (vapor + liquid) phase diagram for this type of system is shown in Figure 6.42 [168]. Two liquid phases are present below the temperature T_1 in the region between lines ad and ce. They are the solubility lines for liquids L_1 and L_2, respectively, with the lines relating the phases at equilibrium. Thus, at a temperature given by point p_1, the two liquids in equilibrium have compositions given by f and g with the ratio (mol of L_1/mol of L_2) equal to the ratio of the lengths $(\overline{p_1 g}/\overline{f p_1})$.

Any composition in the range from 0 to a and c to 1 shows the same behavior on boiling as the homoazeotrope with one major difference. Two liquid layers are formed if the distillate from boiling a liquid such as the one given by point p_2 is condensed. Thus, when the mixture at p_2 is heated until it boils at T_2, the

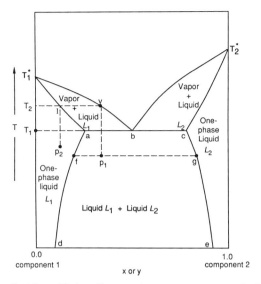

Figure 6.42 (Vapor + liquid) equilibrium diagram of temperature T versus liquid mole fraction x or vapor mole fraction y for a heteroazeotropic system. The liquid phases in equilibrium are represented by L_1 and L_2; and T_1^* and T_2^* are the boiling points of pure components 1 and 2, respectively. Reprinted with permission from J. Laidler and J. H. Meiser, *Physical Chemistry*, Benjamin/Cummings, Menlo Park, CA, 1982, p. 229.

vapor has a composition given by v, which condenses to two liquids when cooled to p_1.

If a two-phase liquid mixture whose overall composition is given by p_1 is heated, it will boil at T_1 and the vapor will have a composition given by point b. This is the heteroazeotropic point described in Figure 6.33. With continued distillation, L_2 will be consumed and the temperature will stay at T_1 until all of L_2 is gone. Once this happens, L_1 is evaporated, the temperature increases, the liquid composition changes along al and the vapor changes along bv. The last drop of liquid disappears when T_2 is reached and only vapor remains.

If the composition of the liquid lies between b and c, distillation again occurs at T_1, with vapor formed at the expense of the L_1 layer. When L_1 is depleted, L_2 distills in a manner similar to that described earlier for L_1.

24.6 Steam Distillation

The extreme case of a heteroazeotropic distillation is one in which the two liquids may be considered completely insoluble in one another. In this case each liquid exerts the same vapor pressure as if it were pure; the total vapor pressure p above the mixture is given by

$$p = p_1^* + p_2^* \tag{53}$$

where p_1^* and p_2^* are the vapor pressures of the pure components. In the distillation, p remains constant at this value until one of the components disappears.

Since the two vapor pressures are additive, the total vapor pressure is constant at a value greater than either p_1^* or p_2^*, and the mixture boils at a constant temperature, which is less than the boiling point of the more volatile component. Furthermore, the relative amounts of components 1 and 2 in the vapor are given by

$$\frac{n_1}{n_2} = \frac{p_1^*}{p_2^*} \tag{54}$$

Thus, as long as both components are present, both temperature and vapor composition remain constant.

Water is often chosen as one of the components in this type of distillation for purifying organic compounds. The process is called steam distillation since it is often accomplished by bubbling steam through the organic liquid. If the organic phase is immiscible in water, the distillate, when condensed and collected, separates into essentially pure aqueous and organic layers, with the relative amounts of the two layers given by (54). This process is especially useful when the organic liquid has a much lower volatility than water and is distilled at a temperature less than 100°C rather than at its much higher boiling point, where it may decompose. High yields in terms of mass or volume of the organic compound are often obtained despite the low volatility of the organic compound because of its high molecular weight compared to water. This can be seen by writing (54) as

$$\frac{g_1}{g_2} = \frac{M_1}{M_2} \frac{p_1^*}{p_2^*} \tag{55}$$

where M_1, p_1^*, and g_1 are the molecular weight, vapor pressure, and grams of organic liquid collected, respectively, and M_2, p_2^*, and g_2 are the corresponding values for water.

25   MEASUREMENT OF BOILING TEMPERATURES

25.1   Boiling Temperature as a Measure of Purity

Organic chemists often purify a chemical by distillation in an apparatus schematically similar to that shown in Figure 6.36, although it often has automatic devices to control the reflux ratio and to monitor the temperature at several points along the distillation column. Variations of the basic design, such as the spinning band column [175, 182], are also often employed. The usual distillation procedure for purification is to collect the distillate while monitoring the condensation temperature at the top of the column. A center or "heart" cut of the distillate is retained and the temperature range corresponding to this cut is reported as a measure of the purity of the distillate.

In this application, mercury thermometers are often employed since accurate measurement and a continuous temperature record is not required. Thermocouples can also be used, and modern commercial units often employ resistance thermometers with a digital readout. Atmospheric pressure is also usually reported since the boiling temperature depends on the pressure, although sometimes the temperature is corrected to a pressure of 1 atm by (40) or by more accurate modifications of this equation.

25.2 Accurate Measurement of Boiling Temperatures for Thermodynamic Applications

Accurate boiling temperatures are usually obtained in a vapor pressure apparatus or in an equilibrium still. The operations of these apparatus are very different. With a vapor pressure apparatus one measures pressure p at a fixed temperature T. Continuous dilution vapor pressure apparatus is now available [183] that allows one to measure the total pressure p as a function of x (liquid composition) accurately over the entire composition region, all at a fixed temperature T. By changing T and repeating the process (T, p, x) data can be obtained from which the normal boiling point is determined as the temperature where $p = 1$ atm.

In the equilibrium still, the boiling temperature at a fixed p is determined as a function of the compositions x and y. Available in the literature are extensive compilations that report (vapor + liquid) equilibria for many systems [158–161]. Other compilations give vapor pressures and normal boiling points for a large variety of pure substances [135, 136, 139]. These vapor pressures are correlated by functional group and in terms of position in a homologous series, so that vapor pressures can be predicted as a function of temperature for a wide variety of substances [140].

A useful application of (vapor + liquid) measurements is the calculation of excess Gibbs free energy G^E in a mixture. The necessary (T, p, x) measurements are made in a vapor pressure apparatus, or (T, p, x, y) measurements, in an equilibrium still. Either method requires accurate measurement of p and T and extreme caution to ensure that equilibrium is obtained in the apparatus. Both methods have disadvantages that affect the accuracy of the G^E calculation. Vapor pressure measurements usually do not give the vapor phase composition, which limits the accuracy with which partial pressures can be calculated; and partial pressures are necessary for calculating G^E. Equilibrium stills give liquid and vapor compositions at a fixed p, but over a range of boiling temperatures. The measurements must be corrected to a common temperature. Often, G^E is wanted at a temperature considerably different from the boiling temperature; and the temperature extrapolation is long. Direct vapor pressure measurements usually give more accurate G^E results. Details of apparatus and procedures for both methods are described in the literature [183, 184].

26   MEASUREMENT OF MOLECULAR WEIGHT FROM BOILING POINT ELEVATION

In dilute solution, the vapor pressure of a solvent (major component) can be approximated by Raoult's law. In fact, in the limit of infinite dilution the solvent in any solution must obey Raoult's law exactly. This behavior can be used to determine the molecular weight of a nonvolatile solute by measuring the rise in boiling point when a known amount of the unknown substance is dissolved in a volatile solvent. For an ideal solution with nonvolatile solute, the boiling temperature is related to pressure by

$$\ln p_1 = \frac{\Delta_{vap} H_1}{R}\left(\frac{T - T_1^*}{T T_1^*}\right) + \ln(1 - x) \tag{56}$$

where T and T_1^* are the boiling temperatures of the solution and pure solvent, respectively, at an atmospheric pressure equal to p_1 (in atmospheres); $\Delta_{vap} H_1$ is the enthalpy of vaporization of the solvent; R is the gas constant; and $(1 - x)$ is the mole fraction of solvent in the solution. By measuring T and T_1^*, x can be calculated, from which M_2, the molecular weight of the solute, can also be calculated.

A simplified form of (56), valid in very dilute solutions, is often employed instead:

$$\Delta T_b = K_b m \tag{57}$$

with

$$K_b = \frac{R(T_1^*)^2 M_1}{\Delta_{vap} H_1} \tag{58}$$

where M_1 is the molecular weight of the solvent (kg/mol), T_1^* is the normal boiling point of the solvent (in kelvins), ΔT_b is the boiling point elevation, and m is the molality. Again, when ΔT_b is measured, m and hence M_2, the unknown molecular weight, can be calculated.

In principle, (56) and (57) provide a simple way of determining M_2; but in practice, the procedure is fraught with difficulties. First of all, the solute must be nonvolatile. But even when this is true, boiling solutions tend to superheat, and variations in atmospheric pressure change the boiling point. All this makes it difficult to measure ΔT_b accurately. Furthermore, it is difficult to achieve equilibrium conditions in the boiling point apparatus and even more difficult to know the composition of the boiling solution, since some of the solvent evaporates and is held up in the condenser, column, and other parts of the apparatus.

Differential ebuilliometers, which measure ΔT_b directly, have been designed, and are used in an attempt to minimize these problems. Their design, which is described in the literature [184], is usually one of two types. A differential ebulliometer of the first kind measures only ΔT_b. It is the simplest to use, but a more useful apparatus is a differential ebulliometer of the second type, which measures T and T_1^* in addition to ΔT_b.

Another attempt to improve the accuracy of the molecular weight determination involves calibrating the apparatus by measuring the boiling point elevation for a standard and comparing it with the boiling point elevation of the unknown instead of relying on (56) or (57). In this case, M_2 is given by

$$M_2 = \frac{(M_2)_s (G_1)_s (\Delta T_b)_s g_2}{G_1 \Delta T_b (g_2)_s} \tag{59}$$

where the subscript s refers to the standard substance. For both standard and unknown, g_2 is the mass of solute introduced, G_1 is the mass of solvent, M_2 is molecular weight, and ΔT_b is the boiling point increase. Even with all the refinements, determination of M_2 by boiling point elevation is difficult and is not often employed. A comparable measurement of M_2 from freezing point lowering is easier, more reliable, and more often used. This procedure is described in Section 36.

27   BINARY (SOLID + LIQUID) PHASE EQUILIBRIUM: PHASE DIAGRAMS FOR IDEAL SOLUTIONS

Many binary (solid + liquid) phase diagrams are given in the literature [185–191].[†] A simple diagram consists of the curves shown in Figure 6.23, predicted for an ideal solution with no solid phase solubility, which are obtained by integrating (60).

$$\frac{\partial \ln x}{\partial T} = \frac{\Delta_{fus} H}{RT^2} \tag{60}$$

In (60), x is the mole fraction of the component that freezes from solution; T is the melting temperature; and $\Delta_{fus} H$ is the enthalpy of fusion for the pure component, again, the one that freezes from the solution. The condition of pure solid freezing from solution often occurs in organic mixtures, although solid solutions sometimes form when the components are very similar. In that case, (60) does not apply.

[†]Wisniak [185] presents a bibliography of the multicomponent (solid + liquid) phase equilibria results reported in the literature. Binary and ternary (solid + liquid) phase diagrams for molten salts and ceramic materials are summarized in [186, 187]. For compilation of binary (solid + liquid) phase diagrams of metal alloys, see [188–190], and a theoretical discussion of multicomponent (solid + liquid phase equilibria is given in Palatnik and Landau [191].

Equation (60) can be integrated, by assuming $\Delta_{fus}H$ is constant with temperature, to give

$$\ln x = \frac{\Delta_{fus}H}{R}\left(\frac{1}{T^*} - \frac{1}{T}\right) \tag{61}$$

where T^* is the melting temperature of the pure substance. In more elaborate treatments, $\Delta_{fus}H$ is expressed as a function of temperature by expressing $\Delta_{fus}C_p$ (the change in heat capacity upon melting for the pure substance) as a polynomial function of temperature and integrating, in which case

$$\Delta_{fus}H = a + bT + cT^2 + dT^3 \tag{62}$$

where a, b, c, and d are constants. Combining (60) and (62) and integrating gives

$$R \ln x = a\left(\frac{1}{T^*} - \frac{1}{T}\right) + b \ln \frac{T}{T^*} + c(T - T^*) + \frac{d}{2}(T^2 - T^{*2}) \tag{63}$$

Figure 6.43 compares a graph of (63) for {benzene (C_6H_6) + p-xylene [1,4-$C_6H_4(CH_3)_2$]} with experimentally measured melting temperatures [192], illustrating how well ideal solution behavior is approximated for this system. Equation (63) must be applied twice to graph the melting lines in Figure 6.43, with the eutectic composition dividing the two calculations.

Equation (61) or (63) applies over the entire composition range from pure substance to the eutectic when the two components of the mixture are similar, so that ideal solutions form. Most binary mixtures do not obey Raoult's law over the entire composition range, but do approach the ideal melting curves [as predicted by (61) or (63)] as a limiting law in dilute solution. It can be shown from thermodynamic principles that in the absence of solid solubility, (61) must apply for the solvent in the limit of $x = 1$.

28   DEVIATION FROM IDEAL SOLUTION BEHAVIOR

The deviations from ideal melting behavior are usually the result of a combination of several effects, but sometimes specific effects dominate and can be used to explain the differences. If we designate A as the solvent, that is, the substance freezing from solution, and B as the solute or substance added to A to lower its melting temperature, then deviations can occur when the following chemical equilibrium processes occur in solution:

$$2B = B_2 \tag{64}$$

$$B = 2C \tag{65}$$

$$A + B = AB \tag{66}$$

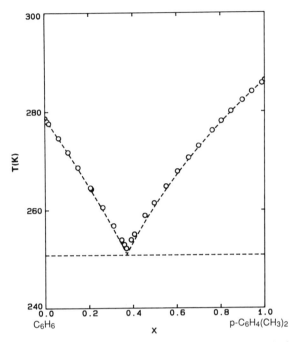

Figure 6.43 (Solid + liquid) phase diagram of temperature T versus mole fraction x for xp-$C_6H_4(CH_3)_2 + (1-x)C_6H_6$ [192]. The broken line is the ideal solution approximation from (62) and the $\bigcirc$ values are the experimental results.

In (64), the solute dimerizes and less solute molecules are present in solution than predicted from the stoichiometric concentration. The ideal freezing point lowering is a colligative property; that is, it depends on the total mole fraction of solute present. When association occurs, the total mole fraction of solute decreases, the mole fraction of solvent increases, and the freezing point lowering is less than the ideal solution prediction. The freezing point lowering in naphthalene when acetic acid is added as a solute (Figure 6.44) [193–195] is an example of this effect. The acetic acid forms a hydrogen-bonded dimer in naphthalene, and the freezing point lowering is approximately one-half as great as when benzene (which forms nearly an ideal solution with naphthalene) is the solute.

In (65), the solute dissociates so that the total number of molecules or moles of solute increase and the freezing point lowering is greater than the ideal prediction. The most common examples are the dissociations of electrolytes causing the well-known abnormal depression of the freezing point of water. In another example, hexaphenylethane dissolved in naphthalene dissociates partially to form the triphenylmethyl radical.

$$(C_6H_5)_3C\!\!-\!\!C(C_6H_5)_3 = 2(C_6H_5)_3C\cdot \tag{67}$$

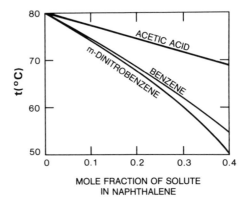

Figure 6.44 Lowering of the freezing temperature t of naphthalene by the different solutes acetic acid, benzene, and m-dinitrobenzene. The graph is reprinted with permission from E. L. Skau and J. C. Arthur, Jr., "Determination of Freezing and Melting Temperatures," in A. Weissberger and B. W. Rossiter, Eds., *Physical Methods of Chemistry*, Part V, *Determination of Thermodynamic and Surface Properties*, Wiley-Interscience, 1971, Chap. 3, p. 116. Skau and Arthur used the data from [194, 195] to construct the figure.

However, a solute that dissociates does not always cause an abnormally large freezing point lowering. In naphthalene, for example, naphthalene picrate dissociates to some extent to form naphthalene and picric acid. This dissociation increases the mole fraction of solvent and the observed freezing point lowering is less than the ideal solution prediction, even though dissociation occurs.

In (66), A and B combine partially to form a third substance AB. The net effect is an exchange of solute B for solute AB with some of the solvent being used up. Hence, the mole fraction of solvent decreases, the concentration of solute increases, and the freezing point lowering is greater than the ideal solution prediction. This effect is shown in Figure 6.44 for m-dinitrobenzene in naphthalene. These two substances combine in solution to form a molecular addition compound.

Solid phase solubility also causes deviations from the ideal solution predictions, and the ideal prediction is not approached in the limit of dilute solution. For example, Figure 6.45 compares the experimental (solid + liquid) phase diagram for $xC_6H_6 + (1 - x)c\text{-}C_6H_{12}$ [196] with the ideal prediction. The deviation at low x, which persists to $x = 0$, is attributed to the formation of a solid phase containing benzene dissolved in the cyclohexane. When this occurs, (61) or (63) does not predict the correct slope for the (solid + liquid) line. Later (Section 36), we describe the use of freezing point lowering to determine the molecular weight of a solute. The procedure is based on the assumption that the solvent obeys (61) in the dilute solution. Care must be taken to avoid solvents where effects such as association, dissociation, and solid solutions can occur.

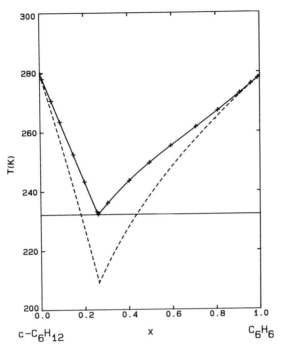

Figure 6.45 (Solid + liquid) phase diagram of temperature T versus mole fraction x for $x C_6 H_6 + (1-x) c\text{-}C_6 H_{12}$ [196]. The + signs are the experimental results and the solid line is the best fit to the experimental results. The dashed line is the ideal solution prediction from (63). The deviation at low x is due to solid solution formation.

29 MEASUREMENT OF (SOLID + LIQUID) PHASE DIAGRAMS

Accurate (solid + liquid) phase measurements are usually made in an apparatus similar to the one used at the National Institute of Science and Technology [197–199]. We have used a similar apparatus for many years [200] and found it to be simple, but effective.[†] The apparatus is shown in Figure 6.46. The sample is placed inside the inner tube of the bottom compartment. The thermometer (usually a resistance type) is inserted into the thermometer well. A coiled stirrer in the bottom of the sample tube moves vertically to stir the sample. A vacuum flask filled with coolant (usually liquid nitrogen) is placed around the apparatus to cool the sample. The apparatus is connected to a vacuum system through the stopcock to vary the gas pressure in the vacuum chamber, which changes the rate of cooling. An inert or dry atmosphere can be maintained above the sample.

[†]A computerized system for data collection for the apparatus described in [200] is reported in [201].

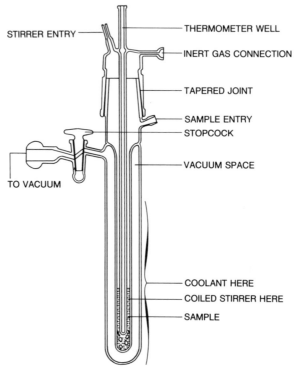

STIRRER ENTRY — THERMOMETER WELL

— INERT GAS CONNECTION

— TAPERED JOINT

— SAMPLE ENTRY
— STOPCOCK

— VACUUM SPACE

TO VACUUM

— COOLANT HERE
— COILED STIRRER HERE
— SAMPLE

Figure 6.46   Apparatus for making accurate melting points measurements on mixtures [200, 201].

30   SIMPLE EUTECTIC SYSTEM

The thermometer of the melting point apparatus [200] is attached to a temperature monitoring and recording system [201], and heat is removed or added at a constant rate to give time-versus-temperature cooling or warming curves. Figure 6.47 shows the curves that result when various mixtures of the simple eutectic system (benzene + p-xylene) are cooled or heated [192]. Curves ab and ba are cooling and warming curves, respectively, for pure benzene. To review a previous discussion, when liquid benzene is cooled, solid forms at T_1 (after some supercooling), and the temperature stays at T_1 until all of the liquid is solidified (cooling curve ab). For the reverse process (warming curve ba), solid benzene melts at T_1 when heated, and a temperature halt occurs at T_1 until all the solid is melted. The major difference between curves ab and ba is that superheating (analogous to supercooling) does not occur in ba. Similar cooling or heating curves are obtained if pure p-xylene is cooled or heated except that the temperature halt occurs at the melting point of p-xylene instead of at that of benzene.

Curve cd results when a liquid mixture with composition corresponding to c

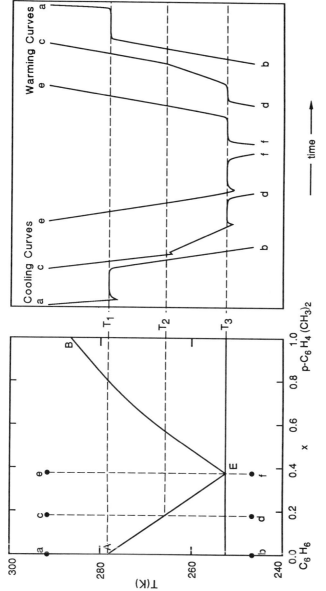

Figure 6.47 (Solid + liquid) phase diagram of temperature T versus mole fraction x for xp-$C_6H_4(CH_3)_2 + (1-x)C_6H_6$ [192], showing time-versus-temperature cooling and warming curves. The labeling of the cooling and warming curves (on the right) correspond to the temperature changes represented by the broken lines in the phase diagram (on the left).

537

is cooled. After supercooling, pure solid benzene crystallizes from the liquid mixture at a temperature which would extrapolate to T_2. At this composition, the cooling curve changes in slope and a temperature halt is not obtained. When solid benzene crystallizes from solution, the remaining liquid is richer in p-xylene and has a lower melting temperature. Continued freezing enriches the liquid mixture in p-xylene, and the temperature continues to fall. In effect, the cooling curve follows AE, the (solid + liquid) equilibrium line for benzene in the phase diagram. When T_3 is reached, p-xylene starts to crystallize from solution (after supercooling). Now both components freeze from solution at rates such that the composition stays at the eutectic composition. No change in composition occurs and the temperature stays at T_3 until all liquid is solidified. The result is a temperature halt in the cooling curve at the eutectic temperature T_3. It is the lowest temperature at which liquid can exist in this system. Point E in the phase diagram is known as the eutectic point.

Curve dc is the warming curve for the same mixture. Melting of the solid mixture starts when the temperature reaches T_3 (the eutectic temperature), and the temperature stays at this value until all the solid p-xylene is melted. After the solid p-xylene is gone, solid benzene starts to melt at T_3 and continues to do so, with the temperature changing until T_2 is reached. The break in the curve at T_2 is at the melting point of the mixture. It occurs at the same temperature as the break in the cooling curve, except that supercooling does not obscure the break in the warming curve. For this reason, time-versus-temperature warming curves are usually used instead of cooling curves to give melting points of solutions. It is important to note that melting of a mixture can occur over a considerable temperature range, in this example, from T_3 to T_2.

Cooling and warming curves similar to cd or dc are obtained all across the composition region, with the break in the curve occurring at the temperature given by the (solid + liquid) equilibrium line. In all the curves, the eutectic halt occurs at the same temperature, T_3. For x less than the eutectic composition, AE is the (solid + liquid) equilibrium line, and benzene is the component that freezes when the liquid is cooled. For x greater than the eutectic composition, p-xylene freezes from solution during cooling, or melts during warming, and BE is the equilibrium line.

Curve ef results when a solution with the eutectic composition is cooled. The only break in the curve is a temperature halt at the eutectic temperature T_3. This curve should not be confused with the cooling curve for a pure substance even though it appears to be the same. The temperature halt occurs because the eutectic is an invariant point with both components crystallizing from solution, and not because a pure solid compound is forming.

31   SYSTEM WITH A SOLID PHASE TRANSITION

Figure 6.48, which is the (solid + liquid) phase diagram for (cyclohexane + n-hexane), is an example of a system with a solid phase transition [192]. Solid cyclohexane changes crystal structure at 186.12 K. Along line AE, n-hexane

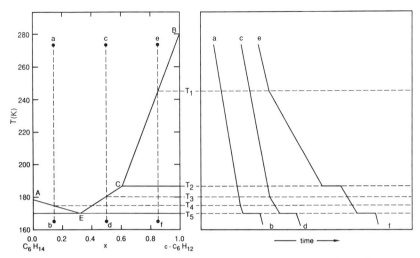

Figure 6.48 (Solid + liquid) phase diagram of temperature T versus mole fraction x for xc-$C_6H_{12} + (1-x)C_6H_{14}$ [192], showing time-versus-temperature cooling curves for a system with a solid phase transition. For simplicity, supercooling effects have not been shown. The labeling of the cooling curves on the right correspond to the temperature changes represented by the broken lines in the phase diagram on the left.

crystallizes from solution; along BC, a form of cyclohexane stable at high temperature crystallizes from solution; and along CE, a form of solid cyclohexane stable at low temperatures is formed. The two solid forms of cyclohexane are in equilibrium at point C which is at the transition temperature of 186.12 K. The phase changes that occur in this system are easiest understood by looking at the time–temperature cooling curves shown in Figure 6.48. (For simplicity, supercooling effects are not shown.)

Curve ab is obtained when a solution rich in n-hexane is cooled along line ab. Behavior in this region is the same as in the simple eutectic system. Solid n-hexane crystallizes from solution at temperature T_4 and the composition of the liquid follows along the (solid + liquid) equilibrium line AE until T_5 is reached. The low-temperature form of solid cyclohexane now starts to crystallize from solution, and a eutectic half is obtained. The temperature stays at T_5 until the system solidifies completely.

When a solution is cooled along line cd, the low-temperature form of cyclohexane crystallizes from solution at T_3 and a eutectic halt occurs at T_5. When a solution is cooled along ef, the high-temperature form of solid cyclohexane crystallizes from solution at T_1 and continues to collect as the temperature and composition of the liquid follow along the equilibrium line BC. At T_2, the high-temperature form of cyclohexane converts to the low-temperature form, and a temperature halt occurs until the conversion is complete. Continued cooling produces more of the low-temperature form, and the temperature and composition follow CE until the eutectic halt is obtained at T_5.

32 SYSTEMS WITH SOLID COMPOUND FORMATION

Phase diagrams can occur in which the components combine together in the solid phase to form a molecular addition compound. The phase diagram is quite different, depending on whether the compound melts congruently or incongruently.

32.1 Congruently Melting Addition Compounds

Figure 6.49 is the phase diagram for (hexafluorobenzene + benzene), an example of a system that forms a congruently melting addition compound [202]. In the figure, AE_1 and BE_2 are the temperature versus mole fraction (T, x) lines for solid C_6H_6 and solid C_6F_6, respectively, in equilibrium with the liquid; E_1CE_2 is the (T, x) line for the solid addition compound in equilibrium with the liquid. The maximum in the curve at C is at the stoichiometric composition of the compound x_c. Since $x_c = 0.5$, the compound is $1:1$, that is $C_6H_6 \cdot C_6F_6(s)$. For a $1:2$ compound (AB_2), $x_c = 0.333$, and $x_c = 0.667$ for a $2:1$ (A_2B) compound. In general, for a compound with the formula A_mB_n, $x_c = m/(m + n)$. When a liquid solution with composition between E_1 and E_2 is cooled, the addition compound $C_6H_6 \cdot C_6F_6(s)$, rather than $C_6H_6(s)$ or $C_6F_6(s)$, crystallizes from solution. Figure 6.49 may be thought of as two simple eutectic phase diagrams side by side, one for $(C_6H_6 + C_6H_6 \cdot C_6F_6)$ and the other for $(C_6H_6 \cdot C_6F_6 + C_6F_6)$.

When a solution is cooled along line ab, cooling curve ab is obtained, with solid C_6H_6 crystallizing from solution at T_4 and a eutectic halt occurring at T_5. When line cd is followed, $C_6H_6 \cdot C_6F_6(s)$ crystallizes from solution at T_2; the liquid solution follows line CE_1 with a eutectic halt again at T_5. Along line gh,

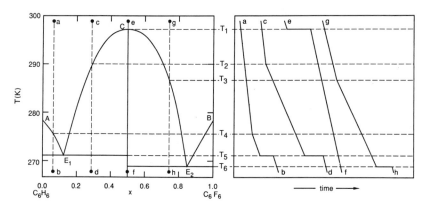

Figure 6.49 (Solid + liquid) phase diagram of temperature T versus mole fraction x for $xC_6F_6 + (1 - x)C_6H_6$, showing time-versus-temperature cooling curves for a system that forms a congruently melting solid addition compound [202]. Supercooling effects have not been shown. The labeling of the cooling curves on the right correspond to the temperature changes represented by the broken lines in the phase diagram on the left.

$C_6H_6 \cdot C_6F_6(s)$ crystallizes at T_3, and the liquid solution follows line CE_2 with a eutectic halt at T_6. The two eutectics differ by having $C_6H_6(s)$, $C_6H_6 \cdot C_6F_6(s)$, and liquid in equilibrium at E_1; and $C_6H_6 \cdot C_6F_6(s)$, $C_6F_6(s)$, and liquid in equilibrium at E_2.

Temperature T_1 is the melting temperature for $C_6H_6 \cdot C_6F_6(s)$. Cooling along line ef results in a temperature halt at T_1 (the same type of cooling curve is obtained as for any other pure substance).

When $C_6H_6 \cdot C_6F_6(s)$ is heated along line fe, melting occurs at T_1. We can represent this process by the chemical reaction.

$$C_6H_6 \cdot C_6F_6(s) = \text{liquid solution} \tag{68}$$

or in general

$$A_m B_n = \text{liquid solution} \tag{69}$$

32.2   Incongruently Melting Addition Compounds

The process in (69) in which a solid completely melts at a given temperature to form liquid is called *congruent melting*. Often, a solid melts at a temperature to form a liquid solution plus another solid, which must be heated to a higher temperature before it melts. This process, which is sometimes thought of as a decomposition, is called *incongruent melting*. An example of incongruent melting is shown in Figure 6.50 for the (methanol + water) system [203]. A solid hydrate

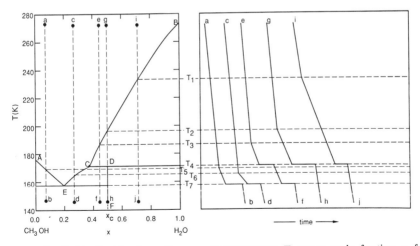

Figure 6.50   (Solid + liquid) phase diagram of temperature T versus mole fraction x for $x H_2O + (1 - x)CH_3OH$ showing time-versus-temperature cooling curves for a system that forms an incongruently melting solid addition compound [203]. Supercooling effects have not been shown. The labeling of the cooling curves on the right correspond to the temperature changes represented by the broken lines in the phase diagram on the left.

(molecular addition compound) forms in this system with the formula $CH_3OH \cdot H_2O(s)$. This $1:1$ hydrate appears in the phase diagram at a composition $x_c = 0.5$ (point D). If the hydrate had been $2:1$ or $1:2$, x_c would equal 0.667 or 0.333, respectively. The vertical line DF represents the formation of this hydrate. If solid hydrate at point h is heated incongruent melting occurs at T_4. The reaction is

$$CH_3OH \cdot H_2O(s) = H_2O(s) + \text{liquid solution} \qquad (70)$$

The composition of the liquid solution is given by C. Point C is known as the peritectic point, and it is an invariant point. It gives the decomposition temperature for the hydrate and represents the highest temperature at which the hydrate can exist. This can be seen by considering the time-versus-temperature cooling curves shown in Figure 6.50. When solution is cooled along ab, solid CH_3OH crystallizes at T_5 and a eutectic halt occurs at T_7. Cooling along cd results in the formation of $CH_3OH \cdot H_2O(s)$ at T_6, followed by a eutectic halt at T_7.

Cooling a solution along ef gives $H_2O(s)$ at T_3. As ice crystallizes from solution, the liquid follows line BC until T_4 is reached. At this temperature, the peritectic reaction, which is the reverse of the peritectic decomposition [reaction (70)], occurs.

$$H_2O(s) + \text{liquid solution} = CH_3OH \cdot H_2O(s) \qquad (71)$$

The temperature stays at T_4 until all the accumulated ice is used up in this reaction. The result is a temperature halt in the cooling curve. Once the peritectic reaction is complete, the composition and temperature follow line CE as solid hydrate crystallizes from solution. At T_7, $CH_3OH(s)$ forms and a eutectic halt occurs until all the liquid is gone. The result of this cooling process is a mixture of $CH_3OH(s)$ and $CH_3OH \cdot H_2O(s)$.

When a solution with composition x_c is cooled (line gh), ice crystallizes from solution at T_2. The liquid follows along BC as ice accumulates. At T_4, the peritectic reaction (71) occurs, and a temperature halt is obtained. In this solution, the stoichiometry is just right for the liquid solution to react completely with the ice, leaving only pure $CH_3OH \cdot H_2O(s)$. Continued cooling produces no further breaks or halts in the cooling curve.

A solution with composition corresponding to point i is richer in water than the stoichiometric composition of the compound. Cooling this solution along ij gives ice at T_1. Continued cooling gives more ice as the liquid follows BC. At T_4 the peritectic reaction (71) again occurs, and a temperature halt is obtained. This time all the liquid reacts, leaving a mixture of ice and $CH_3OH \cdot H_2O(s)$. Once this reaction is complete, the temperature falls, but no further breaks or halts occur in the cooling curve.

33   MORE COMPLICATED PHASE DIAGRAMS

The phase diagrams we have used as examples were kept simple to illustrate a single type of phase behavior, such as a solid phase transition or molecular addition compound formation. A wide variety of combinations are possible that have several types of phase equilibria in a single phase diagram. For example, Figure 6.51 is the phase diagram for (N,N-dimethylformamide + trichloro-bromomethane) [204]. A solid phase transition in the $CBrCl_3$ can be seen in this diagram along with the formation of a congruently melting 1:1 addition compound and an incongruently melting 2:1 compound. Many examples of phase diagrams are available in the literature that show a number of incongruently and/or congruently melting compounds, solid solutions, transitions and other such features. The (magnesium + mercury) system [205], for example, has three congruently melting compounds ($MgHg$, Mg_5Hg_3, and Mg_2Hg), three incongruently melting compounds ($MgHg_2$, Mg_5Hg_2, and Mg_3Hg), and solid solution regions. (Solid + liquid) phase diagrams play an important role in the description of alloys, fused salt systems, and ceramics, as well as organic systems [186–190].

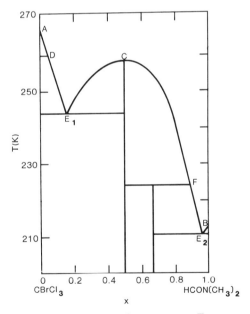

Figure 6.51   (Solid + liquid) phase diagram of temperature T versus mole fraction x for $x\,HCON(CH_3)_2 + (1-x)CBrCl_3$ [204]. Points A and B are the melting points of the pure substances, C is the melting temperature for a 1:1 congruently melting addition compound, F is the peritectic point for the 1:2 incongruently melting addition compound, D is a solid phase transition, and E_1 and E_2 are eutectic points.

34   (SOLID + LIQUID) PHASE DIAGRAMS WITH SOLID SOLUTIONS

Only pure solids (including addition compounds) form in the phase diagrams we have described. We will now consider phase diagrams in which solid phase solubility is present. The nature of the phase diagram depends on the extent of solubility, which can vary from very limited to complete miscibility in the solid phase.

34.1   Limited Solid Solubility

Figure 6.52 is the (solid + liquid) phase diagram for (silver + copper) [206]. It is an example of a phase diagram for a system with limited solid phase solubility. Lines AC and BD are the solid solubility lines. Thus, liquid solutions with compositions given by lines AE are in equilibrium with solids with composition along AC. Likewise, liquids with compositions along BE are in equilibrium with solids of the composition given by BD. If a liquid is cooled along line ab, cooling curve ab is obtained. Solid with the composition given by point s_1 crystallizes from solution at t_3. With continued cooling, the liquid composition follows line AE. At the same time, the solid adjusts in composition along line AC. At t_4, the liquid has a composition given by point E and the solid by point C. At this temperature, a second solid rich in Cu with composition given by point D begins to crystallize from solution, and a eutectic halt is obtained. The temperature stays at t_4 until all of the liquid is used up, leaving a mixture of the two solids. With continued cooling, the solids change along lines CF and DG, although this

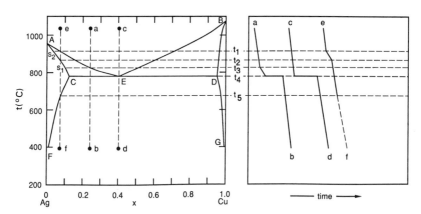

Figure 6.52   (Solid + liquid) phase diagram of temperature t versus mole fraction x for $x\mathrm{Cu} + (1-x)\mathrm{Ag}$ showing time-verus-temperature cooling curves for a simple eutectic system with partial solid phase solubility. Supercooling effects are not shown. The labeling of the cooling curves on the right corresponds to the temperature changes represented by the broken lines in the phase diagram on the left. Phase diagram reprinted with permission from M. Hansen and K. Anderko, *Constitution of Binary Alloys*, 2nd ed., McGraw-Hill, New York, 1958, p. 18.

process is usually very slow and not often observed since it requires diffusion in the solids to readjust the compositions.

Cooling curve cd is obtained when a solution with the eutectic composition is cooled along line cd. A eutectic halt is obtained at t_4, and the conditions differ from those for the simple eutectic system described earlier only in that solid solutions with composition given by points C and D are formed instead of pure Ag and Cu.

When a liquid is cooled along line ef, solid solution with composition s_2 forms at t_1 causing a break in the cooling curve ef. With continued cooling, the liquid follows line AE and the solid line AC. At t_2 the solid has the same composition as the beginning liquid, and the last of the liquid freezes. This causes another break in the cooling curve with an increase in slope. Continued cooling gives no changes in phase or composition until t_5 is reached. At this temperature, the second solid solution given by DG can form, although this may not happen to an appreciable extent because the process is slow. Note that a eutectic halt is not observed during this cooling process. A eutectic forms only when solutions with compositions between C and D are cooled.

34.2   Solid State Miscibility

Three types of systems exhibit complete miscibility in the solid state. Figure 6.53 illustrates these three possibilities with the phase diagrams for (naphthalene + 2-hydroxynaphthalene) (curve I), (d-carvoxime + l-carvoxime) (curve II), and (p-chloroiodobenzene + p-dichlorobenzene) (curve III) [207, 208].

34.2.1   Freezing Points of the Mixture Are Intermediate

In the first (curve I), the freezing temperatures of the mixtures are intermediate between the freezing points of the pure components. The upper curve is known as the liquidus line and the lower curve is the solidus line. At any specified temperature, the liquidus line gives the composition of the liquid in equilibrium with the solid whose composition is given by the solidus line. The region between the curves is a two phase region in which the composition of the phases in equilibrium are connected by a tie line.

Cooling curve ab is obtained when liquid is cooled along line ab. A solid with composition l_1, crystallizes from solution at t_3, causing a break in the cooling curve. The first solid that crystallizes from solution has a composition given by s_1. With continued cooling the liquid solution follows the liquidus line; and solid, whose composition is given by the solidus line, crystallizes from solution. When the liquid composition is given by l_2, the solid has a composition s_2, and the relative amounts of liquid to solid are given by the lever rule [see the (vapor + liquid) discussion in Section 18].

$$\frac{\overline{l_2 p}}{\overline{s_2 p}} = \frac{\text{mol of solid}}{\text{mol of liquid}} \tag{72}$$

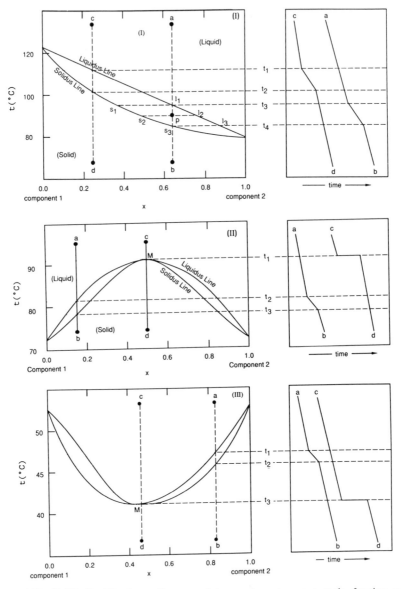

Figure 6.53 (Solid + liquid) phase diagram of temperature t versus mole fraction x for $xC_{10}H_8 + (1-x)2\text{-}C_{10}H_7OH$ (curve I), $xd\text{-}C_{10}H_{15}NO + (1-x)l\text{-}C_{10}H_{15}NO$ (curve II), and $xp\text{-}C_6H_4ICl + (1-x)p\text{-}C_6H_4Cl_2$ (curve III). The time-versus-temperature cooling curves are representative for systems with solid phase miscibility. Supercooling effects have not been shown. The labeling of the cooling curves on the right corresponds to the temperature changes represented by the broken lines in the phase diagram on the left. The (solid + liquid) phase diagrams are reprinted with permission from E. L. Skau and J. C. Arthur, Jr., "Determination of Melting and Freezing Temperatures," in A. Weissberger and B. W. Rossiter, Eds., *Physical Methods of Chemistry*, Part V, *Determination of Thermodynamic and Surface Properties*, Wiley-Interscience, 1971, Chap. 3, p. 123. Skau and Arthur constructed the figures using data from [208].

The last liquid with composition l_3 freezes at t_4, giving a solid with composition s_3. A break in the freezing curve occurs at this point.

The cooling curves at other compositions are similar but displaced in temperature. For example, cooling curve cd results when liquid is cooled from c to d. The first solid forms at t_1, giving a break in the cooling curve; and the last liquid freezes at t_2, causing a second break in the cooling curve.

34.2.2   Freezing Points of the Mixture Pass Through a Maximum

Curve II results when the liquidus (upper line) and solidus (lower line) curves pass through a maximum. This type of behavior is very rare. Optical isomers, such as d- and l-carvoxime, sometimes show this behavior, in which case, the melting temperatures of the pure substances are the same, and the curve is symmetrical. Note that for cooling curve cd, a temperature halt is obtained at a temperature t_1 corresponding to the maximum M. At this mole fraction the solid has the same composition as the liquid.

34.2.3   Freezing Points of the Mixtures Pass Through a Minimum

In curve III, the solidus and liquidus lines pass through a minimum. This type of behavior is not uncommon. In general, time-versus-temperature cooling and warming curves in II and III are similar to I, except at the maximum or minimum, where the solid has the same composition as the liquid; and, therefore, a temperature halt is obtained (cooling curve cd).

35   PURIFICATION BY CRYSTALLIZATION

(Solid + liquid) phase diagrams can be used to predict the procedure to follow to purify a liquid by crystallization. In the simple eutectic system (Figure 6.47) or the system with a phase transition (Figure 6.48), pure component 1 can be obtained by starting with a solution rich in this component and cooling it to a temperature just above the eutectic temperature. The solid, which can be separated from the liquid by filtering or decantation, is essentially pure component 1. Starting with a solution rich in component 2 and following the same procedure gives essentially pure B. If solid solutions do not form, very pure chemical can be obtained, especially if a melting and recrystallization process is repeated several times. This purification procedure is usually not used, however, unless fairly pure starting material is available. Otherwise, it can be quite wasteful, since the impurity starts to crystallize at the eutectic, where a considerable amount of the chemical to be purified is still in the liquid phase and will be lost.

Care must be taken in applying these crystallization techniques to systems in which congruently or incongruently melting addition compounds are formed, such as those in Figures 6.49 and 6.50. In regions of the phase diagram where addition compounds form, they, instead of the pure components, will be obtained upon freezing. In fact, this is the usual procedure for preparing pure addition compounds, for example, salt hydrates.

35.1   Fractional Crystallization

When solid solutions with limited solubility form as in Figure 6.52, a one-step crystallization similar to the process just described will not give pure material since solid solutions freeze from the mixture. When, however, the components are miscible in both the liquid and solid states, a fractional crystallization process can be used, which is in many ways analogous to fractional distillation. If the liquid with composition corresponding to point a in Figure 6.54 is cooled along the line ab to T_1, a mixture of almost equal amounts of liquid with composition l_2 and solid with composition s_2 is obtained (since the length of the lines $\overline{bs_2}$ and $\overline{bl_2}$ are almost equal). The liquid l_2 now is considerably richer in component 1 than was l_1 and the solid s_2 is richer in component 2.

If the liquid l_2 is separated and cooled to T_2, almost equal amounts of l_3 and s_3 are obtained. By repeating this process, almost pure component 1 can soon be obtained. Similarly, if s_2 is separated, melted, and then cooled to T_3, liquid l_4 and solid s_4 will be obtained. By repeating this procedure with the solid, almost pure component 2 can be obtained after a few recrystallizations.

Fractional crystallization will not separate the components completely when a maximum or minimum occurs in the melting curve as in Figure 6.53, curves II and III. In curve II, fractional crystallization gives a final solid with the

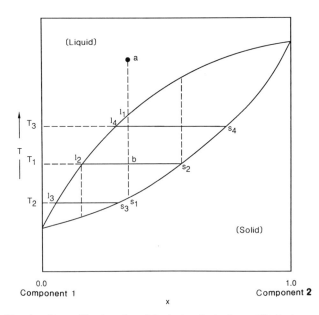

Figure 6.54   Fractional crystallization of a solid solution. In the figure, T is the temperature and x is the mole fraction of component 2.

composition corresponding to the maximum in the curve (point M), and pure component 1 or pure component 2 can be obtained from the liquid, depending on whether the starting mixture is richer in (1) or in (2), respectively, than in M. Thus, after fractional crystallization, a solution with composition given by line ab will yield a solid solution with composition M and pure liquid component 1.

In Figure 6.53, curve III, fractional crystallization will give a liquid with composition corresponding to the minimum (point M) and either pure solid component 1 or component 2, depending on whether the starting solution is richer in (1) or in (2), respectively, than in M. Thus, after fractional crystallization, a solution with composition corresponding to line ab will give a liquid with composition M and pure solid component 2.[†]

35.2 Zone Refining

Fractional crystallization is in many ways similar to fractional distillation, but is not as convenient to apply since it is not as easy to construct an apparatus for fractional crystallization as it is for fractional distillation. One such apparatus, which does purify by fractional crystallization is the zone refiner. The solid to be purified is placed in a vertical cylindrical tube (usually glass) and melted from the bottom by an annular heater that fits around the tube. The heater is slowly moved up the tube, melting solid as it goes and allowing the liquid below to recrystallize. In the process, the impurities are carried to the top of the tube in the liquid layer. The procedure can be repeated several times to increase the purification. When the process is completed, the tube is removed and cut into sections to expose the purified portions.

Zone refining is also used to purify refractory materials and metals at high temperatures. A rod of the substance is placed in an evacuated chamber and heated, for example, by an induction heater or an electron beam, which moves up the rod, causing the melting (and subsequent recrystallization) of a narrow band of the rod as it goes. Surface tension keeps the liquid zone in place and the rod intact. Zone melting has an additional advantage in that large single crystals are produced. It is used to produce ultrahigh purity substances such as silicon or germanium for use in making transistors [215].

Principles, procedures, and apparatus for zone refining and for other techniques based on crystallization procedures can be found in the literature [216–220].

36 MEASUREMENT OF MOLECULAR WEIGHT BY FREEZING POINT LOWERING

Equation (63), which relates the freezing point of a solution to mole fraction, is valid for an ideal solution over the entire composition range or for a real

---

[†]Details of methods for purification of samples by crystallization are given in [209]. Other references describing crystallization techniques are [210–214].

solution at low concentrations, provided dissociation or association as described in Section 26 does not occur and solid solutions do not form, since (63) is
valid only when pure solvent crystallizes from solution.

A version of (63), even simpler than (61), which applies in dilute solution, is
the familiar expression

$$\Delta T_f = K_f m \tag{73}$$

where ΔT_f is the freezing point lowering when a solute is added to a solvent to
give a solution of molality m and K_f is the freezing point lowering constant given
by

$$K_f = -\frac{R(T_1^*)^2 M_1}{\Delta_{fus} H_1} \tag{74}$$

In (74), M_1 is the molecular weight of the solvent and the other physical
constants are as described for (61).

When it applies, (61) can be used to calculate the molecular weight of a solute
from the freezing point lowering. In the procedure followed, the melting
temperature T_1^* of a solvent is measured. Then a small amount of solute of mass
W_2, whose molecular weight M_2 is to be determined, is added to a mass W_1 of
the solvent, whose molecular weight is M_1 and whose enthalpy of fusion is
$\Delta_{fus} H_1$. The melting temperature of the mixture T is measured and M_2 is
calculated from (75), which is derived from (61) with assumptions that apply in
dilute solution

$$M_2 = -\frac{W_2 M_1 R(T_1^*)^2}{W_1 \Delta_{fus} H_1 \Delta T_f} \tag{75}$$

where

$$\Delta T_f = T - T_1^* \tag{76}$$

is the freezing point lowering due to the addition of the solute. This equation can
be obtained more directly by starting with (73) and (74). From (75), we see that
this method is best applied by using a solvent with a small enthalpy of fusion,
since this gives the largest and most easily measured freezing point lowering.
This procedure can also be improved by making freezing point measurements
and calculating M_2 values from solutions of varying concentrations, and then
extrapolating the M_2 values to zero concentration, where (75) applies rigorously.

Table 6.13 gives $\Delta_{fus} H_1$ for some liquids commonly used as a solvent to
determine M_2 from a freezing point lowering. The values in the table were taken
from [134].

Table 6.13 Molecular Weight, Normal Melting Temperature, and Enthalpy of Fusion for Some Common Solvents Used for Molecular Weight Determination From Freezing Point Lowering[a]

Solvent	Formula	M_1 (g/mol)	T^* (K)	$\Delta_{fus}H$ (J/mol)
Acetic acid	CH_3COOH	60.05	289.69	11,720
Benzene	C_6H_6	78.11	278.69	9,866
Camphor	$C_{10}H_{16}O$	152.24	451.5	6,820
Cyclohexane	C_6H_{12}	84.16	279.82	2,677
Hexadecane	$C_{16}H_{34}$	226.44	291.34	53,359
Naphthalene	$C_{10}H_8$	128.17	353.38	19,046
Nitrobenzene	$C_6H_5NO_2$	123.11	278.8	12,121
Tetrachloromethane	CCl_4	153.82	250.3	2,515
Water	H_2O	18.016	273.15	6,010
p-Xylene	$C_6H_4(CH_3)_2$	106.17	286.39	17,113

[a]$\Delta_{fus}H$ and T^* taken from [134].

37 CALCULATION OF PURITY FROM THE CHANGE OF FREEZING POINT WITH FRACTION MELTED

An example of a time-versus-temperature freezing curve for a substance with a small amount of impurity is shown in Figure 6.55. From the change in melting temperature with fraction melted, the amount of impurity can be calculated [197–199, 221]. The procedure for measurement of impurity level that follows can be used for liquids with 99–99.99 mol% purity.

Equation (61), which applies for the solvent in dilute solution with no solid phase solubility, can be used to derive the equation

$$T = T_1^* - \frac{R(T_1^*)^2 y}{\Delta_{fus}H_1}\left(\frac{1}{f}\right) \tag{77}$$

In (77), y is the impurity level in the original liquid in terms of moles of impurity per mole of liquid, and f is the mole fraction melted; R is the gas constant, and T_1^* and $\Delta_{fus}H_1$ are the melting temperature and enthalpy of fusion of the pure liquid. From (77), we see that a graph of melting temperature against $1/f$ should give a straight line, with intercept T_1^* and slope $-R(T_1^*)^2 y/\Delta_{fus}H_1$. To find f, the freezing curve in Figure 6.55 is divided into equal increments of time (perhaps 10), and the corresponding temperatures are read from the graph. By assuming uniform cooling, these increments give the fraction melted. Figure 6.56 shows a typical graph of T versus $1/f$. The intercept gives T_1^*, which can be combined with the slope to give the mole fraction of impurity in the liquid.

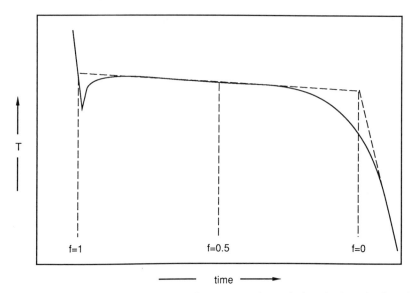

Figure 6.55   Time-versus-temperature T cooling curve used to calculate the impurity level in a substance; T is the temperature and f is the fraction melted.

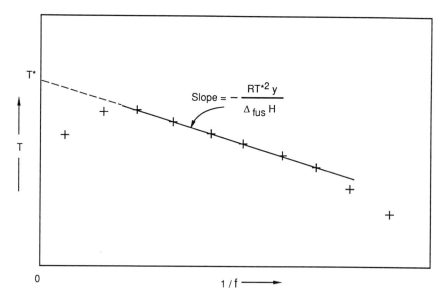

Figure 6.56   Graph of melting temperature T versus the reciprocal of the fraction melted f. The intercept gives the melting point of the pure substance T^*, and the slope can be used to calculate the amount of impurity y. The + signs represent the experimental results.

This procedure can be applied to time–temperature cooling curves as described above, but it works even better with low-temperature calorimetric measurements. In the procedure followed, the sample is first frozen in the calorimeter; then electrical heat is added to melt a fraction of the substance, and this fraction can be calculated precisely from the amount of heat added. The (solid + liquid) mixture is allowed to sit for a time period to ensure temperature and phase equilibrium before the temperature is measured. Repeating the procedure gives an accurate measure of T as a function of fraction melted, from which T^* and y can be calculated from equation (77) [221].

38 THERMODYNAMIC APPLICATIONS OF BINARY (SOLID + LIQUID) PHASE EQUILIBRIA

One important application of (solid + liquid) equilibria information is calculating thermodynamic properties of solutions. The procedure is based on (78), which applies when a pure solid crystallizes from solution[†]

$$\frac{\partial \ln a}{\partial T} = \frac{\Delta_{fus}H}{RT^2} \tag{78}$$

In (78) T is the melting temperature, $\Delta_{fus}H$ is the enthalpy of fusion, and a is the activity at T of the component freezing from solution. By knowing a, we can calculate excess Gibbs free energies of solution [222].

One of the keys to the successful application of (78) is the development of the fitting equation [196]

$$T = T^* \left[1 + \sum_{j=1}^{j=n} A_j(x - x^*)^j \right] \tag{79}$$

In the equation, T is the melting temperature at mole fraction x, T^* is the melting temperature of the pure substance, x^* is the x value at which T^* is taken (0 or 1 for a simple eutectic system), and the A_j values are parameters to be fit by least squares methods. Equation (79) fits the (solid + liquid) results for a wide variety of systems, including those shown in Figures 6.45 and 6.47–6.51.

In another application, (solid + liquid) data can be used to calculate the enthalpies of formation of solid addition compounds [223, 224]; that is, ΔH can be obtained for the process.

$$mA(s) + nB(s) = A_m B_n(s) \tag{80}$$

These ΔH values are useful in understanding molecular interaction in solution.

[†]Equation (78) reduces to (60) when the solution is ideal so that $a = x$.

39   USE OF MELTING POINTS FOR SAMPLE IDENTIFICATION AND PURITY CHECK

The melting point measurement is one of the first tests applied by an organic chemist to a new compound, since the melting point helps to identify the compound if it has been previously synthesized. Tables are available in sources such as the *Handbook of Chemistry and Physics* or *Tables for Identification of Organic Compounds* [225], which list the compounds described in the handbook by increasing melting point. Referring to such a table can help identify possibilities for the compound when the melting temperature has been measured. Also, the size of the melting temperature range serves as an indication of purity. Usually, the sharper the melting point, the more pure is the sample.

The melting point determinations from time–temperature cooling and warming curves give accurate melting points and melting ranges, as well as other useful information, such as transition temperatures, eutectic points, compound formation, and solid solution formation, but the method is often not applicable in organic chemistry research. Organic chemists usually have available a few milligrams to at most a few grams of the compound, and the time–temperature methods using an apparatus such as the one described in Figure 6.46 require much larger samples. Other methods, however, are available for small-sized samples. For example, milligram-sized samples can be studied with differential scanning calorimeters, but the apparatus is complicated, expensive, and difficult to maintain. Differential scanning calorimetry (DSC) techniques are described in detail in Chapter 8. Two procedures commonly employed by organic chemists are the capillary tube and the hot-stage methods described below [226, 227].

39.1   Capillary Tube Method

The capillary tube melting point method is the one most commonly used by organic chemists for measuring the melting temperatures of substances that melt above room temperature. The technique, relatively simple to apply, uses inexpensive and simple equipment; but it is not, in general, a highly accurate method for measuring melting points.

The capillary tubes that contain the sample are approximately 6 cm long by 1 mm in diameter. They can be obtained commercially or made by melting and drawing out clean soft glass tubing, cutting out a section, and sealing one end. A sample of the solid compound is forced into the mouth of the tube and moved to the bottom by tapping or vibration. The top of the tube can then be sealed in a flame to isolate the sample from the atmosphere.

The capillary tube is placed in a melting point apparatus and heated. The melting point, which is taken as the temperature at which the last crystal melts, is obtained from a visual observation. The melting range can sometimes also be obtained by observing the temperature when the first liquid forms.

A simple melting point apparatus is shown in Figure 6.57. A nonvolatile oil such as di-*n*-butyl phthalate is placed in the flask and heated, and the

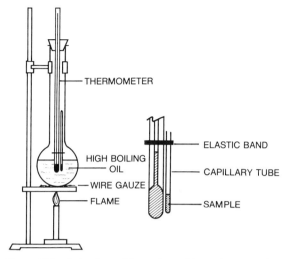

Figure 6.57   A simple capillary tube melting point apparatus.

thermometer is read when the solid melts in the capillary. Three commercial capillary melting point devices are shown in Figures 6.58–6.60. The Thiele apparatus (Figure 6.58) is similar to the apparatus shown earlier except that heating occurs in the sidearm. This sets up convection currents, which stir the oil and keep it at a more uniform temperature.

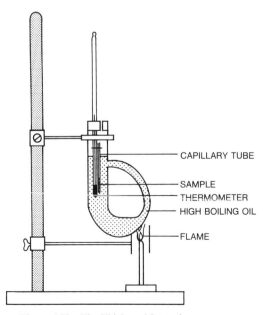

Figure 6.58   The Thiele melting point apparatus.

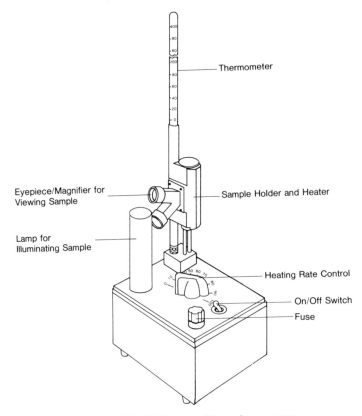

Figure 6.59 The Mel-Temp melting point apparatus.

The Mel-Temp apparatus (Figure 6.59) contains an aluminum block that can hold three capillaries at a time. The block is electrically heated at controlled rates. The sample is illuminated through the lower part and melting is observed through the upper part with a 6-power lens. The Thomas–Hoover Uni-Melt apparatus (Figure 6.60) contains a small beaker of high-boiling silicone oil that is stirred and heated electrically, with the heating rate controlled with a variable transformer. As many as seven capillaries can be observed during a single heating. Reading of the mercury thermometer is accomplished with a traveling periscope device, so that the eye can remain focused on the capillaries.

39.2   Hot Stage Melting Point Method

The hot stage apparatus does not use capillary tubes to hold the sample. Instead, the sample is sandwiched between thin glass windows and placed on a heating block. Hot stages have the advantage of rapid heating, no liquid bath, and ease of operation. They are particularly well suited for substances that melt with decomposition. They can also be equipped with microscopes for studying very

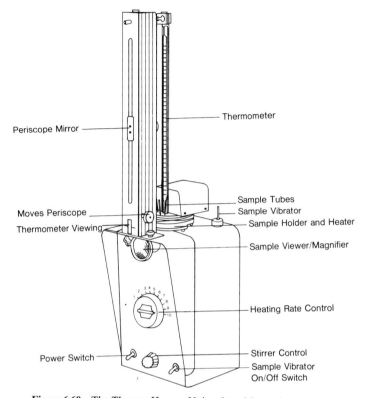

Figure 6.60   The Thomas-Hoover Uni-melt melting point apparatus.

small samples and with polarizing filters. The latter are especially useful in studying melting of liquid crystals. A disadvantage of a hot stage apparatus is that, in general, it lacks the accuracy and precision attainable with a high-quality capillary apparatus. There are exceptions. Kofler[†] hot stages use a microscope to observe the melting and they operate to 350°C; an accuracy in the melting point of ±0.2°C is possible.

The Fisher-Johns apparatus[‡] shown in Figure 6.61 [228] is one of the more commonly encountered hot stage devices. A thermometer is inserted horizontally into the block, which is heated electrically. The sample is placed between two microscope cover slides and set on the block. The melting process is observed from above through a magnifying glass.

39.3   Thermometer Calibration

The mercury thermometer used for the melting point measurements should be calibrated. This is done by comparing the known melting point of standard

[†]Thomas Scientific, P.O. Box 99, Swedesboro, NJ 08085-0099.
[‡]Fisher Scientific, 113 Hartwell Avenue, Lexington, MA 02173-3190.

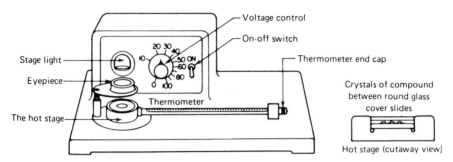

Figure 6.61   The Fisher-Johns hot stage melting point apparatus. Reprinted with permission from J. W. Zubrick, *The Organic Chem Lab Survival Manual: A Student's Guide to Techniques*, Wiley, New York, 1984, p. 39.

substances with the melting point obtained in the apparatus. Commonly used melting point standards are given in Table 6.14. Compounds of high purity must be chosen for the standards or the observed melting points will be low. For the best measurements, stem corrections for the thermometer should be applied to the readings (see Section 10).

40   OTHER AREAS OF PHASE EQUILIBRIA STUDY

We do not have space in this chapter to discuss the many applications of phase equilibria, some of which are interesting and productive areas of research. For example, we have not described ternary and other multicomponent phase equilibria [117–120, 185–191], and this is an important area of study since many alloys, ceramics, and other substances involve three or more components. We also have not described (liquid + liquid) equilibria [229–233], that is, the effect of temperature and pressure on the mutual solubility of liquids in one another. An extension of (liquid + liquid) equilibria is (fluid + fluid) equilibria in

Table 6.14   Compounds for Calibrating A Melting Point Apparatus

Compound	Melting Point (°C)	Compound	Melting Point (°C)
p-Dichlorobenzene	53.2	4-Toluic acid	182
Naphthalene	80.2	Isatin	204
Benzoic acid	122.4	Anthracene	216.2
Urea	132.7	Caffeine	238
Salicylic acid	158.3	Anthraquinone	286
Sulfanilamide	165		

which a supercritical fluid is substituted for one or more of the liquids. A supercritical fluid is obtained by taking a gas at a temperature high enough that it cannot liquefy and subjecting it to high pressure (10–20 MPa or 100–200 atm) so that it has a density approaching that of a liquid [234]. Examples are high temperature–high pressure steam, and CO_2 or C_2H_6 at high pressures and at temperatures a little above ambient.[†]

Supercritical fluids retain some of the useful properties of a gas but also take on some liquidlike properties. For example, they have diffusion coefficients approaching that of a gas but have solvent characteristics similar to a liquid. They find application in a variety of fields, ranging from extraction to supercritical fluid chromatography. Supercritical CO_2, for example, can be used as an extraction fluid in such applications as removing caffeine from coffee and the bitter compounds from the hops in beer, and in tertiary oil recovery. For this and other recent examples of the application of supercritical fluids, see [235].

The superconductivity transition provides another exciting area of research. Until recently, superconductivity was limited to very low temperatures, requiring liquid helium to cool a substance below its transition temperature. Even so, superconducting alloys have been used in such applications as the construction of high-field magnets for use in high-resolution nuclear magnetic resonance spectroscopy and imaging.

Recent developments have identified ceramics such as $YBa_2Cu_3O_x$ with superconducting transitions above 90 K. Attempts are being made to push the transition temperature even higher with the hope that some day, room temperature superconductors will be available. Recent developments in high-temperature superconductors are summarized in [236–238], [239] is a non-technical account of the discovery of the high-temperature superconductor, and [240] is an earlier reference that captures the excitement of the discovery of the high-temperature superconductor.

References

1. *Temperature. Its Measurement and Control in Science and Industry*, Vols. 1–5, American Institute of Physics, New York, 1941–1982.

2. M. H. Aronson, Ed., *Temperature Measurement and Control Handbook*, Instruments Publishing, Pittsburgh, PA, 1964.

3. A. Weissberger and B. W. Rossiter, Eds., *Physical Methods of Chemistry*, Part V, *Determination of Thermodynamic and Surface Properties*, Wiley-Interscience, New York, 1971, Chaps. 1, 3, and 4.

4. T. J. Quinn, *Temperature*, Academic, New York, 1983.

5. J. F. Schooley, *Thermometry*, CRC Press, Boca Raton, FL, 1986.

6. T. D. McGee, *Principles and Methods of Temperature Measurement*, Wiley, New York, 1988.

[†]For a discussion of properties near the critical point, see [234].

7. G. N. Lewis and M. Randall, revised by K. S. Pitzer and L. Brewer, *Thermodynamics*, 2nd ed., McGraw-Hill, New York, 1961, pp. 31–33.

8. T. J. Quinn, *Temperature*, Academic, New York, 1983, pp. 61–120.

9. J. F. Schooley, *Thermometry*, CRC Press, Boca Raton, FL, 1986, pp. 115–138.

10. L. A. Guildner and R. E. Edsinger, *J. Res. Natl. Bur. Stand.*, **80A**, 703 (1976).

11. L. A. Guildner and R. E. Edsinger, "Progress in NBS Gas Thermometry Above 500°C," *Temperature. Its Measurement and Control in Science and Industry*, Vol. 5, American Institute of Physics, New York, 1982, p. 43.

12. I. R. Levine, *Physical Chemistry*, 2nd ed., McGraw-Hill, New York, 1983, pp. 200–202.

13. A. W. Adamson, *A Textbook of Physical Chemistry*, 3rd ed., Academic, Orlando, FL, 1986, pp. 11–31.

14. J. B. Ott, J. R. Goates, and H. T. Hall, Jr., "Comparisons of Equations of State in Effectively Describing pVT Relations," *J. Chem. Educ.*, **48**, 515 (1971).

15. M. W. Kemp, R. E. Thompson, and D. J. Zigrang, "Equations of State with Two Constants," *J. Chem. Educ.*, **52**, 802 (1975).

16. K. K. Shah and G. Thodos, "A Comparison of Equations of State," *Ind. Eng. Chem.*, **57(3)**, 30 (1965).

17. P. A. Rock, *Chemical Thermodynamics*, University Science Books, Mill Valley, CA, 1983, p. 19.

18. R. Battino and S. E. Wood, *Thermodynamics, An Introduction*, Academic, New York, 1968, pp. 199–206.

19. J. A. Beattie and I. Oppenheim, *Principles of Thermodynamics*, Elsevier, Amsterdam, 1979, pp. 167–170.

20. J. R. Goates and J. B. Ott, *Chemical Thermodynamics, an Introduction*, Harcourt Brace Jovanovich, New York, 1971, pp. 25–27.

21. L. A. Guildner and W. Thomas, "The Measurement of Thermodynamic Temperature," *Temperature. Its Measurement and Control in Science and Industry*, Vol. 5, American Institute of Physics, New York, 1982, p. 9.

22. *International Temperature Scale of 1927, ITS-27, Comptes Rendus de la Septieme Conference Generale 1927*, BIPM, Paris, 1927.

23. *International Temperature Scale of 1948, ITS-48, Comptes Rendus de la Neuvieme Conference Generale 1948*, BIPM, Paris 1948.

24. *International Practical Scale of Temperature of 1948, IPTS-48* (amended edition of 1960), *Comptes Rendus de al Onzieme Conference Generale 1960*, BIPM, Paris, 1960; published in English in *J. Res. Natl. Bur. Stand.*, **65A**, 139 (1961).

25. *International Practical Temperature Scale of 1968, IPTS-68, Comptes Rendus de la Treizieme Conference Generale 1968*, BIPM, Paris, 1968; published in English in *Metrologia*, **5**, 35 (1969).

26. *International Practical Temperature Scale of 1968, IPTS-68* (amended edition of 1975), *Comptes Rendus de la Quinzieme Conference Generale 1975*, BIPM, Paris, 1975; published in English in *Metrologia*, **13**, 7 (1976).

27. *The 1976 Provisional 0.5 K to 30 K Temperature Scale, EPT-76*, BIPM. Paris, 1976; published in English in *Metrologia*, **15**, 65 (1979).

28. R. E. Bedford, "The International Practical Temperature Scale of 1968 and Its Probable Future Development," *High Temp. High Pressures*, **11**, 135 (1979).

29. H. Preston-Thomas, "The Origin and Present Status of the IPTS-68," *Temperature. Its Measurement and Control in Science and Industry*, Vol. 4, American Institute of Physics, New York, 1972, p. 3.

30. R. P. Hudson, "Temperature Scales, the IPTS, and Its Future Development," *Temperature. Its Measurement and Control in Science and Industry*, Vol. 5, 1982, p. 1.

31. T. J. Quinn, *Temperature*, Academic, New York, 1983, pp. 381–389.

32. J. F. Schooley, *Thermometry*, CRC Press, Boca Raton, FL, 1986, pp. 94–111.

33. H. Kunz, "Current Status of Applied Precision Radiation Thermometry," *High Temp. High Pressures*, **11**, 193 (1979).

34. T. J. Quinn, *Temperature*, Academic, New York, 1983, pp. 284–367.

35. J. F. Schooley, *Thermometry*, CRC Press, Boca Raton, FL, 1986, pp. 208–222.

36. T. J. Quinn, *Temperature*, Academic, New York, 1983, pp. 387.

37. M. Durieux, D. N. Astrov, W. R. G. Kemp, and C. A. Swenson, "The Derivation and Development of the 1976 Provisional 0.5 K to 30 K Scale," *Metrologia*, **15**, 57 (1979).

38. "The 1976 Provisional 0.5 K to 30 K Temperature Scale," *Metrologia*, **15**, 57 (1979).

39. T. J. Quinn, *Temperature*, Academic, New York, 1983, pp. 50–53.

40. J. F. Schooley and R. J. Soulen, Jr., "Superconductive Thermometric Fixed Points," *Temperature. Its Measurement and Control in Science and Industry*, Vol. 5, American Institute of Physics, New York, 1982, p. 251.

41. R. C. Kemp, W. R. G. Kemp and L. M. Besley, "A Determination of Thermodynamic Temperatures and Measurements of the Second Virial Coefficient of ^{4}He between 13.81 K and 287 K Using a Constant Volume Gas Thermometer," *Metrologia*, **23**, 61 (1986).

42. T. J. Quinn and J. E. Martin, "Radiometric Measurements of the Stefan–Boltzman Constant and Thermodynamic Temperature between $-40°C$ and $+100°C$," *Metrologia*, **20**, 163 (1984); *Phil. Trans. R. Soc. London Ser. A*, **316**, 85 (1985).

43. L. A. Guildner and R. E. Edsinger, "Deviation of International Practical Temperatures in the Temperature Range from 273.16 K to 730 K," *J. Res. Natl. Bur. Stand.*, **80A**, 703 (1976).

44. H. J. Jung, "An Optical Measurement of the Deviation of International Practical Temperatures T_{68} from Thermodynamic Temperatures in the Range from 730 K to 930 K," *Metrologia*, **20**, 67 (1984).

45. H. J. Jung, "A Measurement of Thermodynamic Temperatures between 683 K and 933 K by an Infrared Pyrometer," *Metrologia*, **23**, 19 (1986).

46. J. Bonhoure and R. Pello, Document CCT/84-21, *BIPM Com. Cons. Thermometrie*, **15**, 1984.

47. P. B. Coates, J. W. Andrews, and M. V. Chattle, "Measurement of the Difference between IPTS-68 and Thermodynamic Temperature in the Range 457°C to 630°C," *Metrologia*, **21**, 31 (1985).

48. J. Bonhoure, "Determination Radiometrique des Temperatures Thermodynamiques Comprises entre 940 et 1338 K," *Metrologia*, **11**, 141 (1975).

49. J. W. Andrews and Gu Chuanxin, Document CCT/84-39, *BIPM Com. Cons. Thermometrie*, **15**, 1984.

50. L. Crovini and A. Actis, "Noise Thermometry in the Range 630–962°C," *Metrologia*, **14**, 69 (1978).

51. W. R. Blevin and W. J. Brown, "A Precise Measurement of the Stefan–Boltzman Constant," *Metrologia*, **7**, 15 (1971).

52. T. J. Quinn, T. R. D. Chandler, and M. V. Chattle, "The Departure of IPTS-68 from Thermodynamic Temperatures between 725°C and 1064.43°C," *Metrologia*, **9**, 44 (1973).

53. T. J. Quinn, "Corrections in Optical Pyrometry for the Refractive Index of Air," *Metrologia*, **10**, 115 (1974).

54. R. E. Bedford and C. K. Ma, "Measurement of the Melting Temperature of the Copper 71.9% Silver Eutectic Alloy with a Monochromatic Optical Pyrometer," *Temperature. Its Measurement and Control in Science and Industry*, Vol. 5, American Institute of Physics, New York, 1982, p. 361.

55. B. W. Mangum, "Special Report on the International Temperature Scale of 1990. Report on the 17th Session of the Consultative Committee on Thermometry," *J. Res. Natl. Inst. Stand. Technol.*, **95**, 69 (1990).

56. H. Preston-Thomas, "The International Scale of 1990 (ITS-90)," *Metrologia*, **27**, 3 (1990).

57. M. L. McGlashan, "The International Temperature Scale of 1990 (ITS-90)," *J. Chem. Thermodyn.*, **22**, 653 (1990).

58. *Supplementary Information for the ITS-90*. International Bureau of Weights and Measures: Pavillon de Breteuil, F-92312 Sèvres, France, 1990.

59. *Techniques for Approximating the ITS-90*. International Bureau of Weights and Measures: Pavillon de Breteuil, F-92312 Sèvres, France, 1990.

60. A. R. Colclough, "Primary Acoustic Thermometry: Principles and Current Trends," *Temperature. Its Measurement and Control in Science and Industry*, Vol. 5, American Institute of Physics, New York, 1982, p. 65.

61. H. B. Callen and T. A. Welton, "Irreversibility and Generalized Noise," *Phys. Rev.*, **83**, 34 (1951).

62. D. Gugan and G. W. Michel, "Dielectric Constant Gas Thermometry from 4.2 to 27.1 K," *Metrologia*, **16**, 149 (1980).

63. D. Gugan, "Dielectric Constant Gas Thermometry (DCGT): A New Method of Accurate Thermodynamic Thermometry," *Temperature. Its Measurement and Control in Science and Industry*, Vol. 5, American Institute of Physics, New York, 1982, p. 49.

64. A. R. Colclough, "Systematic Errors in Primary Acoustic Thermometry in the Range 2–20 K," *Metrologia*, **10**, 73 (1974).

65. A. R. Colclough, "A Refractive Index Thermometer for Use at Low Temperatures," *Temperature. Its Measurement and Control in Science and Industry*, Vol. 5, American Institute of Physics, New York, 1982, p. 89.

66. L. Crovini, "Resistance Thermometry," *High Temp. High Pressures*, **11**, 151 (1979).

67. J. L. Riddle, G. T. Furukawa, and H. H. Plumb, "Platinum Resistance Thermometry," *Natl. Bur. Stand. (U.S.) Monogr.*, **126** (1973).

68. T. J. Quinn, *Temperature*, Academic, New York, 1983, pp. 167–240.

69. J. F. Schooley, *Thermometry*, CRC Press, Boca Raton, FL, 1986, pp. 186–206.

70. H. J. Jung and H. Nubbemeyer, "The Stability of Commercially Available High Temperature Resistance Thermometers of a 5Ω Silica Cross Type Up to 961.93°C," *Temperature. Its Measurement and Control in Science and Industry*, Vol. 5, American Institute of Physics, New York, 1982, p. 763.

71. J. P. Evans, "Experiences with High-Temperature Platinum Resistance Thermometers," *Temperature. Its Measurement and Control in Science and Industry*, Vol. 5, American Institute of Physics, New York, 1982, p. 771.

72. L. Guang and T. Hongtu, "Stability of Precision High Temperature Platinum Resistance Thermometers," *Temperature. Its Measurement and Control in Science and Industry*, Vol. 5, American Institute of Physics, New York, 1982, p. 783.

73. J. V. McAllan, "Practical High Temperature Resistance Thermometry," *Temperature. Its Measurement and Control in Science and Industry*, Vol. 5, American Institute of Physics, New York, 1982, p. 789.

74. R. J. Berry, "Oxidation, Stability, and Insulation Characteristics of Rosemount Standard Platinum Resistance Thermometers," *Temperature. Its Measurement and Control in Science and Industry*, Vol. 5, American Institute of Physics, New York, 1982, p. 753.

75. H. Sachse, *Semiconducting Temperature Sensors and their Applications*, Wiley, New York, 1975.

76. F. J. Hyde, *Thermistors*, Iliffe, London, 1971.

77. T. J. Quinn, *Temperature*, Academic, New York, 1983, pp. 220–223.

78. J. F. Schooley, *Thermometry*, CRC Press, Boca Raton, FL, 1986, pp. 202–204.

79. L. G. Rubin, B. L. Brandt, and H. H. Sample, "Cryogenic Thermometry: A Review of Recent Progress: II," *Temperature. Its Measurement and Control in Science and Industry*, Vol. 5, American Institute of Physics, New York, 1982, p. 1333.

80. J. E. Kunzler, T. H. Geballe, and G. W. Hull, "Germanium Resistance Thermometers," *Temperature. Its Measurement and Control in Science and Industry*, Vol. 3, American Institute of Physics, New York, 1962, p. 391.

81. P. Lindenfield, "Carbon and Semiconductor Thermometers for Low Temperatures," *Temperature. Its Measurement and Control in Science and Industry*, Vol. 3, American Institute of Physics, New York, 1962, p. 399.

82. R. L. Rusby, "Resistance Thermometry Using Rhodium-Iron," *Inst. Phys. Conf. Ser.*, **26**, 125 (1975).

83. R. L. Rusby, "The Rhodium–Iron Resistance Thermometer: Ten Years On," *Temperature. Its Measurement and Control in Science and Industry*, Vol. 5, American Institute of Physics, New York, 1982, p. 829.

84. J. F. Schooley, *Thermometry*, CRC Press, Boca Raton, FL, 1986, pp. 205.

85. J. F. Schooley, *Thermometry*, CRC Press, Boca Raton, FL, 1986, pp. 174.

86. L. A. Guildner and G. W. Burns, "Accurate Thermocouple Thermometry," *High Temp. High Pressures*, **11**, 173 (1979).

87. G. W. Burns and W. S. Hurst, "Thermocouple Thermometry," in B. F. Billing and T. J. Quinn, Eds., *Temperature Measurement 1975*, Institute of Physics Conference Series, Volume 26 (London and Bristol: Institute of Physics), 1975, pp. 144–159.

88. T. J. Quinn, *Temperature*, Academic, New York, 1983, pp. 241–280.

89. J. F. Schooley, *Thermometry*, CRC Press, Boca Raton, FL, 1986, pp. 172–183.

90. P. A. Kinzie, *Thermocouple Temperature Measurement*, Wiley-Interscience, New York, 1973.

91. R. L. Powell, W. J. Hall, C. H. Hynik, Jr., L. L. Sparks, G. W. Burns, M. G. Scroger, and H. H. Plumb, "Thermocouple Reference Tables Based on IPTS-68," *Natl. Bur. Stand. (U.S.) Monogr.*, **125** (1974).

92. R. L. Powell and G. W. Burns, "Thermocouple Reference Tables Based on the IPTS-68: Reference Tables in Degrees Fahrenheit for Thermoelements versus Platinum (Pt-67)," *Natl. Bur. Stand. (U.S.) Monogr.*, **125(1)** (1975).

93. T. J. Quinn, *Temperature*, Academic, New York, 1983, pp. 398–408.

94. R. Berman and J. Kopp, "Thermoelectric Power of Dilute Gold–Iron Alloys," *J. Phys. F.*, **1**, 457 (1971).

95. C. P. Cannon, "2200°C Thermocouples for Nuclear Reactor Fuel Centerline Temperature Measurements," *Temperature. Its Measurement and Control in Science and Industry*, Vol. 5, American Institute of Physics, New York, 1982, p. 1061.

96. G. W. Burns, "The Nicrosil Versus Nisil Thermocouple: Recent Developments and Present Status," *Temperature. Its Measurement and Control in Science and Industry*, Vol. 5, American Institute of Physics, New York, 1982, p. 1121.

97. T. P. Wang and C. D. Starr, "Oxidation Resistance and Stability of Nicrosil–Nisil in Air and in Reducing Atmospheres," *Temperature. Its Measurement and Control in Science and Industry*, Vol. 5, American Institute of Physics, New York, 1982, p. 1147.

98. N. A. Burley, R. M. Hess, C. F. Howie, and J. A. Coleman, "The Nicrosil Versus Nisil Thermocouple: A Critical Comparison with the ANSI Standard Letter-Designated Base Metal Thermocouples," *Temperature. Its Measurement and Control in Science and Industry*, Vol. 5, American Institute of Physics, New York, 1982, p. 1159.

99. B. LeNeindre and Y. Garrabos, "Temperature Measurement under Pressure," in B. LeNeindre and B. Vodar, Eds., *Experimental Thermodynamics*, Vol. II, Butterworths, London, 1975, Chap. 3, pp. 87–113.

100. I. C. Getting and G. C. Kennedy, "The Effect of Pressure on the emf of Chromel–Alumel and Platinum–Platinum 10% Rhodium Thermocouples," *J. Appl. Phys.*, **41**, 4552 (1970).

101. J. F. Schooley, *Thermometry*, CRC Press, Boca Raton, FL, 1986, p. 167.

102. J. Busse, "Liquid in Glass Thermometers," *Temperature. Its Measurement and Control in Science and Industry*, Vol. 1, American Institute of Physics, New York, 1941, p. 228.

103. H. F. Stimson, D. R. Lovejoy, and J. R. Clement, "Temperature Scales and Temperature Measurement," in J. P. McCullough and D. W. Scott, Eds., *Experimental Thermodynamics*, Vol. 1, Butterworths, London, 1968, pp. 30–32.

104. J. A. Hall and V. M. Leaver, "Some Experiments in Mercury Thermometry," *Temperature. Its Measurement and Control in Science and Industry*, Vol. 3, American Institute of Physics, New York, 1962, p. 231.

105. J. A. Hall and V. A. Leaver, "The Design of Mercury Thermometers for Calorimetry," *J. Sci. Instrum.*, **36**, 183 (1959).

106. G. T. Furukawa and W. R. Bigge, "Reproducibility of Some Triple Point of Water

Cells," *Temperature. Its Measurement and Control in Science and Industry*, Vol. 5, American Institute of Physics, New York, 1982, p. 291.

107. F. L. Swinton, "The Triple Point of Water," *J. Chem. Educ.*, **44**, 541 (1967).

108. G. T. Furukawa and W. R. Bigge, "Reproducibility of Some Triple Point of Water Cells," *Temperature. Its Measurement and Control in Science and Industry*, Vol. 5, American Institute of Physics, New York, 1982, p. 291.

109. B. W. Mangum, "Triple Point of Gallium as a Temperature Fixed Point," *Temperature. Its Measurement and Control in Science and Industry*, Vol. 5, American Institute of Physics, New York, 1982, p. 299.

110. S. Sawada, "Realization of the Triple Point of Indium in a Sealed Glass Cell," *Temperature. Its Measurement and Control in Science and Industry*, Vol. 5, American Institute of Physics, New York, 1982, p. 343.

111. J. V. McAllan and J. J. Connolly, "The Use of the Cadmium Point to Check Calibrations on the IPTS," *Temperature. Its Measurement and Control in Science and Industry*, Vol. 5, American Institute of Physics, New York, 1982, p. 351.

112. G. T. Furukawa and E. R. Pfeiffer, "Investigation of the Freezing Temperature of Cadmium," *Temperature. Its Measurement and Control in Science and Industry*, Vol. 5, American Institute of Physics, New York, 1982, p. 355.

113. J. V. McAllan and M. M. Ammar, "Comparison of the Freezing Points of Aluminum and Antimony," *Temperature. Its Measurement and Control in Science and Industry*, Vol. 4, American Institute of Physics, New York, 1972, p. 275.

114. G. T. Furukawa, "Investigation of Freezing Temperatures of National Bureau of Standards Aluminum Standards," *J. Res. Natl. Bur. Stand.*, **78A**, 477 (1974).

115. G. T. Furukawa, J. L. Riddle, and W. R. Bigge, "Investigation of Freezing Temperatures of National Bureau of Standards Tin Standards," *Temperature. Its Measurement and Control in Science and Industry*, Vol. 4, American Institute of Physics, New York, 1972, p. 232.

116. J. P. Evans and S. D. Wood, "An Intercomparison of High Temperature Platinum Resistance Thermometers and Standard Thermocouples," *Metrologia*, **7**, 108 (1971).

117. H. A. J. Oonk, *Phase Theory. The Thermodynamics of Heterogeneous Equilibria*, Elsevier, Amsterdam, 1981.

118. A. Reisman, *Phase Equilibria. Basic Principles, Applications, Experimental Techniques*, Academic, New York, 1970.

119. J. E. Ricci, *The Phase Rule and Heterogeneous Equilibrium*, Van Nostrand, New York, 1951.

120. A. Findlay, A. N. Campbell, and N. O. Smith, *The Phase Rule and Its Applications*, 9th ed., Dover, New York, 1951.

121. M. L. McGlashan, *Chemical Thermodynamics*, Academic, New York, 1979, Chaps. 1, 3, 9, 16, and 17.

122. J. Timmermans, "Plastic Crystals: A Historical Review," *J. Phys. Chem. Solids*, **18**, 1 (1961).

123. L. A. K. Staveley, "Plastic Crystals: A Historical Review," *J. Phys. Chem. Solids*, **18**, 46 (1961).

124. J. G. Aston, "Plastic Crystals: A Historical Review," *J. Phys. Chem. Solids*, **18**, 62 (1961).

125. E. F. Westrum, Jr., "Plastic Crystals: A Historical Review," *J. Phys. Chem. Solids*, **18**, 83 (1961).

126. B. Weinstock, "Plastic Crystals: A Historical Review," *J. Phys. Chem. Solids*, **18**, 86 (1961).

127. W. J. Dunning, "The Crystal Structure of some Plastic and Related Crystals," from J. N. Sherwood, Ed., *The Plastically Crystalline State (Orientationally-Disordered Crystals)*, Wiley, New York, 1979, pp. 1–31.

128. G. H. Brown, "Liquid Crystals—The Chameleon Chemicals," *J. Chem. Educ.*, **60**, 900 (1983).

129. J. L. Fergason, "Liquid Crystals," *Sci. Am.*, **211(2)**, 77 (1964).

130. J. W. Goodby, "Melting Phenomena and Liquid-Crystalline Behavior," *Chemalog Hi-lites*, **11**, 3 (1987).

131. F. D. Saeva, Ed., *Liquid Crystals—The Fourth State of Matter*, Dekker, New York, 1979.

132. S. Chandrasekhar, *Liquid Crystals*, Cambridge, New York, 1977.

133. A. Tamir, E. Tamir, and K. Stephan, *Heats of Phase Change of Pure Components and Mixtures, A Literature Guide*, Elsevier, Amsterdam, 1983.

134. E. S. Domalski, W. H. Evans, and E. D. Hearing, "Heat Capacities and Entropies of Organic Compounds in the Condensed Phase," *J. Phys. Chem. Ref. Data*, **13(1)**, 1984.

135. B. J. Zwolinski and R. C. Wilhoit, *Handbook of Vapor Pressures and Heats of Vaporization of Hydrocarbons and Related Compounds*, Thermodynamics Research Center, Texas A & M University, College Station, TX, 1971.

136. S. Ohe, *Computer Aided Data Book of Vapor Pressure*, Data Book Publishing Company, Tokyo, Japan, 1976.

137. P. Ambrose, "Vapor Pressures," in B. LeNeindre and B. Vodar, Eds., *Experimental Thermodynamics*, Vol. II, Butterworths, London, 1975, pp. 607–656.

138. M. L. McGlashan, Ed., *Chemical Thermodynamics*, Vol. 1, The Chemical Society, London, 1973, Chap. 7, pp. 218–261.

139. W. Braker and A. L. Mossman, *Matheson Gas Data Book*, Matheson, Lyndhurst, NJ, 1980.

140. K. L. Nelson, "Correcting Observed Boiling Points to Standard Pressure," *Am. Lab.*, August, 1984, pp. 14–27.

141. J. Timmermans, *Physical-Chemical Constants of Pure Organic Compou..ds*, Vols. I and II, Elsevier, Amsterdam, 1950 and 1965, respectively.

142. R. R. Dreisbach, *Physical Properties of Chemical Compounds, Advances in Chemistry*, Vol. I, Series No. 15, American Chemical Society, Washington, DC, 1955.

143. R. R. Dreisbach, *Physical Properties of Chemical Compounds, Advances in Chemistry*, Vol. II, Series No. 22, American Chemical Society, Washington, DC, 1959.

144. R. R. Dreisbach, *Physical Properties of Chemical Compounds, Advances in Chemistry*, Vol. III, Series No. 29, American Chemical Society, Washington, DC, 1961.

145. J. A. Riddich and W. B. Bunger, "Organic Solvents: Physical Properties and

Methods of Purification," in A. Weissberger, Ed., *Techniques of Chemistry*, Volume I, 3rd ed., Wiley-Interscience, New York, 1970.

146. D. R. Stull and G. C. Sinke, *Thermodynamic Properties of the Elements, Advances in Chemistry*, Series 18, American Chemical Society, Washington, DC, 1956.

147. R. Hultgren, P. D. Desai, D. T. Hawkins, M. Gleiser, K. K. Kelley, and D. G. Wagman, *Selected Values of the Thermodynamic Properties of the Elements*, American Society for Metals, Metals Park, Ohio, 1973.

148. C. W. F. T. Pistorius, "Part 1, Melting Points and Volume Changes Upon Melting," in B. LeNeindre and B. Vodar, Eds., *Experimental Thermodynamics*, Butterworths, London, 1975, Chap. 17, pp. 803–854.

149. Y. Doucet, "Part 2, Liquid–Solid Phase Equilibria II-Cryoscopy," in B. LeNeindre and B. Vodar, Eds., *Experimental Thermodynamics*, Butterworths, London, 1975, Chap. 17, pp. 835–900.

150. J. Timmermans, *Bull. Soc. Chem. Belg.*, **28**, 392 (1919).

151. W. E. Garner, F. C. Madden, and J. E. Rushbrooke, "Alternation in the Heats of Crystallisation of the Normal Monobasic Fatty Acids, Part II," *J. Chem. Soc.*, **1926**, 2498.

152. J. Timmermans, *Les Constantes Physiques des Composés Organiques Crystallisés*, Masson, et Cie, Paris, 1953.

153. J. B. Austin, "A Relation Between the Molecular Weights and Melting Points of Organic Compounds," *J. Am. Chem. Soc.*, **52**, 1049 (1930).

154. E. L. Skau and J. C. Arthur, Jr., "Determination of Melting and Freezing Temperatures," in A. Weissberger and B. W. Rossiter, Eds., *Physical Methods of Chemistry*, Part V, *Determination of Thermodynamic and Surface Properties*, Wiley-Interscience, New York, 1971, Chap. 3, pp. 128–129.

155. J. H. Hildebrand and R. L. Scott, *The Solubility of Nonelectrolytes*, 3rd ed., Reinhold, New York, 1950 or Dover, New York, 1964.

156. J. S. Rowlinson and F. L. Swinton, *Liquids and Liquid Mixtures*, Butterworths, Boston, MA, 1982, Chaps. 4 and 6.

157. J. Timmermans, *Physico-Chemical Constants of Binary Systems*, Vols. 1–4, Interscience, New York, 1959–1960.

158. I. Wichterle, J. Linek, and E. Hála, *Vapor–Liquid Equilibrium Data Bibliography*, Elsevier, Amsterdam, 1973, and Supplements, 1976, 1979, 1982, and 1985.

159. H. Knapp, R. Doring, L. Oellrich, U. Plocker, and J. M. Prausnitz, *Vapor–Liquid Equilibria for Mixtures of Low Boiling Substances*, Dechema, Frankfurt, 1982.

160. J. Gmehling and U. Onken, *Vapor–Liquid Equilibrium Data Collection*, Dechema, Frankfurt, 1977.

161. M. Hirata, S. Ohe, and K. Nigohama, *Computer Aided Data Book of Vapor–Liquid Equilibrium*, Kodanska Limited Elsevier, Tokyo, 1975.

162. A. G. Williamson, "Part 1. Phase Equilibria of Two-Component Systems and Multicomponent Systems," in B. LeNeindre and B. Vodar, Eds., *Experimental Thermodynamics*, Vol. II, Butterworths, London, 1975, Chap. 16, pp. 749–786.

163. G. M. Schneider, "Phase Equilibria of Liquid and Gaseous Mixtures at High Pressures," in B. LeNeindre, and B. Vodar, Eds., *Experimental Thermodynamics*, Vol. II, Butterworths, London, 1975, pp. 787–801.

164. G. M. Schneider, "High-Pressure Phase Diagrams and Critical Properties of Fluid Mixtures," in M. L. McGlashan, Ed., *Chemical Thermodynamics*, Vol. II, The Chemical Society, London, 1978, pp. 105–146.

165. C. L. Young, "Experimental Methods for Studying Phase Behaviour of Mixtures at High Temperatures and Pressures," in M. L. McGlashan, Ed., *Chemical Thermodynamics*, Vol. II, The Chemical Society, London, 1978, pp. 71–104.

166. J. R. Goates and J. B. Ott, *Chemical Thermodynamics, An Introduction*, Harcourt Brace Jovanovich, New York, 1971, pp. 112–113.

167. M. L. McGlashan, "Deviations from Raoult's Law," *J. Chem. Educ.*, **40**, 516 (1963).

168. J. Laidler and J. H. Meiser, *Physical Chemistry*, Benjamin/Cummings, Menlo Park, CA, 1982, pp. 221 and 229.

169. J. R. Anderson, "Determination of Boiling and Condensation Temperatures," in A. Weissberger and B. W. Rossiter, Eds., *Physical Methods of Chemistry*, Part V, *Determination of Thermodynamic and Surface Properties*, Wiley-Interscience, 1971, Chap. 4, pp. 207, 208.

170. W. J. Gaw and F. L. Swinton, "Occurrence of a Double Azeotrope in the Binary System Hexafluorobenzene + Benzene," *Nature (London)*, **212**, 284 (1966).

171. L. H. Horsley, *Azeotropic Data-III, Advances in Chemistry Series 116,* American Chemical Society, Washington, DC, 1973.

172. C. D. Holland, *Fundamentals of Multicomponent Distillation*, McGraw-Hill, New York, 1981.

173. R. Billet, *Distillation Engineering*, Chemical Publishing Company, New York, 1979.

174. E. Krell and E. C. Lumb, *Handbook of Laboratory Distillation*, Elsevier, Amsterdam, 1963.

175. E. A. Coulson and E. F. G. Herington, *Laboratory Distillation Practice*, Interscience, New York, 1958.

176. F. G. Shinskey, *Distillation Control for Productivity and Energy Conservation*, 2nd ed., McGraw-Hill, New York, 1984.

177. W. Malesinski, *Azeotropy and Other Theoretical Problems of Vapor–Liquid Equilibrium*, Interscience, New York, 1965.

178. W. Swietoslawski, *Azeotropy and Polyazeotropy*, Macmillan, New York, 1963.

179. E. J. Hoffman, *Azeotropic and Extractive Distillation*, New York, 1964.

180. J. R. Anderson, "Determination of Boiling and Condensation Temperatures," in A. Weissberger and B. W. Rossiter, Eds., *Physical Methods of Chemistry*, Part V, *Determination of Thermodynamic and Surface Properties*, Wiley-Interscience, 1971, Chap. 4, pp. 213 and 219.

181. L. Sieg, "Vapor–Liquid Equilibria in Binary Systems of Hydrocarbons of Various Types," *Chem. Ing. Tech.*, **22**, 322 (1950).

182. E. Knell, E. C. Lumb, and C. G. Verver, *Handbook of Laboratory Distillation*, Elsevier, Amsterdam, 1963, pp. 209–214.

183. K. N. Marsh, "The Measurement of Thermodynamic Excess Functions of Binary Liquid Mixtures," in M. L. McGlashan, Ed., *Chemical Thermodynamics*, Vol. 2, The Chemical Society, London, 1978, pp. 1–45.

184. J. R. Anderson, "Determination of Boiling and Condensation Temperatures," in A.

Weissberger and B. W. Rossiter, Eds., *Physical Methods of Chemistry*, Part V, *Determination of Thermodynamic and Surface Properties*, Wiley-Interscience, 1971, Chap. 4, pp. 218–232.

185. J. Wisniak, *Phase Diagrams, A Literature Source Book Part A and Part B*, Elsevier, Amsterdam, 1981; Supplement 1, 1986.

186. E. M. Levin, C. F. Robbins, and H. F. McMurdie, *Phase Diagrams for Ceramists*, Vol. 1, The American Ceramic Society, Columbus, OH, 1964; Vol. 2, 1969; Vol. 3, 1975; Vol. 4, 1981; Vol. 5, 1983; Cumulative Index to Vol. 1–5; Vol. 6, 1987.

187. A. M. Alper, Ed., *Phase Diagrams: Materials Science and Technology*, Academic, New York, Vols. 1–3, 1970; Vol. 4, 1976; and Vol. 5, 1978.

188. M. Hansen and K. Anderko, *Constitution of Binary Alloys*, McGraw-Hill, New York, 1958.

189. R. P. Elliott, *Constitution of Binary Alloys*, First Supplement, McGraw-Hill, New York, 1965.

190. R. Hultgren, P. D. Desai, D. T. Hawkins, M. Gleiser, and K. K. Kelley, *Selected Values of the Thermodynamic Properties of Binary Alloys*, American Society for Metals, Metals Park, OH, 1973.

191. L. S. Palatnik and A. I. Landau, *Phase Equilibria in Multicomponent Systems*, Holt, Rinehart, and Winston, New York, 1964.

192. J. R. Goates, J. B. Ott, J. F. Moellmer, and D. W. Farrell, "(Solid + Liquid) Phase Equilibria in *n*-Hexane + Cyclohexane and Benzene + *p*-Xylene," *J. Chem. Thermodyn.*, **11**, 709 (1979).

193. E. L. Skau and J. C. Arthur, Jr., "Determination of Melting and Freezing Temperatures," in A. Weissberger and B. W. Rossiter, Eds., *Physical Methods of Chemistry*, Part V, *Determination of Thermodynamic and Surface Properties*, Wiley-Interscience, 1971, Chap. 3, p. 116.

194. H. L. Ward, "The Solubility Relations of Napthalene," *J. Phys. Chem.*, **30**, 1316 (1926).

195. E. L. Skau, "Compound Formation in the System Naphthalene–*Meta*-Dinitrobenzene," *J. Am. Chem. Soc.*, **52**, 945 (1930).

196. J. B. Ott and J. R. Goates, "(Solid + Liquid) Phase Equilibria in Binary Mixtures Containing Benzene, a Cycloalkane, and an *n*-Alkane or Tetrachloromethane. An Equation for Representing (Solid + Liquid) Phase Equilibria," *J. Chem. Thermodyn.*, **15**, 267 (1983).

197. B. J. Mair, A. R. Glasgow, Jr., and F. D. Rossini, "Determination of Freezing Points and Amounts of Impurity in Hydrocarbons from Freezing and Melting Curves," *J. Res. Natl. Bur. Stand.*, **26**, 591 (1941).

198. A. R. Glasgow, Jr., A. J. Streiff, and F. D. Rossini, "Determination of the Purity of Hydrocarbons by Measurement of Freezing Points," *J. Res. Natl. Bur. Stand.*, **35**, 355 (1945).

199. A. R. Glasgow, Jr., N. C. Krouskop, J. Beadle, G. D. Axilrod, and F. D. Rossini, "Compounds Involved in Production of Synthetic Rubber. Determination of Purity by Measurement of Freezing Points," *Anal. Chem.*, **20**, 410 (1948).

200. J. R. Goates, J. B. Ott, and A. H. Budge, "Solid–Liquid Phase Equilibria and Solid Compound Formation in Acetonitrile–Aromatic Hydrocarbon Systems," *J. Phys. Chem.*, **65**, 2162 (1961).

201. J. B. Ott, B. F. Woodfield, C. Guanquan, J. Boerio-Goates, and J. R. Goates, "(Solid + Liquid) Phase Equilibria in Acetonitrile + Tetrachloromethane, + Trichloromethane, + Trichlorofluoromethane, and + 1,1,1-Trichlorotrifluoroethane," *J. Chem. Thermodyn.*, **19**, 177 (1987).

202. J. R. Goates, J. B. Ott, and J. Reeder, "Solid + Liquid Phase Equilibria and Solid Compound Formation in Hexafluorobenzene + Benzene, + Pyridine, + Furan, and + Thiophene," *J. Chem. Thermodyn.*, **5**, 135 (1973).

203. J. B. Ott, J. R. Goates, and B. A. Waite, "(Solid + Liquid) Phase Equilibria and Solid Hydrate Formation in Water + Methyl, + Ethyl, + Isopropyl, and + Tertiary Butyl Alcohols," *J. Chem. Thermodyn.*, **11**, 739 (1979).

204. J. R. Goates, J. B. Ott, and D. E. Oyler, "Intermolecular Compound Formation in Solutions of N,N-Dimethylformamide with Carbon Tetrachloride and Several Related Substances," *Trans. Faraday Soc.*, **62**, 1511 (1966).

205. M. Hansen and K. Anderko, *Constitution of Binary Alloys*, 2nd ed., McGraw-Hill, New York, 1958, p. 823.

206. M. Hansen and K. Anderko, *Constitution of Binary Alloys*, 2nd ed., McGraw-Hill, New York, 1958, p. 18.

207. E. L. Skau and J. C. Arthur, Jr., "Determination of Melting and Freezing Temperatures," in A. Weissberger and B. W. Rossiter, Eds., *Physical Methods of Chemistry*, Part V, *Determination of Thermodynamic and Surface Properties*, Wiley-Interscience, 1971, Chap. 3, p. 123.

208. *International Critical Tables*, Vol. 4, McGraw-Hill, New York, 1928, pp. 123, 154, and 155.

209. E. L. Skau and J. C. Arthur, Jr., "Determination of Melting and Freezing Temperatures," in A. Weissberger and B. W. Rossiter, Eds., *Physical Methods of Chemistry*, Part V, *Determination of Thermodynamic and Surface Properties*, Wiley-Interscience, 1971, Chap. 3, pp. 178–185.

210. M. Zief and W. R. Wilcox, Eds., *Fractional Solidification*, Vol. 1, Dekker, New York, 1967.

211. M. Zief, Ed., *Purification of Inorganic and Organic Materials*, Dekker, New York, 1969.

212. G. F. Reynolds, "Crystal Growth," in D. Fox, M. M. Labes, and A. Weissberger, Eds., *Physics and Chemistry of the Organic Solid State*, Interscience, New York, 1963, pp. 223–286.

213. E. A. D. White, "The Growth of Single Crystals from the Fluxed Melt," in H. B. Jonassen and A. Weissberger, Eds., *Technique of Inorganic Chemistry*, Vol. 4, Interscience, New York, 1965, pp. 31–64.

214. R. G. Bautista and J. L. Margrave, "High Temperature Techniques," in H. B. Jonassen and A. Weissberger, Eds., *Technique of Inorganic Chemistry*, Vol. 4, Interscience, New York, 1965, pp. 65–135.

215. W. Keller and A. Muhlbayer, *Floating Zone Silicon*, Dekker, New York, 1981.

216. W. R. Wilcox, "Zone Refining," in H. F. Mark, D. F. Othmir, C. G. Overberger, and G. T. Seaborg, Eds., *Encyclopedia of Chemical Technology*, 3rd ed., Vol. 24, Wiley, New York, 1986, pp. 903–917.

217. W. G. Pfann, *Zone Melting*, 2nd ed., Wiley, New York, 1966.

218. H. Schildknecht, *Zone Melting*, Academic, New York, 1966.

219. E. F. G. Herington, *Zone Melting of Organic Compounds*, Wiley, New York, 1963.

220. W. R. Wilcox, R. Friedenberg, and N. Back, "Zone Melting of Organic Compounds," *Chem. Rev.*, **64**, 187 (1964).

221. E. F. Westrum, Jr., "Determination of Purity and Phase Behavior by Adiabatic Calorimetry," in R. S. Porter and J. F. Johnson, Eds., *Analytical Calorimetry*, Plenum, New York, 1968, pp. 231–238.

222. R. L. Snow, J. B. Ott, J. R. Goates, K. N. Marsh, S. O'Shea, and R. N. Stokes, "(Solid + Liquid) and (Vapor + Liquid) Phase Equilibria and Excess Enthalpies for (Benzene + n-Tetradecane), (Benzene + n-Hexadane), (Cyclohexane + n-Tetradecane), and (Cyclohexane + n-Hexadecane) at 293.15, 298.15, and 308.15 K. Comparison of G^E Calculated from (Vapor + Liquid) and (Solid + Liquid) Equilibria," *J. Chem. Thermodyn.*, **18**, 107 (1986).

223. J. Boerio-Goates, S. R. Goates, J. B. Ott, and J. R. Goates, "Enthalpies of Formation of Molecular Addition Compounds in Tetrachloromethane + p-Xylene, + Toluene, and + Benzene from (Solid + Liquid) Phase Equilibria," *J. Chem. Thermodyn.*, **17**, 665 (1985).

224. J. R. Goates, J. Boerio-Goates, S. R. Goates, and J. B. Ott, "(Solid + Liquid) Phase Equilibria for (N,N-Dimethylacetamide + Tetrachloromethane): Enthalpies of Melting of Pure Components and Enthalpies for Formation of Molecular Addition Compounds from Phase Equilibria," *J. Chem. Soc., Faraday Trans. 1*, **83**, 1553 (1987).

225. M. Frankel and S. Patai, *Tables for Identification of Organic Compounds*, The Chemical Rubber Co., Cleveland, OH, 1964.

226. L. F. Fieser and K. L. Williamson, *Organic Experiments*, 6th ed., Heath, Lexington, MA, 1987, pp. 34–38.

227. R. L. Shriner, R. C. Fuson, D. Y. Curtin, and T. C. Morrill, *The Systematic Identification of Organic Compounds*, 6th ed., Wiley, New York, 1980, pp. 37–45.

228. J. W. Zubnick, *The Organic Chem Lab Survival Manual: A Student's Guide to Techniques*, Wiley, New York, 1984, p. 39.

229. A. W. Francis, *Liquid–Liquid Equilibriums*, Interscience, New York, 1963.

230. L. Alders, *Liquid–Liquid Extraction: Theory and Laboratory Practice*, 2nd ed., Elsevier, Amsterdam, 1959.

231. J. Wisniak and A. Tamir, *Liquid–Liquid Equilibrium and Extraction: A Literature Source Book, Parts A and B*, Elsevier, Amsterdam, 1981.

232. J. M. Sorensen and W. Arlt, *Liquid–Liquid Equilibrium Data Collection: Part 1, Binary Systems; Part 2, Ternary Systems; Part 3, Ternary and Quaternary Systems*, Dechema, Frankfurt, Germany, 1979 and 1980.

233. A. Kreglewski, *Equilibrium Properties of Fluids and Fluid Mixtures*, Texas A & M University Press, College Station, TX, 1984.

234. J. H. M. Levelt Sengers, "Thermodynamic Properties Near the Critical Point," in B. LeNeindre and B. Vodar, Eds., *Experimental Thermodynamics*, Vol. II, Butterworths, London, 1975, Chap. 14, pp. 657–724.

235. T. G. Squires and M. E. Paulaitis, Eds., *Supercritical Fluids: Chemical and Engineering Principles and Applications ACS Symposium Series 329*, American

Chemical Society, Washington, DC, 1987.

236. W. E. Hatfield and J. H. Miller, Jr., Eds., *High Temperature Superconducting Materials*, Dekker, New York, 1988.

237. R. Simon and A. Smith, *Superconductors, Conquering Technology's New Frontier*, Plenum, New York, 1988.

238. T. Forester, Ed., *The Materials Revolution, Superconductors, New Materials and the Japanese Challenge*, The MIT Press, Cambridge, MA, 1988.

239. R. M. Hazen, *The Breakthrough: The Race for The Superconductor*, Ballantine, New York, 1989.

240. R. Dagani, "Superconductivity: A Revolution in Electricity is Taking Shape," *Chem. Eng. News*, **65**, 6 (May 11, 1987).

Chapter **7**

CALORIMETRY

John L. Oscarson and Reed M. Izatt

Physical Methods of Chemistry, Second Edition Volume Six: Determination of Thermodynamic Properties   Edited by Bryant W. Rossiter and Roger C. Baetzold
ISBN 0-471-57087-7   Copyright 1992 by John Wiley & Sons, Inc.

1   INTRODUCTION

Calorimeters are instruments used to measure heat changes, and many different calorimeters have been designed. In this chapter, not all of the different types of calorimeters are described in detail, but the main new types of calorimeters, important modifications, and novel uses developed since Sturtevant's chapter in the previous edition of this series [1] are presented. In general, calorimetry can be used to investigate a physical or chemical change where heat is evolved or absorbed. Since nearly all physical and chemical changes are accompanied by heat changes, calorimetry can be used as a tool to investigate most of these processes. Besides the measurement of obvious quantities such as the enthalpy change ΔH_R and heat capacity change ΔC_p values for chemical reactions, pure component and mixture heat capacity C_p values, and latent heats of phase changes ΔH_L, calorimeters are used to measure directly or indirectly many other quantities of interest or importance. The clever application of calorimeters and careful analysis of calorimetric data can give much more information than those quantities directly associated with heat changes.

1.1   Scope

J. M. Sturtevant [1] provided excellent descriptions of many of the applications and types of calorimeters used prior to 1971. In general, the types of calorimeters and their applications as described by him are still in use today. It is not the intent of this chapter to repeat or revise what Sturtevant wrote, but to discuss new types and new or modified uses of calorimeters.

1.2   Units

Some comments should be made about units. The scientific community has agreed to use the Systéme International d'Unités (SI). Therefore, energy is now reported in joules, temperature in kelvins, pressure in pascals, mass in kilograms, distance in meters, and volume in terms of meters cubed. However, the standard mole is still a gram · mole. The appropriate prefixes can be added to the units. As a consequence, concentrations are in units such as moles per kilogram of water as opposed to molal and moles per decimeter cubed instead of molar.

2   OVERVIEW

Calorimetry is a diverse subject, and an introductory short overview of the types and uses of calorimetry is included in this section. Much of the material on the traditional uses mentioned here is given in more detail by Sturtevant [1].

2.1   Recent Review Articles on Calorimetry

Selected review papers that were written over the past several years on calorimetry are referred to here for the interested reader. Parrish [2] has written a review emphasizing the role of calorimetry in obtaining the thermophysical properties of fluids for chemical process design. He described the experimental methods used and the strengths and weaknesses of these methods. He also presents some of the developments that have extended the capability of calorimetry. Gmelin and Brill [3] wrote an article on recent developments in low-temperature calorimetry. Their article is in abstract form, but contains references to pertinent articles. Karlsen and Villadsen [4] wrote review articles on the use of isothermal batch calorimeters in industry and on the treatment of the data collected using these calorimeters [5]. An excellent review on the technique of using data from a flow or titration calorimeter to determine the equilibrium constant for a reaction K_{eq} was written by Eatough and co-workers [6]. Lamprecht [7] and Wadsö [8, 9] wrote excellent reviews on the subject of using calorimetry to investigate biological systems. An article was written about the applications of solution calorimetry to a wide range of problems [10] and another on the changes that have occurred in the field of solution chemistry during the past two or three decades [11]. Review articles are available on the measurement of gas–solid and solution–solid adsorption heats [12], the use of adsorption calorimetry in the study of heterogeneous catalysis reactions [13], and the adsorption at the liquid–solid interface [14]. Hansen and Eatough [15] discussed the sensitivity and detection limits of various calorimeters. They compared different calorimetric methods with regard to the minimum detectable heat effect. Zielenkiewicz and co-workers [16] presented details for mathematically modeling heat conduction calorimeters and Kuessner [17] described the processing of data for measurements made with isoperibol calorimeters.

2.2   Classification of Calorimeters

Although the measurement of heat changes is common to all calorimeters, they differ in how heat changes are detected, how the temperature changes during the process of making a measurement are determined, how the changes that cause heat effects to occur are initiated, what materials of construction are used, what temperature and pressure ranges of operation are used, and so on. Heat changes in calorimeters are detected in five main ways.

1. The temperature change is measured when a chemical reaction, phase change, physical change, or heating of the material of interest occurs. The temperature change is measured by using conventional liquid expansion thermometers, resistance thermometers, thermistors, pyrometers, gas thermometers, or thermocouples–thermopiles. To obtain accurate data, it is important to measure the temperature change as precisely as possible (it is possible to determine the temperature change two or three orders of magnitude more accurately than the absolute temperature). This temperature change can then be converted to energy by use of the appropriate heat capacity, correcting for the heat leaks in and out of the calorimeter and for the power used to mix the contents. Much effort is made to reduce the corrections needed, since a significant amount of error can be introduced by the uncertainty in these corrections. The heat leaks can be reduced by using evacuated, silvered Dewar reaction vessels, as is done in isoperibol calorimeters and is shown schematically in Figure 7.1, or by using adiabatic reaction vessels where the temperature of the jacket surrounding the reaction vessel is controlled to follow the temperature of the reaction vessel, as shown in Figure 7.2.

2. The heat change is detected by using a thermopile to measure the difference between the temperature of the reaction vessel and that of its thermostated surroundings, as shown in Figure 7.3. In this case, the main heat leak path is

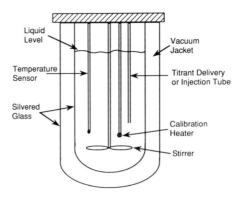

Figure 7.1   Schematic of a typical isoperibol titration calorimeter. This instrument is usually placed inside a temperature bath.

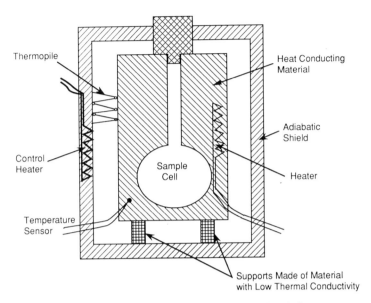

Figure 7.2   Schematic of a typical adiabatic batch calorimeter.

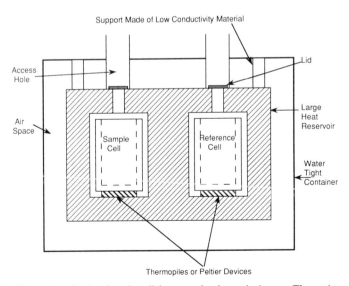

Figure 7.3   Schematic of a batch twin-cell heat conduction calorimeter. The entire ensemble is typically placed in a temperature bath.

through the thermopile. Calorimeters using this heat detection system are called *heat conduction* or *Calvet calorimeters*.

3. In isothermal calorimeters, the energy that must be added to maintain a constant temperature at some strategic location within the instrument is measured. In some of these calorimeters, only heat changes due to endothermic reactions can be measured, while the more versatile types, such as the one shown in Figure 7.4 have a constant heat leak that allows us to measure both exothermic and endothermic changes.

4. The amount of energy absorbed in a well-characterized substance when undergoing a similar temperature rise is measured and compared to that of the reaction of interest, as shown in Figure 7.5.

5. The quantity of fluid that is caused to undergo a phase change by the energy input or output from the calorimeter is measured. The instrument used in this measurement is the same as that depicted in Figure 7.5 except that the reference fluid undergoes a phase change at a constant temperature and pressure and the amount of vapor leaving the calorimeter heat exchanger is measured.

The heat change that occurs in a calorimeter can be initiated in several ways. The method used to initiate this change is determined by the nature of the process to be studied. Combustion calorimeters usually employ an electrical spark to initiate the reaction of the mixture in the calorimeter. Calorimeters used to measure heats of solution of a solid in a liquid or a liquid in another liquid either can have a small ampoule that is broken to initiate the reaction of interest as in Figure 7.6, an ampoule that holds one of the reactants and has a removable plug(s) that is (are) removed at the desired time; a syringe injection device that is connected to the injection tube, as shown in Figures 7.1 and 7.4; or an open container holding one reactant above the main compartment in which the other

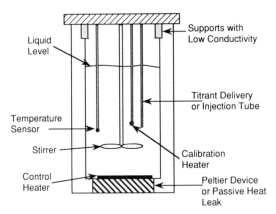

Figure 7.4 Schematic of a typical isothermal titration calorimeter. This instrument is usually placed in a temperature bath.

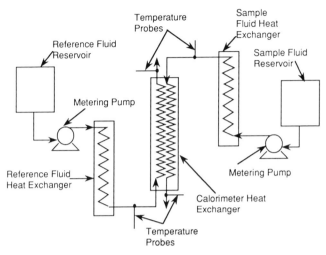

Figure 7.5   Schematic of a flow calorimeter for measuring incremental enthalpy changes.

reactant is located, allowing the two reactants to mix by inverting the calorimeter. Titration calorimeters, which can be isoperibol, adiabatic, isothermal, or heat conduction instruments, initiate a reaction by starting the flow of a titrant. The calorimeters seen in Figures 7.1 and 7.4 are examples of titration calorimeters when the titrant is pumped at a programmed rate to the reaction vessel through the titrant delivery tube. Reaction flow calorimeters initiate heat production (either from a chemical reaction and/or from mixing) by allowing two streams to mix as they flow through the calorimeter as seen in Figure 7.7.

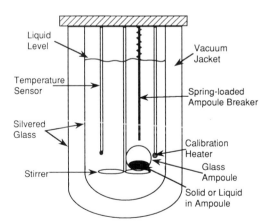

Figure 7.6   Schematic of a simple isoperibol batch calorimeter with glass ampoule. This instrument is usually placed inside a temperature bath.

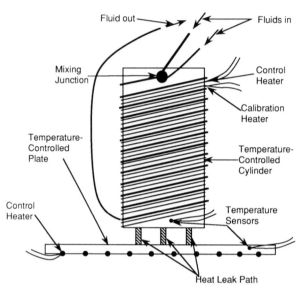

Figure 7.7 Schematic of the mixing cell of an isothermal flow calorimeter used for measuring heat changes associated with mixing fluids.

Calorimeters used to measure heat capacities and transition energies usually have energy added to the material in the calorimeter and the resultant temperature change is monitored. Some calorimeters measure very slow reactions such as those that occur in open-circuit batteries or in drugs sitting on a shelf. The heat given off by these reactions is measured by simply placing the object into the calorimeter.

2.3 Uses of Calorimeters

Calorimeters are used to detect a wide variety of heat effects. The traditional uses are to measure heats of combustion; heat capacities; total enthalpy changes; fusion, transition, and phase change heats; heats of solution and dilution; heats of mixing; heats of adsorption; and heats of chemical reactions. These measured heats can be used in a variety of clever ways to determine other quantities. Sometimes the calorimeter is designed to detect the heat effect in such a way as to allow for the easy interpretation of these quantities. For example, a calorimeter can be used as a titration device to measure concentrations of one or more species in solution [18, 19]. Calorimetric data can also be used to calculate the equilibrium constant of a reaction, measure the rate of a reaction, estimate the shelf life of a product (such as drugs or batteries), predict the potential for a runaway accident to occur in an industrial plant, determine phase equilibria, determine the critical point of a mixture, determine activity coefficients, and test theoretical models such as molecular dynamics.

2.4 Operating Temperatures and Pressures

Bomb calorimeters are designed to operate at high temperatures and pressures. Many titration calorimeters were designed to operate near ambient pressure and close to room temperature. The use of flow calorimeters, where the reaction takes place in a tube, allows high pressures to be obtained. Exploitation of new types of materials and designs has allowed the development of calorimeters that can be operated over wide temperature and pressure ranges.

3 NEW INSTRUMENTATION AND AUTOMATION

The introduction of solid-state electronics and microcomputers has resulted in profound changes in the way calorimeters are operated. These modern tools have allowed the size of the auxiliary equipment needed to operate a calorimeter to be reduced. Large potentiometers used previously in conjunction with galvanometers and controllers have been replaced with small "black boxes" to maintain temperature control. The new instruments are sophisticated, and the average calorimetrist no longer understands them well enough to repair or modify them. The initiation of a reaction, whether it be the turning on of a pump, the breaking of an ampoule, the removal of a plug, the introduction of a spark, or the turning on of a heating current, is usually controlled by a computer, rather than by a manual switch. It is important that the conditions are appropriate before a reaction is initiated; for example, the temperature must be steady and at a predetermined point. When this was done manually, the temperature was usually observed visually on a strip-chart recorder and the operator decided when the temperature was stable enough. Now the decision is made by a computer. Data collection and manipulation are also done with a computer [20–22], allowing the calorimeters to be run with less supervision. Once the data are collected, they are readily reduced by computer techniques. Hence, data can be collected when no one is present. Fewer operators are needed and much less time is required to reduce and interpret the data. The introduction of solid-state electronics also allows better temperature control of the temperature bath, both where the temperature is to be held constant with time [23, 24] and when the bath is to duplicate the temperature of the reaction of interest [25]. This automation comes with a price, however, since the operator may not immediately be aware of problems that arise in the automated equipment. Great care must be exercised when using a fully automated calorimeter, especially in the set up and calibration of the instrument.

4 TYPES OF CALORIMETERS

Different types of calorimeters are discussed in this section. One of the more popular types, the differential scanning calorimeter, is discussed in Chapter 8 on "Differential Thermal Methods" and is not discussed here. The emphasis of this section is on the new types, modifications, and uses of calorimeters, rather than on a complete treatise on each instrument. The classification of calorimeters is difficult and somewhat arbitrary as they can be classified by the type of measurements made, type of mixing, type of temperature control, type of heat detection, and so on. The classification used here is ours and no claim is made that it is the best.

4.1   Bomb Calorimeters

Bomb calorimeters have been in existence for many decades. They are used mainly to measure heats of combustion. The oxidizing agent is either oxygen or fluorine. In recent years, these calorimeters have been used to measure heats of different types of reactions such as that between liquid metals [26]. The precision of these instruments has increased [27], and better instrumentation has been used in their operation [28].

4.2   Batch Calorimeters

As defined in this section, batch calorimeters are those instruments in which there is no flow of mass in or out of the calorimeter during the time that the heat is being measured. Bomb calorimeters fit this definition, but because of their special role they are treated separately. Batch calorimeters differ in the way in which the reactants are mixed, the way the heat change is measured, and the nature of the heat measured.

4.2.1   Ampoule Design

Frequently, batch calorimeters make use of a thin-walled glass ampoule that is broken in order to release its contents to initiate the reaction, as illustrated in Figure 7.6. This method has been used widely, but it has inherent problems. It is difficult to ensure that each ampoule has the same heat capacity. The ampoules have long narrow necks that are flame sealed once the contents are added to the ampoule (see Figure 7.8). This can create some problems if the contents of the ampoule are heat sensitive unless care is taken when the ampoule is sealed. This

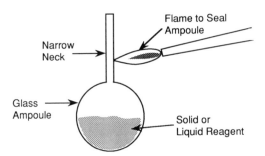

Figure 7.8   Schematic of a typical glass ampoule used in batch calorimeters.

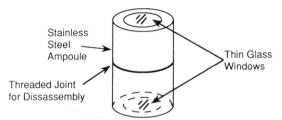

Figure 7.9 Schematic of a stainless steel ampoule with breakable glass windows used in batch calorimeters [29, 30].

long narrow neck makes adding solids difficult. Solids must be ground fine enough to fit through the neck and it is difficult to ensure that none of the fine solid adheres to the neck wall.

Significant developments in batch calorimeters have involved improvement of the type of ampoule used. One such improvement [29, 30], represented schematically in Figure 7.9, is the use of a stainless steel cylinder with breakable glass windows at each end. The ampoule can be taken apart easily and new glass windows can be installed. It can be partially assembled to form a cup to which solid can be added and weighed with ease. Once the solid is added, the final end is attached and the sample is ready. A device of this type is very convenient when mixing a solid with a liquid, such as is done in the measurement of the heat of solution of a solid or the heat of adsorption on the surface of a solid. Another innovation is shown in Figure 7.10 in which the glass ampoule is attached directly to the stirrer [31] thus ensuring good mixing of its contents.

4.2.2   Isoperibol

The need to measure small heats, especially those involving infinite dilution data, has led to the development of new designs that allow these measurements

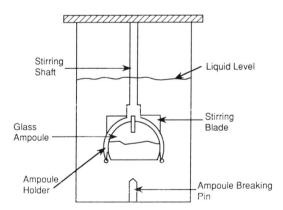

Figure 7.10 Schematic showing a glass ampoule attached to stirrer for good dispersion of ampoule contents [31].

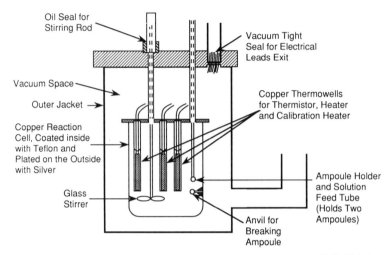

Figure 7.11   Schematic of a batch calorimeter used in measuring small heats [22]. This instrument is usually placed inside a temperature bath.

as well as rapid automatic data acquisition. One such design (see Figure 7.11), reported by Shin and Criss [22], involves a batch calorimeter made of thin copper coated on the inside with Teflon material. This calorimeter has a small vapor space and a silicone oil seal in the stirring shaft in order to minimize evaporation and exclude atmospheric moisture. The data acquisition is fully automated. It also has two sample ampoules so that two successive additions of a second solution can be made for one calorimetric charge.

All calorimeters have some heat exchange with their surroundings. Batch calorimeters usually have stirrers to ensure good mixing. These stirrers add energy, which is converted to heat. The temperature-sensing device is usually a thermistor or resistance thermometer that dissipates heat into the calorimeter. The measured heat change is the sum of all of these effects and can be written as

$$q_m = q_{reaction} + q_{heater} + q_{friction} + q_{resist} - q_{loss} \tag{1}$$

where q_m is the heat measured; $q_{reaction}$ is the heat to be determined (heat of reaction, heat of phase change, or change in sensible heat) plus heat of dilution and heat of extraneous reactions, if any occur; $q_{friction}$ is the heat due to stirring, mixing, or flowing; q_{resist} is the heat due to the electrical heating of the resistance thermometer or thermistor; and q_{loss} is the heat lost to the surroundings. The heat effects other than those desired may be made very small by careful design of the calorimeter; but even when the heat loss and other thermal effects are small, they can be significant if the heat being measured is small. This is true especially when following the heat of a slow reaction. All heat effects other than $q_{reaction}$ can be accounted for either by measuring the heat of a blank run when a fluid is mixed with itself, by use of an electrical calibration when a known amount of electrical energy is added, or by use of a chemical calibration running a well-characterized chemical reaction in the calorimeter. Usually, all three methods are used to calculate the extraneous heat effects. Even if the instrument is calibrated using all three methods, the exact conditions of the calorimetric runs of interest are not duplicated. If the heat measured is large, the errors introduced by the uncertainties of the last four terms of (1) are small compared to $q_{reaction}$; but in the case of slow reactions with small rates of heat generation, the errors may be significant.

One approach used to overcome this difficulty is to store the temperature or heat effects as a function of time for the reaction of interest as the measurement is being made and then to measure the amount of energy it takes to replicate this temperature trace [20]. This operation was not feasible before the availability of computers for automatic control and data collection. A measurement is made by adding one solution to the reaction vessel and the second solution to a syringe. Both the syringe and the vessel are allowed to come to the desired temperature, then the reaction is initiated by injecting the syringe contents into the reaction vessel. The temperature data are collected and stored by the computer. Once the reaction is completed, the vessel and its contents are allowed to return to the initial temperature, then an electronic calorimetric run is made. During this run, the computer controls the energy output of the heater inside the reaction vessel so that the temperature history of the chemical run is duplicated by this electronic run. The amount of electrical energy needed to duplicate the temperature of the chemical run is stored and is assumed to be equal to the heat change due to the earlier chemical reaction.

In the past, operation of batch calorimeters was limited mainly to ambient temperatures and pressures. Many interesting phenomena, however, occur at temperatures and pressures well above ambient conditions. Batch calorimeters are developed that can operate at higher temperatures and pressures. These calorimeters must be made of materials that maintain their strength and resist chemical attack at the elevated temperatures. Even materials that react slowly may be unsuitable since possible corrosion reactions will have a heat effect that may be significant when compared to that being measured. One apparatus is operated at a temperature of 987 K [32] to measure the heats of mixing of high melting point solutions. It is constructed of platinum and uses perforated platinum holders that are stirred throughout the mix. This calorimeter can make measurements on low vapor-pressure substances only. Greiger [33] and Chen [34] developed a large (800-cm³) batch calorimeter made of titanium–0.2% palladium that will withstand both high temperatures and high pressures. One mixes the contents in the vessel by using a magnetically coupled stirrer. During operation, the vapor space for this calorimeter is kept to a minimum and the volume is well characterized as a function of temperature and pressure to allow the precise calculation of the heat due to phase changes. The large size of this calorimeter allows the measurement of small heat changes per unit of reactant, since the total amount of reactant in the calorimeter is large. The materials of construction work well at high temperatures and pressures and with acidic and basic solutions. However, the collection of data in such a calorimeter is time consuming.

4.2.3   Heat Conduction

The detection of very small heat changes has been facilitated by the use of Calvet or heat conduction calorimeters [35, 36]. All Calvet calorimeters make use of reaction cells whose main heat leak path is through thermopiles or Peltier devices to some relatively constant temperature heat reservoir, as seen in Figure 7.3. Originally, the temperatures of both the cell(s) and the reservoir are equal and the voltage across the thermopile(s) is zero. The reactants are then added to the cell creating a temperature difference across the thermopile. This temperature difference causes heat to be conducted across the thermopile and a voltage difference across the thermopile. Usually, Calvet-type calorimeters have twin cells, each thermally connected in a manner identical to the same heat reservoir. The measured voltage is the difference between those produced by the two thermopiles. This method of construction eliminates most of the noise due to the change in reservoir temperature. Suurkuusk and Wadsö [37] made extensive use of this technique in several similar calorimeter designs. These calorimeters have small twin cells each of which is thermally connected by two Peltier devices to two small aluminum blocks, which, in turn, are in contact with two large aluminum cylinders that are held inside a stainless steel can that is in a constant temperature water bath. A simple schematic of this device is shown in Figure 7.12. Many different inserts are used inside the cells, depending on the type of measurement needed. Batch type cells with ampoules are used when batch operations are required. Usually, this type of calorimeter is used to measure the heat change when two materials come in contact with each other, but it is also used to measure the specific heats of reactive materials [38].

Another Calvet-type calorimeter was developed specifically to measure the heat released by batteries when they are in an open circuit mode in order to measure nondestructively the life of a battery [39, 40]. Several designs were developed, depending on the temperature of operation and the size of sample cell needed, but the basic design is virtually the same in all cases and is illustrated in Figure 7.3. The twin cells are connected by the thermopiles to large metal blocks, which are in turn suspended using poorly conducting standoffs inside a metal can. The metal can is suspended in a water or air temperature bath. The electronics, which among other things amplify the voltage output from the thermopiles, are contained within an air temperature bath. These calorimeters are used extensively to check the quality of batteries. Virtually the same design has been used in calorimeters that test the aging of pharmaceuticals at room temperature [41].

Calvet-type calorimeters have been designed to operate at high temperatures [42–44]. One of these calorimeters [42, 43] consists of a central block with twin cells made of Inconel metal, which is surrounded by a high-temperature furnace. The block is supported on ceramic supports, and the thermopiles are made of platinum–platinum rhodium junctions. A Calvet-type calorimeter was designed

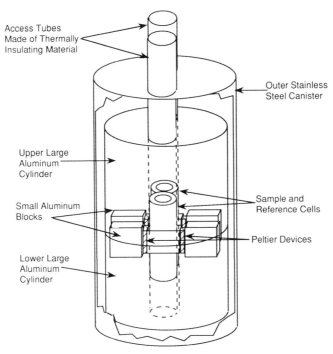

Figure 7.12 Schematic of a twin-cell heat conduction calorimeter that allows the insertion of batch or flow devices in the cells [37]. This instrument is usually placed inside a temperature bath.

to study aerobic growth [45]. This instrument is similar to the others in [45], and reaction cells are in contact through thermopiles with the same large thermal mass. The large thermal mass consists of two cylindrical aluminum blocks, one placed on top of the other, with cylindrical holes drilled in them for insertion of the reaction and reference vessels. The aluminum block is placed inside an inner can, which in turn is placed inside an outer can that is insulated. The entire assembly is placed inside a tank. The unique features of this calorimeter are that it provides for stirring and for the inflow of oxygen to the reaction cell to sustain the aerobic reaction.

4.2.4   Adiabatic

When the temperature rise in a batch calorimeter is measured directly, it is important to quantify the last four terms of (1) and to measure the temperature increase as accurately as possible in order to convert these temperature measurements to the quantity desired. The most difficult of these three terms to measure is the heat loss to the surroundings. This heat loss can be described by Newton's law of cooling:

$$q_{\text{loss}} = UA(T_c - T_s) \tag{2}$$

where U is the overall heat transfer coefficient, A is the area for heat exchange, T_c is the temperature of the sample cell of the calorimeter, and T_s is the temperature of the bath and so on surrounding the sample cell. The approach taken in adiabatic calorimeters is to eliminate q_{loss} by controlling the temperature of a shield around the sample cell so that the shield temperature is the same as that of the sample cell during the time of the reaction so that $T_c - T_s$ is equal to 0. A schematic representation of a typical adiabatic calorimeter is shown in Figure 7.2.

The advent of relatively inexpensive sophisticated computers has led to large improvements in adiabatic calorimeters. Much of the effort in recent years has been to automate and make the calorimeters easier to use. Adiabatic calorimeters have found special use in the measurement of heat capacity changes at temperatures approaching 0 K and in the measurement of the heat associated with phase transitions [46–51]. Almost all adiabatic calorimeters used to measure heat capacities, the temperature of phase transitions, and the enthalpy changes associated with such transitions are designed using the same basic principles. These calorimeters have a sample cell that holds the material whose properties are to be measured inside an adiabatic temperature shield. The temperature difference between the shield and the cell is monitored by a thermopile, and an automatic control system is used to keep the voltage output of the thermopile as close to 0 as possible. The main thermal connection between the shield and the cell is through the thermopile. Connections necessary for support are made using poorly conducting material. A run consists of placing the sample in the sample cell, placing the cell inside the calorimeter, letting the sample come to the desired initial temperature, and then heating the sample using a precisely known amount of electrical energy and monitoring the temperature rise. It is important that any electrical leads do not conduct heat from the sample vessel. Thermal anchors are used to prevent this heat flow.

Designs of such calorimeters were made to improve the cryostat around the calorimeter for low-temperature measurements [46], to improve the automatic control of the shield [46–48, 51], and to decrease the downtime between measurements [50]. In one such calorimeter [49] a pyroelectric thermometer, which has a fast response time, is used thus increasing the resolution of the first derivative of temperature with time. Other adiabatic calorimeters have been developed to follow the heat production of reactions with time as a function of temperature [52–54]. These calorimeters operate in much the same manner as those used to measure heat capacities and phase transitions, except that a reactive sample is placed in the cell and the heating is done only to raise the temperature to the desired initial value. The temperature is then monitored as a function of time to give data that can be interpreted to give the heat production and kinetics of a reaction.

Adiabatic calorimeters are used to make thermal hazard evaluations for industry. An exothermic chemical reaction has the potential of creating a thermal runaway and possibly a serious accident. As the reaction proceeds, the heat released increases the temperature in the reactor and speeds up the chemical reaction, thus increasing the rate of heat release. If sufficient heat transfer capability is not available, the reaction rate and consequent heat release may become so rapid that hazardous pressures and/or temperatures can develop. Special adiabatic calorimeters were developed to evaluate the possibilities of such hazards. The accelerating rate calorimeter (ARC) was developed at Dow Chemical USA for this purpose [55, 56]. Figure 7.13 shows a schematic of this calorimeter. The reaction vessel of the ARC is a spherical bomb designed to withstand high pressures (the pressure transducer measures pressures to 17 MPa and the equipment is designed to withstand even higher pressures) and high temperatures (to 773 K). The reaction vessel is surrounded by a temperature-controlled, nickel-plated, copper jacket. The complete apparatus in an insulated aluminum canister is mounted in a rugged steel containment vessel for safety. The operation of the calorimeter is under computer control. The temperature of the jacket is programmed to follow the temperature of the reaction vessel, as it is in all adiabatic calorimeters. In operation, a sample is added to the cell, which is then heated to some preset temperature. The calorimeter is allowed to come to thermal equilibrium, and the rate of the temperature rise due to the chemical reaction is measured. If the rate of temperature rise is equal to or greater than a preset value, heating is discontinued. However, if the rate of temperature increase is less than the preset value, the heater is turned on and the process is repeated. This heat–wait–search mode of operation is continued until the temperature is sufficiently high that the reaction rate is large enough to produce the desired rate of temperature increase. The chemical reaction is then allowed to proceed at its own adiabatic rate. The

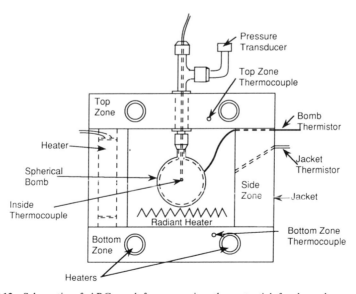

Figure 7.13 Schematic of ARC used for measuring the potential for hazardous conditions occurring in industrial chemical reactors [55, 56].

temperature and pressure are monitored and the pertinent kinetic parameters are calculated by using the temperature and pressure trace. A similar but more versatile calorimeter, developed by Sandoz, Ltd., allows adiabatic, isoperibolic, and scanning modes [57].

A problem exists with the two calorimeters mentioned previously; that is, they have high thermal mass because the sample vessels have thick walls to contain the materials at high pressures. One way to overcome this problem is to place the sample vessel in a thick-walled containment vessel. The pressure difference between the inside and the outside of the sample cell is kept small by regulating the pressure automatically in the containment vessel so that it follows the pressure inside the sample cell [58]. This apparatus is operated in much the same manner as the ARC, but it requires a much smaller correction to the heat data due to the small thermal mass of the sample vessel.

4.2.5 Alternating Current

Sullivan and Seidel [59] developed a new type of calorimeter in the late 1960s called an alternating current (ac) calorimeter. Many calorimeters have been developed using modifications of their design. The basic principle employed in the use of ac calorimeters is that an oscillating heat input is applied to the sample that in turn is connected to a heat sink through a thermal link of some sort as shown schematically in Figure 7.14. The resulting sample temperature is recorded as a function of time. When the heating current is sinusoidal and the temperature is uniform throughout the sample and its container, the temperature of the sample can be represented by the following equation once steady operation is obtained

$$T = T_0 + \frac{V_0^2}{2R}\left\{\frac{1}{K} - \frac{1}{\omega C}\frac{\cos(\omega t - \phi)}{[1+(1/\omega^2\tau^2)]}\right\} \tag{3}$$

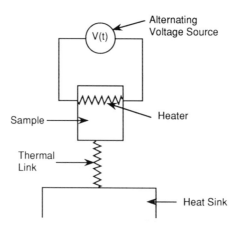

Figure 7.14 Schematic showing principles of operation of an ac calorimeter.

where T is the sample temperature, T_0 is the sink temperature, V_0 is the peak voltage across the heater of resistance R, K is the thermal conductance of the heat link, ω is twice the frequency of the voltage, C is the heat capacity of the sample plus container, τ is the thermal relaxation time and is equal to C/K, t is time, and ϕ is the phase shift and is found by

$$\phi = \sin^{-1}\left(1 + \frac{1}{\omega^2\tau^2}\right)^{-1/2} \tag{4}$$

We can find the heat capacity of the sample by using (3); it is inversely proportional to the amplitude of the temperature oscillations in the sample. Since the measured signal is repeated many times, signal averaging can lead to very precise results. The base or average temperature of the material can be adjusted by changing the temperature of the heat sink and/or by changing the power of the oscillating heat source. This calorimeter works well in measuring the heat capacity and the first- and second-order phase transitions in both solids and liquids [60–63]. The technique is used to measure the heat capacities of solids up to pressures of 300 MPa [64]. The thermal link between the sample and the heat reservoir was the inert gas, argon.

Improvement in the resolution is obtained in some cases by using a square wave heating source instead of the usual sinusoidal energy input [65]. We can use ac calorimeters to measure the thermal conductivities as well as the heat capacities of thin samples by using the correct frequency of oscillation and a more complex analysis of the output data [66, 67]. This technique is used to measure heat capacities at temperatures below 2 K [68]. In this low-temperature calorimeter, the frequency of the energy input is controlled by a computer and the data acquisition is also done automatically by a computer, allowing the operator to set the frequency of the heat input oscillations.

4.2.6   Isothermal

One of the problems with adiabatic or near-adiabatic calorimeters is that the rise in temperature must be converted to its energy equivalent according to the equation

$$q_m = C(T_f - T_i) \tag{5}$$

where C is the sample plus cell heat capacity, T_f is the final temperature, and T_i is the initial temperature. To do this, the heat capacity of the calorimeter and the reacting material must be known throughout the reaction. One way to overcome this problem is to reproduce the temperature trace electrically [20] as already mentioned; however, even then the assumption is implicitly made that the heat capacity does not change during the course of the reaction.

Another way to eliminate the need for heat capacity data is to run the instrument isothermally. This method was applied to batch calorimeters [4, 5].

Most isothermal calorimeters have a constant heat leak to a heat reservoir. A typical batch isothermal calorimeter is shown in Figure 7.15. We can effect the heat leak to the reservoir in an active way by using a Peltier device [a thermopile with a direct current (dc) voltage applied to remove heat] or in a passive way by thermally linking the sample vessel to the reservoir and by keeping the heat sink reservoir at a temperature lower than that of the reaction vessel. We can control the temperature at a constant value by using an automatically controlled heater. The amounts of energy added by the control heater in the absence and presence of a reaction are recorded. The difference between the energy needed to keep isothermal conditions in the absence and presence of a reaction is the amount of heat liberated by the reaction. No conversion of temperature changes to energy changes is required. A typical isothermal batch calorimeter was developed by Schildknecht [69]. The reaction vessel is submerged in a bath that is held at a temperature slightly lower than the desired reaction temperature. The higher the expected heat output from the reaction, the lower is the temperature of the bath that must be selected. The lid of the reaction vessel has different necks allowing dosage, gassing, taking samples, measuring pressure, measuring pH, and so on. The energy added to keep the vessel isothermal is added as dc through a resistance heater, and the amount is monitored and controlled by a computer. An isothermal calorimeter was designed specially to measure the heat flow from high-temperature batteries [70]. This calorimeter consists of three closely fitting chambers. The outermost chamber is a stainless steel box lined on the inside with high-temperature insulation. The outside of the box is in contact with ambient air that acts as the low-temperature heat sink. The middle chamber is a constant temperature oven that consists of a nickel heater sandwiched between stainless steel walls. The inner can is a composite of layered silver coated plates, nickel heaters, and temperature sensors. The battery is placed inside this

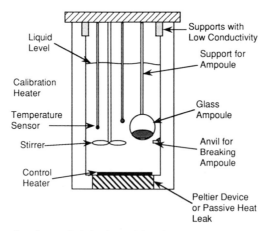

Figure 7.15 Schematic of a typical isothermal batch calorimeter with glass ampoule. This instrument is usually placed inside a temperature bath.

chamber. The calorimeter is constructed of metal and high-temperature insulation allowing the calorimeter to operate at temperatures up to 773 K. It will not operate at near-ambient temperatures since it requires a heat leak to the ambient air.

Zhong [71, 72] developed a new isothermal calorimeter that he calls a *compensation calorimeter*. Both the heating and the cooling are done with Peltier devices. The temperature is controlled automatically by a microcomputer. If the reaction vessel is too hot, the voltage across the Peltier devices is such that they are in the cooling mode. If the reaction vessel is too cold, the polarity of the voltage is reversed and the Peltier devices are put in a heating mode. The outer temperature bath is controlled in this same manner. The current to the Peltier devices is delivered in discrete pulses. Each pulse has the same amount of energy, and the total amount of energy is determined by the number of pulses needed to heat or cool the calorimeter. The energy of the reaction is determined by the difference in the pulses needed to keep isothermal conditions in the presence and absence of a reaction.

4.2.7   Relaxation

Another type of batch calorimeter, the relaxation calorimeter, is used primarily to measure heat capacities. A schematic of this calorimeter is found in Figure 7.16. In these calorimeters, the sample is placed in intimate thermal contact with a small holder. We can link the holder thermally to the surroundings by using a well-characterized path such as a small wire or wires. The sample and holder are heated to a temperature slightly above that of their surroundings. The temperatures of the holder and sample are monitored as the temperature "relaxes" back to the temperature of the surroundings. The temperature decays

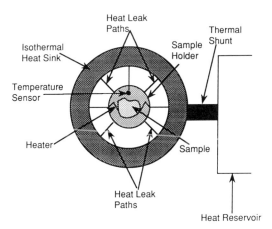

Figure 7.16 Schematic showing principles of operation of relaxation calorimeter. Leads to the temperature sensor and heater also serve as heat leak paths.

exponentially if the heat capacity is constant according to the following equation

$$T = T_0 + \Delta T e^{-t/\tau} \tag{6}$$

where T_0 is the temperature of the heat sink, ΔT is the initial temperature difference between the sample temperature and the heat sink temperature, t is the time after the heating has been terminated, and τ is the characteristic time of the system. The characteristic time of the system τ is equal to C/K, where C is the heat capacity of the sample and sample holder and K is the combined thermal conductance of the heat leak paths.

This calorimeter is used widely to measure heat capacities at low temperatures. One such calorimeter for measuring heat capacities in the 1–35 K range uses a silicon chip as the sample holder [73]. This chip contains two silicon resistance thermometers and a Cr—Ti resistor for heating. Six lead wires from a copper ring to the chip support the chip and the sample. Four of the wires serve as leads to the heaters and two serve as leads to the temperature sensors. These wires also serve as heat leak paths during the relaxation period. The sample is bonded to the chip with a thermally conducting grease. The resistance thermometer is part of an ac Wheatstone bridge. The whole calorimeter is in a cryostat submerged in liquid helium. A relaxation calorimeter for measuring heat capacities at near ambient temperatures was designed by Hatta [74]. The sample chamber is a thick-walled copper block that serves as a thermal bath. The chamber is filled with an appropriate heat-exchange gas, such as helium or air, at pressures of 0.1 MPa or lower. A platelike sample of the material to be analyzed is suspended by fine Chromel–Alumel thermocouple wires. One junction of the thermocouple is attached by varnish to the sample and the other to the thick-walled chamber so that the voltage measured across each thermocouple is a measure of the difference in temperature between the sample and the temperature bath. The sample is placed in the bath and allowed to come to thermal equilibrium. Heat flux for warming the sample is then supplied by light from a stable halogen lamp.

As in many relaxation calorimeters, the heat capacity is calculated from the temperature trace as the sample warms up instead of when it cools down. Griffing and Shivashankar [75] designed a relaxation calorimeter to operate over the 4–380 K range. This extended range is made possible by using a GaSiP light-emitting diode as the temperature sensor. This diode is effective over a wide temperature range, but is not as accurate at temperatures below 10 K. The sample platform is a thin single-crystal sapphire. One side of the platform is polished for good thermal bonding of the sample. The other side contains a gold heater and contact pads that have been evaporated onto the surface. The temperature sensor is indium soldered to one of the contact pads. A low-temperature calorimeter has been designed to measure the capacities of solids at temperatures between 0.020 and 25 K [76]. Its operation and design are similar to those of other low-temperature relaxation calorimeters except that it uses a $(^{3}He + {}^{4}He)$ dilution refrigerator to obtain these low temperatures.

4.2.8   Pulse Heated

A calorimetric method quite similar to the relaxation method for measuring heat capacities is the pulse-heating method. The principle of operation is to heat the sample using a precise amount of power for different lengths of time. Between each pulse of power, the sample is allowed to return to the initial temperature. The longer the time of the pulse of energy, the higher is the temperature. By monitoring the time of the pulse and the final temperature of the sample after each pulse, we can establish the temperature derivative experimentally with respect to time and find the heat capacity by using the equation

$$C = \frac{q_{heater}}{dT/dt} \tag{7}$$

where C is the heat capacity of the sample heated, q_{heater} is the power added to the sample corrected for heat loss, and dT/dt is the temperature derivative with respect to time. As seen in (7) the heat capacity is directly proportional to the power input and inversely proportional to the first derivative of the temperature with time. This scheme is used to measure the heat capacity of electrically conducting materials at high temperatures by passing an electrical current through the material, which is held in a vacuum [77]. Such a calorimeter is limited in measuring heat capacities of electrically conducting materials with low vapor pressures. Heat-pulse calorimeters are also used to measure heat capacities of radioactive samples [78]. Radioactive materials present special problems in that the heat generated by radioactive decay is difficult to separate from other heat effects. The unique feature of this heat-pulse calorimeter is that it has a heat leak path from the sample to a heat reservoir. The temperature of the heat reservoir is adjusted so that the heat leak compensates only for the heat generated by the radioactive decay. In this way the heat effects of the decay process can be eliminated from the calculation of the heat capacity.

4.3   Titration Calorimeters

A titration calorimeter is characterized by the addition of one fluid at a well-defined rate to a reaction vessel that already contains a second fluid. In this way, one experiment or titration is equivalent to many batch experiments that involve adding different ratios of reactants.

4.3.1   Reaction

A titration calorimeter developed by Christensen and co-workers [79] in the early 1960s and its subsequent modifications were used successfully for many years. This calorimeter is based on operation in the isoperibol mode. A schematic of it is given in Figure 7.1. Christensen and co-workers [80] later developed a titration calorimeter that operates in an isothermal mode [(see Figure 7.4). This isothermal calorimeter uses a Peltier device as a constant heat leak and adds energy by a control heater. The heat is added in pulses and the amount of heat is proportional to the number of pulses. A titration calorimeter based on the design of Christenson and co-workers is now available commercially.

Examples of the diverse work being done with such calorimeters are (1) the measurement of the formation constant and enthalpy of reaction of crown ethers and cryptands with cations [81] and (2) the measurement of the boundaries between two-phase and three-phase regions in emulsified systems [82]. The calorimeter developed by Christensen and co-workers [79, 80] has a vapor space above the liquid and a vent through the stirring shaft hole to the atmosphere. This allows evaporation to occur, which is no problem when using oxygen-stable, low-volatility liquids. Spokane and Gill [83] developed a titration calorimeter that eliminates the vapor space and also allows studies to be done on very small quantities of material. Figure 7.17 is a schematic of this calorimeter. The vapor space is eliminated by filling the vessel and allowing the excess fluid to flow out of the vessel into a reservoir as titrant is added. A micrometer driven by a stepping motor is used to deliver the titrant stepwise in discrete volumes through the center of the stirring rod, which is a stainless steel hypodermic needle with a bent Teflon tube attached to the end to act as a paddle. The excess fluid escapes into the reservoir through the annulus between the stirring rod and the filling tube. The principle of this procedure is that each volume of liquid added near the bottom of the cell displaces an equal volume of liquid from the top of the cell before the newly added titrant can mix with the liquid that leaves. In reducing the data, a proper mass balance on the material in the reaction vessel is done. The reaction cell is a 1 cm^3 glass bulb with 0.1-cm^3 capillary tubing used as the filling tube. The cell is embedded in an aluminum cup using Woods metal. A reference block is placed inside the calorimeter below the reaction cell and is connected to the reaction cell with a specially designed thermopile. The reference block and reaction cell are shielded by two adiabatic shields. During operation, a constant current is applied to the heater in the reference block, causing the temperature to rise at a very low rate (about $0.8 \mu\text{K/s}$), and the control causes the reaction cell and the shields to follow this temperature. The difference in the amount of energy added to the reaction cell in the presence and absence of a reaction gives the amount of energy released during the reaction.

Another titration instrument developed specifically to measure small heats is a twin-cell isoperibol calorimeter [84–86]. This calorimeter has twin cells made of thin polycarbonate that are placed in the same temperature bath and are filled with the same solution. The titrants are stored in tubes that are inside the bath so that the titrants are at the initial temperature of the reaction vessel. The solution in one tube (nonreactant titrant) is the same as that in the reaction

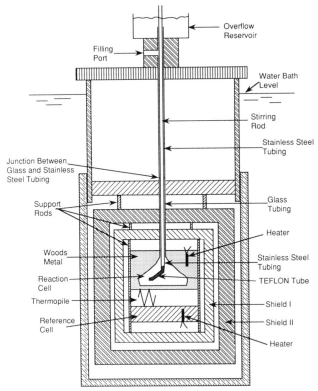

Figure 7.17   Schematic of a calorimeter that employs overflow of solution from reaction cell into reservoir as increments of reactant are titrated into the cell [83].

vessels, while that in the other tube is the reactant solution. During a titration, the reactant and nonreactant solutions are added simultaneously and at equal rates to the two vessels. The temperatures of the solutions in the two cells are monitored by thermistors whose temperature resistances match each other as closely as possible. These two thermistors are the opposing legs in an electrical bridge. The output of the bridge is amplified. This arrangement allows the measurement of very low heat changes as much of the noise is eliminated by the twin-cell arrangement. This calorimeter is fully automated and also has pH probes inserted into the cells for simultaneous pH measurement.

Wadsö [87] developed a perfusion-titration cell to use with a 2277 multi-channel microcalorimeter system, LKB Products Bromma, Sweden. This instrument can be used as a precision titration calorimeter. It is designed specifically for use in biological reactions where samples may be small, resulting in small heat effects. The basic calorimeter is the same as that mentioned in [37]. The cell consists of a sample compartment that is connected to a steel tube that holds the shaft connecting the motor with the stirrer. This steel tube is thermally anchored by brass bolts to the upper aluminum block and water bath. The sample cups can be of different volumes from 1 to 3 cm^3.

Bishop and co-workers [88] developed a combination heat conduction and compensation calorimeter that allows titration with a gas. The instrument has two cells, a reaction cell and a reference cell, thermally connected by thermoelectric devices to a large copper heat sink that is enclosed in a submersible box that is suspended in a constant-temperature water bath. The thermoelectric devices are wired so that the voltages oppose each other. Contrary to most Calvet-type calorimeters, electrical energy is added to these cells, and the amount of energy input that is increased or decreased when a reaction occurs is a measure of the heat change due to the reaction. In actual operation, the change in energy input to the calorimeter accounts for about 97% of the energy of reaction and the final 3% is calculated by the output from the thermoelectric devices. The reaction and reference cells are copper cups held in place with nylon screws with the thermopiles held snugly in place between the cups and the copper block. Glass nuclear magnetic resonance (NMR) sample tubes placed snugly in these cups hold the reacting fluid. To ensure rapid gas exchange with the liquid, small pieces of filter paper are wetted with the solution and placed in the calorimeter. Water-saturated gas is first added to the cell, and then by using a gas dilution valve, the partial pressure of the reactive gas is reduced by diluting with a nonreactive gas in several dilution steps, giving a reverse titration curve.

A large titration calorimeter used in simulating industrial processes has been designed [89]. The calorimeter consists of a cylindrical metal reaction vessel inside three concentric cylindrical cans as shown if Figure 7.18. The innermost can acts as an adiabatic shield and is controlled at the same constant temperature as that of the reaction cell. Heating for this shield is provided by a foil heater and cooling is provided by a gas that enters the annular space between the adiabatic shield and the second can by ducts in a perforated coil. The space between the second can and the outermost can is filled with insulation. The titrant is pumped through a coil in the annular space between the reaction cell and the adiabatic shield so that it reaches the proper temperature before entering the reaction cell. Heat exchange between the bottom of the reaction vessel and a low viscosity silicone oil through a heat exchanger is the only significant heat leak. During operation, the temperatures of the reaction vessel, the adiabatic shield, and the cooling fluid are kept constant. The difference in the amount of energy added to the reaction vessel in the presence and absence of a reaction gives the heat change due to the reaction. A large glass reactor for simulating industrial reactions was developed by Mettler Instruments AG [90]. This calorimeter can be operated in the isothermal, adiabatic, or ramp mode automatically by use of an International Business Machines Corporation (IBM) PC/XT computer. The reaction vessel is surrounded by a shield through which thermostated fluid is circulated. The temperature of this fluid can be changed automatically and rapidly to control the temperature in the reaction vessel. When operating in the isothermal mode, we determine the heat of reaction by calculating the heat loss using Newton's

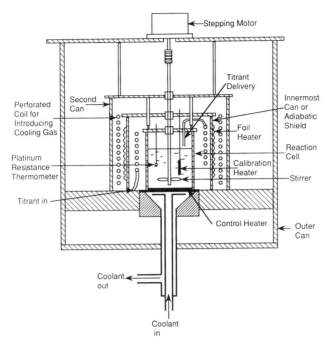

Figure 7.18   Schematic of a titration calorimeter used in simulating industrial conditions [89].

law of heat transfer [see (2)]. The product of the heat transfer coefficient and heat exchange area in this equation is found empirically by using a calibration heater. The reaction vessel is first loaded with a charge of reactant and a second reactant is added in doses. The pH and pressure are monitored as well as the temperatures of the cell and shield.

4.3.2   Adsorption

The adsorption of a fluid on a solid is of interest to those engaged in studies of surface chemistry, heterogeneous catalysis, lubrication, enhanced oil recovery, and so on. We investigate the heats of adsorption and sometimes the kinetics by using calorimetry. We study the heats of adsorption of a liquid on a surface by using conventional batch calorimeters where the ampoule is filled with the solid. Also, we study the heats of adsorption of a fluid (gas or liquid) by using special titration calorimeters where the solid is placed in a cell and incremental amounts of the fluid are added to the cell in one or in several steps. O'Neil and co-workers [91] reported a Calvet type calorimeter with three cells designed specifically to measure the differential heats of adsorption of gases on solids containing high surface areas. The three cells include a reaction cell, a reference cell with an opposing thermopile, and a cell that holds the gas at the proper temperature before dosing. The three cells are connected through high-vacuum stopcocks to a vacuum system that allows the pressure to be lowered down to 10^{-4} Pa. The heat sink is a common large aluminum block encased inside a double box. The reaction cell and the reference cells are thermally connected to the large aluminum block through thermopiles. The cells are made of Pyrex glass. The inner box is held at temperature by adding or removing heat with thermopiles. The useful temperature range is from 273 to 373 K. The solid is added to the reaction vessel under vacuum, allowed to reach thermal equilibrium, and then dosed with successive amounts of gas. The heat for each addition of gas is measured and the amount of gas adsorbed is determined by pressure measurements. An isoperibol calorimeter for the measurement of the heats of adsorption of a gas on a single crystal was designed [92]. A single crystal of Pt (crystal face 111) about 1 cm in diameter is supported by two 0.38-mm tantalum wires in an ultrahigh vacuum cell (vacuum to 4×10^{-9} Pa). Two thermistors, one for temperature measurement and one for heat input for calibration, are inserted in glass capillary tubes that are mounted into holes drilled into the crystal. A Chromel–Alumel thermocouple is spot-welded to the crystal for temperature measurement during the cleaning cycle. An Auger electron spectrometer is mounted into the vacuum cell so that the relative surface coverage can be measured and the cleanliness of the surface can be monitored. The crystal is cleaned by electric heating to 1073 K and by electron bombardment to 1573 K. The thermistor is part of a bridge whose output is monitored by a strip chart recorder. After the crystal has been cleaned, allowed to come to thermal equilibrium, and coated with a certain amount of gas, it is dosed with a gas and the temperature is monitored. The total heat released is proportional to the area under the curve.

4.3.3   Solution of Gases in Liquids

The heats of solution of gases are performed in a special calorimeter designed to titrate gases into a liquid [93]. This calorimeter is of the twin-cell Calvet type and has the output voltages from the cells opposed. The cells (reaction and reference) are made of borosilicate glass. Precisely known quantities of gas are injected into the reaction cell through a sintered glass–mercury one-way valve. We measure the gas quantities by using a mercury manometer.

4.4   Flow Calorimeters

Flow calorimeters are instruments in which at least one stream flows into and at least one stream flows out of the measurement section of the calorimeter. Flow calorimeters are particularly well suited to measurements at high pressures, since the measurement is usually made on a fluid flowing in a tube. The nature of a flow calorimeter also allows for the easy elimination of a vapor space. The disadvantages of these instruments are that relatively large amounts of the fluid are needed, solids are difficult to handle, and the precision of the instrument is limited by the accuracy of the rate of flow produced by the pumping system.

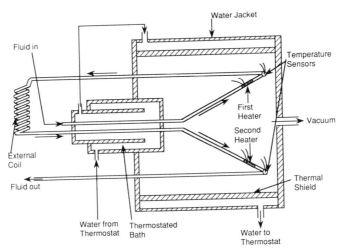

Figure 7.19   Schematic of a flow calorimeter used to measure heat capacities of solutions relative to the heat capacity of the solvent [94]. The thermostated bath section is shown much shorter than actually exists in the calorimeter.

4.4.1   Heat Capacity

Flow calorimeters are used extensively for the measurement of heat capacities of fluids. Picker and co-workers [94] report a calorimeter shown schematically in Figure 7.19 that is capable of measuring the heat capacities of solutions relative to those of the pure solvent. The ratio of the heat capacity of the solution to that of a reference fluid (the solvent) is determined by measuring the amounts of energy required to heat the solution and the reference fluid to the same temperature when they both start at the same temperature and are flowing at the same volumetric flow rate. This heat capacity measurement is accomplished by pumping the solvent from a thermostated bath through a tube in the first heater section to cause a small temperature rise. The solvent in the tube then flows through a coil and then back through the thermostated bath, which brings the temperature of the solvent back to its initial value. It then flows through a second heater and detection system identical to the first heater. The energy input to the second heater is controlled so that the outlet temperature is the same as the outlet temperature in the first heater. The amount of energy input to the second section must be the same as that to the first section since the flow and temperature rise are the same for both sections. Once steady operation with a pure solvent is attained, a small volume of the solution is introduced in the solvent stream. Measurements are then made of the energies required to heat the solution leaving the first heater and the solvent leaving the second heater. The ratio of the heat capacities of the solution and solvent is equal to that of the energies added to the two heaters. As the flow continues, the pure solvent is reintroduced into the flow system and the process is repeated. In this type of calorimeter, the heat capacity ratios can be measured with great accuracy. However, the accuracy of the solution heat capacity calculated from this ratio is limited by the accuracy of the reference heat capacity.

Wood and co-workers [95, 96] developed a calorimeter for the measurement of the heat capacities of aqueous solutions up to 600 K similar in operation to the one just described. This calorimeter, shown in Figure 7.20, consists of two cells made of loops of Hastelloy C-276 tubing that are held in thermal contact with a large copper block, except for those measurement sections where the heaters and temperature sensors are located. The direction of flow is (1) through a part of the tube that is in contact with the copper block, (2) into the first measurement section, (3) through a sample injection valve, (4) back through a part of the tube that is in contact with the copper block, and (5) into the second measurement section. The large copper block is held at a constant temperature and surrounded by an adiabatic shield. Held in a thermostated sample loop is 10 cm³ of the solution, which can be injected into the flow loop between the two measurement sections by switching the sample injection valve. After a steady-state condition is established with water in the entire loop, the sample injection valve is switched so that the solution is injected into the flow loop between the two measurement sections. The measurements made and conversions to heat capacity are essentially the same as those in the Picker calorimeter.

Ogawa and Murakami [97] developed a similar calorimeter to measure liquid heat capacities. The calorimeter is enclosed in a vacuum cell inside a water

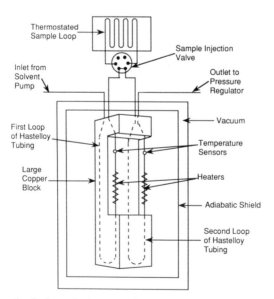

Figure 7.20 Schematic of a flow calorimeter used to measure heat capacities of solutions relative to the heat capacity of the solvent at elevated temperatures and pressures [95, 96].

bath, and both the reference and sample cells have two flow paths allowing greater flexibility. Zegers and co-workers [98] modified a Picker-type calorimeter so that the heat capacities of viscous fluids can be measured. In this instrument, the entering liquids are pressurized sufficiently to permit flow.

4.4.2   Incremental Enthalpy Change

Wormald and Yerlett [99] developed a calorimeter that uses heat exchange with a well-characterized fluid (water) to measure the incremental enthalpy changes that occur when a vapor is cooled and condensed. The construction is similar to that shown in Figure 7.5. The fluid for which the enthalpy change is to be measured is pumped at a precise rate through a heater where it is heated to the desired temperature and then is cooled and condensed with water in a very carefully constructed heat exchanger. From the flow rates and the inlet and outlet temperatures of the water, enthalpy changes can be calculated. This apparatus has been used to measure the enthalpy changes when condensation occurs. The clever design of the heat exchanger in the calorimeter minimizes the heat leaks. Simoni and Chagas [100] developed a labyrinth flow calorimeter to measure enthalpy changes by using the same principles as those used by Wormald and Yerlett [99]. The heat exchange of this calorimeter occurs in a labyrinth, and we measure the temperature rise in the exchange fluid by using calibrated thermistors. This instrument is used to measure small sensible heat changes and is effective for determining the heat capacity.

4.4.3   Mixing of Two Fluids

Christensen and co-workers [101–104] developed several similar isothermal flow calorimeters that are used to measure heats of mixing, reaction, and solution of gases in liquids over a wide range of temperatures and pressures. Figure 7.7 shows the basic design for the reaction cell of these calorimeters. The overall flow path for a high-temperature instrument is shown in Figure 7.21. Each of the two fluids to be mixed is pumped at a rate of approximately 0.005 cm^3/s through approximately 2 m of 1.56-mm outside diameter (o.d.) metal tubing, which is inside a temperature bath so that it reaches the desired temperature before mixing. The two tubes join in a mixing junction on a temperature-controlled plate or cylinder. The fluids mix at this junction and continue to flow through a single tube that is in intimate thermal contact with the temperature-controlled cylinder or plate. Upon leaving the plate or cylinder, the tube carrying the fluid mixture is in thermal contact with the two tubes carrying the incoming fluid to allow heat exchange between the incoming and outgoing fluids. The tube then exists from the calorimeter. Thermal energy is removed from the temperature-controlled plate on the low-temperature calorimeter at a constant rate by a Peltier device [101]. The voltage applied to this device can be set at different values to control the rate of energy removal. The temperature is kept constant by adding just enough heat to compensate for the heat loss. The energy is added in discrete pulses, each pulse having the same

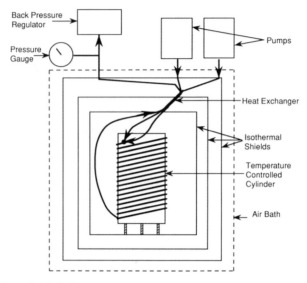

Figure 7.21 Schematic of the flow scheme of a flow calorimeter used for mixing two fluids at elevated temperatures and pressures [101–104]. The reaction cell portion is shown in more detail in Figure 7.7.

amount of energy, so that the frequency of the pulses gives the energy added. The difference between the energy added in the presence and absence of a reaction is equal to the heat change due to mixing of the two fluids. The high-temperature versions [102–104] differ from the low-temperature version in that an air bath is used instead of a water bath, a temperature-controlled cylinder is used instead of a temperature-controlled plate, and the heat is removed by conduction through nickel bolts that connect the cylinder to a plate controlled at a temperature 0.5 to 2.0 K below that of the cylinder. The tubing in the low-temperature calorimeter is made of stainless steel 316, that of one of the high-temperature calorimeters is made of Hastelloy C metal [102], and that of another one is made of tantalum [104] to avoid corrosion. The high-temperature calorimeters have been operated up to 593 K and 15 MPa, although the materials used in their construction are rated to be stable at substantially higher temperatures and pressures.

One of the shortcomings of these calorimeters is that there is no good way to account for the viscous heating terms. Raal and Webley [105] designed a flow calorimeter to eliminate the errors due to viscous heating. The incoming fluids flow through tubes 2 m long that are submerged in a temperature bath. These fluids are then mixed in the mixing cell, which consists of a Teflon tube containing a Nichrome heating element that is twisted helically in one direction and then in the other to ensure complete mixing. After mixing, the fluid is restored to its initial temperature and flows through a reference cell similar in design to the mixing cell, but somewhat shorter. The pressure drop across the reference cell is controlled to be the same as that across the mixing cell with a throttling valve, so that the frictional heat generation is the same through both cells. Each cell contains a thermistor, one matching the other, and these thermistors are opposing legs in an electrical bridge. The heater in the mixing cell is controlled automatically so that the temperature in the mixing cell is kept equal to that in the reference cell to eliminate the effects of frictional heating on the measurement. The heat supplied to the mixing cell is equal to the heat of mixing. Much care was taken in the design of the apparatus to minimize the heat leaks. This apparatus should prove useful in measuring heats of mixing of viscous fluids. Limitations of the instrument are that only heats of endothermic reactions can be measured and the temperature and pressure ranges are limited to those near ambient due to the materials of construction.

Randzio and Tomaszkiewicz [106, 107] developed flow calorimeters that use active temperature control of the incoming fluids. The incoming fluids are pumped through capillaries that are inbedded in heat equilibrators. These heat equilibrators are controlled so that they are at the same temperature as the calorimetric vessel. The fluids are mixed by flowing them either into a mixing cell [106] or into a mixing junction [107]. In either case, the incoming fluids heat exchange with the outgoing mixture in an aneroid type countercurrent heat exchanger. The calorimeters have high-temperature (to 723 K) and high-pressure (to 39 MPa) capabilities. Generally, we use these flow calorimeters to run a heat compensation method in an isothermal mode.

To measure small heat changes that occur in aqueous solutions at elevated temperatures and pressures, workers at Oak Ridge National Laboratories developed a flow calorimeter of the Calvet type [108, 109]. This calorimeter uses two identical heat exchangers each made of platinum + rhodium tubing coiled around a tube and tightly encased inside lower and upper stainless steel containment cylinders. The tubing is coiled such that a countercurrent heat exchange occurs between the entering two fluids and the exiting fluid. The upper cylinder is in thermal contact with the calorimeter block and serves as a heat exchanger to bring the incoming solutions to the block temperature. The lower containment cylinder is in thermal contact with the calorimeter block by a 450-couple thermopile. During a measurement, the liquids to be mixed (an aqueous solution and pure water) are pumped through one cell and an equivalent amount of water is pumped through the reference cell. The thermopile in the reference cell is wired so that its voltage is in opposition to the thermopile in the reaction cell. Ruska differential proportioning and proportioning pumps are used. The tubes are made of platinum + rhodium gold soldered at the junctions and the valves are platinum lined so that they are corrosion resistant. To protect the pumps from corrosion, the pusher device shown in Figure 7.22 is used. The Teflon bag in this device is filled with the corrosive fluid to be studied. The pumps are filled with distilled water. In operation, the water is pumped into the annular space between the stainless steel cylinder and the Teflon bag squeezing the potentially corrosive material into the calorimeter.

Arntz and Gottlieb [110] designed a cell to insert in a Setaram heat-flow

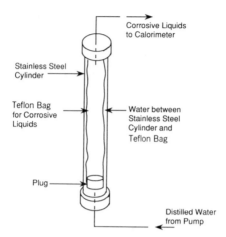

Figure 7.22   Schematic of pusher used to isolate pumps from corrosive material [108].

calorimeter to study heats of reaction at high pressures. The cell consists of a cylinder that encloses a catalyst basket. The reactants and products flow in the inner and outer tubes, respectively, of a concentric pair. This apparatus was used to measure the heat of formation of methyl *t*-butyl ether from methanol and 2-methyl propene. Other modifications of similar types of flow calorimeters were made such as the modifications made on the Beckman 190 microcalorimeter by Weber and Hinz [111]. They replaced the platinum tubing with Teflon tubing and introduced a reference cell identical to the reaction cell so that it could be used for biophysical studies.

4.4.4   Heat of Solution

Gill and Seibold [112] have developed a Calvet flow calorimeter to measure the heats of solution of solids, as shown in Figure 7.23. It is similar to other Calvet type calorimeters in that it consists of twin cells, one of which is a reference cell, both thermally connected to a large heat sink through thermopiles. The twin cells are in a metal box that is submerged in a water bath. The flow insert consists of about 2 m of concentric Teflon tubing that are held firmly in place between two stainless steel cylinders. The flow insert is placed in the sample cell of the Tian–Calvet block. A similar dummy insert is placed in the reference cell. The liquid enters the inner tube of the concentric pair of tubes in the flow insert and exits in the annulus between the inner and the outer tubes. There is a space for an insertion probe at the point in the flow path where the liquid leaves the inner tube and enters the annulus. A solid insert probe consisting of a Teflon tube with a glass capillary tube holding the solid and two sections of stainless steel needlestock, one section to form a tight connection with the inner flow tube

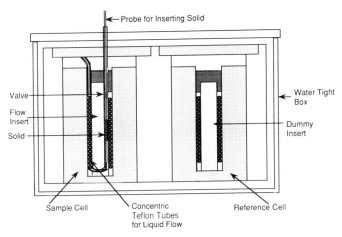

Figure 7.23   Schematic of a heat conduction flow calorimeter used to measure heats of solution of solids [112].

and the other to form a valve to connect the probe to the outer flow tube, is placed in this space. The liquid is pumped through the inner tube, through the insert probe past the solid, and then out the annular space between the inner and outer tubes. The perfusion titration calorimeter developed by Wadsö [87] can be used as a flow cell. Liquid can be pumped into the cell through the center of the stirring shaft and can leave through the annulus between the stirring shaft and the steel tube by which the cell is suspended. Nilsson and Wadsö [113] modified this flow cell so that it can be used to measure the heat of solution of solids. It was designed specifically to measure the heat of solution of slightly soluble solids. The stirrer used in the Nilsson–Wadsö cell was replaced with a dissolution ampoule that consisted of a steel cylinder attached to the stirring shaft in a tight fit so that the incoming fluid flowing down the inside of the stirring shaft flows into the ampoule. The flow in the ampoule is directed down a central tube past the solid to be dissolved on the bottom of the ampoule and out the top of the ampoule through steel tubes. The steel tubes are connected to Teflon tubes that direct the flow to the bottom of the cell. As in the flow cell, the liquid then leaves the cell in the annulus between the stirring shaft and the steel support tube.

Wadsö [114] also designed an insert to put in the calorimeter that allows us to measure dissolution of small quantities of highly or slightly soluble liquids in other liquids. This insert consists of two concentric glass tubes with two heat-conducting regions in thermal contact with the surrounding water thermostat and the calorimeter heat sink. Below these regions, there is a brass cup in thermal contact with the thermopiles. This cup contains a system of spiraled tubes in Woods metal. During a measurement, a constant flow of solvent is pumped through the vessel and brought to reaction temperature as it flows to the dissolution region. The solute is typically added in 5×10^{-3}-cm^3 increments to the dissolution zone by means of a thin metal tube fastened to a microsyringe.

Dec and Gill [115] developed a flow calorimeter for measuring the heats of solution of slightly soluble gases in water. It is of the heat-conduction type with reaction and reference cells in thermal contact by thermopiles to the same large heat sink; the thermopiles are wired so that their voltage outputs oppose each other. The calorimeter is enclosed in a stainless steel box that is submerged in a constant temperature water bath. Inside the stainless steel box is a large aluminum block that acts as the heat sink. Water from a distillation column enters the calorimeter by a stainless steel tube and is first equilibrated with the water in the temperature bath and then with the aluminum block. The water enters at the bottom of a gold-plated copper cup, which forms the bottom half of the reaction zone. This cup is in thermal contact with the block through the thermopile. The gas, after being thermally equilibrated, enters through the top of a gold-plated copper bell, which forms the top boundary of the reaction zone. The gas flow is computer controlled, so that a steady-state flow is maintained and an appropriate material balance is made. A hydrophobic glass ring around the circumference of the reaction zone helps maintain a stable gas–liquid interface.

4.4.5   Biological System

Marison and von Stockar [116] use a BSC-81 calorimeter developed by Ciba–Geigy AG, Basel, Switzerland, to measure the heat generated during the growth of microorganisms. The operation is similar to that of the calorimeter developed by Grob and co-workers [90]. The temperature of the reaction vessel is kept constant by controlling the temperature of the fluid in the cooling jacket around the vessel. In this calorimeter, all terms in (1) except $q_{reaction}$ and q_{loss} are negligible. Therefore, $q_{reaction}$ is equal to q_{loss} and can be calculated from (2). The reaction cell is a 2-dm^3 glass culture cell with a glass jacket for the cooling fluid. Air can be pumped through the cell for aerobic reactions, making this calorimeter a flow calorimeter. Kobayashi and co-workers of Tokyo Riko Co., Ltd., Tokyo, Japan [117], developed a twin cell heat conduction calorimeter for studying biological reactions. This instrument can be used in a batch or flow mode by changing the cells. The heat sink is a large aluminum block held in a constant temperature air bath. The block can be rotated for mixing. Leiseifer and Schleser [118] coupled an LKB Company heat conduction flow calorimeter to a fermenter to measure the heat released by steady-state microbial cultures. Some culture from a thermostated fermenter is pumped through the measuring cell, and sterile solution is pumped through the reference cell. The measured heat was found to be a function of residence time outside the fermenter. Measurements were made as a function of flow rate through the cell, and the results were extrapolated to infinite flow rate or zero residence time.

5 SOME USES OF CALORIMETERS AND CALORIMETRIC DATA

Calorimeters serve as effective tools in investigating many chemical and physical changes. This section discusses some of the new applications of calorimetry and improvements or extensions of earlier applications. It is anticipated that many additional uses will be found for calorimeters and calorimetric data.

5.1   Measuring Equilibrium Constant and Heat of Reaction Values

Titration or flow calorimeter data collected when two reactants are mixed were used to calculate the equilibrium constant K_{eq} as well as the heat of reaction ΔH_R values for chemical reactions [119–123]. This technique was applied more recently to high temperature aqueous solutions [124, 125]. At high temperatures (above 473 K), the data reduction becomes more complex. Most ionic species associate at high temperatures so that reactions other than the one of interest occur. The heat effects due to changing ionic strength during the course of the measurement are large, sometimes up to 50% of the total heat. It is not advisable to carry out the reaction in a solution containing a high concentration of "inert electrolyte," which is intended to keep the ionic strength constant. The anion and cation of the added salt will likely associate with each other as well as with the reactant ions thereby greatly complicating data analysis. The enthalpy departure of the water also contributes significantly to the measured heat. The activity coefficients of the electrolytes are significantly different from unity even at ionic strengths as low as 0.01, especially for multiply charged ions. As a result of these complications, great care must be exercised in the reduction of calorimetric data collected for aqueous reactions at elevated temperatures, and even then the values obtained are dependent on the activity coefficient model chosen. It is usually necessary to collect the data at many temperatures and ionic strengths to calculate the ΔH_R and K_{eq} values.

5.2   Determination of Activity Coefficients for Electrolytes in Solution

Workers at Oak Ridge National Laboratories [108, 109] have used high-temperature flow calorimetry to calculate the parameters in a Pitzer type activity coefficient model for solutions up to temperatures of 673 K. They have collected data by mixing pure water with solutions of HCl and NaCl over wide temperature and concentration ranges. This work requires a sensitive calorimeter since heats are needed at very low concentrations. As in measuring heats of reaction, this calculation process is complicated by the tendency of even strong electrolytes to associate at high temperatures.

5.3   Measuring Heats of Conformational Changes in Solids

Lindenbaum and co-workers [126] have used dissolution calorimetry to determine heats of conformational changes in solids. Since the conformational changes in a solid take place very slowly, it is difficult to measure this change directly. However, the change was determined by measuring the heat of dissolution of the different solid forms in various solvents. Since the dissolved solid is the same irrespective of its original solid form, the difference in the heats of solution gives the heat of the solid conformational change. Lindenbaum, has confirmed this method by obtaining the same conformational heat with different solvents.

5.4   Characterization of Surfaces

Several workers have used calorimetry to study the adsorption of gases or liquids on catalytic surfaces [12–14, 91, 92]. Calorimeters serve to measure the heat of adsorption and the equilibrium concentration of the fluid on the surface. Calorimeters can also be used to measure the rate of adsorption. This has been done on bulk catalytic as well as on single crystal surfaces. These measurements are often done at high-vacuum levels, requiring the use of specialized high-vacuum calorimeters.

5.5   Investigating Biological Reactions

Calorimeters have been used for some time to investigate biological systems. This is an area where the use of calorimeters has grown rapidly during the last decade [7–9]. Calorimetry is used to identify and characterize microorganisms by their distinctive power-time curves during metabolism and growth. Care must be taken in defining the medium that is used, as the composition of the medium strongly influences the metabolism. Calorimeters are also used to test the effect drugs have on microorganisms. The heat generated by a drug-treated culture compared to that generated by a control can give information on the effectiveness of the drug in killing or controlling the growth rate of the microorganism. We can monitor the heat production rate readily during fermentation processes by using calorimetry. This rate is important as the fermentation processes are usually very temperature sensitive and the intelligent design of effective temperature control for industrial fermentation reactors is dependent upon knowing the heat production.

All living systems involve chemical reactions in or with water. Much information about biological systems can be obtained by calorimetric measurements of the interaction of biochemicals or simple model compounds with water or in aqueous solution [127]. Information on the structure is obtained by measuring the heat of dissolution of the solid compounds or the dilution of solutions of these compounds. Studies of biochemical substrate–receptor interactions help in understanding structural properties of biochemicals. Thermodynamic values for the transfer of solutes from the gas phase to solution give much information about solute–solvent interactions. Heat capacity values of solutes in water give information about the hydrophobicity of the solute, which is believed to be important for understanding many biochemical binding reactions. The thermodynamics of protein–ligand binding is a field of great interest. Since purified proteins are difficult to obtain in large quantities, the reactions are usually carried out at the submicromolar level. Very sensitive calorimeters must be used such as those developed by Gill and Seiblod [112] and Wadsö [114]. The information gained by such measurements helps in understanding the complex reactions between proteins and metal ions, proteins and hydrogen ions, and proteins and organic ligands.

Calorimetry is also used to study enzyme activity [128]. This is accomplished when we first determine the heat of a given reaction by using a calorimeter. The rate of heat production in the presence of an enzyme is then measured. This rate of heat production is proportional to the heat of reaction times the activity of the enzyme.

5.6   Nondestructive Determination of the Life of Products

Calorimeters are used in some specialized cases to determine the rate of decomposition of certain products. This information is extremely important in some cases. For example, it is important to know that a particular pacemaker battery that is to be implanted in a person will perform for the prescribed length of time. It is also important to know the shelf life of a specific drug or the deterioration rate of the paper in a book. This information can often be obtained using calorimetric techniques. Sensitive Calvet-type calorimeters were developed to measure the heat released by a battery in an open and a closed circuit [39]. The heat given off with an open circuit is a measure of the rate at which parasitic reactions are taking place and gives an estimate of how long the battery will last. Many battery companies use calorimeters to test their batteries. The shelf life of drugs can also be estimated by the use of these sensitive calorimeters [41]. Such tests can be done in a short time and in a nondestructive manner. A related proposed use for such sensitive calorimeters is the testing of different papers in different environments to estimate how long documents printed on the paper will last.

5.7   Determining the Concentration of Species in Complex Mixtures

Titration calorimeters are used to determine the quantity of a given species in solution [18, 19]. This procedure can be used as long as there is a specific reagent that will react quantitatively with the species in question and not with the other species in the mixture. The end point on a heat–time curve generated from a measurement using a titration calorimeter can give the concentration of a given species as long as the titrant delivery rate is known. The total heat of a calorimetric determination can give the species concentration as long as the heat of reaction is known. Eatough, Hansen, and co-workers [129–134] exploited this technique to determine specific species in airborne pollutants. They found calorimetry to be quicker, less expensive, and easier to use than other types of analytical procedures in determining the species present.

5.8   Determination of Reaction Kinetics

Many of the calorimeters mentioned previously are used to study the kinetics of chemical reactions [4, 5, 55–58, 69, 90]. A calorimeter with a fast time response relative to the half-life of the reaction can be used to find the reaction rate. The rate of heat production is proportional to the reaction rate; and if the heat of reaction is known, the heat–time curve can be used to find the rate of reaction. Jansson and co-workers [135] studied the rate of radical formation under emulsion polymerization conditions using a calorimetric system. Grases and co-workers [136] developed a calorimetric technique to determine certain inorganic species from the reaction rate measurements. This latter technique was tested by determining the amount of iodide in solution by measuring the initial rate of the iodide-catalyzed cerium(IV)–arsenic(III) reaction. They found the method gave results comparable to those found using other analytical methods.

5.9   Measuring the Enthalpies of Evaporation and Sublimation

In principle, calorimeters should be the instruments of choice for measuring the enthalpies of evaporation and sublimation since a direct measurement of the enthalpy change is made. However, to get accurate results great care must be taken in the design of the calorimeter and in data collection and reduction. A review of the experimental techniques and data handling has been written [137]. Since evaporation involves large volume changes, allowance must be made for the expansion or removal of some of the material both experimentally and mathematically. The fact that heat must be added creates the additional problem that significant temperature gradients may exist in the calorimeter. Parisod and Plattner [138] designed a calorimeter to measure the enthalpy of vaporization of water from salt solutions at temperatures to 723 K and pressures to 45 MPa. This calorimeter is isothermal and is contained in a high-pressure autoclave. The high-pressure, 1000-cm^3 cell is made of a titanium alloy containing a wound internal heater that adds the heat necessary to keep the temperature constant. The cell is filled about three-fourths full of liquid and is allowed to equilibrate about 12 h. The amount of heat added to compensate for the heat losses to the environment is then determined. Once this is done, a small amount of vapor is withdrawn resulting in an equivalent amount of liquid being evaporated. The vapor is condensed and weighed and the amount of energy needed to maintain the temperature constant during evaporation corrected for heat leaks is the heat of evaporation of the liquid. Natarajan and Viswanath [139] developed an adiabatic calorimeter to measure the vapor pressure and enthalpy of vaporization of certain coal chemicals. This calorimeter is a steady-flow recycle instrument. The vapor generated is condensed and the flow rate of the resulting liquid is measured. This condensate flows into a temperature-equilibrated reservoir from which it flows into the calorimeter where evaporation takes place, thus completing the cycle. This instrument was calibrated by measuring the heat of vaporization of benzene between 350 and 480 K. A pressure gauge in the vapor line was used to measure the equilibrium vapor pressure. Kusano [140] designed and used a heat conduction calorimeter to measure the heat of evaporation. The calorimeter is of the twin-cell Calvet type. Small glass sample vials containing the liquid are lowered into the two cells. Once thermal equilibration is obtained, a valve leading to a vacuum line is opened in the measurement vial. As the evaporation takes place, the differential voltage between the sample and reference cell thermopiles is recorded. The evaporation is terminated and the temperature is allowed to return to the base temperature. The area between the base-line and measurement voltage curves gives the heat absorbed by the evaporating fluid. The amount of liquid evaporated is determined gravimetrically. Use a heat conduction calorimeter to allow the measurement to be made on small amounts of liquid. Murata and co-workers [141] measured the heats of sublimation of solids using a Calvet-type calorimeter. Pithon and Rouyer [142] used a Setaram C 80 heat conduction calorimeter to measure heats of evaporation as well as vapor pressures and thermal conductivities. The heat of vaporization is measured in much the same manner as that done by Kusano [140]. Sváb and co-workers [143] developed an isothermal calorimeter for the measurement of enthalpies of evaporation at high temperatures (to 600 K) and pressures (to 3 MPa). It is an isothermal type with vapor withdrawal capability and controlled heat flow. Its design and principle of operation are similar to those of the calorimeter designed by Parisod and Plattner [138]. An interesting extension of the use of calorimeters to measure the heat of evaporation is that employed by Reading and Reisner [144]. These workers designed a calorimeter to measure the rates of evaporation of liquids. An open-top tray with a thermocouple embedded in it is placed in a thermostated wind tunnel. The liquid to be studied is poured into the tray and the temperature trace is recorded as evaporation occurs. The temperature curve generated during and after evaporation can be reduced to give the evaporation rate, provided the enthalpy of evaporation is known.

5.10   Determination of Fluid Phase Equilibria

Flow calorimeters can be used to determine the compositions of two fluid phases in equilibrium with each other. This is done in the cases of liquid–liquid equilibrium of a binary mixture [145], the solubility of a reactive gas in a liquid [146], and vapor–liquid equilibrium [147]. The phase compositions are obtained from enthalpy composition curves generated by mixing two fluids that exhibit two-phase behavior at the temperature and pressure of the measurement. The enthalpy composition curve in the two-phase region for a binary mixture is linear, as seen in Figure 7.24. The phase rule states that the compositions of the two phases in equilibrium with each other must remain

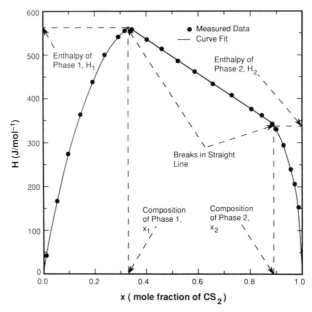

Figure 7.24   Plot of enthalpy against mole fraction of CS_2 for the binary $CS_2 + CH_3OH$ at 293.15 K and 0.531 MPa showing the compositions of the two immiscible phases [145]. The reference enthalpies are those for the pure components at the system temperature and pressure.

constant, independent of the overall composition of the mixture. The enthalpy of the two-phase portion of a binary mixture at constant temperature and pressure is given by the straight line relationship in (8)

$$H = x \left(\frac{H_1 - H_2}{x_1 - x_2} \right) + \frac{x_1 H_2 - x_2 H_1}{x_1 - x_2} \tag{8}$$

where H is the molal enthalpy of the overall mixture, H_1 is the molal enthalpy of phase 1, H_2 is the molal enthalpy of phase 2, x is the mole fraction of one of the components in the total mixture, x_1 is the mole fraction of that component in phase 1, and x_2 is the mole fraction of that component in phase 2. The two end points of the linear portion give the compositions of the two phases as seen in Figure 7.24. This method is convenient since sample removal and analysis are not needed.

5.11   Miscellaneous Uses

The uses of calorimetry are varied and limited only by the imagination of the scientists involved. An example of one use is the determination of the thermal expansion coefficient of a material. Pruzan and co-workers [148] developed a calorimeter that measures the thermal expansion of liquids at pressures to

1000 MPa at temperatures between 253 and 373 K. The calorimeter consists of well-insulated twin measurement and reference cylindrical cells immersed in the same liquid temperature bath. The measurement cell is constructed to withstand high pressures, whereas the reference does not have high-pressure capability. A differential gas thermometer capable of measuring temperature differences as low as 10^{-5} K is used to measure temperature differences between the measurement and reference cells. This small temperature difference is necessary since the heat capacity of the high-pressure cell is large. The measurement cell is filled with the fluid to be studied and is pressurized to the highest desired pressure, and the two cells are brought to the same temperature by adding heat to the appropriate cell. The pressure is then reduced in 10 to 20 MPa increments and the resultant temperature change is recorded. The resulting series of temperature changes are then converted to the corresponding thermal compressibilities by the appropriate equations.

A heat conduction calorimeter used to investigate the hydration of cement was developed by Prosen and co-workers at the National Institute of Standards and Technology [149]. The temperature range of this calorimeter is 275–348 K, and it can be used to make measurements on samples as small as 0.0001 kg over extended periods of time. This calorimeter allows the initial mixing of cement and water within the cell and has a fast time response so that acquisition of data during the early hydration period can be made. It has four chambers so that four experiments can be carried out simultaneously.

6   FUTURE NEEDS IN CALORIMETRY

This section includes some of our thoughts about what needs to be done in the future with calorimetry. A key point made in the Pimentel report [150] is that the number of new compounds produced by chemists is increasing at a rate that is faster than exponential. For example, in 1950 the number of known compounds was 2 million; in 1985, this number was nearly 9 million. One of the pressing needs is to characterize these compounds. Some of the properties that must be measured are thermodynamic quantities, such as heats of formation, heat capacities, and enthalpies of phase changes. The interaction of these compounds, where appropriate, with protons, metal ions, and other ligands must be studied. In most cases, these measurements are best performed in a calorimeter. The temperature and pressure ranges of all types of calorimetry must be extended. Many of the phenomena of interest occur at elevated temperatures and pressures or at very low temperatures and pressures. Some work was done at these extreme ranges, but much more must be done. Many of the measurements involve corrosive substances. More development is needed in the design and construction of calorimeters capable of resisting corrosion at high temperatures and/or with a variety of corrosive materials. Calorimeters have proven very useful in determining equilibrium phase compositions. They should also prove to be useful in measuring solubilities of solids when the dissolution or precipitation is associated with an enthalpy change. A calorimeter and a method must be developed for handling precipating solids, but it must avoid the problems of supersaturation and clogging.

In this chapter, many novel uses of calorimetry are described. The usefulness of calorimetry in making measurements describing chemical systems are demonstrated. It is anticipated that new uses for calorimetry will be developed as more investigators become aware of its power in solving and understanding chemical problems.

References

1. J. M. Sturtevant, "Calorimetry," in A. Weissberger and B. W. Rossiter, Eds., *Physical Methods of Chemistry*, Vol. 1, Wiley-Interscience, New York, 1971, Chap. 7.

2. W. R. Parrish, *Fluid Phase Equilib.*, **29**, 177 (1986).

3. E. Gmelin and W. Brill, *Thermochim. Acta*, **119**, 35 (1987).

4. L. G. Karlsen and J. Villadsen, *Chem. Eng. Sci.*, **42**, 1153 (1987).

5. L. G. Karlsen and J. Villadsen, *Chem. Eng. Sci.*, **42**, 1165 (1987).

6. D. J. Eatough, E. A. Lewis, and L. D. Hansen, "Determination of ΔH_R and K_{eq} Values," in K. Grime, Ed., *Analytical Solution Calorimetry*, Wiley, New York, 1985, Chap. 5.

7. I. Lamprecht, *Thermochim. Acta*, **83**, 81 (1985).

8. I. Wadsö, *Thermochim. Acta*, **88**, 35 (1985).

9. I. Wadsö, *Tibtech*, 45 (Feb. 1986).

10. R. M. Izatt, E. H. Redd, and J. J. Christensen, *Thermochim. Acta*, **64**, 355 (1983).

11. J. L. Oscarson, R. M. Izatt, and J. J. Christensen, *Thermochim. Acta*, **100**, 271 (1986).

12. G. D. Gatta, *Thermochim. Acta*, **96**, 349 (1985).

13. P. C. Gravelle, *Thermochim. Acta*, **96**, 365 (1985).

14. J. Rouquerol, *Thermochim. Acta*, **95**, 337 (1985).

15. L. D. Hansen and D. J. Eatough, *Thermochim. Acta*, **70**, 257 (1983).

16. W. Zielenkiewicz, E. Margas, and J. Hatt, *Thermochim. Acta*, **88**, 387 (1985).

17. A. Kuessner, *Thermochim. Acta*, **119**, 59 (1987).

18. G. A. Vaughan, *Thermometric and Enthalpimetric Titrimetry*, Van Nostrand-Reinhold, London, 1973.

19. H. J. V. Tyrrell and A. E. Beeser, *Thermometric Titrimetry*, Chapman & Hall, London, 1968.

20. R. Abrosetti, N. Ceccanti, and C. Festa, *J. Phys. E*, **16**, 265 (1983).

21. D. L. Martin, *Rev. Sci. Instrum.*, **46**, 1670 (1975).

22. C. Shin and C. M. Criss, *Rev. Sci. Instrum.*, **46**, 1043 (1975).

23. C. J. Wormald and J. M. Eyears, *J. Chem. Thermodyn.*, **19**, 845 (1987).

24. P. A. Baisden, P. M. Grant, and W. F. Kinard, *Rev. Sci. Instrum.*, **58**, 1937 (1987).

25. T. Kotoyori and M. Maruta, *Thermochim. Acta*, **67**, 35 (1983).

26. R. W. Carling, *Rev. Sci. Instrum.*, **49**, 1489 (1978).

27. H. Lenski and D. Böhler, *Rev. Sci. Instrum.*, **51**, 221 (1980).

28. R. Jochems, H. Dekker, and C. Mosselman, *Rev. Sci. Instrum.*, **50**, 859 (1979).

29. R. A. Winnike, D. E. Wurster, and J. K. Guillory, *Thermochim. Acta*, **124**, 99 (1988).

30. D. L. Raschella, R. L. Fellows, and J. R. Peterson, *Rev. Sci. Instrum.*, **50**, 1481 (1979).

31. R. Zimmermann, G. Wolf, and H. A. Schneider, *Colloids Surf.*, **22**, 1 (1987).

32. J. D. Clemens, S. Circone, A. Navrotsky, P. F. McMilland, B. K. Smith, and V. J. Wall, *Geochim. Cosmochim. Acta*, **51**, 2569 (1987).

33. R. D. Greiger, "High-Temperature Pressurized Solution Calorimeter. High-Temperature Thermodynamic Parameters for Aqueous Sodium Flouride and Cesium Chloride," Ph.D. Thesis, Purdue University, West Lafayette, IN, 1973.

34. K. D. Chen, "Thermodynamic Properties of Representative Electrolytes at High Temperatures," Ph.D. Thesis, San Diego State University, San Diego, CA, 1987.

35. A. Tian, *Bull. Soc. Chim. Fr.*, **33**, 427 (1923).

36. E. Calvet and H. Prat, *Recent Progress in Microcalorimetry*, translation by H. A. Skinner, Pergamon, London, 1963.

37. J. Suurkuusk and I. Wadsö, *Chem. Scr.*, **20**, 155 (1982).

38. P. F. Bunyan, *Thermochim. Acta*, **130**, 335 (1988).

39. L. D. Hansen and R. M. Hart, *J. Electrochem. Soc.*, **125**, 842 (1978).

40. L. D. Hansen, E. A. Lewis, and D. J. Eatough, "Instrumentation and Data Reduction," in K. Grime, Ed., *Analytical Solution Calorimetry*, Wiley, New York, 1985, Chap. 3.

41. L. D. Hansen, E. A. Lewis, D. J. Eatough, R. G. Bergstrom, and D. DeGraft-Johnson, *Pharm. Res.*, **6**, 20 (1989).

42. S. C. Mraw and O. J. Kleppa, *J. Chem. Thermodyn.*, **16**, 865 (1984).

43. S. C. Mraw, *Fuel*, **65**, 54 (1986).

44. P. L. Parlouër, J. Mercier, and Francis Pithon, *High Temp. High Pressures*, **17**, 261 (1985).

45. Z. Dermoun, R. Boussand, D. Cotten, and J. P. Belaich, *Biotechnol. Bioeng.*, **27**, 996 (1985).

46. J. C. van Miltenburg, G. J. K. van den Berg, and M. J. van Bommel, *J. Chem. Thermodyn.*, **19**, 1129 (1987).

47. Z. Tan, A. Yin, S. Chen, L. Zhou, F. Li, M. Cai, R. Pan, S. Li, and Y. Xia, *Thermochim. Acta*, **123**, 105 (1988).

48. E. F. Westrum, Jr., *Fluid Phase Equilib.*, **27**, 221 (1986).

49. S. B. Lang, M. Ikura, and J. Brunet, *Ferroelectrics*, **73**, 431 (1987).

50. M. J. M. Van Oort and M. A. White, *Rev. Sci. Instrum.*, **58**, 1239 (1987).

51. J. Zubillaga, A. Lopez-Echarri, and M. J. Tello, *Thermochim. Acta*, **92**, 283 (1985).

52. T. Grewer, *Thermochim. Acta*, **119**, 1 (1987).

53. J. Hakl, *Thermochim. Acta*, **81**, 319 (1984).

54. J. Hakl, *Thermochim. Acta*, **85**, 353 (1985).

55. D. I. Townsend and J. C. Tou, *Thermochim. Acta*, **37**, 1 (1980).

56. L. F. Whiting and J. C. Tou, *J. Therm. Anal.*, **24**, 111 (1982).

57. L. Hub, *Thermochim. Acta*, **85**, 361 (1985).

58. J. C. Leung, H. K. Fauske, and H. G. Fisher, *Thermochim. Acta*, **104**, 13 13 (1986).

59. P. F. Sullivan and G. Seidel, *Phys. Rev.*, **173**, 679 (1968).

60. J. E. Smaardyk and J. M. Mochel, *Rev. Sci. Instrum.*, **49**, 988 (1978).

61. O. S. Tanasijczuk and T. Oja, *Rev. Sci. Instrum.*, **49**, 1545 (1978).

62. C. W. Garland, *Thermochim. Acta*, **88**, 127 (1985).

63. S. Imaizumi, K. Suzuki, and I. Hatta, *Rev. Sci. Instrum.*, **54**, 1180 (1983).

64. J. D. Baloga and C. W. Garland, *Rev. Sci. Instrum.*, **48**, 105 (1977).

65. X. C. Jin, P. H. Hor, M. K. Wu, and W. C. Chu, *Rev. Sci. Instrum.*, **55**, 993 (1984).

66. Y. Sasuga, R. Kato, A. Maesono, and I. Hatta, *Thermochim. Acta*, **92**, 279 (1985).

67. C. C. Huang, J. M. Viner, and J. C. Novak, *Rev. Sci. Instrum.*, **56**, 1390 (1985).

68. G. M. Schmiedeshoff, N. A. Fortune, J. S. Brooks, and G. R. Stewart, *Rev. Sci. Instrum.*, **58**, 1743 (1987).

69. J. Schildknecht, *Thermochim. Acta*, **49**, 87 (1981).

70. L. D. Hansen, R. H. Hart, D. M. Chen, and H. F. Gibbard, *Rev. Sci. Instrum.*, **53**, 503 (1982).

71. G. Zhong, J. Shen, P. Guo, and J. He, *Netsusokutei*, **10**, 91 (1983).

72. G. Zhong, S. Ma, H. Zhao, J. Shen, and Y. Huang, *Thermochim. Acta*, **123**, 93 (1988).

73. R. E. Schwall, R. E. Howard, and G. R. Stewart, *Rev. Sci. Instrum.*, **46**, 1054 (1975).

74. I. Hatta, *Rev. Sci. Instrum.*, **50**, 292 (1979).

75. B. F. Griffing and S. A. Shivashankar, *Rev. Sci. Instrum.*, **51**, 1030 (1980).

76. S. Murakawa, T. Wakamatsu, M. Nakano, M. Sorai, and H. Suga, *J. Chem. Thermodyn.*, **19**, 1275 (1987).

77. M. S. Wire, Z. Fish, and G. W. Webb, *Rev. Sci. Instrum.*, **56**, 1223 (1985).

78. R. J. Trainor, G. S. Knapp, M. B. Brodsky, G. J. Pokorny, and R. B. Snyder, *Rev. Sci. Instrum.*, **46**, 1368 (1975).

79. J. J. Christensen, R. M. Izatt, and L. D. Hansen, *Rev. Sci. Instrum.*, **36**, 779 (1965).

80. J. J. Christensen, H. D. Johnston, and R. M. Izatt, *Rev. Sci. Instrum.*, **39**, 1356 (1968).

81. H. J. Buschmann, *Thermochim. Acta*, **102**, 179 (1986).

82. D. H. Smith and G. C. Allred, *J. Colloid Interface Sci.*, **124**, 199 (1988).

83. R. B. Spokane and S. J. Gill, *Rev. Sci. Instrum.*, **52**, 1728 (1981).

84. R. L. Berger, H. E. Cascio, N. Davids, C. G. Gibson, M. Marini, and L. Thiebault, *J. Biochem. Biophys. Methods*, **10**, 245 (1985).

85. N. Davids, R. L. Berger, and M. A. Marini, *J. Biochem. Biophys. Methods*, **10**, 261 (1985).

86. M. A. Marini, W. J. Evans, and R. L. Berger, *J. Biochem. Biophys. Methods*, **10**, 273 (1985).

87. I. Wadsö, *Thermochim. Acta*, **85**, 245 (1985).

88. G. A. Bishop, A. Parody-Morreale, C. H. Robert, and S. J. Gill, *Rev. Sci. Instrum.*, **58**, 632 (1987).

89. G. W. Stockton, S. J. Ehrlich–Moser, D. H. Chidester, and R. S. Wayne, *Rev. Sci. Instrum.*, **57**, 3034 (1986).

90. B. Grob, R. Riesen, and K. Vogel, *Thermochim, Acta*, **114**, 83 (1987).

91. M. O'Neil, R. Lovrien, and J. Phillips, *Rev. Sci. Instrum.*, **56**, 2312 (1985).

92. D. A. Kyser and R. I. Masel, *Rev. Sci. Instrum.*, **58**, 2141 (1987).

93. P.-C. Maria, J.-F. Gal, L. Elegant, and M. Azzaro, *Thermochim. Acta*, **115**, 67 (1987).

94. P. Picker, P. A. Leduc, P. R. Philip, and J. E. Desnoyers, *J. Chem. Thermodyn.*, 631 (1971).

95. D. Smith-Magowan and R. H. Wood, *J. Chem. Thermodyn.*, **13**, 1047 (1981).

96. R. H. Wood, D. E. White, J. A. Gates, H. J. Albert, D. R. Biggerstaff, and J. R. Quint, *Fluid Phase Equilib.*, **20**, 283 (1985).

97. H. Ogawa and S. Murakami, *Thermochim. Acta*, **88**, 255 (1985).

98. H. C. Zegers, R. Boegschoten, W. Mels, and G. Somsen, *Can. J. Chem.*, **64**, 40 (1986).

99. C. J. Wormald and T. K. Yerlett, *J. Chem. Thermodyn.*, **17**, 1171 (1985).

100. J. de A. Simoni and A. P. Chagas, *Thermochim. Acta*, **113**, 3 (1987).

101. J. J. Christensen, L. D. Hansen, D. J. Eatough, and R. M. Izatt, *Rev. Sci. Instrum.*, **47**, 730 (1976).

102. J. J. Christensen, L. D. Hansen, R. M. Izatt, D. J. Eatough, and R. M. Hart, *Rev. Sci. Instrum.*, **52**, 1226 (1981).

103. J. J. Christensen and R. M. Izatt, *Thermochim. Acta*, **73**, 117 (1984).

104. J. J. Christensen, P. R. Brown, and R. M. Izatt, *Thermochim. Acta*, **99**, 159 (1986).

105. J. D. Raal and P. A. Webley, *AIChE Journal*, **33**, 604 (1987).

106. S. Randzio and I. Tomaszkiewicz, *J. Phys. E*, **13**, 1292 (1980).

107. I. Tomaszkiewicz and S. Randzio, *J. Phys. E*, **18**, 92 (1985).

108. R. H. Busey, H. F. Holmes, and R. E. Mesmer, *J. Chem. Thermodyn.*, **16**, 343 (1984).

109. H. F. Holmes, R. H. Busey, J. M. Simonson, R. E. Mesmer, D. G. Archer, and R. H. Wood, *J. Chem. Thermodyn.*, **19**, 863.

110. H. Arntz and K. Gottlieb, *J. Chem. Thermodyn.*, **17**, 967 (1985).

111. K. Weber and H.-J. Hinz, *Rev. Sci. Instrum.*, **47**, 592 (1976).

112. S. J. Gill and M. L. Seibold, *Rev. Sci. Instrum.*, **47**, 1399 (1976).

113. S.-O. Nilsson and I. Wadsö, *J. Chem. Thermodyn.*, **18**, 1125 (1986).

114. I. Wadsö, *Thermochim. Acta*, **96**, 313 (1985).

115. S. F. Dec and S. J. Gill, *Rev. Sci. Instrum.*, **55**, 765 (1984).

116. I. W. Marison and U. von Stockar, *Thermochim. Acta*, **85**, 493 (1985).

117. K. Kobayashi, T. Fujita, and S. Hagiwara, *Thermochim. Acta*, **88**, 329 (1985).

118. H. P. Leiseifer and G. H. Schleser, *Z. Naturforsch.*, **38c**, 259 (1983).

119. J. J. Christensen, D. P. Wrathall, J. L. Oscarson, and R. M. Izatt, *Anal. Chem.*, **40**, 1713 (1968).

120. J. J. Christensen, D. P. Wrathall, and R. M. Izatt, *Anal. Chem.*, **40**, 175 (1968).

121. J. J. Christensen, J. H. Rytting, and R. M. Izatt, *J. Chem. Soc. A*, 47 (1969).

122. R. M. Izatt, D. J. Eatough, R. L. Snow, and J. J. Christensen, *J. Phys. Chem.*, **72**, 1208 (1968).

123. D. J. Eatough, *Anal. Chem.*, **42**, 635 (1970).

124. J. L. Oscarson, R. M. Izatt, P. R. Brown, Z. Pawlak, S. E. Gillespie, and J. J. Christensen, *J. Solution Chem.*, **17**, 841 (1988).

125. J. L. Oscarson, S. E. Gillespie, J. J. Christensen, R. M. Izatt, and P. R. Brown, *J. Solution Chem.*, **17**, 865 (1988).

126. D. P. Ip, G. S. Brenner, J. M. Brenner, J. M. Stevenson, S. Lindenbaum, A. W. Douglas, S. D. Klein, and J. A. McCauley, *Int. J. Pharm.*, **28**, 183–191 (1986).

127. I. Wadsö, *Pure Appl. Chem.*, **52**, 465 (1980).

128. L. Sica, R. Gilli, C. Briand, and J. C. Sari, *Anal. Biochem.*, **165**, 341 (1987).

129. L. D. Hansen, L. Whiting, D. J. Eatough, T. E. Jensen, and R. M. Izatt, *Anal. Chem.*, **48**, 634 (1976).

130. L. D. Hansen, B. E. Richter, and D. J. Eatough, *Anal. Chem.*, **49**, 1779 (1977).

131. D. J. Eatough, T. Major, J. Ryder, M. Hill, N. F. Mangelson, N. L. Eatough, L. D. Hansen, R. G. Meisnenheimer, and J. W. Fischer, *Atmos. Environ.*, **12**, 263 (1978).

132. C. M. Hilton, J. J. Christensen, D. J. Eatough, and L. D. Hansen, *Atmos. Environ.*, **13**, 601 (1979).

133. D. J. Eatough, J. J. Christensen, N. L. Eatough, M. W. Hill, T. D. Major, N. F. Mangelson, M. E. Post, J. F. Ryder, L. D. Hansen, R. G. Meisnenheimer, and J. W. Fischer, *Atmos. Environ.*, **16**, 1001 (1982).

134. D. J. Eatough and L. D. Hansen, *Sci. Total Environ.*, **36**, 319 (1984).

135. L. Jansson, H. Nilsson, C. Silvergren, and B. Törnell, *Thermochim. Acta*, **118**, 97 (1987).

136. F. Grases, R. Forteza, J. G. March, and V. Cerda, *Talanta*, **32**, 123 (1985).

137. V. Majer, *Fluid Phase Equilib.*, **20**, 93 (1985).

138. Ch. J. Parisod and E. Plattner, *Rev. Sci. Instrum.*, **53**, 54 (1982).

139. G. Natarajan and D. S. Viswanath, *Rev. Sci. Instrum.*, **54**, 1175 (1983).

140. K. Jusano, *Thermochim. Acta*, **88**, 109 (1985).

141. S. Murata, M. Sakiyama, and S. Seiki, *Thermochim. Acta*, **88**, 121 (1985).

142. F. Pithon and M. Rouyer, *Thermochim. Acta*, **114**, 91 (1987).

143. L. Sváb, L. Petros, V. Hynek, and V. Svoboda, *J. Chem. Thermodyn.*, **20**, 545 (1988).

144. C. A. Reading and A. Reisner, *J. Phys. E*, **10**, 1069 (1977).

145. T. A. McFall, M. E. Post, J. J. Christensen, and R. M. Izatt, *J. Chem. Thermodyn.*, **14**, 509 (1982).

146. S. P. Christensen, J. J. Christensen, and R. M. Izatt, *Thermochim. Acta*, **106**, 241 (1986).

147. D. R. Cordray, R. M. Izatt, and J. J. Christensen, *Sep. Sci. Technol.*, **22**, 1169 (1987).

148. P. Pruzan, L. T. Minassian, P. Figuiere, and H. Szwarc, *Rev. Sci. Instrum.*, **47**, 6 (1976).

149. E. J. Prosen, P. W. Brown, G. Frohnsdorff, and F. Davis, *Cem. Concr. Res.*, **15**, 703 (1985).

150. *Opportunities in Chemistry*, National Academy Press, Washington, DC, 1985.

Chapter **8**

DIFFERENTIAL THERMAL METHODS†

Juliana Boerio-Goates and Jane E. Callanan

†Partial contribution of the National Institute of Standards and Technology. Not subject to copyright.

Physical Methods of Chemistry, Second Edition Volume Six: Determination of Thermodynamic Properties   Edited by Bryant W. Rossiter and Roger C. Baetzold
ISBN 0-471-57087-7   Copyright 1992 by John Wiley & Sons, Inc.

1   INTRODUCTION

1.1   Introduction to Thermoanalytical Methods

Thermal analysis has been defined by the International Union of Pure and Applied Chemistry (IUPAC) [1] as "a group of techniques in which a physical property of a substance is measured as a function of temperature whilst the substance is subjected to a controlled temperature programme." More than 100 techniques were explored and developed that satisfy this description. Table 8.1 classifies the most common of these techniques according to the physical property measured and gives accepted abbreviations for the methods.

The 1970s and 1980s brought an explosive growth in the use of thermal

Table 8.1   Classification of Thermoanalytical Methods[a]

Physical Property	Derived Technique(s)	Abbreviation
Mass	Thermogravimetry	TG
	Evolved gas detection	EGD
	Evolved gas analysis	EGA
Temperature	Differential thermal analysis	DTA
Power	Differential scanning calorimetry	DSC
Mechanical characteristics	Thermomechanical analysis	TMA

[a]Adapted with permission from W. W. Wendlant and P. K. Gallagher, "Instrumentation," in E. Turi, Ed., *Thermal Characterization of Polymeric Materials*, Academic, New York, 1981, pp. 1–90.

analysis methods in both research and routine applications in industry and academia. This chapter discusses three of the most commonly used techniques in this field: differential scanning calorimetry (DSC), differential thermal analysis (DTA), and thermogravimetry (TG). The first two techniques involve the measurement of the heat flux, either directly through power measurements or indirectly through temperature measurements, associated with some physical or chemical process of interest. As the two names imply, a differential heat measurement is made: A reference sample is subjected to the same conditions as the material of interest, and the difference between the reference and unknown is measured. The process is usually studied while a programmed temperature change is applied to the system, although isothermal measurements are also possible. In the third technique, mass changes in a sample are followed, again either in a temperature-scanning or isothermal mode.

In Figure 8.1 is given a stylized representation of a DTA or DSC curve, plotted either as the difference between sample and reference temperatures, $T_S - T_R$, or the differential power necessary to keep the sample and reference at the same temperature, versus temperature. The various phenomena that can be detected by these techniques are illustrated in this diagram with the characteristic shapes of the features associated with them. An endothermic process arising from fusion or from a first-order solid–solid phase transition typically appears as a sharp peak, shown in Figure 8.1a. (Endothermic processes are taken by informal convention to extend below the base line; figures reproduced here that use a different convention are noted.) Chemical processes such as rearrangements, polymerizations, or decompositions occur over a wider temperature range than first-order phase transitions and appear as a broad deflection from the base line (Figure 8.1b). An exothermic transition arising from a change from

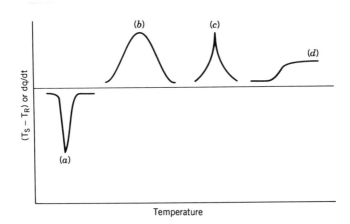

Figure 8.1   Stylized representation of DTA and DSC curves, plots of the temperature difference between sample and reference $T_S - T_R$ or differential power dq/dt versus temperature, for different thermal processes: (a) endothermic phase transition, (b) exothermic chemical reaction, (c) exothermic phase transition, and (d) glass transition.

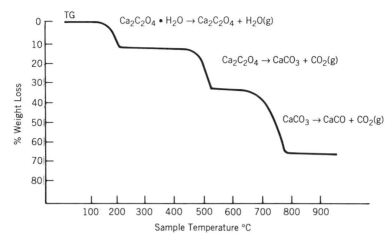

Figure 8.2 Thermogravimetric curve for the dehydration and decomposition of calcium oxalate monohydrate.

a metastable crystalline phase to the stable solid phase is shown in Figure 8.1c. Finally, a glass transition, which appears as a shift in the horizontal base line and which arises from the onset of translational and/or rotational motion in highly disordered materials, is presented in Figure 8.1d. The same types of curves are obtained from either a DSC or DTA instrument; however, the dependent variable recorded by the instrument and plotted as the ordinate of Figure 8.1 varies. The area of the peak is related to the enthalpy of the process associated with each feature.

Chemical or physical processes that result in a mass change of the sample are suitable for study by thermogravimetry. Figure 8.2 shows the thermogravimetric curve for the dehydration and decomposition of calcium oxalate monohydrate. The temperature range as well as the number of molecules evolved during each process can be determined from a TG plot.

1.2   Scope and Organization

The techniques of DTA and DSC are comparable in many respects. These methods differ in the primary physical property measured, but provide similar information, for example, temperatures and enthalpies of transitions. Differential scanning calorimetric measurements are generally limited to temperatures below 1000 K; DTA instruments can go to higher temperatures, but are less accurate for quantitative measurements. Many of the experimental factors affecting the quality of the data are analogous, and we treat the topics of DSC and DTA together. The discussion of TG is done as a separate part of the chapter. Within each part are sections on theory; experimental procedures, including methods of data reduction; and important experimental parameters.

Applications of differential scanning calorimetry and differential thermal analysis (DSC/DTA) and TG methods are diverse, and to cover them exhaustively is beyond the scope of this chapter. They can be summarized as shown in Tables 8.2 and 8.3 [2], respectively. We describe several of the general categories of applications of the DSC/DTA and TG methods shown in these tables within the respective sections. Although applications are specific to either DSC/DTA or TG, the complete resolution of many scientific or technical problems is best accomplished when combination studies with multiple techniques are used. Several of the most common combination techniques are discussed.

A separate section on the application of these methods to kinetics is presented describing the numerous methods of analysis and their limitations. Finally, we conclude the chapter by highlighting specific examples from various scientific disciplines that illustrate the potential of these thermal methods.

The surge in popularity of differential thermal methods has arisen from the appearance of a wide range of commercial units spanning temperature intervals of greatest interest. Few people starting out in the field are concerned with the design and construction of a home-built instrument. Therefore, the emphasis in this chapter is to equip a newcomer in the field with the information necessary to investigate thermal phenomena using a commercial unit: a basic, but not rigorous development of the underlying theoretical principles of the techniques; the important experimental variables; the experimental and calculational procedures; and a review of some applications. Readers with more specialized interests who require extensive information about design principles may wish to consult several of the more comprehensive sources mentioned below.

Table 8.2   Applications of Differential Scanning Calorimetry and Differential Thermal Analysis[a]

Process Enthalpies	Thermophysical Property Measurement	Materials Characterization
Sublimation	Heat capacities	Phase diagrams
Vaporization	Thermal conductivity	Hazards evaluation
Adsorption		Kinetics
Dehydration		"Fingerprint"
Polymerization		identification of
Chemical reaction		compounds
Structural phase transition		
Magnetic phase transition		
Order–disorder phase transition		
Glass transition		
Conformational changes in molecules		

[a]Adapted from W. W. Wendlant and P. K. Gallagher, "Instrumentation," in E. Turi, Ed., *Thermal Characterization of Polymeric Materials*, Academic, New York, 1981, pp. 1–90.

Table 8.3   Applications of Thermogravimetry[a]

Thermal Stabilities and Chemical Reactions

Decompositions of inorganic, organic, and polymeric substances
Corrosion of metals in various atmospheres at high temperatures
Roasting and calcining of minerals
Pyrolysis of coal, petroleum, and wood

Physical Property Measurements

Distillation and evaporation of liquids
Rates of evaporation and sublimation
Vapor pressure determinations and heats of sublimation
Curie temperature and magnetic susceptibility
Adsorption and desorption studies

Quantitative Analysis

Determination of moisture, volatiles, and ash content
Dehydration and hygroscopicity studies
Development of gravimetric analytical procedures

[a]Adapted from W. W. Wendlant and P. K. Gallagher, "Instrumentation," in E. Turi, Ed., *Thermal Characterization of Polymeric Materials*, Academic, New York, 1981, pp. 1–90.

1.3   Sources of Information on Differential Thermal Analysis, Differential Scanning Calorimetry, and Thermogravimetry

Numerous review articles, books, and chapters of books have been written on these techniques. Several monographs, each having a different focus, have been published. The book by Wendlandt [3] covers a wide variety of thermoanalytical techniques including DTA, DSC, and TGA, with numerous citations to the literature. The monographs by Šesták [4] and Hemminger and Höhne [5] are excellent introductions to DTA and DSC. The Šesták book takes a more theoretical approach and highlights the common ground of the two fields, while the Hemminger and Höhne book concentrates on the calorimetric aspects of the thermal analysis field and features a section with details about many commercial instruments. Brown [6] gives an introduction to a broad spectrum of thermal techniques, including several that are not discussed in this chapter.

Biennial reviews of the thermal analysis field are published [7–9] in *Analytical Chemistry*. In its "A" pages *Analytical Chemistry* also publishes occasional feature articles that emphasize current instrumental capabilities in the field (e.g., see [10]). Numerous reviews dealing with applications of these methods to more specialized fields like coal [11], polymers [12], and catalysis [13] are published; others are mentioned in the last section of this chapter. The

manufacturers of many commercial instruments maintain extensive bibliographies of applications and provide them on request.

Two international journals, *Thermochimica Acta* and *Journal of Thermal
Analysis*, are devoted exclusively to research in the various subdisciplines of
thermal analysis and have many articles dealing with these three topics.
According to [14], however, only about 20% of the articles published on thermal
analysis appear in these two journals. That the remainder are found in other,
diverse, publications indicates the widespread interest and applicability of these
methods. It was estimated in 1972 that more than 1000 publications dealing
specifically with thermal analysis appear each year, and there has been a steady
increase since then [15]. A current trend is to incorporate one or more of these
techniques as a small but integral part into a larger research project by scientists
not trained specifically in thermal analysis or calorimetry; these papers may not
even be detected by those counting the numbers of publications.

Several professional associations have been organized to further the development of thermal techniques. These include the International Confederation for
Thermal Analysis (ICTA), whose activities include the development of standards
for nomenclature, experimental procedures, and reference materials; the North
American Thermal Analysis Society (NATAS), which holds annual conferences
at which a wide variety of thermal analysis topics are presented; the American
Society for Testing and Materials ASTM-E37 Committee on Thermal Methods;
and the Calorimetry Conference, at whose annual meeting applications of
scanning calorimetry are presented frequently. Additional organizations exist in
Japan and Europe.

1.4   Commercial Instrumentation

1.4.1   Manufacturers

The major manufacturers (US offices or distributors where available) of thermal
analysis equipment and a general summary of their instrument capabilities are
listed in Table 8.4. Because manufacturers constantly update their product lines,
the information cited here should be taken only as a preliminary guide. Up-to-
date information concerning manufacturers of thermal analysis equipment can
be found in the annual buyers' guides of *Analytical Chemistry, American
Laboratory, Science,* and *Physics Today*. Descriptions of various commercial
units can be found in [2, 3, 5, 6].

1.4.2   Choosing Thermal Analysis Equipment

The most important question to answer when considering the purchase of
thermal analysis equipment is the scope of its use. Will the instrument be used
for repetitive, quality control analyses or for research and development work
where the range of applications is broad and high-quality, high-accuracy results
are desired? If the first is the case, an instrument capable of multiple-sample
measurements may be desirable. When instrument flexibility is necessary, the
ability to operate with a variety of sample sizes, over a wide range of

Table 8.4 Manufacturers of Differential Thermal Analysis Equipment[a]

Manufacturer	Product Line
Cahn–Ventron 16207 South Carmenita Road Cerritos, CA 90701	TG
Hart Scientific 220 North 1300 West Pleasant Grove, UT 84602-0435	DSC
Mettler Instrument Corporation Princeton–Hightstown Road Box 71 Hightstown, NJ 08520	TG, DSC, DTA, TG–DSC
Netzch Incorporated Thermal Analysis Division 119 Pickering Way Exton, PA 19341-1393	DSC, DTA, TG–DTA–DSC, TG–DTA–EGA
Perkin–Elmer Corporation Analytical Instruments Main Avenue (MS-12) Norwalk, CT 06856 USA	DSC, DTA, TG
PL Thermal Sciences 300 Washington Boulevard Mundelein, IL 60060	DSC, TG–DSC, TG–DTA–MS, DSC, TG
Rigaku–Denki Company 9-8, 2-Chome Sotokanda, Chiyoda-Ku Tokyo 101, Japan	DSC, DTA, TG, TG–DTA, TG-DSC
Seiko Instruments USA, Inc. 2990 West Lomita Boulevard Torrance, CA 90505	DSC, TG–DTA
Setaram Astra Scientific International, Inc.[b] 1961 Concourse Drive San Jose, CA 95131	DSC, DTA, TG, TG–DSC
Shimadzu Scientific Instruments, Inc. 7102 Riverwood Drive Columbia, MD 21046	DSC, TG, DTA
T.A. Instruments, Inc. 109 Lukens Drive New Castle, DE 19720 (formerly DuPont Instruments)	TG, DSC, DTA
Ulvac Sinku–Riko, Inc. P.O. Box 799 105 York Street Kennebunk, ME 04043	DSC, TG, DTA

[a]Certain commercial equipment, instruments, or materials are identified in this chapter to specify the theory and the experimental procedures adequately. Such identification does not imply recommendation or endorsement by Brigham Young University or by the National Institute of Standards and Technology, nor does it imply that the materials or equipment identified are necessarily the best available for the purpose.
[b]United States distributor.

temperature and pressure regions, and in isothermal or scanning modes are important factors.

Most commercial instruments are highly automated, both in the acquisition and analysis of data, and so another question of importance concerns the quality and flexibility of the programs supplied with the instrument. Manufacturers emphasize in their product literature the ease of operation, the ability of internal programs to calibrate data as they are collected, and to perform calculations on the data with little operator intervention. Although such systems significantly reduce operator labor involved in producing data, care must be taken to ensure that the data actually represent what the operator believes they do. This requires that the operator have access to information concerning the algorithms used to determine transition temperatures, enthalpies, heat capacities, and kinetic parameters. In a research environment, the operator may also desire the capability to override or modify an algorithm or have access to the raw data for use in user-developed routines.

2   DIFFERENTIAL THERMAL ANALYSIS AND DIFFERENTIAL SCANNING CALORIMETRY

2.1   Historical Development of Differential Thermal Analysis and Differential Scanning Calorimetry

Advances in the ability to measure the two primary quantities, temperature and power, foreshadowed each development in the thermal methods. Mackenzie [16, 17] published accounts of the history of the development of temperature measurement, furnace design, and so on. We merely summarize here the highlights. Theoretical and practical details of current methods of temperature measurement are given in Chapter 6.

Many historical accounts cite the first thermal analysis experiments as those of Le Chatelier in 1887 on clays [18–21], but Mackenzie [17] accords that honor to Rudberg [22] who measured inverse cooling-rate data for several metals and alloys. The experiments consisted of measuring the time taken for the temperature of a sample suspended inside a large vessel to fall by successive intervals of 10 K. The apparatus, which was covered with snow to ensure a constant external temperature, allowed a controlled temperature program to be established and produced freezing temperatures that were surprisingly accurate. Investigations were limited, however, to narrow temperature regions.

Although Le Chatelier's experiments were conducted some 18 years after Rudberg's, Le Chatelier was apparently unaware of the work and arrived at a different experimental arrangement independently [17]. Also, Le Chatelier was responsible for significant improvements in thermocouple design that made the thermocouple a viable temperature measuring device [23].

Roberts-Austen [24, 25] made the first differential thermal measurements by measuring the temperature difference between a sample and reference material placed side by side in the chamber. The device used Pt/Pt–Ir thermocouples connected in opposition with the output taken to a galvanometer and had Pt as the reference material.

In the early designs, instrumental parameters such as heating rate, sample size, nature of the sample holder, and so on were generally ignored. Reference materials were not chosen with any particular care. Applications were qualitative rather than quantitative and primarily involved the characterization of geological and inorganic substances. By 1920, many of the basic advances in the field had taken place and the next decades saw both a refinement in techniques and an increase in the breadth of the applications of the method. The Mackenzie article [17] gives an exhaustive account of these developments.

In a classic paper published in 1955, Boersma [26] reported the next significant advance in the technique. He revised what was then the current practice of immersing the thermocouples directly in the sample and reference by attaching the thermocouples in good thermal contact to the holders containing the sample and reference. This change greatly improved reproducibility and made the technique more quantitative. All modern commercial DTA instruments employ some varient of the Boersma design for placement of the thermocouples.

The first differential scanning calorimeter (DSC) was developed by Watson and co-workers [27] and employed the power-compensation technique. In this DSC temperature is measured with platinum resistance thermometers, and separate heaters maintain the sample and reference at the same temperature. The dependent variable recorded during an experiment is the differential power necessary to keep the two heating or cooling at the same rate. Truly quantitative measurements of both heat capacity and phase transition enthalpies became possible for the first time. The Perkin–Elmer Corporation patented this concept in the United States and is the primary supplier of commercial power-compensated DSC instruments in this country.

Another type of calorimeter, which is referred to in the literature and by manufacturers variously as a DSC, a heat-conduction calorimeter, or a heat-flux calorimeter, was developed by Tian and Calvet (see [28]). In this calorimeter, multiple thermocouples wired in series, called a *thermopile*, connect the sample and a large block enclosing the sample. Also present in the block is an identical chamber with thermopile connections for the reference material. The differential output resulting when the thermopiles are connected in opposition is related to the heat flow between sample and reference and provides a means to obtain quantitative measurements of both heat capacities and enthalpies.

The amazing proliferation of commercial instruments and the concomitant increase in the range of users and applications can be attributed to the great breakthroughs in the electronics industry, both in terms of improved components for the instruments and the greater ease of operation afforded by the addition of computer control of the temperature programming, data acquisition, and data reduction.

2.2   Description of the Instruments and Nomenclature

Figure 8.3 [29] gives schematic illustrations of the fundamental types of instruments that have been developed. A DTA instrument, according to the current nomenclature recommended by IUPAC [1], is one that uses a single thermocouple to measure the temperature difference between a reference and sample. In Figure 8.3a is shown the "classical" DTA arrangement in which a single heat source is used to change the temperature of both the reference and sample simultaneously, and the junctions of the single differential thermocouple are imbedded directly in the two specimens. The Boersma DTA arrangement shown in Figure 8.3b also employs a single heater and a single differential thermocouple, but the thermocouple junctions are placed outside the specimen holders, in good thermal contact with them.

During an experiment, both the sample and reference are subjected to a controlled program of temperature change by applying power to the block heater. When an endothermic phase change or chemical reaction occurs in the sample, the heat absorption causes the sample temperature T_S to lag behind that of the reference T_R and $\Delta T = T_S - T_R$ become negative. The reverse occurs for an exothermic process: The heat generated on the sample side causes T_S to increase above T_R and ΔT is positive. Even during a temperature interval in which no

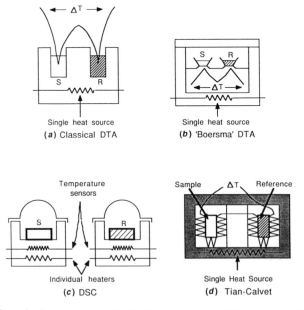

Figure 8.3   Schematic illustrations of the fundamental types of instruments. Thermocouples measure the temperature difference ΔT between sample S and reference R. Parts a, b, and c are reprinted with permission from W. P. Brennan and A. P. Gray, *Thermal Analysis Application Study 9*, Perkin–Elmer Corporation, Instrument Division, Norwalk, CT 06856, 1973.

chemical or physical changes occur, a slight temperature imbalance will develop because of differences in thermal conductivity, heat capacity, and emissivity of the two materials as well as small instrumental deviations.

The output of the differential thermocouple is recorded either as a function of time or as a function of sample temperature. In this latter instance, an additional temperature sensor must be placed in the sample chamber.

The relation of the magnitude of ΔT to units that will permit evaluation of enthalpies or heat capacities depends in a complex fashion on the total thermal resistance to heat flow R, which is temperature dependent. The thermal resistance also varies with the thermal conductivity of the sample and with the degree of thermal contact between the sample and its holder. For this reason, DTA experiments in which quantitative results are desired require careful experimental technique.

In DTA systems, the dependent variable recorded during an experiment is ΔT. In heat-flux DSC instruments, ΔT is brought out electronically as a differential power. Most commercial DTA (and many heat-flux) instruments currently use some variation of the Boersma arrangement.

In the power-compensated DSC arrangement shown in Figure 8.3c, separate heaters are employed for the sample and reference holders. The temperatures of the two holders are compared and the results of the comparison are used to control the power applied to the two heaters. Power to the sample side is increased when the sample temperature falls below that of the reference, as occurs during an endothermic phase transition or chemical reaction. During an exothermic process, power to the reference side is increased. In this manner, the system works to keep the sample and reference at their programmed temperatures. The differential power output, designated dH/dt or dq/dt, necessary to maintain this condition is the dependent variable recorded during a power-compensated DSC experiment.

The advantages of this system over the DTA arrangement are (1) a significant reduction in the thermal resistance of the sample and (2) the ability to produce quantitative information that can be less dependent on sample characteristics and temperature scan rates.

In the Tian–Calvet [see 28] calorimeter (Figure 8.3d), multiple junction thermocouples are used to measure the total heat flux between the sample and a massive block, the temperature of which is controlled. An identical thermopile measures the heat flux between the reference chamber and the block. The output from the two thermopiles connected in opposition is related to the differential heat flux between the sample and reference.

Theoretical analysis shows that similar equations result for instruments employing single or multiple thermocouples placed outside the sample container: The thermocouple connection provides a defined, reproducible pathway for heat conduction and enables measurements related to the differential heat flux between sample and reference chambers to be made. The IUPAC recommendation [1] is to refer to both instruments as *heat-flux calorimeters*.

The term DSC is applied by manufacturers to instruments of both the

Boersma and Tian–Calvet types where the actual measurement is still of a temperature difference between sample and reference. Several instruments are capable of being operated in a DTA or DSC mode. In DTA mode, the output is recorded as a temperature difference; in DSC mode, electronic manipulations are performed internally on the temperature difference, and the output is given as a power difference. This nomenclature is summarized in Table 8.5.

2.3   Theoretical Analysis of Differential Thermal Analysis and Differential Scanning Calorimetry

2.3.1   Principles of Heat Transport

Three types of mechanisms are responsible for the transfer of energy in the calorimetric systems under consideration here: conduction, radiation, and convection. Conductive heat transfer occurs by the transfer of vibrational energy between solid bodies and involves no mass transport. This is the primary means of heat transfer in the systems of interest, especially at low temperatures. Radiative heat transfer occurs by the net absorption of electromagnetic radiation emitted by a hot body and absorbed by a cold one. The contribution to the net heat transport in a calorimetric system from radiative processes increases at high temperature. In convection, energy transfer occurs through molecular collisions of the atmosphere or bath with the walls of the calorimetric system.

The detailed treatment of heat conduction [30] in anisotropic bodies like calorimetric systems is beyond the scope of this chapter. However, several important considerations for both calorimeter and experimental design can be achieved with a less sophisticated analysis. (The treatment presented here follows the development of [5].)

Table 8.5   IUPAC Recommendations for DTA/DSC Nomenclature[a]

Nomenclature	Characteristics of Instrument
DTA	ΔT is measured between sample and reference
	Has a single differential thermocouple
	Output is recorded in units of temperature
DSC	Power compensation maintains sample and reference at the program temperature
	Output is the differential power
Heat-flux	ΔT exists between sample and reference
DSC	Has a differential thermocouple or thermopiles
	Output is differential power related to ΔT

[a]Adapted with permission from R. C. Mackenzie, *Pure Appl. Chem.*, **57**, 1737 (1985). Copyright © 1985 International Union of Pure and Applied Chemistry.

In a simplified, one-dimensional analysis, the heat flux dq/dt across a temperature differential can be represented by

$$\frac{dq}{dt} = A \cdot \kappa \cdot \frac{dT}{dx} \tag{1}$$

where A is the cross-sectional area of the body and dx is the thickness of a slab over which there is the temperature gradient dT. A constant thermal conductivity κ is assumed. A second differential equation relates the temperature gradient to the time rate of change of temperature

$$\frac{\partial^2 T}{\partial x^2} = \frac{\rho \cdot C_p}{\kappa} \left(\frac{\partial T}{\partial t} \right) \tag{2}$$

where ρ is the density of the material in g/cm^3 and C_p is the specific heat capacity in $J/K \cdot g$.

Selection of a particular solution of (2) allows a solution for (1) to be determined. An important effect of heat flow by conduction can be illustrated by consideration of the following simple example. In Figure 8.4a is shown a bar of material having some constant thermal conductivity κ whose temperature can be measured at various positions x as a function of time $tT(x, t)$. If the bar is brought in contact with a block having an infinitely high thermal conductivity and constant temperature T_B at some initial time t_0, the heat flux density $(1/A) \cdot dq/dt$ through cross-sectional areas of the bar at later times has the forms given in Figure 8.4b for two positions x_0 and x_1 within the bar.

The result to be noted from Figure 8.4b is that the instantaneous event of bringing the two bodies, block and bar, together produces a response that is prolonged and broadened. At x_0 there is a gradual decrease in the heat-flux density with time, while at positions x farther along the bar there is a gradual increase in heat flux to a maximum, followed by a gradual decline. This result is a general property of heat conduction and means that the instantaneous generation or absorption of heat in the sample cell will produce a smeared-out response in the detector. The deviation of the measured signal from the true instantaneous heat flow depends on the thermal conductivities, heat capacities, and rate of the thermal event. It can be quite large when fast reactions are measured in calorimetric systems with slow response times and is a concern when kinetic measurements are being made.

For the analyses to be presented in Section 2.3.2, we must also consider the exchange of heat that takes place at the boundaries of two solids in addition to that which takes place within a solid. We approximate the heat-flux relationship as follows

$$\frac{dq}{dt} = \lambda \cdot \Delta T \tag{3}$$

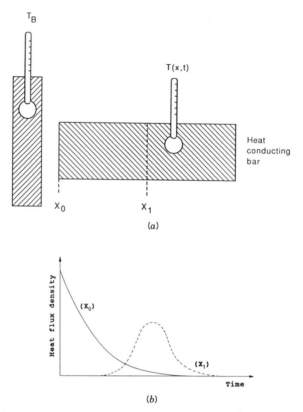

Figure 8.4 Modeling of the effect of thermal conductivity on heat flux for an idealized system (*a*) consisting of a bar with finite, constant heat conductivity in contact with an infinite-conductivity heat sink at a constant temperature T_B, and (*b*) calculated heat flux density at positions x_0 and x_1 in the bar as a function of time. Reproduced with permission, VCH Publishers, Inc. 220 East 23rd St., New York, NY 10010 from W. Hemminger and G. Höhne: *Calorimetry: Fundamentals and Practice*, 1984, pp. 68–69.

where λ is the thermal conductance, the inverse of the thermal resistance R connecting two bodies whose temperature difference is given by ΔT.

Two other mechanisms for heat transfer, radiation and convection, provide additional contributions and are discussed only briefly here.

All matter emits infrared (IR) radiation, the frequency spectrum of the radiation being dependent on the temperature of the body, its emissivity, and some geometrical parameters. A net transfer of energy occurs through radiation between two bodies of unequal temperature: The hotter body emits radiation at a higher rate than the colder body because of its increased temperature. The cold body, emitting radiation itself, will still experience a net increase in energy because it absorbs radiation emitted by the hot body.

The net heat exchange by radiation dq_r/dt can be represented as

$$\frac{dq_r}{dt} = 4 \cdot \sigma_B \cdot A_1 \cdot A_2 \cdot K_{12} \cdot (T_2^4 - T_1^4) \qquad (4)$$

where σ_B is the Stefan–Boltzmann constant, A_1 and A_2 are the surface areas of bodies 1 and 2, K_{12} is an empirical coefficient, and the T values are the temperatures of the two bodies. When $(T_2 - T_1)$ is small, as is usually the case in thermal analysis instruments, (4) can be simplified, as shown in (5)

$$\frac{dq_r}{dt} = l \cdot (T_2 - T_1) \cdot T^3 \qquad (5)$$

where σ_B, the areas, and K_{12} have been incorporated as a single proportionality constant l. From (5) we see that the radiative contribution to the net heat flux, being proportional to the cube of the temperature, is a strongly increasing function of temperature.

Heat exchange by convection occurs through the transport of heat by a flowing gas or liquid from a region of high temperature to one of lower temperature. In the systems of interest here, convective heat transfer may occur through flowing purge gases or pressure-generating fluids that may experience thermal gradients as they pass through the apparatus. A change of purge gas or its flow rate can cause variations in the extent of heat exchange by convection. Thus, the conditions under which calibration measurements are made must be reproduced as closely as possible in the rest of the experiments.

2.3.2 Theoretical Analysis

Theoretical treatments of DTA and DSC have been published at various levels of sophistication. The most complex analyses have usually been geared to the details of specific instruments and are not particularly useful for the general reader. Of the more general approaches, two common lines of attack appear: (1) the assumption of a homogeneous temperature distribution in the calorimetric sample, and (2) the complete incorporation of the effects of temperature gradients in the sample on the resulting heat flux. Within those divisions are also variations concerning assumptions about radiation and convection and about direct conduction between sample and reference.

Our approach here will be, first, to give a treatment of DTA based on the presentations of Gray [31] and Šesták and co-workers [32]. Although this treatment applies strictly to the classical DTA in which the thermocouples are inserted directly in the sample, comparison of the equations with those developed by Baxter [33], which incorporate the additional thermal resistance between sample and sample holder, shows that similar relationships result. We neglect radiative and convective contributions as well as temperature gradients in the sample in the detailed analysis, but then indicate the results of the more complete models that make fewer assumptions. The theoretical basis of the power-compensated technique is then developed according to the paper of Gray [31].

Other theoretical analyses of interest include the following: the first rigorous study of DTA incorporating temperature gradients by Ozawa [34]; studies on power-compensated DSC by O'Neill [35] and by Flynn [36]; studies on the heat-flux calorimeter by Calvet and Prat [28] and by Tanaka [37]; studies on the disk-type, heat-flux calorimeter by Baxter [33], P. Claudy and co-workers [38–40], and Lee and Levy [41]; and a treatment by Mraw [42] that uses a single model for both DTA and DSC. Many additional references are found in the citations of these papers.

The assumptions to be incorporated in the theoretical treatments of both the power-compensated and heat-flux methods are (1) heat exchange occurs only by conduction; radiative and convective effects are ignored; (2) the heat capacities and thermal resistances of the sample and reference are independent of temperature; and (3) there are no temperature gradients within the sample or reference.

Representations of two commercial units, one that uses the Boersma thermocouple placement and one that uses the Tian–Calvet design are given in Figures 8.5a and b, respectively. The two designs are similar in that both provide a heat-conduction path between the sample and reference chambers and the block, the temperature of which is controlled. In Figure 8.5a the constantan disk serves both as one leg of the measuring thermocouple and as the conduction path between cell and block. In Figure 8.5b the multijunction thermopile provides the heat conduit between cell and block. The temperature-measuring device is the primary difference between the two types of instrument. The Tian–Calvet arrangement has the advantage, through the averaging ability of the thermopile, of providing an accurate average temperature difference while instruments with a single thermocouple junction must ensure good thermal conductivity throughout the sample to obtain the average temperature difference. This design also allows for the heating block to encircle the sample and the reference holders more completely and symmetrically. The calorimetric response time tends to be longer for Tian–Calvet instruments, however, and kinetic studies of fast reactions might require deconvolution of the signal for adequate results.

In Figure 8.5c is shown a schematic of the arrangement found in DTA and heat-flux calorimeters that can be applied in the analysis of both types of instruments. The sample and reference compartments are in contact with the block, which is at temperature T_B through thermal resistances R_S and R_R, respectively. The actual sample and reference are in thermal contact with the two holders through thermal resistances R_{SC} and R_{RC}, respectively. In this analysis, it is assumed that R_{SC} and R_{RC} are effectively zero, which means that the temperatures of the sample and reference materials T_S and T_R are equal to that of the respective holders T_{SC} and T_{RC} and that the thermal resistances between the

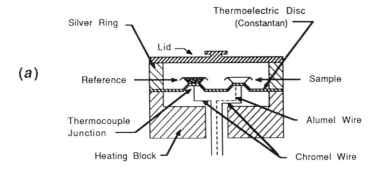

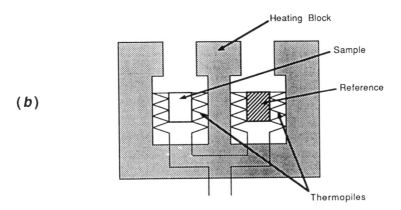

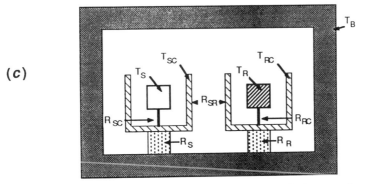

Figure 8.5   Representations of two heat-flux calorimeters showing (*a*) the Boersma thermocouple placement and (*b*) the Tian–Calvet design. (*c*) Schematic diagram appropriate for analysis of the response of both types of calorimeters. The temperature of the block, sample, sample holder, reference, and reference holder are given by T_B, T_S, T_{SC}, T_R, and T_{RC}, respectively. The thermal resistances to heat flow between two surfaces are designated by R values. Part (*a*) has been reprinted with permission from R. A. Baxter, "A Scanning Microcalorimetric Cell Based on a Thermoelectric Disc—Theory and Applications," in R. F. Schwenker and P. D. Garn, Eds., *Thermal Analysis*, Vol. 1, 1969, p. 75.

holders and block govern the heat fluxes between these elements of the system. The thermal resistance R_{SR} provides a heat-conduction path between the sample and reference holders. In some systems this may occur through radiative as well as conductive paths [41].

The heat balance equations for the two holders can be written as

$$C_{P_S}\frac{dT_S}{dt} = \lambda_S(T_B - T_S) + \lambda_{SR}(T_R - T_S) + \frac{dH}{dt} \tag{6}$$

$$C_{P_R}\frac{dT_R}{dt} = \lambda_R(T_B - T_R) + \lambda_{SR}(T_S - T_R) \tag{7}$$

where dH/dt is the rate of heat generated or absorbed in the sample container due to a chemical or physical process and the effective thermal conductances λ are the inverse of the respective thermal resistances. The heat capacities of sample and reference C_{P_S} and C_{P_R}, respectively, are in fact the heat capacities of the sample holder plus sample material and reference holder plus reference material, if any is used. Thermal resistances between the sample and sample holder are assumed to be zero.

If the substitution $\tau_S = C_{P_S}/\lambda_S$ and $\tau_R = C_{P_R}/\lambda_R$ is made, (6) and (7) can be rewritten as

$$\tau_S\left(\frac{dT_S}{dt}\right) = (T_B - T_S) + \frac{\lambda_{SR}}{\lambda_S}(T_R - T_S) + \frac{1}{\lambda_S}\frac{dH}{dt} \tag{8}$$

$$\tau_R\left(\frac{dT_R}{dt}\right) = (T_B - T_R) + \frac{\lambda_{SR}}{\lambda_S}(T_S - T_R) \tag{9}$$

In the simplest approximation $\lambda_{SR} = 0$, which neglects any direct heat flow between sample and reference, and the second terms on the left-hand sides of both (8) and (9) go to zero. If the block temperature is subjected to a linear temperature program with rate α, $T_B = \alpha t + T_{B,0}$, where $T_{B,0}$ is the initial block temperature. Thus (8) and (9) can be written

$$\tau_S\left(\frac{dT_S}{dt}\right) - \frac{1}{\lambda_S}\left(\frac{dH}{dt}\right) + T_S = T_{B,0} + \alpha t \tag{10}$$

$$\tau_R\left(\frac{dT_R}{dt}\right) + T_R = T_{B,0} + \alpha t \tag{11}$$

When $dH/dt = 0$, for example, in studies of heat capacity or before the thermal process begins in the sample cell, the following solutions for T_S and T_R can be obtained

$$T_S = T_{B,0} + \alpha t - \alpha\tau_S + (T_{S,0} - T_{B,0} + \alpha\tau_S)\,e^{(-t/\tau_S)} \tag{12}$$

$$T_R = T_{B,0} + \alpha t - \alpha\tau_R + (T_{R,0} - T_{B,0} + \alpha\tau_R)\,e^{(-t/\tau_R)} \tag{13}$$

where temperatures with subscript 0 indicate initial temperatures and it is assumed that the time constants for both the sample and reference do not change with temperature.

At times t when $t \gg \tau$, T_S and T_R are linear functions of time

$$T_S = T_{B,0} + \alpha t - \alpha \tau_S \tag{14}$$

$$T_R = T_{B,0} + \alpha t - \alpha \tau_R \tag{15}$$

and they lag behind the block temperature by a term equal to the scan rate times the respective time constant. From (14) and (15), it can be seen that $dT_S/dt = dT_R/dt = \alpha$. The difference, $\Delta T = T_S - T_R$, measured by the differential thermocouple or differential connection of the thermopiles, is determined by the scan rate and the difference in the time constants of the sample and reference. With the restrictions assumed to this point, the plot of ΔT versus time in Figure 8.6 is horizontal with magnitude $\alpha(\tau_R - \tau_S)$. This region is referred to as the *base line*. Of course, if $\tau_S = \tau_R$, then ΔT will equal zero.

When a thermal process occurs in the sample cell, dH/dt is no longer zero and dT_S/dt does not equal α. If one assumes that dT_R/dt remains equal to the rate at which T_B is changed, then on subtracting (10) from (11) and rearranging, one can obtain the following equation

$$\frac{1}{\lambda_S}\left(\frac{dH}{dt}\right) = T_S - T_R + \tau_S \left[\frac{d(T_S - T_R)}{dt}\right] + (\tau_S - \tau_R)\frac{dT_R}{dt} \tag{16}$$

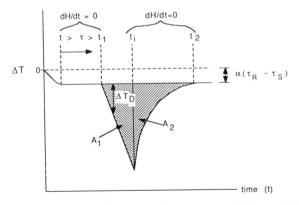

Figure 8.6 Ideal DTA output for a fusion process. The temperature difference ΔT between sample and reference is monitored as a function of time t. During the time interval from t_1 to t_i, dH/dt, the rate of heat absorption in the sample cell, is nonzero. The time interval t_i to t_2 represents the time required for the system to return to the base line following the completion of the thermal process. The sum of the areas A_1 and A_2 is proportional to the integrated enthalpy of the process. Thermal time constants for the reference and sample sides are given by τ_R and τ_S, respectively, and α is the programmed temperature scan rate. The displacement corrected for a nonzero base line is given by ΔT_D.

If the instrument construction is such that $\lambda_S = \lambda_R$, (16) can be rearranged to give

$$\frac{dH}{dt} = \lambda_S(T_S - T_R) + C_{P_S}\left[\frac{d(T_S - T_R)}{dt}\right] + (C_{P_S} - C_{P_R})\frac{dT_R}{dt} \tag{17}$$

Inspection of (17) shows that there is not a direct relationship between the instantaneous power developed in the sample and the measured property $(T_S - T_R)$. Additional terms in the relationship involve the rate of change of the relative temperature displacement and the scan rate dT_R/dt. These last two terms represent correction terms whose magnitude must be evaluated if kinetic studies are of interest, because dH/dt is proportional to the rate of the reaction.

When dH/dt is not zero, ΔT begins to deviate from the constant base line established before the onset of the thermal process. It is useful to consider the temperature displacement relative to the base line ΔT_D rather than $\Delta T = 0$ during this time. The corrected displacement ΔT_D can be expressed as shown in (18) in terms of the measured quantity ΔT and the base line displacement from zero ΔT_B

$$\Delta T_D = \Delta T - \Delta T_B = \Delta T - \alpha(\tau_R - \tau_S) \tag{18}$$

In Figure 8.6, the deviation from the base line is taken to begin at time t_1.

In a first-order phase transition, such as fusion of a pure material, the sample temperature remains constant until fusion is completed. Thus, ΔT_D increases linearly with a slope proportional to the scan rate until $dH/dt = 0$. The signal then decays exponentially [31] back to the base line, as is shown in Figure 8.6.

Substitution of (18) in (17), recognition that $dT_R/dt = \alpha$, and replacement of $(T_S - T_R)$ by ΔT allows (17) to be written as

$$\frac{dH}{dt} = \lambda_S \Delta T_D + C_{P_S}\frac{d(\Delta T_D)}{dt} \tag{19}$$

If the measured signal returns to the base line after the completion of the thermal process at time t_2, (19) can be integrated to give

$$\Delta H = \lambda_S \int_{t_1}^{t_2} \Delta T_D\, dt + C_{P_S} \int_{t_1}^{t_2} d(\Delta T_D) \tag{20}$$

Since times t_1 and t_2 correspond to points on the curve where ΔT_D is zero, the second integral on the left-hand side goes to zero and a proportionality between the enthalpy and the area $A = A_1 + A_2$ shown in Figure 8.6 is evident

$$\Delta H = \lambda_S \cdot A \tag{21}$$

In practice, the proportionality constant λ_S is temperature dependent and must be obtained by calibration experiments conducted over the temperature range of interest. Standard procedures for calibration methods are discussed in Section 2.6. Also encountered in practice is a significant shift in base line when the heat capacity or thermal conductivity of the sample undergoes a large change during the process. Theoretical treatments of the method to calculate the enthalpy under such conditions have been developed [43–47], but they are not presented here.

In the next few paragraphs we discuss the results of more detailed calculations that invoke fewer assumptions than we have made here.

In the treatment of Baxter [33], R_{SC} and R_{RC} are not taken as zero and an equation analogous to (17) was derived in which the coefficient of the second term includes a contribution from R_{SC} as well as R_S.

The assumption of a uniform temperature within the sample or reference is strictly true for well-stirred liquids only. Solid samples develop temperature gradients during heating or cooling, especially when a heat-generating process is under way. Detailed treatments that include the effects of these gradients show that a proportionality still exists between the area of the peak and the enthalpy of the thermal process under study, but the proportionality constant involves geometric details of the sample shape as well as its thermal conductivity.

We have neglected in this treatment the effects of heat transfer directly between sample and reference, radiative heat transfer, and changes in the heat capacities and thermal conductivities of the sample or reference. When λ_{SR} is nonzero, the area under the peak remains proportional to the enthalpy but the proportionality constant includes a term arising from λ_{SR} [34]

$$\Delta H = \lambda_S \left[1 + 2 \left(\frac{\lambda_{SR}}{\lambda_S} \right) \right] \cdot A \tag{22}$$

Thus, for a given enthalpy, the sensitivity of the apparatus is reduced by having a nonzero interaction between the sample and reference.

By including deviations from exact balance of the sample and reference sides of the instrument, radiative contributions, and conduction along the thermocouple wires between the block and reference and sample chambers, Šesták [4] derived

$$\frac{dH}{dt} = K_T(T_S - T_R) + C_{p_S} \frac{d(T_S - T_R)}{dt} + (C_{p_S} - C_{p_R}) \frac{dT_R}{dt} - \Delta K_T \tag{23}$$

The coefficient K_T equals $\lambda_S + 4l_s^{rad} T_R^3 + \tau_S$, where λ_S is the general heat exchange coefficient between the sample and its environment, l_s^{rad} is the proportionality coefficient for radiation between the sample and its environment, and τ_S is the heat conduction coefficient for the thermocouple wire between block and sample cell. The last term in (23), ΔK_T, arises from the differences between sample and reference sides of the instrument and contains contributions with factors including $\lambda_S - \lambda_R$, $l_s^{rad} - l_r^{rad}$, and $\tau_S - \tau_R$. If the two halves of the system are perfectly balanced, ΔK_T goes to zero.

Claudy and co-workers [38] showed that the coupling between reference and sample cells in disk-type instruments is significant and should not be neglected as we have done above. They have presented a model [39, 40] based on the analogy between thermal and electrical resistances and capacitances. Included in the derivation are heat exchange between the reference and sample through the gaseous atmosphere surrounding them and the support disk that connects both of them to the furnace and the effect of the thermal resistance between the sample and its container. This model quantitatively reproduces the behavior observed in the commercial instruments.

Gray [31] developed an analysis of the power-compensation method employing Newton's law of cooling between the sample and sample holder that we present in our treatment of this method, because it highlights some fundamental differences between this mode of thermal analysis and that by heat-flux and classical DTA methods. In his analysis, Gray assumes that temperatures of the sample and sample holder are homogeneous and that the temperature gradient occurs only at the interface between the holder and the sample container.

Figure 8.7 is a schematic diagram of a power-compensated DSC that corresponds to the more representational illustration of Figure 8.3c. The sample (sample container + sample) is at temperature T_S and has a heat capacity C_S, which is taken to be constant with temperature. The reference side is analogously defined with temperature T_R and heat capacity C_R. An electronic control system maintains the temperatures of both the sample and reference containers

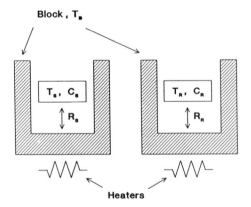

Figure 8.7 Schematic diagram of a power-compensated DSC. The sample temperature, heat capacity, and thermal resistance between the sample and the block are given by T_S, C_S, and R_S, respectively; T_R, C_R, and R_R designate the analogous quantities for the reference; T_B is the block temperature.

at T_B by controlling the power applied to their respective heaters. The difference in power dq/dt is the difference in the energy exchanged per unit time between the sample and its holder from that exchanged between the reference and its holder. The energy exchange between each cell and holder occurs through a thermal resistance R

$$\frac{dq}{dt} = \frac{T_B - T_S}{R} - \frac{T_B - T_R}{R} \tag{24}$$

The heat balance equations for the sample and reference sides of the calorimeter are given in (25) and (26), respectively.

$$\frac{dH}{dt} = C_S \left(\frac{dT_S}{dt}\right) - \frac{T_B - T_S}{R} \tag{25}$$

$$0 = C_R \left(\frac{dT_R}{dt}\right) - \frac{T_B - T_R}{R} \tag{26}$$

The quantity dH/dt represents the heat absorbed or generated in the sample due to some thermal process. Subtracting (26) from (25) gives

$$\frac{dH}{dt} = C_S \left(\frac{dT_S}{dt}\right) - C_R \left(\frac{dT_R}{dt}\right) - \frac{dq}{dt} \tag{27}$$

where (24) has been incorporated. If the assumption is made that the rate of temperature change of the reference cell equals the rate of the block temperature change $dT_B/dt = dT_R/dt$, the second derivative of q with respect to time can be given as

$$\frac{d^2q}{dt^2} = \frac{1}{R}\left(\frac{dT_B}{dt} - \frac{dT_S}{dt}\right) \tag{28}$$

With the incorporation of (25), (27) can be rewritten as

$$\frac{dH}{dt} = -\frac{dq}{dt} - RC_S\left(\frac{d^2q}{dt^2}\right) + (C_S - C_R)\frac{dT_B}{dt} \tag{29}$$

This equation can be compared with (17) developed earlier for DTA. In both, the power developed in the sample cell is the contribution of three terms, the first term is related to the directly measured signal, either $(T_S - T_R)$ or dq/dt; the second, to the first derivative of the measured signal; and the third, to a heat capacity difference between sample and reference cells. There is a major difference, however, because the thermal resistance R $(=1/\lambda)$ appears as a coefficient in front of the second term in (29), while it appears as the proportionality coefficient for the measured signal in (17). This difference results in the ability of DSC instruments to reduce R, thereby decreasing the time constant without decreasing the sensitivity, while in DTA equipment there is a trade-off between the need for a large R for high sensitivity but a low time constant ($\tau = RC_S$) for high resolution. A small R ensures that the lag of the sample temperature behind that of the sample holder is minimized.

Theoretical analysis of enthalpy determinations similar to those described by (18)–(21) can be made. The area under a DSC peak is equal (but opposite) to the integrated enthalpy associated with the thermal event. A calibration is required to convert the area to energy units, but this is an electrical factor not dependent on thermal resistances of the sample or sample holder as is found in DTA. Within the limits of this model, then, the calibration factor in DSC is a constant, independent of temperature. In actual practice, the constant does exhibit some temperature dependence. The relationship for heat capacity determinations can also be derived from (29). When no thermal event occurs in the sample cell, $dH/dt = 0$. At steady-state, $dT_S/dt = dT_R/dt$ and $d^2q/dt^2 = 0$; thus,

$$\frac{dq}{dt} = (C_S - C_R)\frac{dT_B}{dt} \tag{30}$$

Calibration using a substance with known heat capacity, such as sapphire, allows the measurement of unknown heat capacities. If $(dq/dt)_s$ stands for the signal obtained for a substance with a known heat capacity and $(dq/dt)_u$ for an unknown one, the ratio

$$\frac{(dq/dt)_u}{(dq/dt)_s} = \frac{(C_S - C_R)_u}{(C_S - C_R)_s} \tag{31}$$

can be obtained by dividing the equations obtained from (30) for the unknown and standard. The proportionality that results allows the unknown heat capacity to be determined. This topic is discussed in more detail in Section 2.6.5.

Several studies comparing the advantages and disadvantages of the heat-flux versus power-compensated DSC methods were conducted. Höhne and co-workers [48] found comparable capabilities for instruments of both types when measuring enthalpies of several phase transitions within an uncertainty of 1%, but observed that the enthalpies obtained from the heat-flux instrument were consistently low because of the base line correction algorithm used in the computer software.

Randzio [49] found from theoretical considerations that errors arising from the lag time of the calorimetric system, which the author calls "dynamic errors in time space", are comparable in both the heat-flux and power-compensated techniques. However, "parameter–space-induced dynamic errors," which are caused by changes in the time constant of the sample arising either from changes in the heat exchange coefficient or the heat capacity of the sample, are less significant in the power-compensated method.

2.4 Experimental Considerations

2.4.1 Introduction

In the discussion of experimental factors and procedures to follow, both power-compensated DSC and heat-flux DSC instruments are referred to as DSC, and we do not distinguish between those instruments that measure power directly and those configured to convert differential temperatures to power. The term DTA is used here for instruments that are either not capable of determining enthalpies or not configured to do so. In practice, these are often instruments operating at higher temperatures than commonly available with a DSC. We use the terms *pan* and *crucible* to indicate the small container that is in direct contact with the material under study. *Cell* is used to denote the fixed part of the apparatus in which the pan is placed. This nomenclature differs from some usage in which the sample containers are also referred to as cells.

2.4.2 Sample Mass

Samples used in DSC and DTA experiments should be of low mass, typically, 2–20 mg. Large specimens, even of good thermal conductors such as metals, do not achieve a uniform temperature. For poorly conducting organic solids or polymers, a specimen that is 1 mm thick may exhibit temperature gradients of as much as 2 K. Loose, granular material often gives poorly shaped curves because of low thermal conductivity and limited contact with the sample pan. Preparing the sample in a pellet may improve performance; the use of a inert, metallic plate or disk or an inverted pan lid on top of a lumpy specimen helps to form the molten material into a thin layer. When large specimens must be studied, either for reasons of instrument sensitivity or to ensure representative sampling of the material, slower scan rates should be used to allow for internal thermal equilibration of the sample. Extremely small samples can also pose problems, however. Dumas [50] notes that use of liquid specimens of 1 μL or less may result in the formation of metastable phases when the sample solidifies.

When transition or reaction enthalpies are to be determined, specimens must be reweighed after each use to ensure that no mass loss has occurred. The constancy of the transition temperature is relatively independent of the amount of substance present; the enthalpy of transition, however, requires accurate knowledge of the sample mass.

The mass of the specimen should be determined as accurately as possible using a microbalance. For the typical specimen size cited above, a 0.1-mg error in mass represents a 5–0.5% error in the final enthalpy or heat capacity value. The improved sensitivity of modern instrumentation should not be compromised by inaccurate mass determinations.

2.4.3   Sample Pans

A great variety of sample containers are available commercially, and some are interchangeable between different types of instruments. Specialized designs allow for the application of pressure and the addition of a reactant either in a batch or flowing mode. Metallic pans reduce the thermal resistances in the system and minimize thermal gradients between sample and temperature sensors. Quartz or glass liners can be inserted in metallic cup or cradle and used with reactive materials [51–54]; crucibles of graphite, alumina, or boron nitride are also useful with specimens that are incompatible with metallic pans. It is essential that sample pans be in good thermal contact with the temperature sensors and heating elements of the instrument; pans should be discarded when they become deformed or scratched through sealing or use.

Pans or crucibles are used in both the reference and sample sides of the instrument to ensure the equality of the factors, thermal resistances and such, which are assumed to be identical in the theoretical derivation of all differential methods. For this reason, the reference and sample pans should be of the same type (crimped, hermetic, or high pressure) and in the same condition (open or sealed). Ideally, when sealed pans are used, the atmosphere in both pans should be the same. No difficulty will be experienced, however, with a sample sealed in air and standards sealed in nitrogen because of the similar thermal conductivities of air and nitrogen. As long as the pans are of the same type, some flexibility in their materials is permissible. For example, most metallic fusion standards cannot be melted in gold cells because of their tendency to form alloys with the gold, yet gold pans may have to be used for high-temperature measurements on salts. No problems are introduced by using gold and aluminum pans of the same design. The use of a ceramic pan against a metallic one is unsatisfactory, however, because of their different thermal characteristics.

In DSC experiments the specimen pan on the reference side of the instrument is typically left empty, while in DTA experiments a material is placed in this pan. This material, most often alumina, serves to maintain and improve thermal contact between the cell, crucible, and sample on the sample side of the DTA. When alumina is used on the sample side, a like amount should also be used on the reference side to balance the thermal characteristics. When calibrating a DTA instrument, the certified reference material is placed in the sample pan, with or without alumina. Compensated measurements (Section 2.5.3) are an exception to the general rule that the specimen pan in the reference cell of a DSC is left empty.

2.4.4   Scan Rates

The scan rates used for DTA/DSC experiments vary widely; typically, rates from 5 to 20 K/min are recommended. The choice is dependent on several factors: the thermal conductivity of the sample, the reaction rate of the process under study, instrument design, and the purpose of the study. Slow scan rates maintain the sample near equilibrium and should be used for high-accuracy measurements;

their use also facilitates the resolution of closely spaced transitions. A fast scan rate, however, enhances a low-energy thermal process and may aid in its detection.

The enhancement occurs with the faster scan rate because the total energy required by the system must be input in a shorter period of time. This requires a greater energy per unit time and appears as a larger displacement. The optimum scan rate for a particular study must balance these factors. It is sometimes necessary to use a less desirable, fast, scan rate in order to be able to measure effects that would be nearly lost in base line noise at a slower scan rate.

The cooling rate and thermal history of a sample can affect both the temperature and enthalpy of a phase transition, particularly for inorganic salts and organic solids [27]. Once it is ascertained that no irreversible changes occur, it is good practice to heat specimens through the transition and then cool reproducibly, at a reasonably slow rate, to a temperature far enough below the transition to ensure that the specimen has reverted to the low-temperature form before beginning measurements. Fast cooling can result in formation of a metastable state. Suzuki and Wunderlich [55] discuss the influence of thermal history on the reproducibility of heat capacity measurements.

The behavior of amorphous materials is different from the behavior of the crystalline materials just discussed. To measure the glass transition of amorphous or glassy materials, preliminary heating and cooling at the same rate are necessary to prevent hysteresis and to bring the specimen to a defined state of relaxation.

2.4.5   Instrument Preparation

For reliable measurements, routine inspections of the overall condition of the instrument should be performed. Some instruments have provision for the use of lids on the sample and reference cells. Their use helps to maintain a uniform temperature within the sample by reducing the effects of radiation and convection. Misshapen or poorly placed lids are a major source of error in DSC measurements. The common practice of reusing the specimen pan placed in the reference side eventually results in a discoloration of the pan and a mismatch of heat flow conditions in the reference and sample sides of the instrument. The pans should be flat and free of scratches or dents that would reduce the contact with the sample cell.

The base line should be adjusted, according to the manufacturer's instructions, over the temperature region to be used in a series of measurements. Variations in the base line will affect the quantitative evaluation of DSC/DTA results, and the stability should be checked with a regularity sufficient to ensure that significant changes have not occurred.

Calibration procedures appropriate to the type of measurements must be made as discussed in Section 2.6. The frequency with which these must be performed depends on both instrumental and experimental factors. The operator must become familiar with the stability of each aspect of instrument performance and adjust operating procedures so as to produce data whose quality is consistent with the objectives of the experiment. Callanan and Sullivan [56] discuss these factors in more detail.

2.4.6   Purge Gas

For most thermal analytical instrumental systems and for many measurements made with them, it is customary to maintain a flow of inert gas through the measuring system of the instrument. This flow should be fast enough to maintain a constant temperature gaseous atmosphere around the cells. For some instrumental configurations, warming will take place if there is a static atmosphere and the warming will affect the accuracy of the measurements.

If open sample pans are used, the nature of the inert gas may affect the quantity being measured. These same differences can be noted if different gases are in contact, in a sealed pan, with a sample that can react. Nitrogen and argon are most commonly used as nonreactive purge gases; oxygen and carbon dioxide are used for reactive studies.

When liquid nitrogen is used as coolant, these heavier gases are not satisfactory and manufacturers often suggest helium as a purge gas. However, the thermal conductivity of helium is so high by comparison with the other purge gases mentioned that the transition curve obtained may be severely distorted. Mraw and Naas [57] found a commercial mixture of helium and neon eliminated this distortion for transition studies. In a recent comparison of heat capacities measured both by adiabatic calorimetry and by scanning calorimetry, use of this same gas mixture proved very satisfactory [58]. However, Jang and co-workers [59] do not report serious distortion with the use of helium. Since instrument design, heating rate, and gas flow rate affect the thermal measurement, differences in these factors may explain conflicting results reported in the literature.

2.4.7   Comparison With Other Calorimetric Methods

The agreement observed between DSC/DTA results and calorimetric data from high-precision adiabatic and drop calorimetry depends on the care with which the differential measurements are performed, the nature of the material, and the problem under study. Mraw [42] summarized the results of direct C_p comparisons. In the subambient region, agreement of ± 2–3% in heat capacity has been attained, while from room temperature to 700 K agreement can be as good as $\pm 1\%$. Achieving the best agreement requires reducing errors to a minimum by rigorous calibration procedures, careful attention to sample size and scan rates, precise mass determinations, and careful manipulations of samples and the instrument [60]. The use of rapid scans and large samples can preclude obtaining accurate thermodynamic data, because the sample is not maintained near equilibrium. This can be a particular problem in detecting phase transitions, particularly when multiple phase transitions occur in a narrow temperature region [61].

Determining enthalpies of polymerizations and solid-state chemical reactions and of all high-temperature, high-pressure processes are only feasible with DSC/DTA instruments so that direct comparison in these examples is rarely possible.

While the most accurate and precise heat capacity and transition data are obtained from adiabatic and drop calorimeters, there are several disadvantages to these techniques: Large quantities of material (5–50 g) are required, and the experiments are time-consuming. Commercial instruments are not currently available. The maintenance and use of the specialized equipment requires highly trained operators.

The techniques of DSC and DTA have some strong advantages that can compensate for the reduced precision and accuracy inherent in the methods. When objectives of a study do not require thermodynamic data of an exceptionally high precision and accuracy, these advantages can make DSC/DTA the favored approach.

A principal advantage of DSC/DTA is the ability to work with very small samples. The use of small quantities is particularly important for the study of materials that are expensive, difficult to synthesize, or hazardous in large quantities. However, replicate specimens should be measured for high-precision work, and always for the study of heterogeneous materials [62].

The DSC/DTA instruments are capable of operation over a wide range of temperatures, and measurements can be made in both a heating and cooling mode. Because of the use of small samples, rapid cooling can be achieved and one can study metastable, quenched phases. The procedures for loading samples are relatively quick and the scans can be collected over large temperature intervals at the rate of 5–20 K/min so the duration of an experiment is reduced.

2.5 Measurement Procedures

2.5.1 Transition Temperatures

The transition temperature reported from DSC measurements is traditionally obtained as the temperature determined from the intersection of the pretransition base line and the linear portion of the leading edge of the peak. This temperature is referred to as the *extrapolated onset temperature* and is indicated as T_{eo} in Figure 8.8a [63]. The extrapolated onset temperature gives closest agreement with the thermodynamic transition temperature and should be used for DSC measurements. When the presence of impurities or the use of large samples results in a distorted or broadened peak (Figure 8.8b), the instrument-limited slope determined in the discussion of thermal lag (Section 2.6.6) is used to establish the intersection temperature.

The peak temperature T_{pk} is the temperature recorded when the process is complete. In the past, it was common to report T_{pk} rather than T_{eo} for transitions measured by DTA, although better agreement with equilibrium thermodynamic measurements is obtained with T_{eo}. Commercial instruments that can be operated in either a DSC or DTA mode frequently display both T_{eo} and T_{pk}

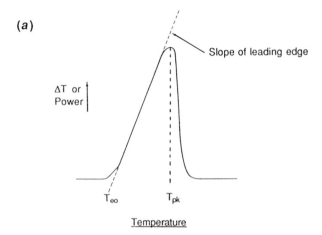

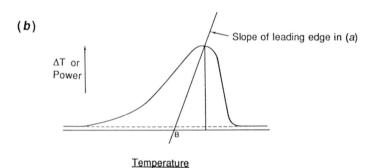

Figure 8.8 Determination of transition temperatures for samples (*a*) without thermal lag and (*b*) with thermal lag. Endothermic direction is upward. Differential temperature ΔT or power is plotted as a function of temperature. In (*a*), T_{eo} is the extrapolated onset temperature and T_{pk} is the temperature at the peak. In (*b*), point *B* represents the transition temperature corrected for thermal lag. Adapted with permission from *Thermal Analysis Newsletter*, No. 5, Perkin–Elmer Corporation, Norwalk, CT 06856.

when in the DSC mode and only T_{pk} when in the DTA mode. The recommended procedure would be to operate in the DSC mode and report T_{eo}; but, in any case, the manner of determining the temperature reported as the transition temperature should be indicated when publishing results.

2.5.2 Enthalpies

The enthalpy of a transition is determined by subtracting the background curve from the transition curve and integrating the resulting curve to obtain an area that is proportional to the enthalpy. The determination of the correct background curve is an important consideration in obtaining accurate values for the enthalpy. Several ways of drawing in this background curve or scanning base line, as it is sometimes called, have been suggested. The simplest approach is to draw a straight line or a sigmoid curve connecting the point at which the transition curve begins to move away from the background curve to the point at which it returns to background. Guttman and Flynn [44] recommend extrapolating the background curves backward and forward to the selected transition temperature; such a technique is particularly important when the heat capacities of phases 1 and 2 differ considerably. The background curve AB shown in Figure 8.9 is a measure of the heat capacity of phase 1; EF is that for phase 2. Heuvel and Lind [64] describe a calculation of the background curve that takes thermal lag into account as well as the differing heat capacities of the two phases. Goldberg and Prosen [45] discuss the theory involved in making these corrections and note that the problem can be minimized by use of small samples.

The choice of base line may be even more problematic when determining the enthalpy of chemical reactions. Differences in the pre- and posttransition base lines are pronounced because the reactants and products will generally have heat capacities that differ more than do two phases of the same material. If the reactions occur over a wide temperature range or take place over a long time interval, instability of the base line may pose additional problems and lead to an increased uncertainty in the measured enthalpy.

The software for the analysis of transition measurements that is incorporated in most commercial instruments realigns the background curve, so that the mismatch of the heat capacity may not be observed. A recorder tracing of the

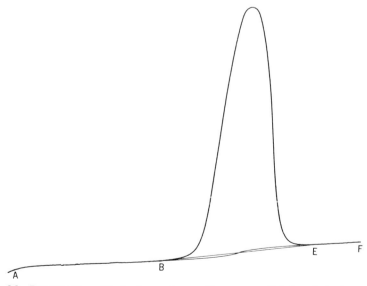

Figure 8.9 Determination of the background curve. The segment AB defines the background curve prior to the onset of the process and EF defines the base line following the completion of the process.

curve with isothermals well matched will alert one to the need for a more sophisticated background estimation.

2.5.3   Heat Capacities

Heat capacity measurements require that an instrument be very stable during the period of measurement because the energies involved are much smaller than those usually associated with transitions. Accurate measurements demand careful attention to many factors, each of which has the potential of affecting the measurements by a small amount. If these errors are not eliminated, their sum leads to unacceptable accuracy and precision.

There are two methods available for heat capacity determination with the DSC: the scanning method and the enthalpic (intermittent heating) method. Both methods require three measurements or "runs": the empty pan, a standard or reference material, and the unknown. The heat capacity of the empty pan is subtracted from both standard and sample heat capacities. When instrumental drift is a problem, an additional measurement on a reference material performed after that of the unknown allows a correction to be calculated [56]. Both the scanning and enthalpic methods can be modified by the addition of a reference material to the reference cell in what we call here *compensated heat capacity measurements.*

The scanning method is more widely used and faster than the enthalpic method. The required runs are made over temperature intervals of 50–200 K and the heat capacity is determined from the relative displacements of the standard and the sample at specific temperatures. Figure 8.10 shows the traces obtained from the empty pan, the sample, and the standard (sapphire) [65]. For manual calculations, the heat capacity of the sample is calculated as

$$C_{samp} = \left(\frac{m_{std} \cdot D_{samp}}{m_{samp} \cdot D_{std}} \right) \cdot C_{std} \tag{32}$$

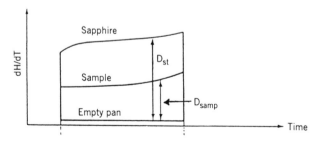

Figure 8.10   Traces used in the determination of heat capacity by the scanning method. The differential power outputs dH/dt for the scans made with the empty pan, (sample + pan), and (sapphire + pan) are shown. Displacements corrected for the base line of the empty pan are shown as D_{samp} and D_{st}. Reprinted with permission from M. J. O'Neill, *Anal. Chem.*, **38**, 1331 (1960). Copyright © 1960 American Chemical Society.

where C_{samp} is the specific heat capacity (J/K-g) of the sample at a given temperature, C_{std} is the literature value of the specific heat capacity of the standard at that temperature, m refers to mass, D is the net ordinate displacement [$D =$ (deflection of the specimen)–(deflection of the empty pan)], and the subscripts identify quantities associated with the sample or standard.

For precise heat capacity determinations, corrections for differences in the masses of the sample pans used for the empty, the standard, and the sample must be applied. A correction to the displacement of sample (standard) D' can be applied manually using the following equation

$$D' = (\Delta m \cdot C_{pan} \cdot S)/R \tag{33}$$

where Δm is the difference in pan mass ($\Delta m = m_s - m_{empty}$) for the pans used in the sample (standard) and empty runs; C_{pan} is the specific heat capacity of the pan material, S is the scan rate, and R is the range control setting. The value ($D' + D$) is then used in (32) to calculate the specific heat capacity of the sample.

In automated systems, C_{samp} may not be determined directly from (32) because of compensations applied internally for base line drift and other factors. Also, the algorithm for the pan weight corrections may not be included in the software and should be added by the user.

One approach, suggested by Luebke and Tria [66], is to correct the measured heat capacity of the sample using

$$C_{cor} = C_{meas} - \frac{C_{pan} \cdot \Delta m}{m_{samp}} \tag{34}$$

where C_{cor} and C_{meas} are the corrected and measured heat capacity, and C_{pan} and Δm are defined above. For accurate work, C_{cor} must be determined at each temperature for which a heat capacity is to be reported and then be further corrected using the calibration procedures indicated in Section 2.6.5.

The enthalpic and compensation methods to be discussed next address some of the instrumental and operating factors of the DSC that reduce the accuracy and precision of heat capacity measurements; these factors are considered in detail in Section 2.6.5.

The enthalpic method mimics the procedures used in adiabatic calorimetry in allowing for equilibration of the system after each step in a series of small step increases in temperature. The area of the resulting curve (the total enthalpy ΔH in the step) is evaluated and divided by the temperature step (ΔT) to give the heat capacity at the midpoint of the temperature step $C(T_{mid})$

$$C(T_{mid}) = \Delta H/\Delta T \tag{35}$$

A series of enthalpy increments is shown in Figure 8.11. The enthalpic method compensates for thermal lag by use of the total area. Because small temperature steps are used, the system comes to equilibrium every few minutes; as a result,

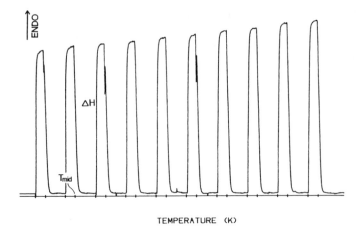

TEMPERATURE (K)

Figure 8.11 Trace obtained in stepwise fashion for heat capacity determination by the enthalpic method. The arrow indicates the direction for an endothermic process and T_{mid} designates the midpoint temperature for the temperature interval.

the drift in any one step is negligible. (The potential for drift from empty to standard to sample, etc. remains.)

Either method can be used in a compensated measurement procedure in which the reference cell contains a reference material such as sapphire. Cassel [67] measured one specimen of talc and two specimens of sapphire that were prepared to have nearly equal heat capacities. One of the sapphire specimens was placed in the reference cell and left there during the entire experiment. The other sapphire specimen and the talc were placed in the sample cell in turn and measured. In this work, the measured quantity is ΔC and (32) becomes

$$C_{samp} = \frac{C_{std} \cdot m_{std} + \Delta C}{m_{samp}} \tag{36}$$

where $\Delta C = (dq_{samp}/dt - dq_{std}/dt)/(dT/dt)$.

Because of the improved balance in heat flow between the reference and sample, a more sensitive measuring scale can be used. The heat capacity of talc was measured by this method from 223 to 773 K. In the temperature region in which adiabatic calorimetric measurements are available for comparison, the average deviation of the DSC data from the adiabatic data is 0.3%.

Cassel [67] also used the compensation procedure in conjunction with enthalpic measurements on 10 μL of D_2O with 10 μL of H_2O in the reference cell. Since the specimen in the reference cell effectively canceled out 90% of the total heat, inaccuracy in the temperature–area measurements of 2% would be expected to lead to an error of 0.2%, but actual results were not reported.

2.5.4   Measurements Made on Cooling

One advantage of the DSC is that measurements can be made while cooling the sample as well as during heating. This ability is useful in demonstrating that measurements of thermodynamic quantities are in fact those of a system at equilibrium; for example, heat capacities from cooling and from heating measurements should agree. Measurements of heat capacity made in the cooling mode are useful also to elucidate the behavior of systems that undergo irreversible changes on heating [68]. The observation of thermal hysteresis, a difference in transition temperature with measurements made in the heating or cooling mode, is a way of identifying a first-order transition from one that is second order.

The heat capacities of the standard and the empty pan must also be measured in a cooling mode as factors will differ from those obtained during heating. The thermal hysteresis associated with many first-order transitions requires that caution be used in the selection of calibrants for transition measurements on cooling.

2.5.5   Summary of Procedures and Precautions

For the convenience of the newcomer to these disciplines a short summary of operating procedures and precautions is included here.

ALL MEASUREMENTS

For all measurements with a DSC or DTA, instrument maintenance procedures, base line adjustment, and instrument calibration are necessary. It is often much easier to adjust enthalpies through a calibration of the data than by instrument adjustments. For DSC, specimens are prepared in appropriate pans that are sealed if possible, with no distortion of the pan bottom, and the mass is determined on a microbalance.

All specimens should be reweighed after use to ensure that no mass loss has occurred. A significant mass loss will invalidate enthalpy or heat capacity measurements; the mass loss that can be tolerated is determined from the mass of the specimen and the desired accuracy. For maximum accuracy and precision, all specimens should be subject to the same thermal history.

For instruments in which the sample and reference holders are heated directly and surrounded by an "isothermal" block, care should be taken that the block temperature is stable by use of a refrigerator, a circulating fluid, or a two-phase or liquid nitrogen bath.

Since operating conditions and manipulation of specimens are major sources of error and variability, the need for careful and reproducible conditions, and especially placement of specimen pans and cell lids, is paramount.

DTA MEASUREMENTS

For measurements with a DTA, it is sometimes helpful to place a metal disk or lid on top of the powdered alumina in the crucible. The lid helps to reduce temperature differences within the crucible that are caused by convection and to maintain good contact between the crucible and the sample. To drive off moisture and adsorbed gases, the alumina should be heated to a temperature above that at which it will be used and stored in a desiccator. If it is necessary to exclude oxygen from a sample to be heated in a DTA, the system should be evacuated and purged with an inert gas; there is no way to exclude air as the crucible is filled and placed in the DTA. Caution must be used in working with alumina crucibles at very high temperatures because of the changes in emissivity.

TRANSITION MEASUREMENTS

Small specimens, 1–3 mg, of unknowns and appropriate bracketing with reference materials are used. It is wise to run a test specimen in the laboratory from day to day to evaluate the stability of the instrument temperature calibration. The reference specimens and the unknowns are run under identical conditions of scan rate and gas flow, and resulting data are calibrated (corrected).

HEAT CAPACITY MEASUREMENTS

The masses of all empty pans used for heat capacity determinations must be recorded so that pan mass corrections can be made. For each determination, measurements are made on the empty, the standard, and the sample(s) and then a repeat measurement is made on the standard. Heat capacities are determined, and the resulting data are calibrated (corrected). Because heat capacity measurements are very sensitive to pan and lid placements in the DSC cell, care should be taken to reproduce their positions as closely as possible.

2.6   Calibration and Correction Procedures

2.6.1   Need for Calibration

Calibration of a DSC/DTA involves adjustment of instrument electronics and manipulation of the data to ensure the accuracy of measured quantities of temperature, heat capacity, or enthalpy. Calibrations are required in these instruments for several reasons. Temperature sensors such as thermocouples or thermistors may experience drifts that affect the mathematical relationship between voltage or resistance and absolute temperature. Also, significant differences between the true internal temperature of a sample with poor thermal conductivity and the temperature recorded by a probe in contact with the sample cup can develop when the sample is subjected to rapid temperature scans. A temperature calibration made under conditions closely approximating the experimental ones allows for correction of these problems.

The other important quantity measured in a DSC/DTA experiment is the differential temperature (DTA) or the differential power (DSC) from which enthalpy or heat capacity information is extracted. There is an obvious need for a calibration of DTA instruments: The proportionality constant in (21) must be determined using a known enthalpy or heat capacity. Since the output of a power-compensated instrument is already in units of power, the need for enthalpy or heat capacity calibration in these instruments is perhaps less obvious. Factors such as mass of the specimen, its form, the gaseous atmosphere, heating rate, location of the heating and sensing elements, constancy of the heat sink temperature, and thermal resistances in the system affect the magnitude of the power levels recorded in both DSC and DTA instruments, however. Placement of the sample within the sample cup or small changes in the shape or placement of the lid can also cause deviations of a few percent in the primary measurements. The correction procedures are designed to compensate for the fact that environmental and manipulative factors distort the proportionality of the primary differential measurement, temperature or power, with enthalpy or heat capacity differences. To compensate properly for these effects, the calibration experiments must be performed using reference materials under identical conditions to those used for the sample of interest.

Some commercial instruments incorporate electrical calibration procedures that provide an evaluation of the performance of the instrument sensors. These procedures do not correct data for the kind of sample-related effects listed above. Also, only instrumental precision can truly be determined from these procedures, and absolute accuracy must be established with a standard reference material.

Several types of recommended procedures for calibration are described in the following sections. The exact procedure and frequency of its use to calibrate the instrument or correct the data depends on the type of instrumentation employed, the stability of the particular instrument, and the accuracy desired from the final results. Experiments designed to yield only qualitative information about the existence of phase transitions require less rigorous and frequent calibration procedures than do studies in which high-quality thermodynamic data are desired. It is extremely important that the calibration be carried out in the same manner and under the same conditions that will be used for measurements on the samples. Callanan and Sullivan [56] discuss the development of standard operating procedures for DSC that incorporate these considerations. We treat here considerations for DSC calibrations since the same procedures for DTA are generally applicable, and, we note deviations required for DTA when they are appropriate.

2.6.2 Reference Materials

Ideally, calibrants should be certified reference materials. A certified reference material is a material for which values of the property in question have been determined for a particular lot of material. The certified reference values can be ascribed only to that lot and not to another lot of the same substance. Properties to be desired in a reference material are discussed by Lindsey and co-workers [69].

Certified values for thermal properties measured in DSC/DTA are determined through precision calorimetric measurements. For example, fusion standards are measured with an adiabatic or drop calorimeter. When certification through precision measurements is not possible, two independent measures of the property can be used to establish the certificate values. For example, the purity of DSC purity standards has been determined by DSC and by chromatographic analysis. If an absolute or a second method is not available, an interlaboratory comparison or round-robin study that involves several laboratories and various types of equipment can be conducted. Proper statistical design of the experiment and evaluation of the results yield meaningful certificate values. When no certified material is available, literature values obtained from absolute measurements on a very pure material can be used, if a correspondingly pure material can be obtained for the calibration experiments.

Table 8.6 lists the certified materials presently available for transition temperature and enthalpy calibrations, their approximate values, and the source from which the material can be obtained. The reference materials from the National Physical Laboratory (NPL) were certified for enthalpy by adiabatic calorimetry; temperatures were determined in a triple point cell. Barnes and co-workers [70] and Callanan and Vecchia [71] reported on the use of these materials. The ICTA reference materials cited in Table 8.6 are certified only for comparisons among instruments and not for calibration of temperatures or enthalpies.

Some of the NIST materials were certified by adiabatic or drop calorimetry; others were certified through a statistically designed program of measurements with a scanning calorimeter. The NIST program was undertaken to supply a series of reference samples of materials with different thermal properties (metals, inorganic salts, and organic powders) at suitably narrow temperature intervals for calibration of both the temperature and enthalpy scales. The literature contains numerous suggestions of additional possibilities for reference materials, for example Breuer and Eysel [72].

Materials certified for use in heat capacity calibrations, their temperature ranges, and suppliers are given in Table 8.7.

2.6.3 Temperature Calibration

The temperature scale of a scanning calorimeter can be calibrated by isothermal or by scanning methods. An isothermal approach is recommended for calibration of the instrument sensors. As mentioned earlier, however, response of the instrument during a scan can be affected by several factors, and calibration of measurements made with the instrument should be carried out in the same way the measurements are made. Therefore, isothermal temperature calibrations are preferred if enthalpically determined heat capacities are to be measured and dynamic ones are to be used for scanning heat capacity measurements. The isothermal procedure, which is discussed in detail by Brennan and co-workers [73], Flynn [74], and Richardson [75], calls for stepwise heating through the transition of a calibrant in small temperature intervals, typically 0.1 K. When using a pure fusion standard, it is possible that the melting will occur completely

Table 8.6 Certified Reference Materials for Transition Properties

| Material | Type[a] | Transition Properties | | Source[c] |
		T (K)	ΔH (J/g)[b]	
Indium	Fusion	430	28	NPL, NIST/ICTA
Mercury	Fusion	234	11	NIST
Tin	Fusion	505	60	NIST, NIST/ICTA
Zinc	Fusion	693	112	NIST
Silver sulfate	S–S	703	NA	NIST/ICTA
Barium carbonate	S–S	1083	NA	NIST/ICTA
Potassium perchlorate	S–S	573	NA	NIST/ICTA
Potassium chromate	S–S	938	NA	NIST/ICTA
Potassium nitrate	S–S	401	NA	NIST/ICTA
Quartz	S–S	846	NA	NIST/ICTA
Strontium carbonate	S–S	1198	NA	NIST/ICTA
Acetanilide	Fusion	388	160	NPL
Benzil	Fusion	368	112	NPL
Benzoic acid	Fusion	396	148	NPL
Biphenyl	Fusion	342	120	NIST
Cyclohexane	S–S	190	NA	NIST/ICTA
Cyclohexane	Fusion	280	NA	NIST/ICTA
1,2-Dichloroethane	Fusion	241	NA	NIST/ICTA
Diphenylacetic acid	Fusion	420	147	NPL
Phenyl ether	Fusion	303	NA	NIST/ICTA
o-Terphenyl	Fusion	331	NA	NIST/ICTA

[a]S–S = Solid–solid phase transition.
[b]NA = Not available.
[c]NIST = National Institute of Standards and Technology, Office of Standard Reference Materials, Gaithersburg, MD 20899; NPL = National Physical Laboratory, Office of Reference Materials, Teddington, Middlesex, TW11 OLW, UK; ICTA = International Confederation for Thermal Analysis; NIST/ICTA = ICTA certified reference materials supplied by NIST.

Table 8.7 Certified Reference Materials for Heat Capacity

Material	Temperature Range (K)	Source
Benzoic acid[a]	10–350	NIST
Copper	1–300	NIST
Molybdenum	273–2800	NIST
Platinum[a]	298–1500	
Poly(styrene)	10–35	NIST
Poly(ethylene)	5–360	NIST
Sapphire	0–2250	NIST
Tungsten	273–1200	NIST

[a]Material certified for other types of analysis but recommended by IUPAC for heat capacity; excellent literature data are available.

in one such interval. Otherwise, the temperature interval in which the greatest proportion of the transition enthalpy occurs is designated the *transition temperature*. In Figure 8.12 the 0.1 K interval marked T_{tr} is that in which most of the transition occurs. The accepted transition temperature of the reference material is assigned to the temperature of this interval. Clearly, the accuracy in temperature measurements is limited in the best case to the calibration interval employed. When less pure standards are used, the fusion peak is broader, and the transition interval may not be well resolved.

Differential scanning calorimetric instruments are more commonly calibrated with scanning temperature procedures. The ASTM [76] recommends a two-point calibration procedure in which two reference materials are chosen whose transition temperatures bracket the region of interest; an assumption is made that the temperature scale of the instrument is linear. The observed transition temperatures of the reference materials (OT_1 and OT_2) along with their literature values (ST_1 and ST_2) are used to establish the slope S and intercept I of the instrumental response

$$S = (ST_1 - ST_2)/(OT_1 - OT_2) \tag{37}$$

$$I = [(OT_1 \cdot ST_2) - (OT_2 \cdot ST_1)]/(OT_1 - OT_2) \tag{38}$$

The corrected specimen temperature T is then obtained from the observed temperature OT by (39)

$$T = OT \cdot S + I \tag{39}$$

This method works well when the standards are chosen to bracket the temperature of the unknown by about 30–50 K on either side and when they have similar characteristics, such as thermal conductivity and heat capacity. For this reason, a variety of materials with well-established transition temperatures are needed as standards: metals, inorganic salts, and organic powders [60].

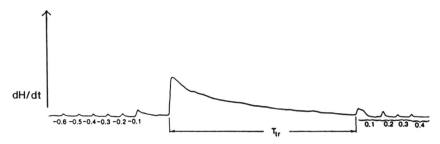

Figure 8.12   Trace obtained in stepwise, isothermal fashion for temperature calibration. The output of the DSC is given by dH/dT, the temperature interval in which the standard melted is designated T_{tr}. The endothermic direction is indicated by the arrow. Reprinted with permission from J. E. Callanan and S. A. Sullivan, *Rev. Sci. Instrum.*, **57**, 2584 (1986).

Experience with the two-point calibration procedure demonstrates the linearity of the temperature scale in both subambient and superambient regions [77].

A one-point calibration procedure [76] for temperature can be used if the slope, as calculated by (37), is close to unity. In this procedure, the intercept is calculated from (40)

$$I = ST_1 - OT_1 \tag{40}$$

A difference of 1% in the slope, that is, $S = 0.99$ or 1.01, yields an error of 1 K in the temperature when the calibration transition occurs 100 K from that of the unknown.

A calibration procedure that can be thought of as a multipoint procedure in analogy with the terminology used for one- and two-point methods is employed in some commercial DSC units. The computer that controls the DSC generates a correction polynomial by fitting values of ST and OT, which are input by the user, for several temperature standards. The measured temperature is fed into the correction polynomial, and the corrected temperature is stored and used for calculations.

2.6.4   Calibrations for Enthalpy Measurements

While the temperature calibration procedures are well established, the enthalpy procedures currently available are not. The simplest enthalpy calibration procedure calls for determination of a correction factor E by comparison of the observed enthalpy of transition of a reference material ΔH_{obsd} with the certified or literature value for that enthalpy ΔH_{lit}

$$E = \Delta H_{lit}/\Delta H_{obsd} \tag{41}$$

The correction factor is assumed to be independent of temperature, and the adjusted value of the enthalpy of the phase transition of the specimen is taken as

$$\Delta H_{adj} = E \cdot \Delta H_{meas} \tag{42}$$

where ΔH_{meas} is the measured enthalpy of the process under study.

A two-point calibration method analogous to that described earlier for temperature was used with some success [56]. The application of this procedure uses (37)–(39) with ΔH substituted for T. A comparison of the enthalpy of fusion of tin foil determined with this procedure using tin and zinc as the bracketing reference standards gave a value of 60.15 ± 0.15 J/g while drop calorimetric measurements yield a value of 60.22 ± 0.18 J/g [71].

In a variation of this two-point enthalpy procedure, the specimen is bracketed with reference materials in order of increasing enthalpy rather than in order of temperature. The linearity of the resulting enthalpy calibration curve is better than that obtained when transition temperatures are used to choose the bracketing materials [78].

The ASTM procedures [79] for determining transition enthalpy call for a modification of the one-point calibration method to incorporate a temperature dependence in the E factor. The value of ΔH_{adj} [see (42)] is obtained at the temperature of the fusion standard, and the heat capacity of a reference is obtained at this temperature and at the temperature of interest for the unknown. A correction factor F is taken as the ratio of the quantities C_{lit}/C_{obsd} at the two temperatures. The corrected enthalpy of fusion is obtained by multiplying ΔH_{adj} by a ratio of correction factors CF for the unknown and for the reference.

$$\Delta H_{cor} = \frac{CF_u}{CF_r} \cdot \Delta H_{adj} \tag{43}$$

This procedure does not appear to give satisfactory agreement with accepted literature values for several materials [60]; studies leading to revision of these procedures are under way at NIST.

2.6.5 Calibrations for Heat Capacity Measurements

The usual procedure for calibration of heat capacity data involves the determination of the heat capacity of a reference material over the temperature range for which C_p measurements are desired for the unknown. (Again, it is important to duplicate the experimental conditions used in the measurement of the unknown as closely as possible.) A temperature-dependent correction factor $F(T)$ is then calculated as a ratio

$$F(T) = C_{lit}(T)/C_{obsd}(T) \tag{44}$$

where $C_{lit}(T)$ is the literature value for the heat capacity of the standard at temperature T; C_{obsd} is the value obtained under the conditions of the experiment and corrected for pan mass differences at the same temperature.

The calibrated specimen heat capacity is then determined by

$$C(T) = F(T) \cdot C_{cor}(T) \tag{45}$$

where C_{cor} is the observed heat capacity of the unknown, corrected for pan mass differences at each temperature T.

It is often assumed that F is constant over the entire range of the DSC and does not change with time. For most instruments, this assumption is not warranted, the factor varies both with temperature and time, and the corrections indicated should be applied. The performance of an individual instrument and the accuracy and precision required of the measurements enter into the decision about the necessity for these corrections.

A correction polynomial analogous to that described in the temperature calibration can be generated and applied internally in computer-controlled instruments. Care must be exercised, however, when such polynomials have been established using reference materials certified for enthalpy measurements and not for heat capacity.

2.6.6   Corrections for Thermal Lag

The term *thermal lag* is used to indicate the difference between the temperature recorded by the instrument display and that actually experienced by the sample. The most serious contribution to thermal lag comes from the interfaces between the calorimeter cell and the specimen pan *A* and the specimen itself *B* as indicated in Figure 8.13. The usual strictures regarding good thermal contact between pan and calorimeter or thermocouple (that is, flat bottom, reproducible placement) help to minimize the first contribution; the form and shape of the specimen influence the second. A further correction is required by the fact that the temperature-sensing element is not in the sample.

In transition measurements, the problem is manifested in the data as a curvature in the leading edge of a transition rather than as a sharp rise (Figure 8.8*b*) and by significant differences in observed results as the scan rate is varied.

To obtain accurate transition temperatures when thermal lag is a problem, the following procedure can be followed [80]. A peak is obtained from a thin sample of a pure material having thermal conductivity comparable to the unknown and using the same scan rate. A sharp peak is observed, and the slope of its leading edge is determined by instrumental factors (Figure 8.8*a*). Corrections to the data obtained from the unknown are made by using this slope, applied at the peak of the transition, to extrapolate downward for the intersection with the base line as illustrated in Figure 8.8*b*. Point *B* in this figure gives the transition temperature when such a correction is applied.

The work of Richardson and co-workers [81, 82] in the treatment of thermal lag in heat capacity measurements is discussed here. Corrections suggested by these authors were incorporated into data processing algorithms of instruments currently on the market.

A simple and rapid way to estimate thermal lag in a scanning heat capacity

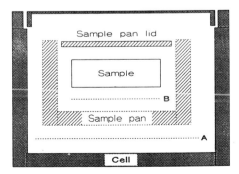

Figure 8.13   Representation of a DSC cell and the resistances that contribute to thermal lag: (*A*) pan-cell resistance and (*B*) sample-pan resistance.

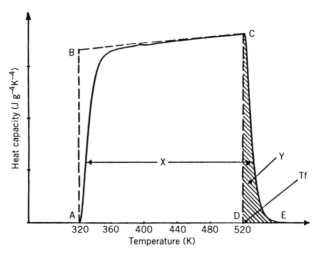

Figure 8.14   Estimation of thermal lag. In the absence of thermal lag, curve $ABCD$ will be obtained. In the presence of thermal lag, curve ACE is followed. The area of curve ACE is given by X while the area Y represents the tail that results after the final temperature T_f has been reached. Reprinted with permission from J. E. Callanan and S. A. Sullivan, *Rev. Sci. Instrum.*, **57**, 2584 (1986).

measurement, illustrated in Figure 8.14 [81], is to compare the area of the tail Y once the scan has reached its final temperature T_f with the total area of the scan X. The temperature lag δT can be evaluated by multiplying the temperature range of the scan $\Delta T = T_f - T_i$ by the ratio of the two areas. The exact formula

$$\delta T = \frac{Y}{X} \cdot \frac{C(T_m)}{C(T_f)} \cdot \Delta T \tag{46}$$

requires that the ratio of the heat capacity at the midpoint of the temperature scan $C(T_m)$ to the heat capacity at T_f $C(T_f)$ be used. If the temperature interval of the scan is small and/or the heat capacity is not changing rapidly with temperature (e.g., outside of a transition region), the correction introduced by use of the heat capacity ratio is insignificant.

The temperatures at which heat capacities are reported are adjusted by (46) or by the simplified form

$$\delta T = Y \cdot \Delta T / X \tag{47}$$

A different, more rigorous treatment for thermal lag corrects the temperature $T(d)$, as read on the DSC digital output, for the instrument calibration $dT(s)$ as well as for the thermal resistance factor. Richardson and Burrington [81] suggest the following correction

$$T_{act} = T(d) + dT(s) + dT(o) + dT(mat) \tag{48}$$

Here $dT(s)$ is the correction arising from the calibration of the calorimeter $(T_S - T_o)$; T_S is the literature value of the calibrant, and T_o is the observed transition temperature; $dT(s)$ can be positive or negative.

The $dT(o)$ term corrects for the effect of the interface between the sample pan and cell. It is a function of heating rate and pan area and condition; details of its evaluation are discussed in [81, 82]. At scan rates of 20 K/min, $dT(o)$ for a thin aluminum sample pan is about 1.5 K; at 5 K/min, the lag is 0.35 K.

The term $dT(mat)$ corrects for the effect of the thermal resistance of the specimen. It is a function of sample mass and form, area in contact with the pan, and factors that affect the contact between the pan and sample. If very thin specimens are used, as is often done in transition measurements, $dT(mat)$ can be insignificant. If large specimens are used, as in heat capacity measurements, or the specimens are poorly shaped, correction for $dT(mat)$ can be significant.

Given the circuitry of modern instruments, (48) is expected to be linear; the linearity should be verified, however, particularly in subambient temperature regions or if the temperature interval used is large.

The effects of the finite thermal resistances on the output signal recording the differential power (DSC) or differential temperature (DTA) become important when the details of peak shape are of great interest, for example, in kinetic analysis. Deconvolution techniques that attempt to produce an instantaneous record of the thermal event are discussed in Section 5.

2.7 Applications

2.7.1 Purity Determinations

One of the earliest applications of DSC/DTA was to the determination of the purity of organic compounds. Although many experimental techniques are available for purity determination, DSC provides a rapid technique that does not require the use of a reference material. Reviews of the method can be found in [6, 83, 84] and in references cited in [85].

Figure 8.15 [63] shows the DSC output obtained from a series of impure benzoic acid samples. It is apparent that the fusion process begins at lower temperatures and the peak gets broader as the concentration of impurity is raised in a solid. Very small impurity levels are seen to have significant effects on the shape of the curves. Quantitative purity determinations are possible from an analysis of the DSC curve.

The procedure for quantitative evaluation of purity is based on the thermodynamic relationship for the depression of the melting point by the presence of impurities. In the derivation of this procedure, it is assumed that (1) the major component (solvent) and its impurities (solute) form ideal solutions upon melting; (2) the solvent and solute are immiscible in the solid phase; (3) $x_2 \approx 0$; that is, the concentration of impurities, expressed as mole fraction of solute, is small; and (4) the enthalpy of fusion of the sample ΔH_f is constant over the range of temperatures employed in the analysis.

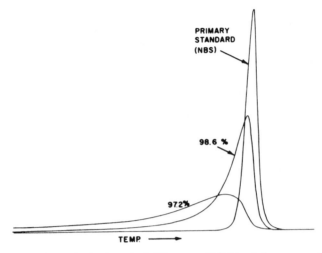

Figure 8.15 Effect of impurity level on the DSC output dH/dT for fusion curves of benzoic acid. Reprinted with permission from *Thermal Analysis Newsletter*, No. 5, Perkin–Elmer Corporation, Norwalk, CT 06856.

If these four assumptions are valid, one can write that

$$x_2^* = \frac{\Delta H_f}{R}\left(\frac{1}{T_m} - \frac{1}{T_o}\right) \tag{49}$$

where T_o is the melting point of the pure material and T_m is the melting point of the impure material with composition x_2^*. Since $T_o \approx T_m$ under the condition of assumption 3, (49) can be written as

$$x_2^* = (\Delta H_f/RT_o^2)\cdot(T_o - T_m) \tag{50}$$

The fraction F of sample melted at temperature T is given by the ratio of the composition of the sample x_2^* to the composition of the liquid x_2 which is in equilibrium with the solid at that temperature.

$$F = x_2^*/x_2 = (T_o - T_m)/(T_o - T) \tag{51}$$

The second equality follows directly from the mole fraction ratio, if (50) holds for both compositions x_2 and x_2^*. With this latter assumption, substitution of the appropriate form of (50) for x_2 into (51) gives

$$F = \frac{RT_o^2 x_2^*}{\Delta H_f(T_o - T)} \tag{52}$$

Upon rearrangement, (52) yields a linear relationship between T and $(1/F)$

$$T = T_o - \left(\frac{x_2^* R T_o^2}{\Delta H_f}\right)\frac{1}{F} \qquad (53)$$

A plot of the sample temperature T versus the reciprocal of the fraction melted $1/F$ should yield a straight line with the slope equal to $-(x_2^* R T_o^2/\Delta H_f)$ and an intercept of T_o.

Values of the fraction melted at various temperatures are obtained from a series of partial integrations of the DSC peak, where

$$F(T) = \Delta H(T)/\Delta H_f \qquad (54)$$

The definition of $\Delta H(T)$ can be made with reference to Figure 8.16 [85]. The top curve in this figure represents the DSC output for the sample undergoing fusion. The middle curve is the background chosen for the calculation of the total enthalpy of the process. The dashed curve at the bottom represents the instrument base line. Integration of the difference of the top curve and the background gives the enthalpy of fusion ΔH_f. The area of the peak up to AB is generally taken as $\Delta H(T)$, where the temperature at point A is obtained as the intersection of line AD with the base line, as discussed in the procedure for thermal lag corrections. Brennan and co-workers [85] also suggest that the area

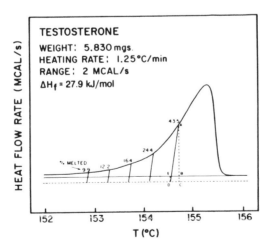

Figure 8.16 Example of the determination of the fraction of a sample melted using the fractional areas of the measured enthalpy change. Copyright ASTM. Reprinted with permission from W. P. Brennan, M. P. DeVito, R. L. Fayans, and A. P. Gray, "An Overview of the Calorimetric Purity Measurement," in R. L. Blaine and C. K. Schoff, Eds., *Purity Determinations by Thermal Methods*, American Society for Testing and Materials, Philadelphia, PA, 1984, pp. 5–15.

EBCD be included as part of $\Delta H(T)$ to compensate for the heat capacity of the sample and pan.

A van't Hoff plot obtained in this manner is shown as the curve marked *original data* in Figure 8.17 [85]. The nonlinearity of this curve is attributed to underestimation of the fraction melted in the early part of the melting process. The movement of the dH/dt curve away from the base line is so gradual that it is difficult to estimate the exact beginning of the melting. The process referred to as *van't Hoff linearization* adds a number y representing the amount first melted, but not included, to both the numerator and denominator in (51). Thus, (51) becomes

$$F = \frac{T_o - T_m + y}{T_o - T + y} \tag{55}$$

In this equation, y is an arbitrary constant chosen to linearize the plot. In Figure 8.17 it can be seen that for these data, a value of 6.6% produces a linear plot.

Sondack [86] eliminates the arbitrariness in this procedure by selecting three points in the 20–50% melt region from which he calculates the correction factor. The application of the correction factor linearizes the data. Garn and co-workers [87] suggest that the use of (NMR) measurements will aid in the detection of the solidus or early melting.

To achieve good results with this procedure, the fusion measurements must

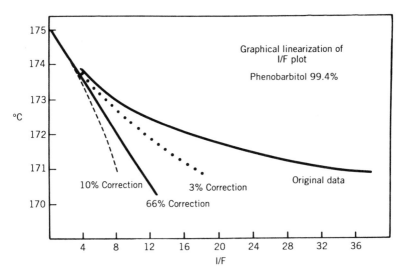

Figure 8.17 Linearization of the van't Hoff curve; F is the fraction of sample melted. Copyright ASTM. Reprinted with permission from W. P. Brennan, M. P. DeVito, R. L. Fayans, and A. P. Gray, "An Overview of the Calorimetric Purity Measurement," in R. L. Blaine and C. K. Schoff, Eds., *Purity Determinations by Thermal Methods*, American Society for Testing and Materials, Philadelphia, PA, 1984, pp. 5–15.

be made on a small specimen (1–2 mg) at a very slow scan rate (under 1 K/min). The specimen should be melted and recrystallized before beginning the fusion scan to ensure that it is in an equilibrium crystalline state prior to and during melting. Because the technique is measuring small differences in pure materials, care during the preparation of the specimen is critical. The measurement can be in scanning mode (dynamic purity) or stepwise; the observed melting point should be corrected for thermal lag. The fraction melted up to successive areas, from about 10 to 50% melted is calculated and plotted. The beginning of the curve is eliminated to avoid putting much weight on the early, poorly defined melting region; the latter part is eliminated because the melting in this region is too rapid for equilibrium to exist.

Two major limitations of DSC purity determinations appear to be the formation of solid solutions, and the applicability of the method to small impurity levels only. Brostow and co-workers [88] addressed the failure of the methods to deal with systems in which solid solutions were formed. They suggest a new equation based on Guggenheim's theory of strictly regular solutions. They applied their equation as well as the van't Hoff equation and that of Mastrangelo and Dornte [89]. It appears from the results that their equation, which assumes that Guggenheim's model applies simultaneously to solid and liquid phases, is valid. By means of a stepwise scan through the melting region, Elder [90] obtained satisfactory results for systems with impurity levels up to 10%; in the analysis of the scan, the heat capacity of the system is not added to each step. When correction is made for the heat capacity contribution from values at the beginning and end of the steps, the corrected plot is linear; excellent values are obtained for the melting point, heat of fusion, and impurity level. Elder [91] also found that when there are two impurities present in a system, their relative proportions do not affect the total purity level. Ramsland [92] modifies (55) by addition of a partition coefficient; again, accurate purity determinations on samples with up to 10% impurity were possible.

2.7.2   Phase Diagrams

The ability of the DSC/DTA instrument to determine temperatures and enthalpies of phase transitions can be exploited to construct phase diagrams for multiple-component systems. A simple binary phase diagram is shown in Figure 8.18 to illustrate the approach. Figure 8.19 shows the DSC/DTA output for cooling a series of mixtures whose compositions correspond to those designated in Figure 8.18. For the pure components A and B and a mixture exactly at the eutectic composition X_A^E, a single sharp peak is observed. Mixtures of intermediate composition will exhibit two features, a broad maximum that marks the onset of the phase reaction

$$\text{Liquid} \rightarrow \text{Liquid} + \text{Solid A (or B)}$$

and a sharp peak at the eutectic temperature. In principle, the peak is observed

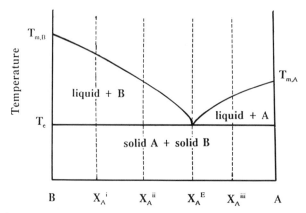

Figure 8.18 Binary phase diagram for a system of immiscible solids A and B with melting temperatures $T_{m,A}$ and $T_{m,B}$, respectively. Compositions of various mole fractions X_A and their melting temperatures T are identified for correlation with the curves of Figure 8.19. Reproduced with permission from M. E. Brown, *An Introduction to Thermal Analysis: Techniques and Applications*, Chapman & Hall, New York, NY, 1988, p. 41.

at the eutectic melting temperature in each mixture. The area under this peak ΔH_{melt} is proportional to the relative amount of sample that melts isothermally at the eutectic temperature. Since this quantity decreases as the composition of the mixture moves away from the eutectic composition, it is frequently difficult to detect the eutectic peak very close to the pure components because of its small size. Thus, differentiating between simple eutectic and other phase behavior such as solid solution formation may be impossible near the end points of the phase

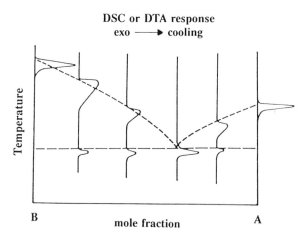

Figure 8.19 Comparison of DSC/DTA traces as a function of temperature T for compositions with the mole fractions identified in Figure 8.18. Reproduced with permission from M. E. Brown, *An Introduction to Thermal Analysis: Techniques and Applications*, Chapman & Hall, New York, NY, 1988, p. 42.

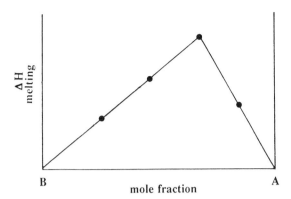

Figure 8.20 Variation of the enthalpy of eutectic melting for binary mixtures of Compounds A and B for the compositions identified in Figure 8.18.

diagram. However, the variation of ΔH_{melt} with composition can be used to determine the eutectic composition without having to make many mixtures around the eutectic. As Figure 8.20 shows, the plot of ΔH_{melt} with composition has two linear regions that intersect in a maximum value for ΔH_{melt} at the eutectic composition.

Very complex phase diagrams can be constructed in this manner with the following generalizations. When a phase boundary, which is required by the Gibbs phase rule to be isothermal, is crossed, a sharp peak is observed. Phase reactions that occur over a range of temperatures are located by the appearance of broad features. A treatment of theoretical and practical considerations for the generation of phase diagrams from thermal methods can be found in [4, 93, 94].

Care must be taken, however, to ensure that the samples are homogeneous mixtures of both components, usually by mixing the pure components together in the molten state, or premelting the mixture before taking measurements. The scan rate and direction of temperature change can also be important factors in obtaining reliable results. Measurements should be made as slowly as is necessary to ensure that the system is in temperature equilibrium. Also, many phases have a tendency to supercool, and operation in a heating mode may be necessary to observe the equilibrium phase temperatures. However, the ability of the DSC to cool quickly and trap metastable phases can be an advantage, if one is interested in studying the metastable state, for example [5], with glassy phases.

2.7.3 Glass Transitions

At the molecular level a glass can be described as a highly disordered material in which the translational and some rotational degrees of molecular freedom are frozen in, and the solid lacks the full long-range order associated with a crystal. The glassy state in polymers arises when motions of segments of the polymer molecules freeze in on cooling. These motions involve torsional oscillations and rotations about bonds within the backbone of the polymer chain and side groups attached to the backbone. In crystals of smaller, more symmetrical molecules, long-range translational order can be achieved but orientational motion ceases before a completely ordered crystal is obtained. Upon heating, the molecules regain internal degrees of kinetic energy and resume a variety of complex internal motions.

In the temperature region at which motional degrees of freedom are frozen in (cooling) or become excited (heating), the material is said to have undergone a *glass transition*. Changes in mechanical, electrical, and thermal properties accompany this molecular phenomenon. The glass transition temperature is a temperature chosen to represent the range over which the glass transition takes place. The two temperatures generally selected from DSC/DTA measurements

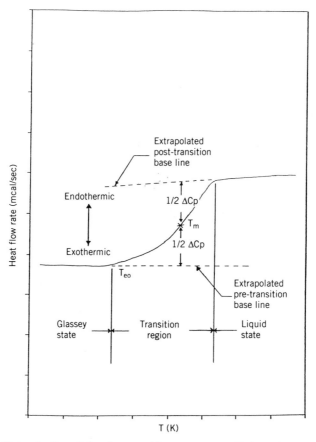

Figure 8.21   Determination of the glass transition temperature T_m from the extrapolated onset temperature T_{eo} and the change in heat capacity ΔC_p. Reprinted with permission from W. P. Brennan, *Thermal Analysis Application Study*, No. 7, Perkin–Elmer Corporation, Norwalk, CT 06856, 1973, p. 13.

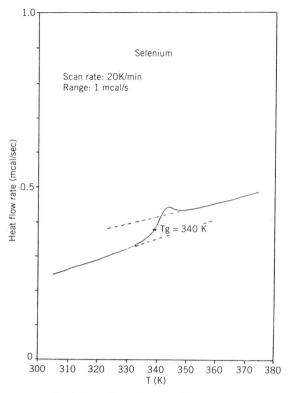

Figure 8.22 Appearance of endotherm in a glass transition at temperature T_g. Reprinted with permission from W. P. Brennan, *Thermal Analysis Application Study*, No. 7, Perkin–Elmer Corporation, Norwalk, CT 06856, 1973, p. 22.

are T_{eo}, the extrapolated onset temperature, or T_m, the point on the thermal curve corresponding to one-half the difference between the extrapolated pre- and posttransition base lines (Figure 8.21) [95]. This latter temperature corresponds more closely with the glass transition temperature determined by other methods. An endotherm (Figure 8.22 [95]) is sometimes observed in the first pass through the material as the heat capacity approaches the posttransition base line. This endotherm is usually related to the thermal history of the sample and is an enthalpy relaxation that occurs when the cooling rate is slower or on the same time scale as the heating scan.

The glass transition has occasionally been classified, mistakenly, as a *second-order phase transition*. A second-order phase transition arises from a continuous change as the low-temperature phase approaches the high-temperature phase on heating. Both phases are thermodynamically stable at the temperatures at which they are observed, and the transition is manifested by discontinuous changes in the second derivatives of the free energy, such as heat capacity or thermal expansion. The glass formed upon cooling through the glass transition is a metastable phase; cooling the high-temperature material in an infinitely slow process allows the internal degrees of freedom to come to equilibrium before they are frozen in, resulting in a crystalline rather than a glassy material. Properties like the heat capacity change continuously as the material is heated through this region.

Because the glass transition is a phenomenon of a metastable state, the temperature at which it is observed varies sensitively with cooling rate. Other factors that influence the measurement of the glass transition temperature include operating conditions such as the nature and flow rate of the purge gas and sample pan contact with the cell, sample properties such as size, shape, thermal conductivity, crystallinity, and purity, and contact between the sample and the sample pan [96, 97]. Therefore, the observed transition temperature should be considered an approximate value, valid only for the particular technique and test conditions. These conditions should be detailed whenever the transition temperature is reported.

Although the glass transition is manifested by a steplike change in the heat capacity, the exact magnitude of the change is not generally of interest. Only the temperature calibration of the instrument need be performed when this is the case.

Before determining the glass transition temperature, the specimen (10–20 mg) is heated at the same scan rate to be used for measurement, to a temperature high enough to erase its previous thermal history. Once it has equilibrated, it is cooled at approximately 20 K/min to at least 50 K below the anticipated glass transition. After the specimen has equilibrated at the lower temperature, it is again heated at the same rate as before until the transition is completed. These measurements can be made in a cooling mode also.

The glass transition serves as an indicator of the degree of cure of a thermoset polymer and aids in the evaluation of the effects of additives and copolymers and in the design of new polymers [98].

2.7.4   Thermal Conductivity

Several attempts to measure thermal conductivity are reported in the literature. By definition, thermal conductivity is the rate of heat transfer per unit temperature gradient per unit area. Knowledge of the thermal conductivity is essential to the design of equipment and processes, but it remains a very difficult property to measure. Even complex equipment in the hands of experts in the field often yields results of no better than $\pm 5\%$ accuracy. The goal of the use of scanning calorimetry for these measurements is to devise a rapid, nonrigorous technique of known accuracy for approximate, preliminary measurements.

A discussion of the theory of thermal conductivity is beyond the scope of this chapter. We summarize here some attempts to measure thermal conductivity with commercial scanning calorimeters and point out some of the difficulties encountered, but not with the kind of critical evaluation one would expect in a treatise on thermal conductivity.

Several investigators used the sample cells of a DSC for the input of energy to the system and to measure the temperature at the base of a cylindrical specimen. Thermocouples or a sensor made of a material with a well-known melting point were used to record the temperature at the other, cooler end of the specimen. In some instances a grease was used to improve sample contact with the detectors; attempts were also made to provide some level of shielding from the environment. Reference materials were measured to provide for calibration. Brennan and co-workers [99] report results for poly(tetrafluorethylene) (PTFE) and poly(styrene) that differ from literature values by 2–10%. Chiu and Fair [100] made isothermal measurements, insulated their system, and evaluated the effects of specimen length on their calibration constant. Boddington and co-workers [101] buried one junction of the thermocouple in the center of a cylinder of a pyrotechnic sample; the other junction was mounted on the outside of the cylinder in the same radial plane. Measurements were made both in scanning and in temperature-jump modes. Hakvoort and van Reijen [102] used a material that melted at an accurately known temperature as sensor for the cooler end. Flynn and Levin [103] used encapsulated sensor materials to evaluate the thermal conductivity of films. LeParlouer [104] worked with two identical cylinders that had a core surrounded by a cylinder of a different material. The cylinders were fabricated from copper and a polymer, one with the metal as core and the other with the metal surrounding the polymer. Thermal conductivities for three polymers calculated from his data agreed with literature values within 4%. Unfortunately, none of these investigators reports results on a reference material with accepted values for thermal conductivity.

2.7.5 Enthalpies of Vaporization and Sublimation

There are several reports in the literature of efforts to measure heats of vaporization or sublimation with commercial calorimeters. Those pertaining to DSC are discussed here.

The principles involved in the measurement of vaporization–sublimation enthalpies are the same as those discussed in Section 2.5.3. Problems arise, however, in preventing vaporization from occurring until the sample has equilibrated and in determining the mass of the material vaporized. Early efforts were directed toward the first of these problems: keeping the sample in the sample pans, without vaporization, until it had equilibrated and measurements were begun. These included use of pans with pinholes and pinholes with restrictions and devices to puncture sealed pans after they had equilibrated [105]. Mita and co-workers [106] modified cells and instruments to allow injection of the sample directly into the pan. Beech and Lintonbon [107] and Brostow and co-workers [108] used sealed pans with fine aluminum powder to fill the space between the specimen and the lid of the pan in an attempt to control convection within the cell and prevent unwanted condensation. As these measurements were made in the scanning mode, no comparison with literature values for isothermal vaporization is possible. Callanan [109] added sample to both sample and reference cells of a combined DSC/TG. Thus, vaporization occurred from both cells during the equilibration period. The mass of sample vaporized was determined with the thermobalance. The resulting enthalpies of vaporization, for isothermal temperatures 5–8 K lower than the known boiling points, were from 1.5 to 4% high.

3   THERMOGRAVIMETRY

3.1   Introduction

Thermogravimetry (TG) is a technique in which the mass changes of a substance are measured as the sample is subjected to a controlled temperature change. These mass changes may arise from chemical processes such as a decomposition of a thermally unstable material; an oxidation or reduction reaction with an ambient gas; or a physical process such as sublimation, vaporization, or desorption. A representation of the output of a TG experiment, where mass is plotted as a function of time or temperature, is shown in Figure 8.23. An idealized process in which mass is lost in both steps of the reaction is depicted here. The beginning temperature at which a mass change is detectable is referred to as T_i, while the temperature at which the mass becomes constant again is referred to as T_f; the difference $(T_f - T_i)$ is described as the reaction interval. The change in the ordinate is proportional to the mass lost during the process and can be reported in absolute mass units or as percent of mass loss.

The quantities T_i, T_f, and the reaction interval are sensitive functions of instrumental parameters such as scan rate, the presence of a static or flowing atmosphere, and the geometry of the sample container and furnace. Particle size and packing density are sample characteristics that also influence the details of a TG curve. Unlike thermodynamic properties, such as the enthalpy of a transition, the TG quantities are not intrinsic characteristics of a material, and

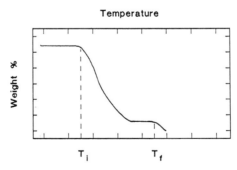

Figure 8.23   Idealized thermogravimetric curve that shows the initial T_i and final T_f temperatures used to define the reaction interval.

care must be taken in the comparison of data obtained under different experimental conditions and with different instruments.

Thermogravimetric experiments can be classified into three types according to the nature of the temperature program. In isothermal thermogravimetry, the sample is held at a fixed temperature until a constant mass is obtained. In dynamic TG, the most common form of the technique, a linear temperature increase is applied to the sample and the mass loss is followed as a function of temperature. A third temperature program, sometimes referred to as *quasiisothermal thermogravimetry*, involves alternate periods of rapid heating and isothermal holds, during which the temperature is held constant until the mass loss ceases.

3.2 Historical Development of Thermogravimetry

As a detailed treatment of the historical aspects of thermogravimetry is presented in Keattch and Dollimore [110], we give only a brief sketch here.

Honda [111] was the first person to apply the term thermobalance to an instrument where a sample could be subjected to a controlled change in its temperature while its mass was monitored simultaneously. Improvements were made through the years with different types of balances and various furnace and sample holders. In the early days, studies were primarily qualitative; Guichard [112, 113] was the first to point out the precautions that had to be taken in interpreting thermal curves because of the influence of sample size, heating rate, and a variety of other factors on the curves. Thermogravimetric studies were instrumental in characterizing and defining the conditions for preparing precipitates for classical wet-chemical inorganic analysis [114].

3.3 Components of a Thermobalance

3.3.1 Introduction

The basic components found in a modern TG system are outlined in Figure 8.24. These elements include a detection mechanism for observing mass changes, an inert container to hold the sample, a furnace that establishes a uniform and controllable temperature zone around the sample, and a means for controlling the gaseous atmosphere in contact with the sample. In the 1980s, many commercial instruments also have, as integral components, complete microcomputer systems and numerous microprocessors to control the apparatus, to acquire data, and to store data for additional numerical processing at the completion of the experiment.

In the next section, we give brief details about these major components of current TG systems. Additional information concerning design criteria and specifics of commercial units can be found in the books by Wendlandt [3] and Brown [6]. Of course, manufacturers should be consulted for the most up-to-date information about currently available commercial units.

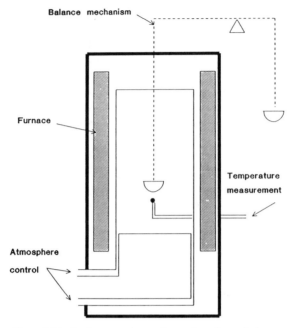

Figure 8.24   Basic components of a modern thermobalance.

3.3.2   Balances

Two important considerations in the selection of a balance are its sensitivity and its accuracy. For sample sizes in the milligram range, a microbalance is often required to achieve an accuracy of better than $\pm 0.01\%$. For versatility, the balance should be capable of operation at this sensitivity level for a range of sample masses. The mass calibration process should be a simple one and should remain stable for time periods over which an experiment is to be conducted. However, if the mass loss is measured as a percent of mass loss, no mass calibration is required because of the relative nature of that quantity, providing the balance response is linear with mass. The balance should also exhibit a high degree of stability and reproducibility. It must have a low noise level with a minimum amount of electronic filtering to maintain a quick response time that will allow for the detection of fast mass changes. This is a particularly important consideration if kinetic studies are to be made since one assumes in such studies that the signal is a measure of the instantaneous change of the sample.

The deflection and the null-point balances meet these criteria and are the most commonly applied to thermobalances. Deflection instruments incorporate one of the following: a beam, cantilever, spring, or torsion wire. Null-balance systems are preferable in TG systems because the sample is maintained in the same location in the furnace, and effects of spatial inhomogeneities in the furnace temperature are minimized.

For studies where extremely high sensitivity is required, the piezoelectric effect can be exploited to measure small changes in mass. This method takes advantage of the sensitivity of the resonant frequency of a piezoelectric crystal to crystal mass. Changes of as little as 10^{-12} g have been detected in microgram quantities of sample adsorbed onto the surface of a quartz crystal oscillator [115]. Because the resonant frequency is also a function of temperature, most studies using this technique are isothermal. A variant can be employed for nonisothermal studies; the sample is subjected to a programmed temperature change and the condensation of gaseous reaction products on the quartz crystal microbalance, maintained at some low temperature, is studied. The development of more conventional nonisothermal studies has been discussed by Glassford [116] for low temperature (< 100 K) applications; Henderson and co-workers [117] reported a system in which a computer is used to subtract out the temperature variation of the oscillator frequency. This apparatus has been used for studies below 843 K, the temperature at which quartz loses its piezoelectric character.

3.3.3   Sample Containers

The primary requirement of the sample container is that it be inert to reactions with the sample and the gaseous atmosphere that may be flowing through the system. It is desirable to have the mass of the container be as small as possible to maximize the sensitivity, since the limits of detectability and the total mass on the beam are related. Good thermal conductivity of the sample holder is also important to facilitate rapid heat transfer into the sample and to minimize the uncertainty in the sample temperature measurement. These two criteria favor the use of metals or alloys for the construction of the holder, but the temperature range of the experiment, nature of the atmosphere, or possible catalytic activity of a metallic holder may dictate glass, quartz, or alumina containers.

The geometrical design of the container can also be a critical factor in the outcome of the experiment. To maintain a uniform temperature throughout the sample, provide good contact of reagent gases with the sample, and allow easy diffusion of product gases out of the sample, the ideal container is a shallow dish or plate containing a thin layer of sample. However, the practicalities of some experiments require other configurations. For example, when inhomogeneous samples are to be studied, large quantities of material may be needed to get representative behavior. Inhomogeneity is best dealt with, however, by use of replicate specimens. A compromise can be to use a multiplate container in which several tiers of thin sample holders are stacked together. The differences that can be observed with the use of different sample containers can be seen in Figure 8.25 [118], which illustrates curves showing the dehydration of $CuSO_4 \cdot 5H_2O$ and the sample containers used in the experiments [118]. Curve 1 was obtained with the crucible shown in Figure 8.25b, while curve 2 was obtained with the multiplate container, also shown in Figure 8.25b.

Some reactions occur with spattering and these may need to be studied in

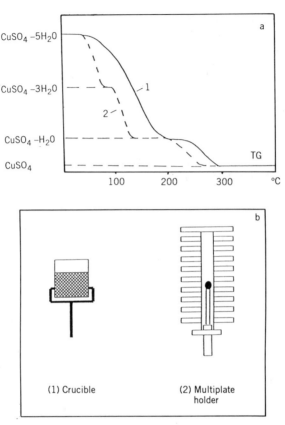

Figure 8.25 (a) Two thermogravimetric (TG) curves obtained with different sample holders: 1, crucible; 2, multiplate holder. (b) Representations of the two sample holders used to obtain the data in (a). Adapted with permission, from J. Paulik, F. Paulik, and L. Erdey, *Anal. Chim. Acta*, **34**, 419 (1966).

crucible-type containers with lids to prevent mechanical loss of sample. In such cases, care should be exercised to minimize thermal and concentration gradients within the sample by proper selection of scan and gas flow rates.

Specialized applications of the technique such as vapor pressure measurements [119, 120] or the use of self-generated atmospheres, in which the sample is maintained in contact with its reaction products [121], impose other design constraints, and the literature should be consulted for additional information on such topics.

3.3.4   Furnaces, Temperature Programmers, and Temperature Measurement

Both horizontal and vertical configurations of the furnace surrounding the sample are used. Most commercial furnaces employ electrical resistive heaters in which the resistance wire is wound noninductively around a hollow ceramic tube. The choice of wire is determined largely by the desired operating range of the instrument. Nichrome wire, for example, is limited to furnaces operating below 1200 K, while tungsten can be used up to 3000 K.

Thermogravimetric furnaces are designed to establish a uniform temperature zone around the sample rapidly. The temperature programmers provide the control necessary to vary scan rates and operating limits. Convection currents can be a problem when dynamic atmospheres are employed, and the furnace design should minimize these effects. The balance mechanism must be isolated from the furnace and protected by convection and radiation through the use of baffles and shields.

Temperatures in a TG system are usually measured with an appropriate thermocouple. In DSC and DTA systems, the device that monitors the sample temperature is in direct thermal contact with the sample holder. Such an arrangement is not typically implemented in a conventional TG system because the thermocouple interferes with the accurate detection of mass changes in the sample. Figure 8.26 shows several possible placements of the thermocouple.

Early systems monitored only the temperature of the furnace (Figure 8.26a) and assumed an equivalence of the sample temperature when, in fact, the sample temperature could lag behind that of the furnace by several tens of degrees, the actual magnitude being a function of the heating rate [122]. In modern

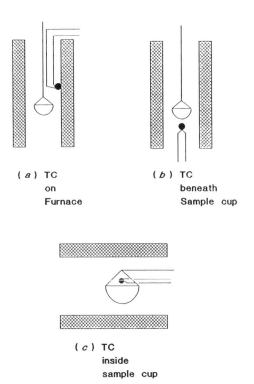

(*a*) TC
on
Furnace

(*b*) TC
beneath
Sample cup

(*c*) TC
inside
sample cup

Figure 8.26 Possible thermocouple (TC) placements in a thermobalance.

instruments, the furnace temperature is monitored independently, and a thermocouple is placed in close proximity to the sample to determine its temperature (Figure 8.26*b,c*). Because the thermocouple is not in direct contact with the sample, there is still an uncertainty about how closely the temperature sensed by the thermocouple corresponds to the true sample temperature. The problem is exacerbated when the system is operated in a vacuum because under such conditions heat transfer is greatly reduced, and radiation becomes the dominant energy transfer mechanism. Under any circumstances, radiation effects dominate the heat transfer above 760 K, and the blackbody radiation absorbing characteristics of the sample and sample holder may introduce errors in sample temperature measurement as well. A deeply colored sample, carbon black, for example, will absorb IR radiation strongly from the furnace and heat up more than a white material. The thermocouple might record the same temperature for these two samples, when in fact their true temperatures may differ by as much as 20 K above 1200 K [123]. Strongly exothermic or endothermic reactions generate heat effects within the sample that may not be detected immediately by a thermocouple located some distance away from the sample. When volatile products are produced, the dynamic atmosphere must be carefully controlled to prevent the buildup of a deposit that can alter subsequent temperature measurements.

The uncertainty in sample temperature because of thermocouple placement is further compounded by the difficulty in calibrating the temperature scale in a TG system. In DSC/DTA instruments, the temperature scale is calibrated using well-characterized samples that undergo some phase transition, usually fusion, reproducibly and over a very narrow temperature range. For TG calibrations, chemical reactions that exhibit mass changes are usually quite sensitive to variables such as scan rate, shape of the sample holder, flow rate, and composition of the purge gas, and they are no longer used as calibration standards.

In 1970 Norem and co-workers [123] introduced the use of magnetic transitions for the temperature calibration of thermobalances. Ferromagnetic materials undergo a phase transition, as the temperature is raised, to a paramagnetic state that exhibits no net magnetization. This transition occurs rapidly and at a sharply defined temperature known as the *Curie point*. When a magnet is placed around the sample holder containing the ferromagnetic material, the balance interprets the force arising from the interaction of the two magnetic fields as a gravitational pull on the mass. As the sample temperature is scanned through the Curie point, the internal magnetization of the sample drops to zero and the balance records an apparent change in mass. Repeating the measurement with several magnetic materials that have different Curie points allows one to establish the temperature scale. Garn and co-workers [124] present the results of an ICTA study to test the interlaboratory correlation of temperatures measured in this manner. From this study, a series of certified reference materials was formulated, and these are available from the United States National Institute of Standards and Technology [125]. The protocol for the ICTA study [124] required the measurement of three temperatures for each material: T_1 the extrapolated onset temperature, T_2 the midpoint temperature, and T_3 the extrapolated endpoint temperature as shown in Figure 8.27. Traditionally, T_3 is taken as the calibration temperature because it corresponds most closely to the true Curie temperature and it exhibits the least variation with the strength of the applied magnetic field [126]. Details of the magnetic calibration procedures are given in [127] for temperature calibration of a thermobalance.

The uncertainties associated with these measurements are larger than one might desire for calibration purposes, particularly if kinetic studies are to be performed. In a study of thermobalance calibration, Gallagher and Gyorgy [126] demonstrate the dependence of the transition temperatures on both heating rate and magnetic field strength. New values for the magnetic transition temperatures were reported by Blaine and Fair [128], who applied corrections to bring the experimental values closer to the true Curie temperatures. Charsley and co-workers [129] calibrated a simultaneous DTA/TG instrument with materials having well-characterized fusion temperatures and then measured the magnetic transition temperatures of the ICTA standards. The ICTA test program results were obtained at scan rates of 1-2 K/min, the measurements of Blaine and Fair [128] at 10 K/min and those of Charsley [129] at 3 and 10 K/min. No strong heating rate dependence was observed. During a study of the temperature-control procedures associated with a commercial thermobalance, Elder [130] found that the magnetic reference materials recommended by Norem [123] show significantly greater linearity than do the ICTA materials. The materials used for magnetic calibration and their approximate transition temperatures are identified in Table 8.8.

A second technique was developed that overcomes the large uncertainties associated with the ICTA standards. This method [131] utilizes several of the

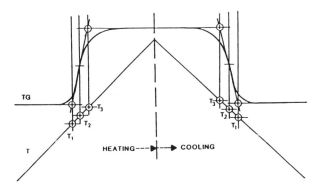

Figure 8.27 The defined points on the thermogravimetric temperature calibration curve: T_1 is the extrapolated onset, T_2 is the midpoint, and T_3 is the extrapolated end point. Reprinted with permission from P. D. Garn, O. Menis, and H. G. Wiedeman, *J. Therm. Anal.*, **20**, p. 191. Copyright © 1981 John Wiley and Sons, Ltd.

Table 8.8   Reference Materials for Thermogravimetry[a]

Material	Transition Temperature (K)	Source
Permanorm 3 metal	532	NIST/ICTA
Nickel	626	NIST/ICTA
Mumetal metal	654	NIST/ICTA
Permanorm 5 metal	727	NIST/ICTA
Trafoperm metal	750	NIST/ICTA

[a]NIST/ICTA = ICTA-certified reference materials supplied by NIST.

calibration standards established for DSC/DTA instruments for which temperatures are well known (Table 8.6). A link of a fusible metal standard is used to suspend a platinum coil above a TG sample holder. The furnace is scanned upward in temperature, the metal link melts at its fusion temperature, and the platinum coil falls either into the balance pan or through an opening cut into this pan. When the coil falls into the balance pan, an action–reaction blip occurs; with automated data acquisition procedures, the data sampling may cause this blip to be missed. However, when the coil falls through an opening cut in the balance pan, a dramatic decrease in the mass that is easily recorded by the TG is observed. The arrangement of the system for a horizontal furnace is shown in Figure 8.28a [131]. A configuration that results in a mass difference is more difficult to achieve for a vertical balance; Figure 8.28b [126] illustrates one configuration that works successfully. The TG output for several standards is shown in Figure 8.29 [10].

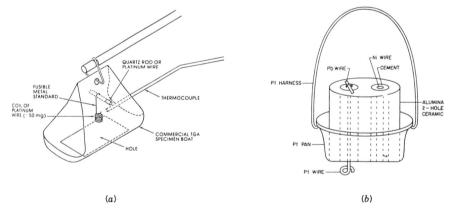

(a)   (b)

Figure 8.28 Use of a fusible link for temperature calibration in (a) a horizontal balance configuration and (b) a vertical balance configuration. Parts (a) and (b) are reprinted with permission from A. R. McGhie, J. Chiu, P. G. Fair, and R. L. Blaine, *Thermochim. Acta*, **67**, 241 (1983) and P. K. Gallagher and E. M. Gyorgy, *Thermochim. Acta*, **109**, 193 (1986), respectively.

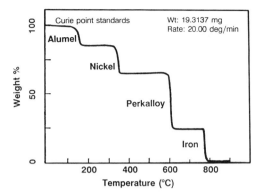

Figure 8.29 Thermal curve for selected Curie point standards (alumel, 163°C; nickel, 354°C; perkalloy, 596°C; and iron, 780°C). Reprinted with permission from C. M. Earnest, *Anal. Chem.*, **56**, 1471A (1984). Copyright © 1984 American Chemical Society.

Gallagher and Gyorgy [126] compared the Curie point and the fusible link techniques and found both to be suitable for the temperature calibration of TG instruments. The particular choice seems to be a matter of convenience or desired temperature range for the experiment.

The results of [126] illustrate one important fact. They observed a difference of about 16 K between the true melting temperature and their measured melting temperature, even with extrapolation to zero heating rate and field strength. A similar difference was observed for the Curie temperature. These differences reinforce the comparative nature of thermogravimetric calibration: The true temperature of a thermal event measured with this system would be different from the indicated temperature by this same amount (16 K). The indicated temperature for that event should be adjusted accordingly.

3.3.5 Atmosphere Control

For many studies, controlling the atmosphere surrounding the sample is a critical part of the experiment. The composition of the ambient gas and its pressure and flow rate must be controlled carefully to obtain reproducible results. A variety of configurations and control capabilities have been in-corporated in commercial units, including the ability to change atmospheric conditions under computer control. Typical operations for which atmosphere control is necessary include preliminary purging of the reaction chamber and balance area under vacuum, the operation of the furnace as well as the reaction chamber under vacuum, or switching from a purge gas to a reactive gas. Provision can also be made for sampling the products by chromatography or mass spectrometry. Further discussion of the combination of TG instruments with other techniques is discussed in Section 4.

3.3.6   Instrumentation Electronics

In commercial TG systems that are produced today, very sophisticated electronic techniques, the detailed discussion of which is beyond the scope of this chapter, are used to regulate the furnace temperature, operate the balancing mechanism, and simultaneously record its output and the sample temperature for further analysis on some storage medium such as a floppy diskette or hard disk. The data can be displayed as they are accumulated on a CRT screen, and hard copies can be obtained from a plotter before or after additional numerical processing.

Thermobalances marketed prior to the availability of integrated automated systems are automated to various levels of sophistication by accessing the analog signals these instruments normally send to recorders. Appropriate amplifiers and analog–digital conversion circuitry are used to obtain signals compatible with the computer's input requirements and timing devices to synchronize the temperature and balance measurements. Wendlandt [3] and Brown [6] summarize the requirements for such applications.

3.4   Experimental Parameters

3.4.1   Scan Rates

The effects of scan rate on the results of TG experiments were studied in considerable detail. In general, T_i and T_f for single-step endothermic processes are functions of the heating rate. Faster heating rates push both temperatures up, although not necessarily to the same extent, so that the reaction interval is also a function of the rate of heating. For example, the effect of the heating rate on the pyrolysis of poly(styrene) in a nitrogen atmosphere is shown in Figure 8.30 [122].

When consecutive reactions are involved or when a reaction occurs reversibly, the effects of scan rate become more complicated and may be interdependent upon other parameters such as atmosphere and sample characteristics. Copper sulfate pentahydrate decomposes according to the following reactions

$$CuSO_4 \cdot 5H_2O \rightarrow CuSO_4 \cdot 3H2O + 2H_2O$$

$$CuSO_4 \cdot 3H_2O \rightarrow CuSO_4 \cdot H2O + 2H_2O$$

The effect of scan rate on this process has been studied by Nagase and co-workers [132] and provides an excellent example of these possibilities, as is illustrated in Figure 8.31.

Typically, increasing the heating rate in a TG experiment causes plateaus to shorten as T_i increases faster than T_f. At a heating rate of 15 K/h in an atmosphere of static air, the two steps are well resolved as shown in Figure 8.31. As the scan rate is increased from 15 K/h to 60 K/h to 300 K/h, the onset temperatures of both reactions increase, but the length of the horizontal plateau decreases and it becomes difficult to resolve the two reactions. When the heating

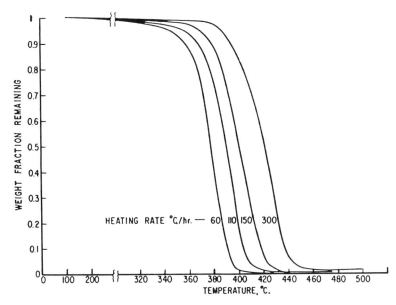

Figure 8.30 Effects of heating rate on the pyrolysis of poly(styrene) in a nitrogen atmosphere. Reprinted with permission from A. E. Newkirk, *Anal. Chem.*, **32**, 1558 (1960). Copyright © 1960 American Chemical Society.

rate is decreased, however, some unusual behavior is observed. The well-defined plateaus are lost and only some poorly resolved inflection points are observable. At these slow scan rates, with water vapor accumulating around the sample because of the static air atmosphere, equilibrium conditions are being approached, and the rates of decomposition depend more on thermodynamic factors than on the kinetic factors that determine the behavior obtained when fast scan rates are used.

A desirable scan rate, therefore, depends on several experimental conditions, usually related to thermal conductivity, and should be chosen to ensure that thermal and concentration gradients within the sample are minimized.

3.4.2 Atmosphere

The atmosphere surrounding the specimen may be inert or reactive, static or flowing. During the developments that led to modern instrumentation, much attention was given to the use of static atmospheres and, later, to self-generated atmospheres [121]. Typical effects for the decomposition of $CaCO_3$ conducted in vacuum, in static air, and in static CO_2 atmospheres are shown in Figure 8.32*a–c* [118], respectively. However, reproducible results are difficult to obtain in static atmospheres. The environment around the sample changes because of the production of volatile products that tend to stay concentrated around the sample. Convection currents sweep these away in an irregular and uncon-

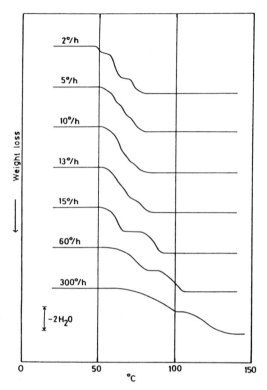

Figure 8.31 Effect of scan rate on the thermal curves for the dehydration of $CuSO_4 \cdot 5H_2O$ in static air. Reprinted with permission from K. Nagase, H. Yokobayashi, M. Kikuchi, and K. Stone, *Thermochim. Acta*, **35**, 99 (1980).

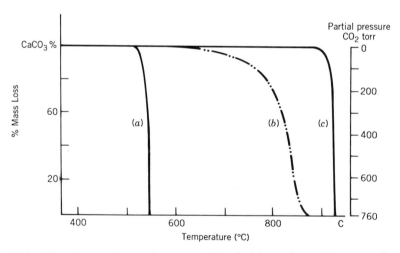

Figure 8.32 Effects of atmosphere on the decomposition of calcium carbonate: (*a*) vacuum, (*b*) static air, and (*c*) static CO_2. Adapted with permission from J. Paulik, F. Paulik, and L. Erdey, *Anal. Chim. Acta*, **34**, 419 (1966).

trollable manner. For these reasons, the use of flowing or dynamic gas atmospheres is encouraged.

There are several additional advantages associated with dynamic atmospheres. The temperature around the specimen is stabilized, and the effects of volatile products and side reactions are minimized. The flowing gas acts as a diluent for corrosive or condensible products that might damage or interfere with the sensitive balance and it sweeps them out of the reaction chamber. With proper design, flow-rate fluctuations and their effect on the noise level of the system can be minimized.

The effect of the nature of the dynamic atmosphere on the observed TG output depends on the process studied and the characteristics of the atmosphere. When inert gases are used, for example, He, N_2, or Ar, the primary influence arises from the different thermal conductivities of the gases. In comparison studies of the decomposition of $CaCO_3$ with the above gases, Caldwell and co-workers [133] found that the trend in reaction rate paralleled that of the thermal conductivities of the gases. The use of a reactive gas, such as CO_2, in this same reaction has the effect of increasing T_i, but decreasing the reaction interval [133].

Operation of the thermobalance under reduced pressure or at pressures greater than 20 kPa, poses some problems not yet discussed. When the sample area is evacuated, radiation becomes the dominant mechanism for heat transfer and significant temperature gradients can develop. Thermomolecular flow, a streaming of molecules from a hot region to a cold region, can result. It is manifested as spurious mass changes since the temperature gradient frequently occurs from the sample holder up through the balance suspension. During the initial heating of a sample decomposing under vacuum, one can also observe effects arising from the collisions of the effluent molecules with the sample holder [134]. The effect, apparently dependent on crucible design and sample characteristics, can be minimized using slower heating rates or increased pumping cross sections.

3.4.3 Sample Characteristics

Parameters of the sample that influence the thermal curve include sample mass, packing characteristics, and particle size. The first two items generally have effects that can be interpreted in terms of thermal gradients and of concentration gradients of reactant or product gases within the sample. In large samples, temperature gradients can arise either from slow heat transfer between the sample and the furnace or because of self-heating or self-cooling effects when strongly exothermic or endothermic reactions are taking place. Powders that are loosely packed have especially poor thermal conductivity and are subject to greater temperature gradients. Compressing the powder, on the other hand, may reduce the ability of reactant or product gases to diffuse uniformly throughout the sample. Sample packing is also difficult to reproduce. The most desirable combination of sample mass, packing density, and scan rate may be a matter of compromise depending on the final goals of the experiment.

The effects of particle size may be due in part to thermal conductivity: Large single crystals conduct heat better than an equivalent mass of powder. Other factors also come into play, however. Finely ground powders have very large surface areas, and the presence of significant surface energy may cause the powder to undergo different reactions, or the same reaction at markedly different rates than would be observed for a bulk crystal.

3.5   Differential Thermogravimetry

The resolution of multistep processes is frequently improved by a variant of the TG method known as *differential thermogravimetry* or DTG. In a DTG experiment, the recorded signal may be the first derivative of the mass loss with respect to time or temperature. Alternatively, the derivative curve may be calculated, after the completion of the experiment, from the thermogravimetric curve. The mass differentiation can be performed numerically on mass versus time data obtained with a conventional thermobalance or electronically in an internal mode of the instrument. In principle, the DTG curve contains exactly the same information as the TG curve. The horizontal plateaus observed with TG correspond to regions where there is no change in mass; therefore, dm/dt (or dm/dT) equals zero. At the inflection point of a TG curve, the change in mass is most rapid, and the DTG signal appears as a maximum. Whereas the change in mass is proportional to the magnitude of the ordinate of a TG curve, the integrated area under the DTG peak is related to the mass change. The correspondence between TG and DTG curves is shown in Figure 8.33.

The primary advantage to the DTG method is in its ability to detect consecutive reactions for which there are only inflection points in the TG curves, as shown in Figure 8.34. Quantitative information about the mass changes in such processes can be obtained by estimating the contribution of each peak in the region in which they overlap.

3.6   Applications

3.6.1   Compositional Analysis

The use of TG to characterize and analyze various precipitates for use in gravimetric analysis was one of its early applications [114]. Today the technique is widely used to characterize the thermal stability and composition of a broad spectrum of complex materials.

Figure 8.35 shows the results of an analysis of a bituminous coal specimen [10] in which the percentage of adsorbed water and other volatiles was determined by analysis with a flowing nitrogen atmosphere, and the amounts of fixed carbon and ash were determined by oxidation in an oxygen atmosphere. The use of TG for composition analysis is greatly assisted when the volatile products can be identified through gas chromatography or mass or infrared

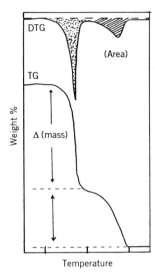

Figure 8.33 Correspondence between thermogravimetry (TG) and differential thermogravimetry (DTG) curves.

spectroscopy. The use of these combination instruments is discussed in Section 4.2.

3.6.2 Vapor Pressure Measurements

The Knudsen effusion equation relates vapor pressure of a substance to the rate of loss of molecules effusing from an orifice of a cell into a vacuum

$$p = \frac{\Delta m}{\Delta T} \cdot \frac{1}{q} \left(\frac{2\pi RT}{M} \right)^{1/2} \tag{56}$$

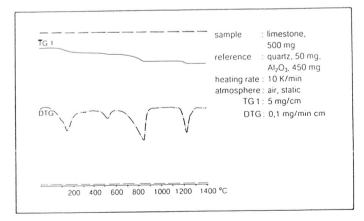

Figure 8.34 Comparison of DTG and TG curves for consecutive reactions where only inflection points are observed by TG. Adapted with permission from "Thermal Analysis Systems," Netzsch Incorporated, Exton, PA, 1984, p. 3.

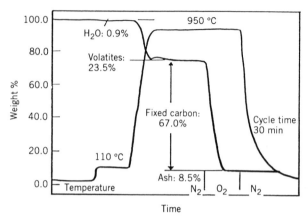

Figure 8.35 Proximate analysis of a bituminous coal in which the sample is subjected to programmed isothermal holds and changes in the atmosphere. Reprinted with permission from C. M. Earnest, *Anal. Chem.*, **56**, 1471A (1984). Copyright © 1984 American Chemical Society.

where p is the vapor pressure in units of dyne/cm^2, R is the ideal gas constant, q is the area of the orifice in cm^2, $\Delta m/\Delta t$ is the rate of mass loss through the orifice, T is the temperature, and M is the molecular weight.

To measure vapor pressures using the Knudsen effusion method, the sample holder of a thermobalance is modified [119, 120] to incorporate the effusion cell. Mass loss rates are determined from the output of the thermobalance in an isothermal, vacuum experiment. Important restrictions on experimental design are that the diameter of the orifice must be smaller than the mean free path of the molecules, the pressure outside the effusion cell must be at least an order of magnitude lower than that inside the cell, and provision must be made to trap the effusing molecules so as not to have condensation on the balance arms. Wiedemann [135] reviewed this method for vapor pressures in the range 0.0001–130 Pa. Seyler [136] evaluated the parameters affecting the determination of vapor pressure by a number of differential thermal methods. Several of these methods are discussed in Section 2.7.5. In addition to these methods, the present availability of variable pressure DSC instruments allows for the determination of the vapor pressures.

4   COMBINATION TECHNIQUES

4.1   Introduction

The current frontier in thermal analysis methods lies in the development and application of combined techniques. These techniques include the whole range of thermal methods: DSC/DTA, TG/DTG, TMA, and DMA; analytical methods such as gas chromatography (GC), mass spectroscopy (MS), Fourier transform infrared (FTIR) and ultraviolet (UV) spectroscopies; and optical methods including X-ray diffraction (XRD), microscopy, and reflectance. Many manufacturers supply instruments in a variety of configurations that incorporate two or more of these methods. In some instruments measurements are made concurrently: separate samples are used for each technique, but they are subjected to the same environment. Simultaneous measurements made on a single sample are possible in other designs.

We indicate here several of the most common types of combined techniques in use and give examples that illustrate the advantages afforded by the combination of multiple methods. One class of combination techniques allows for the correlation of mass effects with heat effects (DSC/DTA/TG), a second class allows for the identification and quantification of the volatile products of a TG experiment (TG/MS and TG/FTIR), and a third class facilitates the identification of energetic processes with solid or liquid phase behavior (DSC/DTA/XRD and DSC/microscopy).

4.2   Differential Scanning Calorimetry/Differential Thermal Analysis Thermogravimetry

Enthalpy changes are ubiquitous side effects of the processes that are suitable for study by TG methods, and one commonly finds studies in which both DSC/DTA and TG techniques are employed simultaneously, either in a single experiment using a single sample, or in parallel experiments with separate thermobalance and DSC/DTA apparatus using two samples. Comparison of the two outputs allows one to distinguish between physical changes within the sample, such as conformational changes or phase transitions, and chemical reactions. For example, a TG/DSC instrument was used to study the stability of a lubricating oil [137]. As Figure 8.36 [137] shows, there is an initial, rapid, mass loss beginning near 70°C associated with an endothermic feature in the DSC. Above 220°C, there is a slower mass loss, and the DSC registers the onset of an exothermic process. The observation of the exothermic peak from the DSC measurements indicates that a new process is associated with the mass loss at this temperature. Such a distinction is not possible, solely on the basis of the TG curve.

In many of the DSC/DTA/TG designs, the sample temperature is now measured directly and some of the problems that can occur because of thermocouple placement in TG instruments can be overcome. Charsley and co-workers [129] report the results of studies on ICTA reference materials with simultaneous TG/DTA where the temperature is calibrated using melting point standards.

4.3   Thermogravimetry/Mass Spectroscopy and Thermogravimetry/Fourier Transform Infrared

Chemical analysis of the volatile products produced during a TG experiment greatly increases the usefulness of the mass loss data. The variety of methods developed for this analysis have come to be known as *evolved gas analysis* or

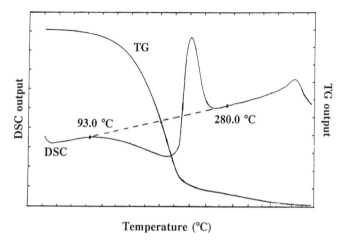

Temperature (°C)

Figure 8.36 Combined TG/DSC curve of a system showing an endothermic phase change (vaporization) followed by a chemical reaction (oxidation). Adapted with permission from P. Le Parlouer, "The Simultaneous TG-DSC 111: A New Digital and Computer Controlled Presentation," *Proceedings of the 17th North American Thermal Analysis Symposium,* Vol. 1, 1988, p. 378.

EGA. Analytical instruments that are coupled to TG equipment for the purpose of gas identification include GC, MS, and FTIR spectrometers.

Mass spectrometry is probably the most commonly used detector in EGA studies today. Numerous references to the details of the development of the combined method can be found in Brown [6], Chiu and Beattie [138, 139] and Behrens [140]. Because the optimum experimental conditions for MS and TG are not directly compatible, some difficulties are involved in coupling the two instruments. Most quadrupolar mass spectrometers are designed to operate at low pressures, and direct connection to a TG requires that the TG be operated in a vacuum. This arrangement has the inherent problems already discussed in Section 3 that are associated with vacuum TG experiments. Also, the valuable flexibility of varying atmosphere conditions is lost and the differences from the thermal curves produced in a TG operating with static or flowing atmospheres complicate the interpretation of the results. Two interface designs were developed that allow for operation of the TG at atmospheric pressure and the MS at reduced pressure. In one arrangement, a small portion of the TG effluent and purge gas is siphoned off through a molecular leak connection for analysis through the MS [141]; the other employs a jet separator in which the carrier gas, hydrogen or helium, is separated from the TG effluent before injection into the MS analysis region [142, 143]. The molecular leak interface can be used with the common gases used in TG experiments, while the operation of the jet requires the use of low molecular weight helium or hydrogen to effect proper separation of carrier and product. A commercial instrument that incorporates a dual inlet section, allowing selection of either a molecular leak or jet separator is described in [144].

The atmospheric pressure chemical ionization/MS (APCI/MS) was also interfaced successfully to a TG [145]. As its name implies, this instrument operates at ambient pressures, and the difficulties associated with the vacuum interface can be avoided. Also, it uses a lower energy ionization process, less fragmentation of the parent ion results, and additional information about product distribution can be obtained [146].

The most recent development in EGA is the interfacing of FTIR and TG instrumentation (e.g., see [147]). The FTIR can be used as a nonspecific detector yielding the total IR absorbance of the gases evolving from the sample as the reaction progresses. This quantity is proportional to the rate of gas evolution. Upon completion of the experiment, a high-resolution spectrum of the effluent as a function of time can be determined. The FTIR has the advantage over some MS systems of being able to identify more readily certain small gaseous molecules and to provide qualitative characterization of the functional groups of more complex products. The complementarity of TG/FTIR and TG/APCI/MS is discussed in [146].

4.4   Differential Scanning Calorimetry/Microscopy and Differential Scanning Calorimetry/X-Ray Diffraction

The combination of DSC and hot-stage microscopy (and other optical techniques) allows one to correlate qualitative morphological changes in the specimen with enthalpy changes. Wiedemann and Bayer [93] describe a commercial DSC/microscopy unit that allows for these techniques to be studied simultaneously on the same sample and they discuss its application to the study of phase diagrams. Additional discussion of such instruments that focuses on design considerations is found in [148]. The importance of this method in the study of liquid crystals is discussed in Section 6.2.

The development of a DSC/XRD instrument that is capable of simultaneous measurements is reported by Fawcett and co-workers [149]. Applications of this method allow correlation of the DSC data with factors such as percent crystallinity, crystallite size, and polymorphic recrystallization. Using this instrument, the authors were able to conclude from the X-ray patterns that two polymorphs of a compound were formed upon slow recrystallization and to interpret the DSC behavior in light of this information.

5   KINETICS

5.1   Introduction

The methods of DSC, DTA, and TG are applied in ever-increasing numbers to the study of reaction kinetics. The detailed treatment of kinetic theory lies outside the scope of this chapter, but excellent reviews of the various models and mathematical treatments can be found in [4, 32, 150]. A comprehensive review of thermoanalytical kinetic studies is given in [32], and a critical discussion of the factors that influence such studies is presented by Garn [151, 152], and Šesták [4, 32].

Kinetic studies that employ thermal methods can be classified according to the nature of the temperature programming of the experiment: isothermal or nonisothermal. Each type has some advantages and disadvantages. In the isothermal method, the sample is quickly heated to a temperature at which the reaction proceeds at a reasonable rate and then maintained at constant temperature. Because of the finite time required to heat the sample, there is some uncertainty associated with the initial time t_0 from which to begin the kinetic analysis. Also, a series of measurements made at different temperatures must be made to obtain kinetic parameters such as the activation energy.

In principle, measurements made with scanning-temperature instrumentation circumvent both problems. The experiment can be initiated and the base line established at some low temperature where the reaction rate is very slow. The use of a changing temperature during the measurement suggests that the kinetic parameters can be evaluated in a single experiment. This was disputed [153], however, and there is still much discussion about this topic [154, 155]. There are examples in the literature where kinetic functions obtained from nonisothermal measurements agree with those obtained from isothermal studies [156, 157], but more common are instances where the two types of measurements yield significant discrepancies [157–159].

5.2   Kinetics of Heterogeneous Reactions

Although the first work to develop a theoretical analysis of the use of thermal methods in kinetic studies was performed on solutions [160], a proliferation of studies on heterogeneous reactions has occurred in the last two decades. Because most of the difficulties in applying thermal methods to kinetics occur when solid phases are present, we concentrate the remainder of the discussion in this area.

The kinetic description of heterogeneous reactions is complicated by the existence of phase boundaries in such reactions, and the overall process may consist of three types of steps [32]: (1) transport of the reactants to the phase boundary, (2) reaction at the phase boundary, and (3) movement of the products away from the reaction zone. Any of the three steps may be a rate-limiting step. In some systems, however, several steps with comparable rates may exist, or the rate-limiting step may change during the course of the reaction.

In the study of kinetics in heterogenous reactions, concentration units used to describe the extent of reactions in solutions are replaced by a quantity α called the fractional degree of conversion, degree of reaction, or fraction decomposed. This quantity is defined according to the specific requirements of the system under study so that α proceeds from 0 to 1 as the reaction goes from initiation to completion. The kinetic rate expressions are commonly given in the form [150]

$$\frac{d\alpha}{dt} = k \cdot f(\alpha) \tag{57}$$

where k, the Arrhenius rate constant, is given by

$$k = A \cdot \exp(-E^*/RT) \tag{58}$$

and $f(\alpha)$ is called the kinetic function whose form varies for a particular mechanism. Many applications of nonisothermal kinetics use this equation in its integrated form [150]

$$g(\alpha) = \int_0^\alpha \frac{d\alpha}{f(\alpha)} = \frac{A}{\beta} \int_{T_0}^T \exp(-E^*/RT) \, dT \tag{59}$$

where β is the heating rate, assumed to be linear, and T_0 is the temperature at which the reaction begins to take place at an appreciable rate. For isothermal experiments, $g(\alpha) = kt$. Some examples of $f(\alpha)$ and $g(\alpha)$ for selected rate-limiting steps are given in Table 8.9 [150].

There is much discussion [150] of whether the frequency factor A and the activation energy E^* have a physical significance in solid-state reactions. These parameters appear to vary with experimental factors such as heating rate and sample size and also are correlated in what is known as the *compensation effect* [161]. Within a related series of experiments, variations in A are approximately compensated by changes in E^*, according to the empirical relationship

$$\log A = B + e \cdot E^* \tag{60}$$

where B and e are constants.

Table 8.9   Forms of $f(\alpha)$ and $g(\alpha)$ for Selected Mechanisms[a]

Reaction Mechanism	$f(\alpha)$	$g(\alpha)$
Power law[b]	α^{1-r}	$(1/r)\alpha^r$
Nucleation and nuclei growth		
Unimolecular decay	$(1-\alpha)$	$-\ln(1-\alpha)$
Avrami–Erofeev[c]	$(1-\alpha)[-\ln(1-\alpha)]^{1-r}$	$(1/r)[-\ln(1-\alpha)]^r$
Prout–Tompkins	$\alpha(1-\alpha)$	$\ln[\alpha/(1-\alpha)]$
Exponential	α	$\ln \alpha$
Phase boundary		
Contracting sphere	$(1-\alpha)^{2/3}$	$3[1-(1-\alpha)^{1/3}]$
Contracting cylinder	$(1-\alpha)^{1/2}$	$2[1-(1-\alpha)^{1/2}]$

[a]Adapted with permission from W. E. Brown, D. Dollimore, and A. K. Galwey, *Comprehensive Chemical Kinetics*, Vol. 22, Elsevier, New York, 1980, Table 6, pp. 90–91.
[b]$r = \frac{1}{4}, \frac{1}{3}, \frac{1}{2}$, or 1.
[c]$r = \frac{1}{4}, \frac{1}{3}, \frac{1}{2}$, or $\frac{2}{3}$.

5.3   Quantitative Methods

5.3.1   Determination of α

A typical kinetics study consists of fitting the experimentally obtained values of α or $d\alpha/dt$ to expressions of the type given in Table 8.9 to seek agreement with a particular mechanism. In TG studies, α is determined from the mass loss and stoichiometry of the reaction; $d\alpha/dt$ can be obtained from DTG experiments or by differentiation of the α versus time plot. In DTA and DSC, the measured signal, which is related to the heat absorption or evolution that accompanies the reaction, can be used to determine either α or $d\alpha/dt$ [162]. If $q(t)$ is the DSC/DTA signal at time t, calibrated according to the enthalpy methods described in Section 2.6.4, α can be obtained as a function of time according to the following equation

$$\alpha(t) = \frac{\displaystyle\int_{t_0}^{t} q(t)\,dt}{\displaystyle\int_{t_0}^{t_{end}} q(t)\,dt} = \frac{Q(t)}{Q_e} \tag{61}$$

where the denominator is the total area of the peak obtained from time t_0, the initiation of the reaction to time t_{end}, when it was completed, and the numerator is the area integrated from time t_0 to time t [162]. The time derivative of α, $d\alpha/dt$, can be obtained from (61) as

$$\frac{d\alpha}{dt} = \frac{q(t)}{Q_e} \tag{62}$$

since $q(t)$, the instantaneous power developed in the calorimeter, is the time derivative of the enthalpy. When a known heating rate is used, time and temperature become interchangeable, so that α can be determined as a function of temperature as well. This derivation assumes that only one process is responsible for the output signal developed in the calorimeter and that the enthalpy change is proportional to α across the entire composition region of interest. This may not always be the case; see, for example [163].

5.3.2   Isothermal Methods

To measure the effect of reaction rate on temperature, a series of isothermal measurements is made at different temperatures. The data are fit to various expressions of the type given in Table 8.9 at each temperature to determine the appropriate form of $f(\alpha)$ or $g(\alpha)$. Brown and Galwey [164] discuss criteria for determining obedience of isothermal data to the various rate equations. There is some indication that the kinetic function of a reaction is best determined by isothermal techniques [153].

Once the kinetic function is determined, the temperature variation of k can be obtained. Plots of $\log k$ versus $1/T$ exhibit linearity for a wide range of solid-state reactions. In fact, deviation from this behavior is frequently taken to indicate that more than one reaction may be taking place or that a different rate-limiting step controls the process with changing temperature. Šesták [165] has noted that non-Arrhenius type behavior can govern an entire solid-state process, for example, in the crystallization of some glasses.

5.3.3   Nonisothermal Methods

In most nonisothermal methods, the goal is to obtain simultaneously the kinetic function, the activation energy, and the frequency factor. A particular expression for $f(\alpha)$ or $g(\alpha)$, the temperature dependence of k, and the experimentally imposed temperature variation are combined to yield an equation for α versus time (or temperature) against which the experimental data can be tested. Various methods have been developed to analyze the data; they can be classified broadly as peak temperature, integral, and derivative methods. Critical reviews of their applications are given in [153, 165, 166].

Kissinger [167] was the first to develop a method to utilize the peak temperature T_m. He considered reactions of the type $f(\alpha) = (1-\alpha)^n$ and derived the result that β/T_m^2 is proportional to $\exp(E^*/RT_m)$, where β is the heating rate. By performing a series of experiments at various heating rates, E^* can be obtained. However, several assumptions of Kissinger are invalid for general values of reaction order n in this type of kinetic function [153]. In general, the value of α at T_m varies with β and is not constant as assumed in the Kissinger method. The method is strictly applicable only when $n = 1$ and may give misleading results for different values of n or when $f(\alpha)$ is not of the type for which the original derivation was made.

The isoconversional method, developed independently by Ozawa [168, 169] and Flynn and Wall [170], with modifications by Doyle [171, 172] and Flynn [173], extracts only the activation energy from nonisothermal data and requires no previous assumption about reaction mechanism. It is similar to the Kissinger method in that experiments are performed at various heating rates. In this method, the variation in temperature at which a particular degree of conversion is reached with different heating rates is exploited to give E^*. A series of experiments is made at a number of heating rates, β_1, $\beta_2 \cdots \beta_j$, and the temperatures T_{kj} are determined for different degrees of conversion α_k. Plots of log (or ln) β_j versus T_{kj} are then made for each α_k. The slope of the line is given, according to the approximation of Doyle [171, 172], by $0.457E^*/R$ for a $\log \beta$ plot or by $1.052E^*/R$ for a $\ln \beta$ plot. Flynn [173] gives correction factors that should be applied when E^*/RT falls outside the range $32 < E^*/RT < 45$. From this method, the constancy of E^* over temperature can be determined from the linearity of the data for each α_k series, and the constancy of the reaction mechanism as α changes can be tested by comparing the values for E^* determined at each α.

The temperature integral in (59) cannot be evaluated analytically, and much of the literature on the integral method of nonisothermal kinetics deals with techniques to approximate it. Many approximate solutions have been obtained and are discussed in [150] and [32]. These include approximation by rational fractions [174], asymptotic expansions [175, 176], and numerical integration [177, 178]. Each approximation is only valid for a limited range of E^*/RT; and because E^* is generally unknown a priori, there is uncertainty about which approximation to apply.

Use of the differential form (57) avoids the errors introduced by the approximation of the temperature integral, but differentiation of the data to obtain $d\alpha/dt$ may propagate additional errors. Because the output from DSC and DTA instruments, when properly corrected, is proportional to $d\alpha/dt$, these methods lend themselves to treatments using the differential form. Expressions for $f(\alpha)$ in terms of the variables that are obtainable directly from the calorimetric experiment are developed in [162]. Thermogravimetric data, on the other hand, yield information more directly about α versus time and must be differentiated for these methods.

The Freeman and Carroll [179] technique is widely used to determine the order of reactions where $f(\alpha)$ is given by $(1-\alpha)^n$. In this method values of $\ln(d\alpha/dt)$, $\ln(1-\alpha)$, and $1/T$ are determined for different values of α. Incremental plots of $\Delta[\ln(d\alpha/dt)]/\Delta[\ln(1-\alpha)]$ versus $\Delta[1/T]/\Delta[\ln(1-\alpha)]$, where Δ means the difference between successive values of the quantity, are expected to yield a straight line with slope E^*/R and an intercept of n for reactions obeying nth-order kinetics. Criado and co-workers [180] show, however, that this method can lead to an erroneous agreement by reactions having kinetic functions not of the form

$$f(\alpha) = (1-\alpha)^n$$

Problems arise in both the integral and derivative methods because the differences between the experimental data and the fit are about the same for several choices for $f(\alpha)$. For example, the dehydration of $Mg(OH)_2$ is studied [181] and, as is shown in Table 8.10 [182], the data give equally good fits to several diverse mechanisms based on the correlation coefficient R obtained for each fit. It is difficult, therefore, to select the best mechanism from these considerations only. There is not yet an accepted method for determining the best fit, although Vyazovkin and Lesnikovich [182] discussed the use of statistical analysis in this problem.

5.3.4   Difficulties With the Methods

We discussed at length that variables, such as heating rate, sample and particle size, and atmospheric conditions, can have a significant effect on the enthalpy change or mass loss detected in the thermal experiment. Consequently, the kinetic parameters E^*, k, and A measured from the thermal experiments can show significant variation from laboratory to laboratory [183].

Table 8.10   Kinetic Parameters for the Decomposition of $Mg(OH)_2$ Derived for Various $g(\alpha)$[a]

$g(\alpha)$	E^* (kJ/mol)	A (s^{-1})	R
$-\ln(1-\alpha)$	210.62	1.3×10^{14}	0.997
$[-\ln(1-\alpha)]^{2/3}$	136.81	1.4×10^8	0.997
$[-\ln(1-\alpha)]^{1/2}$	99.905	1.2×10^5	0.997
$[-\ln(1-\alpha)]^{1/3}$	63.417	8.9×10^3	0.997
$[-\ln(1-\alpha)]^{1/4}$	44.539	<1	0.997
$(1-\alpha)^{-1}-1$	254.03	1.3×10^{14}	0.999
$(1+\alpha)^{-1/4}+1$	231.30	1.3×10^{14}	0.999

[a]Reprinted with permission from S. V. Vyazovakin and A. I. Lesnikovich, *J. Therm. Anal.*, **30**, 832 (1984). Copyright © 1984 John Wiley & Sons, Inc.

The heat generated or absorbed during the reaction can cause local hot or cold spots that will greatly affect the apparent rate of reaction. The use of smaller samples and slow scan rates helps to reduce, but may not completely eliminate, the problem of temperature inhomogeneities. The heating rate must be fast enough to ensure that equilibrium of the chemical reaction is not achieved, however; and minimum sample sizes may be limited by instrument sensitivity, sample homogeneity, and the energetics of the process. Other factors that affect the thermal conductivity need also be considered to reduce this problem. These include sample characteristics, such as particle size and packing density. However, factors such as particle size can introduce additional complications into the kinetic analysis by changing the relative contributions of different processes to the observed reaction rate.

Methods that employ DSC/DTA techniques are subject to a common problem when the heat capacity of the sample undergoes significant changes during the course of the reaction. As discussed in Section 2.5.2, such changes cause base line shifts in the DTA and DSC curves that may interfere with the accurate evaluation of the enthalpy. Increasing the mass of the sample holder may help to minimize the contribution of the sample heat capacity, but it can reduce the instrument sensitivity significantly.

Another important consideration is the magnitude of the time constant of the calorimeter relative to the instantaneous rate of the reaction. The transfer of heat by conduction causes a smearing out of the instantaneous heat absorption as measured by the instrument detector. The slow response of a calorimeter with a large time constant will result in the measured output being deformed significantly from the true instantaneous enthalpy change, the quantity needed to determine the time variation of α.

Various techniques were developed to deconvolute the observed signal, and numerous studies were published on the topic. Representative examples of methods include dynamic optimization [184], harmonic analysis [185], and electronic [186] and numerical [187] filtering to yield the true enthalpy value. These techniques are compared and evaluated by Zielenkiewicz [188]; Randzio and Suurkuusk [189] discuss when such corrections are necessary. More commonly, they are significant for Calvet-type calorimeters, where the time constant may be as large as 40 min and can be ignored for disk-type and power-compensated calorimeters. Šesták [165], however, has pointed out that the basic equation of heat-flux calorimeters (17) always prevents a direct determination of the instantaneous signal from these kinds of instruments.

5.3.5 Conclusion

In conclusion, while thermal techniques are widely used to study kinetics of solid-state reactions, there are still questions about their general validity. A survey of recent literature [155] shows that many different methods are used to calculate E^*, with little or no justification given for the choice. There is only rare agreement between E^* values calculated for the same reaction using different methods. The lack of consensus suggests that caution be used in accepting values for the kinetic parameters that are determined solely from nonisothermal methods. Agreement of a single set of thermal data with a particular kinetic model does not guarantee that the reaction will proceed according to the mechanism on which the model is based. Additional studies exploring geometric factors, using microscopic techniques, for example, are recommended [150].

Garn [152] points out that (57) implies a dependence of the reaction rate only on composition α and not on the experimental parameters of atmosphere, scan rate, or particle size. He contends that when these incidental factors are observed to influence the reaction rate, any relation derived from the fundamental assumption of (57) is likely to be invalid. This may be the reason for the wide range of parameters one finds in the literature for the same reaction. Before one attempts to embark on a fitting procedure discussed above, Garn recommends some simple tests to determine if the reaction rate is truly a function of composition alone.

The ICTA has formed a committee to establish guidelines for the use of thermal analysis methods for the study of kinetics [190]. This committee is preparing a report that defines terms, and compiles kinetic data for solid decomposition reactions and practical applications such as shelf-life predictions and hazards analysis.

6 EXAMPLES OF APPLICATIONS IN SELECTED DISCIPLINES

6.1 High-Pressure Studies

High pressures are of interest both in chemical reactions, where the effects of an active gas under high pressure can be studied, and in the study of nonreacting systems, where pressure versus temperature phase diagrams can be generated.

The pressure-generating medium most commonly used in thermal analysis instrumentation is a gas such as argon, nitrogen, and helium for inert gas applications and oxygen, hydrogen, carbon dioxide, carbon monoxide, and water for reactive gas applications. For pressures above 100 atm, it is generally necessary to use an auxillary pressure generator. Below this pressure, the high-pressure gas can be obtained directly from a gas cylinder. Liquids are also used to generate the pressure. Reports of the use of solid pressure-transmitting media are rare.

Table 8.11 lists representative instruments that are designed for studies at high pressure. The entries are not meant to be exhaustive, but can serve as an introduction to the types of design considerations required. The development of high-pressure thermobalances up to 1976 is reviewed by Dobner and co-workers [196]. Many commercial instruments are currently available for use in a variety of pressure and temperature regimes.

Applications of DSC/DTA and TG techniques under high-pressure gas atmospheres are reviewed in [197]. They include studies of decompositions of inorganic compounds under high pressures of active gases, explosives that are of interest to the rocket industry, and studies dealing with coal gasification and liquifaction.

6.2 Liquid Crystals

A rich variety of phases was discovered in liquid crystals, and thermal methods are very valuable in characterizing them. The capability to perform measurements of the enthalpy and temperature of phase transitions [198] on small samples is an important aspect of the studies in this field. Metastable phases can be detected and studied because the measurements can be made both in heating and cooling modes. In conjunction with optical microscopy and X-ray diffraction measurements, DSC studies are used extensively to delineate phase diagrams for binary mixtures of liquid crystals. A review of thermodynamic data on pure mesogens is presented in [198].

Table 8.11 High-Pressure DSC/DTA and Thermobalances

Instrument	Maximum Temperature Range (K)	Pressure (MPa)	Reference
DTA	Unspecified	60	[191]
DTA	250–470	250	[192]
DTA	Unspecified	300	[193]
DSC	300–600	600	[194]
TG	298–770	50	[195]
TG	298–1370	30	[196]

A disadvantage of the DSC method, however, is its inability to distinguish between first-order and second-order transitions; these fine distinctions can be important in the testing of theoretical models of the phase transitions. Ratna and Chandrasekhar [199] demonstrate that studying the effect of scan rate on the excess enthalpy from a liquid crystalline phase transition may allow such differentiation. In Figure 8.37 [199], the apparent enthalpy of transition for two phase transitions in 4-*n*-octyl-4'-cyanobiphenyl (8CB) is depicted as a function of scan rate. The nematic to smectic-*A* phase of 8CB is known by other methods to be second order; the enthalpy of this transition extrapolates to zero at zero heating rate (Figure 8.37*a*). For the first-order, isotropic liquid to nematic phase transition, there is a nonzero value for the apparent latent heat at zero heating rate (Figure 8.37*b*).

6.3   Polymers

Thermal analysis methods are widely used by polymer scientists. In this field, as in many others, the availability of high-quality commercial instruments, the small samples required, and the rapidity with which measurements can be conducted contribute to the explosive growth in the use of thermal analysis techniques for a wide variety of polymer-related problems. Thousands of publications on their use appear annually. An excellent monograph that is dedicated exclusively to this general topic is [12]; another monograph with more specialized articles is [200].

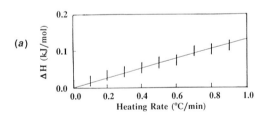

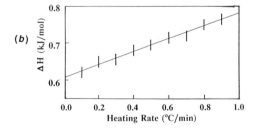

Figure 8.37   The effect of scan rate on the observed enthalpy of transition Δ*H* for (*a*) a second-order transition, and (*b*) a first-order transition in a liquid crystal. Reprinted with permission from B. R. Ratna and S. Chandrasekhar, *Mol. Cryst. Liq. Cryst.*, **162B**, 157 (1989). Courtesy of Gordon & Breach Science Publishers S. A.

Wunderlich [201] performed a systematic analysis of the magnitude of linear macromolecular heat capacities. Using a data bank generated from these studies and adjuvant vibrational spectroscopy studies, he calculates heat capacities for polymers at high temperatures above and below the melt (see [202] for an extensive list of references). He also shows that the heat capacity as well as the glass transition yield quantitative information about the structure of two-phase polymeric systems such as semicrystalline polymers and block copolymers [202].

Studies by TG and DSC contribute to the understanding of the curing of thermosetting resins [203], decomposition and degradation phenomena, characterization of polymer additives [204], and polymer flammability [205]. These techniques find use both on the routine level, for quality control purposes, as well as for research and development applications.

6.4   Biochemistry and Biology

Differential scanning calorimetry has been used to study phase transitions such as the helix to coil transition in DNA; the irreversible denaturation of proteins; and the reversible gel to liquid–crystalline transition that occurs in lipid bilayers and biological membranes. Such studies have the advantage not only of indicating the temperatures at which these phenomena occur, but also of yielding information about the enthalpies and entropies for the process under study. Comparison of these quantities can produce information about the relative stabilities of various states of the biological system under different conditions of pH, ionic strength, or concentration of ions that bind with the substrate. From such studies, for example, the mode of drug–DNA interactions is delineated [206], predictions of the stability of various DNA base pairing arrangements are made [207], and the contribution to the helix stability arising from hydrogen bonding and stacking interactions is determined for different types of DNA [208].

When the resolution of the signal is not instrument-limited, information about the relative cooperativity of a phase transition can be obtained from the breadth of the thermal peak. For example, the width of a transition in pure synthetic phospholipids, where intermolecular cooperativity is high, can be 0.1 K, while in very complex biological membranes, where the transition is less cooperative, the breadth can be as much as 10–15 K [209].

A DSC operated in an isothermal mode can be used as a bioactivity monitor in which heat effects can be used to study metabolic processes. "Fingerprint" curves of heat production versus time in a specific growth medium can be used to identify microbes, in many cases on a faster time scale than with conventional methods. The efficacy of drugs can be tested by observing the reduction in power level, reflecting a reduction in metabolic activity, when the active ingredient is applied to an organism. The feasibility of using the heat output from tomato and carrot cells to monitor their metabolic response to variations in temperature and oxygen concentration was demonstrated [210]. Good correlation between cell culture results and those obtained from tissue sections was observed. This method allows for rapid testing of cellular responses to various kinds of stress and is useful for screening purposes.

Reviews that provide numerous references to biological and biochemical applications of DSC include those of McElhany [211] on studies of the interactions of lipids with proteins in model membranes, and those on biological membranes [209, 212] and other biological macromolecules [213]. Thermodynamic studies of nucleic acids and proteins are reviewed by Privalov [214]. Reviews of the literature dealing with broader topics in biochemistry and biotechnology are found in [215–217]. Schwarz and Kirchhoff [218] discuss factors for testing the performance of calorimeters used in biological studies and recommend ribonuclease A in a glycine buffer as a calibration material.

6.5   Pharmaceuticals

Physicochemical interactions between a drug and an excipient, the medium in which the drug is incorporated, can alter important properties of the drug, such as its shelf life or its in vivo distribution after ingestion. Thermal analysis techniques, primarily DSC, are used to probe these interactions through the establishment of phase diagrams (e.g., see [219, 220]) or signs of chemical reactivity. Predictions of drug-excipient compatibilities are made on the basis of a comparison of the DSC trace of a mixture of drug and excipient to the sum of the traces of the two pure components [221, 222]. Deviation of the trace of the mixture from that expected by additivity can be taken as an indication of some kind of interaction, such as chemical reaction, adsorption, or eutectic or complex formation.

The ability of many organic compounds to exist with different crystallographic structures, known as *polymorphs*, can pose problems for the formulation and processing of drugs because the various polymorphs of an active agent usually have different densities, solubilities, and melting points. These differences can result in significant changes in dosage requirements or stabilities for formulations prepared from different polymorphs. Measurements by DSC can be used to determine the existence of polymorphic phases, to determine the relative stabilities and temperature ranges of each, and to investigate any differences in drug-excipient interactions that may tend to favor the use of one polymorph [221].

Isothermal and nonisothermal kinetics studies of drugs and drug formulations are frequently made for prediction of stability and aging predictions. The general basis for these studies is described in more detail in Section 5, and specific applications to the pharmaceutical industry can be found in [221] and [223].

Another important application is purity determination with DSC (Section 2.7.1), using ASTM Standard E914-87 [127], and with TG to determine the composition of hydrates [224] or the presence of other volatile components [222, 225]. Thermogravimetry is also useful for identifying whether an endothermic phase transition detected by DSC is caused by chemical decomposition or by a polymorphic phase change [225].

7 EPILOGUE

The techniques of DTA, DSC, and TGA are established workhorses in the assembly of modern physical methods. In some instances, one of them may be the only technique available to produce desired results; in other situations, they are used to provide valuable adjuvant data as part of a larger project. It is important to keep in mind, especially since instruments are now more "user-friendly," that the appropriate procedures and attention to detail must still be followed to provide the most reliable, and, therefore, worthwhile results.

References

1. R. C. Mackenzie, *Pure Appl. Chem.*, **57**, 1737 (1985).
2. W. W. Wendlandt and P. K. Gallagher, "Instrumentation," in E. Turi, Ed., *Thermal Characterization of Polymeric Materials*, Academic, New York, 1981, pp. 1–90.
3. W. W. Wendlandt, *Thermal Analysis*, 3rd ed., Wiley, New York, 1986.
4. J. Šesták, "Thermophysical Properties of Solids. Their Measurements and Theoretical Thermal Analysis," in G. Svehla, Ed., *Wilson and Wilson's Comprehensive Analytical Chemistry*, Vol. 12, Part D, Elsevier, New York, 1984.
5. W. Hemminger and G. Höhne, *Calorimetry—Fundamentals and Practice*, Verlag Chemie, Deerfield Beach, FL, 1984.
6. M. E. Brown, *An Introduction to Thermal Analysis: Techniques and Applications*, Chapman and Hall, London, 1988.
7. W. W. Wendlandt, *Anal. Chem.*, **56**, 250R (1984).
8. W. W. Wendlandt, *Anal. Chem.*, **58**, 1R (1986).
9. D. Dollimore, *Anal. Chem.*, **60**, 274R (1988).
10. C. M. Earnest, *Anal. Chem.*, **56**, 1471A (1984).
11. W. A. Kneller, *Thermochim. Acta*, **108**, 357 (1986).
12. E. Turi, Ed., *Thermal Characterization of Polymeric Materials*, Academic, New York, 1981.
13. D. Dollimore, *Thermochim. Acta*, **50**, 123 (1981).
14. G. Liptay, *J. Therm. Anal.*, **25**, 235 (1982).
15. B. Wunderlich, *Thermochim. Acta*, **83**, 35 (1985).
16. R. C. Mackenzie, *Thermochim. Acta*, **73**, 251 (1984).
17. R. C. Mackenzie, *Thermochim. Acta*, **73**, 307 (1984).
18. H. Le Chatelier, *Bull. Soc. Fr. Mineral.*, **10**, 204 (1887).
19. H. Le Chatelier, *Compt. Rend.*, **104**, 1443 (1887).
20. H. Le Chatelier, *Compt. Rend.*, **104**, 1517 (1887).

21. H. Le Chatelier, *Z. Phys. Chem.*, **1**, 396 (1887).

22. F. Rudberg, *K. Sven. Vetenskapsakad. Handl.*, **1829** 157 (1830).

23. H. Le Chatelier, *C.R. Acad. Sci., Paris*, **102**, 819 (1886).

24. W. C. Roberts-Austen, *Proc. Inst. Mech. Eng.*, 543 (1891).

25. W. C. Roberts-Austen, *Proc. Inst. Mech. Eng.*, 35 (1899).

26. S. L. Boersma, *J. Am. Ceram. Soc.*, **38**, 281 (1955).

27. E. S. Watson, M. J. O'Neill, J. Justin, and N. Brenner, *Anal. Chem.*, **36**, 1233 (1964).

28. E. Calvet and H. Prat, *Recent Progress in Microcalorimetry*, Pergamon, New York, 1963.

29. W. P. Brennan and A. P. Gray, *Thermal Analysis Application Study 9*, Perkin–Elmer Corporation, Instrument Division, Norwalk, CT 06856, 1973.

30. H. S. Carslaw and J. C. Jaeger, *Conduction of Heat in Solids*, 2nd ed., Oxford University Press, London, 1959.

31. A. P. Gray, "A Simple Generalized Theory for the Analysis of Dynamic Thermal Measurement," in R. S. Porter and J. F. Johnson, *Analytical Calorimetry*, Vol. 1, Plenum, New York, 1968, pp. 209–218.

32. J. Šesták V. Šatava, and W. W. Wendlandt, *Thermochim. Acta*, **7**, 333 (1973).

33. R. A. Baxter, "A Scanning Microcalorimetry Cell Based on a Thermoelectric Disc—Theory and Applications," in R. F. Schwenker and P. D. Garn, Eds., *Thermal Analysis*, Vol. 1, Academic, New York, 1969, pp. 65–84.

34. T. Ozawa, *Bull. Chem. Soc. Jpn.*, **39**, 2071 (1966).

35. M. J. O'Neill, *Anal. Chem.*, **36**, 1238 (1964).

36. J. H. Flynn, "Theory of Differential Scanning Calorimetry—Coupling of Electronic and Thermal Steps," in R. S. Porter and J. F. Johnson, *Analytical Calorimetry*, Vol. 3, Plenum, New York, 1974, pp. 17–44.

37. S. Tanaka, *Thermochim. Acta*, **61**, 147 (1983).

38. P. Claudy, J. C. Commercon, and J. M. Letoffe, *Thermochim. Acta*, **65**, 245 (1983).

39. P. Claudy, J. C. Commercon, and J. M. Letoffe, *Thermochim. Acta*, **68**, 305 (1983).

40. P. Claudy, J. C. Commercon, and J. M. Letoffe, *Thermochim. Acta*, **68**, 317 (1983).

41. J. D. Lee and P. F. Levy, "Heat Flux Differential Scanning Calorimetry—Theory and Practice," in *Proceedings of the 11th North American Thermal Analysis Symposium* 1981, pp. 215–228.

42. S. C. Mraw, *Rev. Sci. Instrum.*, **53**, 228 (1982).

43. J. J. Kessis, *C. R. Acad. Sci., Paris Ser. C*, **283**, 83 (1976).

44. C. M. Guttman and J. H. Flynn, *Anal. Chem.*, **45**, 408 (1973).

45. R. N. Goldberg and E. J. Prosen, *Thermochim. Acta*, **6**, 1 (1973).

46. Y. Saito, K. Saito, and T. Atake, *Thermochim. Acta*, **104**, 275 (1986).

47. Y. Saito, K. Saito, and T. Atake, *Thermochim. Acta*, **107**, 277 (1986).

48. G. W. H. Höhne, K.-H. Breuer, and W. Eysel, *Thermochim. Acta*, **69**, 145 (1983).

49. S. L. Randzio, *Thermochim. Acta*, **89**, 215 (1985).

50. J. P. Dumas, *J. Phys. C*, **9**, L143 (1976).

51. A. Dyer and T. R. Nowell, *Thermochim. Acta*, **29**, 171 (1979).

52. J. L. Illinger, N. S. Schneider, and F. E. Karasz, *Thermochim. Acta*, **42**, 51 (1980).

53. J. C. Tou and L. F. Whiting, *Thermochim. Acta*, **42**, 21 (1980).

54. L. F. Whiting, M. S. LaBean, and S. S. Eadie, *Thermochim. Acta*, **136**, 231 (1988).

55. H. Suzuki and B. Wunderlich, *J. Therm. Anal.*, **29**, 1369 (1984).

56. J. E. Callanan and S. A. Sullivan, *Rev. Sci. Instrum.*, **57**, 2584 (1986).

57. S. C. Mraw and D. F. Naas, *J. Chem. Thermodyn.*, **11**, 567 (1979).

58. J. E. Callanan, K. M. McDermott, R. D. Weir, and E. F. Westrum, Jr., *J. Chem. Thermodyn.*, **24**, 233 (1992).

59. G.-W. Jang, R. Segal, and K. Rajeshwar, *Anal. Chem.*, **59**, 684 (1987).

60. J. E. Callanan, S. A. Sullivan, and D. F. Vecchia, "Feasibility Study for the Development of Standards Using Differential Scanning Calorimetry", Natl. Bur. Stand. (U.S.) Spec. Publ. 260-99 (1985).

61. M. A. White, *Thermochim. Acta*, **74**, 55 (1984).

62. R. A. McDonald, J. E. Callanan, and K. M. McDermott, *Energy Fuels*, **1**, 535 (1987).

63. *Thermal Analysis Newsletter*, No. 5, Perkin–Elmer Corporation, Norwalk, CT 06856.

64. H. M. Heuvel and K. C. J. B. Lind, *Anal. Chem.*, **42**, 1044 (1970).

65. M. J. O'Neill, *Anal. Chem.*, **38**, 1331 (1966).

66. H. W. Luebke and J. J. Tria, "Improvements in the Perkin–Elmer Specific Heat Capacity Software for the DSC-2 Differential Scanning Calorimeter," in *Proceedings of the 13th North American Thermal Analysis Symposium*, 1984, p. 207.

67. B. Cassel, *New Techniques in DSC: Differential Heat Capacity Determinations for Maximum Accuracy*, Pittsburgh Conference on Analytical Chemistry, 1974.

68. J. G. Hust, J. E. Callanan, and S. A. Sullivan, *Therm. Conduct.*, **19**, 533 (1988).

69. A. S. Lindsey, H. M. Paisley, B. E. Broderick, and J. M. Ellender, *National Physical Laboratory (UK) Report CHEM 64*, Teddington, UK, 1977.

70. P. A. Barnes, E. L. Charsley, J. A. Rumsey, and S. B. Warrington, *Anal. Proc.*, **21**, 5 (1984).

71. J. E. Callanan and D. F. Vecchia, "Scanning Calorimetric Measurements of the Temperature and Enthalpy of Fusion of Tin for Certification as a Reference Material," to be published.

72. K.-H. Breuer and W. Eysel, *Thermochim. Acta*, **57**, 317 (1982).

73. W. P. Brennan, B. Miller, and J. C. Whitwell, "Thermal Resistance Factors in Differential Scanning Calorimetry," in R. S. Porter and J. F. Johnson, *Analytical Calorimetry*, Vol. 2, Plenum, New York, 1970, pp. 441–450.

74. J. H. Flynn, *Thermochim. Acta*, **8**, 69 (1974).

75. M. J. Richardson, *J. Polym. Sci.*, **38**, 25 (1972).

76. ASTM Standard E967-83, *Standard Practice for Temperature Calibration of Differential Scanning Calorimeters and Differential Thermal Analyzers*, Annual Book of Standards, Vol. 14.02, Philadelphia, PA, 1988, p. 583.

77. J. E. Callanan, K. M. McDermott, and S. A. Sullivan, to be published.

78. J. E. Callanan and D. F. Vecchia, "Enthalpy Calibration of Scanning Calorimeters," to be published.

79. ASTM Standard E968-83, *Standard Practice for Heat Flow Calibration of*

Differential Scanning Calorimeters, Annual Book of Standards, Vol. 14.02, American Society for Testing and Materials, Philadelphia, PA, 1988, p. 587.

80. J. L. McNaughton and C. T. Mortimer, "Differential Scanning Calorimetry," in H. A. Skinner, Ed., *International Reviews of Science: Physical Chemistry, Series Two*, Vol. 10, Butterworth, London, 1975, pp. 1–44.

81. M. J. Richardson and P. Burrington, *Thermochim. Acta*, **6**, 345 (1974).

82. M. J. Richardson and N. G. Savill, *Thermochim. Acta*, **12**, 213 (1975).

83. E. E. Marti, *Thermochim. Acta*, **5**, 173 (1972).

84. E. F. Joy, J. D. Bonn, and A. J. Barnard, Jr., *Thermochim. Acta*, **2**, 57 (1971).

85. W. P. Brennan, M. P. DeVito, R. L. Fayans, and A. P. Gray, "An Overview of the Calorimetric Purity Measurement," in R. L. Blaine and C. K. Schoff, Eds., *Purity Determinations by Thermal Methods*, American Society for Testing and Materials, Philadelphia, PA, 1984, pp. 5–15.

86. D. L. Sondack, *Anal. Chem.*, **44**, 888 (1972).

87. P. D. Garn, B. Kawalec, J. J. Houser, and T. F. Habash, "Dynamic Purity Measurements," in B. Miller, Ed., *Thermal Analysis, Proceedings of the 7th International Conference on Thermal Analysis*, Vol. 2, Wiley, New York, 1982, p. 899.

88. W. Brostow, M. A. Macip, M. Sanchez-Rubio, and M. A. Valerdi, *Mater. Chem. Phys.*, **10**, 31 (1984).

89. S. V. R. Mastrangelo and R. W. Dornte, *J. Am. Chem. Soc.*, **77**, 6200 (1955).

90. J. P. Elder, "Purity Analysis by Dynamic and Isothermal Step Differential Scanning Calorimetry," in R. L. Blaine and C. K. Schoff, Eds., *Purity Determinations by Thermal Methods*, American Society for Testing and Materials, Philadelphia, PA, 1984, pp. 50–60.

91. J. P. Elder, *Thermochim. Acta*, **34**, 11 (1979).

92. A. C. Ramsland, *Anal. Chem.*, **52**, 1474 (1980).

93. H. G. Wiedemann and G. Bayer, *J. Therm. Anal.*, **30**, 1273 (1985).

94. R. E. Mills and R. T. Coyle, *Thermochim. Acta*, **124**, 65 (1988).

95. W. P. Brennan, *Thermal Analysis Application Study*, No. 7, Perkin–Elmer Corporation, Norwalk, CT 06856, 1973.

96. J. R. Saffell, *Thermochim. Acta*, **36**, 251 (1980).

97. E. Donoghue, T. S. Ellis, and F. E. Karasz, "The Effect of Sample Temperature Gradients on DSC Thermograms at the Glass Transition Temperature," in J. F. Johnson and P. S. Gill, Eds., *Analytical Calorimetry*, Vol. 5, Plenum, New York, 1984, pp. 325–341.

98. D. M. Leisz, L. W. Kleiner, and P. G. Gertenbach, *Thermochim. Acta*, **35**, 51 (1980).

99. W. P. Brennan, B. Miller, and J. C. Whitwell, *J. Appl. Polym. Sci.*, **12**, 1800 (1968).

100. J. Chiu and P. G. Fair, *Thermochim. Acta*, **34**, 267 (1979).

101. T. Boddington, P. G. Laye, and J. Tipping, *Combust. Flame*, **50**, 139 (1983).

102. G. Hakvoort and L. L. van Reijen, *Thermochim. Acta*, **93**, 317 (1985).

103. J. H. Flynn and D. M. Levin, *Thermochim. Acta*, **126**, 93 (1988).

104. P. LeParlouer, "Rapid Determination of Specific Heat and Thermal Conductivity by DSC," in *Proceedings of the 16th North American Thermal Analysis Symposium*, Washington, DC, September, 1987, p. 5.

105. R. E. Farritor and L. C. Tao, *Thermochim. Acta*, **1**, 297 (1970).

106. I. Mita, I. Imai, and H. Kambe, *Thermochim. Acta*, **2**, 337 (1971).

107. G. Beech and R. M. Lintonbon, *Thermochim. Acta*, **2**, 86 (1970).

108. W. Brostow, D. M. McEachern, and J. A. Valdez, *Mater. Chem.*, **6**, 187 (1981).

109. J. E. Callanan, "Development of Standard Measurement Techniques and Standards Reference Materials for Heat Capacity and Heat of Vaporization of Jet Fuels," National Bureau of Standards (U.S.) Interagency Report 88-3093, NIST, Washington, DC, 1988.

110. C. J. Keattch and D. Dollimore, *An Introduction to Thermogravimetry*, 2nd ed., Heyden, London, 1975.

111. K. Honda, *Sci. Rep. Tohoku Univ.*, **4**, 97 (1915).

112. M. Guichard, *Bull. Soc. Chim. Fr.*, **33**, 258 (1923).

113. M. Guichard, *Bull. Soc. Chim. Fr.*, **37**, 251 (1925).

114. C. Duval, *Inorganic Thermogravimetric Analysis*, 2nd ed., Elsevier, Amsterdam, 1963.

115. A. W. Czanderna and C. Lu, "Introduction, History, and Overview of Applications of Piezoelectricity in Quartz Crystal Microbalances," in A. W. Czanderna and C. Lu, Eds., *Applications of Piezoelectric Quartz Crystal Microbalances*, Elsevier, New York, 1984, pp. 1–17.

116. A. P. Glassford, "Application of the Quartz Crystal Microbalance to Space System Contamination Studies," in A. W. Czanderna and C. Lu, Eds., *Applications of Piezoelectric Quartz Crystal Microbalances*, Elsevier, New York, 1984, pp. 281–350.

117. D. E. Henderson, M. B. DiTaranto, W. G. Tonkin, D. J. Ahlgren, D. A. Gatenby, and T. W. Shum, *Anal. Chem.*, **54**, 2067 (1982).

118. J. Paulik, F. Paulik, and L. Erdey, *Anal. Chim. Acta*, **34**, 419 (1966).

119. S. J. Ashcroft, *Thermochim. Acta*, **2**, 512 (1971).

120. H. G. Wiedemann, *Chem. Ing. Tech. Z.*, **11**, 1105 (1964).

121. A. E. Newkirk, *Thermochim. Acta*, **2**, 1 (1971).

122. A. E. Newkirk, *Anal. Chem.*, **32**, 1558 (1960).

123. S. D. Norem, M. J. O'Neill, and A. P. Gray, *Thermochim. Acta*, **1**, 29 (1970).

124. P. D. Garn, O. Menis, and H. G. Wiedemann, *J. Therm. Anal.*, **20**, 185 (1981).

125. *NIST Standard Reference Materials Catalog 1990–91*, NIST (U.S.) Spec. Publ. 260 (1990).

126. P. K. Gallagher and E. M. Gyorgy, *Thermochim. Acta*, **109**, 193 (1986).

127. ASTM Standard E914-87, *Standard Practice for Evaluating the Temperature Scale for Thermogravimetry Using the ICTA Standards*, *Annual Book of Standards*, Vol. 14.02, American Society for Testing and Materials, Philadelphia, PA, 1988, p. 558.

128. R. L. Blaine and P. G. Fair, *Thermochim. Acta*, **67**, 233 (1983).

129. E. L. Charsley, S. St. J. Warne, and S. B. Warrington, *Thermochim. Acta*, **114**, 53 (1987).

130. J. P. Elder, *Thermochim. Acta*, **52**, 235 (1982).

131. A. R. McGhie, J. Chiu, P. G. Fair, and R. L. Blaine, *Thermochim. Acta*, **67**, 241 (1983).

132. K. Nagase, H. Yokobayashi, M. Kikuchi, and K. Sone, *Thermochim. Acta*, **35**, 99 (1980).

133. K. M. Caldwell, P. K. Gallagher, and D. W. Johnson, Jr., *Thermochim. Acta*, **18**, 15 (1977).

134. H. G. Wiedemann, *Thermochim. Acta*, **6**, 257 (1973).

135. H. G. Wiedemann, *Thermochim. Acta*, **3**, 355 (1972).

136. R. J. Seyler, *Thermochem. Acta*, **17**, 129 (1976).

137. P. Le Parlouer, "The Simultaneous TG-DSCIII: A New Digital and Computer Controlled Presentation," in *Proceeding of the 17th North American Thermal Analysis Symposium*, Vol. 1, Lake Buena Vista, FL, October, 1988, p. 378.

138. J. Chiu and A. J. Beattie, *Thermochim. Acta*, **40**, 251 (1980).

139. J. Chiu and A. J. Beattie, *Thermochim. Acta*, **50**, 49 (1981).

140. R. Behrens, Jr., *Rev. Sci. Instrum.*, **58**, 451 (1987).

141. W.-D. Emmerich and E. Kaiserberger, *J. Therm. Anal.*, **17**, 197 (1979).

142. K. C. Chan, R. S. Tse, and S. C. Wong, *Anal. Chem.*, **54**, 1238 (1982).

143. E. Clarke, *Thermochim. Acta*, **51**, 7 (1981).

144. E. L. Charsley, N. J. Manning, and S. B. Warrington, *Thermochim. Acta*, **114**, 47 (1987).

145. S. M. Dyzel, *Thermochim. Acta*, **61**, 169 (1983).

146. J. Khorami, A. Lemieux, and R. B. Prime, "Characterization of Volatile Products from the Thermal Degradation of Polyolefin Fibers Using TGA/FTIR and TGA/APCI-MS," in *Proceedings of the 17th North American Thermal Analysis Symposium*, Vol. 2, Lake Buena Vista, FL, October, 1988, p. 596.

147. B. Cassel, G. L. McClure, and T. Lever, "A New System for the Measurement of Evolved Gases by TGA/FT-IR," in *Proceedings of the 17th North American Thermal Analysis Symposium*, Vol. 2, Lake Buena Vista, FL, October, 1988, p. 581.

148. D. Schultze, *Thermochim. Acta*, **29**, 233 (1979).

149. T. G. Fawcett, C. E. Crowder, L. F. Whiting, J. C. Tou, W. F. Scott, R. A. Newman, W. C. Harris, F. J. Knoll, and V. J. Caldecourt, *Adv. X-Ray Anal.*, **28**, 227 (1985).

150. W. E. Brown, D. Dollimore, and A. K. Galwey, *Comprehensive Chemical Kinetics*, Vol. 22, Elsevier, New York, 1980.

151. P. D. Garn, *Crit. Rev. Anal. Chem.*, **3**, 65 (1972).

152. P. D. Garn, *Thermochim. Acta*, **135**, 71 (1988).

153. T. B. Tang and M. M. Chaudhri, *J. Therm. Anal.*, **18**, 247 (1980).

154. A. K. Galwey, *Thermochim. Acta*, **96**, 259 (1985).

155. N. J. Carr and A. K. Galwey, *Thermochim. Acta*, **79**, 323 (1984).

156. M. E. Brown and G. M. Swallowe, *Thermochim. Acta*, **49**, 333 (1981).

157. P. M. D. Benoit, R. G. Ferillo, and A. H. Granzow, *J. Therm. Anal.*, **30**, 869 (1985).

158. H. Tanaka, *Thermochim. Acta*, **46**, 139 (1981).

159. G. O. Reddy, V. K. Mohan, B. K. M. Murali, and A. K. Chatterjee, *Thermochim. Acta*, **43**, 61 (1981).

160. H. J. Borchardt and F. Daniels, *J. Am. Chem. Soc.*, **79**, 41 (1957).

161. A. K. Galwey, *Adv. Catal.*, **26**, 247 (1977).

162. S. L. Randzio and J. Boerio-Goates, *J. Phys. Chem.*, **91**, 2201 (1987).

163. J. Boerio-Goates, J. I. Artman, and D. Gold, *J. Phys. Chem. Solids*, **48**, 1185 (1987).

164. M. E. Brown and A. K. Galwey, *Thermochim. Acta*, **29**, 129 (1979).

165. J. Šesták, *J. Therm. Anal.*, **30**, 1223 (1985).

166. J. H. Flynn and L. A. Wall, *J. Res. Natl. Bur. Stand.*, **70A**, 487 (1966).

167. H. E. Kissinger, *Anal. Chem.*, **29**, 1702 (1957).

168. T. Ozawa, *Bull. Chem. Soc. Jpn.*, **38**, 1881 (1965).

169. T. Ozawa, *J. Therm. Anal.*, **2**, 301 (1970).

170. J. H. Flynn and L. A. Wall, *Polym. Lett.*, **4**, 323 (1966).

171. C. D. Doyle, *J. Appl. Polym. Sci.*, **5**, 285 (1961).

172. C. D. Doyle, *J. Appl. Polym. Sci.*, **6**, 639 (1962).

173. J. H. Flynn, *J. Therm. Anal.*, **27**, 95 (1983).

174. M. R. Keenan, *Thermochim. Acta*, **98**, 263 (1986).

175. A. W. Coats and J. P. Redfern, *Nature (London)*, **201**, 68 (1964).

176. C. D. Doyle, *Nature (London)*, **207**, 290 (1965).

177. J. Zsakó, *J. Phys. Chem.*, **72**, 2406 (1968).

178. V. Satava and F. Sávara, *J. Am. Ceram. Soc.*, **52**, 591 (1969).

179. E. S. Freeman and B. Carroll, *J. Phys. Chem.*, **62**, 394 (1958).

180. J. M. Criado, D. Dollimore, and G. R. Heal, *Thermochim. Acta*, **54**, 159 (1982).

181. P. H. Fong and D. T. Y. Chen, *Thermochim. Acta*, **18**, 273 (1977).

182. S. V. Vyazovkin and A. I. Lesnikovich, *J. Therm. Anal.*, **30**, 831 (1985).

183. P. K. Gallagher and D. W. Johnson, Jr., *Thermochim. Acta*, **6**, 67 (1973).

184. J. Gutenbaum, E. Utizig, J. Wiśniewski, and W. Zielenkiewicz, *Bull. Acad. Pol. Sci. Sér. Sci. Chim.*, **24**, 193 (1976).

185. J. Navarro, V. Torra, and E. Rojas, *An. Fis.*, **67**, 367 (1971).

186. J. P. Dubes, M. Barres, E. Boitard, and H. Tachoire, *Thermochem. Acta*, **39**, 63 (1980).

187. E. Cesari, V. Torra, J. L. Macqueron, R. Prost, J. P. Dubes, and H. Tachoire, *Thermochim. Acta*, **53**, 1 (1982).

188. W. Zielenkiewicz, *J. Therm. Anal.*, **29**, 179 (1984).

189. S. Randzio and J. Suurkuusk, "Interpretation of Calorimetric Thermograms and Their Dynamic Correction," in A. Beezer, Ed., *Biological Microcalorimetry*, Academic, London, 1980, pp. 311–401.

190. J. H. Flynn, M. Brown, and J. Šesták, *Thermochim. Acta*, **110**, 101 (1987).

191. J. R. Williams and W. W. Wendlandt, *Thermochim. Acta*, **7**, 269 (1973).

192. M. Kamphausen, *Rev. Sci. Instrum.*, **46**, 668 (1975).

193. A. Wurflinger and G. M. Schnieder, *Ber. Bunsenges. Phys. Chem.*, **77**, 121 (1973).

194. R. Sandrock, *Rev. Sci. Instrum.*, **53**, 1079 (1982).

195. J. R. Williams and W. W. Wendlandt, *Thermochim. Acta*, **7**, 253 (1973).

196. S. Dobner, G. Kan, R. A. Graff, and A. M. Squires, *Thermochim. Acta*, **16**, 251 (1976).

197. Y. Sawada, H. Henmi, N. Mizutani, and M. Kato, *Thermochim. Acta*, **121**, 21 (1987).

198. A. Beguin, J. Billiard, F. Bonamy, J. M. Buisine, P. Curelier, J. C. Dubois, and P. LeBarny, *Mol. Cryst. Liq. Cryst.*, **115**, 1 (1984).

199. B. R. Ratna and S. Chandrasekhar, *Mol. Cryst. Liq. Cryst.*, **162B**, 157 (1989).

200. E. A. Turi, Ed., *Thermal Analysis in Polymer Characterization*, Heyden, London, 1981.

201. B. Wunderlich, *Thermochim. Acta*, **92**, 15 (1985).

202. B. Wunderlich, *J. Therm. Anal.*, **30**, 1217 (1985).

203. R. B. Prime, "Thermosets," in E. A. Turi, Ed., *Thermal Characterization of Polymeric Materials*, Academic, New York, 1981, pp. 435–569.

204. H. E. Bair, "Thermal Analysis of Additives in Polymers," in E. A. Turi, Ed., *Thermal Characterization of Polymeric Materials*, Academic, New York, 1981, pp. 845–909.

205. E. M. Pearce, Y. P. Khanna, and D. Raucher, "Thermal Analysis in Polymer Flammability," in E. A. Turi, Ed., *Thermal Characterization of Polymeric Materials*, Academic, New York, 1981, pp. 793–843.

206. L. A. Marky, J. G. Snyder, D. P. Remeta, and K. J. Breslauer, *J. Biomol. Struct. Dynam.*, **1**, 487 (1983).

207. K. J. Breslauer, R. Frank, H. Blocker, and L. A. Marky, *Proc. Natl. Acad. Sci. USA*, **83**, 3746 (1986).

208. H. Klump, *Thermochim. Acta*, **85**, 457 (1985).

209. R. N. McElhaney, *Chem. Phys. Lipids*, **30**, 229 (1982).

210. R. S. Criddle, R. W. Breidenbach, E. A. Lewis, D. J. Eatough, and L. D. Hansen, *Plant Cell Environ.*, **11**, 695 (1988).

211. R. N. McElhaney, *Biochem. Phys. Acta*, **864**, 361 (1986).

212. D. Bach, "Calorimetric Studies of Model and Natural Biomembranes," in D. Chapman, Ed., *Topics in Molecular and Structural Biology*, Vol. 4, Verlag Chemie, New York, 1984, pp. 1–41.

213. R. L. Biltonen and E. Freire, *CRC Crit. Rev. Biochem.*, **5**, 85 (1978).

214. P. Privalov, "Heat Capacity Studies in Biology," in A. E. Beezer, Ed., *Biological Microcalorimetry*, Academic, New York, 1980, pp. 413–451.

215. I. Wadsö, "Biochemical Thermochemistry," in H. A. Skinner, Ed., *International Review of Science: Physical Chemistry, Series One*, Vol. 10, Butterworth, London, 1975, pp. 1–43.

216. G. Rialdi and R. L. Biltonen, "Thermodynamics and Thermochemistry of Biologically Important Systems," in H. A. Skinner, Ed., *International Review of Science: Physical Chemistry, Series Two*, Vol. 10, Butterworth, London, 1975, pp. 147–189.

217. I. Lamprecht, *Thermochim. Acta*, **83**, 81 (1985).

218. F. P. Schwarz and W. H. Kirchhoff, *Thermochim. Acta*, **128**, 267 (1988).

219. G. P. Bettinetti, C. Caramella, F. Giordano, A. La Manna, C. Margheritis, and C. Sinistri, *J. Therm. Anal.*, **28**, 285 (1983).

220. D. J. W. Grant and I. K. A. Abougela, *Anal. Proc.*, **19**, 559 (1982).

221. M. J. Hardy, "Applications of Thermal Methods in the Pharmaceutical Industry— Part 1," in B. Miller, Ed., *Thermal Analysis, Proceedings of the 7th International Conference on Thermal Analysis*, Vol. 2, Wiley, New York, 1982, p. 876.

222. M. J. Hardy, "Applications of Thermal Methods in the Pharmaceutical Industry— Part 2," in B. Miller, Ed., *Thermal Analysis, Proceedings of the 7th International Conference on Thermal Analysis,* Vol. 2, Wiley, New York, 1982, p. 887.

223. A. Li Wan Po, *Anal. Proc.,* **23**, 391 (1986).

224. D. E. Brown and M. J. Hardy, *Thermochim. Acta,* **85**, 421 (1985).

225. J. A. McCauley, "Application of Thermal Analysis to Analytical and Process Research in the Pharmaceutical Industry," in B. Miller, Ed., *Thermal Analysis, Proceedings of the 7th International Conference on Thermal Analysis,* Vol. 2, Wiley, New York, 1982, p. 893.

INDEX